AF888144

Valeriy K. Zakharov, Timofey V. Rodionov
Sets, Functions, Measures

De Gruyter Studies in Mathematics

Edited by
Carsten Carstensen, Berlin, Germany
Gavril Farkas, Berlin, Germany
Nicola Fusco, Napoli, Italy
Fritz Gesztesy, Waco, Texas, USA
Niels Jacob, Swansea, United Kingdom
Zenghu Li, Beijing, China
Karl-Hermann Neeb, Erlangen, Germany

Volume 68/1

Valeriy K. Zakharov, Timofey V. Rodionov

Sets, Functions, Measures

Volume I: Fundamentals of Set and Number Theory

DE GRUYTER

Mathematics Subject Classification 2010
03-02, 03B10, 03C20, 03C62, 03E10, 03E15, 03E30, 03E35, 03E55, 18A05, 18A15, 54H05

Authors
Prof. Dr. Valeriy K. Zakharov
Lomonosov Moscow State University
Faculty of Mathematics & Mechanics
Department of Mathematical Analysis
Leninskie Gory b.1, GSP-1
119991 Moscow
Russia

Prof. Dr. Timofey V. Rodionov
Lomonosov Moscow State University
Faculty of Mathematics & Mechanics
Department of Mathematical Analysis
Leniniskie Gory b.1, GSP-1
119991 Moscow
Russia

ISBN 978-3-11-055008-5
e-ISBN (PDF) 978-3-11-055094-8
e-ISBN (EPUB) 978-3-11-055021-4
Set-ISBN 978-3-11-055095-5
ISSN 0179-0986

Library of Congress Cataloging-in-Publication Data
A CIP catalog record for this book has been applied for at the Library of Congress.

Bibliographic information published by the Deutsche Nationalbibliothek
The Deutsche Nationalbibliothek lists this publication in the Deutsche Nationalbibliografie; detailed bibliographic data are available on the Internet at http://dnb.dnb.de.

© 2018 Walter de Gruyter GmbH, Berlin/Boston
Typesetting: Compuscript Ltd., Shannon, Ireland
Printing and binding: CPI books GmbH, Leck
♾ Printed on acid-free paper
Printed in Germany

www.degruyter.com

The authors dedicate their book to the centenary of Felix Hausdorff's outstanding book "Set Theory"

Contents

Historical foreword on the centenary after Felix Hausdorff's classic *Set Theory*

From Euclid's "Elements" (III c. BC), the most extensive and influential mathematical treatise of the antiquity, mathematics was presented as the totality of several separate domains such as arithmetic, algebra, and geometry. Until the nineteenth century, these domains (and analysis, appearing in XVII c.) were developed rather independently. There was no general foundation connecting them in any integrity.

This peculiarity distinguished mathematics disadvantageously from the most part of natural sciences, since each of them were united up to this time in some special integrities on the base of some uniting concepts going back to philosophy of the antiquity. For physics, the notion of *atom*, for chemistry the notion of *chemical element*, and for biology the notion of *biological cell* became such uniting concepts.

This situation in mathematics changed cardinally when Georg Cantor (1845–1918) developed the theory of abstract sets consisting of abstract elements, i.e. connected with each other by only one *membership relation*. Unfortunately, such a general idea allowed the use such boundless and indicationless notions as the *set of all sets*. This brought to the discovery of paradoxes in *Cantor's set theory* and induced the distrust to it among mathematicians. But *Set Theory* as a dream has been defended by David Hilbert (1862–1943). In his famous expression, "Aus dem Paradies, das Cantor uns geschaffen, soll uns niemand vertreiben können", or in English, "From the paradise, that Cantor created for us, no-one can expel us" (D. Hilbert, *Grundlagen der Geometry*, Teubner, Leipzig-Berlin, 1930, p.274). Hilbert used the word *paradise* because he excellently understood that some *well-postulated set theory* can be that general foundation, which will give the opportunity to unite arithmetic, algebra, geometry, analysis, and other domains of mathematics in a unique integrity.

Felix Hausdorff (1868–1942) was one of those mathematicians who were occupied with the creating and forming of this *mathematical set-theoretical paradise*.

In his famous book, *Grundzüge der Mengenlehre* [Vien, Leipzig, 1914; 2nd ed. *Mengenlehre*, Walter de Gruyter, Berlin, 1927], F. Hausdorff described the architecture of contemporary mathematics in the form of a tree with the set theory as a trunk and all separate domains of mathematics as its branches. Hausdorff himself laid there the foundations of two such main domains of mathematics, function theory and measure and integration theory. This outstanding book became a model for all subsequent authors who certainly built their books dealing with any branch of mathematics on the basis of the set theory.

Starting at Hausdorff's initial architecture of mathematics, a group of mathematicians acting under the pseudonym Nicolas Bourbaki described the final architecture of contemporary mathematics. On the basis of set theory and formal logic, they introduced a general concept of a *mathematical structure* and a more substantial concept

https://doi.org/10.1515/9783110550948-204

of a *mathematical system* consisting of a principle *carrier set* and a totality of some mathematical structures on this set connected by certain logical axioms [N. Bourbaki, *Eléments de Mathématique. Livre I. Théorie des ensembles. Chapitres 1–4*, Hermann, Paris, 1956–1960]. This allowed presenting every branch of the "mathematical tree" as a mathematical theory studying some separated totality of mathematical systems with their own special structures and axioms.

Thus, the books of Hausdorff and Bourbaki have played the leading role in the consolidation of mathematics as well as mathematicians on the basis of just a few general ideas: element, set, structure, system (with an indispensable and indiscernible involvement of logical tools).

Unfortunately, Bourbaki's book on sets and structures turned out to be so formalized and difficult that it could not eclipse Hausdorff's book and become an acknowledged introduction to contemporary mathematics.

The aforementioned famous book by Hausdorff expounding on the set theory, the theory of (real-valued) functions, and the measure and integration theory as foundations of mathematics plays this role up to our days, of course, along with remarkable later books of other authors such as K. Kuratowski and A. Mostovski [*Set theory*. North-Holland Publishing Company, Amsterdam, 1967], K. Kuratowski [*Topology. Volume 1*, Academic Press, New York-London, 1966], and so on. Surely, for the past centenary after the first edition of Hausdorff's book in 1914, these domains (as the trunk and two main branches) of mathematics developed swiftly. Therefore, in our time, they differ considerably from those in the beginning of the twentieth century.

The discovery of paradoxes in *Cantor's set theory* forced mathematicians to bring the strictness up to a higher level. Therefore, mathematics advanced from the naive set theory expounded by Hausdorff to axiomatic set theories with strict logical language and adjusted axioms restricting the bounds of mathematical creation. The *Zermelo – Fraenkel set theory* and the *Neumann – Bernays – Gödel set theory* became the most well-known axiomatic set theories. At present, there are remarkable texts on this subject, although almost all of them are rather intended for particular specialists than for a wide circle of mathematicians.

Further, using von Neumann's approach to the construction of ordinal and cardinal numbers, mathematicians were able to construct at first the *set of natural numbers* and then the following *sets of integer*, *rational*, and *real numbers* within the aforementioned axiomatic set theories. This achievement allowed overcoming the gaps in Hausdorff's book, where (a) ordinal and cardinal numbers were introduced not as some special sets but on the naive level by means of extended notions such as "thing", "symbol", and others, and (b) number theory was lacking since it was considered as a prolegomenon to the naive set theory.

The enormous virtue of Hausdorff's book is the *general theory of measurable* (real-valued) *functions on descriptive spaces* (a *descriptive space* is a set with a fixed set of its subsets). The extreme importance and naturalness of the family of all measurable functions follows from the famous *Borel – Lebesgue – Hausdorff theorem* asserting that (a) this family is closed under all natural mathematical operations (in particular, addition, multiplication, division, and so on) and uniform convergence and (b) every family of (real-valued) functions on a set with the mentioned properties is some family of all measurable functions on some descriptive space. This family is extraordinary abundant in its concrete forms.

However, it turned out that the concept of a measurable function is not sufficient for the solution of some problems which arose later in function theory. Thus, in 2006, the *family of uniform functions on a prescriptive space* was discovered (a *prescriptive space* is a set with a fixed set of its finite covers). The importance and naturalness of the family of all uniform functions follows from the *characterization theorem* (proved in 2008) asserting that (a) this family is closed under all natural mathematical operations (in particular, addition, multiplication, bounded division, and so on) and uniform convergence and (b) every family of (real-valued bounded) functions on a set with the mentioned properties is some family of all uniform functions on some prescriptive (in particular, descriptive) space. This family also became sufficiently abundant in its concrete forms (for example the *family of all Riemann integrable functions on the real interval* was described as some family of all uniform functions on it).

Further, the most important concrete family of measurable functions considered in Hausdorff's book is the *family of Borel functions on a descriptive space*. The outstanding result about this family presented in his book is the *Lebesgue-Hausdorff classification* describing the family of Borel functions on a metric space by means of the transfinite application of the *Baire operation of addition of the pointwise limits of sequences of functions from the preceding families*. This result essentially uses the remarkable transfinite construction of Borel sets given by William Henry Young and Hausdorff himself. However, these classifications and constructions are not valid for an arbitrary descriptive space.

Thus, in 2002, new more general and more complicated constructions and classifications were created. It is remarkable that the finest classification (2014) uses, in the capacity of the initial functional family, some narrow family of uniform functions.

In the first edition of his book published in 1914, Hausdorff also expounded on an important branch of mathematics, the *theory of the Lebesgue integral*. Naturally, for the past centenary, the measure and integration theory had an enormous development. This is reflected in the large number of excellent books on this domain with different degrees of generality and profundity. From the times of Lebesgue and Young, two parallel points of view were developed in integration theory: the first considers the integral as a special structure over a descriptive space with some measure; the second considers the integral as a superstructure over a functional linear space with some linear functional on it. For many years, the efforts of many outstanding mathematicians were devoted to the proof of the parallelism of these points of view.

For the most popular topological space with a Radon measure, this supposed parallelism is known as the *Riesz – Radon – Fréchet problem of characterization of Radon integrals as linear functionals*. The solution of this problem as well as of the *problem of the general parallelism* in the most general and complete form took up almost one hundred years.

Finally, although the Lebesgue integral substantially darkened the Riemann integral, the latter continued to develop and in result has been generalized onto an arbitrary Tychonoff topological space with some bounded positive Radon measure. It is remarkable that the description of Riemann integrable functions requires involving some family of uniform functions. By the same token, the Riemann integrable functions was characterized in 2006 and this characterization is completely different from the famous Lebesgue characterization (as almost everywhere continuous functions) even for the real interval.

All previously described (along with many other) changes and achievements happening in the centenary after the first edition of Hausdorff's book are reflected (in detail and in up-to-date mathematical language) in the present comprehensive two-volume book *Sets, Functions, Measures* published by Walter de Gruyter in 2018.

The present work expounds set theory, the theory of (real-valued) functions, and the measure and integration theory as the fundamental domains of contemporary mathematics successively built on each other. It may be said that the authors of this book have attempted to solve one hundred years later the problem solved successfully by Hausdorff at the beginning of the twentieth century. In particular, continuing Hausdorff's line, the material of the book is presented in such a way that there is no need for references to other sources.

V. K. Zakharov and T. V. Rodionov
October 2017

Preface

The book's title *Sets, Functions, Measures* shows that it is devoted to the exposition of the most general fundamentals of mathematics. It may be said that the book goes back to the famous *Set Theory* by Felix Hausdorff [1914], where he expounded the set theory, the function theory, and the function theory as the general fundamentals of mathematics. The authors of this book have attempted to solve a hundred years later the problem solved successfully by F. Hausdorff in the beginning of 20th century.

The manner of exposition also goes back to the manner of F. Hausdorff. As in "Set Theory", we set as an object to expound the most general results of the set theory, the function theory, and the measure theory in such a way that there was no need for references to other sources. Since the number theory was not included by F. Hausdorff in his book, we, following the indicated line, considered it necessary to eliminate this minor defect and to expound the theory of natural, integer, rational, and real numbers, deducing it from the set theory itself.

It follows from the above that this book is addressed to a wide range of mathematicians, and it can be useful both to mature mathematicians and to students and young mathematicians, who would like to be acquainted with the fundamentals of the listed theories.

According to its title, the book is divided into three chapters, each leaning on the other subsequently and is supplied with special appendices. The content of the book and the motivations of the authors are explicitly given in the introduction to each chapter. Here, we shall touch only some of the peculiarities of the presented material.

The first chapter is devoted to the theory of classes, sets, and numbers. This theory is expounded in the framework of Neumann – Bernays – Gödel axiomatics with generality, completeness, and thoroughness. It is called the *Neumann – Bernays – Gödel set theory* (NBG). The summit of this chapter is the theory of real and extended real numbers, including the theory of series in the extended real line $[-\infty, \infty]$. The other version of the theory of sets (the *Zermelo – Fraenkel set theory with the choice axiom* (ZF)) is presented in Appendix A (see A.2).

The second chapter is devoted to the theory of functions. It is based on the first chapter, especially on its last section. It contains together with the more or less standard material a lot of new and non-trivial materials such as the theory of uniform functions on prescriptive spaces.

The third chapter is devoted to the general theory of measure and integral. It is based on two preceding chapters, especially on the theory of series in the extended real line. This allows us to consider measures taking their values in $[-\infty, \infty]$ at the very beginning. Note that in the last chapter, the authors widely used ideas and methods expounded in the remarkable books of Fremlin [1974], Jacobs [1978], and Konig [1997]. The particular value of this chapter consists of the solution of the problem of characterization of Lebesgue integrals and the problem of characterization

https://doi.org/10.1515/9783110550948-205

of Radon integrals in the most general cases, and the key tool in the solution of the second problem is some concrete family of uniform functions.

Appendix A is devoted to the characterization of all natural models of the NBG and ZF set theories. These models are extremely important in virtue of their simplicity. For the reader's convenience, the first section of this appendix describes the structure of an arbitrary first-order language (theory) and contains necessary notions from mathematical logic. In the second and third sections, the proper axioms and axiom schemes of the ZF set theory are formulated, the notions of ordinals, cardinals, and inaccessible cardinals are introduced, and properties of cumulative Mirimanov – Neumann sets are described. Thus, these sections are good supplements to the first chapter, giving a more profound presentation about contemporary mathematical logic and axiomatic set theories.

Appendix B is devoted to the local theory of sets giving the solution of Maclane's problem of constructing a new and more flexible axiomatic set theory that could serve as an adequate logical foundation for all the naive category theory.

Appendix C is devoted to the proof of the compactness theorem for some generalized second-order language. The compactness theorem is valid for the first-order language, but it is not valid for the usual second-order language.

Appendix D contains historical notes on the famous Riesz – Radon – Fréchet problem of characterization of Radon integrals as linear functionals.

Each chapter C is divided into sections with two-valued numeration C.P. Each section is divided into subsections with three-valued numeration C.P.S. Important statements in each subsection such as lemmas, propositions, and theorems are numbered in a subsection by natural numbers in the manner Lemma N, Proposition N, Theorem N. When referring in some subsection to statements from another subsection C.P.S, we use number C.P.S in round brackets in the manner N (C.P.S).

The symbol □ is used throughout the text to indicate the end of a proof.

If some notion has several names, then the other names are written in parentheses after the name chosen by the authors as the main name.

To shorten writings, we use the *method of parallel writing* in the following form. A short writing "$A\ \pi\ [\varkappa, \rho, \ldots]\ B\ p\ [q, r, \ldots]\ C$." is equivalent to the following full writing: "(1) $A\ \pi\ B\ p\ C$; (2) $A\ \varkappa\ B\ q\ C$; (3) $A\ \rho\ B\ r\ C$; …", where the capital letters denote some texts and the small letters denote some words, word combinations, or formulae.

By the recommendation of the publisher, the book is divided into two volumes. For the convenience of readers, each volume is equipped by the same index of terms, index of notations, and bibliography.

The authors express their profound gratitude to the Rector of the Lomonosov Moscow State University, Professor Victor Antonovich Sadovnichy for his continuous support during the twenty-year work on this book under the aegis of the University.

1 Fundamentals of the theory of classes, sets, and numbers

Introduction

The first chapter is devoted to the theory of classes, sets, and numbers. It is the basis for all other chapters.

In particular, we need notions of a *class* and a *property of a class*.

In connection with the notion of a class, it is not enough for us to use the theory of sets in Zermelo – Fraenkel's axiomatics (ZF) (see Appendix A, A.2), but it is necessary to use the theory of classes and sets in Neumann – Bernays – Gödel's axiomatics (NBG).

For the axiomatic construction of the theory of classes and sets, we have chosen not the finitary version of NBG, presented for instance in [*Mendelson*, 1997], but a simpler equivalent version, close to the version presented by Kelley [1975]. In this version, several explicit axioms from the finitary version of NBG, claiming the existence of some classes, are substituted by one axiom scheme AS2 (see 1.1.5) using the selecting term $\{x \mid \varphi(x)\}$ with an arbitrary formula φ (see [*Mendelson*, 1964, ch. 4, 1]). The finitary version of NBG and the proof of equivalence between the scheme and finitary versions are given in Appendix B (see B.7.3).

The first chapter begins with a strict inductive definition of *formulas*.

Not having the notion of a natural number at this stage of the theory, we were forced to give rather unaccustomed definitions of *deducibility* and *correctness* because we could not use chains $\sigma_1, \sigma_2, \dots, \sigma_n$, usually used in mathematical logic. For the strict definition of deducibility and correctness, we need the full family of axioms, i. e. not only the proper axioms and axiom schemes about classes and sets, but also the initial *logical axiom schemes* LAS1 – LAS14 (see 1.1.4) and the *rules of deduction* ($\equiv$ *rules of inference*).

The presented logical axiom schemes and rules of deduction are used also directly in proofs of some starting mathematical assertions such as Proposition 1 (1.1.5).

Furthermore, we were forced to introduce in the first chapter the new unaccustomed notion of a *(multivalued) collection of classes* $(A_i \subset A \mid i \in I)$ in addition to the usual notion of a *simple collection* ($\equiv$ *indexed family*) *of sets* $(a_i \in A \mid i \in I)$. These notions reflect our intuitive ideas about *collections of totalities* and *collections of wholenesses*.

In addition, the necessity to have finite suits (pairs, triplets, quadruplets,...) of classes forced us to introduce a new notion of *(multivalued) sequential suits of classes* (A, A'), (A, A', A''), (A, A', A'', A'''),...; and on this base to define *sequential products of classes* $A \times A'$, $A \times A' \times A''$, $A \times A' \times A'' \times A'''$,... (see 1.1.12).

All the above-mentioned topics constitute the main part of the first section of the first chapter. The other three sections are devoted to the theory of ordinal, cardinal, natural, and real numbers.

https://doi.org/10.1515/9783110550948-001

The use of axiom of regularity A6 (see 1.1.11) makes it possible to simplify strongly the definition of ordinal and cardinal numbers. The theory of ordinal and cardinal numbers is given by the method of J. Neumann. In this method, an ordinal is just equal to the class of all preceding ordinals. The operations over ordinal and cardinal numbers are described sufficiently in details.

Natural numbers are defined as some ordinal numbers. The theory of real numbers is based on the progressive extension of the set of natural numbers. The operations over numbers are introduced gradually starting from the operations over cardinal numbers.

Note, that proofs of all basic mathematical assertions of the first section are very detailed. However, further in the book, proofs gain a form more habitual for the most of mathematical texts.

1.1 Classes and sets

The *axiomatic theory of classes and sets* postulates the existence of some undefinable objects called classes and sets and formulates the rules of action with them. Axiomatics of the theory of classes and sets consists of two parts. The first part is *logical* and describes rules of construction of correct conclusions about classes and sets. The second part is proper *mathematical* and describes some primary properties of classes and sets.

1.1.1 Symbols, symbol-strings, and texts of the theory of classes and sets

The theory of classes and sets uses the following *special symbols* (≡ *signs*): $\neg$ (the *negation*); $\wedge$ (sometimes &) (the *conjunction*); $\vee$ (the *disjunction*); $\Rightarrow$ (the *implication*); $\forall$ (the *quantifier of generality*); $\exists$ (the *quantifier of existence*); $\{|\}$ (the *selector*); $\in$ (the *belonging*).

The symbols $\neg$, $\wedge$, $\vee$, $\Rightarrow$, $\forall$, $\exists$, and $\{|\}$ are called *logical*. The symbols $\neg$, $\wedge$, $\vee$, $\Rightarrow$ are called the *logical propositional connectives*. The symbol $\neg$ is *unary*; the symbols $\wedge$, $\vee$, $\Rightarrow$, and $\{|\}$ are *binary*.

The symbol $\in$ is the single *proper* special symbol. It is binary as well.

These special symbols have the following sense: $\neg \ldots$ (*It is not...*); $\cdots \wedge \ldots$ (*...and...*); $\cdots \vee \ldots$ (*...or...*); $\ldots \Rightarrow \ldots$ (*...implies...; If..., then...*); $\forall \ldots$ (*For all...; For any...*); $\exists \ldots$ (*There is...; There exists...*); $\{\cdots \mid \ldots\}$ (*All...with the property...*); $\cdots \in \ldots$ (*...belongs to...; ...is an element of...; ...is a member of...*).

In the capacity of *general symbols* (≡ *signs*), the theory of classes and sets uses the *letters* of Latin, Greek, Gothic, and other alphabets, Arabic and Roman numerals, the comma ",", the point ".", the colon ":", the prime "′", covers "$\bar{}$, $\hat{}$, $\check{}$, $\tilde{}$,...", circles "$\circ$, $\otimes$, $\oplus$,...", the round, curly, square, broken, and angular brackets "(,), {,}, [,], ⌊, ⌋, ⟨, ⟩", the blank symbol (␣), and so on.

The special and general symbols compose the *initial alphabet of the theory of classes and sets* (note that it is neither a class nor a set).

A *symbol-string* (≡ *expression*) *of the theory of classes and sets* is a sequence of symbols of the initial alphabet of this theory except the blank symbol, written one after another. More strictly a *symbol-string* is defined by induction in the following way: 1. every symbol α is a *symbol-string*; 2. if σ and ρ are symbol-strings, then $\sigma\rho$ and $\rho\sigma$ are *symbol-strings*. A *letter-string* is a symbol-string every symbol of which is a letter.

The usage of only symbol-strings brings to insuperable difficulties. Therefore, further new *designating* (≡ *reducing, shortening*) *symbols* and *symbol-strings* will be introduced. A *designating symbol-string* σ for a symbol-string ρ is introduced in the form of the symbol-string $\sigma \equiv \rho$ or $\rho \equiv \sigma$ (σ is a *designation for* ρ). Examples of designating symbol-strings are the following: $A \subset B$, $A = B$, $A \cup B$, $\varnothing$, $\mathbb{N}$, $\prod(A_i \subset A \mid i \in I)$, the *real line*, the *function* exp, and so on.

On the level of epilogic and epimathematics (i. e. before carrying out formal descriptions) the following *initial opportunities of a mathematician for reasoning about symbol-strings* are assumed: 1. the opportunity to insert one symbol-string into another symbol-string (in particular to write alongside); 2. the opportunity to distinguish a part of a symbol-string in the capacity of a new symbol-string.

If a symbol-string ρ is a part of a symbol-string σ staying in one of the three following positions: $\ldots\rho$, $\rho\ldots$, $\ldots\rho\ldots$, then we say that ρ *occurs in* σ (or else ρ is an *occurrence in* σ).

A sequence of symbol-strings, written one after another with the blank symbol $\sqcup$ between them, is called a *text of the theory of classes and sets*. More strictly *text* is defined by induction in the following way: 1. every symbol-string σ is a *text*; 2. if Φ and Ψ are texts, then $\Phi_\sqcup\Psi$ and $\Psi_\sqcup\Phi$ are *texts*.

We say that a text Φ *occurs in a text* Σ (or else Φ is an *occurrence in* Σ), if Φ is a part of the text Σ, staying in one of the three following positions: $\ldots_\sqcup\Phi$, $\Phi_\sqcup\ldots$, $\ldots_\sqcup\Phi_\sqcup\ldots$.

The following *initial opportunities of a mathematician for reasoning about texts* are assumed: (1) the opportunity to insert one text Φ into another text Ψ (in particular to write alongside), inserting every occurrence Ω in the text Φ in the capacity of an occurrence in the text Ψ; (2) the opportunity to distinguish a part of a text Ψ in the capacity of a new text Φ so that every occurrence Ω in the text Φ is also an occurrence in the initial text Ψ.

Note that in practice the blank symbol $\sqcup$ is simply omitted or substituted by the usual point, if there is no confusion in understanding.

Not all of possible symbol-strings and texts can be used in the theory of classes and sets. In the next subsection, we shall describe the admissible ones.

1.1.2 Formulas and terms

At first, we shall describe those symbol-strings which are admissible in the theory of classes and sets. The exacting reader can find the more formalized exposition of this material in Appendix A (see A.1).

Admissible symbol-strings are divided into two types: terms and formulas. Intuitively, terms are symbol-strings representing objects, and formulas are symbol-strings representing statements about these objects.

For *variables*, we shall use the letters of Latin, Greek, Gothic, and other alphabets.

Now, we shall give the inductive definition of *terms* and *formulas* and *free* and *connected occurrences of variables* (≡ *arguments*) in them.

1) Every variable x is a *term* with the single *free occurrence of the variable* x. Such a term is called a *letter term.*
2) Every symbol-string $(x \in y)$ for any letter terms x and y is a *formula* with the two *free occurrence of the variables* x *and* y. Such a formula is called *simplest relational.*
3) Every symbol-string $(\neg\varphi)$, $(\varphi \wedge \psi)$, $(\varphi \vee \psi)$ and $(\varphi \Rightarrow \psi)$ for any formulas φ and ψ is a *formula*. Such a formula is called *derivative logical.*
   Every free occurrence of some variable in the formula φ is called a *free occurrence of this variable in the formula* $(\neg\varphi)$. Every free occurrence of some variable in the formulas φ and ψ is called a *free occurrence of this variable in the formulas* $(\varphi \wedge \psi)$, $(\varphi \vee \psi)$ *and* $(\varphi \Rightarrow \psi)$.
4) Every symbol-string $(\forall x\varphi)$ and $(\exists x\varphi)$ for any formula φ is a *formula*. Such a formula is called *derivative quantified.*
   Every free occurrence of some (except x) variable in the formula φ is called a *free occurrence of this variable in the formulas* $(\forall x\varphi)$ *and* $(\exists x\varphi)$. Every occurrence of the variable x in the formulas $(\forall x\varphi)$ *and* $(\exists x\varphi)$ is called the *connected occurrence.*

A variable x is called a *free* [*connected*] *variable of a term or a formula* ζ if there is at least one free [connected] occurrence of x in ζ. If there is no free variable for a symbol-string ζ, then ζ is called *closed.*

If a variable x is a free variable of a formula φ [term τ], then this is denoted by $\varphi(x)$ [$\tau(x)$]. If another variable y is a free variable of the formula φ, then this is denoted by $\varphi(x)(y)$, $\varphi(y)(x)$, $\varphi(x, y)$ or $\varphi(y, x)$; the notation $\varphi(x, y, z)$ is defined in a similar way, and so on. Every not closed formula φ has *lists* $\vec{l}$ *of its free variables.* We can divide a list $\vec{l}$ of free variables of φ into two different lists $\vec{x}$ and $\vec{p}$ and use the *parametric list* $\vec{x}, \vec{p}$, where $\vec{x}$ is some list of *basic* free variables of φ and $\vec{p}$ is some list of *auxiliary* free variables of φ. All the said statements are valid also for the term τ.

If at least one of the free variables of the formula φ [term τ] occurs in the symbol-string x, y, then we shall write $\varphi|x, y|$ [$\tau|x, y|$]. If at least one of the free variables of the formula φ [term τ] occurs in the symbol-string x, y, z, then we shall write $\varphi|x, y, z|$ [$\tau|x, y, z|$], and so on.

If ζ is a term or a formula and τ is a term, then the symbol-string obtained by the substitution of every free occurrence of a variable x in the symbol-string ζ by the term τ is denoted by $\zeta(x \parallel \tau)$. In this case, we say also that τ *substitutes for free occurrences of* x *in* ζ. If x does not occur freely in ζ, then $\zeta(x \parallel \tau)$ is ζ itself.

A letter term y is called *free for a variable x in the symbol-string ζ* if every free occurrence of x in ζ is not a free occurrence in some quantified formulas $\forall y\psi$ and $\exists y\psi$ occurring in ζ.

If ζ is a term [formula] and a term τ is free for a variable x in ζ, then the symbol-string $\zeta(x \parallel \tau)$ is a term [formula]. Every free occurrence of some variable y (except x) in ζ and every free occurrence of some variable z in the term τ are free occurrences of these variables in the symbol-string $\zeta(x \parallel \tau)$.

If the term τ is free for the variable x in the symbol-string ζ, then along with $\zeta(x \parallel \tau)$ we shall write also $\zeta(\tau)$.

Further, instead of the formula $((\varphi \Rightarrow \psi) \wedge (\psi \Rightarrow \varphi))$, we shall use the designation $(\varphi \Leftrightarrow \psi)$ and say that *φ is equivalent to ψ*. Also, along with the formulas $(\forall x((x \in y) \Rightarrow \varphi(x)))$ and $(\exists x((x \in y) \wedge \varphi(x)))$, we shall sometimes write $((\forall x \in y)\varphi)$ and $((\exists x \in y)\varphi)$, respectively. Such the quantifiers are called *bounded.*

Formulas and terms contain round brackets. However, for the facilitation of notations, some round brackets are omitted frequently; in particular, the exterior pair of brackets are omitted. When omitting the brackets the agreement is used that the symbol $\neg$ is stronger that the symbols $\wedge$ and $\vee$, the symbols $\wedge$ and $\vee$ have the equal status and both are stronger than the symbol $\Rightarrow$, which is stronger than $\Leftrightarrow$. The quantifiers $\forall$ and $\exists$ have the equal status and both are stronger than every previous logical symbol. The symbol $\in$ is stronger than every logical symbol.

1.1.3 Axioms, deducibility, and theorems

After that as symbol-strings, terms, formulas were defined, it is necessary to select some basic symbol-strings.

Along with the initial language of the theory of classes and sets, we shall use some broader language (*epilanguage*) for this theory. In this epilanguage, we need variables for formulas of the theory of classes and sets. Now, we shall give the inductive definition of *formula schemes*:

1) if f is a variable for formulas, then f is a formula scheme;
2) if ρ and σ are formula schemes, then $(\neg\varphi)$, $(\varphi \wedge \psi)$, $(\varphi \vee \psi)$, and $(\varphi \Rightarrow \psi)$ are formula schemes;
3) if ρ is a formula scheme and x is a variable of the initial language, then $(\forall x\rho)$ and $(\exists x\rho)$ are formula schemes.

The following property is easily proved by induction:

If f is a variable for formulas, $\gamma(f)$ is a formula scheme, φ is a formula, then $\gamma(f \parallel \varphi)$ is a formula.

This formula $\gamma(f \parallel \varphi)$ is called a *formula generated by the formula scheme γ.*

Some fixed text Γ is called an *axiom text* if every symbol-string γ occurring in Γ is either a formula or a formula scheme. If γ is a formula in this text, then it is called an *explicit axiom*. If γ is a formula scheme in this text, then it is called an *axiom scheme*. Every formula generated by the axiom scheme γ is called an *implicit axiom*.

Let Γ be a fixed axiom text, Ψ be some text, such that every symbol-string occurring in it is a formula, and δ be a formula.

The text Ψ is called a *deduction from the condition* δ *and the axiom text* Γ if for every formula ψ occurring in the text Ψ at least one of the following conditions is fulfilled:

D^{c}1. ψ is a rewriting of the formula δ or some explicit axiom occurring in the text Γ;

D^{c}2. ψ is some implicit axiom generated by some axiom scheme occurring in the text Γ;

D^{c}3. some formulas φ and $(\varphi \Rightarrow \psi)$ occurs in the text Ψ, such that φ precedes $(\varphi \Rightarrow \psi)$ and $(\varphi \Rightarrow \psi)$ precedes ψ (the *rule of implication*);

D^{c}4. some formula $\varphi(x)$ occurs in the text Ψ such that φ precedes ψ, x is not a free variable of formula δ, and ψ is the formula $(\forall x \varphi)$ (the *rule of generalization with the condition* δ).

The final formula ρ occurring in the deduction Ψ from the condition δ and the axiom text Γ is called the *result of the deduction* Ψ (or *deduced by the deduction* Ψ) *from the condition* δ *and the axiom text* Γ.

The text Ψ is called a *deduction from the axiom text* Γ if for every formula ψ occurring in the text Ψ at least one of the following conditions is fulfilled:

D1. ψ is a rewriting of some explicit axiom occurring in the text Γ;

D2. ψ is some implicit axiom generated by some axiom scheme occurring in the text Γ;

D3. some formulas φ and $(\varphi \Rightarrow \psi)$ occurs in the text Ψ such that φ precedes $(\varphi \Rightarrow \psi)$ and $(\varphi \Rightarrow \psi)$ precedes ψ (the *rule of implication* or *modus ponens* (*MP*));

D4. some formula $\varphi(x)$ occurs in the text Ψ such that φ precedes ψ and ψ is the formula $(\forall x \varphi)$ (the *rule of (unconditional) generalization* (*Gen*)).

The final formula ρ occurring in the deduction Ψ from the axiom text Γ is called the *result of the deduction* Ψ (or *deduced by the deduction* Ψ) *from the axiom text* Γ.

Any formula is called *true* (*in the sense of deducibility from the axiom text* Γ) if it is a result of some deduction from the axiom text Γ. A formula φ is called *false* if the formula $\neg\varphi$ is true.

Usually, the indication of the axiom text Γ in all the definitions mentioned above is omitted because Γ is fixed.

The text $\delta \vdash \rho.\Psi$ consisting of the base δ for the deduction Ψ, the *symbol of deduction* $\vdash$, the result ρ of the deduction Ψ, and the deduction Ψ written one after another is called a *theorem of conditional deduction*. The text $\delta \vdash \rho$ is called the *statement* ($\equiv$ *assertion, formulation*), the formula δ is called the *condition*, the formula ρ is called the *conclusion*, and the deduction Ψ is called the *proof of the theorem of conditional deduction* $\delta \vdash \rho.\Psi$.

If formulas π and ρ occur in some theorems of conditional deduction $(\delta \wedge \pi) \vdash \rho.\Phi$ and $(\delta \wedge \rho) \vdash \pi.\Psi$ with the conditions $\delta \wedge \pi$ and $\delta \wedge \rho$, respectively, then the conclusions π and ρ are called *mutually deducible* or *equivalent under the condition* δ. This situation will be denoted by $\delta \vdash (\pi \sim \rho)$.

The text $\delta \vdash (\pi \sim \rho).\Phi.\Psi$, composed of the parts of the theorems of conditional deduction $(\delta \wedge \pi) \vdash \rho.\Phi$ and $(\delta \wedge \rho) \vdash \pi.\Psi$, is called a *theorem of conditional equivalence*. The text $\delta \vdash (\pi \sim \rho)$ is called the *statement*, the formula δ is called the *condition*, the formulas π and ρ are called the *conclusions*, and the deductions Φ and Ψ are called the proof of the *theorem of conditional equivalence* $\delta \vdash (\pi \sim \rho).\Phi.\Psi$.

Theorems of conditional deduction and theorems of conditional equivalence are usually called simply *conditional theorems*.

The theorem of conditional deduction $\delta \vdash \rho.\Psi$ is usually written in the following form:

Theorem. *Let* δ. *Then,* ρ.

Proof. Ψ. □

The theorem of conditional equivalence $\delta \vdash (\pi \sim \rho).\Phi.\Psi$ is usually written in the following form:

Theorem. *Let* δ. *Then, the following conclusions are equivalent:*
1) π;
2) ρ.

Proof. 1) $\vdash$ 2). Φ.
2) $\vdash$ 1). Ψ. □

The text $\vdash \rho.\Psi$ consisting of the symbol of deduction $\vdash$, the result ρ of the deduction Ψ and the deduction Ψ written one after another is called a *theorem of unconditional deduction*. The text $\vdash \rho$ is called the *statement*, the formula ρ is called the *conclusion*, and the deduction Ψ is called the proof of the *theorem of unconditional deduction* $\vdash \rho.\Psi$.

If formulas π and ρ occur in some theorems of conditional deduction $\pi \vdash \rho.\Phi$ and $\rho \vdash \pi.\Psi$ with the conditions π and ρ, respectively, then the conclusions π and ρ are called *mutually deducible* or *equivalent*. This situation will be denoted by $\vdash (\pi \sim \rho)$.

The text $\vdash (\pi \sim \rho).\Phi.\Psi$, composed of the parts of the theorems of conditional deduction $\pi \vdash \rho.\Phi$ and $\rho \vdash \pi.\Psi$, is called a *theorem of unconditional equivalence*. The text $\vdash (\pi \sim \rho)$ is called the *statement*, the formulas π and ρ are called the *conclusions*, and the deductions Φ and Ψ are called the *proof of the theorem of unconditional equivalence* $\vdash (\pi \sim \rho).\Phi.\Psi$.

Theorems of unconditional deduction and theorems of unconditional equivalence are usually called simply *unconditional theorems*.

The theorem of unconditional deduction $\vdash \rho.\Psi$ is usually written in the following form:

Theorem. ρ.

Proof. Ψ. □

The theorem of unconditional equivalence $\vdash (\pi \sim \rho).\Phi.\Psi$ is usually written in the following form:

Theorem. *The following conclusions are equivalent:*
1) π;
2) ρ.

Proof. 1) $\vdash$ 2). Φ.
2) $\vdash$ 1). Ψ. □

Sometimes, in statements of these theorems, some explanatory texts about some terms and formulas occurring in these theorems can be used. In this case, the form of these theorems can be slightly modified.

In some cases, along with the form "*Let* δ. *Then*, ρ." the forms "*If* δ, *then* ρ.", "*Suppose* δ. *Then*, ρ.", and others are used.

Sometimes, along with the short initial phrases "*Let* ρ.", "*If* ρ,", "*Suppose* ρ.", and others we shall use the more expanded initial phrases "*Let we are given* ρ.", "*If we are given* ρ,", "*Suppose we are given* ρ.", and others.

In the theorem of conditional equivalence, the forms "*Then*, π *if and only if* ρ." and "*Then, for* π *it is necessary and sufficient* ρ." are also used. Along with the words "*if and only if*", the shorter variant "*iff*" is used. In the theorem of unconditional equivalence, these forms without the word "*Then*" are also used.

Note that along with the word "*theorem*" the words "*proposition*", "*lemma*", "*corollary*" and others are used for the designation of less important results.

Some important properties obtain usually special names in the following form. A text "τ is called T if $\varphi(\tau)$" including a term τ, a formula $\varphi(\tau)$ and a text T is called a *definition* (*of the property of the term* τ *by means of the formula* φ). The text T is called a *name of the property* $\varphi(\tau)$.

With the usage of the definition "τ is called T if $\varphi(\tau)$" the theorem $\vdash \varphi(\tau).\Psi$ asserting that the term τ possesses the property $\varphi(\tau)$ is usually written in the following form:

Theorem. τ *is* T.

Proof. Ψ. □

The following *logical epitheorem of deduction* (the *deduction theorem*) is valid.

Theorem. *Let δ, ρ be formulas. Suppose the theorem $\delta \vdash \rho.\Phi$ of conditional deduction holds, where the deduction Φ is constructed without applying the rule of generalization D^c4 to the free variables of the formula δ. Then, there exists a deduction Ψ such that the theorem $\vdash (\delta \Rightarrow \rho).\Psi$ of unconditional deduction holds.*

This epitheorem is used when one needs in the formula $\delta \Rightarrow \rho$, but it is very difficult to deduce it. In this case, the simpler deduction Φ of ρ from δ is constructed, and by the epitheorem of deduction, it is possible to conclude that the formula $\delta \Rightarrow \rho$ is deducible.

Further, we begin to fix some concrete *axiom text of the theory of classes and sets.*

1.1.4 Logical axiom schemes of the theory of classes and sets

Axioms and axiom schemes of the theory of classes and sets are diveded in two classes: *logical* and *proper* (*non-logical*).

In this subsection, we shall formulate some logical axiom schemes of the theory of classes and sets which use only the special logical symbols from 1.1.1.

Further, φ, ψ, χ denote variables for formulas and τ denotes a term.

LAS1. $(\varphi \Rightarrow (\psi \Rightarrow \varphi))$.
LAS2. $(\varphi \Rightarrow \psi) \Rightarrow ((\varphi \Rightarrow (\psi \Rightarrow \chi)) \Rightarrow (\varphi \Rightarrow \chi))$.
LAS3. $((\varphi \wedge \psi) \Rightarrow \varphi)$.
LAS4. $((\varphi \wedge \psi) \Rightarrow \psi)$.
LAS5. $((\varphi \Rightarrow \psi) \Rightarrow ((\varphi \Rightarrow \chi) \Rightarrow (\varphi \Rightarrow (\psi \wedge \chi))))$.
LAS6. $(\varphi \Rightarrow (\varphi \vee \psi))$.
LAS7. $(\psi \Rightarrow (\varphi \vee \psi))$.
LAS8. $((\varphi \Rightarrow \chi) \Rightarrow ((\psi \Rightarrow \chi) \Rightarrow ((\varphi \vee \psi) \Rightarrow \chi)))$.
LAS9. $((\varphi \Rightarrow \neg\psi) \Rightarrow (\psi \Rightarrow \neg\varphi))$.
LAS10. $((\neg(\neg\varphi)) \Rightarrow \varphi)$.
LAS11. $((\forall x\varphi) \Rightarrow \varphi(x \parallel \tau))$ if τ is free for x in φ.
LAS12. $(\varphi(x \parallel \tau) \Rightarrow (\exists x\varphi))$ if τ is free for x in φ.
LAS13. $((\forall x(\psi \Rightarrow \varphi(x))) \Rightarrow (\psi \Rightarrow (\forall x\varphi)))$ if x is not a free variable of ψ.
LAS14. $((\forall x(\varphi(x) \Rightarrow \psi)) \Rightarrow ((\exists x\varphi) \Rightarrow \psi))$ if x is not a free variable of ψ.

Using axiom schemes LAS13 and LAS14 and rules of deduction D3 and D4 in 1.1.3 we obtain the following *derivative rules of (unconditional) deduction*:

D5. there is in the deduction Ψ some formula $(\chi \Rightarrow \varphi(x))$, such that χ does not contain the free variable x of φ, $(\chi \Rightarrow \varphi(x))$ precedes ψ, and ψ is the formula $(\chi \Rightarrow (\forall x\varphi))$;

D6. there is in the deduction Ψ some formula $(\varphi(x) \Rightarrow \chi)$, such that χ does not contain the free variable x of φ, $(\varphi(x) \Rightarrow \chi)$ precedes ψ, and ψ is the formula $((\exists x\varphi) \Rightarrow \chi)$.

The *derivative rules of (conditional) deduction* D^c5 and D^c6 are obtained by adding to D5 and D6 the condition that x is not a free variable of the formula δ.

The rules of deduction D6 and D^c6 are used often in the following case. Suppose that we deduced the formula $\exists x\varphi$ and we need to deduce the formula $(\exists x\varphi) \Rightarrow \chi$. Then, it is sufficient to deduce the simpler formula $\varphi(x) \Rightarrow \chi$.

Remark. We can take initially only axiom schemes LAS1 – LAS12 and rules D1, D2, D3, D5, and D6. Then, the rule D4 can be obtained as a derivative rule of deduction.

This means that the deducibility under axiom schemes LAS1 – LAS14 and rules D1 – D4 can be interchanged by the deducibility under axiom schemes LAS1 – LAS12 and rules D1, D2, D3, D5, and D6. The same is valid for the (conditional) deducibility under axiom schemes LAS1 – LAS14 and rules D^c1 – D^c4 and the (conditional) deducibility under axiom schemes LAS1 – LAS12 and rules D^c1, D^c2, D^c3, D^c5, and D^c6, respectively.

1.1.5 First non-logical axioms and axiom schemes of the theory of classes and sets

"Intuitively, $\in$ is to be thought of as the membership relation and the values of the variables are to be thought of as classes. ... The axioms will reveal more about what we have mind. They will provide us with the classes we need in mathematics and appear modest enough so what contradictions are not derivable from them" [*Mendelson*, 1997, ch. 4, § 1].

A class A is called a *set* if $\exists x(A \in x)$. A class A is called a *proper class* if $\neg(\exists x(A \in x))$.

A class B is called a *subclass of a class A* if $\forall x(x \in B \Rightarrow x \in A)$. This formula is denoted by $B \subset A$. In this case, we also say that B is *contained in A*, *B is a part of A*, *A contains B*. Classes A and B are called *equal* if $A \subset B$ and $B \subset A$, i. e. $\forall x(x \in A \Leftrightarrow x \in B)$. This formula is denoted by $A = B$. The formula $\neg(A = B)$ is denoted by $A \neq B$. The formula $\neg(x \in A)$ is denoted by $x \notin A$.

A1. (The *extensionality axiom.*) $(A = B) \Rightarrow (A \in C \Leftrightarrow B \in C)$.

A formula φ is called *predicative* ($\equiv$ *such that every connected variable of φ is a variable for sets*) if all symbol-strings $\forall x$ and $\exists x$, occurring in the formula φ, are situated only in positions of the following kind: $\forall x((\exists X(x \in X)) \Rightarrow \ldots)$ and $\exists x((\exists X(x \in X)) \wedge \ldots)$.

AS2. (The *full comprehension axiom scheme.*) Let $\varphi(x, \vec{p})$ be a predicative formula, such that X is not a free variable of φ. Then, $\exists X\forall x((x \in X) \Leftrightarrow ((\exists Y(x \in Y)) \wedge \varphi(x, \vec{p})))$.

This axiom scheme postulates the existence of the unique class depending on the *parameter* $\vec{p}$ and denoted by $\tau(\vec{p}) \equiv \{x \mid \varphi(x, \vec{p})\}$. It will be said that the class $\tau(\vec{p})$ *is selected by the property* $\varphi(x, \vec{p})$.

Having the equality of classes, we can introduce some convenient designations. A class $\{x \mid \exists a((\exists A(a \in A)) \wedge (x = \tau(a)) \wedge \varphi(a))\}$, where τ is any class with a parameter a, and φ is any formula freely containing the variable a, will be denoted shortly by $\{\tau(a) \mid \varphi(a)\}$. A class $\{x \mid \exists a((\exists A(a \in A)) \wedge \exists b((\exists B(b \in B)) \wedge (x = \tau(a, b)) \wedge \varphi(a, b)))\}$ will be denoted also by $\{\tau(a, b) \mid \varphi(a, b)\}$, and so on.

It follows from AS2 that if B is a class and φ is a formula as in AS2, then $\{x \mid (x \in B) \wedge \varphi\}$ is a subclass of B. It is denoted also by $\{x \in B \mid \varphi\}$.

By means of axiom A1 and axiom scheme AS2, we can construct some classes from others.

Let A and B be classes. The class $\{x \mid x \in A \vee x \in B\}$ is called the (*binary*) *union of the classes A and B* and is denoted by $A \cup B$. The class $\{x \mid x \in A \wedge x \in B\}$ is called the (*binary*) *intersection of the classes A and B* and is denoted by $A \cap B$.

Lemma 1.
1) $A \cup A = A, A \cap A = A$;
2) $A \cup B = B \cup A, A \cap B = B \cap A$ (the *commutativity of union and intersection*);
3) $(A \cup B) \cup C = A \cup (B \cup C), (A \cap B) \cap C = A \cap (B \cap C)$ (the *associativity of union and intersection*);
4) $A \cup (B \cap C) = (A \cup B) \cap (A \cup C), A \cap (B \cup C) = (A \cap B) \cup (A \cap C)$ (the *distributivity of union with respect to intersection and intersection with respect to union*).

The proof of these equalities follows directly from the definitions and axiom scheme AS2.

The class $\{x \mid x \notin A\}$ is called the *complement of the class A* and is denoted by A^c.

Lemma 2.
1) $(A^c)^c = A$;
2) $(A \cup B)^c = A^c \cap B^c$;
3) $(A \cap B)^c = A^c \cup B^c$.

The proof of these equalities also follows directly from the definitions.

The class $B \cap A^c$ is called the *complement of the class A in the class B* or the *difference of the classes B and A* and is denoted by $B \setminus A$.

The class $\{x \mid x \neq x\}$ is called the *empty* (≡ *void, vacuous*) *class* and is denoted by $\varnothing$.

Classes A and B are called *disjoint* if $A \cap B = \varnothing$.

The class $\{x \mid x = x\}$ is called *universal* and is denoted by $\mathfrak{U}$.

Lemma 3. *The following conclusions are equivalent:*
1) *A is a set;*
2) $A \in \mathfrak{U}$.

Proof. 1) $\vdash$ 2). Let A be a set. Take the formula $x = x$. Then, by axiom scheme AS2, the equality $A = A$ implies $A \in \mathfrak{U}$.

2) $\vdash$ 1). This follows from axiom scheme AS2. □

The following assertions are used very often without the exact indication.

Lemma 4.
1) $\varnothing \subset A, \varnothing \cup A = A, \varnothing \cap A = \varnothing$, *and* $\varnothing^c = \mathfrak{U}$;
2) $A \subset \mathfrak{U}, A \cup \mathfrak{U} = \mathfrak{U}, A \cap \mathfrak{U} = A$, *and* $\mathfrak{U}^c = \varnothing$;
3) *If* $A \subset B$ *and* $B \subset C$, *then* $A \subset C$;
4) $A \subset B$ *is equivalent to* $A \cup B = B$;
5) $A \subset B$ *is equivalent to* $A \cap B = A$.

Proof. We shall check only the inclusion $\varnothing \subset A$. Let $X \in \varnothing$. It was mentioned above that $X \notin \varnothing$. Therefore, by LAS1 (1.1.4) $(X \notin \varnothing) \Rightarrow ((X \notin A) \Rightarrow (X \notin \varnothing))$. By rule of implication D3 (1.1.3), we get $(X \notin A) \Rightarrow \neg(X \in \varnothing)$. By LAS9 (1.1.4), we get $((X \notin A) \Rightarrow \neg(X \in \varnothing)) \Rightarrow ((X \in \varnothing) \Rightarrow \neg\neg(X \in A))$. Again by rule of implication D3 (1.1.3), we get $(X \in \varnothing) \Rightarrow \neg\neg(X \in A)$. By the same rule, this implication and the formula $X \in \varnothing$ imply $\neg\neg(X \in A)$. By LAS10 (1.1.4), $\neg\neg(X \in A) \Rightarrow (X \in A)$. Finally, again by the same rule, we conclude that $X \in A$.

All the other assertions follow directly from the corresponding definitions. □

Let A be a class. The class $\{x \mid x \subset A\}$ will be called the *complete* (*full*) *ensemble* (≡ *class of all parts*) *of the class* A and will be denoted by $\mathcal{P}(A)$.

Lemma 5.
1) $\mathcal{P}(\mathfrak{U}) = \mathfrak{U}$.
2) $X \in \mathcal{P}(A)$ *iff* X *is a set and* $X \subset A$.

Proof. 1. Let X be a class and $X \in \mathcal{P}(\mathfrak{U})$. Then, X is a set and so $X \in \mathfrak{U}$ by Lemma 3. Conversely, if $X \in \mathfrak{U}$, then X is a set and by Lemma 4 we have $X \subset \mathfrak{U}$. Therefore, $X \in \mathcal{P}(\mathfrak{U})$.

2. Consider $\varphi(x)$, such that $x \subset A$. Let $X \in \mathcal{P}(A)$. Then, X is a set and the condition $X \in \{x \mid \varphi(x)\}$ implies by axiom scheme AS2 that $X \subset A$. Conversely, let X be a set and $X \subset A$. Then, $\varphi(X)$ implies by AS2 that $X \in \{x \mid \varphi(x)\} = \mathcal{P}(A)$. □

Proposition 1. $A = \varnothing$ *iff* $\forall x(x \notin A)$.

Proof. Let $A = \varnothing$. Take any class x. Suppose that $x \in A$. Since $A = \varnothing$, we get $x \in \varnothing$. Then, by axiom scheme AS2 $x \neq x$. But this contradicts the obvious equality $x = x$. It follows from this contradiction that $x \notin A$. By rule of deduction D4 (1.1.3), we get $\forall x(x \notin A)$.

Conversely, let $\forall x(x \notin A)$. By Lemma 4, we get $\varnothing \subset A$. Let $y \in A$. Take any class z. By axiom scheme LAS11 (1.1.4), we have $(\forall x(x \notin A)) \Rightarrow (z \notin A)$. By rule of deduction D3 (1.1.3), $z \notin A$. Suppose that $y = z$. Then, by axiom A1 $(y = z) \Rightarrow ((y \in A) \Leftrightarrow (z \in A))$. By rule of deduction D3 (1.1.3), we conclude that $(y \in A) \Leftrightarrow (z \in A)$ and by the same reason, $z \in A$. However,this contradicts the formula $z \notin A$. From this contradiction, we infer that $y \neq z$. By rule of deduction D4 (1.1.3), we get $\forall z(y \neq z)$. Now, by axiom scheme LAS11 (1.1.4), we have $(\forall z(y \neq z)) \Rightarrow (y \neq y)$. Again, by rule of deduction D3 (1.1.3), $y \neq y$. Finally, by axiom scheme AS2, we get $y \in \varnothing$. This means that $A \subset \varnothing$. As a result, we get $A = \varnothing$. □

Remark. Axiom scheme AS2 is the single axiom scheme in our axiomatic family A1, AS2, and A3 – A8, which uses an arbitrary formula φ. That is why AS2 is very powerful creative axiom scheme. However, axiom scheme AS2 can be replaced by several weak explicit axioms, which are specific cases of AS2. The detailed exposition of this material is given in B.7.3.

1.1.6 First axioms of existence of sets

A3. (The *axiom of the full ensemble.*) For every set A, there exists a set P, such that for every class X the condition $X \subset A$ implies $X \in P$.

Lemma 1. *Let A be a set and $X \subset A$. Then, X is a set.*

Proof. By axiom A3, $X \subset A$ implies $X \in P$. Consequently, X is a set. □

Corollary 1. *If A is a set and φ is a formula as in AS2, then $\{x \in A \mid \varphi\}$ is a subset of the set A.*

Lemma 2. *Let A be a set. Then:*
1) *$\mathcal{P}(A)$ is a set;*
2) *$X \subset A$ iff $X \in \mathcal{P}(A)$.*

Proof. Consider the set P from axiom A3. Let $X \in \mathcal{P}(A)$. Then, X is a set. Therefore, by axiom scheme AS2 (1.1.5) $X \subset A$. Consequently, by axiom A3, $X \in P$. This means that $\mathcal{P}(A) \subset P$. Then, by Lemma 1 we get $\mathcal{P}(A)$ is a set.

Now, let $X \subset A$. By Lemma 1, we see that X is a set. By AS2, $X \in \mathcal{P}(A)$. □

According to this lemma, for a set A, the class $\mathcal{P}(A)$ may be called the *set of all parts of the set A* or the *set of all subsets of the set A*.

Proposition 1. *The class $\mathfrak{U}$ is not a set, i. e. it is a proper class.*

Proof. Consider the class $R \equiv \{x \mid x \notin x\}$. Suppose that R is a set. Then, by virtue of axiom scheme AS2 (1.1.5), $R \in R$ is equivalent to $R \notin R$, but this is impossible. It follows from this contradiction that our supposition is not true, and so R is not a set.

By Lemma 4 (1.1.5), $R \subset \mathfrak{U}$. Now, by Lemma 1, we conclude that $\mathfrak{U}$ is not a set. □

Let A be a class. The class $\{x \mid x = A\}$ is called the *solitary class of the class A* and is denoted by $\{A\}$.

Lemma 3. *Let A be a set. Then:*
1) $X \in \{A\}$ *iff* $X = A$;
2) $\{A\} \subset \mathcal{P}(A)$;
3) $\{A\}$ *is a set.*

Proof. 1. Let $X \in \{A\}$. Then, X is a set, and so by axiom scheme AS2 (1.1.5), $X = A$. Conversely, if $X = A$, then X is a set and again by AS2 $X \in \{A\}$.

2. If $X \in \{A\}$, then by Proof 1, $X = A$. Thus, X is a set and $X \subset A$. Now, by AS2 $X \in \mathcal{P}(A)$. As a result, $\{A\} \subset \mathcal{P}(A)$.

3. By Lemma 2, we see that $\mathcal{P}(A)$ is a set. Therefore, by Proof 2 and Lemma 1, we have that $\{A\}$ is a set as well. □

A4. (The *axiom of the binary union.*) If A and B are sets, then $A \cup B$ is a set as well.

Non-ordered and ordered pairs of classes

Let A and B be classes. The class $\{A\} \cup \{B\}$ is called the *non-ordered pair of the classes A and B* and is denoted by $\{A, B\}$.

Lemma 4. *Let A and B be sets. Then:*
1) $\{A, B\}$ *is a set;*
2) $X \in \{A, B\}$ *iff* $X = A$ *or* $X = B$.

Proof. 1. It follows from Lemma 3 and axiom A4 that $\{A, B\}$ is a set.

2. If $X \in \{A, B\}$, then X is a set and so by axiom scheme AS2 (1.1.5) $(X \in \{A\}) \vee (X \in \{B\})$. Hence, by Lemma 3 $(X = A) \vee (X = B)$. Conversely, if $(X = A) \vee (X = B)$, then again by Lemma 3 $(X \in \{A\}) \vee (X \in \{B\})$. In both cases, X is a set. Now, by axiom scheme AS2 $X \in \{A\} \cup \{B\} = \{A, B\}$. □

Let A and B be classes. The class $\{\{A\}, \{A, B\}\}$ is called the *coordinate* (≡ *ordered*) *pair* (≡ *pair of Kuratowski*) *of the classes A and B with the left member A and the right member B* and is denoted by $\langle A, B\rangle$. Along with the name "member", the name "coordinate" and others are used.

Lemma 5. *If A and B are sets, then* $\langle A, B\rangle$ *is a set.*

Proof. By Lemmas 3 and 4, we see that $\{A\}$ and $\{A, B\}$ are sets. Therefore, by axiom A4, $\langle A, B\rangle$ is a set as well. □

Proposition 2. *Let A, B, C, and D be sets and* $\langle A, B\rangle = \langle C, D\rangle$. *Then,* $A = C$ *and* $B = D$.

Proof. By Lemmas 3 and 4, we have that $\{A\}$, $\{B\}$, $\{C\}$, $\{D\}$, $\{A, B\}$, and $\{C, D\}$ are sets. From $\{C\} \in \langle A, B\rangle$ by virtue of axiom scheme AS2 (1.1.5), we get either the equality (I) $\{C\} = \{A\}$ or the equality (II) $\{C\} = \{A, B\}$. From $\{C, D\} \in \langle A, B\rangle$, we get either the equality (III) $\{C, D\} = \{A\}$ or the equality (IV) $\{C, D\} = \{A, B\}$. Equality (II) is fulfilled if and only if $A = C = B$. In this case, (III) and (IV) coincide and give $C = D = A$. Thus, $A = C = D = B$. In the case, when we have Equality (III), the arguments are the same.

Now, suppose that we have equality (I) or (IV). Then, $C = A$ and either $C = B$ or $D = B$. If $C = B$, then we get the case (II). If $D = B$, then $A = C$ and $B = D$. □

Corollary 1. *Let A and B be sets and* $\langle A, B\rangle = \langle B, A\rangle$. *Then,* $A = B$.

Let A and B be classes. The class $\{x \mid \exists a \exists b\ ((a \in A) \wedge (b \in B) \wedge (x = \langle a, b\rangle))\}$, consisting of all coordinate pairs $\langle a, b\rangle$, will be called the *coordinate* (≡ *direct, Cartesian*) *product of the classes A and B* and will be denoted by $A * B$.

Proposition 3.
1) $A * (B \cup C) = (A * B) \cup (A * C)$, $(A \cup B) * C = (A * C) \cup (B * C)$ (the *distributivity of union with respect to multiplication*);
2) $A * (B \cap C) = (A * B) \cap (A * C)$, $(A \cap B) * C = (A * C) \cap (B * C)$ (the *distributivity of intersection with respect to multiplication*);
3) $(A \cap C) * (B \cap D) = (A * B) \cap (C * D)$.

All of these equalities are checked by the direct application of the definitions.

Proposition 4. *Let A and B be sets. Then,* $A * B$ *is a set as well.*

Proof. Let $a \in A$ and $b \in B$. Then, $\{a\} \subset A$ and $\{b\} \subset B$. Consequently, $\{a\} \subset A \cup B$, $\{b\} \subset A \cup B$, and $\{a, b\} \subset A \cup B$. By axiom A4, $A \cup B$ is a set. Therefore, by Lemma 2, we get $\{a\} \in \mathcal{P}(A \cup B)$ and $\{a, b\} \in \mathcal{P}(A \cup B)$. By Lemma 3, we see that $\{a\}$ is a set. By Lemma 4, we see that $\{a, b\}$ is also a set. Therefore, by virtue of Lemma 4, we get $\langle a, b\rangle = \{\{a\}, \{a, b\}\} \subset \mathcal{P}(A \cup B)$. By Lemma 2, we have that $\mathcal{P}(A \cup B)$ is a set and $\langle a, b\rangle \in \mathcal{P}(\mathcal{P}(A \cup B))$. Again, by Lemma 2, we obtain that $\mathcal{P}(\mathcal{P}(A \cup B))$ is a set. As a result, $A * B$ is contained in the set $\mathcal{P}(\mathcal{P}(A \cup B))$. By Lemma 1, we have that $A * B$ is a set. □

1.1.7 Correspondences

Let A and B be classes. A subclass $u \subset A * B$ of the product $A * B$ is called a *correspondence* (≡ *multivalued mapping*) *from the class A into the class B* and is denoted by $u\colon A \longrightarrow\!\!\!\prec B$.

Let $u\colon A \longrightarrow\!\!\!\prec B$ be a correspondence. If $a \in A$, $b \in B$ and $\langle a, b\rangle \in u$, then we shall say that the *element b* is a *value of the correspondence u on* (or *at*) *the argument a*. The subclass $\{b \in B \mid \langle a, b\rangle \in u\}$ of the class B is called the *class of values of the correspondence u on* (or *at*) *the argument* $a \in A$ and is denoted by $u\langle a\rangle$. It is clear that the class $u\langle a\rangle$ may be empty for some elements a. Along with $\langle a, b\rangle \in u$, we shall write sometimes $b \in u\langle a\rangle$.

The class $\{a \in A \mid \exists b \in B\ (b \in u\langle a\rangle)\}$ is called the *class of assignment* (≡ *domain of definition*) *of the correspondence u* and is denoted by $\operatorname{dom} u$. The class $\{b \in B \mid \exists a \in A\ (b \in u\langle a\rangle)\}$ is called the *class of values of the correspondence u* and is denoted by $\operatorname{rng} u$.

The correspondence $u\colon A \longrightarrow\!\!\!\prec B$ will be called *total*, if $\operatorname{dom} u = A$, i. e. $u\langle a\rangle \neq \varnothing$ for every $a \in A$. The correspondence u will be called *single-valued*, if for every element $a \in \operatorname{dom} u$, the corresponding class $u\langle a\rangle$ contains a single element of the class B in that sense that $b, b' \in u\langle a\rangle$ implies $b = b'$. This single element $b \in u\langle a\rangle$ is called the *value of the correspondence u on* (or *at*) *the argument a* and is denoted by $u(a)$ or simply by ua.

The class $\{x \mid \exists u\,((u\colon A \longrightarrow\!\!\!\prec B) \wedge (x = u))\}$ of all correspondences from the class A into the class B, which are sets, will be denoted by $\operatorname{Cor}(A, B)$. Its subclass of all correspondences $u\colon A \longrightarrow\!\!\!\prec B$, such that for every $a \in A$ the class $u\langle a\rangle$ is a set will be denoted by $\operatorname{Cor}_s(A, B)$.

Let X be a subclass of the class A. The class $\{b \in B \mid \exists a \in X\ (b \in u\langle a\rangle)\}$ is called the *image of the subclass X of the class A under the correspondence u* and is denoted by $u[X]$. It is clear that $u[\{a\}] = u\langle a\rangle$ and $u[A] = \operatorname{rng} u$.

For X, consider the class $v \equiv \{\langle a, b\rangle \in u \mid a \in X\}$. The correspondence $v\colon X \longrightarrow\!\!\!\prec B$ is called the *restriction of the correspondence* $u\colon A \longrightarrow\!\!\!\prec B$ *on the subclass X* and is denoted by $u|X$ or $\operatorname{rest}_X u$. In this case, the correspondence u is called an *extension of the correspondence v*. For X, consider also class $w \equiv \{\langle a, b\rangle \in u \mid (a \in X) \wedge (b \in u[X])\}$. The correspondence $w\colon X \longrightarrow\!\!\!\prec u[X]$ is called the *strict restriction of the correspondence* $u\colon A \longrightarrow\!\!\!\prec B$ *on the subclass* and is denoted by $u\|X$.

The correspondence $u\colon A \longrightarrow\!\!\!\prec B$ is called *surjective* (≡ a *correspondence onto*, a *surjection*) if for every $b \in B$ there exists $a \in A$, such that $b \in u\langle a\rangle$. The correspondence u is called *injective* (≡ mutually *one-valued*, an *injection*), if from $b \in u\langle a\rangle$ and $b \in u\langle a'\rangle$ it follows that $a = a'$. The correspondence u is called *bijective* (≡ a *bijection*) if it is surjective and injective simultaneously.

Let A be a class. The particular correspondence $u\colon A \longrightarrow\!\!\!\prec A$, such that $u\langle a\rangle = \{a\}$ for every $a \in A$ is called the *identical correspondence from the class A into the class A* and is denoted by $\operatorname{Id}_A\colon A \longrightarrow\!\!\!\prec A$. It is called also the *diagonal of the product* $A * A$. If

X is a subclass of the class A, then the particular correspondence $\mathrm{Id}_A\,|X$ is called the *identical* (≡ *canonical*) *correspondence from the subclass X into the class A* and will be denoted by $\mathrm{Id}_{X,A}\colon X \longrightarrow\!\!\prec A$.

Let $u\colon A \longrightarrow\!\!\prec B$ be a correspondence. The correspondence $\{\langle b, a\rangle \in B * A \mid \langle a, b\rangle \in u\}$ from the class B into the class A is called *inverse to the correspondence u* and is denoted by $u^{-1}\colon B \longrightarrow\!\!\prec A$.

Let Y be a subclass of the class B. The subclass $u^{-1}[Y] \equiv \{a \in A \mid \exists b \in Y\ (b \in u\langle a\rangle)\}$ of the class A is called the *preimage of the subclass Y under the correspondence u*. It is clear that $u^{-1}[\{b\}] = u^{-1}\langle b\rangle$ and $u^{-1}[B] = \mathrm{dom}\, u$.

Let $u\colon A \longrightarrow\!\!\prec B$ and $v\colon B \longrightarrow\!\!\prec C$ be correspondences. The correspondence $\{\langle a, c\rangle \in A * C \mid \exists b \in B\ ((b \in u\langle a\rangle) \wedge (c \in v\langle b\rangle))\}$ from the class A into the class C is called the *composition* (≡ *product*) *of the correspondences u and v* and is denoted by $v \circ u$ or simply by vu.

Lemma 1. *Let $u\colon A \longrightarrow\!\!\prec B$ be a correspondence and X be a subclass of the class A. Then, $u|X = u \circ \mathrm{Id}_{X,A}$.*

Proof. It is clear that $u|X = \{\langle a, b\rangle \in u \mid a \in X\} = \{\langle x, b\rangle \in X * B \mid \exists a \in A\ ((a \in \mathrm{Id}_{X,A}\langle x\rangle) \wedge (b \in u\langle a\rangle))\} = u \circ \mathrm{Id}_{X,A}$. □

Proposition 1. *Let $u\colon A \longrightarrow\!\!\prec B$, $v\colon B \longrightarrow\!\!\prec C$ and $w\colon C \longrightarrow\!\!\prec D$ be correspondences. Then:*
1) $w \circ (v \circ u) = (w \circ v) \circ u$;
2) $(v \circ u)^{-1} = u^{-1} \circ v^{-1}$;
3) $(u^{-1})^{-1} = u$;
4) $u \circ \mathrm{Id}_A = \mathrm{Id}_B \circ u = u$.

All of these equalities are checked by the direct application of the definitions.

Lemma 2. *Let $u\colon A \longrightarrow\!\!\prec B$ and $v\colon A \longrightarrow\!\!\prec C$ be correspondences. Then, the following conclusions are equivalent:*
1) $u = v$;
2) $u\langle a\rangle = v\langle a\rangle$ *for every* $a \in A$.

Proof. 1) ⊢ 2). Let $b \in u\langle a\rangle$. Then, $\langle a, b\rangle \in u = v$ implies $b \in v\langle a\rangle$. Thus, $u\langle a\rangle \subset v\langle a\rangle$. Similarly, $v\langle a\rangle \subset u\langle a\rangle$.

2) ⊢ 1). Let $p \in u$. Then, there are $a \in A$ and $b \in B$, such that $p = \langle a, b\rangle$. From $b \in u\langle a\rangle = v\langle a\rangle$, we infer that $p = \langle a, b\rangle \in v$. Thus, $u \subset v$. Similarly, $v \subset u$. □

1.1.8 Mappings

Here we shall consider the most important kind of correspondences.

Let A and B be classes. A total single-valued correspondence $u\colon A \longrightarrow\!\!\!\!\prec B$ is called a *mapping* (≡ *simple correspondence, function, transformation, operator* and so on) *from the class A into the class B* and is denoted by $u\colon A \to B$.

It is clear that the properties from the definition of a mapping u are equivalent to the properties $\mathrm{Id}_A \subset u^{-1} \circ u$ and $u \circ u^{-1} \subset \mathrm{Id}_B$. The mapping u is denoted also by $a \in A \mapsto u(a) \in B$, $a \mapsto u(a)$ $(a \in A)$, $b = u(a)$ $(a \in A)$ and so on with a suitable modification and simplification. The class $\operatorname{rng} u$ of values of the mapping u has the following description: $\operatorname{rng} u = \{b \in B \mid \exists a \in A\ (b = u(a))\}$.

Remark. The class $\{\langle a, b\rangle \in A * B \mid b = u(a)\}$ is called sometimes the *graph of the mapping* $u\colon A \to B$. But it is clear that this class is equal to the class u itself. Therefore, there is no difference between a mapping and its graph.

The class $\{x \mid \exists u\ ((u : A \to B) \wedge (x = u))\}$ of all mappings from the class A into the class B which are sets will be denoted by $\mathrm{Map}(A, B)$. It is also called the *degree of the class B with the exponent A* and is denoted by B^A.

Proposition 1. *Let A and B be sets. Then,* $\mathrm{Map}(A, B)$ *and* $\mathrm{Cor}(A, B)$ *are sets as well.*

Proof. Let $u \in \mathrm{Map}(A, B)$. By the definition, $u \subset A * B$. By Proposition 4 (1.1.6), $A * B$ is a set. Therefore, by Lemma 2 (1.1.6) $u \in \mathcal{P}(A * B)$. This means that $\mathrm{Map}(A, B) \subset \mathcal{P}(A * B)$. Now, from Lemma 2 (1.1.6) and Lemma 1 (1.1.6), it follows that $\mathrm{Map}(A, B)$ is a set. For $\mathrm{Cor}(A, B)$, the proof is the same. □

Let $u\colon A \to B$ be a mapping. If X is a subclass of the class A, then the restriction $u|X\colon X \longrightarrow\!\!\!\!\prec B$ and the strict restriction $u\|X\colon X \longrightarrow\!\!\!\!\prec u[X]$ from 1.1.7 are also mappings. Therefore, they will be denoted by $u|X\colon X \to B$ and $u\|X\colon X \to u[X]$, respectively.

For the notation of an injective, surjective or bijective mapping $u\colon A \to B$ we shall write $u\colon A \rightarrowtail B$, $u\colon A \twoheadrightarrow B$ and $u\colon A \rightarrowtail\!\!\!\!\!\twoheadrightarrow B$, respectively. Classes A and B are called *equivalent* or *equipollent* $(A \sim B)$ if there exists some bijective mapping $u\colon A \rightarrowtail\!\!\!\!\!\twoheadrightarrow B$.

If A is a class, then the identical correspondence $\mathrm{Id}_A\colon A \longrightarrow\!\!\!\!\prec A$ from 1.1.7 is a bijective mapping, such that $\mathrm{Id}_A\langle a\rangle = \{a\}$ for every $a \in A$. Therefore, it will be called the *identical mapping from the class A onto the class A* and as the mapping will be denoted by $\mathrm{id}_A\colon A \rightarrowtail\!\!\!\!\!\twoheadrightarrow A$. Thus, by the definition, $\mathrm{id}_A(a) = a$ for every $a \in A$.

If X is a subclass of the class A, then the identical correspondence $\mathrm{Id}_{X,A}\colon X \longrightarrow\!\!\!\!\prec A$ from the subclass X into the class A from 1.1.7 is an injective mapping, such that $\mathrm{Id}_{X,A}\langle x\rangle = \{x\}$ for every $x \in X$. Therefore, as the mapping, it will be denoted by $\mathrm{id}_{X,A}\colon X \rightarrowtail A$. Thus, by the definition, $\mathrm{id}_{X,A}(x) = x$ for every $x \in X$.

Lemma 1. *Let A, B and C be classes and* $u\colon A \to B$ *and* $v\colon B \to C$ *be mappings. Then, the composition correspondence* $v \circ u$ *is a mapping as well.*

Proof. Denote $v \circ u$ by w. By Proposition 1 (1.1.7), we have $w^{-1} \circ w = u^{-1} \circ (v^{-1} \circ v) \circ u \supset u^{-1} \circ \mathrm{id}_B \circ u = u^{-1} \circ u \supset \mathrm{id}_A$ and $w \circ w^{-1} = v \circ (u \circ u^{-1}) \circ v^{-1} \subset v \circ \mathrm{id}_B \circ v^{-1} \subset v \circ v^{-1} \subset \mathrm{id}_C$. These inclusions means that w is a mapping. □

Lemma 2. *Let A and B be classes and $u\colon A \to B$ and $v\colon B \to A$ be mappings. Then,*
1) *if u is bijective, then $u^{-1} \circ u = \mathrm{id}_A$ and $u \circ u^{-1} = \mathrm{id}_B$;*
2) *if $v \circ u = \mathrm{id}_A$ and $u \circ v = \mathrm{id}_B$, then u and v are bijective, $v = u^{-1}$ and $u = v^{-1}$.*

Proof. 1. Since u is a mapping, we have $u^{-1} \circ u \supset \mathrm{id}_A$ and $u \circ u^{-1} \subset \mathrm{id}_B$. Besides, u^{-1} is a mapping as well. Moreover, by Proposition 1 (1.1.7), $(u^{-1})^{-1} = u$. Therefore, $u \circ u^{-1} = (u^{-1})^{-1} \circ u^{-1} \supset \mathrm{id}_B$ and $u^{-1} \circ u = u^{-1} \circ (u^{-1})^{-1} \subset \mathrm{id}_A$. From these inclusions, we get the necessary equalities.
2. Let $\langle b, a\rangle \in v$. Then, for $a \in A$ there is $c \in B$, such that $\langle a, c\rangle \in u$. Therefore, $\langle b, c\rangle \in u \circ v = \mathrm{id}_B$ implies $c = b$. Consequently, $\langle a, b\rangle \in u$ gives $\langle b, a\rangle \in u^{-1}$. Thus, $v \subset u^{-1}$.

Let $\langle b, a\rangle \in u^{-1}$, i. e. $\langle a, b\rangle \in u$. For $b \in B$, there is $d \in A$, such that $\langle b, d\rangle \in v$. Hence, $\langle a, d\rangle \in v \circ u = \mathrm{id}_A$ implies $d = a$. Thus, $\langle b, a\rangle \in v$ gives $u^{-1} \subset v$. Finally, $v = u^{-1}$.
It follows from here that u^{-1} is a mapping. Hence, u is bijective.
Analogously, $u = v^{-1}$ implies that v is bijective. □

Lemma 3. *Let A, B and C be classes, and $u\colon A \to B$ and $v\colon A \to C$ be mappings. Then, the following conditions are equivalent:*
1) *$u = v$;*
2) *$u(a) = v(a)$ for every $a \in A$.*

Proof. 1) $\vdash$ 2). From $\langle a, v(a)\rangle \in v = u$, we conclude that $v(a) \in u\langle a\rangle$. Since $u\langle a\rangle$ consists of the single element $u(a)$, we get $v(a) = u(a)$.

2) $\vdash$ 1). Let $p \in u$. Then, there is $a \in A$, such that $p = \langle a, u(a)\rangle$. Consequently, $p = \langle a, v(a)\rangle \in v$. Thus, $u \subset v$. Analogously, $v \subset u$. □

The following properties of mappings are used very often and practically without special references.

Lemma 4. *Let A and B be classes; X, X', and X'' be subclasses of the class A; Y, Y', and Y'' be subclasses of the class B and $u\colon A \longrightarrow\!\!\!\!\prec B$ be a correspondence. Then:*
(i) $u[X' \cup X''] = u[X'] \cup u[X'']$;
(ii) $u[X' \cap X''] \subset u[X'] \cap u[X'']$.

If besides the correspondence u is a mapping $u\colon A \to B$, then
(iii) $u[X' \setminus X''] \supset u[X'] \setminus u[X'']$;
(iv) $u^{-1}[Y' \cup Y''] = u^{-1}[Y'] \cup u^{-1}[Y'']$;

(v) $u^{-1}[Y' \cap Y''] = u^{-1}[Y'] \cap u^{-1}[Y'']$;
(vi) $u^{-1}[Y' \setminus Y''] = u^{-1}[Y'] \setminus u^{-1}[Y'']$;
(vii) $X \subset u^{-1}[u[X]]$;
(viii) $u[u^{-1}[Y]] = Y \cap u[A]$;
(ix) $u[X \cap u^{-1}[Y]] = u[X] \cap Y$.

If, moreover, the mapping u is injective, then in the assertions (ii) *and* (iii), *we have the equalities.*

Proof. The formulas (i) – (viii) are proved by the direct checking. Therefore, we shall prove only the formula (ix). Let $y \in u[X] \cap Y$. Then, there is $x \in X$, such that $y = ux \in Y$. Consequently, $x \in X \cap u^{-1}[Y]$ and $y \in u[X \cap u^{-1}[Y]]$. Conversely, according to the formulas (ii) and (viii), $u[X \cap u^{-1}[Y]] \subset u[X] \cap u[u^{-1}[Y]] = u[X] \cap Y$. □

The following statement will be used several times in the sequel.

Lemma 5. *Let A, B be classes and U be a class of mappings $u \subset A * B$, such that* $\operatorname{dom} u \subset A$, $\operatorname{rng} u \subset B$ *and either $u \subset v$ or $v \subset u$ for every $u, v \in U$. Let $w \equiv \{z \mid \exists u \in U\ (z \in u)\}$. Then:*
(i) *w is a mapping, such that* $\operatorname{dom} w \subset A$ *and* $\operatorname{rng} w \subset B$;
(ii) $\operatorname{dom} w = \{x \mid \exists u \in U\ (x \in \operatorname{dom} u)\}$;
(iii) $a \in \operatorname{dom} w$ *implies $w(a) = u(a)$ for each $u \in U$, such that* $a \in \operatorname{dom} u$, *so that* $w | \operatorname{dom} u = u$;
(iv) $\operatorname{rng} w = \{y \mid \exists u \in U\ (y \in \operatorname{rng} u)\}$.

Proof. Let $p \in w$. Then, $p \in u \subset A * B$ for some $u \in U$. Therefore, $w \subset A * B$. Let now $\langle a, b\rangle \in w$ and $\langle a, c\rangle \in w$. Then, there exist u and v in U, such that $\langle a, b\rangle \in u$ and $\langle a, c\rangle \in v$. We know that $u \subset v$ or $v \subset u$; say $u \subset v$. Then, $\langle a, b\rangle \in v$ and $(a, c) \in v$ imply $b = c$. Thus, w is a single-valued correspondence.

Let $a \in \operatorname{dom} w$. Then, $\langle a, b\rangle \in w$ for some $b \in B$. Thus, $\langle a, b\rangle \in u$ for some $u \in U$, where $a \in \operatorname{dom} u$. By axiom scheme AS2 (1.1.5), $a \in \{x \mid \exists u \in U\ (x \in \operatorname{dom} u)\}$.

Conversely, let $z \in \{x \mid \exists u \in U\ (x \in \operatorname{dom} u)\}$. Then, by AS2, $z \in \operatorname{dom} u$ for some $u \in U$. Thus, $\langle z, b\rangle \in u$ for some $b \in B$. Hence, $\langle z, b\rangle \in w$, i. e. $z \in \operatorname{dom} w$. As a result, we get Equality (ii). Consequently, $\operatorname{dom} u \subset \operatorname{dom} w$ for every $u \in U$.

Let $a \in \operatorname{dom} w$ and $a \in \operatorname{dom} u$. Then, $\langle a, u(a)\rangle \in u \subset w$ implies $w(a) = u(a)$ because w is single-valued. Let $a' \in \operatorname{dom} u$. Then, by Equality (ii), $a' \in \operatorname{dom} w$. By the proved property, $w(a') = u(a')$. This means that $w | \operatorname{dom} u = u$.

Let $b \in \operatorname{rng} w$. Then, $\langle a, b\rangle \in w$ for some $a \in A$. Thus, $\langle a, b\rangle \in u$ for some $u \in U$, where $b \in \operatorname{rng} u$. By AS2, $b \in \{y \mid \exists u \in U\ (y \in \operatorname{rng} u)\}$.

Conversely, let $z \in \{y \mid \exists u \in U\ (y \in \operatorname{rng} u)\}$. Then, by AS2 $z \in \operatorname{rng} u$ for some $u \in U$. Thus, $\langle a, z\rangle \in u$ for some $a \in A$. Hence, $\langle a, z\rangle \in w$, i. e. $z \in \operatorname{rng} w$. As a result, we get Equality (iv). □

Lemma 6. *Let A, B and C be classes. Then:*

1) $A * B \sim B * A$ *with respect to the bijection* $\langle a, b\rangle \mapsto \langle b, a\rangle$ (the *commutativity of the coordinate product*);
2) $(A * B) * C \sim A * (B * C)$ *with respect to the bijection* $\langle\langle a, b\rangle, c\rangle \mapsto \langle a, \langle b, c\rangle\rangle$ (the *associativity of the coordinate product*).

Proof. 1. Denote $A * B$ by P and $B * A$ by Q. Consider the correspondence $u \equiv \{\langle x, y\rangle \in P * Q \mid \exists\alpha \in A\, \exists b \in B\, ((x = \langle\alpha, \beta\rangle) \wedge (y = \langle\beta, \alpha\rangle))\} \subset P * Q$. It is clear that $\operatorname{dom} u = P$. Suppose that $\langle p, q\rangle \in u$ and $\langle p, r\rangle \in u$. Then, $p = \langle a, b\rangle$ for some $a \in A$ and $b \in B$. By axiom scheme AS2 (1.1.5), we get $q = \langle b, a\rangle$ and $r = \langle b, a\rangle$. Consequently, $q = r$. This means that the correspondence u is single-valued. Thus, u is a mapping.

Let $u(p) = u(p')$ for some $p = \langle a, b\rangle$ and $p' = \langle a', b'\rangle$. Then, by AS2 $\langle p, u(p)\rangle \in u$ and $\langle p', u(p')\rangle \in u$ imply that $u(p) = \langle b, a\rangle$ and $u(p') = \langle b', a'\rangle$. Therefore, by virtue of Proposition 2 (1.1.6), the equality $\langle b, a\rangle = \langle b', a'\rangle$ gives $b = b'$ and $a = a'$. In result $p = p'$. This means that u is injective.

Now, let $q = \langle b, a\rangle \in Q$. Take $p \equiv \langle a, b\rangle \in P$. By Lemma 5 (1.1.6), p and q are sets, where $\langle p, q\rangle$ is a set as well. Therefore, the formula $\exists\alpha \in A\, \exists b \in B\, ((p = \langle\alpha, \beta\rangle) \wedge (q = \langle\beta, \alpha\rangle))$ means by axiom scheme AS2 that $\langle p, q\rangle \in u$, i. e. $q = u(p)$. This means that u is surjective.

2. Denote $(A * B) * C$ by P and $A * (B * C)$ by Q. Consider the correspondence $u \equiv \{\langle x, y\rangle \in P * Q \mid \exists\alpha \in A\, \exists\beta \in B\, \exists\gamma \in C\ ((x = \langle\langle\alpha, \beta\rangle, \gamma\rangle) \wedge (y = \langle\alpha, \langle\beta, \gamma\rangle\rangle))\} \subset P * Q$. In the similar manner as above, it is checked that u is a bijective mapping from P onto Q. □

Finally, we shall introduce three important mappings which will be constantly used further.

Projections and derivative mapping

Let A and B be non-empty classes. Consider the correspondences $\mathrm{pr}_A \equiv \{\langle\langle a, b\rangle, c\rangle \in (A * B) * A \mid a = c\}$ and $\mathrm{pr}_B \equiv \{\langle\langle a, b\rangle, c\rangle \in (A * B) * B \mid b = c\}$.

Lemma 7. *The correspondences* pr_A *and* pr_B *are surjective mappings* $\mathrm{pr}_A\colon A * B \twoheadrightarrow A$ *and* $\mathrm{pr}_B\colon A * B \twoheadrightarrow B$, *such that* $\mathrm{pr}_A(a, b) = a$ *and* $\mathrm{pr}_B(a, b) = b$ *for every* $\langle a, b\rangle \in A * B$.

Proof. For each element $\langle a, b\rangle \in A * B$, we have $\langle\langle a, b\rangle, a\rangle \in \mathrm{pr}_A$. Thus, $\operatorname{dom} \mathrm{pr}_A = A * B$. Let $\langle\langle a, b\rangle, c\rangle \in \mathrm{pr}_A$ and $\langle\langle a, b\rangle, d\rangle \in \mathrm{pr}_A$. Then, $a = c$ and $a = d$ imply $c = d$. Hence, pr_A is single-valued. Thus, pr_A is a mapping from $A * B$ into A, such that $\mathrm{pr}_A(a, b) = a$.

Let $a_0 \in A$. Since B is non-empty, there is some element $b_0 \in B$. Now, from the equality $\mathrm{pr}_A(a_0, b_0) = a_0$ we conclude that the mapping pr_A is surjective.

For pr_B, the arguments are the same. □

The mappings $\mathrm{pr}_A\colon A * B \twoheadrightarrow A$ and $\mathrm{pr}_B\colon A * B \twoheadrightarrow B$ are called the *projections onto the factors A and B*, respectively.

Let A, B, A', and B' be classes and $u\colon A \to B$ and $u'\colon A' \to B'$ be mappings. Consider the correspondence $\pi \equiv \{\langle\langle a, a'\rangle, \langle b, b'\rangle\rangle \in (A * A') * (B * B') \mid (b = u(a)) \wedge (b' = u'(a'))\}$.

Lemma 8. *The correspondence π is a mapping $\pi\colon A*A' \to B*B'$, such that $\pi(\langle a, a'\rangle) = \langle u(a), u'(a')\rangle$ for every $\langle a, a'\rangle \in A * A'$.*

Proof. For each element, $\langle a, y\rangle \in A * A'$ we have $\langle\langle a, a'\rangle, \langle u(a), u'(a')\rangle\rangle \in \pi$. Thus, $\operatorname{dom}\pi = A * A'$. Let $\langle\langle a, a'\rangle, \langle b, b'\rangle\rangle \in \pi$ and $\langle\langle a, a'\rangle, \langle c, c'\rangle\rangle \in \pi$. Then, $b = u(a)$, $b' = u'(a')$, $c = u(a)$ and $c' = u'(a')$ imply $b = c$ and $b' = c'$, where $\langle b, b'\rangle = \langle c, c'\rangle$. Thus, π is single-valued. Thus, π is a mapping from $A * A'$ into $B * B'$, such that $\pi(a, a') = \langle u(a), u'(a')\rangle$. □

The mapping $\pi\colon A * A' \to B * B'$ will be called the *derivative mapping of the mappings $u\colon A \to B$ and $u'\colon A' \to B'$ with respect to the coordinate products $A * A'$ and $B * B'$* and will be denoted by $(u\colon A \to B) *_m (u'\colon A' \to B')$ or simply by $u *_m u'$.

Let A and B be classes and $u\colon A \to B$ be a mapping.

Define a mapping $\alpha\colon \mathcal{P}(A) \to \mathcal{P}(B)$, setting $\alpha(X) \equiv u[X]$ for every subset X of the class A. The mapping α will be called the *derivative mapping of the mapping $u\colon A \to B$ with respect to the ensembles $\mathcal{P}(A)$ and $\mathcal{P}(B)$* and will be denoted by $\mathcal{P}^m(u\colon A \to B)$ or simply $\mathcal{P}^m(u)$.

Also, for a class I, define a mapping $\delta\colon A^I \to B^I$ setting $\delta(f) \equiv u \circ f$ for every $f \in A^I$. The mapping δ will be called the *derivative mapping of the mapping $u\colon A \to B$ with respect to the degrees A^I and B^I* and will be denoted by $(u\colon A \to B)^I_m$ or simply $(u)^I_m$.

1.1.9 Multivalued and simple collections

Consider now the important parallel terminology for correspondences and mappings.

Let I be a fixed class. A correspondence $u\colon I \longrightarrow\!\!\!\!\prec A$ from the class I into a class A will be called also a (*multivalued*) *collection of subclasses and subsets of the class A, indexed by the class I*, and will be denoted also by $u \equiv \lbrack A_i \subset A \mid i \in I\rbrack\colon I \longrightarrow\!\!\!\!\prec A$, where $A_i \equiv u\langle i\rangle$ or in a shorter form by $u \equiv \lbrack A_i \subset A \mid i \in I\rbrack$ or by $\lbrack A_i \subset A \mid i \in I\rbrack$. The class A_i will be called the *component of the collection u with the index $i \in I$*. The class I is called the *class of indices of the collection u*. If for every $i \in I$ the class A_i is a set, then u will be called a (*multivalued*) *collection of subsets of the class A*.

If the correspondence u is total, i. e. $A_i \neq \varnothing$ for every $i \in I$, then the collection $u \equiv \lbrack A_i \subset A \mid i \in I\rbrack$ will be called *total* as well.

A collection $u \equiv [A_i \subset \mathfrak{U} \mid i \in I]$ will be called also a *collection of classes and sets* and will be denoted also by $u \equiv [A_i \mid i \in I]$. If for every $i \in I$ the class A_i is a set, then u will be called a *collection of sets*.

It is clear that any mapping $u\colon I \to A$ is a multivalued collection $u \equiv [A_i \subset A \mid i \in I]\colon I \mathrel{-\!\!\!<} A$, such that $A_i \equiv \{a_i\}$ and $a_i \equiv u(i)$. This multivalued collection will be called a (*simple*) *collection of elements of the class A* and will be denoted by $u \equiv (a_i \in A \mid i \in I)$ or in a shorter form by $u \equiv (a_i \in A \mid i \in I)$ or by $(a_i \in A \mid i \in I)$. Thus, the notion of the simple collection $u \equiv (a_i \in A \mid i \in I)$ is another form of the notion of the mapping $u\colon I \to A$. The set a_i will be called the *member* (≡ *coordinate*) *of the collection u with the index $i \in I$*. The class $I \equiv \operatorname{dom} u$ is called the *class of indices of the collection u*. The class $\operatorname{rng} u \equiv \{x \in \mathfrak{U} \mid \exists i \in I\,(x = a_i)\}$ will be called the *class of members of the collection u* and will be denoted by $\{a_i \mid i \in I\}$.

A simple collection $u \equiv (a_i \subset \mathfrak{U} \mid i \in I)$ will be called also a *simple collection of sets* and will be denoted also by $u \equiv (a_i \mid i \in I)$.

Let $u \equiv [A_i \subset A \mid i \in I]$ be a collection and X be a subclass of the class I. Then, the correspondence $v \equiv u|X\colon X \mathrel{-\!\!\!<} A$ is a collection $v \equiv [A_x \subset A \mid x \in X]$. It will be called the *restriction of the collection u on the subclass X of the class I*. And the collection u will be called an *extension of the collection v*. The same terminology is valid for simple collections $u \equiv (a_i \subset A \mid i \in I)$.

The identical correspondence $\mathrm{Id}_I\colon I \mathrel{-\!\!\!<} I$ from 1.1.7 is a multivalued collection, which will be called the *identical collection of single element subsets of the class I* and will be denoted also by $\mathrm{Id}_I \equiv [\{i\}_i \subset I \mid i \in I]$, where $\{i\}_i \equiv \{i\}$ for every $i \in I$.

Similarly, the identical mapping $\mathrm{id}_I\colon I \rightarrowtail\!\!\!\!\twoheadrightarrow I$ from 1.1.8 is a collection, which will be called also the *identical collection of elements of the class I* and will be denoted also by $\mathrm{id}_I \equiv (i_i \in I \mid i \in I)$, where $i_i \equiv i$ for every $i \in I$.

If X is a subclass of the class I, then the identical correspondence $\mathrm{Id}_{X,I}\colon X \mathrel{-\!\!\!<} I$ from 1.1.7 is a multivalued collection, which will be called the *identical collection of single element subsets of the subclass X of the class I* and will be denoted also by $\mathrm{Id}_{X,I} \equiv [\{x\}_x \subset I \mid x \in X]$, where $\{x\}_x \equiv \{x\}$ for every $x \in X$.

Similarly, the identical mapping $\mathrm{id}_{X,I}\colon X \rightarrowtail I$ from 1.1.8 is a collection, which will be called also the *identical collection of elements of the subclass X of the class I* and will be denoted also by $\mathrm{id}_{X,I} \equiv \{x_x \in I \mid x \in X\}$, where $x_x \equiv x$ for every $x \in X$.

Lemma 1. *Let $u \equiv [A_i \subset A \mid i \in I]$ and $v \equiv [B_i \subset B \mid i \in I]$ be multivalued collections. Then, the following conclusions are equivalent:*

1) $u = v$;
2) $A_i = B_i$ *for every* $i \in I$.

The same statement is valid for simple collections.

Proof. This statement is a particular case of Lemma 2 (1.1.7). □

Lemma 2. *Let $u \equiv [A_i \subset A \mid i \in I]\colon I \longrightarrow\!\!\!\!\prec A$ be a collection of subsets of a class A. Then, there exists a unique mapping $v\colon I \to \mathcal{P}(A)$, such that $v(i) = A_i$ for every $i \in I$.*

Proof. By the definition from 1.1.5, we see that $A_i \in \mathcal{P}(A)$ for every $i \in I$. Therefore, we can consider the correspondence $v \equiv \{\langle i, x\rangle \in I * \mathcal{P}(A) \mid x = A_i\}$. If $i \in I$, then $A_i = A_i$ means by axiom scheme AS2 (1.1.5) that $\langle i, A_i\rangle \in v$. Thus, $\operatorname{dom} v = I$. Suppose that $\langle i, p\rangle \in v$ and $\langle i, q\rangle \in v$. Then, by axiom scheme AS2 $p = A_i$ and $q = A_i$. Therefore, $p = q$. This means that the correspondence v is single-valued. Thus, v is a mapping and $v(i) = A_i$. The uniqueness of v follows from Lemma 3 (1.1.8). □

Corollary 1. *Let $u \equiv [A_i \subset A \mid i \in I]$ be a collection of sets of a class A. Then, there exists the unique simple collection $v = (A_i \in \mathcal{P}(A) \mid i \in I)$.*

Corollary 2. *Let $u \equiv [A_i \subset \mathfrak{U} \mid i \in I]$ be a collection of sets. Then, there exists the unique simple collection $v \equiv (A_i \subset \mathfrak{U} \mid i \in I)$.*

Proof. It follows from Corollary 1 and Lemma 5 (1.1.5). □

Consider the correspondence $\varphi \equiv \{\langle u, v\rangle \in \operatorname{Cor}(I, A) * \operatorname{Map}(I, \mathcal{P}(A)) \mid \forall i \in I\ (v(i) = u\langle i\rangle)\}$.

Lemma 3. *Let $I \neq \varnothing$. Then, φ is a bijective mapping $\varphi\colon \operatorname{Cor}(I, A) \rightarrowtail\!\!\!\!\twoheadrightarrow \operatorname{Map}(I, \mathcal{P}(A))$, such that $\varphi[A_i \subset A \mid i \in I] = (A_i \in \mathcal{P}(A) \mid i \in I)$ for every collection $[A_i \subset A \mid i \in I]$ of subsets of the class A.*

Proof. By virtue of Corollary 2 to Lemma 2, we get $\operatorname{dom} \varphi = \operatorname{Cor}(I, A)$. Let $\langle u, v\rangle \in \varphi$ and $\langle u, w\rangle \in \varphi$. Then, $v(i) = u\langle i\rangle$ and $w(i) = u\langle i\rangle$ imply $v(i) = w(i)$ for every $i \in I$. Therefore, by Lemma 3 (1.1.8) $v = w$. Thus, φ is single-valued. Thus, the correspondence φ is a mapping, such that $(\varphi u)(i) = u\langle i\rangle$ for every $i \in I$.

Suppose that $\varphi u = \varphi u'$ for some $u, u' \in \operatorname{Cor}(I, A)$. Then, by Lemma 3 (1.1.8), $(\varphi u)(i) = (\varphi u')(i)$ for every $i \in I$. Therefore, $u\langle i\rangle = u'\langle i\rangle$ for every $i \in I$ implies by Lemma 1 that $u = u'$. This means that φ is injective.

Let now $v \equiv (A_i \in \mathcal{P}(A) \mid i \in I) \in \operatorname{Map}(I, \mathcal{P}(A))$. Consider the class $u \equiv \{x \mid \exists i \in I\ (\exists a \in A_i\ (x = \langle i, a\rangle))\}$. If $y \in u$, then by axiom scheme AS2 (1.1.5) $\exists i \in I\ (\exists a \in A_i\ (y = \langle i, a\rangle))$. Then, $y = \langle i, a\rangle$ for some $i \in I$ and $a \in A_i$. Since A_i is a set, by axiom scheme AS2 $A_i \subset A$. Therefore, $y \in I * A$. This means that $u \subset I * A$. Thus, u is a multivalued collection.

Take any $l \in I$. If $d \in u\langle l\rangle$, then $p \equiv \langle l, d\rangle \in u$. Besides, by axiom scheme AS2 (1.1.5), $p = \langle m, e\rangle$ or some $m \in I$ and $e \in A_m$. From the equality $\langle l, d\rangle = \langle m, e\rangle$, we infer by virtue of Proposition 2 (1.1.6) that $m = l$ and $e = d$. As a result, $d \in A_l$, where $u\langle l\rangle \subset A_l$. Conversely, if $f \in A_l \subset A$, then we can take the element $\langle l, f\rangle \in I * A$. Using axiom schemes LAS12 (1.1.4) and AS2 (1.1.5), we deduce that $\langle l, f\rangle \in u$, i. e. $f \in u\langle l\rangle$, where $A_l \subset u\langle l\rangle$. Thus, we get the equality $u\langle l\rangle = A_l$ for every $l \in I$. This means that u is a multivalued collection $u \equiv (A_i \subset A \mid i \in I)$. By the definition of φ, we have $\varphi u = v$. Consequently, φ is surjective. □

Corollary 1. *Let I be a non-empty class. Then, for the universal class $\mathfrak{U}$ the correspondence φ is a bijective mapping $\varphi\colon \operatorname{Cor}(I, \mathfrak{U}) \rightarrowtail\!\!\!\twoheadrightarrow \operatorname{Map}(I, \mathfrak{U})$, such that $\varphi(A_i \mid i \in I) = (A_i \mid i \in I)$ for every collection of sets $(A_i \mid i \in I)$.*

It follows from Lemma 3 and Corollary 1 to it that working only with sets we could use only simple collections. But the necessity to have collections of classes for the future use forced us to introduce the new unaccustomed notion of a (*multivalued*) *collection*.

1.1.10 The union and intersection of a multivalued collection

Let $u \equiv (A_i \subset \mathfrak{U} \mid i \in I)$ be a (multivalued) collection of classes and sets, indexed by the (non-empty) class I.

The class $\operatorname{rng} u = \{x \mid \exists i\ (i \in I \wedge x \in A_i)\}$ is called the *union of the* (*multivalued*) *collection* u and is denoted by $\bigcup(A_i \mid i \in I)$. The class $\{x \mid \forall i\ ((i \in I) \Rightarrow (x \in A_i))\}$ is called the *intersection of the* (*multivalued*) *collection* u and is denoted by $\bigcap(A_i \mid i \in I)$. It is clear that $A_i \subset \bigcup(A_i \mid i \in I)$ and $\bigcap(A_i \mid i \in I) \subset A_i$ for every $i \in I$. Besides, $\bigcap(A_i \mid i \in I) \subset \bigcup(A_i \mid i \in I)$.

The collection u is called a *cover of a class D*, if $D = \bigcup(A_i \mid i \in I)$. The collection u is called *pairwise disjoint*, if $A_i \cap A_j = \varnothing$ for every element $i \neq j$ from I. The collection u is called a *partition* (≡ *dissection*) *of a class D*, if u is a pairwise disjoint cover of D.

Let $u \equiv (A_i \mid i \in I)$ be a cover of a class D. Any restriction $u|J \equiv (A_i \mid i \in J)$ of the collection u on the subclass $J \subset I$, such that $u|J$ is a cover of the class D as well is called a *subcover of the cover u*.

A cover $w \equiv (C_k \mid k \in K)$ of the class D is called a *refinement of the cover u* if for every $k \in K$ there is an index $i \in I$, such that $C_k \subset A_i$. It is clear that every subcover of the cover u is a refinement of u.

Lemma 1. *Let $u \equiv (A_i \mid i \in I)$ be a collection of subclasses of a class A. Then:*
1) $A \setminus \bigcup(A_i \mid i \in I) = \bigcap(A \setminus A_i \mid i \in I)$;
2) $A \setminus \bigcap(A_i \mid i \in I) = \bigcup(A \setminus A_i \mid i \in I)$.

Proof. 1. Let $x \in A \setminus \bigcup(A_i \mid i \in I)$. Since $A_i \subset \bigcup(A_i \mid i \in I)$, we get $x \notin A_i$, i. e. $x \in A \setminus A_i$ for every $i \in I$. Consequently, $x \in \bigcap(A \setminus A_i \mid i \in I)$. Conversely, let $x \in \bigcap(A \setminus A_i \mid i \in I)$. Then, $x \notin A_i$ for every $i \in I$, i. e. $x \notin \bigcup(A_i \mid i \in I)$. Therefore, $x \in A \setminus \bigcup(A_i \mid i \in I)$.

Conclusion 2 is checked in the similar way. □

Let $u \equiv (A_i \mid i \in I)$ be a (multivalued) collection. The class $\bigcup(A_i * \{i\} \mid i \in I)$ is called the *disjoint union of the* (*multivalued*) *collection* u and will be denoted by $\bigcup_d(A_i \mid i \in I)$.

For every $i \in I$, there is the canonical injection $\mathrm{inj}_i\colon\ A_i \rightarrowtail \bigcup_d(A_i \mid i \in I)$, such that $\mathrm{inj}_i(a) = \langle a, i\rangle$ for every $a \in A_i$.

Denote the class $\bigcup(A_i \mid i \in I)$ by P and the class $\bigcup_d(A_i \mid i \in I)$ by Q. Consider the correspondence $\alpha \equiv \{\langle x, y\rangle \mid (x \in P) \wedge (y \in Q) \wedge (\exists i \in I\,((x \in A_i) \wedge (y = \langle x, i\rangle)))\} \subset P * Q$.

Lemma 2. *Let $(A_i \mid i \in I)$ be a pairwise disjoint collection. Then, the correspondence α is a bijective mapping $\alpha\colon\ \bigcup(A_i \mid i \in I) \rightarrowtail\!\!\!\!\twoheadrightarrow \bigcup_d(A_i \mid i \in I)$, such that $\alpha(p) = \langle p, i\rangle$ for every $p \in A_i$ and $i \in I$.*

Proof. Let $p \in P$. Then, $p \in A_j$ for some index $j \in J$. Consider the element $q \equiv \langle p, j\rangle \in Q$. Then, we have the formula $(p \in A_j) \wedge (q = \langle p, j\rangle)$. By axiom scheme LAS12 (1.1.4), we have the formula $(p \in A_j) \wedge (q = \langle p, j\rangle) \Rightarrow \exists i \in I\,((p \in A_i) \wedge (q = \langle p, i\rangle))$. By condition 3 from 1.1.3, we get the formula $\exists i \in I\,((p \in A_i) \wedge (q = \langle p, i\rangle))$. Now, by axiom scheme AS2 (1.1.5), we conclude that $\langle p, q\rangle \in \alpha$. Thus, $\operatorname{dom}\alpha = P$.

Suppose that $\langle p, q\rangle \in \alpha$ and $\langle p, r\rangle \in \alpha$. Then, by axiom scheme AS2, $\exists i \in I\,(p \in A_i \wedge q = \langle p, i\rangle)$ and $\exists i \in I\,(p \in A_i \wedge r = \langle p, i\rangle)$. Therefore, $p \in A_k$ and $q = \langle p, k\rangle$ for some $k \in I$ and $p \in A_l$ and $r = \langle p, l\rangle$ for some $l \in I$. From $p \in A_k \cap A_l$, we conclude that $k = l$. As a result, $q = r$. This means that the correspondence α is single-valued. Thus, α is a mapping, such that $\alpha(p) = \langle p, i\rangle$ for every $p \in A_i$.

Let $\alpha(p) = \alpha(p')$ for some $p, p' \in P$. Then, $\langle p, \alpha(p)\rangle \in \alpha$ and $\langle p', \alpha(p')\rangle \in \alpha$ imply by AS2, that $p \in A_m$ and $\alpha(p) = \langle p, m\rangle$ for some $m \in I$ and $p' \in A_n$ and $\alpha(p') = \langle p', n\rangle$ for some $n \in I$. By virtue of Proposition 2 (1.1.6), we infer from these conditions that $p = p'$. This means that α is surjective.

Now, let $q \in Q$. Then, $q \in A_\varkappa * \{\varkappa\}$ for some index $\varkappa \in I$, i. e. $q = \langle p, \varkappa\rangle$ for some $p \in A_\varkappa \subset P$. By Lemma 5 (1.1.6), $\langle p, q\rangle$ is a set. As well as above, we deduce the formula $\exists i \in I\,((p \in A_i) \wedge (q = \langle p, i\rangle))$, and by axiom scheme AS2, we conclude that $\langle p, q\rangle \in \alpha$, i. e. $q = \alpha(p)$. This means that α is surjective. □

Proposition 1. *Let $(A_i \mid i \in I)$ be a collection of classes. Then:*
1) *if $u\colon\ K \twoheadrightarrow I$ is a surjective mapping, then $\bigcup(A_i \mid i \in I) = \bigcup(A_{u(k)} \mid k \in K)$ and $\bigcap(A_i \mid i \in I) = \bigcap(A_{u(k)} \mid k \in K)$ (the* general commutativity of union and intersection*);*
2) *if $I = \bigcup(I_m \mid m \in M)$ for some collection $(I_m \mid m \in M)$, then $\bigcup(A_i \mid i \in I) = \bigcup(\bigcup(A_i \mid i \in I_m) \mid m \in M)$ and $\bigcap(A_i \mid i \in I) = \bigcap(\bigcap(A_i \mid i \in I_m) \mid m \in M)$ (the* general associativity of union and intersection*).*

The proof of these equalities follows directly from the definitions.

Some useful special form of the general associativity will be derived in the end of 1.1.13 (see Proposition 1 (1.1.13) and its corollary).

Corollary 1. *Let $(A_i \mid i \in I)$ be a collection of classes. Then:*

1) *if* $u\colon K \rightarrowtail\!\!\!\twoheadrightarrow I$ *is a bijective mapping, then* $\bigcup_d(A_i \mid i \in I) = \bigcup_d(A_{u(k)} \mid k \in K)$;
2) *if a collection* $(I_m \mid m \in M)$ *is a partition of the class* I, *then there exists a bijection* $\beta\colon \bigcup_d(A_i \mid i \in I) \rightarrowtail\!\!\!\twoheadrightarrow \bigcup_d(\bigcup_d(A_i \mid i \in I_m) \mid m \in M)$, *such that* $\beta(\langle x, i\rangle) = \langle\langle x, i\rangle, m\rangle$ *for every* $m \in M$, $i \in I_m$ *and* $x \in A_i$.

Corollary 2. *Let* $(A_i \mid i \in I)$ *be a collection of classes,* $j \in I$ *and* $K \equiv I \setminus \{j\} \neq \varnothing$. *Then,* $\bigcup(A_i \mid i \in I) = A_j \cup \bigcup(A_i \mid i \in K)$ *and* $\bigcap(A_i \mid i \in I) = A_j \cap \bigcap(A_i \mid i \in K)$.

Corollary 3. *Let* $(A_i \mid i \in I)$ *be a collection of classes,* $I = \{j, k\}$ *and* $j \neq k$. *Then,* $\bigcup(A_i \mid i \in I) = A_j \cup A_k$ *and* $\bigcap(A_i \mid i \in I) = A_j \cap A_k$.

Lemma 3. *Let* $(A_i \subset A \mid i \in I)$ *and* $(B_j \subset B \mid j \in J)$ *be collections of corresponding subclasses, and* $u\colon A \longrightarrow\!\!\!\prec B$ *be a correspondence. Then:*
(i) $u[\bigcup(A_i \mid i \in I)] = \bigcup(u[A_i] \mid i \in I)$;
(ii) $u[\bigcap(A_i \mid i \in I)] \subset \bigcap(u[A_i] \mid i \in I)$.

If in addition the correspondence u *is a mapping* $u\colon A \to B$, *then*
(iii) $u^{-1}[\bigcup(B_j \mid j \in J)] = \bigcup(u^{-1}[B_j] \mid j \in J)$;
(iv) $u^{-1}[\bigcap(B_j \mid j \in J)] = \bigcap(u^{-1}[B_j] \mid j \in J)$.

If besides the mapping, u *is injective, then* (ii) *becomes an equality.*

All of these formulas are proved by the direct checking.

Let $(A_i \mid i \in I)$ and $(B_i \mid i \in I)$ be (multivalued) collections of classes and $(u_i \mid i \in I)$ be a (multivalued) collection of mappings $u_i\colon A_i \to B_i$. Define a mapping $\delta\colon \bigcup_d(A_i \mid i \in I) \to \bigcup_d(B_i \mid i \in I)$ setting $\delta(\langle a, i\rangle) \equiv \langle u_i(a), i\rangle$ for every $i \in I$ and $a \in A_i$. The mapping δ will be called a *derivative mapping of the (multivalued) collection of mappings* $(u_i\colon A_i \to B_i \mid i \in I)$ *with respect to the disjoint unions* $\bigcup_d(A_i \mid i \in I)$ *and* $\bigcup_d(B_i \mid i \in I)$ and will be denoted by $\bigcup_{dm}(u_i\colon A_i \to B_i \mid i \in I)$ or simply by $\bigcup_{dm}(u_i \mid i \in I)$.

Lemma 4.
1) *If all the mappings* u_i *are injective [surjective, bijective], then the mapping* $\bigcup_{dm}(u_i \mid i \in I)$ *is injective [surjective, bijective] as well.*
2) *If* $(u_i\colon A_i \to B_i \mid i \in I)$ *and* $(v_i\colon B_i \to C_i \mid i \in I)$ *are two collections of mappings, then* $\bigcup_{dm}(v_i \circ u_i \mid i \in I) = \bigcup_{dm}(v_i \mid i \in I) \circ \bigcup_{dm}(u_i \mid i \in I)$.

All of these conclusions are proved by the direct checking because of the pairwise disjointness of components of the collections $(A_i * \{i\} \mid i \in I)$, $(B_i * \{i\} \mid i \in I)$ and $(C_i * \{i\} \mid i \in I)$.

1.1.11 The other axioms of existence of sets

In this subsection, we shall introduce the other axioms of the theory of classes and sets except the axiom of choice which will be introduced in 1.1.12.

A5. (The *axiom of the general union.*) Let $[A_i \subset \mathfrak{U} \mid i \in I)$ be a multivalued collection of sets, indexed by the set I. Then, $\bigcup[A_i \mid i \in I)$ is a set as well.

Lemma 1.
1) *Let* $u \equiv [A_i \mid i \in I)$ *be a multivalued collection of sets, indexed by a set I. Then, u is a set.*
2) *Let* $u\colon A \to B$ *be a mapping from a set A into a class B. Then,* rng u *and u are sets.*

Proof. 1. By axiom A5, $A \equiv \bigcup[A_i \mid i \in I)$ is a set. Let $p \in u$. Then, by axiom scheme AS2 (1.1.5) $p = \langle i, x\rangle$ for some $i \in I$ and $x \in \mathfrak{U}$. From $\langle i, x\rangle \in u$, we conclude that $x \in u\langle i\rangle \equiv A_i$, where $x \in A$. By virtue of axiom scheme LAS12 (1.1.4) and the rule of deduction D3 (1.1.3), we get the formula $\exists i \in I\,(\exists x \in A\,(p = \langle i, x\rangle))$. Now, by axiom scheme AS2 (1.1.5) $p \in I * A$. As a result, $u \subset I * A$. By Proposition 4 (1.1.6), $I * A$ is a set. Now, by Lemma 1 (1.1.6), u is a set.

2. Let now $u\colon A \to B$ be a mapping from a set A into a class B. By the definition from 1.1.8, u is a multivalued collection $u \equiv [B_a \mid a \in A)$, such that $B_a \equiv u\langle a\rangle = \{u(a)\}$ for every $a \in A$. Since $u(a) \in B$, $u(a)$ is a set. Then, by Lemma 3 (1.1.6) B_a is a set as well. Now, by statement 1, u is a set. Besides, by axiom A5 and the definition from 1.1.10, rng $u \equiv \bigcup[B_a \mid a \in A)$ is a set. □

A6. (The *axiom of regularity* ($\equiv$ *foundation*).) Let A be a class and $A \neq \varnothing$. Then, there exists an element $a \in A$, such that $a \cap A = \varnothing$.

Lemma 2. $A \notin A$.

Proof. Suppose that $A \in A$. Then, A is a non-empty set. Besides A is a single element of the class $\{A\}$. By axiom A6, there is $\alpha \in \{A\}$, such that $\alpha \cap \{A\} = \varnothing$. Since $\alpha = A$, we get $A \cap \{A\} = \varnothing$ and $\alpha \in A$. As a result, $\alpha \in A \cap \{A\} = \varnothing$. It follows from this contradiction that $A \notin A$. □

Corollary 1. *Let A be a set. Then,* $A \neq \{A\}$.

Proof. Suppose that $A = \{A\}$. Then, $A \in A$. But this contradicts Lemma 2. □

Lemma 3. *If* $A \in B$, *then* $B \notin A$.

Proof. Suppose that $A \in B$ and $B \in A$. Then, A and B are sets. By Lemma 4 (1.1.6), $\{A, B\}$ is a set and $x \in \{A, B\}$ is equivalent to either $x = A$ or $x = B$. By axiom A6, there is $\alpha \in \{A, B\}$, such that $\alpha \cap \{A, B\} = \varnothing$. If $\alpha = A$, then $B \in \alpha$. As a result, $B \in \alpha \cap \{A, B\} = \varnothing$. If $\alpha = B$, then $A \in \alpha$. As a result, $A \in \alpha \cap \{A, B\} = \varnothing$. In both cases, we get the contradiction. Thus, $B \notin A$. □

A7. (The *axiom of infinity*.) There exists a set A, such that $\varnothing \in A$, and $a \in A$ implies $a \cup \{a\} \in A$.

It follows from the axiom that $\varnothing$ is a set. It will be denoted further also by 0. By Lemma 3 (1.1.6), $\{0\}$ is also a set. By axiom A4 (1.1.6), the class $0 \cup \{0\}$ is also a set. This set will be denoted by 1. Analogously, the class $1 \cup \{1\}$ is a set. It will be denoted by 2. In the similar manner we define the sets $3 \equiv 2 \cup \{2\}$, $4 \equiv 3 \cup \{3\}$ and so on.

Lemma 4.
1) $0 \subset 1 \subset 2 \subset 3 \subset 4$ *and so on.*
2) $0 \neq 1 \neq 2 \neq 3 \neq 4$ *and so on.*
3) $0 \in 1$; $0 \in 2$ *and* $1 \in 2$; $0 \in 3$, $1 \in 3$, *and* $2 \in 3$; $0 \in 4$, $1 \in 4$, $2 \in 4$, *and* $3 \in 4$; *and so on.*

Proof. 1. All of these inclusions follow from the definitions.

2. Suppose that $1 = 2 = 1 \cup \{1\}$. Then, $\{1\} \subset 1$. Since 1 is a set, we get by Lemma 3 (1.1.6) that $1 \in 1$. But this contradicts Lemma 2. Thus, $1 \neq 2$. The similar arguments are used for the other inequalities.

3. All the belongings follow from the definitions. □

Let A and A' be sets. Consider the correspondence $v\colon 2 \longrightarrow \mathfrak{U}$, such that $v \equiv \{\langle 0, A\rangle, \langle 1, A'\rangle\}$.

Proposition 1. *Let A and A' be sets. Then:*
1) $\operatorname{dom} v = 2$ *and* $\operatorname{rng} v = \{A, A'\}$;
2) v *is a collection* $(X_i \in \{A, A'\} \mid i \in 2)$, *such that* $X_0 = A$ *and* $X_1 = A'$.

Proof. 1. By Lemma 5 (1.1.6), $\langle 0, A\rangle$ and $\langle 1, A'\rangle$ are sets. Therefore, by Lemma 4 (1.1.6), v is a set and $x \in v$ iff either $x = \langle 0, A\rangle$ or $x = \langle 1, A'\rangle$. Consequently, $v = 2$, $A \in \operatorname{rng} v$ and $A' \in \operatorname{rng} v$. By virtue of Lemma 4 (1.1.6), $\{A, A'\} \subset \operatorname{rng} v$. Conversely, if $y \in \operatorname{rng} v$, then by axiom scheme AS2 (1.1.5) $\exists i \in 2\,(\langle i, y\rangle \in v)$. Hence, $\langle j, y\rangle \in v$ for some $j \in 2$. If $j = 0$, then $\langle 0, y\rangle \in v$. Thus, either $\langle 0, y\rangle = \langle 0, A\rangle$ or $\langle 0, y\rangle = \langle 1, A'\rangle$. In the first case by Proposition 2 (1.1.6), $y = A$. The second case is impossible in virtue of Lemma 4 (1.1.6) and Proposition 2 (1.1.6). As a result, $y = A \in \{A, A'\}$. If $j = 1$, then in the similar way $\langle 1, y\rangle \in v$ implies $y = A' \in \{A, A'\}$. This means that $\operatorname{rng} v \subset \{A, A'\}$. Thus, $\operatorname{rng} v = \{A, A'\}$.

2. Now, by the definition from 1.1.9, v is the collection $(v\langle i\rangle \subset \{A, A'\} \mid i \in 2)$. Suppose that $X, X' \in v\langle 0\rangle$. Then, $\langle 0, X\rangle \in v$ and $\langle 0, X'\rangle \in v$. Since $0 \neq 1$, we infer that $\langle 0, X\rangle = \langle 0, A\rangle$ and $\langle 0, X'\rangle = \langle 0, A\rangle$. By virtue of Proposition 2 (1.1.6), $X = A = X'$. This means that the subclass $v\langle 0\rangle \subset \{A, A'\}$ consists of the single element $X_0 = A$. By the same arguments, we check that the subclass $v\langle 1\rangle$ consists of the single element $X_1 = A'$. Therefore, by the definition from 1.1.9 $v = (X_i \in \{A, A'\} \mid i \in 2)$, where $X_0 = A$ and $X_1 = A'$. □

Sequential pairs, triplets, and suits of sets

This collection $v \equiv (X_i \in \{A, A'\} \mid i \in 2)$, such that $v(0) \equiv X_0 \equiv A$ and $v(1) \equiv X_1 \equiv A'$ will be called the (*simple*) *sequential pair of the sets A and A′ with the zero member A and the first member A′* and will be denoted by (A, A'). Along with the name "member" the name "coordinate" and others will be used.

Let $A, A', A'', A''', \ldots$ be sets. The collection $(X_i \in \mathfrak{U} \mid i \in 3)$, such that $X_0 \equiv A$, $X_1 \equiv A'$ and $X_2 \equiv A''$ will be called the (*simple*) *sequential triplet of the sets A, A′, and A″* and will be denoted by (A, A', A''). The collection $(X_i \in \mathfrak{U} \mid i \in 4)$, such that $X_0 \equiv A$, $X_1 \equiv A'$, $X_2 \equiv A''$ and $X_3 \equiv A'''$ will be called the (*simple*) *sequential quadruplet of the sets A, A′, A″ and A‴* and will be denoted by (A, A', A'', A'''), and so on. The sets (A, A'), (A, A', A''), (A, A', A'', A'''),... will be called (*simple*) *sequential suits of sets*.

The sequential pair (A, A') has the best properties of the coordinate pair $\langle A, A' \rangle$.

Lemma 5. *Let A, A', B and B' be sets. If $(A, A') = (B, B')$, then $A = B$ and $A' = B'$. The similar properties are valid for corresponding triplets, quadruplets, and so on.*

Proof. By the definition, $(A, A') = (X_i \in \{A, A'\} \mid i \in 2)$, where $X_0 \equiv A$ and $X_1 \equiv A'$, and $(B, B') = (Y_i \in \{B, B'\} \mid i \in 2)$, where $Y_0 \equiv B$ and $Y_1 \equiv B'$. By virtue of Lemma 1 (1.1.9), $\{X_i\} = \{Y_i\}$ for every $i \in 2$. Since X_i and Y_i are sets, we get $X_i = Y_i$ for every i. Therefore, $A \equiv X_0 = Y_0 \equiv B$ and $A' \equiv X_1 = Y_1 \equiv B'$. □

(Multivalued) sequential pairs, triplets, and suits of classes

Now, we shall introduce another pair suitable also for classes. Let A and A' be classes. Consider the correspondence $u \equiv \{x \mid \exists i \in 2\ (\exists s \in \mathfrak{U}\ ((x = \langle i, s \rangle) \wedge ((i = 0) \Rightarrow (s \in A)) \wedge ((i = 1) \Rightarrow (s \in A'))))\}$.

Proposition 2.

1) $\operatorname{dom} u = 2$ *and* $\operatorname{rng} u = A \cup A'$;
2) *u is a collection* $[X_i \subset A \cup A' \mid i \in 2]$, *such that* $X_0 = A$ *and* $X_1 = A'$.

Proof. All the assertions follow from the equalities $u = \{\langle i, s \rangle \in 2 * \mathfrak{U} \mid ((i = 0) \Rightarrow (s \in A)) \wedge ((i = 1) \Rightarrow (s \in A'))\} = \{\langle 0, a \rangle \mid a \in A\} \cup \{\langle 1, a' \rangle \mid a' \in A'\}$. □

This multivalued collection $u \equiv [X_i \subset A \cup A' \mid i \in 2]$, such that $u\langle 0 \rangle = X_0 = A$ and $u\langle 1 \rangle = X_1 = A'$ will be called the (*multivalued*) *sequential pair of the classes A and A′ with the zero component A and the first component A′* and will be denoted by $[A, A')$.

Let $A, A', A'', A''', \ldots$ be classes. The (multivalued) collection $[X_i \subset \mathfrak{U} \mid i \in 3]$, such that $X_0 \equiv A$, $X_1 \equiv A'$ and $X_2 \equiv A''$ will be called the (*multivalued*) *sequential triplet of the classes A, A′ and A″* and will be denoted by $[A, A', A'')$. The (multivalued) collection $[X_i \subset \mathfrak{U} \mid i \in 4]$, such that $X_0 \equiv A$, $X_1 \equiv A'$, $X_2 \equiv A''$ and $X_3 \equiv A'''$

will be called the *(multivalued) sequential quadruplet of the classes* A, A', A'' *and* A''' and will be denoted by (A, A', A'', A'''), and so on. The classes (A, A'), (A, A', A''), (A, A', A'', A'''),... will be called *(multivalued) sequential suits of classes and sets.*

Lemma 6. *If* $(A, A') = (B, B')$, *then* $A = B$ *and* $A' = B'$. *The similar properties are valid for corresponding triplets, quadruplets, and so on.*

Proof. By the definition, $(A, A') = (X_i \subset A \cup A' \mid i \in 2)$, where $X_0 = A$ and $X_1 = A'$, and $(B, B') = (Y_i \subset B \cup B' \mid i \in 2)$, where $Y_0 = B$ and $Y_1 = B'$. By virtue of Lemma 1 (1.1.9), $X_i = Y_i$ for every $i \in 2$. Therefore, $A \equiv X_0 = Y_0 \equiv B$ and $A' \equiv X_1 = Y_1 \equiv B'$. □

Since (A, A') is a multivalued collection, we can consider its union and intersection. The classes $\bigcup(A, A') \equiv \bigcup(X_i \subset A \cup A' \mid i \in 2)$ and $\bigcap(A, A') \equiv \bigcap(X_i \subset A \cup A' \mid i \in 2)$ will be called the *union* and the *intersection of the (multivalued) sequential pair* (A, A'), respectively. Since 2 is a set, then for every set A and A' by virtue of axiom A6, the class $\bigcup(A, A')$ is a set. This implies by Lemma 1 (1.1.6) that $\bigcap(A, A')$ is a set as well.

Lemma 7. $\bigcup(A, A') = A \cup A'$ *and* $\bigcap(A, A') = A \cap A'$.

Proof. 1. By the definition from 1.1.10 $\bigcup(A, A') \equiv \{x \mid \exists i \in 2\,(x \in X_i)\}$. Let $p \in \bigcup(A, A')$. Then, by axiom scheme AS2 (1.1.5), $\exists i \in 2\ (p \in X_i)$. Therefore, for some $j \in 2$ we have $p \in X_j$. If $j = 0$, then $p \in X_0 = A$. If $j = 1$, then similarly, $p \in X_1 = A'$. In both cases, $p \in A \cup A'$. This means that $\bigcup(A, A') \subset A \cup A'$.

Conversely, let $q \in A \cup A'$. Then, by the definition from 1.1.5, $q \in A \vee q \in A'$. Therefore, $q \in X_0 \vee q \in X_1$, i. e. $(k \in 2) \wedge (q \in X_k)$. Now, by axiom scheme LAS12 (1.1.4) and rule of deduction D3 (1.1.3), we get the formula $\exists i\ ((i \in 2) \wedge (q \in X_i))$. Consequently, by virtue of axiom scheme AS2 (1.1.5), we infer that $q \in \bigcup(A, A')$. This means that $A \cup A' \subset \bigcup(A, A')$.

2. Analogously, by the definition from 1.1.10, $\bigcap(A, A') \equiv \{x \mid \forall i\ ((i \in 2) \Rightarrow (x \in X_i))\}$. Let $p \in \bigcap(A, A')$. Then, $\forall i \in 2\,(p \in X_i)$ implies $p \in X_0 \cap X_1 = A \cap A'$. As a result, $\bigcap(A, A') \subset A \cap A'$.s

Conversely, let $q \in A \cap A'$. Then, by the definition from 1.1.5, $q \in A \wedge q \in A'$. Therefore, $q \in X_0 \wedge q \in X_1$. Consequently, we have the formula $(i \in 2) \Rightarrow (q \in X_i)$. Now, by rule of deduction D4 (1.1.3), we get the formula $\forall i\ ((i \in 2) \Rightarrow (q \in X_i))$. By axiom scheme AS2 (1.1.5), this implies $q \in \bigcap(A, A')$. As a result, $A \cap A' \subset \bigcap(A, A')$. □

Let $A, A', A'', A''', \ldots$ be classes. The classes $\bigcup(A, A', A'')$, $\bigcup(A, A', A'', A''')$,... and the classes $\bigcap(A, A', A'')$, $\bigcap(A, A', A'', A''')$,... will be called the *unions*, and correspondingly, the *intersections of the (multivalued) sequential triplet* (A, A', A''), *quadruplet* (A, A', A'', A'''),... and will be denoted by $A \cup A' \cup A''$, $A \cup A' \cup A'' \cup A'''$,... and correspondingly, by $A \cap A' \cap A''$, $A \cap A' \cap A'' \cap A'''$,....

Let $A, A', A'', A''', \ldots$ be sets. The set $\mathrm{rng}(A, A', A'')$ (see 1.1.9 and axiom A5) of members of the sequential triplet (A, A', A'') will be called the *non-ordered triplet of* *the sets A, A' and A''* and will be denoted by $\{A, A', A''\}$. The set $\operatorname{rng}(A, A', A'', A''')$ of members of the single-valued sequential quadruplet (A, A', A'', A''') will be called the *non-ordered quadruplet of the sets A, A', A'' and A'''* and will be denoted by $\{A, A', A'', A'''\}$, and so on. The sets $\{A, A'\}$, $\{A, A', A''\}$, $\{A, A', A'', A'''\}$,... will be called *non-ordered suits of sets.*

Lemma 8. *Let $X, A, A', A'', A''', \ldots$ be sets. Then:*
1) $X \in \{A, A', A''\}$ *iff* $X = A \vee X = A' \vee X = A''$;
2) $X \in \{A, A', A'', A'''\}$ *iff* $X = A \vee X = A' \vee X = A'' \vee X = A'''$;
3) *and so on.*

Proof. By the definition, $\{A, A', A''\} = \{x \mid \exists i\,(i \in 3 \wedge x = X_i)\}$, where $X_0 \equiv A$, $X_1 \equiv A'$ and $X_2 \equiv A''$. If $X \in \{A, A', A''\}$, then by axiom scheme AS2 (1.1.5) $\exists i\,(i \in 3 \wedge X = X_i)$. Consider the formula $i \in 3 \wedge X = X_i$. If $i = 0$, then $X = X_0 = A$; if $i = 1$, then $X = X_1 = A'$; if $i = 2$, then $X = X_2 = A''$. Thus, $X = A \vee X = A' \vee X = A''$.

Conversely, let $X = A \vee X = A' \vee X = A''$. This implies $((X = X_0) \wedge (0 \in 3)) \vee ((X = X_1) \wedge (1 \in 3)) \vee ((X = X_2) \wedge (2 \in 3))$, where $(k \in 3) \wedge (X = X_k)$. Now, by axiom scheme LAS12 (1.1.4) and by rule of deduction D3 (1.1.3), we get the formula $\exists i\,((i \in 3) \wedge (X = X_i))$. By axiom scheme AS2, $X \in \{A, A', A''\}$.

For other cases, the arguments are the same. □

Remark. Lemma 5 shows that, in fact, the pairs of sets $\langle A, A'\rangle$ and (A, A') are similar objects. But for suits of sets with several (three, four,...) members, the situation becomes completely different. The sequential suits (A, A', A''), (A, A', A'', A'''),... are easily defined. Moreover, generalization of the coordinate pair $\langle A, A'\rangle$ requires the use of the angular brackets (for example in the form $\langle\langle A, A'\rangle, A''\rangle$, $\langle\langle\langle A, A'\rangle, A''\rangle, A'''\rangle$,...). But for the bracket form, it is very difficult to formulate the property of associativity of the products $(A * A') * A''$, $((A * A') * A'') * A'''$,.... That is why we were forced to introduce the unaccustomed notion of (*simple*) *sequential suits of sets* (A, A'), (A, A', A''), (A, A', A'', A'''),....

Moreover, the necessity to have suits of classes forced us to introduce the other unaccustomed notion of (*multivalued*) *sequential suits of classes* $[A, A']$, $[A, A', A'']$, $[A, A', A'', A''']$.... In particular, we use them for the definition of the *sequential products of classes* $A * A'$, $A * A' * A''$, $A * A' * A'' * A'''$,... in 1.1.12.

Let $A, A', A'', \ldots$ be classes. Then, $[A, A']$, $[A, A', A'']$,... are the corresponding multivalued collections. The classes $\bigcup_d [A, A']$, $\bigcup_d [A, A', A'']$,... will be called the *disjoint unions of the* (*multivalued*) *sequential pair* $[A, A']$, *triplet* $[A, A', A'']$,... and will be denoted by $A \cup_d A'$, $A \cup_d A' \cup_d A''$,.... By the definitions of the disjoint union, the multivalued sequential pair and the union we have $A \cup_d A' = \bigcup_d [X_i \mid i \in 2] = \bigcup [X_i * \{i\} \mid i \in 2] = (A * \{0\}) \cup (A' * \{1\})$. In the similar manner it is checked that $A \cup_d A' \cup_d A'' = (A * \{0\}) \cup (A' * \{1\}) \cup (A'' * \{2\})$, and so on.

Besides, let $B, B', B'', \ldots$ be classes and $u\colon A \to B$, $u'\colon A' \to B'$, $u''\colon A'' \to B'', \ldots$ be corresponding mappings. Then, (B, B'), $(B, B', B''), \ldots$ and (u, u'), $(u, u', u''), \ldots$ are the corresponding multivalued collections. The derivative mapping $\bigcup_{dm} (u, u')\colon A \cup_d A' \to B \cup_d B'$ from 1.1.10 of the multivalued sequential pair (u, u') with respect to the disjoint unions $A \cup_d A'$ and $B \cup_d B'$ will be denoted by $u \cup_{dm} u'$. The derivative mapping $\bigcup_{dm}(u, u', u'')\colon A \cup_d A' \cup_d A'' \to B \cup_d B' \cup_d B''$ from 1.1.10 of the multivalued sequential triplet (u, u', u'') with respect to the disjoint unions $A \cup_d A' \cup_d A''$ and $B \cup_d B' \cup_d B''$ will be denoted by $u \cup_{dm} u' \cup_{dm} u''$, and so on. It is clear that $(u \cup_{dm} u')(\langle a, 0\rangle) = \langle u(a), 0\rangle$ and $(u \cup_{dm} u')(\langle a', 1\rangle) = \langle u'(a'), 1\rangle$ for every $a \in A$ and $a' \in A'$. Similarly, $(u \cup_{dm} u' \cup_{dm} u'')(\langle a, 0\rangle) = \langle u(a), 0\rangle$, $(u \cup_{dm} u' \cup_{dm} u'')(\langle a', 1\rangle) = \langle u'(a'), 1\rangle$ and $(u \cup_{dm} u' \cup_{dm} u'')(\langle a'', 2\rangle) = \langle u''(a''), 2\rangle$ for every $a \in A$, $a' \in A'$ and $a'' \in A''$, and so on.

Remark. The axiom A5 is equivalent to the conjunction of the two following weaker axioms A5′ and A5″ presented in [*Kelley*, 1975].

A5′. (The *axiom of values*.) If $u\colon A \to B$ is a mapping from a set A into a class B, then the class $\operatorname{rng} u$ of values of the mapping u is a set.

A5″. (The *axiom of the union*.) If A is a set and $(a_a \subset \mathfrak{U} \mid a \in A) = \varphi^{-1}(\mathrm{id}_A)$ is the multivalued collection from Corollary 1 to Lemma 3 (1.1.9), such that $a_a \equiv a$ for every $a \in A$, then $\bigcup(a_a \mid a \in A)$ is a set.

It is clear that A5′ and A5″ are the particular cases of A5. Now, we shall prove that A5 is deduced from A5′ and A5″.

Let $u \equiv (A_i \subset \mathfrak{U} \mid i \in I)$ be a multivalued collection of sets, indexed by a set I. Consider the class $X \equiv \bigcup(A_i \mid i \in I)$. Take the collection $v \equiv \varphi(u) = (A_i \in \mathfrak{U} \mid i \in I)$ from Corollary 1 to Lemma 3 (1.1.9). By axiom A5′, $R \equiv \operatorname{rng} v \equiv \{A_i \mid i \in I\}$ is a set, and by axiom A5″, $Y \equiv \bigcup(r_r \mid r \in R)$ is a set as well.

If $x \in X$, then $x \in A_i$ for some $i \in I$. Since $r \equiv A_i \in R$, we get $x \in r_r$, where $x \in Y$. As a result, $X \subset Y$. Conversely, if $y \in Y$, then $y \in r_r = r$ for some $r \in R$. But $r = A_i$ for some $i \in I$. Therefore, $y \in A_i$ implies $x \in X$. As a result, $Y \subset X$. Thus, $X = Y$.

By Lemma 3 (1.1.5), $Y \in \mathfrak{U}$. Now, using axiom A1 (1.1.5) and rule of deduction D3 (1.1.3), we deduce that $X \in \mathfrak{U}$. Thus, X is a set.

1.1.12 The product of a multivalued collection. The axiom of choice

Let $u = (A_i \subset \mathfrak{U} \mid i \in I)$ be a (multivalued) collection of classes, indexed by a (nonempty) class I. Consider the class $A \equiv \bigcup(A_i \mid i \in I)$.

The class $\{e \equiv (a_i \in A \mid i \in I) \in \operatorname{Map}(I, A) \mid \forall i \in I\, (a_i \in A_i)\}$, consisting of all simple collections $e = (a_i \in A \mid i \in I)$ of elements of the class A, such that $a_i \in A_i$ for every $i \in I$, is called the (*indexed, direct or Cartesian*) *product of the multivalued collection u* and is denoted by $\prod(A_i \mid i \in I)$. The class A_i is called the *factor of the product* $\prod(A_i \mid i \in I)$ *with the index i*.

From the preceding axioms, we can not conclude that if a collection $(A_i \mid i \in I)$ is total and $I \neq \varnothing$, then $\prod(A_i \mid i \in I) \neq \varnothing$. To prove such a property, it is necessary to introduce another axiom.

Lemma 1. *Let $(A_i \mid i \in I)$ be a (multivalued) collection of sets indexed by a set I. Then, $\prod(A_i \mid i \in I)$ is also a set.*

Proof. Denote $\bigcup(A_i \mid i \in I)$ by A. By the definition, $\prod(A_i \mid i \in I) \subset \mathrm{Map}(I, A)$. By axiom A5 (1.1.11), A is a set. Therefore, by Proposition 1 (1.1.8), $\mathrm{Map}(I, A)$ is a set as well. Now, by virtue of Lemma 1 (1.1.6), $\prod(A_i \mid i \in I)$ is a set. □

A8. (The *axiom of choice.*) Let A be a non-empty set. Then, there exists a mapping $p\colon \mathcal{P}(A) \setminus \{\varnothing\} \to A$, such that $p(P) \in P$ for every non-empty subset P of the set A.

Any such a mapping is called a *choice mapping for the set A.*

Theorem 1 (Russell). *The following statements are equivalent:*
1) *the axiom of choice;*
2) *if $(A_i \mid i \in I)$ is a total (multivalued) collection of sets, indexed by a non-empty set I, then the set $\prod(A_i \mid i \in I)$ is non-empty.*

Proof. 1) $\vdash$ 2). Consider the class $A \equiv \bigcup(A_i \mid i \in I)$. By axiom A5 (1.1.11), A is a set. By axiom A8, there is a choice mapping, $p : \mathcal{P}(A) \setminus \{\varnothing\} \to A$. Since $A_i \in \mathcal{P}(A) \setminus \{\varnothing\}$ for every $i \in I$, we can consider the correspondence $e \equiv \{\langle i, a\rangle \in I * \mathfrak{U} \mid a = p(A_i)\}$. It is clear that $\mathrm{dom}\, e = I$. Let $\langle i, a\rangle \in e$ and $\langle i, b\rangle \in e$. Then, $a = p(A_i)$ and $b = p(A_i)$ imply $a = b$. Thus, e is single-valued. Thus, e is a mapping $e\colon I \to \mathfrak{U}$, such that $e(i) = p(A_i) \in A_i$. Therefore, $e \in \prod(A_i \mid i \in I)$.

2) $\vdash$ 1). Let A be a non-empty set. By Lemma 2 (1.1.6), $\mathcal{P}(A)$ is also a set. By Proposition 1 (1.1.5), there exists a set a, such that $a \in A$. By Lemma 3 (1.1.6), $\{a\} \subset A$. From the equality $a = a$ by axiom scheme AS2 (1.1.5), we get $a \in \{a\}$. By Proposition 1 (1.1.5), this implies $\{a\} \neq \varnothing$. By Lemma 3 (1.1.6), $\{a\}$ is a set. Suppose that $\{a\} \in \{\varnothing\}$. Then, by axiom scheme AS2 (1.1.5), $\{a\} = \varnothing$. It follows from this contradiction that $\{a\} \notin \{\varnothing\}$. By Lemma 2 (1.1.6), $\{a\} \in \mathcal{P}(A)$. As a result, $\{a\} \in \mathcal{P}(A) \setminus \{\varnothing\}$. Again by Proposition 1 (1.1.5) this means that $I \equiv \mathcal{P}(A) \setminus \{\varnothing\} \neq \varnothing$. Consider the identical collection $\mathrm{id}_I \equiv (i_i \in I \mid i \in I)$ from 1.1.9. According to Lemma 3 (1.1.9), we can take the multivalued collection $(i_i \subset A \mid i \in I) = \varphi^{-1}(\mathrm{id}_I)$. By the condition, $\prod(i_i \mid i \in I) \neq \varnothing$. By virtue of Proposition 1 (1.1.5), there exists a mapping $p\colon I \to \bigcup(i_i \mid i \in I)$, such that $p \in \prod(i_i \mid i \in I)$. If $a \in A$, then it was checked above that $a \in \{a\} \in I$. Therefore, $A \subset \bigcup(i_i \mid i \in I)$. Conversely, if $b \in \bigcup(i_i \mid i \in I)$, then $b \in j$ for some $j \in I$. By Lemma 5 (1.1.5), $j \subset A$. Thus, $b \in A$. This means that $A = \bigcup(i_i \mid i \in I)$. As a result, p is a mapping from I into A, such that $p(i) \in i_i = i$. Consequently, p is a choice mapping for the set A. □

Now, we shall connect the product of a collection, introduced here, and the binary coordinate product, introduced in 1.1.6.

Theorem 2. *Let $(A_i \mid i \in I)$ be a collection of sets, indexed by a set I, $j \in I$ and $K \equiv I \setminus \{j\} \neq \varnothing$. Then, there exist bijective mapping $\beta\colon \prod(A_i \mid i \in I) \rightarrowtail\!\!\!\twoheadrightarrow A_j * \prod(A_i \mid i \in K)$ and $\gamma\colon \prod(A_i \mid i \in I) \rightarrowtail\!\!\!\twoheadrightarrow (\prod(A_i \mid i \in K)) * A_j$, such that $\beta(a_i \mid i \in I) = \langle a_j, (a_i \mid i \in K)\rangle$ and $\gamma(a_i \mid i \in I) = \langle (a_i \mid i \in K), a_j\rangle$ for every collection $(a_i \mid i \in I) \in \prod(A_i \mid i \in I)$.*

Proof. Denote $\prod(A_i \mid i \in I)$ by E and $\prod(A_i \mid i \in K)$ by D. Consider the correspondence $\beta \equiv \{\langle e, \langle a, d\rangle\rangle \in E * (A_j * D) \mid (a = e(j)) \wedge (d = e|K)\}$. It is clear that $\operatorname{dom}\beta = E$. Suppose that $\langle e, \langle a, d\rangle\rangle \in \beta$ and $\langle e, \langle a', d'\rangle\rangle \in \beta$. Then, $a = e(j)$, $d = e|K$, $a' = e(j)$ and $d' = e|K$. Hence, $a = a'$ and $d = d'$. As a result, $\langle a, d\rangle = \langle a', d'\rangle$. This means that β is single-valued. Thus, β is a mapping from E into $A_j * D$, such that $\beta(e) = \langle e(j), e|K\rangle$.

Let $\beta(e) = \beta(f)$ for some $e, f \in E$. Then, the equality $\langle e(j), e|K\rangle = \langle f(j), f|K\rangle$ implies by Proposition 2 (1.1.6) that $e(j) = f(j)$ and $e|K = f|K$. By Lemma 3 (1.1.8), $e(k) = f(k)$ for every $k \in K$. Therefore, $e(i) = f(i)$ for every $i \in I$. Again, by this Lemma 3, $e = f$. This means that β is injective.

Denote $\bigcup(A_i \mid i \in I)$ by A and $\bigcup(A_k \mid k \in K)$ by C. Let $\langle b, d\rangle \in A_j * D$. By the definition, $d \in \operatorname{Map}(K, C)$ and $d(k) \in A_k$. Consider the correspondence $e \equiv \{\langle i, x\rangle \in I * A \mid ((i \in K) \Rightarrow (x = d(i))) \wedge ((i = j) \Rightarrow x = b)\}$. It is clear that $\operatorname{dom} e = I$. Let $\langle i, x\rangle \in e$ and $\langle i, y\rangle \in e$. If $i \in K$, then $x = d(i)$ and $y = d(i)$ imply $x = y = d(i) \in A_i \subset A$. If $i = j$, then $x = b$ and $y = b$ imply $x = y = b \in A_j \subset A$. This means that e is single-valued. Hence, $e \in \operatorname{Map}(I, A)$ and $e(i) \in A_i$ for every $i \in I$, i. e. $e \in E$. From $e|K = d$ and $e(j) = b$, it follows that $\beta(e) = \langle b, d\rangle$. This means that β is surjective.

For γ, the arguments are the same. □

Corollary 1. *Let $(A_i \mid i \in I)$ be a collection of sets, such that $I = \{j, k\}$ and $j \neq k$. Then, there exist bijective mappings $\beta'\colon \prod(A_i \mid i \in I) \rightarrowtail\!\!\!\twoheadrightarrow A_j * A_k$ and $\gamma'\colon \prod(A_i \mid i \in I) \rightarrowtail\!\!\!\twoheadrightarrow A_k * A_j$, such that $\beta'(a_i \mid i \in I) = \langle a_j, a_k\rangle$ and $\gamma'(a_i \mid i \in I) = \langle a_k, a_j\rangle$ for every collection $(a_i \mid i \in I) \in \prod(A_i \mid i \in I)$.*

Proof. Consider the mappings $u\colon A_j * \prod(A_i \mid i \in K) \to A_j * A_k$ and $v\colon (\prod(A_i \mid i \in K)) * A_j \to A_k * A_j$, such that $u(\langle a, (a_i \mid i \in K)\rangle) = \langle a, a_k\rangle$ and $v(\langle (a_i \mid i \in K), a\rangle) = \langle a_k, a\rangle$, where $K \equiv I \setminus \{j\} = \{k\}$. It is easy to check that u and v are bijective. Then, $\beta' \equiv u \circ \beta$ and $\gamma' \equiv v \circ \gamma$ are necessary bijections. □

Product of multivalued pairs, triplets, ...

Let A and A' be classes. Consider the multivalued sequential pair (A, A') from 1.1.11. Since (A, A') is a collection, we can consider its product. The class $\prod(A, A') \equiv \prod(X_i \subset A \cup A' \mid i \in 2)$ will be called the (*sequential*) *product of the multivalued sequential pair* (A, A') and will be denoted by $A \times A'$.

For classes $A, A', A'', A''', \dots$, the classes $\prod(A, A', A'')$, $\prod(A, A', A'', A''')$,... will be called the *(sequential) products of the multivalued sequential triplet* (A, A', A''), *multivalued sequential quadruplet* (A, A', A'', A'''),... and will be denoted by $A \times A' \times A''$, $A \times A' \times A'' \times A'''$,....

Let A be a class. Consider the sets $2, 3, 4, \dots$ according to 1.1.8 $A^2, A^3, A^4, \dots$, denote the degrees of the class A with the exponents $2, 3, 4, \dots$, respectively.

Consider also the sequential products $A \times A$, $A \times A \times A$, $A \times A \times A \times A$,.... It is easy to check that $A^2 = A \times A$, $A^3 = A \times A \times A$, $A^4 = A \times A \times A \times A$,....

The binary products $A * A'$ and $A \times A'$ are different objects. But there is the following similarity between them.

Lemma 2. *Let $A, A', A'', A''', \dots$ be classes. Then, $A \times A' = \{(a, a') \mid a \in A \wedge a' \in A'\}$, $A \times A' \times A'' = \{(a, a', a'') \mid a \in A \wedge a' \in A' \wedge a'' \in A''\}$, $A \times A' \times A'' \times A''' = \{(a, a', a'', a''') \mid a \in A \wedge a' \in A' \wedge a'' \in A'' \wedge a''' \in A'''\}$,....*

Proof. By the definition from 1.1.11, $(A, A') \equiv (X_i \subset A \cup A' \mid i \in 2)$, where $X_0 = A$ and $X_1 = A'$. Denote $A \times A'$ by P and $\{(a, a') \mid a \in A \wedge a' \in A'\}$ by Q. By Lemma 7 (1.1.11), $\bigcup(A, A') = A \cup A'$. Therefore, by the definition, $P = \{p \in \text{Map}(2, A \cup A') \mid \forall i \in 2\,(p(i) \in X_i)\}$. Let $p \in P$. Then, $p(0) \in X_0 = A$ and $p(1) \in X_1 = A'$. Denote $p(0)$ by a and $p(1)$ by a'. Since $\operatorname{rng} p = \{a, a'\} \subset A \cup A'$, we conclude that $p \in \text{Map}(I, \{a, a'\})$, i. e. $p = (x_i \in \{a, a'\} \mid i \in 2)$. Besides, $p(0) = a$ and $p(1) = a'$ means that $p = (a, a') \in Q$. Thus, $P \subset Q$.

Conversely, let $q \equiv (b, b') \in Q$. By the definition, $q = (y_i \in \{b, b'\} \mid i \in 2)$, where $y_0 \equiv b \in A \equiv X_0$ and $y_1 \equiv b' \in A' \equiv X_1$. Since $\{b, b'\} \subset A \cup A'$, we conclude that $q = (y_i \in A \cup A' \mid i \in 2) \in \text{Map}(2, A \cup A')$ and $q(i) = y_i \in X_i$ for every $i \in 2$. Thus, $q \in P$. Thus, $Q \subset P$.

As a result, $P = Q$. For the other cases, the arguments are the same. □

The equality $A \times A' = \{(a, a') \mid a \in A \wedge a' \in A'\}$ is resembling the equality $A * A' = \{\langle a, a' \rangle \mid a \in A \wedge a' \in A'\}$ from 1.1.6.

Lemma 3. *Let A and A' be non-empty sets. Then, there exists a bijective mapping $\beta' : A \times A' \rightarrowtail\!\!\!\twoheadrightarrow A * A'$, such that $\beta'(a, a') = \langle a, a' \rangle$ for every $(a, a') \in A \times A'$.*

Proof. According to 1.1.11 $(A, A') \equiv (X_i \subset A \cup A' \mid i \in 2)$, where $X_0 \equiv A$ and $X_1 \equiv A'$. Therefore, $X_0 * X_1 = A * A'$. By virtue of Lemma 4 (1.1.6) and Lemma 4 (1.1.11), $2 = \{0, 1\}$. By Corollary 1 to Theorem 2, there is a bijective mapping $\beta' : \prod(X_i \mid i \in 2) \rightarrowtail\!\!\!\twoheadrightarrow X_0 * X_1$, such that $\beta'(x_i \mid i \in 2) = \langle x_0, x_1 \rangle$ for every collection $(x_i \mid i \in 2) \in \prod(X_i \mid i \in 2)$. It follows from above that β' is a bijection $\beta' : A \times A' \rightarrowtail\!\!\!\twoheadrightarrow A * A'$. If $(a, a') \in A \times A'$, then by the definition from 1.1.11 $(a, a') \equiv (x_i \in \{a, a'\} \mid i \in 2)$, where $x_0 = a$ and $x_1 = a'$. Therefore, $\beta'(a, a') = \beta'(x_i \mid i \in 2) = \langle x_0, x_1 \rangle = \langle a, a' \rangle$. □

Remark. Lemmas 2 and 3 show that, in fact, the binary products $A * A'$ and $A \times A'$ are similar objects. But for the products with several (three, four,...) factors, the situation becomes completely different. The sequential products $A \times A' \times A''$, $A \times A' \times A'' \times A'''$,... are easily defined. However, generalization of the binary coordinate product $A * A'$ requires the use of the round brackets (for example in the form of $(A * A') * A''$, $((A * A') * A'') * A'''$,...). But for the bracket form, it is very difficult to formulate the property of associativity of the product. As to the indexed product $\prod(A_i \mid i \in I)$ (in particular, the sequential products $A \times A' \times A''$, $A \times A' \times A'' \times A'''$,...), the property of associativity can be formulated and prove in large generality (see Theorem 3 below).

By this reason, we shall use further as a rule the sequential products $A \times A'$, $A \times A' \times A''$, $A \times A' \times A'' \times A'''$,.... The single case, where we are forced to use the binary coordinate product $X * Y$, is the assignment of correspondences, mappings, and collections in the form $u \subset X * Y$.

Correspondences of several arguments (variables)

Let A, B be classes and $u\colon A \longrightarrow\!\!\!\!< B$ be a correspondence. Let $X, X', X'', \ldots$ be classes. If $A \subset X \times X'$, then u will be called also a *correspondence of the two arguments* (≡ *variables*) x *and* x'. If $A \subset X \times X' \times X''$, then u will be called also a *correspondence of the three arguments* (≡ *variables*) x, x', *and* x'', and so on. Let $(X_i \mid i \in I)$ be a multivalued collection of classes, indexed by a class I. If $A \subset \prod(X_i \mid i \in I)$, then u will be called a *correspondence of the arguments* x_i *for* $i \in I$. The similar terminology is used for mappings $u\colon A \to B$.

Thus, for correspondences (and in particular mappings) of several arguments we have, by the definition, the following inclusions: $u \subset (X \times X') * B$, $u \subset (X \times X' \times X'') * B$, and so on.

Similarly, let $u \equiv (A_i \subset \mathfrak{U} \mid i \in I)$ be a (multivalued) collection indexed by a class I. Let $X, X', X'', \ldots$ be classes. If $I = X \times X'$, then u will be called also a (*multivalued*) *collection, indexed by elements of the classes* X *and* X'. If $I = X \times X' \times X''$, then u will be called also a (*multivalued*) *collection, indexed by elements of the classes* X, X', *and* X'', and so on. In these cases, along with the notations $u \equiv (A_{(x,x')} \mid (x, x') \in X \times X')$, $u \equiv (A_{(x,x',x'')} \mid (x, x', x'') \in X \times X' \times X'')$,..., we shall use also the notations $u \equiv (A_{xx'} \mid x \in X,\ x' \in X')$, $u \equiv (A_{xx'x''} \mid x \in X,\ x' \in X',\ x'' \in X'')$,.... The similar terminology and notations are used also for simple collections $u \equiv (A_i \mid i \in I)$ of sets A_i.

Thus, for collections of several arguments, we have by the definition the following inclusions: $u \subset (X \times X') * \mathfrak{U}$, $u \subset (X \times X' \times X'') * \mathfrak{U}$, and so on.

Commutativity and associativity of the product

Theorem 3. *Let* $(A_i \mid i \in I)$ *be a total collection of sets, indexed by non-empty set* I. *Then:*

1) *if* $u: K \rightarrowtail\!\!\!\twoheadrightarrow I$ *is a bijective mapping, then there exists a bijection* β: $\prod(A_i \mid i \in I) \rightarrowtail\!\!\!\twoheadrightarrow \prod(A_{u(k)} \mid k \in K)$, *such that* $\beta e = e \circ u$, *i. e.* $\beta(a_i \mid i \in I) = (a_{u(k)} \mid k \in K)$ *for every* $e = (a_i \mid i \in I) \in \prod(A_i \mid i \in I)$ (the *commutativity of the product*);
2) *if a collection* $(I_m \subset I \mid m \in M)$ *is a partition of the set* I *for some set* $M \neq \varnothing$ (see 1.1.10), *then there exists a bijection* β: $\prod(A_i \mid i \in I) \rightarrowtail\!\!\!\twoheadrightarrow \prod(\prod(A_i \mid i \in I) \mid m \in M)$, *such that* $\beta(a_i \mid i \in I) = ((a_i \mid i \in I_m) \mid m \in M)$ *for every* $(a_i \mid i \in I) \in \prod(A_i \mid i \in I)$ (the *associativity of the product*).

Proof. 1. Denote $A_{u(k)}$ by B_k. It is evident that $e \circ u \in \prod(B_k \mid k \in K)$ for every $e \in \prod(A_i \mid i \in I)$. Consequently, β is a correctly defined mapping. Let $\beta e = \beta f$ for some $e, f \in \prod(A_i \mid i \in I)$. Then, for every $k \in K$ we get $e(u(k)) = f(u(k))$. Since u is a bijection, we conclude by virtue of Lemma 3 (1.1.8) that $e = f$. Thus, β is injective.

Let $h \in \prod(B_k \mid k \in K)$. Consider the mapping $e \equiv h \circ u^{-1}: I \to A$. If $i \in I$ and $k \equiv u^{-1}(i)$, then $e(i) = h(k) \in B_k = A_{u(k)} = A_i$. Consequently, $e \in \prod(A_i \mid i \in I)$. As a result, $\beta(e) = e \circ u = h \circ u^{-1} \circ u = h$. This means that β is surjective.

2. Denote $\prod(A_i \mid i \in I)$ by E and $\prod(A_i \mid i \in I_m)$ by B_m. By Theorem 1, these sets are non-empty. Consider the multivalued collection $\varkappa \equiv (B_m \mid m \in M)$. Again, by Theorem 1, the set $F \equiv \prod(B_m \mid m \in M)$ is non-empty. Let $e \equiv (a_i \mid i \in I) \in E$. Then, $e_m \equiv e|I_m = (a_i \mid i \in I_m) \in B_m$. Therefore, $f \equiv (e_m \mid m \in M) \in F$. Consequently, β is correctly defined mapping.

Let $\beta(e) = \beta(e')$. Then, $e|I_m = e'|I_m$ for every $m \in M$ implies $e(i) = e'(i)$ for every $i \in I_m$. Thus, $e = e'$. Thus, β is injective.

Let $f \in F$. Then, $f(m) \in B_m$. Therefore, $f(m)(i) \in A_i$ for every $i \in I_m$. Since $(I_m \mid m \in M)$ is a partition of I, we can correctly define a collection $e \equiv (a_i \mid i \in I) \in E$, setting $a_i \equiv f(m)(i)$ for every $i \in I_m$. Then, $e|I_m = f(m)$ for every $m \in M$. Hence, $\beta(e) = (e|I_m \mid m \in M) = (f(m) \mid m \in M) = f$. This means that β is surjective. □

Projections into the subproduct and into the factor

Let $(A_i \mid i \in I)$ be a multivalued collection of classes, indexed by a class I. Let J be a non-empty subclass of I. Then, the mapping $(a_i \mid i \in I) \mapsto (a_i \mid i \in J)$ from $\prod(A_i \mid i \in I)$ into $\prod(A_i \mid i \in J)$ is called the *projection into the subproduct* $\prod(A_i \mid i \in J)$ and is denoted by p_J. Similarly, the mapping $(a_i \mid i \in I) \mapsto a_j$ from $\prod(A_i \mid i \in I)$ into A_j is called the *projection into the factor* A_j and is denoted by pr_{A_j} or simply by pr_j.

Lemma 4. *Let* $(A_i \mid i \in I)$ *be a total collection of sets indexed by a non-empty set* I. *Then, for every index* $j \in I$ *and for every non-empty subset* $J \subset I$ *the projections* p_J *and* pr_j *are surjective.*

Proof. If $J = I$, then the assertion is evident. Therefore, suppose that $K \equiv I \setminus J \neq \varnothing$. Let $f \in \prod(A_j \mid j \in J)$. By Theorem 1, there is a collection $g \in \prod(A_k \mid k \in K)$. Consider the collection $e \in \prod(A_i \mid i \in I)$, such that $e(j) \equiv f(j)$ for every $j \in J$ and $e(k) \equiv g(k)$ for every $k \in K$. Then, $\mathrm{p}_J(e) = e|J = f$.

For the other projection, the arguments are analogous. □

Derivative mapping with respect to the product

Let $(A_i \mid i \in I)$ and $(B_i \mid i \in I)$ be two collections of classes with the same class of indices. Let $(u_i \mid i \in I)$ be a corresponding collection of mappings $u_i\colon A_i \to B_i$. Define a mapping $\pi\colon (A_i \mid i \in I) \to \prod(B_i \mid i \in I)$ setting $\pi((a_i \mid i \in I)) \equiv (u_i(a_i) \mid i \in I)$ for every $e \equiv (a_i \mid i \in I) \in \prod(A_i \mid i \in I)$. The mapping π will be called the *derivative mapping of the collection of mappings* $(u_i\colon A_i \to B_i \mid i \in I)$ *with respect to the product* $\prod(A_i \mid i \in I)$ *and* $\prod(B_i \mid i \in I)$ and will be denoted by $\prod_m(u_i\colon A_i \to B_i \mid i \in I)$ or simply by $\prod_m(u_i \mid i \in I)$.

In the particular case, let I be the set $2, 3, \ldots$; $A, A', A'', \ldots$ and $B, B', B'', \ldots$ be classes and $u\colon A \to B$, $u'\colon A' \to B'$, $u''\colon A'' \to B''$,... be corresponding mappings. Then, (A, A'), (A, A', A''),..., (B, B'), (B, B', B''),..., and (u, u'), (u, u', u''),... are the corresponding collections. The derivative mapping $\prod_m(u, u')\colon A \times A' \to B \times B'$ of the sequential pair (u, u') with respect to the products $A\times A'$ and $B\times B'$ will be denoted by $u \times_m u'$. The derivative mapping $\prod_m(u, u', u'')\colon A\times A'\times A'' \to B\times B'\times B''$ of the sequential triplet (u, u', u'') with respect to the products $A\times A'\times A''$ and $B\times B'\times B''$ will be denoted by $u \times_m u' \times_m u''$, and so on. It is clear that $(u \times_m u')(a, a') = (ua, u'a')$ for every $(a, a') \in A\times A'$, $(u\times_m u' \times_m u'')(a, a', a'') = (ua, u'a', u''a'')$ for every $(a, a', a'') \in A \times A' \times A''$, and so on.

If $(A_i \mid i \in I)$ is a collection of classes indexed by a class I, such that $A_i = A$ for some class A and every index $i \in I$, then the product $\prod(A_i \mid i \in I)$ of the multivalued collection $(A_i \mid i \in I)$ and the degree A^I of the class A with the exponent I coincide.

Therefore, if $(B_i \mid i \in I)$ is a collection, such that $B_i = B$ for some class B and every index $i \in I$, and $(u_i \mid i \in I)$ is a collection of mappings $u_i\colon A_i \to B_i$, such that $u_i = u$ for some mapping $u\colon A \to B$ and every index $i \in I$, then the mappings $\prod_m(u_i \mid i \in I)\colon \prod(A_i \mid i \in I) \to \prod(B_i \mid i \in I)$ and $(u)^I_m\colon A^I \to B^I$ coincide (see 1.1.8).

Lemma 5. *Let* $(A_i \mid i \in I)$, $(B_i \mid i \in I)$, *and* $(C_i \mid i \in I)$ *be collections of classes, indexed by a class* I, *and* $(u_i\colon A_i \to B_i \mid i \in I)$ *and* $(v_i\colon B_i \to C_i \mid i \in I)$ *be collections of corresponding mappings. Then:*

1) $\prod_m(v_i \mid i \in I) \circ \prod_m(u_i \mid i \in I) = \prod_m(v_i \circ u_i \mid i \in I)$;
2) *if all the mappings* u_i *are injective, then the mapping* $\prod_m(u_i \mid i \in I)$ *is injective as well.*

Moreover, if $(A_i \mid i \in I)$ *and* $(B_i \mid i \in I)$ *are collections of sets, indexed by the set* I, *then:*

3) *if all the mappings* u_i *are surjective, then the mapping* $\prod_m(u_i \mid i \in I)$ *is surjective as well;*
4) *if all the mappings* u_i *are bijective, then the mapping* $\prod_m(u_i \mid i \in I)$ *is bijective as well.*

Proof. The statements 1 and 2 follow from the definitions and Lemma 1.

3. Let $(b_i \in B_i \mid i \in I) \in \prod(B_i \mid i \in I)$. Consider the non-empty sets $D_i \equiv u_i^{-1}(b_i) \subset A_i$. By Theorem 1 and Proposition 1 (1.1.5), there exists a collection $(d_i \subset D \mid i \in I) \in \prod(D_i \mid i \in I)$. From the equality $\prod_m(u_i \mid i \in I)(d_i \mid i \in I) = (u_i(d_i) \mid i \in I) = (b_i \mid i \in I)$, we conclude now that the mentioned mapping is surjective.

4. It follows from 2 and 3. □

Corollary 1. *The corresponding conclusions are valid for corresponding suits* (A, A'), (A, A', A''),..., (B, B'), (B, B', B''),..., (C, C'), (C, C', C''),..., (u, u'), (u, u', u''),..., *and* (v, v'), (v, v', v''),....

1.1.13 Formulas of the distributivity for union, intersection, and product of a multivalued collection

Theorem 1. *Let* $(I_m \mid m \in M)$ *be a collection of sets and* $(\varkappa_m \mid m \in M)$ *be a collection of collections of sets* $\varkappa_m \equiv (A_{mi} \mid i \in I_m)$, *indexed by non-empty sets* M *and* I_m. *Consider the set* $U \equiv \prod(I_m \mid m \in M)$. *Then:*

1) $\bigcup(\bigcup(A_{mi} \mid i \in I_m) \mid m \in M) = \bigcup(\bigcup(A_{mu(m)} \mid m \in M) \mid u \in U)$ (the *general distributivity of union);*
2) $\bigcap(\bigcap(A_{mi} \mid i \in I_m) \mid m \in M) = \bigcap(\bigcap(A_{mu(m)} \mid m \in M) \mid u \in U)$ (the *general distributivity of intersection);*
3) $\bigcap(\bigcup(A_{mi} \mid i \in I_m) \mid m \in M) = \bigcup(\bigcap(A_{mu(m)} \mid m \in M) \mid u \in U)$ (the *general distributivity of intersection with respect to union);*
4) $\bigcup(\bigcap(A_{mi} \mid i \in I_m) \mid m \in M) = \bigcap(\bigcup(A_{mu(m)} \mid m \in M) \mid u \in U)$ (the *general distributivity of union with respect to intersection);*
5) $\prod(\bigcup(A_{mi} \mid i \in I_m) \mid m \in M) = \bigcup(\prod(A_{mu(m)} \mid m \in M) \mid u \in U)$ (the *general distributivity of product with respect to union);*
6) $\prod(\bigcap(A_{mi} \mid i \in I_m) \mid m \in M) = \bigcap(\prod(A_{mu(m)} \mid m \in M) \mid u \in U)$ (the *general distributivity of product with respect to intersection).*

Proof. Consider the sets $E_m \equiv \bigcup(A_{mi} \mid i \in I_m)$, $F_m \equiv \bigcap(A_{mi} \mid i \in I_m)$, $G_u \equiv \bigcup(A_{mu(m)} \mid m \in M)$, $H_u \equiv \bigcap(A_{mu(m)} \mid m \in M)$ and $Q_u \equiv \prod(A_{mu(m)} \mid m \in M)$.

1. Denote the left part of Equality (1) by E and the right part by G. Let $e \in E$. Then, $e \in A_{mi}$ for some $m \in M$ and $i \in I_m$. By Lemma 4 (1.1.12), $\mathrm{pr}_m[U] = I_m$. Therefore, there is $u \in U$, such that $i = \mathrm{pr}_m(u) = u(m)$. Consequently, $e \in A_{mu(m)} \subset G_u \subset G$. As a result, $E \subset G$. Conversely, let $g \in G$. Then, $g \in A_{mu(m)}$ for some $u \in U$ and $m \in M$. By the definition of the product, $u(m) \in I_m$. Therefore, $g \in A_{mu(m)} \subset E_m \subset E$. As a result, $G \subset E$. Thus, $E = G$.

2. Denote the left part of Equality (2) by F and the right part by H. Let $f \in F$. Then, $f \in A_{mi}$ for every $m \in M$ and $i \in I_m$. Take any $u \in U$ and $m \in M$. Since $u(m) \in I_m$, we have $f \in A_{mu(m)}$. By rule of deduction D4 (1.1.3), we get the formula $\forall u \in U$

$(f \in A_{mu(m)})$. By virtue of axiom scheme AS2 (1.1.5), $f \in H_u$. Similarly, the formula $\forall u \in U\ (f \in H_u)$ implies $f \in H$. As a result, $F \subset H$. Conversely, let $h \in H$. Then, $h \in A_{mu(m)}$ for every $u \in U$ and $m \in M$. Take any $m \in M$ and $i \in I_m$. Since $\mathrm{pr}_m[U] = I_m$, there is $u \in U$, such that $i = u(m)$. Consequently, $h \in A_{mi}$. As above, this implies $h \in F_m$ and then $h \in F$, where $H \subset F$. Thus, $F = H$.

3. Denote the left part of Equality (3) by E and the right part by H. Let $e \in E$. Then, $e \in A_{mi}$ for every $m \in M$ and some $i \in I_m$. As above, there is $u \in U$, such that $i = u(m)$. Therefore, $e \in A_{mu(m)}$ for every $m \in M$ implies $e \in H_u$. By virtue of axiom scheme LAS12 (1.1.4) and rule of deduction D3 (1.1.3), we get the formula $\exists u \in U\ (e \in H_u)$. By axiom scheme AS2 (1.1.5), we get $e \in H$. As a result, $E \subset H$. Conversely, let $h \in H$. Then, $h \in A_{mu(m)}$ for some $u \in U$ and every $m \in M$. Since $u(m) \in I_m$, we get the formula $\exists i \in I_m\ (h \in A_{mi})$, where $h \in E_m$ for every $m \in M$. This implies $h \in E$. As a result, $H \subset E$. Thus, $E = H$.

4. Denote the left part of Equality (4) by F and the right part by G. Let $f \in F$. Then, $f \in A_{mi}$ for some $m \in M$ and every $i \in I_m$. Take any $u \in U$. Since $u(m) \in I_m$, we have $f \in A_{mu(m)}$. As above, this implies $f \in G_u$. By rule of deduction D4 (1.1.3), we get the formula $\forall u \in U\ (f \in G_u)$, where $f \in G$. As a result, $F \subset G$. Conversely, let $g \in G$. Then, $g \in A_{mu(m)}$ for every $u \in U$ and some $m \in M$. Take any $i \in I_m$. Since $\mathrm{pr}_m[U] = I_m$, there is $u \in U$, such that $i = u(m)$. Consequently, $g \in A_{mi}$. By rule of deduction D4 (1.1.3), we get the formula $\forall i \in I_m\ (g \in A_{mi})$, where $g \in F_m$. As above, this implies $g \in F$. As a result, $G \subset F$. Thus, $F = G$.

5. Denote the left part of Equality (5) by P and the right part by Q. Let $p \in P$. Then, $p = (e_m \in E_m \mid m \in M)$. Take any $m \in M$. It follows from $e_m \in E_m$, that $e_m \in A_{mi}$ for some $i \in I_m$. Therefore, the subset $J_m \equiv \{i \in I_m \mid e_m \in A_{mi}\}$ of the set I_m is non-empty. By rule of deduction D4 (1.1.3), we get the formula $\forall m \in M\ (J_m \neq \varnothing)$. Consider the set $I \equiv \bigcup(I_m \mid m \in M)$ and the correspondence $\pi \equiv \{\langle m, x\rangle \in M * I \mid x \in J_m\}$. It is clear that $\pi\langle m\rangle = J_m$. By virtue of the mentioned formula, $\mathrm{dom}\,\pi = M$. Thus, π is a multivalued collection $(J_m \subset I \mid m \in M)$. By virtue of Theorem 1 (1.1.12), $\prod(J_m \mid m \in M) \neq \varnothing$. Therefore, by Proposition 1, there exists a mapping $u\colon M \to \bigcup(J_m \mid m \in M)$, such that $u(m) \in J_m$ for every $m \in M$. It is clear that $u \in U$.

If $m \in M$, then $u(m) \in J_m$ implies $e_m \in A_{mu(m)}$. Thus, we get the formula $\forall m \in M\ (e_m \in A_{mu(m)})$. By axiom scheme AS2 (1.1.5), this means that $p = (e_m \mid m \in M) \in Q_u$. By virtue of axiom scheme LAS12 (1.1.4) and rule of deduction D3 (1.1.3), we get the formula $\exists u \in U\ (p \in Q_u)$. By axiom scheme AS2 (1.1.5), we get $p \in Q$. As a result, $P \subset Q$.

Conversely, let $q \in Q$. Then, $q \in Q_u$ for some $u \in U$, where $q \in \mathrm{Map}(M, G_u)$ and $q(m) \in A_{mu(m)}$ for every $m \in M$. From $u(m) \in I_m$, we conclude that $q(m) \in E_m$ for every $m \in M$. Thus, $q \in \mathrm{Map}(M, \bigcup(E_m \mid m \in M))$ and $q(m) \in E_m$ for every $m \in M$, where $q \in P$. As a result, $Q \subset P$. Finally, $P = Q$.

6. Denote the left part of Equality (6) by P and the right part by Q. Let $p \in P$. Then, $p = (f_m \in F_m \mid m \in M)$. Take any $m \in M$. It follows from $f_m \in F_m$ that $f_m \in A_{mi}$ for every $i \in I_m$. Take any $u \in U$. From $u(m) \in I_m$, we conclude that $f_m \in A_{mu(m)}$. By rule of

deduction D4 (1.1.3), we get the formula $\forall m \in M\,(f_m \in A_{mu(m)})$. This gives $p \in Q_u$. Similarly, we get the formula $\forall u \in U\,(p \in Q_u)$, where $p \in Q$. As a result, $P \subset Q$.

Conversely, let $q \in Q$. Then, $q \in Q_u$ for every $u \in U$. Thus, $q(m) \in A_{mu(m)}$ for every $m \in M$. Take any $m \in M$ and $i \in I_m$. Then, as in the proof of 1, there is $u \in U$, such that $i = u(m)$. Consequently, $q(m) \in A_{mi}$. This gives the formula $\forall i \in I_m\,(q(m) \in A_{mi})$, where $q(m) \in F_m$. Similarly, this gives the formula $\forall m \in M\,(q(m) \in F_m)$, where $q \in P$. As a result, $Q \subset P$. Thus, $P = Q$. □

Corollary 1. *In the conditions of Theorem 1, there is some bijection* $\delta\colon \prod(\bigcup_d(A_{mi} \mid i \in I_m) \mid m \in M) \rightarrowtail\!\!\!\twoheadrightarrow \bigcup_d(\prod(A_{mum(m)} \mid m \in M) \mid u \in U)$.

Proof. Denote the left part of this mapping by R and the right part by S. Since $A_{mu(m)} \neq \varnothing$ for every $m \in M$, we can consider the non-empty sets $Q_u \equiv \prod(A_{mu(m)} \mid m \in M)$, $R_u \equiv \prod(A_{mu(m)} * \{u(m)\} \mid m \in M)$ and $S_u \equiv Q_u * \{u\}$ for every $u \in U$. Then, $(R_u \mid u \in U)$ and $(S_u \mid u \in U)$ are multivalued collections.

By formula 5 of Theorem 1, $R = \prod(\bigcup(A_{mi} * \{i\} \mid i \in I_m) \mid m \in M) = \bigcup(R_u \mid u \in U)$. It is easy to check that the mapping $\beta_u\colon R_u \to S_u$, such that $\beta_u(\langle a_{mu(m)}, u(m)\rangle \mid m \in M) = \langle(a_{mu(m)} \mid m \in M), u\rangle$ is a bijection for every $u \in U$. The multivalued collection $(R_u \mid u \in U)$ is pairwise disjoint. In fact, let $u, v \in U$ and $u \neq v$. Suppose that there exists $r \in R_u \cap R_v$. By the condition, $r = (x_m \in A_{mu(m)} * \{u(m)\} \mid m \in M)$ and $r = (y_m \in A_{mv(m)} * \{v(m)\} \mid m \in M)$. Take any $m \in M$. Then, $x_m = \langle a, u(m)\rangle$ and $y_m = \langle b, v(m)\rangle$ for some $a \in A_{mu(m)}$ and $b \in A_{mv(m)}$. In virtue of Lemma 1 $x_m = y_m$. Therefore, by virtue of Proposition 2, $u(m) = v(m)$. By rule of deduction D4 (1.1.3), we get the formula $\forall m \in M\,(u(m) = v(m))$. Again, by Lemma 1 (1.1.9), we conclude that $u = v$. From the obtained contradiction, it follows that $R_u \cap R_v = \varnothing$. That is why we can correctly define a mapping δ from $R = \bigcup(R_u \mid u \in U)$ into $S = \bigcup(S_u \mid u \in U)$ setting $\delta(r) \equiv \beta_u(r)$ for every $r \in R_u$. It is easy to check that δ is a bijection. □

Corollary 2. *Let* $(X_j \mid j \in J)$ *and* $(Y_k \mid k \in K)$ *be multivalued collections of sets, indexed by non-empty sets* J *and* K. *Then:*
1) $\bigcup(X_j \mid j \in J) \cup \bigcup(Y_k \mid k \in K) = \bigcup(X_j \cup Y_k \mid (j, k) \in J \times K)$;
2) $\bigcap(X_j \mid j \in J) \cap \bigcap(Y_k \mid k \in K) = \bigcap(X_j \cap Y_k \mid (j, k) \in J \times K)$;
3) $\bigcup(X_j \mid j \in J) \cap \bigcup(Y_k \mid k \in K) = \bigcup(X_j \cap Y_k \mid (j, k) \in J \times K)$;
4) $\bigcap(X_j \mid j \in J) \cup \bigcap(Y_k \mid k \in K) = \bigcap(X_j \cup Y_k \mid (j, k) \in J \times K)$;
5) $\bigcup(X_j \mid j \in J) \times \bigcup(Y_k \mid k \in K) = \bigcup(X_j \times Y_k \mid (j, k) \in J \times K)$;
6) $\bigcap(X_j \mid j \in J) \times \bigcap(Y_k \mid k \in K) = \bigcap(X_j \times Y_k \mid (j, k) \in J \times K)$.

Proof. Take $M \equiv 2$, $I_0 \equiv J$, $I_1 \equiv K$, $A_{0i} \equiv X_i$ for every $i \in J$, $A_{1i} \equiv Y_i$ for every $i \in K$, $\varkappa_0 \equiv (X_j \mid j \in J) = (A_{0i} \mid i \in I_0)$, $\varkappa_1 \equiv (Y_k \mid k \in K) = (A_{1i} \mid i \in I_1)$ and $U \equiv J \times K = \prod(I_m \mid m \in M)$.

Then, all the necessary equalities are particular cases of the corresponding equalities of Theorem 1. □

The formulas of this corollary are called the *formulas of the binary distributivity.*

Formula 5 gives us the ability to obtain some special form of the general associativity.

Proposition 1. *Let $(J_i \mid i \in I)$ be a collection of non-empty sets indexed by a non-empty set I and $(A_k \mid k \in K)$ be a collection of sets indexed by the set $K \equiv \bigcup(\{i\} \times J_i \mid i \in I)$. Then,*

$$\bigcup(A_k \mid k \in K) = \bigcup(\bigcup(A_{ij} \mid j \in J_i) \mid i \in I) \text{ and}$$

$$\bigcap(A_k \mid k \in K) = \bigcap(\bigcap(A_{ij} \mid j \in J_i) \mid i \in I).$$

Proof. Consider the sets $K_i \equiv \{i\} \times J_i$. Then, $(K_i \mid i \in I)$ is a partition of the set K. Therefore, according to assertion 2 of Proposition 1 (1.1.10), we get the equality $B \equiv \bigcup(A_k \mid k \in K) = \bigcup(\bigcup(A_k \mid k \in K_i) \mid i \in I)$. Consider the bijective mappings $u_i : J_i \rightarrowtail\!\!\!\twoheadrightarrow K_i$ such that $u_i(j) = (i, j)$ for every $j \in J_i$. Then, assertion 1 of Proposition 1 (1.1.10) implies $\bigcup(A_k \mid k \in K_i) = \bigcup(A_{ij} \mid j \in J_i)$. Thus, $B = \bigcup(\bigcup(A_{ij} \mid j \in J_i) \mid i \in I)$.

For the intersection, the arguments are the same. □

Corollary 1. *Let I and J be non-empty sets and $((A_{ij} \mid j \in J) \mid i \in I)$ be a collection of collections of sets. Then:*

1) $\bigcup(\bigcup(A_{ij} \mid j \in J) \mid i \in I) = \bigcup(\bigcup(A_{ij} \mid i \in I) \mid j \in J) = \bigcup(A_{ij} \mid (i, j) \in I \times J)$;
2) $\bigcap(\bigcap(A_{ij} \mid j \in J) \mid i \in I) = \bigcap(\bigcap(A_{ij} \mid i \in I) \mid j \in J) = \bigcap(A_{ij} \mid (i, j) \in I \times J)$.

Proof. Consider the collections $(J_i \mid i \in I)$, where $J_i \equiv J$ and $(\iota_i \mid i \in I)$, where $\iota_i \equiv \{i\}$. Then, $\bigcup(\iota_i \mid i \in I) = I$ and $\bigcup(J_i \mid i \in I) = J$. Take also the sets $K_i \equiv \{i\} \times J_i$ and $K \equiv \bigcup(K_i \mid i \in I)$. Since the mapping $u : I \times I \to I$ such that $u(i, i') = i$ is surjective, assertion 1 of Proposition 1 (1.1.10) and assertion 5 of Corollary 2 to Theorem 1 imply that $K = \bigcup(\{i\} \times J_i \mid i \in I) = \bigcup(\iota_i \times J_{i'} \mid (i, i') \in I \times I) = \bigcup(\iota_i \mid i \in I) \times \bigcup(J_{i'} \mid i' \in I) = I \times J$. According to Proposition 1, we infer that $\bigcup(\bigcup(A_{ij} \mid j \in J) \mid i \in I) = \bigcup(A_k \mid k \in K) = \bigcup(A_{ij} \mid (i, j) \in I \times J)$.

Similarly, $\bigcup(\bigcup(A_{ij} \mid i \in I) \mid j \in J) = \bigcup(A_{ij} \mid (j, i) \in J \times I)$. Since the mapping $v : J \times I \to I \times J$ such that $v(i, j) = (j, i)$ is bijective, by assertion 1 of Proposition 1 (1.1.10) we get $\bigcup(A_{ij} \mid (i, j) \in I \times J) = \bigcup(A_{ij} \mid (j, i) \in J \times I)$.

The second assertion is proved in the same way. □

1.1.14 Binary relations. Equivalence relations. Preorder and order relations

Let A be a fixed class. A subclass $\theta \subset A^2 = A \times A$ of the sequential product $A \times A$ is called a *(binary) relation on the class A*. If $(a, b) \in \theta$, then we shall say that *a is in the relation θ with b* and shall write also $a\theta b$.

If B is a subclass of the class A, then the relation $\theta_B \equiv \theta \cap (B \times B)$ is called the *relation on the subclass B, induced by the relation* θ, or the *restriction of the relation* θ *on the subclass B*.

The relation θ is called *symmetric*, if $(a, a') \in \theta$ implies $(a', a) \in \theta$. It is called *antisymmetric*, if $(a, a') \in \theta$ and $(a', a) \in \theta$ imply $a = a'$. It is called *reflexive*, if $(a, a) \in \theta$ for every $a \in A$. It is called *transitive*, if $(a, a') \in \theta$ and $(a', a'') \in \theta$ imply $(a, a'') \in \theta$. It is called *total* ($\equiv$ *connecting*), if either $(a, a') \in \theta$, or $(a', a) \in \theta$, or $a = a'$. A subclass B of the class A is called a *chain with respect to the relation* θ, if the relation θ_B is total on B.

An element $a \in A$ is called *minimal [maximal] with respect to the relation* θ, if $a' \in A$ and $(a', a) \in \theta$ $[(a, a') \in \theta]$ imply $a' = a$. An element b of a subclass B of the class A is called a *minimal [maximal] element of the subclass B with respect to the relation* θ, if b is a minimal [maximal] in the class B with respect to the relation θ_B, i. e. $b' \in B$ and $(b', b) \in \theta$ $[(b, b') \in \theta]$ imply $b' = b$.

Equivalence relations, factor-classes, and factor-correspondences

A reflexive, symmetric, and transitive relation ε on the class A is called an *equivalence relation* or simply an *equivalence on A*. In this case, along with $(a, a') \in \varepsilon$, we shall write $a \sim a'$ or $a = a' (\mathrm{mod}\ \varepsilon)$ and shall say that *a is equivalent to a' with respect to* $\sim \equiv \varepsilon$ or *a is equal to a' modulo* ε.

Let ε be an equivalence relation on the class A. If $a \in A$, then the subclass $\{a' \in A \mid (a, a') \in \varepsilon\}$ of the class A is called the *equivalence class of the element a, with respect to* ε and is denoted by εa, $\bar{a} \bmod \varepsilon$ or simply by $\bar{a}$. It is clear that $\bar{a} \neq \varnothing$ for every $a \in A$, and if $a, a' \in A$, then either $\bar{a} = \bar{a}'$ or $\bar{a} \cap \bar{a}' = \varnothing$.

Let n be some fixed neutral element of A. The equivalence class of the element n will be called the *neutral subclass of A with respect to n and* ε and will be denoted by A^n. The common examples are $n = 0$ and $n = 1$.

The subclass $\{x \mid \exists a \in A\ (x = \bar{a})\}$ of the class $\mathcal{P}(A) \setminus \{\varnothing\}$, consisting of all equivalence classes $\bar{a}$ which are sets, is called the *factor-class of the class A with respect to the equivalence relation* ε and is denoted by A/ε or simply by $\bar{A}$.

Consider the correspondence $p \equiv \{\langle a, x\rangle \in A * (A/\varepsilon) \mid a \in x\}$.

Lemma 1. *Let A be a class and* ε *be an equivalence on A. Then, the correspondence* $p\colon A \longrightarrow\!\!\!\!\!\prec A/\varepsilon$ *is single-valued and surjective. If besides A is a set, then* A/ε *is a set and p is a surjective mapping* $p\colon A \twoheadrightarrow A/\varepsilon$.

Proof. If $\langle a, x\rangle \in p$ and $\langle a, y\rangle \in p$, then $a \in x$ and $a \in y$ imply $x = y$. Thus, p is single-valued. If $x \in \bar{A}$, then $x \neq \varnothing$ and so there is $a \in A$, such that $a \in x$. Then, $\langle a, x\rangle \in p$. This means that p is surjective.

Now, let A be a set. Then, by Lemma 1 (1.1.6) $\bar{a}$ is a set as well for every $a \in A$. Therefore, $\langle a, \bar{a}\rangle \in p$ for every $a \in A$. Thus, $\mathrm{dom}\, p = A$. By virtue of Lemmas 2 and 1 (1.1.6), $\bar{A}$ is a set. As a result, p is a surjective mapping from the set A onto the set $\bar{A}$. □

The correspondence $p\colon A \longrightarrow A/\varepsilon$ is called the *factor-correspondence from the class A onto the factor-class A/ε with respect to the equivalence ε*. If A is a set, then the mapping $p\colon A \twoheadrightarrow A/\varepsilon$ is called the *factor-mapping from the set A onto the factor-set A/ε with respect to the equivalence ε*.

Consider the identical collection $\mathrm{id}_{A/\varepsilon} \equiv (x_x \in \mathcal{P}(A)\backslash\{\varnothing\} \mid x \in A/\varepsilon)$ from 1.1.9, such that $x_x = x$. By virtue of Lemma 3 (1.1.9), there exists the multivalued collection $[x_x \subset A \mid x \in A/\varepsilon] = \varphi^{-1}(\mathrm{id}_{A/\varepsilon})$.

Lemma 2. *Let A be a set and ε be an equivalence on A. Then, $[x_x \subset A \mid x \in A/\varepsilon]$ is a partition of the set A.*

Proof. If $a \in A$, then by Lemma 1 (1.1.6), $x \equiv \bar{a}$ is a set. Therefore, $x \in \bar{A}$ and $a \in x = x_x$. Consequently, $A = \bigcup[x_x \mid x \in \bar{A}]$. Besides, $x \neq y$ implies $x_x \cap y_y = \varnothing$. □

This multivalued collection $[x_x \subset A \mid x \in A/\varepsilon]$ is called the *partition of the set A determined by the equivalence ε*.

Lemma 3. *Let a collection of sets $[A_i \mid i \in I]$, indexed by a set I, be a partition of a set A. Then, there is an equivalence ε on A and a bijection $u\colon A/\varepsilon \rightarrowtail\!\!\!\twoheadrightarrow I$, such that $x_x = A_{u(x)}$ for every component x_x of the partition of A determined by ε.*

Proof. Consider the relation $\varepsilon \equiv \{(a, a') \in A \times A \mid \exists i \in I\,(a \in A_i \wedge a' \in A_i)\}$. If $a \in A$, then there is $i \in I$, such that $a \in A_i$. Therefore, $(a, a) \in \varepsilon$. This means that ε is reflexive. Let $(a, a') \in \varepsilon$ and $(a', a'') \in \varepsilon$. Then, there are $i, j \in I$, such that $a, a' \in A_i$ and $a', a'' \in A_j$. From $a' \in A_i \cap A_j$, we infer that $i = j$. Therefore, $(a, a'') \in \varepsilon$. Thus, ε is transitive. Finally, let $(a, a') \in \varepsilon$. Then, $a, a' \in A_i$ for some i. Therefore, $(a', a) \in \varepsilon$, i. e. ε is symmetric. Thus, ε is an equivalence relation.

Consider the set $\bar{A} \equiv A/\varepsilon$ and the correspondence $u \equiv \{\langle x, i\rangle \in \bar{A} * I \mid x = A_i\}$. Let $x \in \bar{A}$. Then, $x = \bar{a}$ for some $a \in A$. By the definition, $A = \bigcup[A_i \mid i \in I]$. Therefore, $a \in A_i$ for some $i \in I$. If $a' \in x$, then $(a, a') \in \varepsilon$ implies $a, a' \in A_j$. From $a \in A_i \cap A_j$, we infer that $i = j$. Thus, $a' \in A_i$, where $x \subset A_i$. Conversely, if $a'' \in A_i$, then $(a, a'') \in \varepsilon$ implies $a'' \in \bar{a} = x$, where $A_i \subset x$. Thus, $x = A_i$ and $\langle x, i\rangle \in u$. This means that $\operatorname{dom} u = \bar{A}$.

If $\langle x, i\rangle \in u$ and $\langle x, j\rangle \in u$, then $x = A_i$ and $x = A_j$. Therefore, $i = j$. This means that u is single-valued. As a result, u is a mapping $u\colon \bar{A} \to I$, such that $x = A_{u(x)}$.

If $u(x) = u(y)$, then $x = A_{u(x)} = A_{u(y)} = y$ means that w is injective. Let $i \in I$. Take $a \in A_i$ and $x \equiv \bar{a} \in \bar{A}$. It was deduced above that these conditions imply $\langle x, i\rangle \in u$, where $i = u(x)$. This means that u is surjective. As a result, u is bijective and $x_x = A_{u(x)}$. □

Preorder and order relations

A reflexive and transitive relation θ on the class A is called a *preorder relation* or simply a *preorder on A*. In this case, we say also that *θ preorders A*. Also, in this case, along with $(a, a') \in \theta$ we shall write $a \leqslant a'$ or $a' \geqslant a$ and shall say that *a is smaller than or equal to* a' or a' *is greater than or equal to a with respect to* $\leqslant \equiv \theta$. If besides $a \neq a'$, then we shall write $a < a'$ or $a' > a$ and shall say that *a is smaller than* a' or a' *is greater than a*.

If θ is a preorder on the class A and B is a subclass of A, then the restriction θ_B is a preorder on the class B.

An antisymmetric preorder relation on A is called an *order relation* or simply an *order on A*. A total order relation on A is called a *linear order on A*. If θ is an order [a linear order] on the class A and B is a subclass of A, then the restriction θ_B is an order [a linear order] on B.

If θ is a transitive relation on the class A, then the relation $\theta \cup \{(a, a') \in A \times A \mid a = a'\}$ is a preorder on A. If θ is a preorder on A, then the relation $\eta \equiv \{(a, a') \in A \times A \mid (a', a) \in \theta\}$ is called the *preorder on A opposite to the preorder* θ.

The preorder and order relations are so important in mathematics that there is an extensive row of notions connected with them. Therefore, all the necessary notions connected with the preorder and order relations used in this book, we shall consider in the separate subsection 1.1.15 (and also in the following sections).

1.1.15 Basic notions connected with preorder and order relations

A class A with a preorder $\leqslant$ on A will be called a *preordered class* and will be denoted by $(A, \leqslant)$ or sometimes simply by A. If $\leqslant$ is an order, then $(A, \leqslant)$ is called an *ordered class*.

Monotonicity

Let $(A, \leqslant)$ and $(B, \leqslant)$ be preordered classes and $u\colon A \to B$ be a mapping.

The mapping u is called *increasing* [*decreasing*] if $a \leqslant a'$ implies $u(a) \leqslant u(a')$ [$u(a) \geqslant u(a')$]. An increasing mapping is called also *order preserving* or *monotone*; a decreasing mapping is called also *order changing* or *antimonotone*.

The mapping u is called *strictly increasing* [*strictly decreasing*] if $a < a'$ implies $u(a) < u(a')$ [$u(a) > u(a')$].

The increasing mapping u will be called *isotone* if $u(a) \leqslant u(a')$ implies $a \leqslant a'$. The decreasing mapping u will be called *antiisotone* if $u(a) \geqslant u(a')$ implies $a \leqslant a'$.

Preordered classes $(A, \leqslant)$ and $(B, \leqslant)$ are called *order equivalent* ($(A, \leqslant) \approx (B, \leqslant)$) if there exists some isotone bijective mapping $u\colon A \rightarrowtail\!\!\!\twoheadrightarrow B$.

Since the mapping $u\colon A \to B$ can be considered as the collection $u \equiv (b_a \mid a \in A)$, where $b_a \equiv u(a)$ for every $a \in A$, the given terminology can be easily reformulated for the collection $(b_a \mid a \in A)$.

Lemma 1. *Let $(A, \leqslant)$ and $(B, \leqslant)$ be ordered classes and $u\colon A \to B$ be a mapping. Then:*

1) *if u is isotone, then u is injective and strictly increasing;*
2) *if u is isotone and surjective, then u is bijective, and inverse mapping* $u^{-1}\colon B \to A$ *is also isotone.*

Proof. 1. Let a and a' be arbitrary elements, such that $ua = ua'$. Then, $ua \leqslant ua'$ implies $a \leqslant a'$ and $ua' \leqslant ua$ implies $a' \leqslant a$. As a result, $a = a'$. Thus, u is injective.

Let $a < a'$. Then, $ua \leqslant ua'$ and $ua \neq ua'$. Thus, $ua < ua'$, i.e. u is strictly increasing.

2. By Proof 1, u is bijective. If $b \leqslant b'$, then $u(u^{-1}b) \leqslant u(u^{-1}b')$ implies $u^{-1}b \leqslant u^{-1}b'$. If $u^{-1}b \leqslant u^{-1}b'$, then applying u we get $b = u(u^{-1}b') \leqslant u(u^{-1}b') = b'$. Thus, u^{-1} is isotone. □

Intervals

Let $(A, \leqslant)$ be a preordered class and $a, b \in A$. The subclass $\{c \in A \mid a < c < b\}$ of the class A is called the *open interval with the beginning a and the end b* and is denoted by $]a, b[$. The subclass $\{c \in A \mid a \leqslant c \leqslant b\}$ of the class A is called the *closed interval with the beginning a and the end b* and is denoted by $[a, b]$. In the similar way the *half-open* (≡ *half-closed*) *intervals* $]a, b]$ and $[a, b[$ are defined. Any subclass X of the class A, such that $]a, b[\subset X \subset [a, b]$ will be called an *interval* (*of the general kind*) *with the beginning a and the end b* and will be denoted by $|a, b|$. Sometimes, a is called the *left end* and b is called the *right end* of the corresponding interval.

A subclass B of A is called *convex* if $[b', b''] \subset B$ for every $b', b'' \in B$.

The subclass $\{c \in A \mid c < b\}$ of the class A is called the *open initial interval with the end b* and is denoted by $]\leftarrow, b[$. The subclass $\{c \in A \mid c \leqslant b\}$ of the class A is called the *closed initial interval with the end b* and is denoted by $]\leftarrow, b]$. In the similar way, the *open* $]a, \to[$ and the *closed* $[a, \to[$ *final intervals with the beginning a* are defined.

Let $(A, \leqslant, \{n\})$ be an ordered class with the fixed neutral element $n \in A$. The subclass $A_n \equiv [n, \to [\equiv \{a \in A \mid a \geqslant n\}$ will be called the *main part of A with respect to the neutral element n*. The common examples are $n = 0$ and $n = 1$.

Let X be a subclass of an preordered class A. The subclass X is called *initial in the class A*, if $]\leftarrow, b] \subset X$ for every $b \in X$. The subclass X is called *final in the class A*, if $[a, \to[\subset X$ for every $a \in X$.

The subclass X is called *coinitial to the class A*, if $]\leftarrow, b] \cap X \neq \varnothing$ for every $b \in A$. The subclass X is called *cofinal to the class A*, if $[a, \to[\cap X \neq \varnothing$ for every $a \in A$.

An element $y \in X$ is called the *greatest element of the subclass X* if $X \subset]\leftarrow, y]$. An element $x \in X$ is called the *smallest element of the subclass X* if $X \subset [x, \to[$. If $\leqslant$ is an order, then the greatest and the smallest elements of the subclass X are unique and they will be denoted respectively by gr X and sm X.

Supremum and infimum, greatest and smallest members

Let $(a_i \in A \mid i \in I)$ be a collection of elements of the class A, indexed by some class I. We consider here simple collections $(a_i \in A \mid i \in I)$. But in the particular case of the identical collection $(i_i \in I \subset A \mid i \in I)$ (see 1.1.9), we shall use along with the collection terminology also the set terminology for the set I.

An element $a \in A$ is called an *upper* [a *lower*] *bound of the collection* $(a_i \mid i \in I)$ and the collection $(a_i \mid i \in I)$ is called (*order*) *bounded above* [*below*] *by the element* a if $a_i \leqslant a$ $[a_i \geqslant a]$ for every $i \in I$. The collection $(a_i \mid i \in I)$ is called (*order*) *bounded above* [*below*], if there is in A an upper [a lower] bound of the collection $(a_i \mid i \in I)$. It is called (*order*) *bounded*, if it is bounded above and below.

The preordered class $(A, \leqslant)$ is called *upward* [*downward*] *directed* ($\equiv$ *filtering*) if every sequential pair (a, a') of elements of A is bounded above [below].

Let $(A, \leqslant_A)$ and $(B, \leqslant_B)$ be preordered classes. A mapping $u : A \to B$ is called (*order*) *bounded* if for every (order) bounded set $E \subset A$ the set $u[E]$ is (order) bounded in B.

An element $a \in A$ is called the *greatest lower bound* or *infimum of the collection* $(a_i \mid i \in I)$ if (1) a is a lower bound of $(a_i \mid i \in I)$ and (2) $a \geqslant a'$ for every lower bound a' of $(a_i \mid i \in I)$. If $\leqslant$ is an order, then this element is unique and is denoted by $\inf(a_i \mid i \in I)$. If $I \subset A$, then we use also the notation $\inf I \equiv \inf (i_i \mid i \in I)$.

An element $a \in A$ is called the *smallest upper bound* or *supremum of the collection* $(a_i \mid i \in I)$ if (1) a is an upper bound of $(a_i \mid i \in I)$ and (2) $a \leqslant a'$ for every upper bound a' of $(a_i \mid i \in I)$. If $\leqslant$ is an order, then this element is unique and is denoted by $\sup(a_i \mid i \in I)$. If $I \subset A$, then we use also the notation $\sup I \equiv \sup (i_i \mid i \in I)$.

Both the supremum and the infimum of the collection (set) will be called the *exact bounds* of this collection (set).

It is evident that for every collection $\alpha \equiv (a_i \in A \mid i \in I)$ and $\beta \equiv (b_i \in A \mid i \in I)$ and element $a, b \in A$ such that $a_i \leqslant b_i$ for every $i \in I$, $a = \sup \alpha$, $b = \sup \beta$, we have $a \leqslant b$. The similar property is valid for the greatest lower bound.

Let $(A, \leqslant)$ and $(B, \leqslant)$ be ordered classes and $u\colon A \to B$ be a mapping. The mapping u is called *preserving exact upper* [*lower*] *bounds* if $a = \sup E$ $[a = \inf E]$ implies $ua = \sup u[E]$ $[ua = \inf u[E]]$ for any $a \in E$ and $E \subset A$. The mapping is called *preserving any exact bounds* if it preserves both the exact upper and lower bounds.

Lemma 2. *Let $(A, \leqslant)$, and $(B, \leqslant)$ be ordered classes and $u : A \to B$ be a surjective and isotone mapping. Then, u and $u^{-1} : B \to A$ (see Lemma* 1) *preserve any exact bounds.*

Proof. Let $a \in A$, $E \subset A$, and $a = \sup E$. Then, $ua \geqslant ue$ for every $e \in E$. Let $b \in B$ and $b \geqslant ue$ for every $e \in E$. Then, $b = ua'$ for some $a' \in A$. By condition, $ua' \geqslant ue$ implies $a' \geqslant e$ for every $e \in E$. Hence, $a' \geqslant a$ and $b \geqslant ua$. This means that $ua = \sup u[E]$. For the infimum, the arguments are similar.

According to Lemma 1, the inverse mapping u^{-1} is also surjective and isotone. Therefore, it preserves any exact bounds. □

In the particular case, let I be the sets 2, 3, ... and $a, a', a'', \ldots$ be elements of A. Then, (a, a'), (a, a', a''),... are the corresponding collections. Therefore, we can consider the elements $\sup(a, a')$, $\sup(a, a', a'')$,..., and $\inf(a, a')$, $\inf(a, a', a'')$,....

The elements $\sup(a, a')$, $\sup(a, a', a'')$, ... will be called the *supremums of the simple sequential pair* (a, a'), *triplet* (a, a', a''), ... and will be denoted also by $a \vee a'$, $a \vee a' \vee a''$, ... By the definition of the simple sequential pair from 1.1.11, we have $a \vee a' = \sup(x_i \mid i \in 2)$, where $x_0 \equiv a$ and $x_1 \equiv a'$. In the similar manner, $a \vee a' \vee a'' = \sup(x_i \mid i \in 3)$, where $x_0 \equiv a$, $x_1 \equiv a'$, and $x_2 \equiv a''$, and so on.

The elements $\inf(a, a')$, $\inf(a, a', a'')$, ... will be called the *infimums of the simple sequential pair* (a, a'), *triplet* (a, a', a''), ... and will be denoted also by $a \wedge a'$, $a \wedge a' \wedge a''$, ...

For the preordered class $(A, \leqslant)$ an element $a_j \in \{a_i \mid i \in I\}$ is called the *greatest* [*smallest*] *member of the collection* $(a_i \in A \mid i \in I)$ if $a_i \leqslant a_j$ [$a_i \geqslant a_j$] for every $j \in I$. If $\leqslant$ is an order, then the greatest and the smallest member of the collection $(a_i \mid i \in I)$ are unique and will be denoted by $\mathrm{gr}\,(a_i \mid i \in I)$ and $\mathrm{sm}\,(a_i \mid i \in I)$, respectively.

As above, the elements $\mathrm{gr}(a, a')$, $\mathrm{gr}(a, a', a'')$, $\mathrm{gr}(a, a', a'', a''')$,... will be called the *greatest member of the pair* (a, a'), *triplet* (a, a', a''), *quadruplet* (a, a', a'', a'''),... and will be denoted also by $a \veebar a'$, $a \veebar a' \veebar a''$, $a \veebar a' \veebar a'' \veebar a'''$, ...

The elements $\mathrm{sm}(a, a')$, $\mathrm{sm}(a, a', a'')$, $\mathrm{sm}(a, a', a'', a''')$,... will be called the *smallest members of the pair* (a, a'), *triplet* (a, a', a''), *quadruplet* (a, a', a'', a'''),... and will be denoted also by $a \barwedge a'$, $a \barwedge a' \barwedge a''$, $a \barwedge a' \barwedge a'' \barwedge a'''$, ...

For the preordered class $(A, \leqslant)$ if a_j is the greatest [smallest] member of the collection $(a_i \mid i \in I)$, then a_j is a supremum [infimum] of $(a_i \mid i \in I)$.

For the preordered class $(A, \leqslant)$ an element $a_j \in \{a_i \mid i \in I\}$ is called the *maximal* [*minimal*] *member of the collection* $(a_i \in A \mid i \in I)$ if $a_i \geqslant a_j$ [$a_i \leqslant a_j$] implies $a_i = a_j$. If $\leqslant$ is an order and a_j is the greatest [smallest] member of the collection $(a_i \mid i \in I)$, then a_j is a unique maximal [minimal] member of $(a_i \mid i \in I)$.

The ordered class $(A, \leqslant)$ is called *upward* [*downward*] *lattice-ordered* if for every sequential pair (a, a') of elements of A there is the $\sup(a, a')$ [$\inf(a, a')$]. It is called *lattice-ordered* if it is upward and downward lattice-ordered simultaneously. In this case, the order $\leqslant$ is called *latticed*.

Let $(A, \leqslant, \{n\})$ be a lattice-ordered class with some fixed neutral element $n \in A$. The elements $a_+ \equiv a \vee n$ and $a_- \equiv a \wedge n$ will be called the *positive* and the *negative parts of an element* $a \in A$ (*with respect to the neutral element* n). Respectively, the subclasses $A_+ \equiv \{a_+ \in A \mid a \in A\}$ and $A_- \equiv \{a_- \in A \mid a \in A\}$ will be called the *positive* and the *negative parts of the class* A (*with respect to the neutral element* n). Note that in this case, $A_+ = A_n$, i. e. the positive part and the main part coincide.

Let $U \equiv (A, \leqslant_A)$ be an ordered class. An ordered class $V \equiv (B, \leqslant_B)$ is called *completely* (*order*) *closed in the ordered class* U if the following properties hold:

1) $B \subset A$;
2) $\leqslant_B \; = \; \leqslant_A |B * B$;

3) if $a \in A$ and $a = \sup_A (b_i \mid i \in I)$ for some simple collection $(b_i \in B \mid i \in I)$, then $a \in B$;
4) if $a \in A$ and $a = \inf_A (b_i \mid i \in I)$ for some simple collection $(b_i \in B \mid i \in I)$, then $a \in B$.

Let $U \equiv (A, \leqslant_A)$ be a lattice-ordered class. A lattice-ordered class $V \equiv (B, \leqslant_B)$ is called a *lattice-ideal* ($\equiv$ an *l-ideal*) *of the lattice-ordered class* U if the following properties hold:
1) $B \subset A$;
2) $\leqslant_B \; = \; \leqslant_A |B * B$;
3) $\vee_B = \vee_A|(B * B) * B$;
4) $\wedge_B = \wedge_A|(B * B) * B$;
5) $\forall\, a \in A\, \forall\, b', b'' \in B\, (b' \leqslant_A a \leqslant_A b'' \Rightarrow a \in B)$.

The ordered class $(A, \leqslant)$ is called *upward* [*downward*] *(order) complete* if for every collection of elements of A there is its supremum [infimum]. It is called *(order) complete* if it is upward and downward complete.

The ordered class $(A, \leqslant)$ is called *upward* [*downward*] *Dedekind complete* if for every bounded above [below] collection of elements of A there is its supremum [infimum]. It is called *Dedekind complete* if it is upward and downward Dedekind complete.

Proposition 1. *Let $(A, \leqslant)$ be an ordered class, I and M be classes, $(a_i \in A \mid i \in I)$ and $(x_m \in A \mid m \in M)$ be collections and $a \in A$. Then:*
1) *if $u\colon K \twoheadrightarrow I$ is a surjective mapping, then $a = \sup(a_i \mid i \in I)$ iff $a = \sup(a_{u(k)} \mid k \in K)$; analogously, $a = \inf(a_i \mid i \in I)$ iff $a = \inf(a_{u(k)} \mid k \in K)$ (the* general commutativity of supremum and infimum*);*
2) *if $I = \bigcup(I_m \mid m \in M)$ for some multivalued collection $(I_m \mid m \in M)$ and $x_m = \sup(a_i \mid i \in I_m)$, then $a = \sup(a_i \mid i \in I)$ iff $a = \sup(x_m \mid m \in M)$; analogously, if $x_m = \inf(a_i \mid i \in I_m)$, then $a = \inf(a_i \mid i \in I)$ iff $a = \inf(x_m \mid m \in M)$ (the* general associativity of supremum and infimum*).*

Proof. Conclusion 1 follows directly from the definitions.

2. Let $a = \sup(a_i \mid i \in I)$. Then, $a \geqslant x_m$ for every $m \in M$. Let $y \in A$ and $y \geqslant x_m$ for every m. Then, $y \geqslant a_i$ for every $i \in I_m$ and $m \in M$ implies $y \geqslant a$. Thus, $a = \sup(x_m \mid m \in M)$. Conversely, let $a = \sup(x_m \mid m \in M)$. Then, $a \geqslant x_m \geqslant a_i$ for every $i \in I_m$ and $m \in M$. Let $y \in A$ and $y \geqslant a_i$ for every $i \in I$. Then, $y \geqslant x_m$ for every $m \in M$ implies $y \geqslant a$. Thus, $a = \sup(a_i \mid i \in I)$.

The checking for the infimum is analogous. □

Corollary 1. *Let $(A, \leqslant)$ be an ordered class and $x, y, a, a', a'', a''' \in A$. Then:*
1) $\sup(a, a) = a$ *and* $\inf(a, a) = a$;

2) $a = \sup(a', a'')$ *iff* $a = \sup(a'', a')$; *analogously,* $a = \inf(a', a'')$ *iff* $a = \inf(a'', a')$ (the *commutativity of the supremum and the infimum);*
3) *if* $x = \sup(a', a'')$ *and* $y = \sup(a'', a''')$, *then* $a = \sup(a', a'', a''')$ *iff* $a = \sup(x, a''')$ *and also iff* $a = \sup(a', y)$; *analogously, if* $x = \inf(a', a'')$ *and* $y = \inf(a'', a''')$, *then* $a = \inf(a', a'', a''')$ *iff* $a = \inf(x, a''')$ *and also iff* $a = \inf(a', y)$ (the *associativity of the supremum and the infimum).*

The following result represents some special form of the general associativity.

Proposition 2. *Let* $(A, \leqslant)$ *be an ordered class,* I *be a non-empty set. Let* $(J_i \mid i \in I)$ *be a collection of non-empty sets,* $(x_i \in A \mid i \in I)$ *be a collection,* $(a_k \in A \mid k \in K)$ *be a collection indexed by the set* $K \equiv \bigcup(\{i\} \times J_i \mid i \in I)$, *and* $a \in A$. *Then:*
1) *if* $x_i = \sup\left(a_{ij} \mid j \in J_i\right)$, *then* $a = \sup(x_i \mid i \in I)$ *iff* $a = \sup(a_k \mid k \in K)$;
2) *if* $x_i = \inf\left(a_{ij} \mid j \in J_i\right)$, *then* $a = \inf(x_i \mid i \in I)$ *iff* $a = \inf(a_k \mid k \in K)$.

Proof. Consider the sets $K_i \equiv \{i\} \times J_i$. Then, $(K_i \mid i \in I)$ is a partition of the set K. Suppose $a = \sup(a_k \mid k \in K)$. Then, according to assertion 2 of Proposition 1 (1.1.15), we get the equality $a = \sup(\sup(a_k \mid k \in K_i) \mid i \in I)$. Consider the bijective mappings $u_i : J_i \rightarrowtail\!\!\!\!\twoheadrightarrow K_i$ such that $u_i(j) = (i, j)$ for every $j \in J_i$. Then, assertion 1 of Proposition 1 (1.1.15) implies $\sup(a_k \mid k \in K_i) = \sup\left(a_{ij} \mid j \in J_i\right) = x_i$. Thus, $a = \sup(x_i \mid i \in I)$.

Conversely, if $a = \sup(x_i \mid i \in I)$, then by the above, we have the equality $a = \sup(\sup(a_k \mid k \in K_i) \mid i \in I) = \sup(a_k \mid k \in K)$.

The second assertion is proved in the same way. □

Corollary 1. *Let* $(A, \leqslant)$ *be an ordered class,* I *and* J *be non-empty sets,* $(x_i \in A \mid i \in I)$ *be a collection,* $\left(\left(a_{ij} \in A \mid j \in J\right) \mid i \in I\right)$ *be a collection of collections, and* $a \in A$. *Then:*
1) *if* $x_i = \sup\left(a_{ij} \mid j \in J\right)$ *and* $y_j = \sup\left(a_{ij} \mid i \in I\right)$, *then* $a = \sup\left(a_{ij} \mid (i, j) \in I \times J\right)$ *iff* $a = \sup(x_i \mid i \in I) = \sup\left(y_j \mid j \in J\right)$;
2) *if* $x_i = \inf\left(a_{ij} \mid j \in J\right)$ *and* $y_j = \inf\left(a_{ij} \mid i \in I\right)$, *then* $a = \inf\left(a_{ij} \mid (i, j) \in I \times J\right)$ *iff* $a = \inf(x_i \mid i \in I) = \inf\left(y_j \mid j \in J\right)$.

Proof. Consider the collections $(J_i \mid i \in I)$, where $J_i \equiv J$ and $(\iota_i \mid i \in I)$, where $\iota_i \equiv \{i\}$. Then, $\bigcup(\iota_i \mid i \in I) = I$ and $\bigcup(J_i \mid i \in I) = J$. Take also the sets $K_i \equiv \{i\} \times J_i$ and $K \equiv \bigcup(K_i \mid i \in I)$. Since the mapping $u : I \times I \to I$ such that $u(i, i') = i$ is surjective, assertion 1 of Proposition 1 (1.1.10) and assertion 5 of Corollary 2 to Theorem 1 (1.1.13) imply that $K = \bigcup(\{i\} \times J_i \mid i \in I) = \bigcup(\iota_i \times J_{i'} \mid (i, i') \in I \times I) = \bigcup(\iota_i \mid i \in I) \times \bigcup(J_{i'} \mid i' \in I) = I \times J$.

Suppose $a = \sup(x_i \mid i \in I)$. Then, according to Proposition 2, we infer that $a = \sup(a_k \mid k \in K) = \sup\left(a_{ij} \mid (i, j) \in I \times J\right)$.

Conversely, if $a = \sup\left(a_{ij} \mid (i, j) \in I \times J\right)$, then by the above, we have $a = \sup(a_k \mid k \in K) = \sup(\sup(a_k \mid k \in K_i) \mid i \in I) = \sup(x_i \mid i \in I)$.

Similarly, we prove that $\sup\left(y_j \mid j \in J\right) = \sup\left(a_{ij} \mid (j, i) \in J \times I\right)$. Since the mapping $v : J \times I \to I \times J$ such that $v(i, j) = (j, i)$ is bijective, by assertion 1 of Proposition 1 (1.1.10) we get $\sup\left(a_{ij} \mid (i, j) \in I \times J\right) = \sup\left(a_{ij} \mid (j, i) \in J \times I\right)$.

The second assertion is proved in the same way. □

Note that neither the analogue of the properties of the general distributivities 3 and 4 from Theorem 1 (1.1.13) nor the analogue of the property of distributivity 4 from Lemma 1 (1.1.5) are valid for the supremum and the infimum.

Here we shall prove the analogue of the properties of the general distributivities 1 and 2 from Theorem 1 (1.1.13).

Theorem 1. *Let $(A, \leqslant)$ be an ordered class, $(I_m \mid m \in M)$ be a total multivalued collection of sets indexed by a non-empty set M and $U \equiv \prod(I_m \mid m \in M)$. Let $a \in A$, $(\varkappa_m \mid m \in M)$ be a collection of collections of elements $\varkappa_m \equiv (a_{mi} \in A \mid i \in I_m)$, and $(e_m \in A \mid m \in M)$ and $(g_u \in A \mid u \in U)$ be collections of elements. Then:*

1) *if $e_m = \sup(a_{mi} \mid i \in I_m)$ and $g_u = \sup(a_{mu(m)} \mid m \in M)$, then $a = \sup(e_m \mid m \in M)$ iff $a = \sup(g_u \mid u \in U)$ (the general distributivity of supremum);*
2) *if $e_m = \inf(a_{mi} \mid i \in I_m)$ and $g_u = \inf(a_{mu(m)} \mid m \in M)$, then $a = \inf(e_m \mid m \in M)$ iff $a = \inf(g_u \mid u \in U)$ (the general distributivity of infimum).*

Proof. 1. Let $a = \sup(e_m \mid m \in M)$. Then, $a \geqslant a_{mi}$ for every $m \in M$ and $i \in I_m$. If $u \in U$, then $u(m) \in I_m$ implies $a \geqslant a_{mu(m)}$. Thus, $a \geqslant g_u$ for every $u \in U$. Let $x \in A$ and $x \geqslant g_u$ for every $u \in U$. Then, $x \geqslant a_{mu(m)}$ for every $u \in U$ and every $m \in M$. Let $i \in I_m$. By Lemma 4 (1.1.12), $\mathrm{pr}_m[U] = I_m$. Therefore, there is $u \in U$, such that $i = \mathrm{pr}_m(u) = u(m)$. Consequently, $x \geqslant a_{mi}$ for every $i \in I_m$. Thus, $x \geqslant e_m$ for every $m \in M$. This implies $x \geqslant a$. As a result, $a = \sup(g_u \mid u \in U)$.

Conversely, let $a = \sup(g_u \mid u \in U)$. Then, $a \geqslant a_{mu(m)}$ for every $u \in U$ and $m \in M$. If $i \in I_m$, then as above $i = u(m)$ for some $u \in U$. Therefore, $a \geqslant a_{mi}$ for every $i \in I_m$ implies $a \geqslant e_m$ for every $m \in M$. Let $y \in A$ and $y \geqslant e_m$ for every $m \in M$. Then, $y \geqslant a_{mi}$ for every $i \in I_m$ and every $m \in M$. If $u \in U$, then $u(m) \in I_m$ implies $y \geqslant a_{mu(m)}$ for every $m \in M$. Thus, $y \geqslant g_u$ for every $u \in U$ implies $y \geqslant a$. This means that $a = \sup(e_m \mid m \in M)$.

Conclusion 2 is checked completely in the same manner. □

Corollary 1. *Let $(A, \leqslant)$ be an ordered class and J and K be non-empty sets. Let $(x_j \in A \mid j \in J)$, $(y_k \in A \mid k \in K)$ and $(z_{(j,k)} \in A \mid (j, k) \in J \times K)$ be collections of elements and $a, x, y \in A$. Then:*

1) *if $x = \sup(x_j \mid j \in J)$, $y = \sup(y_k \mid k \in K)$ and $z_{(j,k)} = \sup(x_j, y_k)$, then $a = \sup(x, y)$ iff $a = \sup(z_{(j,k)} \mid (j, k) \in J \times K)$;*
2) *if $x = \inf(x_j \mid j \in J)$, $y = \inf(y_k \mid k \in K)$ and $z_{(j,k)} = \inf(x_j, y_k)$, then $a = \inf(x, y)$ iff $a = \inf(z_{(j,k)} \mid (j, k) \in J \times K)$.*

Proof. Take $M = 2$, $I_0 \equiv J$, $I_1 \equiv K$, $a_{0i} \equiv x_i$ for every $i \in J$, $a_{1i} \equiv y_i$ for every $i \in K$, $\varkappa_0 \equiv (x_j \mid j \in J) = (a_{0i} \mid i \in I_0)$, $\varkappa_1 \equiv (y_k \mid k \in K) = (a_{1i} \mid i \in I_1)$ and $U \equiv J \times K = \prod(I_m \mid m \in M)$.

Then, all the necessary conclusions are particular cases of the corresponding conclusions of Theorem 1. □

The formulas of Corollary 1 are called the formulas of the *binary distributivity of the supremum and the infimum.*

Nets and order-convergence

Let, for a moment, A be any class. Any collection $(a_\mu \in A \mid \mu \in M)$ of elements of A indexed by the principal set M of an upward directed preordered set $(M, \leqslant)$ is called a *net in the class A*. A net $y \equiv (b_\nu \in A \mid \nu \in N)$ is called a *subnet of the net* $x \equiv (a_\mu \in A \mid \mu \in M)$, if there exists a mapping $u \equiv (\mu_\nu \in M \mid \nu \in N)\colon\ N \to M$, such that:

1) for every $\mu \in M$, there exists $\nu \in N$, such that $\varkappa \in N$ and $\varkappa \geqslant \nu$ imply $\mu_\varkappa \geqslant \mu$;
2) $x \circ u = y$, i.e. $a_{\mu_\nu} = b_\nu$ for every $\nu \in N$.

A net in the class A is also called a *net of elements of A*. In the similar way, we can define a *net* $x \equiv (A_\mu \subset A \mid \mu \in M)$ *of subclasses of the class A* and its subnet $y \equiv (B_\nu \subset A \mid \nu \in N)$.

The mapping u is called *thinning the net x out.*

Lemma 3. *Let $(M, \leqslant)$ and $(N, \leqslant)$ be upward directed preordered sets and $u \equiv (\mu_\nu \in M \mid \nu \in N)\colon\ N \to M$ be an increasing mapping, such that the subset $u[N]$ is cofinal to the set M. Then, for every class A and every net $x \equiv (a_\mu \in A \mid \mu \in M)$, the composition $x \circ u = (a_{\mu_\nu} \in A \mid \nu \in N)$ is a subnet of the net x.*

Proof. By the condition for every $\mu \in M$, there is $\nu \in N$, such that $\mu_\nu \geqslant \mu$. Therefore, if $\varkappa \in N$ and $\varkappa \geqslant \nu$, then $\mu_\varkappa \geqslant \mu_\nu \geqslant \mu$. □

Now, again, let $(A, \leqslant)$ be a preordered class. If $(a_\mu \in A \mid \mu \in M)$ is an increasing [decreasing] net, then we shall sometimes write $(a_\mu \mid \mu \in M)\uparrow$ [$(a_\mu \mid \mu \in M)\downarrow$]. A net $(a_\mu \mid \mu \in M)$ is called *increasing to an element* $a \in A$, if $(a_\mu)\uparrow$ and $a = \sup(a_\mu \mid \mu \in M)$. A net $(a_\mu \mid \mu \in M)$ is called *decreasing to an element* $a \in A$, if $(a_\mu)\downarrow$ and $a = \inf(a_\mu \mid \mu \in M)$. In these cases, we shall write $(a_\mu \mid \mu \in M) \uparrow a$ and $(a_\mu \mid \mu \in M) \downarrow a$, respectively, or more simple $a_\mu \uparrow a$ and $a_\mu \downarrow a$.

A net $(a_\mu \mid \mu \in M)$ is called *order-convergent to an element* $a \in A$ and a is called an *order-limit of the net* (a_μ), if there exist nets $(b_\mu \in A \mid \mu \in M) \uparrow a$ and $(c_\mu \in A \mid \mu \in M) \downarrow a$, such that $b_\mu \leqslant a_\mu \leqslant c_\mu$ for every $\mu \in M$. A net can converge to many elements; the class of all order-limits of the net (a_μ) is denoted by o-lim$(a_\mu \mid \mu \in M)$. When the net (a_μ) has exactly one order-limit a, then we shall write $a = \text{o-lim}(a_\mu \mid \mu \in M)$. If $\leqslant$ is an order, then every net (a_μ) can have exactly one order-limit.

An element $a \in A$ is called the *limit inferior of a net* $(a_\mu \in A \mid \mu \in M)$, if there exists a net $(b_\mu \in A \mid \mu \in M) \uparrow a$, such that $b_\mu = \inf(a_\nu \mid \nu \in M,\ \nu \geqslant \mu)$. It is denoted by $\underline{\lim}(a_\mu \mid \mu \in M)$. An element $a \in A$ is called the *limit superior of the net* $(a_\mu \in A \mid \mu \in M)$, if there exists a net $(c_\mu \in A \mid \mu \in M) \downarrow a$, such that $c_\mu = \sup(a_\nu \mid \nu \in M,\ \nu \geqslant \mu)$. It is denoted by $\overline{\lim}(a_\mu \mid \mu \in M)$. It is obvious that if these limits exist for the net $\left(a_\mu \mid \mu \in M\right)$, then $\underline{\lim}\, a_\mu \leqslant \overline{\lim}\, a_\mu$.

Lemma 4. *Let $(A, \leqslant)$ be an ordered class, $a \in A$ and $(a_\mu \in A \mid \mu \in M)$, $(b_\mu \in A \mid \mu \in M) \uparrow$, $(c_\mu \in A \mid \mu \in M) \downarrow$ be nets, such that $b_\mu = \inf(a_\nu \mid \nu \in M,\ \nu \geqslant \mu)$ and $c_\mu = \sup(a_\nu \mid \nu \in M,\ \nu \geqslant \mu)$. Then, the following conclusions are equivalent:*
1) $a = \text{o-lim}(a_\mu \mid \mu \in M)$;
2) $a = \underline{\lim}(a_\mu \mid \mu \in M) = \overline{\lim}(a_\mu \mid \mu \in M)$.

Proof. 1) $\vdash$ 2). By the condition, there exist nets $(v_\mu \in A \mid \mu \in M) \uparrow a$ and $(w_\mu \in A \mid \mu \in N) \downarrow a$, such that $v_\mu \leqslant a_\mu \leqslant w_\mu$ for every $\mu \in M$. Let λ and μ be arbitrary elements of M. If $\nu \geqslant \lambda$, then $v_\lambda \leqslant v_\nu \leqslant a_\nu$ implies $v_\lambda \leqslant \inf(a_\nu \mid \nu \in M,\ \nu \geqslant \lambda) = b_\lambda \leqslant b_\nu$. If $\nu \geqslant \mu$, then $w_\mu \geqslant w_\nu \geqslant a_\nu$ implies $w_\mu \geqslant \sup(a_\nu \mid \nu \in M,\ \nu \geqslant \mu) = c_\mu \geqslant c_\nu$.

Since M is upward directed, there is $\nu \in M$, such that $\lambda \leqslant \nu$ and $\mu \leqslant \nu$. Therefore, $v_\lambda \leqslant b_\lambda \leqslant b_\nu \leqslant a_\nu \leqslant c_\nu \leqslant c_\mu \leqslant w_\mu$. From these inequalities, we deduce that $b_\lambda \leqslant \inf(w_\mu \mid \mu \in M) = a$ and $c_\mu \geqslant \sup(v_\lambda \mid \lambda \in M) = a$. Let $x \in A$ and $x \geqslant b_\lambda$ for every $\lambda \in M$. Then, $b_\lambda \geqslant v_\lambda$ implies $x \geqslant \sup(v_\lambda \mid \lambda \in M) = a$. Thus, $a = \sup(b_\lambda \mid \lambda \in M)$. Let now $y \in A$ and $y \leqslant c_\mu$ for every $\mu \in M$. Then, $c_\mu \leqslant w_\mu$ implies $y \leqslant \inf(w_\mu \mid \mu \in M) = a$. Thus, $a = \inf(c_\mu \mid \mu \in M)$. As a result, $a = \underline{\lim}(a_\mu \mid \mu \in M)$ and $a = \overline{\lim}(a_\mu \mid \mu \in M)$.

2) $\vdash$ 1). By the conditions $b_\mu \uparrow a$, $c_\mu \downarrow a$ and $b_\mu \leqslant a_\mu \leqslant c_\mu$ for every $\mu \in M$. This means $a = \text{o-lim}(a_\mu \mid \mu \in M)$. □

Finally, let $(U_i \mid i \in I)$ be a collection of preordered classes $U_i \equiv (A_i, \leqslant_{A_i})$, indexed by a class I. Define a binary relation θ on the product $E \equiv \prod(A_i \mid i \in I)$, setting $(e', e'') \in \theta$ for elements $e' \equiv (a'_i \mid i \in I)$ and $e'' \equiv (a''_i \mid i \in I)$ of the class E, if $a'_i \leqslant_{A_i} a''_i$ for every $i \in I$. It is evident that θ is a preorder on the class E. The preordered class (E, θ) is called the *product of the collection of preordered classes* $(U_i \mid i \in I)$ and will be denoted by $\prod_0(U_i \mid i \in I)$ or simply by $\prod_0(A_i \mid i \in I)$.

Let $U \equiv (A, \leqslant_A)$ and $U' \equiv (A', \leqslant_{A'})$ be preordered classes. Consider the corresponding collection $(U, U') \equiv (X_i \mid i \in 2)$ of preordered classes from 1.1.11, such that $X_0 \equiv U$ and $X_1 \equiv U'$. The preordered class $\prod_0(U, U') \equiv \prod_0(X_i \mid i \in 2)$ will be called the *product of the pair* (U, U') and will be denoted by $U \times_0 U'$ or simply by $A \times_0 A'$.

In the similar way, for preordered classes $U \equiv (A, \leqslant_A)$, $U' \equiv (A', \leqslant_{A'})$, $U'' \equiv (A'', \leqslant_{A''})$, $U''' \equiv (A''', \leqslant_{A'''})$,.... the preordered classes $\prod_0(U, U', U'')$, $\prod_0(U, U', U'', U''')$,... will be called the *products of the triplet* (U, U', U''), quadruplet (U, U', U'', U'''),... and will be denoted by $U \times_0 U' \times_0 U''$, $U \times_0 U' \times_0 U'' \times_0 U'''$,....

1.2 Ordinals and ordinal numbers

This section is devoted to the theory of ordinals and ordinal numbers. We follow J. Neumann's definition of an ordinal, which may seem strange to those habituated to G. Cantor's point of view (where an ordinal is the "order type" of a well-ordered set). In J. Neumann's approach, an ordinal is just equal to the class of all preceding ordinals.

1.2.1 The property of minimality. The principle of induction

A relation θ on a class A (see 1.1.14) is called a *relation with the property of minimality* [*maximality*], if every non-empty subclass B of the class A has a minimal [maximal] element $b \in B$ with respect to the relation θ (see 1.1.14).

A linearly ordered class $(A, \leqslant)$ is called *well-ordered* if the order relation $\leqslant$ has the property of minimality or equivalently if every non-empty subclass B of the class A has the unique smallest element $\operatorname{sm} B \in B$. If $(A, \leqslant)$ is a well-ordered class and B is a subclass of the class A, then $(B, \leqslant)$ is also a well-ordered class.

Lemma 1. *Let A be a class and θ be a total relation on A with the property of minimality. Then:*
1) *the relation θ is antisymmetric and transitive* (see 1.1.14);
2) *the relation $\leqslant \equiv \theta \cup \{(a, a') \in A \times A \mid a = a'\}$ is a linear order on A;*
3) *the linearly ordered class $(A, \leqslant)$ is well-ordered.*

Proof. 1. Let $a, b \in A$, $(a, b) \in \theta$ and $(b, a) \in \theta$. By the condition, the set $\{a, b\}$ has a minimal element x. Suppose that $a \neq b$. If $x = a$, then $b \neq x$ implies $(b, a) = (b, x) \notin \theta$. If $x = b$, then $a \neq x$ implies $(a, b) = (a, x) \notin \theta$. But this contradicts the condition. It follows from this contradiction that our supposition is not true. Thus, $a = b$. This means that θ is antisymmetric.

Let $(a, b) \in \theta$ and $(b, c) \in \theta$. Suppose that $(a, c) \notin \theta$. Since θ is total, we conclude that either $a = c$ or $(c, a) \in \theta$. If $a = c$, then $(a, b) \in \theta$ and $(b, a) \in \theta$ imply $a = b$, where $b = c$. As a result, $(a, c) \in \theta$, but this contradicts $(a, c) \notin \theta$.

Suppose now that $a \neq c$. Then, $(c, a) \in \theta$. Besides, $(a, b) \in \theta$, $(b, c) \in \theta$ and $(a, c) \notin \theta$ imply $b \neq c$ and $a \neq b$. By the condition, the set $\{a, b, c\}$ has a minimal element x. If $x = a$, then $x = a \neq c$ implies $(c, a) = (c, x) \notin \theta$ by virtue of the property of minimality. Similarly, if $x = b$, then $x = b \neq a$ implies $(a, b) = (a, x) \notin \theta$. Finally, if $x = c$, then $x = c \neq b$ implies $(b, c) = (b, x) \notin \theta$. In all the cases, we got the contradiction. Thus, our supposition that $(a, c) \notin \theta$ is not true, i. e. $(a, c) \in \theta$. This means that θ is transitive.

2. Since θ is transitive, the relation $\leqslant \equiv \theta \cup \{(a, a') \in A \times A \mid a = a'\}$ is a preorder on A. Since θ is antisymmetric, the relation $\leqslant$ is an order. Finally, since θ is total, the order $\leqslant$ is linear.

3. Let $\varnothing \neq B \subset A$. Then, by the condition B has a minimal element $b \in B$ with respect to θ. Let $b' \in B$ and $b' \leqslant b$. Suppose that $(b', b) \in \theta$. By the property of minimality for θ, we conclude that $b' = b$. If $(b', b) \notin \theta$, then from the definition of the relation $\leqslant$ we conclude that $b' = b$. This means that the order relation $\leqslant$ has the property of minimality. □

Theorem 1 (The *principle of induction*). *Let $(A, \leqslant)$ be an ordered class and the relation of order $\leqslant$ has the property of minimality. If B is a non-empty subclass of the class A such that the conditions $a \in A$ and $]\leftarrow, a[\subset B$ imply $a \in B$, then $B = A$.*

Proof. Suppose that $C \equiv A \setminus B \neq \varnothing$. Then, C has a minimal element $x \in C$. Consider the interval $Y \equiv]\leftarrow, x[$. Suppose that $Y = \varnothing$. Then, $Y \subset B$ implies $x \in B$, but this is impossible. Suppose now that $Y \neq \varnothing$. If $y \in Y$, then $y < x$ implies $y \notin C$, because x is a minimal element of C. Thus, $y \in B$. This means that $Y \subset B$. By the condition, this implies $x \in B$, but this is impossible. In both the cases, we get the contradiction. Thus, $B = A$. □

Corollary 1. *Let $(A, \leqslant)$ be an ordered class and the relation of order $\leqslant$ has the property of minimality. Let $\varphi(x)$ be a formula such that every connected variable of φ is a variable for sets. Consider the subclass $B \equiv \{x \in A \mid \varphi\}$ (see 1.1.5) of the class A. If B is a non-empty subclass such that the conditions $a \in A$ and $]\leftarrow, a[\subset B$ imply $a \in B$, then $B = A$.*

1.2.2 The relation of Neumann on the universal class. Ordinals

The relation $\rho \equiv \{(x, y) \in \mathfrak{U} \times \mathfrak{U} \mid x \in y \vee x = y\}$ on the universal class $\mathfrak{U}$ is called the *relation of J. Neumann*. Along with the relation ρ, consider also the correspondence $\sigma \equiv \{\langle a, b\rangle \in \mathfrak{U} * \mathfrak{U} \mid a \in b \vee a = b\}$.

Lemma 1. *There is the canonical bijection $u\colon \rho \rightarrowtail\!\!\!\!\twoheadrightarrow \sigma$ such that $u(x, y) = \langle x, y\rangle$ for every $(x, y) \in \rho$.*

Proof. Consider the correspondence $u \equiv \{\langle (x, y), \langle a, b\rangle\rangle \in \rho * \sigma \mid a = x \wedge b = y\}$. From $\langle (x, y), \langle x, y\rangle\rangle \in u$, we conclude that $\operatorname{dom} u = \rho$. Let $\langle (x, y), \langle a, b\rangle\rangle \in u$ and $\langle (x, y), \langle c, d\rangle\rangle \in u$. Then, $a = x$, $b = y$, $c = x$ and $d = y$ imply $\langle a, b\rangle = \langle c, d\rangle = \langle x, y\rangle$. Hence u is single-valued. Thus, u is a mapping from ρ into σ such that $u(x, y) = \langle x, y\rangle$. If $\langle a, b\rangle \in \sigma$, then $u(a, b) = \langle a, b\rangle$ means that u is surjective. Finally, if $u(x, y) = u(x', y')$, then $\langle x, y\rangle = \langle x', y'\rangle$ implies $x = x'$ and $y = y'$ in virtue of Proposition 2 (1.1.6). Hence $(x, y) = (x', y')$, i. e. u is injective. □

Proposition 1. *The classes ρ and σ are proper classes.*

Proof. Suppose at first that σ is a set. Then, by Lemma 3 (1.1.6), $\{\sigma\}$ is a set and $\sigma \in \{\sigma\}$. By Lemma 3 (1.1.5), σ and $\{\sigma\}$ belong to $\mathfrak{U}$, where $\langle\sigma, \{\sigma\}\rangle \in \mathfrak{U} * \mathfrak{U}$. Thus, $\langle\sigma, \{\sigma\}\rangle \in \sigma$. As a result, we get the following chain $\sigma \in \{\sigma\} \in \langle\sigma, \{\sigma\}\rangle \in \sigma$.

Consider the class $A \equiv \{x \mid x = \sigma \vee x = \{\sigma\} \vee x = \langle\sigma, \{\sigma\}\rangle\}$. By axiom A6 (1.1.11), there exists an element $a \in A$ such that $a \cap A = \varnothing$. By the definition $a \cap A = \{z \mid (z = \sigma \vee z = \{\sigma\} \vee z = \langle\sigma, \{\sigma\}\rangle) \vee z \in a\}$.

If $a = \sigma$, then $\langle\sigma, \{\sigma\}\rangle \in \sigma$ implies $z = \langle\sigma, \{\sigma\}\rangle \in a \cap A$. If $a = \{\sigma\}$, then $\sigma \in \{\sigma\}$ implies $z = \sigma \in a \cap A$. If $a = \langle\sigma, \{\sigma\}\rangle$, then $\{\sigma\} \in \langle\sigma, \{\sigma\}\rangle$ implies $z = \{\sigma\} \in a \cap A$. In all the three cases, we get the contradiction to the equality $a \cap A = \varnothing$.

This means that our supposition is not valid, and σ is a proper class. Suppose now that ρ is a set. Then, by Lemma 1 and axiom A5′ (1.1.11) $\sigma = \operatorname{rng} u$ is a set, but this is not true. Thus, ρ is a proper class. □

Let A be a class. By Lemma 4 (1.1.5) $A \subset \mathfrak{U}$, where $A \times A \subset \mathfrak{U} \times \mathfrak{U}$. Therefore, we can consider the restriction $\theta_A \equiv \rho \cap (A \times A)$ of J. Neumann's relation ρ on the subclass A.

A class A is called *transitive* or *complete* if $x \in a \in A$ implies $x \in A$. A transitive class A, connected by J. Neumann's relation θ_A, so that $a, b \in A$ implies $a \in b$, $b \in a$, or $a = b$, is called an *ordinal (in the sense of J. Neumann)*. An ordinal, which is a set, is called an *ordinal number*. According to 1.1.11, we obtain that $0 \equiv \varnothing$ is an ordinal number.

Proposition 2. *Let α be an ordinal. Then, θ_α is a linear order on α, and the linearly ordered class (α, θ_α) is well-ordered.*

Proof. Let $\varnothing \neq x \subset \alpha$. By axiom A6 (1.1.11), there exists $a \in x$ such that $a \cap x = \varnothing$. Take any element $b \in x$ such that $(b, a) \in \theta_\alpha$. From $a \cap x = \varnothing$, it follows that $b \notin a$. Thus, $b = a$. This means that a is a minimal element of x. Thus, θ_α is a total relation on α with the property of minimality. Besides, $\theta_\alpha = \theta_\alpha \cup \{(a, a') \in \alpha \times \alpha \mid a = a'\}$. Therefore, by virtue of Lemma 1 (1.2.1), θ_α is a linear order on α, and the linearly ordered class (α, θ_α) is well-ordered. □

According to this proposition, if α is an ordinal and $a, b \in \alpha$, then along with $a \in b \vee a = b$ we can write $a \leqslant b$.

Corollary 1. *Let α be an ordinal and $a, b \in \alpha$. Then, $a < b$ iff $a \in b$.*

Proof. This equivalence follows from the definitions and Lemma 2 (1.1.11). □

According to this, along with $\alpha \in \beta$, we can write $\alpha < \beta$ for ordinal numbers α and β.

1.2.3 Properties of ordinals

Lemma 1. *Let α be an ordinal, $\beta \subset \alpha$, $\beta \neq \alpha$, and the class β is transitive. Then, $\beta \in \alpha$.*

Proof. By virtue of Proposition 2 (1.2.2), the subclass $\alpha \setminus \beta$ has a minimal element y. Check that $\beta = y$. Let $x \in y$. Since α is transitive, we conclude that $x \in \alpha$. Then, by Corollary 1 to Proposition 2 (1.2.2), $x \in y$ implies $x < y$. Thus, $x \notin \alpha \setminus \beta$, where $x \in \beta$. As a result, $y \subset \beta$. Conversely, let $x \in \beta$. From $y \notin \beta$, we conclude that $y \neq x$. Suppose that $y \in x$. Since β is transitive, we conclude that $y \in \beta \cap (\alpha \setminus \beta) = \varnothing$. It follows from this contradiction that $x \in y$. As a result, $\beta \subset y$, where $\beta = y \in \alpha$. □

Lemma 2. *Let α and β be ordinals. Then, either $\alpha \subset \beta$ or $\beta \subset \alpha$.*

Proof. The class $\alpha \cap \beta$ is transitive. By Lemma 1, either $\alpha \cap \beta = \alpha$ or $\alpha \cap \beta \in \alpha$. In the first case, $\alpha \subset \beta$. In the second case, $\alpha \cap \beta \notin \beta$. In fact, if $\alpha \cap \beta \in \beta$, then $\alpha \cap \beta \in \alpha \cap \beta$, but this is impossible in virtue of Lemma 2 (1.1.11). Now, by virtue of Lemma 1, we deduce that $\alpha \cap \beta = \beta$, where $\beta \subset \alpha$. □

Corollary 1. *Let α and β be ordinals. Then, $\alpha \in \beta$, $\beta \in \alpha$, or $\alpha = \beta$.*

Lemma 3. *Let α be an ordinal and $\beta \in \alpha$. Then, β is an ordinal number.*

Proof. Let $x, y \in \beta$. Then, $x, y \in \alpha$. By virtue of Proposition 2 (1.2.2) we get $x \in y$, $y \in x$, or $x = y$. Thus, the relation θ_β is connecting on β.

Let $y \in b \in \beta$. Then, $b \in \beta \in \alpha$ implies $y \in b \in \alpha$. This again implies $y \in \alpha$. Thus, β, b and y belong to α. Therefore, by Corollary 1 to Proposition 2 (1.2.2) $y \in b$ and $b \in \beta$ imply $y < b$ and $b < \beta$. Thus, $y < \beta$, where $y \in \beta$. This means that β is transitive. Thus, β is an ordinal. □

By virtue of axiom scheme AS2 (1.1.5), the class $\mathrm{Ord} \equiv \{x \mid x \text{ is an ordinal}\}$ consists of all ordinal numbers. Consider J. Neumann's relation $\theta \equiv \rho \cap (\mathrm{Ord} \times \mathrm{Ord})$ on the class Ord.

Theorem 1. *The class* Ord *is an ordinal, but not an ordinal number.*

Proof. By virtue of Lemma 3, Ord is transitive. By virtue of Corollary 1 to Lemma 2, θ is connecting on Ord. Therefore, Ord is an ordinal. Suppose that Ord is an ordinal number. Then, $\mathrm{Ord} \in \mathrm{Ord}$. But by Lemma 2 (1.1.11), this is impossible. □

Corollary 1. Ord *is the unique ordinal, which is not an ordinal number.*

Proof. The conclusion follows from Theorem 1 and Corollary 1 to Lemma 2. □

Corollary 2. *The class* (Ord, θ) *is well-ordered.*

Proof. The conclusion follows from Theorem 1 and Proposition 2 (1.2.2). □

Lemma 4. *Let* α *be an ordinal and* $a, b \in \alpha$. *Then,* $a \leqslant b$ *iff* $a \subset b$.

Proof. By Lemma 3, a and b are ordinals. Let $a \leqslant b$, i. e. $a \in b \vee a = b$. If $a = b$, then $a \subset b$. If $a \in b$, then $x \in a \in b$ implies $x \in b$, where $a \subset b$. Conversely let $a \subset b$. If $a = b$, then $a \leqslant b$. If $a \neq b$, then by Lemma 1 we get $a \in b$, where $a \leqslant b$. □

Lemma 5. *Let* α *be an ordinal number. Then,* $\alpha = \{x \mid x \in \text{Ord} \wedge x < \alpha\} \subset \text{Ord}$.

Proof. Let $\beta \in \alpha$. Then, by Lemma 3, we obtain that β is an ordinal number, i. e. $\beta \in$ Ord. Besides, $\alpha \in$ Ord. Therefore, from Theorem 1 and Corollary 1 to Proposition 2 (1.2.2) we deduce that $\beta < \alpha$. This means that $\beta \in \{x \mid x \in \text{Ord} \wedge x < \alpha\}$. Conversely let $\beta \in \{x \mid x \in \text{Ord} \wedge \beta < \alpha\}$. Then, $\beta \in$ Ord and $\beta < \alpha$. By the same reason, we conclude that $\beta \in \alpha$. □

Corollary 1. *Let* $\alpha, \beta \in$ Ord *and* $\alpha \leqslant \beta$. *Then,* $\alpha = \{x \mid x \in \beta \wedge x < \alpha\} \equiv]\leftarrow, \alpha[$ *in* β.

Lemma 6. *Let* $A \subset$ Ord *be some class of ordinal numbers and* $(a_a \subset \mathfrak{U} \mid a \in A) = \varphi^{-1}(\text{id}_A)$ *be the multivalued collection from Corollary 1 to Lemma 3 (1.1.9) such that* $a_a \equiv a$ *for every* $a \in A$. *Then, the class* $\alpha \equiv \bigcup(a_a \mid a \in A)$ *is an ordinal.*

Proof. Let $x \in y \in \alpha$. Then, $y \in b$ for some $b \in A$. Therefore, $x \in y \in b$ implies $x \in b$. Thus, $x \in \alpha$. This means that the class α is transitive.

Now, let $x, y \in \alpha$. Then, $x \in a$ and $y \in b$ for some $a, b \in A$. By Lemma 3, we have that x and y are ordinal numbers. By Corollary 1 to Lemma 2, $x \in y$, $y \in x$, or $x = y$. This means that θ_α connects α. □

For every ordinal number $\alpha \in$ Ord, the class $\alpha \cup \{\alpha\}$ is denoted by $\alpha + 1$.

Proposition 1. *Let* $\alpha \in$ Ord. *Then,* $\alpha + 1$ *is an ordinal number,* $\alpha < \alpha + 1$, *and* $\alpha + 1$ *is a minimal element of the subclass* $\{y \in \text{Ord} \mid \alpha < y\}$ *of the class* Ord *with respect to the relation* θ.

Proof. Let $x \in a \in \alpha + 1 \equiv \alpha \cup \{\alpha\}$. If $a \in \alpha$, then $x \in \alpha$. If $a \in \{\alpha\}$, then by Lemma 3 (1.1.6), $a = \alpha$, where $x \in \alpha$. Thus, $x \in \alpha + 1$. This means that the class $\alpha + 1$ is transitive.

Let $a, b \in \alpha+1$. If $a, b \in \alpha$, then $a \in b$, $b \in a$, or $a = b$. If $a, b \in \{\alpha\}$, then $a = \alpha = b$. If $a \in \alpha$ and $b \in \{\alpha\}$, then $b = \alpha$ implies $a \in b$. Finally, if $a \in \{\alpha\}$ and $b \in \alpha$, then $a = \alpha$ implies $b \in a$. This means that the class $\alpha + 1$ is connected by the relation $\theta_{\alpha+1}$.

Thus, $\alpha + 1$ is an ordinal. By Lemma 3 (1.1.6), $\{\alpha\}$ is a set. Therefore, by axiom A4 (1.1.6), $\alpha + 1$ is a set as well. Thus, $\alpha + 1$ is an ordinal number.

By Lemma 3 (1.1.6), $\alpha \in \{\alpha\}$. Therefore, $\alpha \in \alpha + 1$. Using Theorem 1 and Corollary 1 to Proposition 2 (1.2.2), we conclude that $\alpha < \alpha + 1$.

Thus, $\alpha+1 \in B \equiv \{y \in \mathrm{Ord} \mid \alpha < y\}$. Let $\beta \in B$ and $\beta \leqslant \alpha+1$. Suppose that $\beta \neq \alpha+1$. Using the same arguments as above, we conclude that $\alpha < \beta$ implies $\alpha \in \beta$ and $\beta < \alpha+1$ implies $\beta \in \alpha + 1$. If $\beta \in \alpha$, then by virtue of Lemma 3 (1.1.11), this contradicts $\alpha \in \beta$. If $\beta \in \{\alpha\}$, then $\beta = \alpha$ by virtue of Lemma 2 (1.1.11) contradicts $\alpha \in \beta$. This means that $\alpha+1$ is a minimal element of the subclass B. ☐

Corollary 1. *Let* $\alpha \in \mathrm{Ord}$. *Then,* $\alpha + 1 \subset \mathrm{Ord}$ *and* $\bigcup(a_a \mid a \in \alpha + 1) = \alpha$.

Proof. Let $x \in \alpha$. Take $a \equiv \alpha$. Then, $x \in \alpha \in \{\alpha\} \subset \alpha+1$ means that $x \in a_a$ and $a \in \alpha+1$. Thus, $x \in \beta \equiv \bigcup(a_a \mid a \in \alpha + 1)$.

Conversely, let $x \in \beta$. Then, $x \in a_a$ for some $a \in \alpha + 1$. If $a \in \alpha$, then $x \in a \in \alpha$ implies $x \in \alpha$. If $a \in \{\alpha\}$, then $a = \alpha$ implies $x \in \alpha$. Thus, $\beta = \alpha$. ☐

An ordinal α is called a *limit ordinal* if $\alpha \neq \varnothing$ and $\alpha \neq \beta+1$ for every ordinal number β.

Corollary 2. Ord *is a limit ordinal.*

Lemma 7. *Let* $A \subset \mathrm{Ord}$ *be some set of ordinal numbers. Then, there exists* $\alpha \in \mathrm{Ord}$ *such that* $\alpha \geqslant a$ *for every* $a \in A$.

Proof. By Lemma 6, the class $\alpha \equiv \bigcup(a_a \mid a \in A)$ is an ordinal. By axiom A5″ (1.1.11), α is a set. Thus, $\alpha \in \mathrm{Ord}$. If $a \in A$, then $a \subset \alpha$. Therefore, by Lemma 4 we get $a \leqslant \alpha$. ☐

Lemma 8. *Let* $A \subset \mathrm{Ord}$ *be some set of ordinal numbers. Then, the class* $\mathrm{Ord} \setminus A$ *is non-empty and has a minimal element.*

Proof. By virtue of Theorem 1, we get $B \equiv \mathrm{Ord} \setminus A \neq \varnothing$. By virtue of Corollary 2 to Theorem 1, B has a minimal element. ☐

Lemma 9. *Let* $\alpha, \beta \in \mathrm{Ord}$ *and* α *is a maximal element of the subset* $\beta \subset \mathrm{Ord}$. *Then,* $\beta = \alpha + 1$.

Proof. By Corollary 1 to Proposition 2 (1.2.2), $\alpha < \beta$. It follows from Proposition 1 that $\alpha + 1$ is a minimal element of the subclass $Y \equiv \{y \in \mathrm{Ord} \mid \alpha < y\}$. Suppose that $\alpha + 1 < \beta$, i. e. $\alpha + 1 \in \beta$. Then, $\alpha + 1 \leqslant \alpha$ because α is a maximal element. But by Proposition 1, we get $\alpha < \alpha + 1$. It follows from this contradiction that $\alpha + 1 \geqslant \beta$. Since $\beta \in Y$ and $\alpha + 1$ is a minimal element of Y, we infer that $\alpha + 1 \leqslant \beta$. As a result, $\alpha + 1 = \beta$. ☐

Lemma 10. *Let α be a limit ordinal. Then, $\alpha = \bigcup(a_a \mid a \in \alpha)$.*

Proof. Let $x \in \alpha$. Then, $x + 1 \leqslant \alpha$. Since α is a limit ordinal, we infer that $a \equiv x + 1 < \alpha$. Therefore, by virtue of Proposition 1 we see that $x \in a \in \alpha$ implies $x \in \bigcup(a_a \mid a \in \alpha)$. Conversely, if $x \in \bigcup(a_a \mid a \in \alpha)$, then $x \in a$ for some $a \in \alpha$. By virtue of the property of transitivity, $x \in \alpha$. As a result, we get the necessary equality. □

1.2.4 Relations between well-ordered sets

Recall that all the necessary definitions can be found in 1.1.14 and 1.1.15.

Lemma 1. *Let $(A, \leqslant)$ be a well-ordered class and $u\colon A \to A$ be a strictly increasing mapping. Then, $a \leqslant u(a)$ for every $a \in A$.*

Proof. Suppose that there is $a \in A$ such that $u(a) < a$. Then, the non-empty subclass $B \equiv \{a \in A \mid u(a) < a\}$ has a minimal element $b \in B$. Therefore, $u(b) < b$ implies $u(u(b)) < u(b)$. Thus, $u(b) \in B$. But the inequality $u(b) < b$ contradicts the minimality of b. □

Proposition 1. *Let $(A, \leqslant)$ be a well-ordered class, $a, b \in A$ and $a \neq b$. Then:*
1) *the well-ordered classes $(A, \leqslant)$ and $(]\!\leftarrow, a[, \leqslant)$ are not order equivalent;*
2) *the well-ordered classes $(]\!\leftarrow, a[, \leqslant)$ and $(]\!\leftarrow, b[, \leqslant)$ are not order equivalent.*

Proof. Denote $]\!\leftarrow, a[$ by A_a and $]\!\leftarrow, b[$ by A_b.

1. Suppose that there exists an isotone bijection $u\colon A \rightarrowtail\!\!\!\!\twoheadrightarrow A_a$. Then, we can consider u as an isotone mapping $u\colon A \to A$. By virtue of Lemma 1 (1.1.15), u is strictly increasing. Therefore, it follows from Lemma 1 that $a \leqslant u(a)$. But this inequality contradicts the inequality $u(a) < a$.It follows from this contradiction that $(A, \leqslant) \not\approx (A_a, \leqslant)$.

2. If $a < b$, then $a \in A_b$ and A_a is an open initial interval in the well-ordered class $(A_b, \leqslant)$. Therefore, by virtue of conclusion 1, $(A_b, \leqslant) \not\approx (A_a, \leqslant)$. If $b < a$, then the arguments are the same. □

Lemma 2. *Let $(A, \leqslant)$ and $(B, \leqslant)$ be well-ordered classes, and $u\colon A \rightarrowtail\!\!\!\!\twoheadrightarrow B$ and $v\colon A \rightarrowtail\!\!\!\!\twoheadrightarrow B$ be isotone bijective mappings. Then, $u = v$.*

Proof. By virtue of Lemma 1 (1.1.15), the mappings u, v, u^{-1} and v^{-1} are strictly increasing. Therefore, the mapping $w \equiv u^{-1} \circ v\colon A \to A$ is also strictly increasing. Then, by Lemma 1, we get $a \leqslant w(a)$ for every $a \in A$. Applying u, we get $u(a) \leqslant v(a)$ for every $a \in A$. Interchanging the roles of u and v in this argument, we obtain $v(a) \leqslant u(a)$ for every $a \in A$. From these inequalities, we conclude that $u(a) = v(a)$ for every $a \in A$, where $u = v$. □

Lemma 3. *Let* $(A, \leqslant)$ *and* $(B, \leqslant)$ *be well-ordered classes,* $a', a'' \in A$, $b', b'' \in B$ *and* $u':]\leftarrow, a'[\rightarrowtail\!\!\twoheadrightarrow]\leftarrow, b'[$ *and* $u'':]\leftarrow, a''[\rightarrowtail\!\!\twoheadrightarrow]\leftarrow, b''[$ *be isotone bijective mappings. Then, either* $u' \subset u''$ *or* $u'' \subset u'$.

Proof. Denote $]\leftarrow, a'[$ by A', $]\leftarrow, a''[$ by A'', $]\leftarrow, b'[$ by B' and $]\leftarrow, b''[$ by B''. It is clear that either $A' \subset A''$ or $A'' \subset A'$. Suppose that $X \equiv \{a \in A' \cap A'' \mid u'a \neq u''a\} \neq \varnothing$. Then, X has a minimal element x. Thus, $u'x \neq u''x$. If $u'x < u''x \in B''$, then $u'x \in B''$. Therefore, there exists $a \in A''$ such that $u'x = u''a$. Then, $u''a < u''x$ implies $a < x$. Since x is a minimal element, we infer that $a \notin X$, i. e. $u'a = u''a = u'x$. But then $a = x$ contradicts $a < x$. It follows from this contradiction that $X = \varnothing$. If $u'x > u''x$, then the arguments are analogous. □

1.2.5 The correspondence between well-ordered sets and ordinal numbers

Theorem 1. *Let* $(A, \leqslant)$ *be a well-ordered set. Then, there are the unique ordinal number* α *and the unique isotone bijective mapping* $u: A \rightarrowtail\!\!\twoheadrightarrow \alpha$ *such that the well-ordered sets* $(A, \leqslant)$ *and* $(\alpha, \leqslant)$ *are order equivalent with respect to* u.

Proof. Take a minimal element m of the class A and the ordinal number $0 \equiv \varnothing$. Then, we have the unique isotone bijection $\varnothing:]\leftarrow, m[\rightarrowtail\!\!\twoheadrightarrow 0$.

Consider the class U of all isotone mappings $u: A_a \rightarrowtail\!\!\twoheadrightarrow \alpha$ for some $a \in A$ and $\alpha \in \mathrm{Ord}$. It follows from the first indentation that $U \neq \varnothing$. By the definition, $\mathrm{dom}\, u = A_a \subset A$. By Lemma 5 (1.2.3), $\alpha =]\leftarrow, \alpha[$ in the well-ordered class $(\mathrm{Ord}, \leqslant)$. Thus, $\mathrm{rng}\, u = \alpha \subset \mathrm{Ord}$.

Let $u, v \in U$, i. e. u and v are isotone bijective mappings $u: A_a \rightarrowtail\!\!\twoheadrightarrow \alpha$ and $v: A_b \rightarrowtail\!\!\twoheadrightarrow \beta$ for some $a, b \in A$ and $\alpha, \beta \in \mathrm{Ord}$. Then, $\alpha =]\leftarrow, \alpha[$ and $\beta =]\leftarrow, \beta[$ in the well-ordered class $(\mathrm{Ord}, \leqslant)$. Therefore, by Lemma 3 (1.2.4), we have either $u \subset v$ or $v \subset u$.

Consider the classes $C \equiv \{x \mid \exists u \in U\, (x \in \mathrm{dom}\, u)\}$ and $\gamma \equiv \{y \mid \exists u \in U\, (y \in \mathrm{rng}\, u)\}$. Let $\sigma \in y \in \gamma$. Then, $\sigma \in y \in \mathrm{rng}\, u \in \mathrm{Ord}$ for some $u \in U$ implies $\sigma \in \mathrm{rng}\, u$, and so $\sigma \in \gamma$. This means that the class γ is transitive. Let $y, z \in \gamma$. Then, $y \in \mathrm{rng}\, u$ and $z \in \mathrm{rng}\, v$ for some $u, v \in U$. By Lemma 3 (1.2.3), y and z are ordinal numbers. Therefore, by Corollary 1 to Lemma 2 (1.2.3), $y \in z$, $z \in y$, or $y = z$. This means that θ_γ connects γ (see 1.2.2). Thus, γ is an ordinal.

Consider also the class $w \equiv \{z \mid \exists u \in U\, (z \in u)\}$. According to Lemma 5 (1.1.8), w is the surjective mapping $w: C \twoheadrightarrow \gamma$ such that $w|\,\mathrm{dom}\, u = u$ for every $u \in U$. Let $x, y \in C$ and $x \leqslant y$. Then, $y \in \mathrm{dom}\, u$ for some $u \in U$. Therefore, $x \in \mathrm{dom}\, u$. Consequently, $w(x) = u(x) < u(y) = w(y)$. Conversely let $w(x) \leqslant w(y)$. Then, $u(x) = w(x) \leqslant w(y) = u(y)$ implies $x \leqslant y$. This means that w is isotone. Therefore, by Lemma 1 (1.1.15), w is injective. Thus, w is isotone bijective mapping from the well-ordered class $(C, \leqslant)$ onto the well-ordered class $(\gamma, \leqslant)$. According to Lemma 2 (1.2.4), such a mapping w is unique.

Suppose that $(C, \leqslant)$ is order equivalent to $(\beta, \leqslant)$ for some $\beta \in \mathrm{Ord}$ with respect to an isotone bijective mapping $v\colon\ C \rightarrowtail\!\!\!\!\twoheadrightarrow \beta$. By virtue of Corollary 2 to Theorem 1 (1.2.3), $\beta < \gamma$, $\gamma < \beta$, or $\beta = \gamma$. Consider the case $\beta < \gamma$. By Corollary 1 to Lemma 5 (1.2.3), $\beta = \,]\!\leftarrow, \beta[$ in γ. By virtue of Lemma 1 (1.1.15), the mapping w^{-1} is isotone. Consequently, the bijective mapping $v \circ w^{-1}\colon\ \gamma \rightarrowtail\!\!\!\!\twoheadrightarrow \beta$ is also isotone. Thus, $(\gamma, \leqslant) \approx (]\!\leftarrow, \beta[, \leqslant)$. But this contradicts condition 1 of Proposition 1 (1.2.4). In the case $\gamma < \beta$, the arguments are the same. It follows from these contradictions that $\beta = \gamma$, i. e. the ordinal γ for the class $(C, \leqslant)$ is unique.

Thus, if $C = A$, then the theorem is proven. Assume now that $C \neq A$. Then, the class $A \setminus C$ has a minimal element x. Consider the initial interval A_x. Let $a \in A_x$. Then, $a \in C$, i. e. $A_x \subset C$. Conversely, let $c \in C$. Suppose that $x \leqslant c$. Since $c \in A_a$ for some $a \in A$, we conclude that $x \in A_a \subset C$, but this is not so. It follows from this contradiction that $c < x$, i. e. $c \in A_x$. This means that $C = A_x$.

Consider the new class $B \equiv C \cup \{x\}$. Define a mapping $v\colon\ B \to \gamma{+}1$, setting $v|C \equiv w$ and $v(x) \equiv \gamma$. Let $b \in B$ and $b < x$. Then, $v(b) = w(b) \in \gamma$ implies $v(b) < \gamma = v(x)$. Thus, v is monotone. Conversely, let $v(a) \leqslant v(b)$ for $a, b \in B$. If $a, b \in C$, then $w(a) \leqslant w(b)$ implies $a \leqslant b$. If $a \in C = A_x$ and $b = x$ then $a < x = b$. Finally, if $b \in C$ and $a = x$, then $w(b) = v(b) \geqslant v(a) = v(x) = \gamma$ and $w(b) \in \gamma$, i. e. $w(b) < \gamma$, but this is impossible. As a result, $v(a) \leqslant v(b)$ implies $a \leqslant b$. This means that v is isotone. By virtue of Lemma 1 (1.1.15), v is injective. Since w is surjective, v is also surjective. Thus, v is an isotone bijection.

Suppose that $A \setminus C \neq \{x\}$. Then, the non-empty class $A \setminus B$ has a minimal element y. Suppose that $y \leqslant x$. Then, either $y = x \in B$ or $y < x$. But the latter case means that $y \notin A \setminus C$, i. e. $y \in C \subset B$. It follows from this contradiction that $y > x$. Therefore, $b \in B$ implies either $b \in C = A_x$, i. e. $b < x < y$, or $b = x < y$, where $b \in A_y$. Conversely, $a \in A_y$ implies $a \notin A \setminus B$ by virtue of the minimality of y, i. e. $a \in B$. This means that $B = A_y$. Thus, $v \in U$. This implies $\gamma + 1 = \mathrm{rng}\, v \subset \gamma$. By virtue of Lemma 4 (1.2.3), $\gamma + 1 \leqslant \gamma$ in $(\mathrm{Ord}, \leqslant)$. But this contradicts the inequality $\gamma + 1 > \gamma$ from Proposition 1 (1.2.3). It follows from this contradiction that $A \setminus C = \{x\}$, i. e. $A = B$ and $v\colon\ A \rightarrowtail\!\!\!\!\twoheadrightarrow \gamma + 1$ is the necessary isotone bijection.

Entirely in the same manner as was made above for the mapping w and the ordinal number γ, it is checked that for the class A the mapping v and the ordinal number γ are unique. □

This unique ordinal number α from Theorem 1 is called *the order type of the well-ordered set* $(A, \leqslant)$ and will be denoted by $\mathrm{ord}(A, \leqslant)$.

Proposition 1. *Let* $(A, \leqslant)$ *and* $(B, \leqslant)$ *be well-ordered sets. Then,*
1) $\mathrm{ord}(A, \leqslant) < \mathrm{ord}(B, \leqslant)$ *iff* $(A, \leqslant) \approx (]\!\leftarrow, b[, \leqslant)$ *for some* $b \in B$;
2) $\mathrm{ord}(A, \leqslant) > \mathrm{ord}(B, \leqslant)$ *iff* $(]\!\leftarrow, a[, \leqslant) \approx (B, \leqslant)$ *for some* $a \in A$;
3) $\mathrm{ord}(A, \leqslant) = \mathrm{ord}(B, \leqslant)$ *iff* $(A, \leqslant) \approx (B, \leqslant)$.

Proof. 1. Consider the isotone bijections $u\colon A \rightarrowtail\!\!\!\!\twoheadrightarrow \alpha$ and $v\colon B \rightarrowtail\!\!\!\!\twoheadrightarrow \beta$ from Theorem 1. Let $\alpha < \beta$. Since $\alpha \in \beta$, we can take an element $b \equiv v^{-1}(\alpha)$. Let $a \in A$. Then, $u(a) \in \alpha$ implies $u(a) < \alpha$. By virtue of Lemma 1 (1.1.15), the bijective mapping v^{-1} is isotone and strictly increasing. Therefore, $v^{-1}(u(a)) < v^{-1}(\alpha) = b$ implies $v^{-1}(u(a)) \in B_b \equiv]\leftarrow, b[$. Now, let $c \in B_b$. Then, $c < b$ implies $v(c) < v(b) = \alpha$, i. e. $v(c) \in \alpha$. Take an element $a \equiv u^{-1}(v(c))$. Then, $v^{-1}(u(a)) = c$. This means that $v^{-1} \circ u$ is a surjective mapping from A onto B_b. It is evident that $v^{-1} \circ u$ is isotone and bijective. Thus, $(A, \leqslant) \approx (B_b, \leqslant)$.

Conversely, let $(A, \leqslant) \approx (B_b, \leqslant)$ with respect to an isotone bijection $t\colon A \rightarrowtail\!\!\!\!\twoheadrightarrow B_b$. Consider the ordinal number $\gamma \equiv v(b)$ and the mapping $s \equiv v \circ t \circ u^{-1}\colon \alpha \to \beta$. By virtue of Lemma 1 (1.1.15), the mappings u^{-1}, v and t are isotone, bijective, and strictly increasing. If $x \in \alpha$, then $t(u^{-1}(x)) \in B_b$ implies $t(u^{-1}(x)) < b$, where $s(x) < v(b) = \gamma$, i. e. $s(x) \in \gamma$. If $y \in \gamma$, then $y < \gamma$ implies $v^{-1}(y) < b$, i. e. $v^{-1}(y) \in B_b$. Therefore, we can take an element $x = (u \circ t^{-1} \circ v^{-1})(y) \in \alpha$. Then, $sx = y$. This means that s is a surjective mapping from α onto γ. Thus, s is an isotone bijection from α onto γ, i. e. $(\alpha, \leqslant) \approx (\gamma, \leqslant)$.

By virtue of Corollary 2 to Theorem 1 (1.2.3), $\alpha < \gamma$, $\alpha > \gamma$, or $\alpha = \gamma$. If $\alpha < \gamma$, then by virtue of Corollary 1 to Lemma 5 (1.2.3), $\alpha =]\leftarrow, \alpha[$ in γ. Thus, the well-ordered sets $(\gamma, \leqslant)$ and $(\alpha, \leqslant)$ are not order equivalent by virtue of Proposition 1 (1.2.4). It follows from this contradiction that the cases $\alpha < \gamma$ or $\alpha > \gamma$ can not take place. Therefore, $\alpha = \gamma \in \beta$ implies $\alpha < \beta$.

2. Conclusion 2 is simply another version of conclusion 1.

3. In the notation from the proof of conclusion 1, if $\alpha = \beta$, then $v \circ u^{-1}\colon A \rightarrowtail\!\!\!\!\twoheadrightarrow B$ is an isotone bijection. Conversely, if there is an isotone bijection $w\colon A \rightarrowtail\!\!\!\!\twoheadrightarrow B$, then $v \circ w \circ u^{-1}\colon \alpha \rightarrowtail\!\!\!\!\twoheadrightarrow \beta$ is an isotone bijection. □

Corollary 1. *Let $(A, \leqslant)$ and $(B, \leqslant)$ be well-ordered sets. Then, only the following three conclusions take place:*

1) *either there are the unique element $b \in B$ and the unique isotone bijection $f\colon A \rightarrowtail\!\!\!\!\twoheadrightarrow]\leftarrow, b[$;*
2) *or there are the unique element $a \in A$ and the unique isotone bijection $g\colon]\leftarrow, a[\rightarrowtail\!\!\!\!\twoheadrightarrow B$;*
3) *or there is the unique isotone bijection $h\colon A \rightarrowtail\!\!\!\!\twoheadrightarrow B$.*

Proof. Consider the isotone bijections $u\colon A \rightarrowtail\!\!\!\!\twoheadrightarrow \alpha$ and $v\colon B \rightarrowtail\!\!\!\!\twoheadrightarrow \beta$. By virtue of Corollary 2 to Theorem 1 (1.2.3), $\alpha < \beta$, $\alpha > \beta$, or $\alpha = \beta$.

If $\alpha < \beta$, then by virtue of Proposition 1 there exist an element $b \in B$ and an isotone bijection $f\colon A \rightarrowtail\!\!\!\!\twoheadrightarrow]\leftarrow, b[$. Suppose that there exist an element $c \in B$ and an isotone bijection $g\colon A \rightarrowtail\!\!\!\!\twoheadrightarrow]\leftarrow, c[$. Then, using Lemma 1 (1.1.15) we can consider the isotone bijection $g \circ f^{-1}\colon]\leftarrow, b[\rightarrowtail\!\!\!\!\twoheadrightarrow]\leftarrow, c[$. Now, using conclusion 2, of Proposition 1 (1.2.4), we deduce that $b = c$, and the uniqueness of f follows from Lemma 2 (1.2.4).

If $\alpha > \beta$, then using the analogous arguments, we deduce conclusion 2.

Finally, if $\alpha = \beta$, then by virtue of Proposition 1, there exists an isotone bijection $h\colon A \rightarrowtail\!\!\!\!\twoheadrightarrow B$. The uniqueness of h follows from Lemma 2 (1.2.4). □

1.2.6 Natural numbers. Multivalued and simple sequences

An ordinal number α is called a *natural number*, if J. Neumann's relation θ_α on α (see 1.2.2) has the property of maximality, i. e. if every non-empty subset $X \subset \alpha$ has a maximal element x with respect to θ_α, i. e. such that $x' \in X$ and $x \in x' \vee x = x'$ imply $x' = x$. In the other notations from 1.2.2, the latter condition means that $x' \in X$ and $x \leqslant x'$ imply $x' = x$, i. e. x is the unique greatest element $\operatorname{gr} X$ of the subset X.

The class $\{x \in \text{Ord} \mid x \text{ is a natural number}\}$ of all natural numbers is denoted by ω or by ω_0. It is clear that $0 \equiv \varnothing$ is a natural number.

Lemma 1. *Let $n \in \omega$ and $x \in n$. Then, $x \in \omega$.*

Proof. By virtue of Theorem 1 (1.2.3), $x \in \text{Ord}$. Besides, $x \subset n$. If $x = \varnothing$, then $x \in \omega$. Let $x \neq \varnothing$. Take any non-empty subset $X \subset x$. Then, $X \subset n$. Therefore, X has a maximal element α with respect to θ_n. But then α is a maximal element of X with respect to θ_x. Thus, $x \in \omega$. □

Corollary 1. *Let $n \in \omega$, $\alpha \in \text{Ord}$ and $\alpha < n$. Then, $\alpha \in \omega$.*

Lemma 2. *If $n \in \omega$, then $n + 1 \in \omega$.*

Proof. By Proposition 1 (1.2.3), $n + 1 \in \text{Ord}$. Let $\varnothing \neq X \subset n + 1$. If $X \subset n$, then X has the greatest element. If $X \cap (n + 1 \setminus n) \neq \varnothing$, then $n \in X$. Let $x \in X$. If $x \in X \cap n$, then $x < n$. If $x \in X \setminus n$, then $x \in \{n\}$, i. e. $x = n$. Thus, $n = \operatorname{gr} X$. □

Lemma 3. *Let $n \in \omega$. Then, $0 \leqslant n < n + 1$.*

Proof. From $0 \subset n$, we conclude by Lemma 4 (1.2.3) that $0 \leqslant n$. Now, by Proposition 1 (1.2.3), $n < n + 1$. □

The natural number 0 is called the *null element of the class ω* or *zero*. The natural number $1 \equiv 0 \cup \{0\} = \{0\}$ is called the *first element of the class ω* or *one*. The natural number $2 \equiv 1 \cup \{1\} = \{0, 1\}$ is called the *second element of the class ω* or *two*. The natural number $3 \equiv 2 \cup \{2\} = \{0, 1, 2\}$ is called the *third element of the class ω* or *three*, and so on.
A natural number $n \in \omega$ such that $n \neq 0$ will be called a *strictly natural number*. The class $\{n \in \omega \mid n \neq 0\}$ of all strictly natural numbers is denoted by $\mathbb{N}$.

Lemma 4. *Let $m, n \in \omega$ and $m + 1 = n + 1$. Then, $m = n$.*

Proof. By Corollary 1 to Proposition 1 (1.2.3), $m = \bigcup(a_a \mid a \in m + 1) = \bigcup(a_a \mid a \in n + 1) = n$. □

The following statement is extremely important in mathematics.

Theorem 1 (the *principle of natural induction*). *Let $X \subset \omega$ and $0 \in X$. If $n \in X$ implies $n+1 \in X$, then $X = \omega$.*

Proof. Suppose that $X \neq \omega$. Then, the non-empty subclass $Y \equiv \omega \setminus X$ of the class Ord has the smallest element β. By the condition, $\beta \neq 0$. By Lemma 2, we get $\beta + 1 \in \omega$. Therefore, the non-empty subset β of the set $\beta + 1$ has the greatest element $\alpha \in \beta$. By Lemma 9 (1.2.3), $\beta = \alpha + 1$. From $\alpha < \beta$, we conclude that $\alpha \notin Y$. Besides, $\alpha \in \beta \in \omega$ implies by Lemma 1 that $\alpha \in \omega$. Therefore, $\alpha \in X$. By the condition of the theorem, we get $\beta = \alpha + 1 \in X$. It follows from this contradiction that $X = \omega$. □

Lemmas 2 – 4 and Theorem 1 represent the *Peano axioms for the natural numbers.*

Theorem 2. $\omega \in$ Ord.

Proof. By virtue of Lemma 1, the class ω is transitive. By virtue of Corollary 1 to Lemma 2 (1.2.3), ω is connected by θ_ω. Thus, ω is an ordinal.

By axiom A7 (1.1.11), there exists set A such that $0 \in A$, and $a \in A$ implies $a \cup \{a\} \in A$. Consider the set $X \equiv A \cap \omega$. Then, $0 \in X$, and $n \in X$ implies $n+1 \in A$. By Lemma 2, we get $n+1 \in \omega$. Thus, $n+2 \in X$. Now, by Theorem 1, we infer that $X = \omega$. Therefore, $\omega \subset A$. By Lemma 1 (1.1.6), ω is a set. As a result, $\omega \in$ Ord. □

Corollary 1. *ω is a limit ordinal number.*

Proof. Suppose $\omega = \beta+1$ for some $\beta \in$ Ord. By Proposition 1 (1.2.3), $\omega > \beta$, where $\beta \in \omega$. By Lemma 2, we get $\omega = \beta + 1 \in \omega$. But this contradicts Lemma 2 (1.1.11). Therefore, $\omega \neq \beta + 1$ for every ordinal number β. □

Lemma 5. *Let $m, n \in \omega$ and $m + 1 < n + 1$. Then, $m < n$.*

Proof. It is clear that $m \neq n$. Suppose that $m > n$. Then, $n \in m$ implies $\{n\} \subset \{m\}$. Since m transitive, we conclude that $n \subset m$. Consequently, $n+1 \subset m$ implies $n+1 \leqslant m$. As a result, $m+1 < n+1 \leqslant m$, but this contradicts Proposition 1 (1.2.3). Thus, $m < n$. □

Lemma 6. *Let $n \in \mathbb{N}$. Then, there exists the unique natural number $m \in \omega$ such that $n = m + 1$.*

Proof. Consider the greatest element m of the non-empty set n. By Lemma 9 (1.2.3), $n = m + 1$. By Lemma 1, we have $m \in \omega$. The uniqueness of m follows from Lemma 4. □

This unique natural number is denoted by $n - 1$.

Lemma 7. *Let $(x_n \in \omega \mid n \in \omega)$ be a collection such that $x_n < x_{n+1}$ for every $n \in \omega$. Then, for every number $x \in \omega$, there is an index n such that $x \leqslant x_n$.*

Proof. Consider the set $X \equiv \{x \in \omega \mid \exists n \in \omega\ (x \leqslant x_n)\}$. By Lemma 3, we get $0 \leqslant x_0$. Therefore, $0 \in X$. Let $x \in X$. Then, $x \leqslant x_n < x_{n+1}$ implies $x_{n+1} \in \{y \in \mathrm{Ord} \mid x < y\}$. By virtue of Proposition 1 (1.2.3), $x + 1 \leqslant x_{n+1}$. Thus, $x + 1 \in X$. By Theorem 1, we have $X = \omega$. □

A set X is called *finite*, if there exist a natural number m and a bijection $u\colon\ m \rightarrowtail\!\!\!\twoheadrightarrow X$. A set X is called *infinite* if it is not finite.

Theorem 3.

1) *Let $(A, \leqslant)$ be an upward directed [an upward lattice-ordered, a linearly ordered] set. Then, every non-empty finite subset of the set A has an upper bound [the smallest upper bound, the greatest element].*
2) *Let $(A, \leqslant)$ be a downward directed [a downward lattice-ordered, a linearly ordered] set. Then, every non-empty finite subset of A has a lower bound [the greatest lower bound, the smallest element].*

Proof. 1. Suppose at first that $(A, \leqslant)$ is upward directed. Consider the set X of all numbers $n \in \omega$ such that if B is a non-empty subset of A and $u\colon\ n+1 \rightarrowtail\!\!\!\twoheadrightarrow B$ is a bijection, then B has an upper bound.

Let $\varnothing \neq B \subset A$ and $u\colon\ 0+1 \rightarrowtail\!\!\!\twoheadrightarrow B$. Take any $b \in B$ and consider $x \equiv u^{-1}b \in 0+1 \equiv 0 \cup \{0\} = \{0\}$. Then, $x = 0$ implies $b \leqslant u(0)$ for every $b \in B$. This means that $0 \in X$.

Suppose now that $n \in X$. Denote $n + 1$ by m. Let $0 \neq B \subset A$ and $u\colon\ m + 1 \rightarrowtail\!\!\!\twoheadrightarrow B$. Then, $m + 1 = m \cup \{m\}$ means that we can consider the set $C \equiv u[m] \subset B \subset A$ and the bijection $v \equiv u|m$. By the supposition, there exists $d \in A$ such that $d \geqslant c$ for every $c \in C$. Let $b \in B \setminus C$. Then, $b = u(k)$ for some $k \in m + 1$. Suppose that $k \in m$. Then, $b = u(k) \in C$. It follows from this contradiction that $k \in (m + 1) \setminus m = \{m\}$. Therefore, $k = m$. Thus, $b = u(m)$. By the condition, there exists $a \in A$ such that $a \geqslant d \geqslant c$ and $a \geqslant u(m) \geqslant b$ for every $c \in C$ and $b \in B \setminus C$. This means that a is an upper bound of B. Thus, $n + 1 \equiv m \in X$. By Theorem 1, we get $X = \omega$.

Suppose now that $(A, \leqslant)$ is upward lattice-ordered. Then, we need to make only some small changes in the given proof. It is clear that $u(0) = \sup B$. By the supposition, there exists $d \in A$ such that $d = \sup C$. By the condition, there exists $a \in A$ such that $a = \sup\{d, u(m)\}$. Let $x \in A$ and $x \geqslant b$ for every $b \in B$. Then, $x \geqslant c$ for every $c \in C$ implies $x \geqslant d$. Besides, $x \geqslant u(m)$. Thus, $x \geqslant a$. This means that $a = \sup B$.

Finally, suppose that $(A, \leqslant)$ is linearly ordered. In this case, the changes in the proof are the following ones. It is clear that $u(0) = \mathrm{gr}\, B$. By the supposition, there exists $d \in C$ such that $d = \mathrm{gr}\, C$. If $d \leqslant u(m)$, then $u(m) = \mathrm{gr}\, B$. If $d \geqslant u(m)$, then $d = \mathrm{gr}\, B$. Conclusion 2 is checked in the similar way. □

Corollary 1. *Let $(A, \leqslant)$ be a non-empty finite linearly ordered set. Then, $(A, \leqslant)$ is well-ordered, and the set A has the smallest and the greatest elements.*

A set X is called *denumerable* if there exists a bijection $u\colon \omega \rightarrowtail\!\!\!\twoheadrightarrow X$. A set X is called *countable*, if it is either finite or denumerable. A set X is called *uncountable* if it is not countable.

Lemma 8. *Every denumerable set is infinite.*

Proof. Suppose that a denumerable set X is finite. Then, there exist a natural number m and bijections $v\colon m \rightarrowtail\!\!\!\twoheadrightarrow X$ and $w\colon \omega \rightarrowtail\!\!\!\twoheadrightarrow X$. The bijection $w^{-1} \circ v\colon m \rightarrowtail\!\!\!\twoheadrightarrow \omega$ means that the set ω is finite. Therefore, by Corollary 1 to Theorem 3 the set ω has the greatest element n. By virtue of Lemmas 2 and 3, we have $n < n + 1 \in \omega$. This means that n is not the greatest element. It follows from this contradiction that X is infinite. □

Let A be a class. A collection $(A_i \subset A \mid i \in I)$ and a simple collection $(a_i \in A \mid i \in I)$ are called *finite* [*countable*], if I is a finite [countable] set.

Sequences

Let $(I, \leqslant)$ be an ordered subset of the ordered set $(\omega, \leqslant)$ with the order, induced by the order on the set ω (see 1.1.14). Any (multivalued) collection $(A_i \subset A \mid i \in I)$ of subclasses of the class A indexed by the principal set I of the ordered set $(I, \leqslant)$ will be called a (*multivalued*) *sequence of subclasses of the class* A. Similarly, any simple collection $(a_i \in A \mid i \in I)$ of elements of the class A will be called a (*simple*) *sequence of elements of the class* A. According to 1.1.15, a simple sequence is a net.

A sequence $t \equiv (B_j \subset A \mid j \in J)$ is called a *subsequence of a sequence* $s \equiv (A_i \subset A \mid i \in I)$, if there exists a mapping $u \equiv (i_j \in I \mid j \in J)\colon J \to I$ such that:

1) for every number $i \in I$, there exists a number $j \in J$ such that $k \in J$ and $k \geqslant j$ imply $i_k \geqslant i$;
2) $s \circ u = t$, i. e. $A_{i_j} = B_j$ for every $j \in J$.

The mapping u is called *thinning the sequence s out*. The similar definition take place also for simple sequences.

A simple sequence can have subnets in the sense of 1.1.15, which are not simple subsequences, because N in the subnet $y \equiv (y_\nu \mid \nu \in N)$ of the sequence $s \equiv (a_i \in A \mid i \in I)$ may be an arbitrary ordered set and J in the subsequence $t \equiv (b_j \in A \mid j \in J)$ must be a subset of ω.

Now, we have the notion of sequential suits (A, A'), (A, A', A''), (A, A', A'', A'''), …, introduced in 1.1.11, and the notion of sequences $(A_i \mid i \in n)$. There is the following connection between these two important notions.

Lemma 9. *Let A be a class, $n \in \omega$, $[A_i \subset A \mid i \in n]$ be a sequence of subclasses and $(a_i \in A \mid i \in n)$ be a simple sequence of elements of the class A. If $n = 2$, then $[A_i \subset A \mid i \in 2] = [A_0, A_1]$ and $(a_i \in A \mid i \in 2) = (a_0, a_1)$ (see 1.1.11). If $n = 3$, then $[A_i \subset A \mid i \in 3] = [A_0, A_1, A_2]$ and $(a_i \in A \mid i \in 3) = (a_0, a_1, a_2)$ (see 1.1.11), and so on.*

Proof. Denote a_0 by a and a_1 by a'. Then, by the definition from 1.1.11, we see that $(a, a') \equiv (x_i \mid i \in 2)$, where $x_0 \equiv a$ and $x_1 \equiv a'$. Thus, $x_0 = a_0$ and $x_1 = a_1$. By Lemma 1 (1.1.9), $(x_i \mid i \in 2) = (a_i \mid i \in 2)$. As a result, $(a_0, a_1) = (a, a') \equiv (x_i \mid i \in 2) = (a_i \mid i \in 2)$.

For all other cases, the checking is the same. □

This lemma shows that for $n \in \omega \setminus 3$ we can denote a sequence $[A_i \subset A \mid i \in n]$ and a simple sequence $(a_i \in A \mid i \in n)$ also by $[A_0, \dots, A_{n-1}]$ and $(a_0, \dots, a_{n-1})$, respectively.

For a simple sequence $(a_0, \dots, a_{n-1})$ of sets indexed by a set $n \in \omega \setminus 3$ along with the notation $\text{rng}(a_0, \dots, a_{n-1})$ we shall use the notation $\{a_0, \dots, a_{n-1}\}$. If $a_i = a$ for all $i \in n$, then $\{a_0, \dots, a_{n-1}\} = \{a\}$.

For a sequence $[A_0, \dots, A_{n-1}]$ of classes and sets indexed by a set $n \in \omega \setminus 3$ along with the notations $\bigcap[A_i \mid i \in n]$, $\bigcup[A_i \mid i \in n]$, $\bigcup_d[A_i \mid i \in n]$ and $\prod[A_i \mid i \in n]$, we shall use also the notations $A_0 \cap \dots \cap A_{n-1}$, $A_0 \cup \dots \cup A_{n-1}$, $A_0 \cup_d \dots \cup_d A_{n-1}$ and $A_0 \times \dots \times A_{n-1}$ respectively. If $A_i = A$ for all $i \in n$, then $A_0 \times \dots \times A_{n-1} = A^n$.

Note that now we can not assert that every sequence and every simple sequence are finite or countable. It will be followed from 1.3.3 and 1.3.9.

Lemma 10. *Let $n \in \omega \setminus 2$. Then, $n + 1 = \{0, \dots, n\}$.*

Proof. Consider the class $X \equiv \{0, 1\} \cup \{n \in \omega \setminus 2 \mid n + 1 = \{0, \dots, n\}\}$. It is clear that $0, 1 \in X$.

Since $2 + 1 \equiv 2 \cup \{2\} = \{0, 1\} \cup \{2\} = \{0, 1, 2\}$ (see 1.1.11), we infer that $1 \in X$ implies $1+1 \in X$. If $n \in (\omega \setminus 2) \cap X$, then $(n+1)+1 \equiv (n+1) \cup \{n+1\} = \{0, \dots, n\} \cup \{n+1\} = \{0, \dots, n+1\}$ implies $n + 1 \in X$.

By Theorem 1, $X = \omega$, and therefore, $\{n \in \omega \setminus 2 \mid n + 1 = \{0, \dots, n\}\} = \omega \setminus 2$. □

1.2.7 The construction of mappings by natural induction

Now, we shall study two main methods of construction of mappings with the help of the principle of natural induction from 1.2.6.

The following theorem is called the *scheme of construction of mappings by natural induction with the passage from n to $n + 1$ with respect to the productive mapping V*. In this scheme, the value $u(n + 1)$ depends only on the value $u(n)$.

Theorem 1. *Let B be a class, $b_0 \in B$ and $V\colon B \times \omega \to B$ be a mapping. Then, there is the unique mapping $u\colon \omega \to B$ such that $u(0) = b_0$ and $u(n + 1) = V(u(n), n)$ for every $n \in \omega$.*

Proof. First, check the uniqueness. Suppose that a mapping $v\colon \omega \to B$ possesses also all the necessary properties. Consider the set $K \equiv \{n \in \omega \mid u(n) = v(n)\}$. Then, $0 \in K$. If $n \in K$, then $u(n+1) = V(u(n), n) = V(v(n), n) = v(n+1)$. Hence, $n+1 \in K$. By the principle of natural induction, we get $K = \omega$. Consequently, $u = v$.

Now, define a mapping $u_0\colon 1 \to B$ setting u. Consider the subset X of ω consisting of all $n \in \omega$, such that for n, there is the unique mapping $u_n\colon n+1 \to B$, such that: (1) $u_n(0) = b_0$ and (2) $u_n(m+1) = V(u_n(m), m)$ for every $m \in n$. It is clear that $0 \in X$.

Let $p, q \in X$, $p < q$ and $u_p\colon p+1 \to B$, $u_q\colon q+1 \to B$ be the corresponding mappings. Consider the mapping $u'_p \equiv u_q|p+1$. The mapping $u'_p\colon p+1 \to B$ possesses the properties 1) and 2) listed above. Since such a mapping is unique, we conclude that $u'_p = u_p$. Thus, $u_q|p+1 = u_p$, i. e. $u_q(m) = u_p(m)$ for every $m \in p+1$.

Let $n \in X$ and $u_n\colon n+1 \to B$ be the corresponding mapping. Define a mapping $u_{n+1}\colon (n+1)+1 \to B$ setting $u_{n+1} \equiv u_n \cup \{(u+1, V(u_n(n), n))\}$. We have $u_{n+1}(0) = u_n(0) = b_0$ and $u_{n+1}(m+1) = u_n(m+1) = V(u_n(m), m) = V(u_{n+1}(m), m)$ for every $m \in n$. If $m = n$, then $u_{n+1}(n+1) = V(u_n(n), n) = V(u_{n+1}(n), n)$. Consequently, the mapping u_{n+1} possesses the properties 1) and 2) listed above.

Check now the uniqueness of u_{n+1}. Suppose that a mapping $v\colon (n+1)+1 \to B$ also has the same properties. Consider the set $Y \equiv \{m \in (n+1)+1 \mid u_{n+1}(m) = v(m)\} \cup (\omega \setminus ((n+1)+1))$. Since $u_{n+1}(0) = b_0 = v(0)$, we have $0 \in Y$. Suppose that $m \in Y$. If $m+1 \in (n+1)+1$, then $m \in (n+1)+1$ implies $u_{n+1}(m) = v(m)$. Besides, $u_{n+1}(m+1) = V(u_{n+1}(m), m) = V(v(m), m) = v(m+1)$. Thus, $m+1 \in Y$. If $m+1 \notin (n+1)+1$, then $m+1 \in \omega \setminus ((n+1)+1) \subset Y$. In both of the cases, $m+1 \in Y$. By the principle of natural induction, $Y = \omega$. This means that $u_{n+1} = v$.

Define now a mapping $u\colon \omega \to B$, setting $u(0) \equiv b_0$ and $u(n+1) \equiv u_{n+1}(n+1)$ for every $n \in \mathbb{N}$. As proven above, $u_{n+1}(n) = u_n(n)$. Therefore, $u(n+1) = V(u_{n+1}(n), n) = V(u_n(n), n) = V(u(n), n)$ for every $n \in \omega$. □

Denote by C the class of all simple sequences of the kind $s \equiv (b_i \in B \mid i \in n \in \mathbb{N})$. A mapping $V\colon C' \times \omega \to B$ will be called *compatible with the subclass* $C' \subset C$ if for every $n \in \mathbb{N}$ and for every sequence $s \equiv (b_i \in B \mid i \in n) \in C'$, we have $s \cup \{(n, V(s, n-1))\} \in C'$.

The following theorem is called the *scheme of construction of mappings by natural induction with the passage from all $m \leqslant n$ to $n+1$ with respect to the productive mapping* V. In this scheme, the value $u(n+1)$ depends on all the values $u(0), u(1), \ldots, u(n)$.

Theorem 2. *Let B be a class, $b_0 \in B$, C' be a subclass of the class C, such that the simple sequence $(b_0) \equiv (b_i \in B \mid i \in 1)$ with the single member b_0 belongs to C', and $V\colon C' \times \omega \to B$ be a mapping, compatible with C'. Then, there is the unique mapping $u\colon \omega \to B$ such that $u|n \in C'$ for every $n \in \mathbb{N}$, $u(0) = b_0$ and $u(n+1) = V(u|(n+1), n)$ for every $n \in \omega$.*

Proof. First, check the uniqueness. Let a mapping $v\colon\ \omega \to B$ possess also all the necessary properties. Consider the set $K \equiv \{n \in \omega \mid \forall m \in n+1\ (u(m) = v(m))\}$. Then, $0 \in K$. If $n \in K$, then $u|n{+}1 = v|n{+}1$ and $u(n{+}1) = V(u|(n{+}1), n) = V(v|(n{+}1), n) = v(n{+}1)$. Hence, $n+1 \in K$. By the principle of natural induction from 1.2.6, we get $K = \omega$. Consequently, $u = v$.

Now, define a mapping $u_0\colon\ 1 \to B$, setting $u_0 \equiv \{(0, b_0)\}$. By the condition, $u_0 \in C'$. Consider the subset X of ω, consisting of all $n \in \omega$ such that for n there is the unique sequence $S_n \equiv (u_m \mid m \in n+1)$ of mappings $u_m\colon\ m+1 \to B$ such that: 1) $u_m \in C'$ for every $m \in n{+}1$; 2) $u_m(0) = b_0$ for every $m \in n{+}1$; 3) $u_m|k{+}1 = u_k$ for every $k \in m \in n{+}1$ and 4) $u_m(k+1) = V(u_m|(k+1), k)$ for every $k \in m \in n+1$. It is clear that $0 \in X$.

Let $p, q \in X$, $p < q$, $S_p = (v_m \mid m \in p+1)$ and $S_q = (w_m \mid m \in q+1)$. Consider the sequence $S'_p \equiv S_q|p+1$. The sequence S'_p possesses the properties 1 – 4 listed above. Since such a sequence is unique, we conclude that $S'_p = S_p$. Thus, $S_q|p+1 = S_p$, i. e. $w_m = v_m$ for every $m \in p+1$.

Let $n \in X$ and $S_n \equiv (u_m \mid m \in n+1)$. Define a mapping $u_{n+1}\colon\ (n+1)+1 \to B$, setting $u_{n+1} \equiv u_n \cup \{(n+1, V(u_n, n))\}$. Then, we have $u_{n+1} \in C'$, $u_{n+1}(0) = u_n(0) = b_0$, $u_{n+1}|k{+}1 = u_n|k{+}1 = u_k$ for every $k \in n{+}1$ and $u_{n+1}(k{+}1) = u_n(k{+}1) = V(u_n|k{+}1, k) = V(u_{n+1}|k{+}1, k)$ for every $k \in n$. If $k \in n$, then $u_{n+1}(n{+}1) = V(u_n, n) = V(u_{n+1|n+1,n})$. Consequently, the sequence $S_{n+1} \equiv (u_m \mid m \in n+2)$ possesses the properties 1 – 4 listed above. Besides, $S_{n+1}|n+1 = S_n$.

Check now the uniqueness of S_{n+1}. Suppose that a sequence $T \equiv (v_m \mid m \in (n+1)+1)$ of mappings $v_m\colon\ m+1 \to B$ also has the same properties. Consider the set $Y \equiv \{m \in (n+1)+1 | u_m = v_m\} \cup (\omega \setminus ((n+1)+1))$. Since $u_0(0) = b_0 = v_0(0)$, we have $u_0 = v_0$, where $0 \in Y$. Suppose that $m \in Y$. If $m{+}1 \in (n{+}1){+}1$, then $m \in (n{+}1){+}1$ implies $u_m = v_m$. Besides, $u_{m+1}|m+1 = u_m = v_m = v_{m+1}|m+1$ and $u_{m+1}(m+1) = V(u_{m+1}|m+1, m) = V(v_{m+1}|m{+}1, m) = v_{m+1}(m{+}1)$ means that $u_{m+1} = v_{m+1}$. Thus, $m{+}1 \in Y$. If $m{+}1 \notin (n+1)+1$, then $m+1 \in \omega \setminus ((n+1)+1) \subset Y$. In both of the cases $m+1 \in Y$. By the principle of natural induction, $Y = \omega$. This means that $u_m = v_m$ for every $m \in (n{+}1){+}1$, i. e. $S_{n+1} = T$.

It follows from the properties proven above that $n+1 \in X$. By the principle of natural induction, $X = \omega$.

Now, define a mapping $u\colon\ \omega \to B$ in the following way. For 0, set up $u(0) \equiv b_0$. For $n \in \mathbb{N}$, take the unique sequence $S_n \equiv (u_m \mid m \in n+1)$ and set up $u(n) \equiv u_n(n)$.

We have $u|1 = (u(0)) = (b_0) \in C'$. Let $n \in \mathbb{N}$ and $k \in n$. Take the unique sequence $S_k \equiv (v_j \mid j \in k+1)$. As proven above, $S_m = S_n|m+1$, i. e. $v_k = u_k$. Therefore, $u(k) = v_k(k) = u_k(k) = u_n(k)$. Thus, $u|n+1 = u_n \in C'$. Besides, $u(n+1) \equiv u_{n+1}(n+1) = V(u_n, n) = V(u|n+1, n)$ for every $n \in \omega$. □

Corollary 1. *Let B be a class, $b_0 \in B$ and $V\colon C \times \omega \to B$ be a mapping. Then, there is the unique mapping $u\colon\ \omega \to B$ such that $u(0) = b_0$ and $u(n+1) = V(u|(n+1), n)$ for every $n \in \omega$.*

Let D be a class, $\mathcal{B}$ be a subclass of the class $\mathcal{P}(D)$, and C be the class of all simple sequences of the kind $s \equiv (B_i \in \mathcal{B} \mid i \in n \in \mathbb{N})$. Consider the subclass C_{disj} of the class C, consisting of all pairwise disjoint sequences $(B_i \in \mathcal{B} \mid i \in n \in \mathbb{N})$, i. e. such that $B_i \cap B_j = \varnothing$ for every $i \neq j$.

Corollary 2. *Let D be a class, $\mathcal{B}$ be a subclass of the class $\mathcal{P}(D)$, $B_0 \in \mathcal{B}$, C_{disj} be the class of all pairwise disjoint sequences $(B_i \in \mathcal{B} \mid i \in n \in \mathbb{N})$ and $V\colon C_{\text{disj}} \times \omega \to \mathcal{B}$ be a mapping, compatible with a class C_{disj}. Then, there is the unique mapping $u\colon \omega \to \mathcal{B}$ such that $u(m) \cap u(n) = \varnothing$ for every $m \neq n$, $u(0) = B_0$ and $u(n+1) = V(u|(n+1), n)$ for every $n \in \omega$.*

Proof. Let $m < n$. By Theorem 2, we get $u|n+1 \in C_{\text{disj}}$. Therefore, $u(m) \cap u(n) = ((u|n+1)(m)) \cap ((u|n+1)(n)) = \varnothing$. □

Note that the indicated schemes are not absolute, but are only “the examples for imitation”.

The following scheme is a generalization of the scheme from Theorem 1.

Theorem 3. *Let A and B be classes, $U\colon A \to B$ and $V\colon B \times \omega \times A \to B$ be mappings. Then, there is the unique mapping $u\colon \omega \times A \to B$ such that $u(0, a) = U(a)$ for every $a \in A$ and $u(n+1, a) = V(u(n, a), n, a)$ for every $n \in \omega$ and $a \in A$.*

The proof of this statement is completely similar to the proof of Theorem 1.

Theorem 3 is called the *scheme of construction of mappings by natural induction with a parameter $a \in A$ and the passage from n to $n+1$ with respect to the productive mapping V*. In this scheme, the value $u(n+1, a)$ depends only on the value $u(n, a)$.

The following scheme is a generalization of the scheme from Theorem 2.

Let A and B be fixed classes. Denote by E the class of all functional sequences of the kind $s \equiv (f_i \in \operatorname{Map}(A, B) \mid i \in n \in \mathbb{N})$. A mapping $V\colon E' \times \omega \times A \to B$ will be called *compatible with the subclass $E' \subset E$*, if for every $n \in \mathbb{N}$ and for every sequence $s \equiv (f_i \in \operatorname{Map}(A, B) \mid i \in n) \in E'$, we have $s \cup \{(n, \{(a, V(s, n-1, a)) \mid a \in A\})\} \in E'$.

Theorem 4. *Let A and B be classes, $U\colon A \to B$ be a mapping, E' be a subclass of the class E such that every mapping $(0, a) \mapsto U(a)$ from $1 \times A$ into B belongs to E'. Let $V\colon E' \times \omega \times A \to B$ be a mapping compatible with E'. Then, there is the unique mapping $u\colon \omega \times A \to B$ such that $u|n \times A \in E'$ for every $n \in \mathbb{N}$, $u(0, a) = U(a)$ for every $a \in A$ and $u(n+1, a) = V(u|(n+1) \times A, n, a)$ for every $n \in \omega$ and $a \in A$.*

The proof of this statement is completely similar to the proof of Theorem 2.

Theorem 4 is called the *scheme of construction of mappings by natural induction with a parameter $a \in A$ and the passage from all $m \leqslant n$ to $n+1$ with respect to the* *productive mapping* V. In this scheme, the value $u(n+1, a)$ depends on all the values $u(0, a), u(1, a), \dots, u(n, a)$.

In conclusion, we shall consider an example of using Theorem 1.

A simple sequence $s \equiv (a_i \in \mathfrak{U} \mid i \in I)$ (see 1.2.6) will be called *finally constant*, if there exists $j \in I$ such that $a_k = a_j$ for every $k \geqslant j$. The sequence s will be called *finally non-constant*, if for every $j \in I$ there exists $k > j$ such that $a_k \neq a_j$.

Proposition 1. *Let $s \equiv (a_i \mid i \in I)$ be a finally non-constant sequence. Then, there exists a subsequence $(a_{i_k} \mid k \in \omega)$ (see 1.2.6) such that $k \leqslant i_k < i_{k+1}$ and $a_{i_k} \neq a_{i_{k+1}}$.*

Proof. Consider $i_0 \equiv \mathrm{sm}\, I$. Since s is finally non-constant, for every $i \in I$ and every $j \geqslant i$ there exists $k > j$ such that $a_k \neq a_i$. In fact, in the opposite case, there exist i and $j \geqslant i$ such that $a_k = a_i$ for every $k > j$. Then, $a_k = a_{j+1}$ for every $k \geqslant j+1$, i. e. s is finally constant, but this is not so. Thus, the set $I_{ik} \equiv \{l \in I \mid (l > \mathrm{gr}\{i, k\}) \wedge (a_l \neq a_i)\}$ is non-empty. Therefore, we can define correctly a mapping $V\colon I \times \omega \to I$, setting $V(i, k) \equiv \mathrm{sm}\, I_{ik}$. By Theorem 1, there exists the unique mapping $u\colon \omega \to I$ such that $u(0) = i_0$ and $u(k+1) = V(u(k), k) = \mathrm{sm}\{l \in I \mid (l > \mathrm{gr}\{u(k), k\}) \wedge (a_l \neq a_{u(k)})\}$. It is clear, that $u(k+1) > u(k)$, $u(k+1) \geqslant k+1$ and $a_{u(k+1)} \neq a_{u(k)}$. Since the mapping $u \equiv (i_k \in I \mid k \in \omega)\colon \omega \to I$ is increasing and the subset $u[\omega]$ is cofinal to the set I, we conclude by Lemma 3 (1.1.15) that the composition $s \circ u = (a_{i_k} \mid k \in \omega)$ is a subsequence of the sequence s. □

1.2.8 The principle of transfinite induction. The constructions of mappings by transfinite induction

The principle of induction (Theorem 1) from 1.2.1 in the application to well-ordered classes is called the *principle of transfinite induction*. In the application to ordinals, the principle of transfinite induction takes the following form.

Proposition 1. *Let α be an ordinal. If B is a non-empty subclass of the class α such that $a \in \alpha$ and $a \subset B$ imply $a \in B$, then $B = \alpha$.*

Proof. By Proposition 2 (1.2.2), the relation of order on the ordinal α has the property of minimality. Let $a \in \alpha$ and $]\leftarrow, a[\, \equiv \{x \mid x \in \alpha \wedge x < a\} \subset B$. By Lemma 3 (1.2.3), a is an ordinal number. Therefore, by Lemma 5 (1.2.3), $a = \{x \mid x \in \mathrm{Ord} \wedge x < a\}$. By the property of transitivity from 1.2.2 and Corollary 1 to Proposition 2 (1.2.2), $x < a \in \alpha$ implies $x \in \alpha$. Thus, by virtue of Lemma 3 (1.2.3), $a =]\leftarrow, a[\, \subset B$. By the condition, this implies $a \in B$. Consequently, by Theorem 1 (1.2.1) $B = \alpha$. □

This principle is the key component in the following *scheme of construction of mappings by transfinite induction*.

Let α be an ordinal number, B be a class, and $V\colon \bigcup(\mathrm{Map}(\beta, B) \mid \beta \in \alpha) \to B$ be a mapping. Since $0 \times B = 0$, the inclusion $0 \subset 0 \times B$ means that $0 \in \mathrm{Map}(0, B)$. Consequently, $0 \in \bigcup(\mathrm{Map}(\beta, B) \mid \beta \in \alpha)$.

Theorem 1. *Let α be an ordinal, B be a class and $V\colon (\mathrm{Map}(\beta, B) \mid \beta \in \alpha) \to B$ be a mapping. Then, there is the unique mapping $u\colon \alpha + 1 \to B$, such that $u(0) = V(0)$ and $u(\beta) = V(u|\beta)$ for every $\beta \in \alpha + 1$.*

Proof. First, check the uniqueness. Let a mapping $v\colon \alpha+1 \to B$ possess all the necessary properties. Consider the set $X \equiv \{x \in \alpha + 1 \mid u|x + 1 = v|x + 1\}$. Then, $0 \in X$. Take any $\beta \in \alpha + 1$ such that $\beta \subset X$. Let $y \in \beta + 1$. If $y < \beta$, then $y \in \beta$ implies $y \in X$. Therefore, $u(y) = v(y)$, where $u|\beta = v|\beta$. If $y = \beta$, then $u(y) = V(u|\beta) = V(v|\beta) = v(y)$. Thus, $u|\beta + 1 = v|\beta + 1$ implies $\beta \in X$. By Proposition 1, we have $X = \alpha + 1$. Therefore, $\alpha \in X$ implies $u = v$.

Now, define a mapping $u_0\colon 1 \to B$, setting $u_0 \equiv \{(0, V(0))\}$. Consider the subset X of $\alpha+1$, consisting of all $x \in \alpha+1$ such that for x there is the unique mapping $u_x\colon x+1 \to B$ such that $u_x(0) = V(0)$ and $u_x(y) = V(u_x|y)$ for every $y \in x + 1$. Then, $0 \in X$.

Let $p, q \in X$ and $p < q$. Consider the corresponding mappings u_p and u_q and the mapping $u'_p \equiv u_q|p + 1$. Then, $u'_p(0) = u_p(0) = V(0)$ and $u'_p(y) = u_q(y) = V(u_q|y) = V(u'_p|y)$ for every $y \in p + 1$, since by Lemma 4 (1.2.3), $y \subset p + 1$. This means that the mapping u'_p possesses all the necessary properties. Since such a mapping is unique, we conclude that $u'_p = u_p$. Thus, $u_q|p + 1 = u_p$.

Let $z \in \alpha + 1$ and $z \subset X$. If $x, y \in z$ and $x < y$, then $u_y|x + 1 = u_x$. Therefore, we can define correctly a mapping $v\colon z \to B$ setting $v(x) \equiv u_y(x) = u_x(x)$ for every $x \leqslant y$ in z. Define a mapping $w\colon z + 1 \to B$, setting $w|z = v$ and $w(z) \equiv V(v)$. Then, $w(0) = v(0) = u_0(0) = V(0)$. Let $y \in z + 1$. If $y < z$, then $y \in z$ implies $y \subset z$ and $w(y) = v(y) = u_y(y) = V(u_y|y) = V(v|y) = V(w|y)$. If $y = z$, then $w(y) = V(v) = V(w|y)$. Consequently, the mapping w possesses all the necessary properties, listed above. Besides, w is unique. In fact, suppose that a mapping $w'\colon z + 1 \to B$ also has the same properties. Consider the set $Y \equiv \{y \in z + 1 \mid w'|y + 1 = w|y + 1\}$. Then, $w'(0) = V(0) = w(0)$ means that $w'|0 + 1 = w|0 + 1$, where $0 \in Y$. Take any $r \in z + 1$ such that $r \geqslant Y$. Let $\beta \in r + 1$. If $\beta < r$ then $\beta \in r$ implies $\beta \in Y$. Therefore, $w'(\beta) = w(\beta)$, where $w'|r = w|r$. If $\beta = r$, then $w'(\beta) = V(w'|r) = V(w|r) = w(\beta)$. Thus, $w'|r + 1 = w|r + 1$ implies $r \in Y$. By Proposition 1, we get $Y = z + 1$. Therefore, $z \in Y$ implies $w' = w$.

It follows from the properties proven above that $z \in X$. By Proposition 1, we get $X = \alpha+1$. Since $\alpha \in X$, we can take the mapping $u_\alpha\colon \alpha+1 \to B$. It has all the necessary properties. □

In applications of this theorem, the mapping V is usually assigned by the following three formulas. The first one assigns the value $V(0)$. The second one assigns the value $V(v)$ for a mapping $v\colon \beta \to B$ defined on a non-limit ordinal $\beta \in \alpha$ (see 1.2.3). The third formula assigns the value $V(v)$ for a mapping $v\colon\ \beta \to B$ defined on a limit ordinal $\beta \in \alpha$.

1.2.9 The ordered disjoint union of well-ordered sets. The addition of ordinal numbers

Let $u \equiv (A_i \subset \mathfrak{U} \mid i \in I)$ be a collection of well-ordered sets, indexed by a well-ordered set I. Define an order $\leqslant$ on the disjoint union $\bigcup_d(A_i \mid i \in I) \equiv \bigcup(A_i * \{i\} \mid i \in I)$ of the collection u (see 1.1.10), setting $\langle x, i\rangle \leqslant \langle y, j\rangle$ iff either $i < j$ or $i = j$ and $x \leqslant y$. The set $\bigcup_d(A_i \mid i \in I)$ with this order will be called the *ordered disjoint union of the collection* u and will be denoted by $\bigcup_{do}(A_i \mid i \in I)$.

Let $A, A', A'', \ldots$ be well-ordered sets. Then, (A, A'), (A, A', A''),... are the corresponding multivalued collections (see 1.1.11). The ordered sets $\bigcup_{do}(A, A')$, $\bigcup_{do}(A, A', A'')$,... will be called the *ordered disjoint unions of the sequential pair* (A, A'), *triplet* (A, A', A''),... and will be denoted also by $A \cup_{do} A'$, $A \cup_{do} A' \cup_{do} A''$,... (see 1.1.11). By the definition from 1.1.11, we have $A \cup_d A' \equiv \bigcup_d(A, A') \equiv \bigcup_d(X_i \subset A \cup A' \mid i \in 2) = \bigcup(X_i * \{i\} \mid i \in 2) = (A * \{0\}) \cup (A' * \{1\})$. Thus, $\langle x, i\rangle \leqslant \langle y, j\rangle$ in $A \cup_{do} A'$ iff either $i = 0$ and $j = 1$, or $i = j = 0$ and $x \leqslant y$ in A, or $i = j = 1$ and $x \leqslant y$ in A'.

Lemma 1. *Let $(A_i \mid i \in I)$ be a collection of well-ordered sets, indexed by a well-ordered set I. Then, the ordered set $\bigcup_{do}(A_i \mid i \in I)$ is well-ordered.*

Proof. Denote this ordered set by S. It is clear that S is linearly ordered. Take any set $\varnothing \neq P \subset S$ and consider the sets $P_i \equiv P \cap (A_i * \{i\})$ and $J \equiv \{i \in I \mid P_i \neq \varnothing\} \neq \varnothing$. Take the smallest element j of the set J. Since $P_j \neq \varnothing$, we can take the smallest element y of the set A_j. Then, $q \equiv \langle y, j\rangle$ is the smallest element of P. In fact, if $p \in P$, then $p = \langle x, i\rangle \in P_i$ for some $i \in J$. Consequently, $j \leqslant i$. If $j = i$, then $y \leqslant x$ implies $q \leqslant p$. If $j < i$, then automatically $q \leqslant p$. □

Let I be a well-ordered set, $(I_m \mid m \in M)$ be a collection of well-ordered sets, indexed by a well-ordered set M, and $(I_m \mid m \in M)$ be a partition of the set I. The collection $(I_m \mid m \in M)$ is called an *ordered partition of the well-ordered set* I if: (1) the order on every set I_m is induced by the order on the set I; and (2) $i \leqslant j$ in I for some $i \in I_m$ and $j \in I_n$ iff either $m < n$ or $m = n$ and $i \leqslant j$ in I_m.

The ordered disjoint union is associative in the following sense.

Proposition 1. *Let $(A_i \mid i \in I)$ be a collection of well-ordered sets, indexed by a well-ordered set I and a collection $(I_m \mid m \in M)$ of well-ordered sets, indexed by a well-ordered set M, be an ordered partition of the ordered set I. Then, the mapping $\beta\colon \bigcup_{do}(A_i \mid i \in I) \rightarrowtail\!\!\!\!\twoheadrightarrow \bigcup_{do}(\bigcup_{do}(A_i \mid i \in I_m) \mid m \in M)$ such that $\beta(\langle a, i\rangle) \equiv \langle\langle a, i\rangle, m\rangle$ for every $i \in I_m$ and $a \in A_i$ is bijective and isotone.*

Proof. Denote the first set by S and the second one by T. Let $p \equiv \langle x, i\rangle < q \equiv \langle y, j\rangle$ in S. Then, $i \in I_m$ and $j \in I_n$ for some m and n. If $m = n$, then $\beta p \equiv \langle p, m\rangle < \langle q, m\rangle \equiv \beta q$. If $m \neq n$, then $i \neq j$ and $p < q$ imply $i < j$ and so $m < n$. Therefore, again $\beta p \equiv \langle p, m\rangle < \langle q, n\rangle \equiv \beta q$. This means that β is strictly increasing (see 1.1.15). Since all the orders are linear, we infer that β is isotone. Let $t \in T$. Then, $t = \langle p, m\rangle$ for some m and $p \in \bigcup_{do}(A_i \mid i \in I_m)$. Therefore, $p = \langle a, i\rangle$ for some $i \in I_m \subset I$ and $a \in A_i$. From $\beta p = t$, we infer that β is surjective. Now, by Lemma 1 (1.1.15) β is bijective. □

Lemma 2. *Let A, B, and C be well-ordered sets. Then, $A \cup_{do} B \cup_{do} C \approx (A \cup_{do} B) \cup_{do} C \approx A \cup_{do} (B \cup_{do} C)$ (see 1.1.15).*

The proof of Lemma 2 is analogous to the proof of Proposition 1.

Lemma 3. *Let $(A_i \mid i \in I)$ be a collection of well-ordered sets, indexed by a well-ordered set I. Then, $\operatorname{ord} \bigcup_{do}(A_i \mid i \in I) = \operatorname{ord} \bigcup_{do}(\operatorname{ord} A_i \mid i \in I)$ (see 1.2.5).*

Proof. Denote $\bigcup_{do}(A_i \mid i \in I)$ by S and $\bigcup_{do}(\operatorname{ord} A_i \mid i \in I)$ by T. According to Theorem 1 (1.2.5), for every $i \in I$, there is the unique isotone bijection $u_i\colon A_i \rightarrowtail\!\!\!\twoheadrightarrow \alpha_i$ where $\alpha_i \equiv \operatorname{ord} A_i$. If $r \in S$, then $r = \langle a, i\rangle$ for the unique elements $i \in I$ and $a \in A_i$. Therefore, we can define correctly a mapping $u\colon S \to T$ setting $ur \equiv \langle u_i a, i\rangle$. Let $s = \langle b, j\rangle \in S$ and $r < s$. If $i < j$, then $ur \equiv \langle u_i a, i\rangle < \langle u_j b, j\rangle \equiv us$. If $i = j$, then $a < b$ implies by virtue of Lemma 1 (1.1.15) $u_i a < u_i b$, where $ur < us$.

Thus, u is strictly increasing. Since all the orders are linear, we infer that u is isotone. Let $t \in T$. Then, $t = \langle x, i\rangle$ for some i and $x \in \alpha_i$. Take $a \equiv u_i^{-1}x \in A_i$ and $r \equiv \langle a, i\rangle \in S$. From $ur = t$, we infer that u is surjective. Now, by Lemma 1 (1.1.15) u is bijective. Finally, using Theorem 1 (1.2.5) we conclude that $\operatorname{ord} S = \operatorname{ord} T$. □

Let $u \equiv (\alpha_i \mid i \in I)$ be a simple collection of ordinal numbers, indexed by a well-ordered set I. Take the collection $(\alpha_i \mid i \in I) \equiv \varphi^{-1}(\alpha_i \mid i \in I)$ from Corollary 1 to Lemma 3 (1.1.9). The ordinal number $\operatorname{ord} \bigcup_{do}(\alpha_i \mid i \in I)$ is called the *ordinal sum of the collection u* and is denoted by $\sum_o(\alpha_i \mid i \in I)$.

Let $\alpha, \alpha', \alpha'', \ldots$ be ordinal numbers. Then, (α, α'), $(\alpha, \alpha', \alpha'')$,... are the corresponding simple collections (see 1.1.11). The ordinal sums $\sum_o(\alpha, \alpha')$, $\sum_o(\alpha, \alpha', \alpha'')$,... will be called the *ordinal sums of the simple sequential pair* (α, α'), *triplet* $(\alpha, \alpha', \alpha'')$,... and will be denoted also by $\alpha +_o \alpha'$, $\alpha +_o \alpha' +_o \alpha''$,.... By the definition from 1.1.11, we have $\alpha +_o \alpha' \equiv \sum_o(\alpha, \alpha') \equiv \sum_o(x_i \mid i \in 2) = \operatorname{ord} \bigcup_{do}(x_i \mid i \in 2)$, where $x_0 \equiv \alpha$ and $x_1 \equiv \alpha'$.

The ordinal sum is associative in the following sense.

Proposition 2. *Let $(\alpha_i \mid i \in I)$ be a simple collection of ordinal numbers, indexed by a well-ordered set I and the collection $(I_m \mid m \in M)$ of well-ordered sets, indexed by a well-ordered set M, be an ordered partition of the ordered set I. Then, $\sum_o(\alpha_i \mid i \in I) = \sum_o(\sum_o(\alpha_i \mid i \in I_m) \mid m \in M)$.*

Proof. By virtue of Proposition 1 and Lemma 3, we get the equalities $\sum_o(\alpha_i \mid i \in I) \equiv \operatorname{ord}\bigcup_{do}(\alpha_i \mid i \in I) = \operatorname{ord}\bigcup_{do}(\bigcup_{do}(\alpha_i \mid i \in I_m) \mid m \in M) = \operatorname{ord}\bigcup_{do}(\operatorname{ord}\bigcup_{do}(\alpha_i \mid i \in I_m) \mid m \in M) = \operatorname{ord}\bigcup_{do}(\sum_o(\alpha_i \mid i \in I_m) \mid m \in M) = \sum_o(\sum_o(\alpha_i \mid i \in I_m) \mid m \in M)$. □

Lemma 4. *Let α, β and γ be ordinal numbers. Then, $\alpha +_o \beta +_o \gamma = (\alpha +_o \beta) +_o \gamma = \alpha +_o (\beta +_o \gamma)$.*

The proof of Lemma 4 is analogous to the proof of Proposition 2.

Proposition 3. *Let α be an ordinal number. Then, $\alpha + 1 = \alpha +_o 1$ (see 1.2.3).*

Proof. Assume that there exists $x \in \alpha \cap \{\alpha\}$. Then, $x \in \alpha$ and $x = \alpha$ imply $\alpha \in \alpha$, but this contradicts Lemma 2 (1.1.11). Consequently, $\alpha \cap \{\alpha\} = \varnothing$. By the definition from 1.2.3 and by Lemma 7 (1.1.11), $\alpha + 1 \equiv \alpha \cup \{\alpha\} = \bigcup(\alpha, \{\alpha\}) = \bigcup(x_i \mid i \in 2)$, where $x_0 = \alpha$ and $x_1 \equiv \{\alpha\}$. Therefore, by Lemma 2 (1.1.10), there is a bijective mapping $u\colon \bigcup(x_i \mid i \in 2) \rightarrowtail\!\!\!\twoheadrightarrow \bigcup_d(x_i \mid i \in 2)$ such that $up = \langle p, i\rangle$ for every $i \in 2$ and $p \in x_i$. Let $q \in \bigcup(x_i \mid i \in 2)$ and $p < q$. Then, $q \in x_j$ for some j. If $i = j$, then $p < q$ implies $up < uq$. If $i < j$, then automatically $up < uq$. This means that u is strictly increasing (see 1.1.15). Since all the orders are linear, we infer that u is an isotone mapping from the ordered set $\alpha + 1$ into the ordered set $\bigcup_{do}(x_i \mid i \in 2)$.

By the definition $\alpha +_o 1 = \bigcup_{do}(y_i \mid i \in 2)$, where $y_0 \equiv \alpha$ and $y_1 \equiv 1 = \{0\}$. Define a mapping $v\colon \bigcup_d(x_i \mid i \in 2) \to \bigcup_d(y_i \mid i \in 2)$, setting $v(\langle p, 0\rangle) \equiv \langle p, 0\rangle \in \alpha * \{0\} = y_0 * \{0\}$ for every $p \in x_0 \equiv \alpha$ and $v(\langle \alpha, 1\rangle) \equiv \langle 0, 1\rangle \in 1 * \{1\}$ for $\alpha \in x_1 \equiv \{\alpha\}$. It is clear that v is an isotone bijection from the ordered set $\bigcup_{do}(x_i \mid i \in 2)$ onto the ordered set $\bigcup_{do}(y_i \mid i \in 2)$. As a result, the mapping $v \circ u$ is an isotone bijection from the ordered set $\alpha + 1$ onto the ordered set $\alpha +_o 1$. By virtue of Theorem 1 (1.2.5), $\alpha + 1 = \alpha +_o 1$. □

Corollary 1. *Let α and β be ordinal numbers. Then, $(\alpha +_o \beta) + 1 = \alpha +_o (\beta + 1)$.*

Proof. By Proposition 3 and Lemma 4, we get $(\alpha +_o \beta) + 1 = (\alpha +_o \beta) +_o 1 = \alpha +_o (\beta +_o 1) = \alpha +_o (\beta + 1)$. □

1.2.10 The connection between ordinal and natural numbers

Theorem 1. *Let $\alpha \in \mathrm{Ord} \setminus \omega$. Then, there are the unique limit ordinal $\gamma \geqslant \omega$ and the unique natural number n such that $\alpha = \gamma +_o n$.*

Proof. Consider the set B, consisting of all natural numbers and all ordinal numbers $\alpha \geqslant \omega$, representable in the form $\alpha = \gamma +_o n$ for some limit ordinal $\gamma \geqslant \omega$ and some natural number n. It is clear that all limit ordinals belong to B. By Corollary 1 to Theorem 2 (1.2.6), $\omega \in B$.

Let $\alpha \in \mathrm{Ord}$ and $\alpha \subset B$. If α is a limit ordinal number, then $\alpha \in B$. If α is a non-limit ordinal number, then $\alpha = \beta + 1$ for some ordinal number β. From $\beta \in \alpha \subset B$, we infer that either $\beta \in \omega$ or $\beta \geqslant \omega$ and $\beta = \gamma +_o m$ for some $\gamma \geqslant \omega$ and m. In the first case, by Lemma 2 (1.2.6), $\alpha = \beta+1 \in \omega \subset B$, where $\alpha \in B$. In the second case, $\beta = \mathrm{ord}(\gamma \cup_{do} m)$. By virtue of the definition from 1.2.5 and Theorem 1 (1.2.5), there is an isotone bijection $u\colon \gamma \cup_d m \rightarrowtail\!\!\twoheadrightarrow \beta$. According to 1.2.9, we see that $\gamma \cup_d m = (\gamma * \{0\}) \cup (m * \{1\})$ and $\gamma \cup_d (m+1) = (\gamma * \{0\}) \cup ((m+1) * \{1\})$. According to 1.2.3, we get $\beta \subset \beta+1$ and $m \subset m+1$. Therefore, we can define correctly a mapping $v\colon \gamma \cup_d (m+1) \to \beta+1$, setting $v|\gamma \cup_d m \equiv u$ and $v(\langle m, 1\rangle) \equiv \beta$. It is clear that v is bijective. If $x \in \gamma \cup_d m$, then $v(x) = u(x) < \beta = v(\langle m, 1\rangle)$. This means that v is strictly increasing. Since all the orders are linear, we infer that v is isotone. Consequently, by Theorem 1 (1.2.5) $\alpha = \beta+1 = \mathrm{ord}(\gamma \cup_{do} (m+1)) = \gamma +_o (m + 1) \in B$. In all the cases, $\alpha \in B$. Thus, by Proposition 1 (1.2.8) $B = \mathrm{Ord}$.

Check now the uniqueness of this representation. Let $\alpha = \beta +_o m = \gamma +_o n$. If $m = n = 0$, then $\gamma = \beta$. At first, suppose that $m \neq 0$. Then, there exists an isotone bijection $u\colon \beta \cup_d m \rightarrowtail\!\!\twoheadrightarrow \gamma \cup_d n$. According to 1.2.9 we have that $\beta \cup_d m = (\beta * \{0\}) \cup (m * \{1\})$ and $\gamma \cup_d n = (\gamma * \{0\}) \cup (n * \{1\})$. Assume that there exists $l \in m$ such that $u(\langle l, 1\rangle) \in \gamma * \{0\}$. Consider the non-empty subset B of m, consisting of all $k \in m$ such that $u(\langle k, 1\rangle) \in \gamma * \{0\}$. Let $a \in m$ and $a \subset B$. If $a = 0$, then $a \leqslant l$ implies $u(\langle a, 1\rangle) \leqslant u(\langle l, 1\rangle) \in \gamma * \{0\}$. Therefore, $u(\langle a, 1\rangle) \in \gamma * \{0\}$. This means that $a \in B$. If $a > 0$, then by Lemma 6 (1.2.6) $a = k + 1$ for some $k \in \omega$. Then, $k \in a \subset B$. Thus, for $p \equiv \langle k, 1\rangle \in m * \{1\}$, we have $\varkappa \equiv u(p) \in \gamma * \{0\}$, i.e. $\varkappa = \langle x, 0\rangle$ for some $x \in \gamma$. Since γ is a limit ordinal, we infer that $x + 1 < \gamma$, i.e. $x + 1 \in \gamma$. Consider the elements $\rho \equiv \langle x + 1, 0\rangle \in \gamma * \{0\}$ and $q \equiv u^{-1}(\rho)$. From $u(p) = \varkappa < \rho = u(q)$, we conclude that $p < q$, where $q \in m * \{1\}$.

Since $k + 1 = a \in m$, we can take the element $r \equiv \langle k + 1, 1\rangle \in \beta \cup_d m$. Consider the bijection $\varepsilon\colon m \rightarrowtail\!\!\twoheadrightarrow m * \{1\}$ such that $\varepsilon k = \langle k, 1\rangle$ for every $k \in m$. From $p < q$, we infer that $k = \varepsilon^{-1}p < \varepsilon^{-1}q$. Therefore, $k < k + 1 \leqslant \varepsilon^{-1}q$ implies $p < \langle k + 1, k\rangle \leqslant q$. Hence $\varkappa < ur \leqslant \rho$. Assume that $ur \notin \gamma * \{0\}$. Then, $ur \in n * \{1\}$ implies $\rho < ur$. From this contradiction, it follows that $ur \in \gamma * \{0\}$ and consequently, $a = k + 1 \in B$. As a result, by virtue of Proposition 1 (1.2.8), we conclude that $B = m$.

Thus, $u[m * \{1\}] \subset \gamma * \{0\}$. If $s \equiv \langle x, 0\rangle \in \beta * \{0\}$, then $s < \langle m - 1, 1\rangle$ implies $us < u(\langle m-1, 1\rangle) \in \gamma * \{0\}$. Therefore, $us \in \gamma * \{0\}$. This means that $(\gamma * \{0\}) \cup (n * \{1\}) = \mathrm{rng}\, u \subset \gamma * \{0\}$. If $n \neq 0$, then this inclusion is impossible. If $n = 0$, then u is an isotone bijection from $\beta \cup_d m$ onto $\gamma * \{0\}$. The ordered set $\beta \cup_{do} m$ has the greatest element $\langle m - 1, 1\rangle$. Therefore, $u(\langle m - 1, 1\rangle)$ is the greatest element in $\gamma * \{0\}$. But this is impossible, since γ is a limit ordinal number. Thus, in all the cases we came to the contradictions. This means that our assumption is not valid, i.e. $u[m * \{1\}] \subset n * \{1\}$. Since $m \neq 0$, this inclusion implies $n \neq 0$. But then, using the similar arguments for the isotone bijection u^{-1}, we can deduce that $u^{-1}[n * \{1\}] \subset m * \{1\}$. As a result, we get $u[m * \{1\}] = n * \{1\}$. Therefore, $m \approx m * \{1\} \approx n * \{1\} \approx n$.

It follows from the proven inclusions that $u[\beta * \{0\}] = \gamma * \{0\}$. Therefore, $\beta \approx \beta * \{0\} \approx \gamma * \{0\} \approx \gamma$. By virtue of Theorem 1 (1.2.5), we conclude that $m = n$ and $\beta = \gamma$.

If $n \neq 0$, then the arguments are the same. □

1.2.11 The other forms of the axiom of choice

The terminology of ordered classes will often be applied to the full ensemble $\mathcal{P}(A)$ of a class A. Any non-empty subclass $\mathcal{S}$ of the class $\mathcal{P}(A)$ will be called an *ensemble on the class* A.

On the full ensemble, $\mathcal{P}(A)$ define the *order by inclusion* setting $S \leqslant T$ iff $S \subset T$ for $S, T \in \mathcal{P}(A)$. This order induced the *order by inclusion* on every ensemble $\mathcal{S}$ on the class A. With respect to the order by inclusion, an element $S \in \mathcal{S}$ is a *maximal member of the ensemble* $\mathcal{S}$ if S is a proper subset of no other element of $\mathcal{S}$.

If A is an order class, then we can consider the ensemble of all chains ($\equiv$ linearly ordered subclasses) in A. With respect to the order by inclusion on this ensemble, a chain C in A is *maximal* if C is a proper subclass of no other chain in A.

Consider the universal class $\mathfrak{U}$. By Lemma 5 (1.1.5), $\mathcal{P}(\mathfrak{U}) = \mathfrak{U}$. An ensemble $\mathcal{S}$ on the universal class $\mathfrak{U}$ will be called an *ensemble of sets*.

An ensemble of sets $\mathcal{S}$ is called an *ensemble of sets of finite character* if any set $X \in \mathfrak{U}$ belongs to $\mathcal{S}$ iff each finite subset of the set X belongs to $\mathcal{S}$.

There are a lot of situations in mathematics to which axiom of choice A8 from 1.1.12 is not immediately applicable, but to which one or another equivalent form of this axiom is applicable at once. We next list four such equivalent forms. The names "lemma", "theorem", and "principle" are attached to them only for historical reasons.

Using the axiom of choice, we can prove the following

Proposition 1. *Let $\mathcal{F}$ be an ensemble of sets of finite character, $\mathcal{F}$ be a set and $\mathcal{B}$ be a chain in $\mathcal{F}$. Then, $\bigcup(B_B \mid B \in \mathcal{B}) \in \mathcal{F}$.*

Proof. Denote the considered set by X. Let Y be a non-empty finite subclass of X. Then, there exist a number $m \in \mathbb{N}$ and a bijection $u\colon\ m \rightarrowtail\!\!\!\!\twoheadrightarrow Y$. Define a mapping $v\colon\ Y \to \mathcal{P}(\mathcal{B}) \setminus \{\varnothing\}$, setting $v(y) \equiv \{B \in \mathcal{B} \mid y \in B\} \neq \varnothing$. Take any choice mapping $p\colon\ \mathcal{P}(\mathcal{B}) \setminus \{\varnothing\} \to \mathcal{B}$ from axiom A8 (1.1.12). Consider the mapping $w \equiv p \circ v \circ u \equiv (B_i \in \mathcal{B} \mid i \in m)\colon\ m \to \mathcal{B}$.

Consider the set K of all numbers $k \in m$, such that the set $\{B_i \in \mathcal{B} \mid i \in k+1\}$ has the greatest element. Let $X \equiv K \cup (\omega \setminus m)$. Then, $0 \in X$. Let $n \in K$. If $n = m-1$, then $n+1 = m \in X$. Therefore, further we can assume that $n < m-1$. Supposing that $n+1 = m$, we get $n = m-1$. Supposing that $n+1 > m = (m-1)+1$, we infer by Lemma 5 (1.2.6) that $n > m-1$. In both of the cases, we get contradictions. Thus, $n+1 < m$, i. e. $n+1 \in m$. Since $n \in K$, the set $\{B_i \mid i \in n+1\}$ has the greatest element $A \in \mathcal{B}$. If $A \leqslant B_{n+1}$, then B_{n+1} is the greatest element of the set $\{B_i \mid i \in (n+1)+1\}$. If $A \geqslant B_{n+1}$, then A is the greatest element of this set. This means that $n+1 \in K \subset X$. By the principle of natural induction, $X = \omega$. Consequently, $K = m$. Thus, the set $\{B_i \in \mathcal{B} \mid i \in m\}$ has the greatest element B.

If $y \in Y$, then $y = u(i)$ for some $i \in m$. Since $pv(y) \in v(y)$, we infer that $y \in pv(y) = w(i) \equiv B_i \subset B$. Thus, $Y \subset B \in \mathcal{F}$. By the condition, $Y \in \mathcal{F}$. This implies $X \in \mathcal{F}$. □

Theorem 1. *The following conditions are equivalent:*
1) (the *axiom of choice*) *for every non-empty set A, there exists a mapping $p: \mathcal{P}(A) \setminus \{\varnothing\} \to A$ such that $p(P) \in P$ for every non-empty subset P of the set A;*
2) (the *Tukey lemma*) *every non-empty ensemble of sets of finite character, which is a set, has a maximal member;*
3) (the *Hausdorff maximality principle*) *every non-empty ordered set contains a maximal chain;*
4) (the *Kuratowski – Zorn lemma*) *every non-empty ordered set, in which every chain has an upper bound, has a maximal element;*
5) (the *Zermelo well-ordering theorem* or the *Zermelo principle*) *every set can be well-ordered.*

Proof. (1) $\vdash$ (2). Assume that (2) is false. Then, there exists a non-empty ensemble of sets of finite character $\mathcal{F}$, which is a set and which has no maximal member. Since $\mathcal{F}$ has no maximal element, the ensemble $\mathcal{A}_F \equiv \{E \in \mathcal{F} \mid E \supset F \wedge E \neq F\}$ is non-empty for every $F \in \mathcal{F}$. By Theorem 1 (1.1.12), for the collection $(\mathcal{A}_F \mid F \in \mathcal{F})$, there exists an element $u \in \prod(\mathcal{A}_F \mid F \in \mathcal{F})$, i. e. $F \subset u(F) \in \mathcal{F}$ and $F \neq u(F)$ for every $F \in \mathcal{F}$.

A subensemble $\mathcal{G}$ of $\mathcal{F}$ will be called *u-inductive* if it has the following three properties: (1) $\varnothing \in \mathcal{G}$; (2) $A \in \mathcal{G}$ implies $u(A) \in \mathcal{G}$; (3) if $\mathcal{B}$ is a chain in $\mathcal{G}$, then $\bigcup(B_B \mid B \in \mathcal{B}) \in \mathcal{G}$. By virtue of Proposition 1, the ensemble $\mathcal{F}$ is u-inductive. Consider the ensemble of sets $\mathcal{G}_0 \equiv \{A \in \mathcal{F} \mid A \in \mathcal{G}$ for every u-inductive ensemble $\mathcal{G} \subset \mathcal{F}\}$. It is clear that $\mathcal{G}_0$ is the smallest u-inductive ensemble.

Consider the subensemble $\mathcal{H} \equiv \{A \in \mathcal{G}_0 \mid \forall B \in \mathcal{G}_0\ ((B \subset A \ \wedge\ B \neq A) \Rightarrow u(B) \subset A)\}$. We assert that if $A \in \mathcal{H}$ and $C \in \mathcal{G}_0$, then either $C \subset A$ or $u(A) \subset C$. To prove this assertion, take $A \in \mathcal{H}$ and consider $\mathcal{G}_A \equiv \{C \in \mathcal{G}_0 \mid C \subset A \ \vee\ u(A) \subset C\}$. It suffices to show that $\mathcal{G}_A$ is u-inductive. Since $\varnothing \in \mathcal{G}_0$ and $\varnothing \subset A$, (1) is satisfied. Let $C \in \mathcal{G}_A$. Then, we have $C \subset A$ and $C \neq A$, $C = A$, or $u(A) \subset C$. If $C \subset A$ and $C \neq A$, then $u(C) \subset A$ because $A \in \mathcal{H}$. If $C = A$, then $u(A) \subset u(C)$. If $u(A) \subset C$, then $u(A) \subset u(C)$, because $C \in \mathcal{F}$ implies $C \subset u(C)$. Thus, in every case, $u(C) \in \mathcal{G}_A$ and (2) is satisfied. Next, let $\mathcal{B}$ be a chain in $\mathcal{G}_A$. Then, either $C \subset A$ for each $C \in \mathcal{B}$ or there exists $C \in \mathcal{B}$, such that $u(A) \subset C$. In the first case, $G \equiv \bigcup(B_B \mid B \in \mathcal{B}) \subset A$, and in the second case, $u(A) \subset C \subset G$. Thus, $G \in \mathcal{G}_A$ and (3) is satisfied. Thus, $\mathcal{G}_A$ is u-inductive. Therefore, $\mathcal{G}_A = \mathcal{G}_0$.

We next assert that $\mathcal{H} = \mathcal{G}_0$. We prove this by showing that $\mathcal{H}$ is u-inductive. Since $\varnothing$ has no proper subset, $\mathcal{H}$ satisfied (1). Next, let $A \in \mathcal{H}$ and $B \in \mathcal{G}_0$ be such that $B \subset u(A)$ and $B \neq u(A)$. Since $B \in \mathcal{G}_0 = \mathcal{G}_A$, we have $B \subset A$. If $B \neq A$, the definition of $\mathcal{H}$ yields $u(B) \subset A \subset u(A)$. If $B = A$, then $u(B) \subset u(A)$. In either case, the inclusion $u(B) \subset u(A)$ is valid, so $u(A) \in \mathcal{H}$ and (2) holds for $\mathcal{H}$. Finally, let $\mathcal{B}$ be a chain in $\mathcal{H}$ and let $B \in \mathcal{G}_0$ have the property that $B \subset G \equiv \bigcup(B_B \mid B \in \mathcal{B})$ and $B \neq G$. Since $B \in \mathcal{G}_0 = \mathcal{G}_A$ for every $A \in \mathcal{B}$, we have either $B \subset A$ for some $A \in \mathcal{B}$ or $u(A) \subset B$ for every $A \in \mathcal{B}$. If the latter alternative were true, we would have $B \subset G \subset \bigcup(u(A) \mid A \in \mathcal{B}) \subset B$, which is impossible. Thus, there is some $A \in \mathcal{B}$ such that $B \subset A$. If $B \neq A$, then $u(B) \subset A \subset G$, since $A \in \mathcal{H}$. If $B = A$, then $B \in \mathcal{H}$ and $G \in \mathcal{G}_0 = \mathcal{G}_B$. This implies that $u(B) \subset G$, since

$G \subset B$ is impossible. Thus, in either case, we have $u(B) \subset G$ and so $G \in \mathcal{H}$. This proves that $\mathcal{H}$ satisfies (3). Therefore, $\mathcal{H}$ is u-inductive. As a result, $\mathcal{H} = \mathcal{G}_0$.

We conclude from the above arguments that if $A \in \mathcal{G}_0 = \mathcal{H}$ and $B \in \mathcal{G}_0 = \mathcal{G}_A$, then either $B \subset A$ or $A \subset u(A) \subset B$. Thus, $\mathcal{G}_0$ is a chain. Since $\mathcal{G}_0$ is u-inductive, we infer that $G \equiv \bigcup(A_A \mid A \in \mathcal{G}_0) \in \mathcal{G}_0$. Applying (2), we get $G \subset u(G) \in \mathcal{G}_0$ and $G \neq u(G)$. This contradiction establishes the fact that (1) implies (2).

(2) $\vdash$ (3). Let P be a non-empty ordered set. Consider the ensemble $\mathcal{C}$ of all chains in P. It is clear that $\varnothing \in \mathcal{C}$ and $\{x\} \in \mathcal{C}$ for every $x \in P$. By virtue of Lemmas 2 and 1 from 1.1.6, we see that $\mathcal{C}$ is a set. Besides, $\mathcal{C}$ is an ensemble of sets of finite character. By Tukey's lemma, $\mathcal{C}$ has a maximal member.

(3) $\vdash$ (4). Let P be a non-empty ordered set in which every chain has an upper bound. By (3) there is a maximal chain $M \subset P$. Let m be an upper bound for M. Then, m is a maximal element of P. In fact, if there is $x \in P$ such that $m \leqslant x$ and $m \neq x$, then $M \cup \{x\}$ is a chain, which properly contains M. But this contradicts the maximality of M.

(4) $\vdash$ (5). Let S be a non-empty set. Consider the ensemble $\mathcal{Z}$ of all well-ordered sets $(W, \leqslant)$ such that $W \subset S$. Introduce an order on $\mathcal{Z}$, setting $(W_1, \leqslant_1) \leqslant (W_2, \leqslant_2)$ iff either $W_1 = W_2$ and $\leqslant_1 = \leqslant_2$ or there exists $a \in W_2$ such that $W_1 = \{x \in W_2 \mid x \leqslant_2 a \wedge x \neq a\}$ and $\leqslant_1 \subset \leqslant_2$. Take any non-empty chain $\mathcal{C}$. Consider the set $V \equiv \{x \in S \mid \exists (W, \leqslant) \in \mathcal{C}\ (x \in W)\}$ and the binary relation $\theta \equiv \{(x, y) \in V * V \mid \exists (W, \leqslant) \in \mathcal{C}\ (((x, y) \in W * W) \wedge (x \leqslant y))\}$. It is easy to check that θ is a linear order on V. Let A be a non-empty subset of V. Then, there exists $(W_1, \leqslant_1) \in \mathcal{C}$ such that $A \cap W_1 \neq \varnothing$. Therefore, there is an element $a_1 \in A \cap W_1$ such that $a_1 \leqslant x$ for every $x \in A \cap W_1$. Suppose that there is an element $a_2 \in A$ such that $(a_2, a_1) \in \theta$. Then, there is $(W_2, \leqslant_2)$ such that $(a_2, a_1) \in W_2 * W_2$ and $a_2 \leqslant_2 a_1$. Since $\mathcal{C}$ is a chain, we have either $(W_1, \leqslant_1) \leqslant (W_2, \leqslant_2)$ or $(W_2, \leqslant_2) \leqslant (W_1, \leqslant_1)$.

Consider the first case. Suppose that $W_1 = W_2$ and $\leqslant_1 = \leqslant_2$. Then, $a_2 \in A \cap W_1$ implies $a_1 \leqslant_2 a_2$. As a result, $a_1 = a_2$. Suppose now that there exists $b \in W_2$ such that $W_1 = \{x \in W_2 \mid x \leqslant_2 b \wedge x \neq b\}$ and $\leqslant_1 \subset \leqslant_2$. If $a_2 \leqslant_2 b$ and $a_2 \neq b$, then $a_2 \in A \cap W_1$ implies $a_1 \leqslant_1 a_2$, where $a_1 \leqslant_2 a_2$. As a result, $a_1 = a_2$. If $a_2 \geqslant_2 b$, then $a_1 \in W_1$ implies $a_2 \geqslant_2 a_1$, where again $a_1 = a_2$. In all the cases, we infer that a_1 is a minimal element of A in (V, θ). Thus, (V, θ) is a well-ordered set. Consequently, it belongs to $\mathcal{Z}$ and is an upper bound for $\mathcal{C}$.

By Zorn's lemma, $\mathcal{Z}$ has a maximal element $(W_0, \leqslant_0)$. If $W_0 = S$, then (5) is proven. Assume that $W_0 \neq S$. Take some $s \in S \setminus W_0$. Consider the set $W \equiv W_0 \cup \{s\}$ and the order $\leqslant \equiv \leqslant_0 \cup \bigcup\{(x, s) \mid x \in W\}$, i. e. we place s after everything in W_0. Then, $(W, \leqslant) \in \mathcal{Z}$. This contradicts the maximality of $(W_0, \leqslant_0)$.

(5) $\vdash$ (1). Let A be a non-empty set. By the condition, there exists an order $\leqslant$ on A such that $(A, \leqslant)$ is a well-ordered set. Define a choice mapping $p\colon\ \mathcal{P}(A) \setminus \{\varnothing\} \to A$, setting $p(P) \equiv \operatorname{sm} P \in P$ for every $P \subset \mathcal{P}(A) \setminus \{\varnothing\}$. □

Remark. There exist some other equivalent forms of the axiom of choice. In particular, Tukey's lemma has several equivalent variants.

1.3 Cardinal numbers

This section is devoted to the theory of cardinal numbers. We follow J. Neumann's definition of a cardinal number as an ordinal number with some exclusive properties. Cardinal numbers have some properties of natural numbers, and natural numbers constitute a part of the class of all cardinal numbers. Thus, cardinal numbers are a generalization of natural numbers.

1.3.1 The definition of cardinal numbers. The cardinality of natural numbers. The first denumerable cardinal number

According to 1.1.8 classes, A and B are called equivalent or equipollent ($A \sim B$) if there exists some bijective mapping $u\colon A \rightarrowtail\!\!\!\twoheadrightarrow B$.

An ordinal number α is called a *cardinal number* if for every ordinal number β the conditions $\beta \leqslant \alpha$ and $\beta \sim \alpha$ imply $\beta = \alpha$. The class of all cardinal numbers will be denoted by Card. According to Corollary 2 to Theorem 1 (1.2.3) the class Card with the order, induced from the class Ord, is well-ordered.

Lemma 1. *Let $x, y \in \omega$ and $x + 1 \sim y + 1$. Then, $x \sim y$.*

Proof. By the condition, there is a bijection $u\colon x+1 \rightarrowtail\!\!\!\twoheadrightarrow y+1$. Take a mapping $v\colon x+1 \twoheadrightarrow y+1$ such that $v \equiv (u \setminus \{\langle x, u(x)\rangle, \langle u^{-1}(y), y\rangle\}) \cup \{\langle u^{-1}(y), u(x)\rangle, \langle x, y\rangle\}$. Then, $v(x) = y$ and $v|x$ is a bijective mapping from x onto y. □

Proposition 1. $\omega \subset \mathrm{Card}$.

Proof. Consider the set $X \equiv \omega \cap \mathrm{Card}$. It is clear that $0 \in X$. Let $m \in X$ and assume that $x \equiv m + 1 \notin X$. Then, x is not a cardinal number. Therefore, there exists an ordinal number $y < x$ such that $y \sim x$. By Corollary 1 to Lemma 1 (1.2.6), $y \in \omega$. By Lemma 6 (1.2.6), $y = n + 1$ for some $n \in \omega$. By virtue of Lemma 1, we get $m \sim n$. By virtue of Lemma 5 (1.2.6), $n < m$. Since m is a cardinal number, this is impossible. This means that $m + 1 \in X$. Consequently, by the principle of natural induction, $X = \omega$. □

Lemma 2. $\omega \in \mathrm{Card}$.

Proof. Let $\beta \in \mathrm{Ord}$ and $\beta < \omega$. Then, $\beta \in \omega$. By Lemma 8 (1.2.6), ω is infinite. Therefore, $\beta \nsim \omega$. □

Lemma 3. *ω is the smallest of all denumerable ordinal numbers.*

Proof. Consider the class $D \equiv \{\alpha \in \mathrm{Ord} \mid \alpha \sim \omega\}$. By virtue of Corollary 2 to Theorem 1 (1.2.3), D has the smallest element β. Since $\omega \in D$, we have $\beta \leqslant \omega$. Suppose that $\beta < \omega$. By Corollary 1 to Lemma 1 (1.2.6), $\beta \in \omega$. By Lemma 8 (1.2.6), ω is infinite. Therefore, $\beta \nsim \omega$, i. e. $\beta \notin D$. It follows from this contradiction that $\beta = \omega$. □

The cardinal number ω is called the *first denumerable cardinal number* (*in the sense of Lemma* 3).

1.3.2 The power of sets

Using the axiom of choice, we can prove the following

Proposition 1. *Let A be a set. Then:*
1) *the ordinal number $\alpha \equiv \mathrm{sm}\{\beta \in \mathrm{Ord} \mid \beta \sim A\}$ is a cardinal number;*
2) *if α' is a cardinal number and $\alpha' \sim A$, then $\alpha' = \alpha$.*

Proof. According to Theorem 1 (1.2.11), the set A can be well-ordered. Therefore, by virtue of Theorem 1 (1.2.5), there is an ordinal number β such that $A \sim \beta$. By Corollary 2 to Theorem 1 (1.2.3), the class Ord is well-ordered. Consequently, the non-empty subclass $\{\beta \in \mathrm{Ord} \mid \beta \sim A\}$ of this class has a minimal element α, which is also the smallest element of this subclass. Take any ordinal number $\gamma \leqslant \alpha$ such that $\gamma \sim \alpha$. Then, $\gamma \sim A$ implies $\alpha \leqslant \gamma$, where $\gamma = \alpha$. This means that α is a cardinal number.

From the definition of α, we infer that $\alpha \leqslant \alpha'$. Besides $\alpha \sim A \sim \alpha'$ implies $\alpha \sim \alpha'$. Since α' is a cardinal number, we conclude that $\alpha = \alpha'$. □

Consider the class $P \equiv \{\langle A, \alpha\rangle \in \mathfrak{U} * \mathrm{Card} \mid A \sim \alpha\}$.

Corollary 1. *The class P is a mapping from the universal class $\mathfrak{U}$ into the class* Card.

Corollary 2. *Let A and B be sets. Then, $A \sim B$ iff $P(A) = P(B)$.*

Corollary 3. *Let A be a set. Then, $P(P(A)) = P(A)$.*

Corollary 4. *Let A be an infinite set. Then, there exists an injective mapping $u\colon \omega \rightarrowtail A$.*

Proof. By Proposition 1, there is a bijection $v\colon \alpha \rightarrowtail\!\!\!\!\twoheadrightarrow A$. Assume that $\alpha < \omega$. Then, by Corollary 1 to Proposition 2 (1.2.2), $\alpha \in \omega$. The definition from 1.2.6 implies that A is finite. It follows from this contradiction that $\alpha \geqslant \omega$. Now, by Lemma 4, (1.2.3) and Theorem 1 (1.2.3) $\omega \subset \alpha$. Consequently, the mapping $u \equiv v|\omega$ is a necessary injection. □

For a set A, the cardinal number $P(A)$ is called the *power* or *cardinality of the set A*. It will be denoted also by card A or $|A|$.

Proposition 2. *Let* $\alpha \in \mathrm{Ord}$ *and* $\beta \subset \alpha$. *Then,* $\mathrm{card}\,\beta \leqslant \alpha$.

Proof. It is clear that β is a well-ordered set. Therefore, by Corollary 1 to Proposition 1 (1.2.5), only the following three cases take place: (1) there are the unique element $b \in \beta$ and the unique isotone bijection $f\colon \alpha \rightarrowtail\!\!\!\twoheadrightarrow]\leftarrow, b[$; (2) there are the unique element $a \in \alpha$ and the unique isotone bijection $g\colon]\leftarrow, a[\rightarrowtail\!\!\!\twoheadrightarrow \beta$; (3) there is the unique isotone bijection $h\colon \alpha \rightarrowtail\!\!\!\twoheadrightarrow \beta$.

At first, suppose that the first case takes place. Then, $f(b) < b$. By Lemma 1 (1.1.15), the mapping f is strictly increasing. Thus, the mapping $u \equiv f|\beta\colon \beta \to \beta$ is also strictly increasing. Therefore, by Lemma 1 (1.2.4) $b \leqslant u(b) = f(b) < b$. It follows from this contradiction that the first case is impossible.

In the second case, $a \in \alpha$ implies by Lemma 3 (1.2.3) that a is an ordinal number. Therefore, by Corollary 1 to Proposition 2 (1.2.2) and Theorem 1 (1.2.3), $a < \alpha$. Now, by Corollary 1 to Lemma 5 (1.2.3), $a =]\leftarrow, a[$ in α. Consequently, g is a bijection between a and β. Therefore, by Proposition 1, we get $\mathrm{card}\,\beta \leqslant a < \alpha$.

Finally, in the third case, $\alpha \sim \beta$ implies by Proposition 1 that $\mathrm{card}\,\beta \leqslant \alpha$. □

Corollary 1. *Let* A *be a set and* $B \subset A$. *Then,* $\mathrm{card}\,B \leqslant \mathrm{card}\,A$.

Proof. By Proposition 1, there is a bijection $u\colon A \rightarrowtail\!\!\!\twoheadrightarrow P(A)$. Consider the subset $\beta \equiv u[B]$ of the ordinal number $P(A)$. Then, by Proposition 2, $P(\beta) \leqslant P(A)$. By Corollary 2 to Proposition 1, $B \sim \beta$ implies $P(B) = P(\beta) \leqslant P(A)$. □

Theorem 1 (the Schröder – Cantor – Bernstein theorem). *Let* A *and* B *be sets,* $X \subset A$, $Y \subset B$, $A \sim Y$ *and* $B \sim X$. *Then,* $A \sim B$.

Proof. By Corollary 1 to Proposition 2 and Corollary 2 to Proposition 1, $P(A) = P(Y) \leqslant P(B) = P(X) \leqslant P(A)$ implies $P(A) = P(B)$, where $A \sim B$. □

This theorem can also be proven without using the axiom of choice.

Lemma 1. *Let* A *be a set,* B *be a class and* $u\colon A \to B$ *be a mapping. Then,* $\mathrm{card}(\mathrm{rng}\,u) \leqslant \mathrm{card}\,A$.

Proof. By Lemma 1 (1.1.11), $C \equiv \mathrm{rng}\,u$ is a set. Take a choice mapping $p\colon \mathcal{P}(A) \setminus \{\varnothing\} \to A$ from axiom A8 (1.1.12). Define a mapping $v\colon C \to A$, setting $v(c) \equiv p(u^{-1}(c))$. If $v(c_1) = v(c_2)$, then $u^{-1}(c_1) \cap u^{-1}(c_2) \neq \varnothing$ implies $c_1 = c_2$. Thus, v is injective, where $C \sim \mathrm{rng}\,v$. Now, by Corollary 2 to Proposition 1 and Corollary 1 to Proposition 2, we get $P(C) = P(\mathrm{rng}\,v) \leqslant P(A)$. □

Theorem 2 (the Cantor theorem on the cardinality of the set of all subsets). *Let* A *be a set. Then,* $\mathrm{card}\,A < \mathrm{card}\,\mathcal{P}(A)$.

Proof. Define a mapping $u\colon A \to \mathcal{P}(A)$, setting $u(a) \equiv \{a\}$. It is clear that u is injective. Therefore, by Corollary 2 to Proposition 1 and Corollary 1 to Proposition 2, $P(A) = P(\operatorname{rng} u) \leqslant P(\mathcal{P}(A))$. Suppose that $P(A) = P(\mathcal{P}(A))$. Then, there exists a bijection $v\colon A \rightarrowtail\!\!\!\twoheadrightarrow \mathcal{P}(A)$. Since $v(a) = \varnothing$ for some a, the set $P \equiv \{b \in A \mid b \notin v(b)\}$ is not empty. Therefore, $P = v(c)$ for some $c \in A$ such that $c \neq a$. If $c \in P = v(c)$, then $c \notin v(c)$. If $c \notin v(c) = P$, then $c \in P$. It follows from this contradiction that our supposition is not valid. As a result, $P(A) < P(\mathcal{P}(A))$. □

Corollary 1. *Let α be an ordinal number. Then, there is a cardinal number β such that $\alpha < \beta$.*

Proof. Take the cardinal number $\beta \equiv \operatorname{card} \mathcal{P}(\alpha)$. Since β is an ordinal number, we have the opportunities $\alpha < \beta$ and $\alpha \geqslant \beta$.

If $\alpha \geqslant \beta$, then by Lemma 4 (1.2.3), $\beta \subset \alpha$. Hence, by Corollary 1 to Proposition 2, $\operatorname{card} \beta \leqslant \operatorname{card} \alpha$. Now, Corollary 3 to Proposition 1 and Theorem 2 imply $\beta = \operatorname{card} \beta \leqslant \operatorname{card} \beta < \operatorname{card} \mathcal{P}(\alpha) \equiv \beta$. □

Theorem 3. *The class* Card *of all cardinal numbers is not a set, i. e. it is a proper class.*

Proof. Assume that $C \equiv \mathrm{Card}$ is a set. Then, by axiom A5 (1.1.11), $D \equiv \bigcup(c_c \mid c \in C)$ is a set as well. Consider the element $c \equiv P(\mathcal{P}(D)) \in C$. If $x \in c$, then $x \in D$, where $c \subset D$. Consequently, by Corollary 3 to Proposition 1 and Corollary 1 to Proposition 2, $P(\mathcal{P}(D)) \equiv c = P(c) \leqslant P(D)$. But this inequality contradicts Theorem 2. □

1.3.3 Properties of finite sets

Lemma 1. *A set A is finite iff* $\operatorname{card} A \in \omega$.

Proof. The assertion follows from the definition of a finite set in 1.2.6, Proposition 1 (1.3.1) and Proposition 1 (1.3.2). □

Proposition 1. *A set A is finite iff there exists an order relation θ on A such that the sets (A, θ) and (A, θ^{-1}) are well-ordered.*

Proof. Let A be a finite set, i. e. there exist a number $n \in \omega$ and a bijection $u\colon n \rightarrowtail\!\!\!\twoheadrightarrow A$. Define a mapping $w\colon n \times n \to A \times A$, setting $w(i, j) \equiv (u(i), u(j))$. It is clear that w is a bijection.

By Proposition 2 (1.2.2), the linearly ordered set (n, θ_n) is well-ordered. According to the definition from 1.2.6, the linearly ordered set (n, θ_n^{-1}) is also well-ordered. Consider the binary relation $\theta \equiv w[\theta_n]$ on A. Then, the sets (A, θ) and (A, θ^{-1}) are well-ordered.

Conversely, let θ be an order relation on A such that (A, θ) and (A, θ^{-1}) are well-ordered sets. By Theorem 1 (1.2.5), there are on ordinal number α and an isotone bijective mapping $u\colon A \rightarrowtail\!\!\!\twoheadrightarrow \alpha$. Suppose that $\omega \leqslant \alpha$. Then, by virtue of Theorem 1 (1.2.3), Theorem 2 (1.2.6) and Lemma 4 (1.2.3), $\omega \subset \alpha$. Thus, we can take the set $B \equiv u^{-1}[\omega]$. Since (A, θ^{-1}) is well-ordered, the set B has the smallest element b. Then, b is the greatest element of B with respect to the order θ^{-1}. Therefore, $m \equiv u(b)$ is the greatest element of $\omega = u[B]$ in $(A, \leqslant)$. But this contradicts Lemma 3 (1.2.6). It follows from this contradiction that $\omega > \alpha$. By virtue of Corollary 1 to Proposition 2 (1.2.2), $\alpha \in \omega$. By the definition from 1.2.6, A is finite. □

Lemma 2. *Let A and B be finite sets. Then, the set $A \cup B$ is finite.*

Proof. By Proposition 1, there exist order relations η and θ such that (A, η), (A, η^{-1}), (B, θ) and (B, θ^{-1}) are well-ordered sets. Consider the set $C \equiv B \setminus A$. Define a binary relation $\varkappa$ on $A \cup B = A \cup C$, setting $\varkappa \equiv \eta \cup (\theta \cap (C \times C)) \cup (A \times C)$. Then, $(A \cup C, \varkappa)$ and $(A \cup C, \varkappa^{-1})$ are well-ordered sets. Therefore, by Proposition 1 we obtain that $A \cup C$ is a finite set. □

Lemma 3. *Let $(A_i \subset \mathfrak{U} \mid i \in I)$ be a collection of finite sets, indexed by a finite set I. Then, $\bigcup(A_i \mid i \in I)$ is a finite set as well.*

Proof. Consider the subset X of ω consisting of all natural numbers n such that if $(A_i \mid i \in I)$ is a collection of finite sets, indexed by a set I of the power n, then $\bigcup(A_i \mid i \in I)$ is a finite set. It is clear that $0 \in X$. Suppose that $n \in X$. Take any collection $(A_i \mid i \in I)$ of finite sets such that $P(I) = n + 1$. By Proposition 1 (1.3.2), there is a bijection $u\colon n + 1 \to I$. Consider the sets $J \equiv u[n]$ and $K \equiv \{k\}$, where $k \equiv u(n)$. From $n + 1 = n \cup \{n\}$, we infer that $I = J \cup K$. By the supposition $U \equiv \bigcup_{,} (A_i \mid i \in J)$ is a finite set. Therefore, by Lemma 2, the set $\bigcup(A_i \mid i \in I) = U \cup A_k$ is finite. This means that $n+1 \in X$. By the principle of natural induction, $X = \omega$. □

Lemma 4. *Let A and B be finite sets. Then, the sets $A * B$ and $A \times B$ are finite.*

Proof. By the condition, there are a natural number n and a bijection $u\colon n \to B$. Then, the mapping $v_a\colon B \to \{a\} * B$ such that $v_a(b) \equiv \langle a, b\rangle$ for every $b \in B$ is a bijection for every $a \in A$. Since $v_a \circ u$ is a bijection for every $a \in A$, $(\{a\} * B \mid a \in A)$ is a collection of finite sets, indexed by the finite set A. Consequently, by Lemma 3 the set $A * B$ is finite. Now, by virtue of Lemma 3 (1.1.12), the set $A * B = \bigcup(\{a\} * B \mid a \in A)$ is finite as well. □

Lemma 5. *Let $(A_i \subset \mathfrak{U} \mid i \in I)$ be a collection of finite sets, indexed by a finite set I. Then, $\prod(A_i \mid i \in I)$ is a finite set as well.*

Proof. Consider the subset X of ω consisting of all natural numbers n such that if $(A_i \mid i \in I)$ is a collection of finite sets, indexed by a set I of the power n, then $\prod(A_i \mid i \in I)$ is a finite set. It is clear that $0 \in X$. Suppose that $n \in X$. Take any collection $(A_i \mid i \in I)$ of finite sets such that $P(I) = n + 1$. By Proposition 1 (1.3.2), there is a bijection $u\colon\ n + 1 \to I$. Consider the element $j \equiv u(n)$ and the set $K \equiv I \setminus \{j\}$. From $n + 1 = n \cup \{n\}$, we infer that $K = u[n]$. By the supposition, the set $E \equiv \prod(A_i \mid i \in K)$ is finite. Therefore, by Lemma 4, the set $A_j * E$ is finite.

By virtue of Theorem 2 (1.1.12), $F \equiv \prod(A_i \mid i \in I) \sim A_j * E$. Consequently, the set F is finite. This means that $n + 1 \in X$. By the principle of natural induction, $X = \omega$. □

Lemma 6. *Let A be a finite set. Then, the set $\mathcal{P}(A)$ is finite.*

Proof. Consider the subset X of ω consisting of all natural numbers n such that if A is a set of the power n, then $\mathcal{P}(A)$ is a finite set. It is clear that $0 \in X$. Suppose that $n \in X$. Take any set A such that $P(A) = n+1$. By Proposition 1 (1.3.2), there is a bijection $u\colon\ n + 1 \to A$. Consider the sets $B \equiv u[n]$ and $C \equiv \{c\}$, where $c \equiv u(n)$. From $n + 1 = n \cup \{n\}$, we infer that $A = B \cup C$ and $B \cap C = \varnothing$. Consider the set $Q \equiv \{P \cup \{c\} \mid P \in \mathcal{P}(B)\}$. From $Q \sim \mathcal{P}(B)$, we infer that the set Q is finite. Since $\mathcal{P}(A) = \mathcal{P}(B) \cup Q$, we conclude by virtue of Lemma 2 that the set $\mathcal{P}(A)$ is finite. This means that $n+1 \in X$. By the principle of natural induction, $X = \omega$. □

Lemma 7. *Let A be a finite set and $B \subset A$. Then, B is a finite set as well.*

Proof. By Corollary 1 to Proposition 2 (1.3.2), $\beta \equiv P(B) \leqslant P(A) \equiv \alpha$. By Lemma 1, we get $\alpha \in \omega$. If $\beta = \alpha$, then $\beta \in \omega$. If $\beta < \alpha$, then by Corollary 1 to Lemma 1 (1.2.6), $\beta \in \omega$. Now, by Lemma 1 B is finite. □

Lemma 8. *Let A and B be finite sets. Then, the sets* $\mathrm{Cor}(A, B)$ *and* $\mathrm{Map}(A, B)$ *are finite.*

Proof. It is clear that $\mathrm{Map}(A, B) \subset \mathrm{Cor}(A, B) \subset \mathcal{P}(A * B)$. Therefore, the conclusion follows from Lemmas 4, 6 and 7. □

Lemma 9. *Let A be a finite set, B be a class and $u\colon\ A \to B$ be a mapping. Then,* $\mathrm{rng}\, u$ *and u are finite sets.*

Proof. By Lemma 1 (1.3.2), $P(\mathrm{rng}\, u) \leqslant P(A)$. As in the proof of Lemma 7, we check that the set $\mathrm{rng}\, u$ is finite.

Define a mapping $v\colon\ A \to u$, setting $v(a) \equiv \langle a, u(a)\rangle$ for every $a \in A$. It is clear that v is surjective. Let $v(a) = v(a')$. Then, by Proposition 2 (1.1.6) $a = a'$. Thus, v is injective. Thus, v is bijective. Since A is finite, we conclude that u is finite as well. □

Proposition 2. *Let A be a finite set, $B \subset A$ and $B \neq A$. Then,* $\operatorname{card} B < \operatorname{card} A$.

Proof. Consider the subset X of ω, consisting of all natural numbers n such that if A is a set of the power $n+1$, $A' \subset A$ and $A' \neq A$, then $P(A') < P(A)$. Let $A \sim 0+1 = \{0\}$. Then, $A = \{a\}$ for some $a \in A$. If $A' \subset A$ and $A' \neq A$, then $A' = \varnothing$. Therefore, $P(A') = 0 < 0+1 = P(A)$ by virtue of Lemma 3 (1.2.6). This means that $0 \in X$.

Suppose that $n \in X$. Take any set A such that $P(A) = (n+1)+1$. By Proposition 1 (1.3.2), there is a bijection $u\colon (n+1)+1 \to A$. Consider the sets $B \equiv u[n+1]$ and $C \equiv \{c\}$, where $c \equiv u(n+1)$. From $(n+1)+1 = (n+1) \cup \{n+1\}$, we infer that $A = B \cup C$ and $B \cap C = \varnothing$.

Let $A' \subset A$ and $A' \neq A$. Consider the sets $B' \equiv A' \cap B$ and $C' \equiv A' \cap C$. If $C' = \varnothing$, then $A' \subset B$ implies $P(A') \leqslant P(B) = n+1 < (n+1)+1 = P(A)$ by virtue of Corollary 1 to Proposition 2 (1.3.2) and Lemma 3 (1.2.6). If $C' \neq \varnothing$, then $C' = C$ implies $B' \neq B$. Thus, by the supposition, $y \equiv P(B') < P(B) \equiv x$. By virtue of Proposition 1 (1.2.3), $y+1 = \operatorname{sm}\{z \in \mathrm{Ord} \mid y < z\}$. Therefore, $y < x$ implies $y+1 \leqslant x$.

By Proposition 1, there exist bijections $v\colon B \to x$ and $w\colon B' \to y$. Define a mapping $f\colon A \to x+1$, setting $f \mid B \equiv u$ and $f(c) \equiv x$. Similarly, define a mapping $g\colon A' \to y+1$, setting $g \mid B' \equiv w$ and $g(c) \equiv y$. Since the mappings f and g are bijective, we infer that $P(A) = x+1$ and $P(A') = y+1$. Now, by virtue of Lemma 3 (1.2.6), $P(A') = y+1 \leqslant x < x+1 = P(A)$. In both of the cases, we got the necessary inequality. This means that $n+1 \in X$. By the principle of natural induction, $X = \omega$. □

Corollary 1. *Let A be a finite set, $B \subset A$ and $B \neq A$. Then, $B \nsim A$.*

Proof. The assertion follows from Proposition 2 and Corollary 2 to Proposition 1 (1.3.2). □

The last property is characteristic for finite sets.

Lemma 10. *Let A be an infinite set. Then, there exists a subset B of the set A such that $B \neq A$ and $B \sim A$.*

Proof. Lemma 1 implies that $\varkappa \equiv P(A) \notin \omega$. By virtue of Corollary 2 to Theorem 1 (1.2.3) and Corollary 1 to Proposition 2 (1.2.2), we get $\varkappa \geqslant \omega$. Therefore, by Lemma 4 (1.2.3), $\omega \subset \varkappa$.

By virtue of Lemma 2 (1.2.6), we can define correctly a mapping $u\colon \varkappa \to \varkappa$, setting $u(x) \equiv x+1 \in \omega \subset \varkappa$ for every $x \in \omega$ and $u(x) \equiv x$ for every $x \in \varkappa \setminus \omega$. Let $u(x) = u(y)$. If $x, y \in \omega$, then $x+y = y+1$ implies by Lemma 4 (1.2.6) $x = y$. If $x \in \omega$ and $y \in \varkappa \setminus \omega$, then $x+1 = y$ implies by Lemma 2 (1.2.6) $y \in \omega$. It follows from this contradiction that this case is impossible. Similarly, the case $x \in \varkappa \setminus \omega$ and $y \in \omega$ is also impossible. Finally, if $x, y \in \varkappa \setminus \omega$, then automatically $x = y$. This means that u is injective. Let $x \in \varkappa \setminus \{0\} \equiv X$. If $x \in \varkappa \setminus \omega$, then $x = u(x)$. If $x \in \omega$, then by Lemma 6 (1.2.6) $x = y+1$ for some $y \in \omega$. Therefore, $x = u(y)$. This means that $\operatorname{rng} u = X \neq \omega$.

By Proposition 1 (1.3.2), there exists a bijection $v\colon \varkappa \to A$. Consider the set $B \equiv v[X] \neq A$. Then, $B \sim X \sim \varkappa \sim A$. □

Lemma 11. *Let* $\alpha \in \mathrm{Ord} \setminus \omega$. *Then,* $\mathrm{card}\,\alpha = \mathrm{card}(\alpha + 1)$.

Proof. Since $\alpha \subset \alpha+1$, we infer by Corollary 1 to Proposition 2 (1.3.2) that $P(\alpha) \leqslant P(\alpha+1)$. By Lemma 10, there exists a subset X of α such that $X \neq \alpha$ and $X \sim \alpha$ with respect a bijection $u\colon \alpha \to X$. Consider the set $Y \equiv \alpha \setminus X$ and take some element $y \in Y$. Define a mapping $v\colon \alpha+1 \to \alpha$, setting $v \mid \alpha \equiv u$ and $v(\alpha) \equiv y$. The mapping is injective. Consequently, $P(\alpha + 1) = P(\mathrm{rng}\, v) \leqslant P(\alpha)$. □

1.3.4 The first uncountable cardinal number. The enumeration of infinite cardinal numbers

Consider the class $\Omega \equiv \{\alpha \in \mathrm{Ord} \mid \mathrm{card}\,\alpha \leqslant \omega\}$, consisting of all countable ordinal numbers. By virtue of Lemma 2 (1.3.1) and Theorem 2 (1.3.2), $\omega = \mathrm{card}\,\omega < \mathrm{card}\,\mathcal{P}(\omega)$. Therefore, the subclass $\mathrm{Ord} \setminus \Omega$ is non-empty and so it has the smallest element ω_1.

Theorem 1.
1) ω_1 *is a cardinal number.*
2) $\omega_1 > \omega$.
3) $\omega_1 = \Omega$.
4) *If* $\varkappa \in \mathrm{Card}$ *and* $\varkappa > \omega$, *then* $\varkappa \geqslant \omega_1$.

Proof. 1. If $\alpha < \omega_1$, then $\alpha \in \Omega$. Consequently, by Corollary 1 to Proposition 2 (1.2.2) and Theorem 1 (1.2.3), $\omega_1 \subset \Omega$.

Suppose that $\alpha \sim \omega_1$. Then, $P(\omega_1) = P(\alpha) \leqslant \omega$ implies $\omega_1 \in \Omega$. But this contradicts the definition of ω_1. Thus, $\alpha \nsim \omega_1$. This means that ω_1 is a cardinal number.

2. Since $\omega_1 \notin \Omega$, we infer that $\omega_1 = P(\omega_1) > \omega$.

3. Let $\alpha \in \Omega$. Suppose that $\alpha \geqslant \omega_1$. By virtue of Lemma 4 (1.2.3) and Proposition 2 (1.3.2), $\omega_1 \subset \alpha$ implies $\omega_1 = P(\omega_1) \leqslant P(\alpha) \leqslant \omega$, but this contradicts assertion 2. Consequently, $\alpha < \omega_1$, where $\alpha \in \omega_1$. This means that $\Omega \subset \omega_1$. As a result, $\omega_1 = \Omega$.

4. Let $\varkappa$ be a cardinal number and $\varkappa > \omega$. Suppose that $\omega_1 > \varkappa$. Then, $\varkappa \in \omega_1$ implies $\varkappa = P(\varkappa) \leqslant \omega$ by virtue of equality 3, but this contradicts the condition. Consequently, $\omega_1 \leqslant \varkappa$. □

It follows from this theorem that the cardinal number ω_1 can be called the *first uncountable cardinal number* (compare with 1.3.1).

Thus, we have two infinite cardinal numbers: ω_0 and ω_1. Define now the enumeration of all infinite cardinal numbers.

Theorem 2. *There is unique bijective isotone mapping* $\aleph\colon \operatorname{Ord} \rightarrowtail\!\!\!\twoheadrightarrow \operatorname{Card}\backslash\omega$ *such that* $\aleph(0) = \omega_0$ *and* $\aleph(1) = \omega_1$.

Proof. Denote Ord by A and Card $\backslash\omega$ by X. Consider the class U of all bijective isotone mappings $u\colon B \to Y$, where B is an initial subclass of the class A and Y is an initial subclass of the class X (see 1.1.15). If $v \in U$, then by Lemma 3 (1.2.4) either $u \subset v$ or $v \subset u$.

Consider the correspondence $w \equiv \{\langle a, x\rangle \in A * X \mid \exists u \in U\ (a \in \operatorname{dom} u \ \wedge\ x = u(a))\}$. Denote dom w by C and rng w by Z. Since $w = \{p \mid \exists u \in U\ (p \in u)\}$, we conclude by virtue of Lemma 5 (1.1.8) that w is a mapping from C onto Z such that $C = \{a \mid \exists u \in U\ (a \in \operatorname{dom} u)\}$, $Z = \{x \mid \exists u \in U\ (x \in \operatorname{rng} u)\}$ and $w \mid \operatorname{dom} u = u$ for every $u \in U$. Consequently, C is an initial subclass of the class A and Z is an initial subclass of the class X. Besides, the mapping w is bijective and isotone.

Suppose that $C \neq A$ and $Z \neq X$. Consider the elements $a \equiv \operatorname{sm}(A\backslash C)$ and $x \equiv \operatorname{sm}(X\backslash Z)$. Take any $b \in C$ and suppose that $b \geqslant a$. Since $b \in \operatorname{dom} u$ for some $u \in U$, we infer that $a \in \operatorname{dom} u \subset C$, but this is not true. It follows from this contradiction that $b < a$, i. e. $C \subset]\!\leftarrow, a[$. Take now any $b \in]\!\leftarrow, a[$ and suppose that $b \notin C$. Then, $a \leqslant b$, but this is not true. It follows from this contradiction that $]\!\leftarrow, a[\subset C$. In result $C =]\!\leftarrow, a[$. Analogously, $Z =]\!\leftarrow, x[$.

Consider the correspondence $w' \equiv w \cup \{\langle a, x\rangle\} \subset A * X$. Then, $C' \equiv \operatorname{dom} w' = C \cup \{a\}$ and $Z' \equiv \operatorname{rng} w' = Z \cup \{x\}$. It follows from the previous indentation that C' and Z' are initial subclasses in A and X correspondingly. Besides, w is a mapping from C' onto Z' such that $w' \mid C = w$ and $w'(a) = x$. Therefore, the mapping w' is bijective and isotone. Consequently, $w' \in U$. This implies $a \in C' \subset C$, but this is not true. It follows from the obtained contradiction that only the following three cases are possible: (1) $C = A$ and $Z \neq X$; (2) $C \neq A$ and $Z = X$; (3) $C = A$ and $Z = X$. By Lemma 5 (1.2.3), $a =]\!\leftarrow, a[= C$ and $Z \subset [0, x[= x$. Since a and x are sets, we conclude that C and Z are sets. By Theorem 1 (1.2.3) and Theorem 3 (1.3.2), the classes C and Z are not sets. Therefore, the first and the second cases are not possible.

It follows from Lemma 2 (1.2.4) that the isotone bijective mapping w is unique.

Take $B = 2$, $Y = \{\omega_0, \omega_1\}$ and $u\colon B \to Y$ such that $u(0) \equiv \omega_0$ and $u(1) \equiv \omega_1$. Then, $u \in U$. Therefore, $w(0) = u(0) = \omega_0$ and $w(1) = u(1) = \omega_1$. □

The uniquely defined isotone bijection $\aleph$ from Theorem 2 is written usually in the form of the simple collection $\aleph \equiv (\omega_\alpha \in \operatorname{Card}\backslash\omega \mid \alpha \in \operatorname{Ord})$. Thus, we have the transfinite sequence of cardinal numbers $0, 1, 2, \ldots, \omega_0, \omega_1, \omega_2, \ldots$.

1.3.5 Derivative cardinal numbers

Cardinal sum and product

Let $(\alpha_i \in \operatorname{Card} \mid i \in I)$ be a simple collection of cardinal numbers, indexed by the (nonempty) set I. Consider the corresponding multivalued collection $(\alpha_i \mid i \in I) \equiv \varphi^{-1}(\alpha_i \mid i \in I)$ from Corollary 1 to Lemma 3 (1.1.9). It follows from Lemma 3 (1.1.6), Proposition 4

(1.1.6), axiom A5 (1.1.11) and the definition of disjoint union from 1.1.10 that the class $\bigcup_d(\alpha_i \mid i \in I)$ is a set. The cardinal number card $\bigcup_d(\alpha_i \mid i \in I)$ of the set $\bigcup_d(\alpha_i \mid i \in I)$ is called the *(cardinal) sum of the simple collection* $(\alpha_i \mid i \in I)$ and is denoted by $\sum(\alpha_i \mid i \in I)$.

By Lemma 1 (1.1.12), the class $\prod(\alpha_i \mid i \in I)$ is a set. The cardinal number card $\prod(\alpha_i \mid i \in I)$ of the set $\prod(\alpha_i \mid i \in I)$ is called the *(cardinal) product of the simple collection* $(\alpha_i \mid i \in I)$ and is denoted by $P(\alpha_i \mid i \in I)$.

Lemma 1. *Let $(A_i \mid i \in I)$ be a multivalued collection of sets, indexed by the set I. Then,* card $\bigcup_d(A_i \mid i \in I) = \sum(\text{card}\, A_i \mid i \in I)$ *and* card $\prod(A_i \mid i \in I) = P(\text{card}\, A_i \mid i \in I)$.

Proof. By the definition from 1.3.2, we see that $A_i \sim \text{card}\, A_i$. Therefore, by Lemma 4 (1.1.10), $\bigcup_d(A_i \mid i \in I) \sim \bigcup_d(\text{card}\, A_i \mid i \in I)$. Similarly, by the assertion 4 of Lemma 5 (1.1.12), $\prod(A_i \mid i \in I) \sim \prod(\text{card}\, A_i \mid i \in I)$. Now, the assertions follow from Proposition 1 (1.3.2). □

Corollary 1. *Let $(A_i \mid i \in I)$ be a multivalued collection of pairwise disjoint sets, indexed by the set I. Then,* card $\bigcup(A_i \mid i \in I) = \sum(\text{card}\, A_i \mid i \in I)$.

Proof. By Lemma 2 (1.1.10), $\bigcup(A_i \mid i \in I) \sim \bigcup_d(A_i \mid i \in I)$. By Corollary 2 to Proposition 1 (1.3.2), card $\bigcup(A_i \mid i \in I) = \text{card} \bigcup_d(A_i \mid i \in I)$. Now, the necessary equality follows from Lemma 1. □

Theorem 1. *Let $(\alpha_i \mid i \in I)$ be a simple collection of cardinal numbers, indexed by the set I. Then,*

1) *if K is a set and $u: K \rightarrowtail\!\!\!\twoheadrightarrow I$ is a bijective mapping, then $\sum(\alpha_i \mid i \in I) = \sum(\alpha_{u(k)} \mid k \in K)$ and $P(\alpha_i \mid i \in I) = P(\alpha_{u(k)} \mid k \in K)$* (the *general commutativity of the sum and the product*);
2) *if a collection $(I_m \mid m \in M)$ is a partition on the set I, indexed by the set $M \neq \varnothing$, then $\sum(\alpha_i \mid i \in I) = \sum(\sum(\alpha_i \mid i \in I_m) \mid m \in M)$ and $P(\alpha_i \mid i \in I) = P(P(\alpha_i \mid i \in I_m) \mid m \in M)$* (the *general associativity of the sum and the product*).

Proof. 1. By Corollary 1 to Proposition 1 (1.1.10) $\bigcup_d(\alpha_i \mid i \in I) = \bigcup_d(\alpha_{u(k)} \mid k \in K)$. Thus, $\sum(\alpha_i \mid i \in I) = \sum(\alpha_{u(k)} \mid k \in K)$. By Theorem 3 (1.1.12) $\prod(\alpha_i \mid i \in I) \sim \prod(\alpha_{u(k)} \mid k \in K)$. By Corollary 2 to Proposition 1 (1.3.2), $P(\alpha_i \mid i \in I) = P(\alpha_{u(k)} \mid k \in K)$.

2. By Corollary 1 to Proposition 1 (1.1.10) $\bigcup_d(\alpha_i \mid i \in I) \sim \bigcup_d(\bigcup_d(\alpha_i \mid i \in I) \mid m \in M)$. Therefore, by Corollary 2 to Proposition 1 (1.3.2) and Lemma 1 $\sum(\alpha_i \mid i \in I) = \text{card} \bigcup_d(\bigcup_d(\alpha_i \mid i \in I_m) \mid m \in M) = \sum(\text{card} \bigcup_d(\alpha_i \mid i \in I_m) \mid m \in M) = \sum(\sum(\alpha_i \mid i \in I_m) \mid m \in M)$.

By Theorem 3 (1.1.12), $\prod(\alpha_i \mid i \in I) \sim \prod(\prod(\alpha_i \mid i \in I_m) \mid m \in M)$. Therefore, as above $P(\alpha_i \mid i \in I) = \text{card} \prod(\prod(\alpha_i \mid i \in I_m) \mid m \in M) = P(\text{card} \prod(\alpha_i \mid i \in I_m) \mid m \in M) = P(P(\alpha_i \mid i \in I_m) \mid m \in M)$. □

Theorem 2. *Let $(I_m \mid m \in M)$ be a collection of sets and $(\varkappa_m \mid m \in M)$ be a simple collection of simple collections $\varkappa_m \equiv (\alpha_{mi} \mid i \in I_m)$ of cardinal numbers, indexed by the non-empty sets M and I_m. Consider the set $U \equiv \prod(I_m \mid m \in M)$. Then, $P(\sum(\alpha_{mi} \mid i \in I_m) \mid m \in M) = \sum(P(\alpha_{mu(m)} \mid m \in M) \mid u \in U)$ (the general distributivity of the product with respect to the sum).*

Proof. By Corollary 1 to Theorem 1 (1.1.13), $\prod(\bigcup_d(\alpha_{mi} \mid i \in I_m) \mid m \in M) \sim \bigcup_d(\prod(\alpha_{mu(m)} \mid m \in M) \mid u \in U)$. Therefore, by Lemma 1 and Corollary 2 to Proposition 1 (1.3.2), $P(\sum(\alpha_{mi} \mid i \in I_m) \mid m \in M) = \operatorname{card} \prod(\bigcup_d(\alpha_{mi} \mid i \in I_m) \mid m \in M) = \operatorname{card} \bigcup_d(\prod(\alpha_{mu(m)} \mid m \in M) \mid u \in U) = \sum(P(\alpha_{mu(m)} \mid m \in M) \mid u \in U)$. □

Let $\alpha, \alpha', \alpha'', \alpha''', \ldots$ be cardinal numbers. Then, (α, α'), $(\alpha, \alpha', \alpha'')$, $(\alpha, \alpha', \alpha'', \alpha''')$,... are the corresponding simple collections (see 1.1.11).

The cardinal numbers $\sum(\alpha, \alpha')$, $\sum(\alpha, \alpha', \alpha'')$, $\sum(\alpha, \alpha', \alpha'', \alpha''')$,... will be called the *(cardinal) sums of the simple sequential pair* (α, α'), *triplet* $(\alpha, \alpha', \alpha'')$, *quadruplet* $(\alpha, \alpha', \alpha'', \alpha''')$,... and will be denoted also by $\alpha + \alpha'$, $\alpha + \alpha' + \alpha''$, $\alpha + \alpha' + \alpha'' + \alpha'''$,... By the definition of the simple sequential pair from 1.1.11, we have $\alpha + \alpha' = \sum(x_i \mid i \in 2) \equiv \operatorname{card} \bigcup_d(x_i \mid i \in 2) = \operatorname{card}(\alpha \cup_d \alpha')$, where $x_0 \equiv \alpha$ and $x_1 \equiv \alpha'$. In the similar manner, $\alpha + \alpha' + \alpha'' = \sum(x_i \mid i \in 3) \equiv \operatorname{card} \bigcup_d(x_i \mid i \in 3) = \operatorname{card}(\alpha \cup_d \alpha' \cup_d \alpha'')$, where $x_0 \equiv \alpha$, $x_1 \equiv \alpha'$ and $x_2 \equiv \alpha''$, and so on.

Analogously, the cardinal numbers $P(\alpha, \alpha')$, $P(\alpha, \alpha', \alpha'')$, $P(\alpha, \alpha', \alpha'', \alpha''')$,... will be called the *(cardinal) products of the simple sequential pair* (α, α'), *triplet* $(\alpha, \alpha', \alpha'')$, *quadruplet* $(\alpha, \alpha', \alpha'', \alpha''')$, ... and will be denoted by $\alpha\alpha'$, $\alpha\alpha'\alpha''$, $\alpha\alpha'\alpha''\alpha'''$,... As above, $\alpha\alpha' = P(x_i \mid i \in 2)$, $\alpha\alpha'\alpha'' = P(x_i \mid i \in 3)$, $\alpha\alpha'\alpha''\alpha''' = P(x_i \mid i \in 4)$,..., where $x_0 \equiv \alpha$, $x_1 \equiv \alpha'$, $x_2 \equiv \alpha''$, $x_3 \equiv \alpha'''$,...

For a simple sequence $(\alpha_0, \ldots, \alpha_{n-1}) \equiv (\alpha_i \in \mathrm{Card} \mid i \in n)$ (see 1.2.6) of cardinal numbers, indexed by a set $n \in \omega \setminus 3$, along with the notations $\sum(\alpha_i \mid i \in n)$ and $P(\alpha_i \mid i \in n)$ we shall use also the notations $\alpha_0 + \cdots + \alpha_{n-1}$ and $\alpha_0 \ldots \alpha_{n-1}$.

Lemma 2.

1) *Let $(\alpha_i \mid i \in \{p\})$ be a simple collection of cardinal numbers, indexed by the set $\{p\}$. Then, $\sum(\alpha_i \mid i \in \{p\}) = \alpha_p$ and $P(\alpha_i \mid i \in \{p\}) = \alpha_p$.*

2) *Let $(\alpha_i \mid i \in \{p, q\})$ be a simple collection of cardinal numbers, indexed by the set $\{p, q\}$ with different elements $p \neq q$. Then, $\sum(\alpha_i \mid i \in \{p, q\}) = \alpha_p + \alpha_q$ and $P(\alpha_i \mid i \in \{p, q\}) = \alpha_p\alpha_q$.*

Proof. We shall prove only assertion 2. Consider the simple collection $(x_i \mid i \in 2)$ such that $x_0 \equiv \alpha_p$ and $x_1 \equiv \alpha_q$. Consider the sets $I \equiv 2$ and $K \equiv \{p, q\}$ and the bijective mapping $u\colon K \rightarrowtail\!\!\!\!\twoheadrightarrow I$ such that $u(p) \equiv 0$ and $u(q) \equiv 1$. Then, by Theorem 1, we get $\alpha_p + \alpha_q = \sum(x_i \mid i \in I) = \sum(x_{u(k)} \mid k \in K)$. From $x_{u(p)} = x_0 \equiv \alpha_p$ and $x_{u(q)} = x_1 \equiv \alpha_q$, we conclude that $(x_{u(k)} \mid k \in K) = (\alpha_k \mid k \in K)$. As a result, $\alpha_p + \alpha_q = \sum(\alpha_k \mid k \in K)$.

For the product, the arguments are the same. □

Theorem 3. *Let x, y and z be cardinal numbers. Then:*
1) $x + y = y + x$ *and* $xy = yx$ (the *commutativity of the sum and the product);*
2) $x + y + z = x + (y + z) = (x + y) + z$ *and* $xyz = x(yz) = (xy)z$ (the *associativity of the sum and the product);*
3) $x(y + z) = xy + xz$ (the *distributivity of the product with respect to the sum).*

Proof. 1. Take the sets $I \equiv 2$ and $K \equiv 2$. Consider the simple collections $(\alpha_i \in \{x, y\} \mid i \in I)$ and $(\beta_k \in \{x, y\} \mid k \in K)$ such that $\alpha_0 \equiv x$, $\alpha_1 \equiv y$, $\beta_0 \equiv y$ and $\beta_1 \equiv x$. Take the bijective mapping $u\colon K \rightarrowtail\!\!\!\twoheadrightarrow I$ such that $u(0) \equiv 1$ and $u(1) \equiv 0$. From $\alpha_{u(0)} = \alpha_1 \equiv y \equiv \beta_0$ and $\alpha_{u(1)} = \alpha_0 \equiv x \equiv \beta_1$, we conclude that $(\alpha_{u(k)} \mid k \in K) = (\beta_k \mid k \in K)$. As a result, by virtue of Theorem 1, we get the chain of equalities $x + y = \sum(\alpha_i \mid i \in I) = \sum(\alpha_{u(k)} \mid k \in K) = \sum(\beta_k \mid k \in K) = y + x$.

For the product, the arguments are the same.

2. Take the sets $I \equiv 3$, $M \equiv 2$, $I_0 \equiv 1$, and $I_1 \equiv \{1, 2\}$. Then, the collection $(I_m \mid m \in M)$ is a partition on the set I. Consider the simple collections $(\alpha_i \mid i \in I)$, $(\alpha_i \mid i \in I_0)$, and $(\alpha_i \mid i \in I_1)$, such that $\alpha_0 \equiv x$, $\alpha_1 \equiv y$ and $\alpha_2 \equiv z$. Then, by Theorem 1 and Lemma 2, we get the chain of equalities $x + y + z = \sum(\alpha_i \mid i \in I) = \sum(\sum(\alpha_i \mid i \in I_m) \mid m \in M) = \sum(\alpha_i \mid i \in I_0) + \sum(\alpha_i \mid i \in I_1)$.

Further, by Lemma 2, we have $\sum(\alpha_i \mid i \in I_0) = \alpha_0 \equiv x$ and $\sum(\alpha_i \mid i \in I_1) = \alpha_1 + \alpha_2 = y + z$. As a result, we get $x + y + z = x + (y + z)$.

In a similar way, we prove the equality $x + y + z = (x + y) + z$.

For the product, the arguments are the same.

3. Take the sets $M \equiv 2 = \{0, 1\}$, $I_0 \equiv 1$, and $I_1 \equiv 2$. Consider the simple collection $\varkappa_0 \equiv (\alpha_{0i} \mid i \in I_0)$ and $\varkappa_1 \equiv (\alpha_{1i} \mid i \in I_1)$ such that $\alpha_{00} \equiv x$, $\alpha_{10} \equiv y$ and $\alpha_{11} \equiv z$. Then, using Lemma 2, we get the chain of equalities $x(y + z) = \alpha_{00}(\alpha_{10} + \alpha_{11}) = \sum(\alpha_{0i} \mid i \in I_0) \sum(\alpha_{1i} \mid i \in I_1) = P(\sum(\alpha_{mi} \mid i \in I_m) \mid m \in M)$. Consider the set $U \equiv \prod(I_m \mid m \in M)$. Then, using Theorem 2, we come to the equality $x(y + z) = \sum(P(\alpha_{mu(m)} \mid m \in M) \mid u \in U)$. If $u \in U$, then $u(0) \in I_0 = 1$ and $u(1) \in I_1 = 2$. Thus, we have only two opportunities: either $u(0) = 0$ and $u(1) = 0$ or $u(0) = 0$ and $u(1) = 1$. Denote the first mapping by p and the second by q. Then, $U \subset \{p, q\}$. It is clear that $U = \{p, q\}$. Therefore, by Lemma 2 we obtain $x(y+z) = P(\alpha_{mp(m)} \mid m \in M) + P(\alpha_{mq(m)} \mid m \in M) = \alpha_{0p(0)}\alpha_{1p(1)} + \alpha_{0q(0)}\alpha_{1q(1)} = xy + xz$. □

Less general than the property of distributivity in Theorem 2 and more general than the property of distributivity in Theorem 3 is the following *property of distributivity of the binary product with respect to the general sums.*

Proposition 1. *Let $(x_j \mid j \in J)$ and $(y_k \mid k \in K)$ be simple collections of cardinal numbers, indexed by non-empty sets J and K. Then, $(\sum(x_j \mid j \in J))(\sum(y_k \mid k \in K)) = \sum(x_j y_k \mid (j, k) \in J \times K)$.*

Proof. Take $M \equiv 2$, $I_0 \equiv J$, $I_1 \equiv K$, $\alpha_{0i} \equiv x_i$ for every $i \in J$, $\alpha_{1i} \equiv y_i$ for every $i \in K$, $\varkappa_0 \equiv (x_j \mid j \in J) = (\alpha_{0i} \mid i \in I_0)$, $\varkappa_1 \equiv (y_k \mid k \in K) = (\alpha_{1i} \mid i \in I_1)$, and $U \equiv J \times K = \prod[I_m \mid m \in M] \neq \varnothing$.

Then, the necessary equality is a particular case of the corresponding equality in Theorem 2. □

Corollary 1. *Let x and y be cardinal numbers and $(x_j \mid j \in J)$ and $(y_k \mid k \in K)$ be simple collections of cardinal numbers, indexed by non-empty sets J and K. Then, $x \sum(y_k \mid k \in K) = \sum(xy_k \mid k \in K)$ and $(\sum(x_j \mid j \in J))y = \sum(x_jy \mid j \in J)$.*

Proof. To check the first equality we can take $J \equiv 1$ and $x_0 \equiv x$ and apply the first assertion of Lemma 2. Then, this equality is a particular case of the equality of Proposition 1. To check the second equality, we can take $K \equiv 1$ and $y_0 \equiv y$. □

Lemma 3. *Let $(\alpha_i \mid i \in I)$ be a simple collection of cardinal numbers, indexed by a non-empty set I, and J be a non-empty subset of I. Then:*
1) *if $\alpha_i = 0$ for every $i \in I \setminus J$, then $\sum(\alpha_i \mid i \in I) = \sum(\alpha_i \mid i \in J)$;*
2) *if $\alpha_i = 1$ for every $i \in I \setminus J$, then $P(\alpha_i \mid i \in I) = P(\alpha_i \mid i \in J)$.*

Proof. 1. If $\alpha_i = 0 \equiv \varnothing$, then $\alpha_i * \{i\} = \varnothing * \{i\} = \varnothing$. Therefore, $\bigcup_d[\alpha_i \mid i \in I] = \bigcup[\alpha_i * \{i\} \mid i \in I] = \bigcup[\alpha_i * \{i\} \mid i \in J] = \bigcup_d[\alpha_i \mid i \in J]$ implies the necessary equality.

2. Denote the sets $\prod[\alpha_i \mid i \in I]$ and $\prod[\alpha_i \mid i \in J]$, respectively, by P and Q. By Lemma 4 (1.1.12), the projection $u \equiv p_J\colon P \to Q$ is surjective. Take any elements $e \equiv (x_i \mid i \in I)$ and $f \equiv (y_i \mid i \in I)$ of the set P and suppose that $ue = uf$. Then, $x_i = y_i$ for every $i \in J$. By the definition of product $x_i \in \alpha_i$ and $y_i \in \alpha_i$ for every $i \in K \equiv I \setminus J$. Since $\alpha_i = 1 = \{0\}$ we infer that $x_i = 0 = y_i$ for every $i \in K$. Thus, $e = f$, i. e. the mapping u is injective. This means that $P \sim Q$. By Corollary 3 to Proposition 1 (1.3.2), this implies the necessary equality. □

Corollary 1. *Let α be a cardinal number. Then, $\alpha + 0 = \alpha 1 = \alpha$.*

Corollary 2. *Let α and β be cardinal numbers and $(\alpha_i \mid i \in I)$ and $(\gamma_i \mid i \in I)$ be simple collections of cardinal numbers, indexed by the set I, such that $\alpha_i = \alpha$ and $\gamma_i = 1$ for every $i \in I$ and $I \sim \beta$. Then, $\alpha\beta = \sum(\alpha_i \mid i \in I)$ and $\beta = \sum(\gamma_i \mid i \in I)$.*

Proof. It is evident that $\beta \sim I = \bigcup[\{i\} \mid i \in I] \sim \bigcup[\{\gamma_i\} * \{i\} \mid i \in I] \equiv \bigcup_d[\gamma_i \mid i \in I]$. By virtue of Proposition 1 (1.3.2), we infer that $\beta = \sum(\gamma_i \mid i \in I)$. Therefore, using Corollary 1 to Proposition 1 and Corollary 1 to Lemma 3, we get $\alpha\beta = \alpha \sum(\gamma_i \mid i \in I) = \sum(\alpha\gamma_i \mid i \in I) = \sum(\alpha_i \mid i \in I)$. □

Corollary 3. *Let $(\alpha_i \mid i \in I)$ be a simple collection of cardinal numbers, indexed by a non-empty set I. Then:*

1) *if* $\alpha_i = 0$ *for every* $i \in I$, *then* $\sum(\alpha_i \mid i \in I) = 0$;
2) *if* $\alpha_i = 1$ *for every* $i \in 1$, *then* $P(\alpha_i \mid i \in I) = 1$;

Proof. 1. By the condition, there is $j \in I$. Denote $\{j\}$ by J. Then, by Lemma 3 and Lemma 2, we have $\sum(\alpha_i \mid i \in I) = \sum(\alpha_i \mid i \in J) = \alpha_j = 0$.

2. The second assertion is checked in the same way. □

Lemma 4. *Let* $(\alpha_i \mid i \in I)$ *be a simple collection of cardinal numbers, indexed by the (non-empty) set* I. *Then, the following conditions are equivalent:*
1) $\alpha_i \neq 0$ *for every* $i \in I$;
2) $P(\alpha_i \mid i \in I) \neq 0$.

Proof. (1) ⊢ (2). By virtue of Theorem 1 (1.1.12), $\prod(\alpha_i \mid i \in I) \neq \varnothing$. This gives (2).

(2) ⊢ (1). Condition 2 implies $\prod(\alpha_i \mid i \in I) \neq \varnothing$. This means that there exists a mapping $u \in \prod(\alpha_i \mid i \in I)$. Then, $u(i) \in \alpha_i$ implies $\alpha_i \neq 0$ for every $i \in I$. □

The following assertion is a generalization of Lemma 4 (1.2.6).

Lemma 5. *Let* α *and* β *be cardinal numbers and* $\alpha + 1 = \beta + 1$. *Then,* $\alpha = \beta$.

Proof. It follows from the definition that $\alpha + 1 = \sum(x_i \mid i \in 2)$ and $\beta + 1 = \sum(y_i \mid i \in 2)$ where $x_0 \equiv \alpha$, $x_1 \equiv 1$, $y_0 \equiv \beta$, and $y_1 \equiv 1$. Therefore, $\sum(x_i \mid i \in 2) = \alpha + 1 = \beta + 1 = \sum(y_i \mid i \in 2)$ implies $\alpha \equiv x_0 = y_0 \equiv \beta$. □

Cardinal degree

Let α and β be cardinal numbers. The cardinal number $\operatorname{card}\operatorname{Map}(\beta, \alpha)$ of the set $\operatorname{Map}(\beta, \alpha)$ of all mappings from the set β into the set α is called the (*cardinal*) *degree of the cardinal* α *with the cardinal exponent* β and is denoted by α^β. Note that we have already used this notation in 1.1.8. Therefore, it is necessary to distinguish the cardinal degree from the degree of the set α with the exponent β as a set.

Lemma 6. *Let* A *and* B *be sets and* $\alpha \equiv \operatorname{card} A$ *and* $\beta \equiv \operatorname{card} B$. *Then,* $\operatorname{card}\operatorname{Map}(B, A) = \alpha^\beta$.

Proof. By the condition, there exist some bijective mappings $u\colon A \to \alpha$ and $v\colon \beta \to B$. Define a mapping $w\colon \operatorname{Map}(B, A) \to \operatorname{Map}(\beta, \alpha)$, setting $wf \equiv u \circ f \circ v$ for every $f\colon B \to A$. It is easy to check that w is bijective. By Corollary 2 to Proposition 1 (1.3.2), this implies the necessary equality. □

Lemma 7. *Let* α *and* β *be cardinal numbers and* $(\alpha_i \mid i \in I)$ *be a simple collection of cardinal numbers, indexed by the set* I *such that* $\alpha_i = \alpha$ *for every* $i \in I$ *and* $I \sim \beta$. *Then,* $\alpha^\beta = P(\alpha_i \mid i \in I)$.

Proof. As in the proof of Lemma 6, it is checked that $\mathrm{Map}(\beta,\alpha) \sim \mathrm{Map}(I,\alpha) = \prod(\alpha_i \mid i \in I)$. By Corollary 2 to Proposition 1 (1.3.2), this implies the necessary equality. □

Corollary 1. *Let α be a cardinal number and $(\beta_i \mid i \in I)$ be a simple collection of cardinal numbers, indexed by the non-empty set I. Then, $\alpha^{\sum(\beta_i|i\in I)} = P(\alpha^{\beta_i} \mid i \in I)$*

Proof. Denote the left part of this equality by L and the right part by R. Consider the set $K \equiv \bigcup_d(\beta_i \mid i \in I)$ and the collections $(K_i \mid i \in I)$ and $(\alpha_k \mid k \in K)$ such that $K_i \equiv \beta_i * \{i\}$ for every $i \in I$ and $\alpha_k \equiv \alpha$ for every $k \in K$.

Then, by Lemma 6, Theorem 1, and Lemma 7, we have the equalities $L \equiv \mathrm{card\,Map}(\mathrm{card}\,K,\alpha) = \mathrm{card\,Map}(K,\alpha) = \mathrm{card}\prod(\alpha_k \mid k \in K) \equiv P(\alpha_k \mid k \in K) = P(P(\alpha_k \mid k \in K_i) \mid i \in I) = R$. □

Corollary 2. *Let α, γ and δ be cardinal numbers. Then, $\alpha^{\gamma+\delta} = \alpha^\gamma\alpha^\delta$.*

Proof. Consider the simple collection $(\beta_i \mid i \in 2)$ such that $\beta_0 \equiv \gamma$ and $\beta_1 \equiv \delta$. Then, by Corollary 1 $\alpha^{\gamma+\delta} = \alpha^{\sum(\beta_i|i\in 2)} = P(\alpha^{\beta_i} \mid i \in 2) = \alpha^\gamma\alpha^\delta$. □

Corollary 3. *Let α, β and γ be cardinal numbers. Then, $\alpha^{\beta\gamma} = (\alpha^\beta)^\gamma$.*

Proof. Consider the simple collection $(\beta_i) \mid i \in \gamma$ such that $\beta_i \equiv \beta$ for every $i \in \gamma$. Then, by Corollary 2 to Lemma 3, we get $\beta\gamma = \sum(\beta_i \mid i \in \gamma)$. Therefore, by Corollary 1 $\alpha^{\beta\gamma} = P(\alpha^{\beta_i} \mid i \in \gamma) = \mathrm{card\,Map}(\gamma,\alpha^\beta) = (\alpha^\beta)^\gamma$. □

Lemma 8. *Let $(\alpha_i \mid i \in I)$ be a simple collection of cardinal numbers, indexed by the non-empty set I, and β be a cardinal number. Then, $(P(\alpha_i \mid i \in I))^\beta = P(\alpha_i^\beta \mid i \in I)$.*

Proof. Denote the left part of this equality by L and the right part by R. Consider the set $A \equiv I \times \beta$ and the collections $(A_i \mid i \in I)$, $(A_x \mid x \in \beta)$, $(\alpha_a \mid a \in A)$, $(\pi_i \mid i \in I)$, and $(\pi_x \mid x \in \beta)$, such that $A_x \equiv I \times \{x\}$ for every $x \in \beta$, $A_i \equiv \{i\} \times \beta$ for every $i \in I$, $\alpha_a \equiv \alpha_i$ for every $a \equiv (i,x) \in A$, and $\pi_x \equiv P(\alpha_i \mid i \in I)$ for every $x \in \beta$.

Define the collection $(u_x \mid x \in \beta)$ of bijective mappings $u_x\colon A_x \to I$ setting $u_{(i,x)} \equiv i$ for every $(i,x) \in A_x$. Then, by the assertion 1) of Theorem 3 (1.1.12) $\prod(\alpha_i \mid i \in I) \sim \prod(\alpha_{u(a)} \mid a \in A_x) = \prod(\alpha_i \mid (i,x) \in A_x) = \prod(\alpha_a \mid a \in A_x)$. Thus, by Corollary 2 to Proposition 1 (1.3.2), $\pi_x = P(\alpha_a \mid a \in A_x)$.

Therefore, by Lemma 7 and Theorem 1, we get the equalities $L = P(\pi_x \mid x \in \beta) = P(P(\alpha_a \mid a \in A_x) \mid x \in \beta) = P(\alpha_a \mid a \in A) = P(P(\alpha_a \mid a \in A_i) \mid i \in I) = R$. □

Lemma 9. *Let α be a cardinal number. Then, $\alpha^0 = 1$, $\alpha^1 = \alpha$, and $1^\alpha = 1$. Besides, if $\alpha \neq 0$, then $0^\alpha = 0$.*

Proof. The first formula follows from the equality $\mathrm{Map}(\varnothing, \alpha) = \{\varnothing\}$. Since $\mathrm{Map}(1, \alpha) \sim \alpha$, we infer by Proposition 1 (1.3.2) that $\alpha^1 = \alpha$. The third formula follows from the equality $\mathrm{Map}(\alpha, 1) = \{\alpha * \{0\}\}$. Finally, if $\alpha \neq \varnothing$, then $\mathrm{Map}(\alpha, 0) \subset \alpha * 0 = 0$ implies the forth formula. □

Note that $0^0 = 1$.

Proposition 2. *Let A be a set and α be its cardinal number. Then,* $\mathrm{card}\, \mathcal{P}(A) = 2^\alpha$.

Proof. For every subset B of the set A consider the function $g\colon A \to 2$ such that $g(b) \equiv 1$ for every $b \in B$ and $g(a) \equiv 0$ for every $a \in A \setminus B$. Define a mapping $\chi\colon \mathcal{P}(A) \to \mathrm{Map}(A, 2)$, setting $\chi(B) \equiv g$. Let $B \neq C$, and suppose that there exists $b \in B \setminus C$. Then, $\chi(B)(b) = 1$ and $\chi(C)(b) = 0$ means that $\chi(B) \neq \chi(C)$. Therefore, χ is injective. Let $g \in \mathrm{Map}(A, 2)$. Then, for the subset $B \equiv \{a \in A \mid g(a) = 1\}$, we have $\chi(B) = g$. This means that χ is surjective. Consequently, χ is bijective. Using Corollary 2 to Proposition 1 (1.3.2), we get the necessary equality. □

Lemma 10. *Let α and β be cardinal numbers. Then, the following conclusions are equivalent:*
1) $\alpha \leqslant \beta$;
2) *there is a cardinal number γ such that $\beta = \alpha + \gamma$.*

Proof. (1) ⊢ (2). Consider the set $C \equiv \beta \setminus \alpha$ and its cardinal number γ. Then, we have the following chain of equivalences: $\beta = \alpha \cup C \sim \alpha \cup_d C \sim \alpha \cup_d \gamma$. Using the Corollary 3 to Proposition 1 (1.3.2), we infer that $\beta = \mathrm{card}(\alpha \cup_d \gamma) = \alpha + \gamma$.

(2) ⊢ (1). By Corollary 1 to Proposition 2 (1.3.2), $\alpha = \mathrm{card}(\alpha * \{0\}) \leqslant \mathrm{card}((\alpha * \{0\}) \cup (\gamma * \{1\})) = \mathrm{card}(\alpha \cup_d \gamma) = \alpha + \gamma = \beta$. □

Lemma 11. *Let $(\alpha_i \mid i \in I)$ and $(a_i \mid i \in I)$ be simple collections of cardinal numbers, indexed by the non-empty set I such that $\alpha_i \leqslant a_i$ for every $i \in I$. Then, $\sum(\alpha_i \mid i \in I) \leqslant \sum(a_i \mid i \in I)$ and $P(\alpha_i \mid i \in I) \leqslant P(a_i \mid i \in I)$.*

Proof. It follows from the condition, Theorem 1 (1.2.3) and Lemma 4 (1.2.3) that $\alpha_i \subset a_i$. Consider the identical mapping $u_i \equiv \mathrm{id}_{\alpha_i, a_i}\colon \alpha_i \rightarrowtail a_i$ (see 1.1.8). By Lemma 5 (1.1.12), the mapping $v \equiv \prod_m(u_i \mid i \in I)\colon \prod(\alpha_i \mid i \in I) \to \prod(a_i \mid i \in I)$ is injective. By Lemma 4 (1.1.10), the mapping $u \equiv \bigcup_{dm}(u_i \mid i \in I)\colon \bigcup_d(\alpha_i \mid i \in I) \to \bigcup_d(a_i \mid i \in I)$ is injective. Thus, by Corollary 2 to Proposition 1 (1.3.2) and Corollary 1 to Proposition 2 (1.3.2), $\sum(\alpha_i \mid i \in I) = \mathrm{card}(\mathrm{rng}\, u) \leqslant \sum(a_i \mid i \in I)$ and $P(\alpha_i \mid i \in I) = \mathrm{card}(\mathrm{rng}\, v) \leqslant P(a_i \mid i \in I)$. □

Corollary 1. *Let β, γ, b, and c be cardinal numbers such that $\beta \leqslant b$ and $\gamma \leqslant c$. Then, $\beta + \gamma \leqslant b + c$ and $\beta\gamma \leqslant bc$.*

Proof. By the definition $\beta + \gamma = \sum(\alpha_i \mid i \in 2)$ and $b + c = \sum(a_i \mid i \in 2)$ where $\alpha_0 \equiv \beta$, $\alpha_1 \equiv \gamma$, $a_0 \equiv b$ and $a_1 \equiv c$. Therefore, the inequalities $\alpha_0 \leqslant a_0$ and $\alpha_1 \leqslant a_1$ imply by Lemma 11 the inequality $\beta + \gamma \leqslant b + c$.

The second inequality is checked in a similar way. □

Corollary 2. *Let β be a cardinal number. Then, $\beta \leqslant \beta + 1$.*

Corollary 3. *Let x, y and z be cardinal numbers such that $x = y$. Then, $x + z = y + z$ and $xz = yz$.*

Proof. The equalities follow from Corollary 1 because the order relation is antisymmetric. □

Corollary 4. *Let $(a_i \mid i \in I)$ be a simple collection of cardinal numbers, indexed by the set I, and J be a subset of the set I. Then, $\sum(a_i \mid i \in J) \leqslant \sum(a_i \mid i \in I)$. If, besides, $a_i \neq 0$ for every $i \in I \setminus J$, then $P(a_i \mid i \in J) \leqslant P(a_i \mid i \in I)$.*

Proof. Define a simple collection $(\alpha_i \mid i \in I)$, setting $\alpha_i \equiv a_i$ for every $i \in J$ and $\alpha_i \equiv 0$ (respectively $\alpha_i \equiv 1$) for every $i \in I \setminus J$. Now, apply Lemma 11. □

Corollary 5. *Let β, γ, b and b be cardinal numbers such that $\beta \leqslant b$, $\gamma \leqslant c$ and $b > 0$. Then, $\beta^\gamma \leqslant b^c$.*

Proof. By Lemma 7, we have $\beta^\gamma = P(\beta_i \mid i \in \gamma)$ and $b^\gamma = P(b_i \mid i \in \gamma)$, where $\beta_i \equiv \beta$ and $b_i \equiv b$. Then, Lemma 11 implies $\beta^\gamma \leqslant b^\gamma$. In a similar way, $b^c = P(b_i \mid i \in c)$. It follows from the condition, Theorem 1 (1.2.3) and Lemma 4 (1.2.3) that $\gamma \subset c$. Therefore, by Corollary 4 $b^\gamma \leqslant b^c$. □

Corollary 6. *Let x, y and z be cardinal numbers such that $x = y$. Then, $x^z = y^z$ and $z^x = z^y$.*

Proof. The equalities follow from Corollary 5 because the order relation is antisymmetric. □

1.3.6 Derivative natural numbers

Now, we have the definition of ordinal sum in 1.2.9 and the definition of cardinal sum in 1.3.5. These sums may be different in general.

To distinguish for a cardinal number α the ordinal number $\alpha+1$, introduced in 1.2.3, from the cardinal sum of these numbers, we shall denote sometimes this cardinal sum by $\alpha +_c 1$.

Lemma 1. *Let m and n be natural numbers. Then,* $m +_o n = n + m$.

Proof. Denote the ordinal number in the left part of this equality by α and the cardinal number in the right part by β. Then, $\alpha \approx m +_o n \sim m \cup_d n \sim \beta$ implies $\operatorname{card} \alpha = \beta$. According to Proposition 1 (1.3.2), $\beta \leqslant \alpha$. Therefore, by Lemma 4 (1.2.3) $\beta \subset \alpha$. By the definition from 1.2.6, the sets $m * \{0\} \sim m$ and $n * \{1\} \sim n$ are finite. Thus, by Lemma 2 (1.3.3), the set $m \cup_d n$ is finite as well. Consequently, the set α is finite. Supposing that $\beta \neq \alpha$, we deduce by Proposition 2 (1.3.3) that $\beta = \operatorname{card} \beta < \operatorname{card} \alpha$, but this contradicts the previous equality. As a result, we conclude that $\beta = \alpha$. □

Thus, for natural numbers the binary ordinal and cardinal sums coincide. Moreover, if m is a natural number, then $m + 1 = m +_o 1 = m +_c 1$.

Theorem 1. *Let α be a cardinal number. Then, the following conclusions are equivalent:*
1) *α is a natural number;*
2) $\alpha < \alpha +_c 1$.

Proof. (1) ⊢ (2). By Lemma 3 (1.2.6), Proposition 3 (1.2.9), and Lemma 1, we get $\alpha < \alpha +_o 1 = \alpha +_c 1$.

(2) ⊢ (1). Let $\alpha < \alpha +_c 1$. Suppose that $\alpha \notin \omega$. By the Proposition 3 (1.2.9) and Lemma 11 (1.3.3), $\alpha = \operatorname{card}(\alpha +_o 1)$. Consequently, $\alpha \sim \alpha +_o 1 \approx \alpha \cup_{do} 1$ implies $\alpha \sim \alpha \cup_d 1$. Therefore, by Proposition 1 (1.3.2), $\alpha = \operatorname{card}(\alpha \cup_d 1) = \alpha +_c 1$, but this contradicts the initial inequality. Thus, $\alpha \in \omega$. □

Lemma 2. *Let α be a cardinal number. Then, the following conclusions are equivalent:*
1) *α is a natural number;*
2) *$\alpha +_c 1$ is a natural number.*

Proof. (1) ⊢ (2). By Theorem 1, Lemma 1, Proposition 3 (1.2.9), and Lemma 3 (1.2.6), we obtain $\alpha < \alpha +_c 1 = \alpha +_o 1 = \alpha + 1 < (\alpha + 1) + 1 = (\alpha +_c 1) +_c 1$. Therefore, by Theorem 1, we see that $\alpha +_c 1$ is a natural number.

(2) ⊢ (1). Suppose that $\alpha = \alpha +_c 1$. Then, Theorem 1 implies $\alpha = \alpha +_c 1 < (\alpha +_c 1) +_c 1 = \alpha +_c 1 = \alpha$. It follows from this contradiction that our supposition is not valid. Thus, by Corollary 2 to Lemma 11 (1.3.5), we get $\alpha < \alpha +_c 1$. Now, by Theorem 1, we see that α is a natural number. □

Sum, product, and raising to a degree for natural numbers

Lemma 3. *Let $(n_i \mid i \in I)$ be a simple collection of natural numbers, indexed by a finite set I. Then, the cardinal numbers $\sum(n_i \mid i \in I)$ and $P(n_i \mid i \in I)$ are natural.*

Proof. By the definition from 1.2.6, the sets $n_i * \{i\} \sim n_i$ are finite. Thus, by Lemma 2 (1.3.3), the set $X \equiv \bigcup_d(n_i \mid i \in I)$ is finite as well, where $X \sim m$ for some $m \in \omega$. By virtue of Proposition 1 (1.3.2), this equivalence implies the first assertion of this lemma.

By Lemma 5 (1.3.3), the set $\prod(n_i \mid i \in I)$ is finite. As above, this implies the second assertion. □

Corollary 1. *Let $\alpha, \alpha', \alpha'', \alpha''', \ldots$ be natural numbers. Then, the cardinal numbers $\alpha + \alpha', \alpha + \alpha' + \alpha'', \alpha + \alpha' + \alpha'' + \alpha''', \ldots$ and $\alpha \cdot \alpha', \alpha \cdot \alpha' \cdot \alpha'', \alpha \cdot \alpha' \cdot \alpha'' \cdot \alpha''', \ldots$ are natural.*

Note that usually the symbol of multiplication "·" is omitted.

Corollary 2. *Let m and n be natural numbers. Then, the cardinal number m^n is natural.*

Proof. The assertion follows from Lemma 7 (1.3.5) and Lemma 3. □

Now, we shall formulate especially for natural numbers some properties of addition, multiplication, and raising to a degree, proven for cardinal numbers in 1.3.5.

Theorem 2. *Let $(n_i \mid i \in I)$ be a simple collection of natural numbers, indexed by the finite set I. Then:*
1) *if K is a finite set and $u\colon K \rightarrowtail\!\!\!\twoheadrightarrow I$ be a bijective mapping, then $\sum(n_i \mid i \in I) = \sum(n_{u(k)} \mid k \in K)$ and $P(n_i \mid i \in I) = P(n_{u(k)} \mid k \in K)$* (the *general commutativity of the sum and the product*);
2) *if a collection $(I_m \mid m \in M)$ is a finite partition of the set I, indexed by the finite non-empty set M, then $\sum(n_i \mid i \in I) = \sum(\sum(n_i \mid i \in I_m) \mid m \in M)$ and $P(n_i \mid i \in I) = P(P(n_i \mid i \in I_m) \mid m \in M)$* (the *general associativity of the sum and the product*).

Proof. These assertions are direct consequences of Theorem 1 (1.3.5) and Lemma 3. □

Theorem 3. *Let $(I_m \mid m \in M)$ be a collection of finite sets and $(\varkappa_m \mid m \in M)$ be a simple collection of simple collections $\varkappa_m \equiv (n_{mi} \mid i \in I_m)$ of natural numbers, indexed by the non-empty finite sets M and I_m. Consider the finite set $U \equiv \prod(I_m \mid m \in M)$. Then, $P(\sum(n_{mi} \mid i \in I_m) \mid m \in M) = \sum(P(n_{mu(m)} \mid m \in M) \mid u \in U)$* (the *general distributivity of the product with respect to the sum*).

Proof. This assertion is a direct consequence of Theorem 2 (1.3.5) and Lemma 3. □

Lemma 4.
1) *Let $(n_i \mid i \in \{p\})$ be a simple collection of natural numbers, indexed by the set $\{p\}$. Then, $\sum(n_i \mid i \in \{p\}) = n_p$ and $P(n_i \mid i \in \{p\}) = n_p$.*

2) *Let $(n_i \mid i \in \{p, q\})$ be a simple collection of natural numbers, indexed by the set $\{p, q\}$ with different elements $p \neq q$. Then, $\sum(n_i \mid i \in \{p, q\}) = n_p + n_q$ and $P(n_i \mid i \in \{p, q\}) = n_p n_q$.*

Proof. This lemma is simply a special case of Lemma 2 (1.3.5). □

Theorem 4. *Let x, y and z be natural numbers. Then:*
1) $x + y = y + x$ *and* $xy = yx$ (the *commutativity of the sum and the product);*
2) $x + y + z = x + (y + z) = (x + y) + z$ *and* $xyz = x(yz) = (xy)z$ (the *associativity of the sum and the product);*
3) $x(y + z) = xy + xz$ (the *distributivity of the product with respect to the sum*).

Proof. All the assertions are direct consequences of Theorem 3 (1.3.5) and Corollary 1 to Lemma 3. □

Proposition 1. *Let $(x_j \mid j \in J)$ and $(y_k \mid k \in K)$ be simple collections of natural numbers, indexed by non-empty finite sets J and K. Then, $(\sum(x_j \mid j \in J))(\sum(y_k \mid k \in K)) = \sum(x_j y_k \mid (j, k) \in J \times K)$.*

Proof. This assertion is a direct consequence of Proposition 1 (1.3.5), Lemma 3, and Corollary 1 to it. □

Further, we shall consider inequalities for derivative natural numbers. Before that, we shall recall some basic inequalities.

By the definition of the first natural numbers from 1.2.6, we have $0 \in \{0\} = 1$. By Theorem 2 (1.2.6) and Corollary 1 to Proposition 2 (1.2.2), this means that $0 < 1$.

By the same reason, if $n \in \omega$ and $n > 0$, then $0 \in n$. Therefore, $1 = \{0\} \subset n$ by Lemma 4 (1.2.3) implies $n \geqslant 1$. Conversely, if $n \in \omega$ and $n \geqslant 1$, then $n \geqslant 1 > 0$.

Theorem 5. *Let m and n be natural numbers. Then, the following assertions are equivalent:*
1) $m < n$;
2) *there is a natural number $k > 0$ such that $n = m + k$.*

Proof. (1) ⊢ (2). By Lemma 10 (1.3.5), there is a cardinal number k, such that $n = m + k$. It follows from Corollary 1 to Lemma 3 (1.3.5) that $k > 0$. Suppose that k is not a natural number. Then, by Corollary 2 to Lemma 11 (1.3.5) $k \leqslant k +_c 1$. From Theorem 1, we infer that $k = k +_c 1$. Therefore, by Theorem 3 (1.3.5), $n +_c 1 = (m + k) +_c 1 = m + (k +_c 1) = m + k = n$. From Theorem 1, we infer that n is not a natural number. It follows from this contradiction that k is a natural number.

(2) ⊢ (1). Let $n = m + k$ and $k \geqslant 1$. By Theorem 1, $m < m + 1$. Therefore, by Corollary 1 to Lemma 11 (1.3.5), $m < m + 1 \leqslant m + k = n$. □

Proposition 2. *Let $(m_i \mid i \in I)$ and $(n_i \mid i \in I)$ be simple collections of natural numbers, indexed by the non-empty finite set I, such that $m_i \leqslant n_i$ for every $i \in I$. If $m_i < n_i$ at least for one index, then $\sum(m_i \mid i \in I) < \sum(n_i \mid i \in I)$. If, besides, $n_i > 0$ for every $i \in I$, then $P(m_i \mid i \in I) < P(n_i \mid i \in I)$.*

Proof. Let $m_j < n_j$ for some j. Consider the set $K \equiv I \setminus \{j\}$. By Theorem 5, $n_j = m_j + l$ for some $l > 0$. Using Theorem 2, Lemma 4, Lemma 11 (1.3.5), and Theorem 5, we can deduce that $\sum(n_i \mid i \in I) = n_j + \sum(n_i \mid i \in K) \geqslant l + m_j + \sum(m_i \mid i \in K) = l + \sum(m_i \mid i \in I) > \sum(m_i \mid i \in I)$.

Denote $\sum(n_i \mid i \in K)$ by z. Then, as above, using in addition Theorem 5 we can deduce that $P(n_i \mid i \in I) = n_j z = m_j z + kz \geqslant m_j P(m_i \mid i \in K) + kz = P(m_i \mid i \in I) + kz$. By Lemma 11 (1.3.5), Lemma 7 (1.3.5), and Lemma 9 (1.3.5) $z \geqslant 1^{\varkappa} = 1$, where $\varkappa \equiv \operatorname{card} K$. Therefore, by Corollary 1 to Lemma 11 (1.3.5) and Corollary 1 to Lemma 3 (1.3.5) $kz \geqslant k1 = k > 0$. As a result, by Theorem 5, we get the second necessary inequality. □

Corollary 1. *Let p, q, r and s be natural numbers such that $p < q$ and $r \leqslant s$. Then, $p+r < q+s$. If, besides, $s > 0$, then $pr < qs$.*

Corollary 2. *Let $(n_i \mid i \in I)$ be a finite simple collection of natural numbers, indexed by the non-empty finite set I, such that $n_i \geqslant 0$ for every $i \in I$. If $n_i > 0$ at least for one index, then $\sum(n_i \mid i \in I) > 0$. If, besides, $n_i > 0$ for every $i \in I$, then $P(n_i \mid i \in I) > 0$.*

Proof. Consider the simple collection $(m_i \mid i \in I)$ such that $m_i \equiv 0$ for every $i \in I$. Then, by Corollary 3 to Lemma 3 (1.3.5), $\sum(m_i \mid i \in I) = 0$. By Lemma 4 (1.3.5), $P(m_i \mid i \in I) = 0$. Now, it is sufficient to apply Proposition 2. □

Corollary 3. *Let m, n and k be natural numbers such that $m < n$ and $k > 0$. Then, $m^k < n^k$.*

Proof. The inequality follows from Lemma 7 (1.3.5) and the second assertion of Proposition 2. □

Corollary 4. *Let n and k be natural numbers such that $n > 0$ and $k > 0$. Then, $n^k > 0$.*

Proof. Take $m = 0$. Then, by Lemma 9 (1.3.5), $m^k = 0$. Now, the inequality follows from Corollary 3. □

Corollary 5. *Let m, n and k be natural numbers such that $m < n$ and $k > 1$. Then, $k^m < k^n$.*

Proof. By Theorem 5, $n = m + l$ for some $l > 0$. Therefore, by Corollary 2 to Lemma 7 (1.3.5), $k^n = k^m k^l$. From $k > 1$ and $l > 0$, we infer by Corollary 2 and Lemma 9 (1.3.5) that $k^l > 1^l = 1$. From $m \geqslant 0$, we infer by Corollary 5 to Lemma 11 (1.3.5) and Lemma 9 (1.3.5) that $k^m \geqslant 1^0 = 1 > 0$. Now by Corollary 1, $k^m k^l > k^m 1 = k^m$. □

Corollary 6. *Let m, n, and k be natural numbers. If $m + k = n + k$, then $m = n$. If $k > 0$ and $mk = nk$, then $m = n$.*

Proof. All the assertions follows from Corollary 1 by means of the proof from an opposite assumption. □

Corollary 7. *Let m and n be natural numbers such that $m \leqslant n$. Then, there is a unique natural number k such that $n = m + k$.*

Proof. If $m = n$, then by Corollary 1 to Lemma 3 (1.3.5) $n = m + 0$. If $m < n$, then by Theorem 5, $n = m + k$ for some $k > 0$. Suppose that $n = m + l$ for some l. Then, by Corollary 6, we have $l = k$. □

If m and n are natural numbers and $m \leqslant n$, then the unique natural number from Corollary 7 such that $n = m + k$ is called the *difference of the numbers n and m* and is denoted by $n - m$.

Corollary 8. *Let k, l, m, and n be natural, numbers such that $k \leqslant l$ and $m \leqslant n$. Then, $(l - k) + (n - m) = (l + n) - (k + m)$.*

Proof. By Corollary 1 to Lemma 11 (1.3.5), $k + m \leqslant l + n$. By the definition of difference $l = k + p$ and $n = m + q$ for $p \equiv l - k$ and $q \equiv n - m$. Then, using Theorem 4 several times, we deduce that $l + n = (k + p) + (m + q) = k + (p + (m + q)) = k + ((p + m) + q) = (k + (m + p)) + q + ((k + m) + p) + q = (k + m) + (p + q)$. Thus, $p + q = (l + n) - (k + m)$. □

Now, we can prove an important version of the principle of natural induction from 1.2.6.

Theorem 6 (the *general principle of natural induction*). *Let $X \subset [m, \rightarrow[\subset \omega$ and $m \in X$. If $n \in X$ implies $n + 1 \in X$, then $X = [m, \rightarrow[$.*

Proof. Consider the set $Y \equiv \{x - m \mid x \in X\}$. Then, $0 \in Y$. If $n \in Y$, then $n = x - m$ for some $x \in X$. By the condition, $x + 1 \in X$. Therefore, by Corollary 8 of Proposition 2 $n + 1 = (x - m) + (1 - 0) = (x + 1) - (m + 0) = (x + 1) - m \in Y$. By Theorem 1 (1.2.6), $Y = \omega$. Take any $n \in [m, \rightarrow[$. By Corollary 7 of Proposition 2, $n = m + (n - m)$. Therefore, $n - m \in \omega = Y$ implies $n - m = x - m$ for some $x \in X$. By the definition of difference $x = m + (n - m)$. Thus, $n = x \in X$. This means that $X = [m, \rightarrow[$. □

Division of natural numbers

Theorem 7 (the *Euclidean division*). *Let m and n be natural numbers such that $n > 0$. Then, there exist unique natural numbers q and r such that $m = nq + r$ and $r < n$.*

Proof. If $m < n$, then $q = 0$ and $r = m$. If $m = n$, then $q = 1$ and $r = 0 < n$. Therefore, further we shall assume that $m > n$. Consider the set $P \equiv \{x \in \omega \mid m < nx\}$. By Theorem 1, $m < m+1$. Besides, $1 \leqslant n$. Thus, by Corollary 1 to Lemma 3 (1.3.5) and Corollary 1 to Proposition 2, $m = 1m < n(m+1)$. This means that $m+1 \in P$, where $P \neq \varnothing$. Consider the smallest element p of the set P. Since $p \in \mathbb{N}$, we infer from Lemma 6 (1.2.6) that $p = q + 1$ for some q. By Theorem 1, we have $q < q + 1 = p$. Thus, $q \notin P$, i. e. $nq \leqslant m$. Besides, $m < np = n(q + 1) = nq + n$. Now, by Corollary 7 of Proposition 2, we have $m = nq + r$ for a unique number r.

Suppose that $r \geqslant n$. Then, by Corollary 1 to Proposition 2, $m = nq+r \geqslant nq+n > m$. It follows from this contradiction that $r < n$.

Now, suppose that there exist numbers $\varkappa$ and ρ such that $m = n\varkappa + \rho$ and $\rho < n$. Then, by Corollary 1 to Proposition 2, $n\varkappa \leqslant m = n\varkappa + \rho < n\varkappa + n = n(\varkappa + 1)$. Therefore, $\varkappa + 1 \in P$, where $q + 1 = p \leqslant \varkappa + 1$. Suppose that $q > \varkappa$. Then, by Corollary 1 to Proposition 2 $q+1 > \varkappa+1$, but this is not so. Now, suppose that $q < \varkappa$. Then, $\varkappa = q+k$ for some $k > 0$. From $1 \leqslant k$, we infer by Corollary 1 to Lemma 11 (1.3.5) that $q + 1 \leqslant q + k = \varkappa$. By the same, reason $m < n(q + 1) \leqslant n\varkappa \leqslant m$, but this is impossible. Consequently, $q = \varkappa$. Finally, $\rho = m - n\varkappa = m - nq = r$. □

The number r from Theorem 7 is called the *remainder at the division of the number m by the number n*. If $r = 0$, then the number m is called the *multiple of the number n*, and the number n is called the *divisor of the number m*. In this case, we also say that *m is divided by n* and *n divides m*. If $r = 0$, then the number q is called the *quotient at the division of the number m by the number n* and is denoted by m/n.

Natural numbers that are multiples of the number 2 are called *even*. All the others are called *odd*. By Theorem 7. even [odd] numbers m are described by the formula $m = 2q$ [$m = 2q + 1$].

By Theorem 1 (1.2.10), for every ordinal number $\alpha \in \mathrm{Ord} \setminus \omega$, there are a unique limit ordinal number $\gamma \geqslant \omega$ and a unique natural number n such that $\alpha = \gamma +_0 n$. If the natural number n is even [odd], then the ordinal number α is called *even* [*odd*]. Since $n = 0$ is even, all limit ordinal numbers are even as well.

1.3.7 Ordered sets of natural numbers

Let I be a subset of the set ω (see 1.2.6). Since ω is an ordered set, we can induce an order on I, setting $i \leqslant j$ for $i, j \in I$ if $i \leqslant j$ in ω. This order is called a *natural order on the subset $I \subset \omega$*.

Theorem 1. *Let I be a subset of ω, taken with its natural order, and $m \in \omega$. If I is finite, then there are a unique number $n \in \omega$ and a unique isotone* (see 1.1.5) *bijection $u\colon\ n \setminus m \rightarrowtail\!\!\!\twoheadrightarrow I$. If I is denumerable, then there is a unique isotone bijection $u\colon\ \omega \setminus m \rightarrowtail\!\!\!\twoheadrightarrow I$.*

Proof. At first, note that by virtue of Theorem 2 (1.2.6) every non-empty subset K of ω has the smallest element $\operatorname{sm} K$. By virtue of Theorem 3 (1.2.6), every non-empty finite subset, L of ω has the greatest element $\operatorname{gr} L$. Consider the number $i_0 \equiv \operatorname{sm} I$.

At first, assume that I is denumerable. Define an isotone injection $u_0\colon\ (m+1) \setminus m \rightarrowtail I$, setting $u_0(m) \equiv i_0$. Consider the subset X of ω, consisting of all $k \in \omega$ such that for k there is a unique isotone injection $u_k\colon\ (m+k+1) \setminus m \rightarrowtail I$ such that $u_k(m) \equiv i_0$ and $u_k(i+1) = \operatorname{sm}(I \setminus u_k[(i+1) \setminus m])$ for all $i \in (m+k+1) \setminus m$. It is clear that $0 \in X$.

Let $k \in X$. Supposing that $I = \operatorname{rng} u_k$, we conclude that I is finite. According to Lemma 8 (1.2.6), this contradicts the condition that I is denumerable. Therefore, $I \setminus \operatorname{rng} u_k \neq \varnothing$. Hence, we can define a mapping $u_{k+1}\colon\ (m+(k+1)+1) \setminus m \rightarrowtail I$, setting $u_{k+1} \equiv u_k \cup \{(m+k+1,\ \operatorname{sm}(I \setminus \operatorname{rng} u_k))\}$. The mapping u_{k+1} has the two necessary properties.

Suppose that a mapping $v\colon\ (m+(k+1)+1) \rightarrowtail I$ also has the same properties. Consider the set $Y \equiv \{y \in k+2 \mid v|(m+y+1) \setminus m = u_{k+1}|(m+y+1) \setminus m\} \cup (\omega \setminus (k+2))$. It is clear that $0 \in Y$. Suppose that $y \in Y$. If $y+1 \in k+2$, then $y \in k+2$ implies $v|(m+y+1) \setminus m = u_{k+1}|(m+y+1) \setminus m$. Besides, $v(m+y+1) = \operatorname{sm}(I \setminus v[(m+y+1) \setminus m]) = \operatorname{sm}(I \setminus u_{k+1}[(m+y+1) \setminus m]) = u_{k+1}(m+y+1)$. This means that v and u_{k+1} coincide on the set $(m+(y+1)+1) \setminus m$. Thus, $y+1 \in Y$. If $y+1 \notin k+2$, then $y+1 \in \omega \setminus (k+2) \subset Y$. In both the cases, $y+1 \in Y$. By the principle of natural induction $Y = \omega$. This means that we can take $y = k+1$. Then, $v|(m+(k+1)+1) \setminus m = u_{k+1}|(m+(k+1)+1) \setminus m$ means that $v = u_{k+1}$. This proves the uniqueness of u_{k+1}.

It follows from the properties proven above that $k+1 \in X$. By the principle of natural induction $X = \omega$. Now, we shall check that u_l is an extension of u_k for every $l > k \geqslant 0$. Fix k and l and consider the set $Z \equiv \{z \in k+1 \mid u_l|(m+z+1) \setminus m = u_k|(m+z+1) \setminus m\} \cup (\omega \setminus (k+1))$. Since $u_l(m) = u_k(m)$, we have $0 \in Z$. Suppose that $z \in Z$. If $(z+1) \in k+1$, then $z \in k+1$ implies $u_l|(m+z+1) \setminus m = u_k|(m+z+1) \setminus m$. Besides, $u_l(m+z+1) = \operatorname{sm}(I \setminus u_l[(m+z+1) \setminus m]) = \operatorname{sm}(I \setminus u_k[(m+z+1) \setminus m]) = u_k(m+z+1)$. This means that u_l and u_k coincide on the set $(m+(z+1)+1) \setminus m$. Thus, $z+1 \in Z$. If $z \in \omega \setminus (k+1)$, then $z+1 \in \omega \setminus (k+1) \subset Z$. In both the cases, $z+1 \in Z$. By the principle of natural induction, $Z = \omega$. This means that we can take $z = k$. Then, $u_l|(m+k+1) \setminus m = u_k|(m+k+1) \setminus m$ means that u_l is an extension of u_k.

Thus, we can define correctly a mapping $u\colon\ \omega \setminus m \rightarrowtail I$, setting $u|(m+k+1) \setminus m = u_k = u_l|(m+k+1) \setminus m$ for every $l > k$ in ω.

Let now $q > p$ in $\omega \setminus m$. Then, $q = m+l$ and $p = m+k$ for some $l > k \geqslant 0$. Therefore, $u(q) = u_{l-1}(m+l) > u_{l-1}(m+k) = u_k(m+k)$. If $k = 0$, then $u(q) > u_0(m) = u(m+k) = u(p)$. If $k > 0$, then $u(q) > u_k(m+k) = u(p)$. It follows from this property that u is isotone and injective.

Suppose that u is not surjective and take $i_1 \equiv \mathrm{sm}(I \setminus \mathrm{rng}\, u)$. By the construction, $i_0 < i_1$. Therefore, $J \equiv \{i \in I \mid i_0 \leqslant i < i_1\} \subset \mathrm{rng}\, u$. Consider the number $i_2 \equiv \mathrm{gr}\, J$. Then, $i_2 = u(m + r)$ for some $r \in \omega$. If $s \in r$, then $i_0 \leqslant u(m + s) < u(m + r) = i_2 < i_1$ implies $u(m + s) \in J$, where $u[(m + r + 1) \setminus m] \subset J$. If there exists $j \in J \setminus \{i_0, i_2\}$, then $j = u(m + s)$ for some $s \in \omega$. Thus, $u(m + s) = j < i_2 = u(m + r)$ implies $m + s < m + r$, i. e. $m + s \in (m + r + 1) \setminus m$, where $j \in u[(m + r + 1) \setminus m]$. Thus, $u[(m + r + 1) \setminus m] = J$.

Consequently, $u_{r+1}[(m + r + 1) \setminus m] = J$. Therefore, $u(m + r + 1) = u_{r+1}(m + r + 1) = \mathrm{sm}(I \setminus u_{r+1}[(m + r + 1) \setminus m]) = \mathrm{sm}(I \setminus J) = i_1$, where $i_1 \in \mathrm{rng}\, u$.

It follows from this contradiction that u is surjective. Finally, u is bijective. Suppose that there exists another isotone bijective mapping $w\colon\ \omega \setminus m \rightarrowtail\!\!\!\!\rightarrow I$ such that $w(m + k) \neq u(m + k)$ for some $k \in \omega$. Consider the mapping $w' \equiv w|(m + k + 1) \setminus m$. Then, we have $w'(m + k) \neq u(m + k) = u_k(m + k)$, but this contradicts the uniqueness of u_k. Consequently, u is unique.

At last, assume that I is finite. Take the number $i_3 \equiv \mathrm{gr}\, I$ and consider the countably infinite set $I' \equiv I \cup (\omega \setminus (i_3 + 1))$. As proven above, there exists an isotone bijection $u'\colon\ \omega \setminus m \rightarrowtail\!\!\!\!\rightarrow I'$. Take n such that $u'(n - 1) = i_3$. If $i \in I$, then $i = u'(m + k)$ for some $k \in \omega$. Thus, $u'(m + k) \leqslant i_3 = u'(n - 1)$ implies $m + k \leqslant n - 1$, where $I \subset u'[n \setminus m]$. On the other hand, if $m + k \in n \setminus m$, then $m \leqslant m + k \leqslant n - 1$ implies $i_0 \leqslant u'(m + k) \leqslant i_3$, i. e. $u'(m + k) \notin \omega \setminus (i_3 + 1)$, where $u'(m + k) \in I$. Thus, $I = u'[n \setminus m]$. Take now $u \equiv u'|n \setminus m$. Then, $u\colon\ n \setminus m \rightarrowtail\!\!\!\!\rightarrow I$ is the necessary isotone bijection.

Suppose that there exists another isotone bijection $v\colon\ n \setminus m \rightarrowtail\!\!\!\!\rightarrow I$ such that $v(m+k) \neq u(m+k)$ for some $k \in \omega$. Define a mapping $v'\colon\ \omega \setminus m \to I'$, setting $v'|n \setminus m \equiv v$ and $v'(n + l) \equiv i_3 + l + 1$. It is clear that v' is isotone bijection and $v' \neq u'$. This is in contradiction to the uniqueness of u'. Consequently, u is unique.

Suppose that there exist a natural number p and an isotone bijection $v\colon\ p \setminus m \rightarrowtail\!\!\!\!\rightarrow I$. Then, $v^{-1} \circ u\colon\ n \setminus m \rightarrowtail\!\!\!\!\rightarrow p \setminus n$ is an isotone bijection. Suppose that $n > p$. Then, $p \setminus m$ is the initial interval $]\leftarrow, p[$ in the well-ordered set $(n \setminus m, \leqslant)$. But then, we get the contradiction with conclusion 1 of Proposition 1 (1.2.4). Supposing that $n < p$, we come to the contradiction in the similar manner. Thus, $n = p$, i. e. the number n is unique. □

1.3.8 Properties of infinite cardinal numbers

To prove the properties of infinite cardinal numbers, we introduce some order on the class $R \equiv \mathrm{Ord} \times \mathrm{Ord}$.

First, we shall give an intuitive description of this order for the set $\omega \times \omega$. We shall represent $\omega \times \omega$ by points on a plane. An ordering will be assigned by means of the upper and right sides of squares in the following manner: $a < b < c < d < e < f < h$.

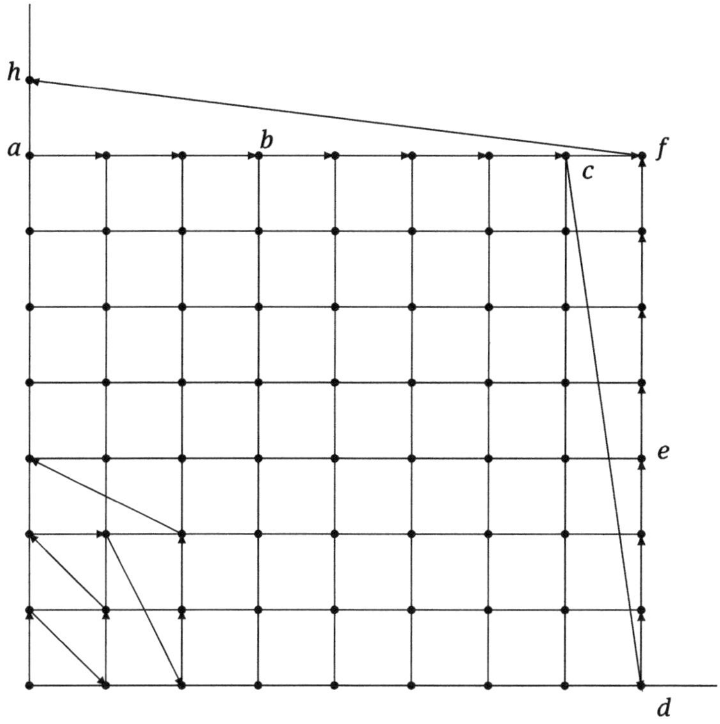

Now, we shall give the strict definition. If α and β are ordinals, then by Lemma 4 (1.2.3) $\mathrm{gr}\{\alpha, \beta\} = \alpha \cup \beta$. Consider a relation θ on R such that $((u, v), (x, y)) \in \theta$ iff either $\mathrm{gr}\{u, v\} < \mathrm{gr}\{x, y\}$, or $\mathrm{gr}\{u, v\} = \mathrm{gr}\{x, y\}$ and $u < x$, or $\mathrm{gr}\{u, v\} = \mathrm{gr}\{x, y\}$, $u = x$ and $v < y$.

Lemma 1. *The relation θ is total and with the property of minimality.*

Proof. Take in R any points $p \equiv (u, v)$ and $q \equiv (x, y)$. Since the class Ord is linearly ordered, we have only the following cases. In the first case, $\mathrm{gr}\{u, v\} < \mathrm{gr}\{x, y\}$, where $(p, q) \in \theta$. In the second case, $\mathrm{gr}\{u, v\} > \mathrm{gr}\{x, y\}$, where $(q, p) \in \theta$. In the third case, $\mathrm{gr}\{u, v\} = \mathrm{gr}\{x, y\}$. Then, we have also the following alternative cases. If $u < x$, then $(p, q) \in \theta$. If $u > x$, then $(q, p) \in \theta$. Finally, let $u = x$. Then, $v < y$ implies $(p, q) \in \theta$; $v > y$ implies $(q, p) \in \theta$; and $v = y$ implies $p = q$.

Take any $P \subset R$. Consider in R for every ordinal number x the square $Q_x \equiv (x+1) \times (x+1)$ with the right side $S_x \equiv \{p \in R \mid \exists \beta (\beta \in \mathrm{Ord} \wedge \beta \leqslant x \wedge p = (x, \beta))\}$ and the upper side $T_x \equiv \{q \in R \mid \exists \alpha (\alpha \in \mathrm{Ord} \wedge \alpha \leqslant x \wedge q = (\alpha, x))\}$. Consider the set $X \equiv \{x \in \mathrm{Ord} \mid Q_x \cap P \neq \varnothing\}$. By virtue of Corollary 2 to Theorem 1 (1.2.3), it has the smallest element z. Consider the square Q_z and its vertex $r \equiv (z, z)$. It is clear that $P \cap Q_z \subset S_z \cup T_z$.

If $P \cap T_z$ contains a point $t \neq z$, then consider the set $A \equiv \{\alpha \in \mathrm{Ord} \mid \exists t\, (t \in P \cap T_z \wedge t \neq r \wedge t = (\alpha, z))\}$, its smallest element u, and the point $q \equiv (u, z)$. By definition,

$(q, t) \in \theta$ for every $t \in P \cap T_z$, $(q, s) \in \theta$ for every $s \in P \cap S_z$ and $(q, p) \in \theta$ for every $p \in P \setminus Q_z$. Thus, q is a minimal element in P.

If $P \cap T_z$ contains only the vertex r and $P \cap S_z$ contains a point $s \neq r$, then consider the set $B \equiv \{\beta \in \mathrm{Ord} \mid \exists s\ (s \in P \cap S_z \wedge s \neq r \wedge s = (z, \beta))\}$, its smallest element v, and the point $p \equiv (z, v)$. By definition, $(p, s) \in \theta$ for every $s \in P \cap S_z$ and $(p, q) \in \theta$ for every $q \in P \setminus Q_z$. Thus, p is a minimal element in P.

Finally, if $P \cap T_z$ and $P \cap S_z$ contain only the vertex r, then r is a minimal element in P. □

Corollary 1. *The relation* $\theta \cup \mathrm{id}_{\mathrm{Ord}}$ *is well-ordering.*

Proof. The assertion follows from Lemma 1 and Lemma 1 (1.2.1). □

Denote the relation from this Corollary by $\leqslant$.

Lemma 2. *Let* u, v, x, y *be ordinal numbers such that* $(u, v) \leqslant (x, y)$. *Then,* $(u, v) \in (\mathrm{gr}\{x, y\} + 1) \times (\mathrm{gr}\{x, y\} + 1)$.

Proof. It is clear that $\mathrm{gr}\{u, v\} \leqslant \mathrm{gr}\{x, y\}$. Thus, by Proposition 1 (1.2.3), $u \leqslant \mathrm{gr}\{x, y\} < \mathrm{gr}\{x, y\} + 1$. By Corollary 1 to Proposition 2 (1.2.2) $u \in \mathrm{gr}\{x, y\} + 1$. The similar property is valid for v. □

Cardinal numbers from the class $\mathrm{Card} \setminus \omega$ are called *infinite* or *transfinite.*

Theorem 1. *Let* $\alpha \in \mathrm{Card} \setminus \omega$. *Then,* $\mathrm{card}(\alpha \times \alpha) = \alpha$. *In particular,* $\mathrm{card}(\omega \times \omega) = \omega$.

Proof. Consider the class $\mathcal{A} \equiv \mathrm{Card} \setminus \omega$ and its subclass $\mathcal{B} \equiv \{\alpha \in \mathcal{A} \mid \mathrm{card}(\alpha \times \alpha) = \alpha\}$. Take any $x \in \mathcal{A}$ and assume that $[\omega, x[\subset \mathcal{B}$. By Theorem 1 (1.2.5) and Corollary 1 to Lemma 1 for the well-ordered set $A \equiv x \times x$, there are an ordinal number α and an isotone bijection $f\colon A \rightarrowtail\!\!\!\twoheadrightarrow \alpha$. Since $x \neq \varnothing$, there is $u \in x$. It is evident that $x \sim x \times \{u\} \subset A$. Therefore, $x = \mathrm{card}(x \times \{u\}) \leqslant \mathrm{card}\, A = \alpha$.

Let $a \equiv (u, v) \in A$, i. e., $u < x$ and $v < x$. Consider the interval $I \equiv [(0, 0), a]$ in A. By Lemma 2, this interval is contained in the square $Q \equiv (\mathrm{gr}\{u, v\} + 1) \times (\mathrm{gr}\{u, v\} + 1)$. Let $\beta \in \alpha$ and assume that $\beta \in J \equiv [0, f(a)]$. Then, $\beta = f(b)$ for some $b \in A$. It follows from $f(b) \leqslant f(a)$ that $b \leqslant a$, where $b \in I$. Therefore, $\beta \in f[I]$. Consequently, $J = f[I]$, i. e., $J \sim I$.

At first, assume that $x = \omega$. Then, $\mathrm{gr}\{u, v\} \in \omega$ implies $\mathrm{gr}\{u, v\} + 1 \in \omega$ by virtue of Lemma 2 (1.2.6). By Lemma 4 (1.3.3), the set Q is finite. By Lemma 7 (1.3.3), the set I is also finite. As a result, the set J is finite. But by Lemma 8 (1.2.6), the set ω is infinite. Consequently, $\omega \notin J$, where $f(a) < \omega = x$.

Now, assume that $x > \omega$. If $u, v \in \omega$, then in the similar manner $f(a) < \omega < x$. Finally, if $\mathrm{gr}\{u, v\} \notin \omega$, then by Lemma 11 (1.3.3) $y \equiv \mathrm{card}(\mathrm{gr}\{u, v\} + 1) = \mathrm{card}$

$\mathrm{gr}\{u,v\} < x$. It follows from this inequality that $\mathrm{card}(y \times y) = y$. Therefore, $\mathrm{gr}\{u,v\}+1 \sim y$ implies $Q \sim y \times y \sim y$, i. e., $\mathrm{card}\, Q = y$. Therefore, $\mathrm{card}\, I \leqslant y$ implies $\mathrm{card}\, J \leqslant y < x$. Suppose that $f(a) \geqslant x$. Then, $x \in J$ implies $x \subset J$, where $x = \mathrm{card}\, x \leqslant \mathrm{card}\, J < x$. It follows from this contradiction that $f(a) < x$.

Thus, in all the cases, $f(a) \in x$. Therefore, $\alpha = f[A] \subset x$ implies $\alpha \leqslant x$. As a result, we get that $\mathrm{card}\, A = \alpha = x$. This means that $x \in \mathcal{B}$. By the principle of induction from 1.2.1, $\mathcal{B} = \mathcal{A}$. □

Corollary 1. *Let α be an infinite cardinal number. Then, $\alpha\alpha = \alpha$.*

Corollary 2. *Let α and β be cardinal numbers, α be infinite, and $0 < \beta \leqslant \alpha$. Then, $\mathrm{card}(\alpha \times \beta) = \alpha$.*

Proof. From $0 \in \beta$, we infer that $\alpha \sim \alpha \times \{0\} \subset \alpha \times \beta \subset \alpha \times \alpha$. Consequently, $\alpha \leqslant \mathrm{card}(\alpha \times \beta) \leqslant \mathrm{card}(\alpha \times \alpha) = \alpha$. □

Corollary 3. *Let α and β be cardinal numbers, α be infinite, and $0 < \beta \leqslant \alpha$. Then, $\alpha\beta = \alpha$.*

Corollary 4. *Let α be an infinite cardinal number. Then, $\alpha + \alpha = \alpha$.*

Proof. By Corollary 2 to Lemma 3 (1.3.5), $2\alpha = \alpha + \alpha$. By the preceding corollary, $\alpha = \mathrm{card}(\alpha \times 2) = \mathrm{card}(2 \times \alpha) \equiv 2\alpha$. □

Corollary 5. *Let α and β be cardinal numbers, α be infinite, and $\beta \leqslant \alpha$. Then, $\alpha + \beta = \alpha$.*

Proof. By virtue of Corollary 1 to Lemma 11 (1.3.5), $\alpha \leqslant \alpha + \beta \leqslant \alpha + \alpha = \alpha$. □

Corollary 6. *Let α be an infinite cardinal number and n be a non-zero natural number. Then, $\alpha^n = \alpha$.*

Proof. Consider the set $X \equiv \{m \in \omega \mid \alpha^{m+1} = \alpha\}$. It is clear that $0 \in X$. Assume that $m \in X$. By Corollary 2 to Lemma 7 (1.3.5), $\alpha^{(m+1)+1} = \alpha^{m+1}\alpha = \alpha\alpha = \alpha$. This means that $m+1 \in X$. By virtue of the principle of natural induction from 1.2.6 $X = \omega$. □

Proposition 1. *Let A be an infinite set of the power α. Then, the set of all finite subsets of the set A has also the power α.*

Proof. Denote the set of all subsets of A by F. From $A \sim \alpha$, we infer that $A^n \sim \alpha^n = \alpha$, i. e. $\mathrm{card}\, A^n = \alpha$ for every non-zero natural number. By definition, A^n consists of all sequences $(a_i \in A \mid i \in n)$. Consider also the sets $F_n \equiv \{B \subset A \mid \mathrm{card}\, B = n\}$. It is clear that $F_m \cap F_n = \varnothing$ and $A_m \cap A_n = \varnothing$ for every $m \neq n$ from $\mathbb{N}$. Consider the surjective mapping u_n from A^n onto F^n such that $u_n(a_i \in A \mid i \in n) = \{a_i \mid i \in n\}$ (see 1.1.9). Since

$F = \bigcup(F^n \mid n \in \mathbb{N})$, we can define correctly a surjective mapping u from $S \equiv \bigcup(A^n \mid n \in \mathbb{N})$ onto F setting $us \equiv u_n s$ for every $s \in A^n$. Then, by Lemma 1 (1.3.2), $\operatorname{card} F \leqslant \operatorname{card} S$. Consider also the mapping $v\colon A \rightarrowtail F$ such that $va \equiv \{a\}$. Then, by Corollary 1 to Proposition 2 (1.3.2), $\alpha = \operatorname{card} v[A] \leqslant \operatorname{card} F$.

By Corollary 1 to Lemma 1 (1.3.5) $\operatorname{card} S = \sum(\operatorname{card} A^n \mid n \in \mathbb{N})$. Since $\operatorname{card} A^n = \alpha$, we infer by virtue of Corollary 2 to Lemma 3 (1.3.5) that $\operatorname{card} S = \alpha\omega$. By Corollary 3 to Theorem 1, $\alpha\omega = \alpha$. As a result, we get the inequality $\alpha \leqslant \operatorname{card} F \leqslant \operatorname{card} S = \alpha$, which give the necessary equality. □

1.3.9 Properties of countable sets

Now, we can prove some basic properties of countable sets defined in 1.2.6.

Lemma 1. *Any subset of countable set is countable.*

Proof. Let A be countable and $B \subset A$. Then, by Corollary 1 to Proposition 1 (1.3.2) $\operatorname{card} B \leqslant \operatorname{card} A \leqslant \omega$. □

Lemma 2. *Let A be a countable set, B be a class, and $u\colon A \to B$. Then,* $\operatorname{rng} u$ *is a countable set.*

Proof. By Lemma 1 (1.3.2) $\operatorname{card}(\operatorname{rng} u) \leqslant \operatorname{card} A \leqslant \omega$. □

Lemma 3. *Let A and B be countable sets. Then, the sets $A \times B$ and $A * B$ are countable.*

Proof. By definition, there are injections $u\colon A \rightarrowtail \omega$ and $v\colon B \rightarrowtail \omega$. Then, the mapping $w \equiv (u, A, \omega) \times_m (v, B, \omega)$ from 1.1.12 is also an injection from $A \times B$ into $\omega \times \omega$. By Theorem 1 (1.3.8), the set $\omega \times \omega$ is countable. Thus, by Lemma 1, the set $C \equiv w[A \times B]$ is countable. Since w is a bijection from $A \times B$ onto C, we infer that $A \times B$ is also countable. Now, by virtue of Lemma 3 (1.1.12), the set $A * B$ is countable. □

Proposition 1. *Let $(A_i \mid i \in I)$ be a finite collection of countable sets. Then, the set $\prod(A_i \mid i \in I)$ is countable.*

Proof. Consider the set X of all natural numbers n such that the assertion of this proposition is valid for any set I with the power $n+1$. If $\operatorname{card} I = 1$, then there is a bijection $u\colon 1 \rightarrowtail\!\!\!\!\twoheadrightarrow I$. Take the element $i_0 \equiv u(0)$. Since $1 = \{0\}$, we infer that $I = \{i_0\}$. Therefore, $\prod(A_i \mid i \in I) \sim A_{i_0}$. This implies that $0 \in X$.

Assume that $n \in X$ and take any collection $(A_i \mid i \in I)$ such that $\operatorname{card} I = (n+1)+1$. By definition, there is a bijection $v\colon (n+1)+1 \rightarrowtail\!\!\!\!\twoheadrightarrow I$ and $(n+1)+1 = (n+1) \cup \{n+1\}$. Consider the element $j \equiv v(n+1)$ and the set $K \equiv I \setminus \{j\} \neq \varnothing$. By Theorem 2 (1.1.12),

$P \equiv \prod(A_i \mid i \in I) \sim A_j * Q$, where $Q \equiv \prod(A_i \mid i \in K)$. According to our assertion the set Q is countable. Thus, by Lemma 3, the set $A_j * Q$ is countable. Then, the set P is also countable. This means that $n + 1 \in X$. By the principle of natural induction from 1.2.6 $X = \omega$. □

Theorem 1. *Let $(A_i \mid i \in I)$ be a countable collection of countable sets. Then, the set $\bigcup(A_i \mid i \in I)$ is countable.*

Proof. Denote $\bigcup(A_i \mid i \in I)$ by S. By definition, there are countable ordinal numbers $\alpha_i \equiv \omega$ and $\beta \equiv \omega$ and bijections $u_i\colon\ \alpha_i \rightarrowtail\!\!\!\twoheadrightarrow A_i$ and $v\colon\ \beta \rightarrowtail\!\!\!\twoheadrightarrow I$. Consider the collection $(B_k \mid k \in \beta)$ of the pairwise disjoint sets $B_0 \equiv A_{v(0)}$, $B_1 \equiv A_{v(1)} \backslash B_0$, $B_2 \equiv A_{v(2)} \backslash B_0 \cup B_1, \dots,$ $B_{n+1} \equiv A_{v(n+1)} \setminus \bigcup(B_k \mid k \in n+1 \in \beta)$. It is clear that $T \equiv \bigcup(B_k \mid k \in \beta) \subset S$. Take any $x \in S$. Then, $x \in A_i$ for some $i \in I$. Consider the number $k \equiv v^{-1}(i)$. If $k = 0$, then $x \in A_{v(0)} \equiv B_0$. If $k \geqslant 1$, then $x \in A_{v(k)}$ implies either $x \in B_k$ or $x \in A_{v(k)} \cap \bigcup(B_j \mid j \in k)$. In all the cases, $x \in T$. Thus, $T = S$.

Define an injective mapping $U\colon\ T \rightarrowtail \bigcup_d(\alpha_{v(k)} \mid k \in \beta)$ setting $U(b) \equiv \langle u_{v(k)}^{-1}(b), \alpha_{v(k)}\rangle \in \alpha_{v(k)} * \{\alpha_{v(k)}\}$ for every $b \in B_k \subset A_{v(k)}$. Define also an injective mapping $V\colon\ \bigcup_d (\alpha_{v(k)} \mid k \in \beta) \rightarrowtail \omega \times \omega$ setting $V(\langle x, \alpha_{v(k)}\rangle) = (x, k)$. Then, the mapping $W \equiv V \circ U$ is an injection from S into $\omega \times \omega$. Using Theorem 1 (1.3.8) and Lemma 1, we infer that the set $W[S]$ is countable. Hence, the set S is also countable. □

1.3.10 Properties of the class of all countable ordinal numbers

Note that according to Theorem 1 (1.3.4), the cardinal number ω_1 consists of all countable ordinal numbers.

Lemma 1. *Let $(\alpha_k \in \omega_1 \mid k \in K)$ be a countable simple collection of countable ordinal numbers. Then, the set $\alpha \equiv \bigcup(\alpha_k \mid k \in K)$ has the following properties:*

1) *$\alpha \in \omega_1$, i. e. α is also a countable ordinal number;*
2) *if $\beta \in \mathrm{Ord}$ and $\beta \geqslant \alpha_k$ for every $k \in K$, then $\beta \geqslant \alpha$, i. e. , $\alpha = \sup(\alpha_k \mid k \in K)$ in Ord and ω_1.*

Proof. 1. By Theorem 1 (1.3.9), α is countable. It is clear that the set α is transitive in the sense of 1.2.2. Take any elements a, b in α. Then, $a \in \alpha_j$ and $b \in \alpha_k$ for some indices. By Lemma 3 (1.2.3), a and b are ordinal numbers. Thus, by Corollary 1 to Lemma 2 (1.2.3), $a \in b$, $b \in a$, or $a = b$. According to 1.2.2, this means that α is an ordinal.

2. Take any element $a \in \alpha$. Then, $a \in \alpha_j$ for some j. By Lemma 4 (1.2.3), $\alpha_j \subset \beta$. Thus, $a \in \beta$. This means that $\alpha \subset \beta$, where $\alpha \leqslant \beta$. Besides, $\alpha_k \subset \alpha$ implies that $\alpha_k \leqslant \alpha$ for every k. □

Lemma 2. *Let $(\alpha_k \in \omega_1 \mid k \in \omega)$ be a strictly increasing sequence of countable ordinal numbers. Then, the countable ordinal number $\alpha \equiv \bigcup(\alpha_k \mid k \in \omega)$ from Lemma 1 is a limit ordinal number (in the sense of* 1.2.3).

Proof. Since the sequence is strictly increasing, $\alpha_k \neq 0$ for some k. Thus, $\alpha \supset \alpha_k \neq \varnothing$. Assume that $\alpha = \beta+1$ for some ordinal number β. If $\beta \geqslant \alpha_k$ for every k, then by Lemma 1, $\beta \geqslant \alpha = \beta + 1$, but this is impossible. Therefore, $\beta < \alpha_k$ for some k. Then, by Proposition 1 (1.2.3), $\alpha = \beta + 1 \leqslant \alpha_k < \alpha_{k+1} \leqslant \alpha$, but this is also impossible. It follows from this contradiction that $\alpha \neq \beta + 1$ For every β. □

Lemma 3. *Let α and β be countable ordinal numbers. Then, the ordinal number $\alpha +_o \beta$ is countable.*

Proof. By definition from 1.2.9, $\alpha +_o \beta \equiv \operatorname{ord} \bigcup_{do}(x_i \mid i \in 2)$, where $x_0 \equiv \alpha$ and $x_1 \equiv \beta$. By Theorem 1 (1.3.9), the set $S \equiv \bigcup_d(x_i \mid i \in 2) \equiv \bigcup(x_i * \{i\} \mid i \in 2)$ is countable. Thus, the set $\alpha +_o \beta \sim S$ is also countable. □

Corollary 1. *Let α be a countable ordinal number. Then, $\alpha + 1$ is also a countable ordinal number.*

Proof. The assertion follows from Lemma 3 and Proposition 3 (1.2.9). □

The following theorem is called the *principle of induction for countable ordinal numbers*.

Theorem 1. *Let A be a subclass of the class* Ord *with the following properties:*
1) $0 \in A$;
2) $\alpha \in A$ *implies* $\alpha + 1 \in A$;
3) *if* $(\alpha_k \in A \mid k \in \omega)$ *is a strictly increasing sequence, then* $\bigcup(\alpha_k \mid k \in \omega) \in A$.

Then, $\omega_1 \subset A$.

Proof. Suppose that the class $\omega_1 \setminus A$ is non-empty. According to Proposition 2 (1.2.2), it has the smallest element β. By property 1, $\beta \neq 0$. Suppose that β is not a limit ordinal, i. e. $\beta = \gamma + 1$ for some ordinal number. Then, $\gamma < \beta \in \omega_1$ implies $\gamma \in \omega_1$, and so $\gamma \in A$. By property 2, $\beta \in A$. It follows from this contradiction that β is a limit ordinal.

Since β is countable, there exists a bijection $u\colon \omega \rightarrowtail\!\!\!\!\twoheadrightarrow \beta$. Consider the ordinal numbers $b_m \equiv u(m)$ and suppose that there exists $a \in \beta$ such that $a \geqslant b_m$ for every m. But $a = b_n$ for some n. Therefore, $b_m \leqslant b_n < b_n + 1$ implies $b_m \in b_n + 1$ for every m, where $\beta \subset b_n + 1$. As a result, $\beta \leqslant b_n + 1 \leqslant \beta$, but this is impossible because β is a limit ordinal. Thus, for every $a \in \beta$, the set $S_a \equiv \{m \in \omega \mid a < b_m\}$ is non-empty.

Take the smallest element $m(a)$ of the set S_a. Define a mapping $V\colon \beta \times \omega \to \beta$ setting $V(a, n) \equiv b_{m(a)}$. By Theorem 1 (1.2.7), there is a unique mapping $v\colon \omega \to \beta$, such that $v(0) = b_0$ and $v(n+1) = V(v(n), n) = b_{m(v(n))}$, where $m(v(n))$ is the smallest element of the set $S_{v(n)} \equiv \{m \in \omega \mid v(n) < b_m\}$. Consequently, $v(n+1) = b_{m(v(n))} > v(n)$. In addition, the inclusion $S_{v(n)} \equiv \{m \in \omega \mid v(n-1) < v(n) < b_m\} \subset \{m \in \omega \mid v(n-1) < b_m\} \equiv S_{v(n-1)}$ implies $m(v(n)) \geqslant m(v(n-1))$. Suppose that $m(v(n)) = m(v(n-1))$. Then, $v(n+1) = b_{m(v(n))} = b_{m(v(n-1))} = v(n)$, but this is false. Thus, $m(v(n)) > m(v(n-1))$. Thus, the sequences $(m(v(n)) \in \omega \mid n \in \omega)$ and $(v(n) \in \beta \mid n \in \omega)$ are strictly increasing.

It is clear that $X \equiv \bigcup(v(n) \mid n \in \omega) \subset \beta$. If $b \in \beta$, then $b = b_k$ for some k. Therefore, by Lemma 7 (1.2.6), there is n such that $k < m(v(n))$. Suppose that $b_k > v(n)$. Then, $m(v(n)) \leqslant k$. It follows from this contradiction that $b = b_k \leqslant v(n) < v(n+1)$, where $b \in v(n+1) \subset X$. Thus, $X = \beta$. From $v(n) \in \beta \in \omega_1$, we infer that $v(n) < \beta$ and $v(n) \in \omega_1$. Thus, $v(n) \in A$ for every n. Therefore, by property 3, $\beta \in A$, but this is false. We conclude from this contradiction that $\omega_1 \subset A$. □

Corollary 1. *Let the set $A \subset \omega_1$ have properties* 1 – 3. *Then, $A = \omega_1$.*

Note that by virtue of Lemma 1, the set ω_1 itself has properties 1 – 3.

1.4 Real numbers

In this section, we set forth basic information about constructions and properties of integers, rational numbers, real numbers, and extending real numbers starting from the set ω of natural numbers (see 1.2.6 and 1.3.6).

1.4.1 Integers

Define on the set $\omega\times\omega$ a binary relation θ setting $((m,p),(n,q)) \in \theta$ iff $m+q = n+p$. We assert that θ is an equivalence relation. In fact, θ is obviously reflexive and symmetric. Let $(l,p)\theta(m,q)$ and $(m,q)\theta(n,r)$, i. e. $l+q = m+p$ and $m+r = n+q$. Then, $(l+r)+q = (l+q)+r = (m+p)+r = (m+r)+p = (n+q)+p = (n+p)+q$ imply by Corollary 6 to Proposition 2 (1.3.6) that $l+r = n+p$. This means that $(l,p)\theta(n,r)$. Thus, θ is transitive.

Consider the factor-set $\mathbb{Z} \equiv (\omega\times\omega)/\theta$ consisting of equivalence classes $x \equiv \theta(m,p)$ of all pairs $(m,p) \in \omega\times\omega$ (see 1.1.14). Elements of the set $\mathbb{Z}$ are called *integers*; and the set $\mathbb{Z}$ is called the *set of all integers*.

Consider the factor-mapping $f\colon\ \omega\times\omega \to \mathbb{Z}$ from 1.1.14. By Lemma 1 (1.1.14) f is surjective. Therefore, using Theorem 1 (1.3.8) and Lemma 2 (1.3.9) we conclude that the set $\mathbb{Z}$ is countable.

Associate with every natural number $m \in \omega$ the integer $\hat{m} = \theta(m,0) \in \mathbb{Z}$, and consider the mapping e from ω into $\mathbb{Z}$ such that $em \equiv \hat{m}$. This mapping is injective. Consider the set $\mathbb{Z}_+ \equiv \{\hat{m} \mid m \in \omega\}$. Since $\omega = \operatorname{card}\omega = \operatorname{card}\mathbb{Z}_+ \leqslant \operatorname{card}\mathbb{Z} \leqslant \omega$, we infer that $\operatorname{card}\mathbb{Z} = \omega$, i. e. the set $\mathbb{Z}$ is denumerable.

Sum of integers

Let $(x_i \in \mathbb{Z} \mid i \in I)$ be a simple collection of integers $x_i \equiv \theta(m_i, p_i)$ indexed by a finite set I. The integer $\theta(\sum(m_i \mid i \in I), \sum(p_i \mid i \in I))$ is called the *sum of the simple collection* $(x_i \in \mathbb{Z} \mid i \in I)$ and is denoted by $\sum(x_i \mid i \in I)$. If $I = n+1$ for $n \in \omega \setminus 2$, then along with $\sum(x_i \mid i \in n+1)$ we shall use the notation $x_0 + \cdots + x_n$.

It is clear that $e(\sum(m_i \in \omega \mid i \in I)) = \sum(em_i \in \mathbb{Z} \mid i \in I)$.

Let $x, x', x'', x''', \ldots$ be integers. Then, (x, x'), (x, x', x''), (x, x', x'', x'''), ... are the corresponding simple collections (see 1.1.11).

The integers $\sum(x, x')$, $\sum(x, x', x'')$, $\sum(x, x', x'', x''')$, ... will be called the *sums of the simple sequential pair* (x, x'), *triplet* (x, x', x''), *quadruplet* (x, x', x'', x'''),... and will be denoted also by $x + x'$, $x + x' + x''$, $x + x' + x'' + x'''$, ... By the definition of the simple sequential pair from 1.1.11 we have $x + x' = \sum(a_i \mid i \in 2)$, where $a_0 \equiv x$ and $a_1 \equiv x'$. In the similar manner, $x + x' + x'' = \sum(a_i \mid i \in 3)$, where $a_0 \equiv x$, $a_1 \equiv x'$ and $a_2 \equiv x''$, and so on.

Theorem 1. *Let $(x_i \in \mathbb{Z} \mid i \in I)$ be a simple collection indexed by a finite set I. Then:*

1) *if K is a finite set and u is a bijective mapping from K onto I, then $\sum(x_i \mid i \in I) = \sum(x_{u(k)} \mid k \in K)$ (the general commutativity of the sum);*
2) *if a collection $[I_m \subset I \mid m \in M]$ is a partition of the set I indexed by a finite non-empty set M, then $\sum(x_i \mid i \in I) = \sum(\sum(x_i \mid i \in I_m) \mid m \in M)$ (the general associativity of the sum).*

Proof. We shall denote the left parts of these equalities by L. Let $x_i \equiv \theta(m_i, p_i)$.

1. Using assertion 1 of Theorem 2 (1.3.6), we get $L \equiv \theta(\sum(m_i \mid i \in I), \sum(p_i \mid i \in I)) = \theta(\sum(m_{u(k)} \mid k \in K), \sum(p_{u(k)} \mid k \in K)) \equiv \sum(\theta(m_{u(k)}, p_{u(k)}) \mid k \in K) \equiv \sum(x_{u(k)} \mid k \in K)$.

2. Analogously, using assertion 2 of the same theorem, we get $L \equiv \theta(\sum(m_i \mid i \in I), \sum(p_i \mid i \in I)) = \theta(\sum(\sum(m_i \mid i \in I_m) \mid m \in M), \sum(\sum(p_i \mid i \in I_m) \mid m \in M)) \equiv \sum(\theta(\sum(m_i \mid i \in I_m), \sum(p_i \mid i \in I_m)) \mid m \in M) \equiv \sum(\sum(\theta(m_i, p_i) \mid i \in I_m) \mid m \in M) \equiv \sum(\sum(x_i \mid i \in I_m) \mid m \in M)$. □

Lemma 1.

1) *Let $(x_i \in \mathbb{Z} \mid i \in \{p\})$ be a simple collection indexed by a set $\{p\}$. Then, $\sum(x_i \mid i \in \{p\}) = x_p$.*
2) *Let $(x_i \in \mathbb{Z} \mid i \in \{p, q\})$ be a simple collection indexed by a set $\{p, q\}$ with different elements $p \neq q$. Then, $\sum(x_i \mid i \in \{p, q\}) = x_p + x_q$.*

Proof. We shall prove only assertion 2. Consider the simple collection $(a_i \mid i \in 2)$ such that $a_0 \equiv x_p$ and $a_1 \equiv x_q$. Consider the sets $I \equiv 2$ and $K \equiv \{p, q\}$ and the bijective mapping $u\colon K \to I$ such that $u(p) \equiv 0$ and $u(q) \equiv 1$. Then, by Theorem 1, $x_p + x_q = \sum(a_i \mid i \in I) = \sum(a_{u(k)} \mid k \in K)$. From $a_{u(p)} = a_0 \equiv x_p$ and $a_{u(q)} = a_1 \equiv x_q$, we infer that $(a_{u(k)} \mid k \in K) = (x_k \mid k \in K)$. As a result, $x_p + x_q = \sum(x_k \mid k \in K)$. □

Corollary 1. *Let $(x_i \in \mathbb{Z} \mid i \in I)$ and $(y_i \in \mathbb{Z} \mid i \in I)$ be simple collections indexed by a finite non-empty set I. Then, $\sum(x_i \mid i \in I) + \sum(y_i \mid i \in I) = \sum(x_i + y_i \mid i \in I)$.*

Proof. Consider the sets $M \equiv 2 \times I$, $M_i \equiv 2 \times \{i\}$, and $M_k \equiv \{k\} \times I$. Consider also the elements $a_m \equiv x_i$ for $m \equiv (0, i)$ and $a_m \equiv y_i$ for $m \equiv (1, i)$. By assertion 2 of Lemma 1, $x_i + y_i = \sum(a_m \mid m \in \{(0, i), (1, i)\}) = \sum(a_m \mid m \in M_i)$. Using assertion 2 of Theorem 1, we get $\sum(x_i + y_i \mid i \in I) = \sum(\sum(a_m \mid m \in M_i) \mid i \in I) = \sum(a_m \mid m \in M)$ because $(M_i \mid i \in I)$ is a partition of M. Analogously, since $(M_k \mid k \in 2)$ is also a partition of M, we get $\sum(a_m \mid m \in M) = \sum(\sum(a_m \mid m \in M_k) \mid k \in 2) = \sum(a_m \mid m \in M_0) + \sum(a_m \mid m \in M_1) = \sum(a_{(0,i)} \mid i \in I) + \sum(a_{(1,i)} \mid i \in I) = \sum(x_i \mid i \in I) + \sum(y_i \mid i \in I)$. □

Theorem 2. *Let x, y and z be integers. Then:*
1) $x + y = y + x$ (the *commutativity of the sum*);
2) $x + y + z = x + (y + z) = (x + y) + z$ (the *associativity of the sum*).

Proof. 1. Take the sets $I \equiv 2$ and $K \equiv 2$. Consider simple collections $(a_i \in \{x, y\} \mid i \in I)$ and $(b_k \in \{x, y\} \mid k \in K)$ such that $a_0 \equiv x$, $a_1 \equiv y$, $b_0 \equiv y$, and $b_1 \equiv x$. Take a bijective mapping $u\colon K \to I$ such that $u(0) \equiv 1$ and $u(1) \equiv 0$. From $a_{u(0)} = a_1 \equiv y \equiv b_0$ and $a_{u(1)} = a_0 \equiv x \equiv b_1$, we infer $(a_{u(k)} \mid k \in K) = (b_k \mid k \in K)$. As a result, by virtue of Theorem 1, we get the chain of equalities $x + y = \sum(a_i \mid i \in I) = \sum(a_{u(k)} \mid k \in K) = \sum(b_k \mid k \in K) = y + x$.

2. Take the sets $I \equiv 3$, $M \equiv 2$, $I_0 \equiv 1$, and $I_1 \equiv \{1, 2\}$. Then, the collection $(I_m \subset I \mid m \in M)$ is a partition of the set I. Consider simple collections $(a_i \mid i \in I)$, $(a_i \mid i \in I_0)$, and $(a_i \mid i \in I_1)$ such that $a_0 \equiv x$, $a_1 \equiv y$, and $a_2 \equiv z$. Then, by Theorem 1 and Lemma 1, we get the chain of equalities $x + y + z = \sum(a_i \mid i \in I) = \sum(\sum(a_i \mid i \in I_m) \mid m \in M) = \sum(a_i \mid i \in I_0) + \sum(a_i \mid i \in I_1)$.

Further, by Lemma 1 $\sum(a_i \mid i \in I_0) = a_0 \equiv x$ and $\sum(a_i \mid i \in I_1) = a_1 + a_2 = y + z$. As a result, we get $x + y + z = x + (y + z)$.

In a similar way, we prove the equality $x + y + z = (x + y) + z$. □

The element $\hat{0}$ is called the *zero element* in $\mathbb{Z}$. For every integer x, we have the equality $\hat{0} + x = x + \hat{0} = x$.

The element $\theta(p, m)$ is called the *opposite element to the element* $x \equiv \theta(m, p)$ and is denoted by $-x$. It is clear that $-(-x) = x$. The zero and opposite elements are connected by the equality $x + (-x) = -x + x = \hat{0}$. Further, along with $x + (-y)$, we shall write also $x - y$; this number is called the *difference of the numbers x and y*.

Consider the sets $\mathbb{Z}_- \equiv \{-\hat{m} \mid m \in \omega\}$ and $\mathbb{Z}^* \equiv \mathbb{Z} \setminus \{\hat{0}\}$.

Lemma 2.
1) $\mathbb{Z} = \mathbb{Z}_+ \cup \mathbb{Z}_-$ *and* $\mathbb{Z}_+ \cap \mathbb{Z}_- = \{\hat{0}\}$.
2) *For every $x \in \mathbb{Z}$, there exist $y \in \mathbb{Z}_+$ and $z \in \mathbb{Z}_-$ such that $x = y + z$.*

Proof. 1. Let $x \equiv \theta(m, p) \in \mathbb{Z}$. If $m = p$, then $x = \hat{0}$. If $m > p$, then, by Theorem 5 (1.3.6), there is a natural number $k > 0$ such that $m = p + k$. Therefore, $x = \hat{k}$. If $m < p$, then there is a natural number $l > 0$ such that $m + l = p$. Therefore, $x = -\hat{l}$.

If $x \in \mathbb{Z}_+ \cap \mathbb{Z}_-$, then $x = \hat{m}$ and $x = -\hat{n}$ for some $m, n \in \omega$. Therefore, $((m, 0), (0, n)) \in \theta$ implies $m + n = 0 + 0 = 0$, where $m = n = 0$. Thus, $x = \hat{0}$.

2. The assertion follows from 1. □

It follows from Lemma 2 that for every integer x, we can define correctly its *modulus* $|x| \in \mathbb{Z}_+$ setting $|x| \equiv x$ if $x \in \mathbb{Z}_+$ and $|x| \equiv -x$ if $x \in \mathbb{Z}_-$.

Product of integers

Let $(\hat{m}_i \in \mathbb{Z}_+ \mid i \in I)$ be a simple collection indexed by a finite set I. The integer $e(P(m_i \in \omega \mid i \in I)) \in \mathbb{Z}_+$ is called the *product of the simple collection* $(\hat{m}_i \in \mathbb{Z}_+ \mid i \in I)$ and is denoted by $P(\hat{m}_i \mid i \in I)$. If $I = n+1$ for $n = \omega\backslash 2$, then along with $P(\hat{m}_i \mid i \in n+1)$, we shall use the notation $\hat{m}_0 \ldots \hat{m}_n$.

The products $P(\hat{m}, \hat{m}')$, $P(\hat{m}, \hat{m}', \hat{m}'')$, ... of the simple collections $(\hat{m}, \hat{m}')$, $(\hat{m}, \hat{m}', \hat{m}'')$, ... composed of elements $\hat{m}, \hat{m}', \hat{m}'', \ldots$ of $\mathbb{Z}_+$ will be denoted also by $\hat{m}\hat{m}', \hat{m}\hat{m}'\hat{m}'', \ldots$.

Lemma 3. *Let $(\hat{m}_i \in \mathbb{Z}_+ \mid i \in I)$ be a simple collection indexed by a finite set I. Then:*
1) *if K is a finite set and u is a bijective mapping from K onto I, then $P(\hat{m}_i \mid i \in I) = P(\hat{m}_{u(k)} \mid k \in K)$;*
2) *if a collection $(I_m \subset I \mid m \in M)$ is a partition of the set I indexed by a finite nonempty set M, then $P(\hat{m}_i \mid i \in I) = P(P(\hat{m}_i \mid i \in I_m) \mid m \in M)$.*

Proof. We shall denote the left and right parts of these equalities by L and R, respectively.

1. Using assertion 1 of Theorem 2 (1.3.6), we get $L \equiv e(P(m_i \in \omega \mid i \in I)) = e(P(m_{u(k)} \mid k \in K)) \equiv R$.

2. Analogously, using assertion 2 of the same theorem, we get $L = e(P(m_i \in \omega \mid i \in I)) = e(P(P(m_i \mid i \in I_m) \mid m \in M)) \equiv P(e(P(m_i \mid i \in I_m)) \mid m \in M) \equiv R$. □

Corollary 1.
1) *Let $(\hat{m}_i \in \mathbb{Z}_+ \mid i \in \{p\})$ be a simple collection indexed by the set $\{p\}$. Then, $P(\hat{m}_i \mid i \in \{p\}) = \hat{m}_p$.*
2) *Let $(\hat{m}_i \in \mathbb{Z}_+ \mid i \in \{p, q\})$ be a simple collection indexed by the set $\{p, q\}$ with different elements $p \neq q$. Then, $P(\hat{m}_i \mid i \in \{p, q\}) = \hat{m}_p\hat{m}_q$.*

The proof is analogous to the proof of Lemma 1.

Let $\varkappa \equiv (x_i \in \mathbb{Z} \mid i \in I)$ be a simple collection indexed by a finite set I. Define the *product* $P\varkappa \equiv P(x_i \mid i \in I) \in \mathbb{Z}$ *of the simple collection* $\varkappa$ setting $P\varkappa \equiv P(|x_i| \in \mathbb{Z}_+ \mid i \in I)$ if the power of the set $I_\varkappa^- \equiv \{i \in I \mid x_i \in \mathbb{Z}_- \setminus \{\hat{0}\}\}$ is an even natural number, and

$P\varkappa \equiv -P(|x_i| \in \mathbb{Z}_+ \mid i \in I)$ if the power of this set is an odd natural number. If $I = n+1$ for $n \in \omega \setminus 2$, then along with $P(x_i \mid i \in n+1)$ we shall use also the notation $x_0 \ldots x_n$.

Let $x, x', x'', x''', \ldots$ be integers. As above, the integers $P(x, x')$, $P(x, x', x'')$, $P(x, x', x'', x''')$, ... will be called the *products of the simple sequential pair* (x, x'), *triplet* (x, x', x''), *quadruplet* (x, x', x'', x'''), ... and will be denoted also by xx', $xx'x''$, $xx'x''x'''$, ...

Lemma 4. *Let $m, n \in \omega$. Then, $(-\hat{m})\hat{n} = -(\hat{m}\hat{n})$, $\hat{m}(-\hat{n}) = -(\hat{m}\hat{n})$, and $(-\hat{m})(-\hat{n}) = \hat{m}\hat{n}$.*

Proof. By definition, $(-\hat{m})\hat{n} \equiv P(-\hat{m}, \hat{n}) \equiv P(x_i \mid i \in 2)$ for the collection $\varkappa \equiv (x_i \in \mathbb{Z} \mid i \in 2)$ such that $x_0 \equiv -\hat{m}$ and $x_1 \equiv \hat{n}$. Since $\operatorname{card} 2^-_\varkappa = \operatorname{card}\{0\} = 1$, we infer that $P(x_i \mid i \in 2) \equiv -P(|x_i| \mid i \in 2) = -(\hat{m}\hat{n})$. The other equalities are checked in a similar way. □

Proposition 1. *Let $(x_i \in \mathbb{Z} \mid i \in I)$ be a simple collection indexed by a finite set I, K be a finite set, and u be a bijective mapping from K onto I. Then, $P(x_i \mid i \in I) = P(x_{u(k)} \mid k \in K)$ (the* general commutativity of the product).

Proof. Denote the collections $(x_i \mid i \in I)$ and $(x_{u(k)} \mid k \in K)$ by π and $\varkappa$, respectively. Then, $u[K^-_\varkappa] = I^-_\pi$ implies the equality of the powers of these sets. Thus, these powers are both odd or both even simultaneously. In the first case, using assertion 1 of Lemma 3, we get $P\pi \equiv -P(|x_i| \in \mathbb{Z}_+ \mid i \in I) = -P(|x_{u(k)}| \mid k \in K) \equiv P\varkappa$. In the second case, the argument is the same. □

Corollary 1.
1) *Let $(x_i \in \mathbb{Z} \mid i \in \{p\})$ be a simple collection indexed by the set $\{p\}$. Then, $P(x_i \mid i \in \{p\}) = x_p$.*
2) *Let $(x_i \in \mathbb{Z} \mid i \in \{p, q\})$ be a simple collection indexed by the set $\{p, q\}$ with different elements $p \neq q$. Then, $P(x_i \mid i \in \{p, q\}) = x_p x_q$.*

The proof is analogous to the proof of Lemma 1.

Lemma 5. *Let $(x_i \in \mathbb{Z} \mid i \in I)$ be a simple collection indexed by a finite set I, $(I_m \subset I \mid m \in M)$ be a partition of the set I indexed by a set M such that $M = \{p, q\}$ and $p \neq q$. Then, $P(x_i \mid i \in I) = P(P(x_i \mid i \in I_m) \mid m \in M) = P(x_i \mid i \in I_p)P(x_i \mid i \in I_q)$.*

Proof. Denote the collections $(x_i \mid i \in I_p)$, $(x_i \mid i \in I_q)$, and $(x_i \mid i \in I)$ by $\varkappa_p$, $\varkappa_q$, and $\varkappa$, respectively. It is clear that the collection $(I^-_{m\varkappa_m} \subset I^-_\varkappa \mid m \in M)$ is a partition of the set $I^-_\varkappa$. Consider the natural numbers $n_p \equiv \operatorname{card} I^-_{p\varkappa_p}$, $n_q \equiv \operatorname{card} I^-_{q\varkappa_q}$, and $n \equiv \operatorname{card} I^-_\varkappa$. According to Corollary 1 to Lemma 1 (1.3.5), $n = n_p + n_q$.

At first, consider the case when the number n is even. Then, n_p and n_q are either both even or both odd. In the first case, we have $P\varkappa_p \equiv P(|x_i| \mid i \in I_p)$ and $P\varkappa_q \equiv P(|x_i| \mid i \in I_q)$. Besides, $P\varkappa = P(|x_i| \mid i \in I)$. Using assertion 2 of Lemma 3 and Corollary 1 to this lemma, we get $P(|x_i| \in \mathbb{Z}_+ \mid i \in I) = P(P(|x_i| \mid i \in I_m) \mid m \in M) = P(|x_i| \mid i \in I_p)P(|x_i| \mid i \in I_q)$. As a result, $P\varkappa = P(P\varkappa_m \mid m \in M) = P\varkappa_p P\varkappa_q$. In the second case, we have $P\varkappa_p \equiv -P(|x_i| \mid i \in I_p)$ and $P\varkappa_q \equiv -P(|x_i| \mid i \in I_q)$. Therefore, by Lemma 4 $P\varkappa \equiv P(|x_i| \mid i \in I) = P(|x_i| \mid i \in I_p)P(|x_i| \mid i \in I_q) = (-P(|x_i| \mid i \in I_p))(-P(|x_i| \mid i \in I_q)) = P\varkappa_p P\varkappa_q$. By Corollary 1 to Proposition 1, $P\varkappa_p P\varkappa_q = P(P\varkappa_m \mid m \in M)$.

Now, consider the case when the number n is odd. Then, either n_p is even and n_q is odd or n_p is odd and n_q is even. In the first case by Lemma 4, we get $P\varkappa \equiv -P(|x_i| \mid i \in I) = -(P(|x_i| \mid i \in I_p)P(|x_i| \mid i \in I_q)) = P(|x_i| \mid i \in I_p)(-P(|x_i| \mid i \in I_q)) = P\varkappa_p P\varkappa_q = P(P\varkappa_m \mid m \in M)$. In the second case, the argument is the same. □

Finally, we can prove the property of general associativity for the product of integers.

Theorem 3. *Let $(x_i \in \mathbb{Z} \mid i \in I)$ be a simple collection indexed by a finite set I and $(I_m \subset I \mid m \in M)$ be a partition of the set I indexed by a finite non-empty set M. Then, $P(x_i \mid i \in I) = P(P(x_i \mid i \in I_m) \mid m \in M)$* (the *general associativity of the product*).

Proof. Consider the set X of all natural numbers n such that for every collection $\varkappa \equiv (x_i \in \mathbb{Z} \mid i \in I)$ and every partition $\pi \equiv (I_m \subset I \mid m \in M)$ with card $M = n + 2$ we have the property $P\varkappa = P(P(x_i \mid i \in I_m) \mid m \in M)$. By Lemma 5, $0 \in X$.

Let $n \in X$. Take any $\varkappa$ and π such that card $M = (n+1)+2$. Fix some element $m_0 \in M$ and consider the sets $M_0 \equiv \{m_0\}$ and $M_1 \equiv M \setminus M_0$. It is clear that $M_1 = n + 2$. Consider also the sets $J_0 \equiv I_{m_0}$ and $J_1 \equiv I \setminus I_0$. Then, $\pi_1 \equiv (I_m \mid m \in M_1)$ is a partition of J_1 and $(J_k \mid k \in 2)$ is a partition of I. Therefore, for the collection $\varkappa_1 \equiv (x_i \mid i \in J_1)$ and the partition π_1 we have $P\varkappa_1 = P(P(x_i \mid i \in I_m) \mid m \in M_1)$. By Corollary 1 to Proposition 1 for the collection $\varkappa_0 \equiv (x_i \mid i \in J_0)$, we have $P\varkappa_0 = P(x_i \mid i \in I_{m_0}) = P(P(x_i \mid i \in I_m) \mid m \in M_0)$.

By Lemma 5 $P\varkappa = P(P\varkappa_k \mid k \in 2) = P(P(P(x_i \mid i \in I_m) \mid M_k) \mid k \in 2) = P(P(x_i \mid i \in I_m) \mid m \in M)$ because $(M_k \mid k \in 2)$ is a partition of the set M. This means that $n+1 \in X$. By the principle of natural induction from 1.2.6, we infer that $X = \omega$. □

Corollary 1. *Let $(x_i \in \mathbb{Z} \mid i \in I)$ and $(y_i \in \mathbb{Z} \mid i \in I)$ be simple collections indexed by a finite non-empty set I. Then, $P(x_i \mid i \in I)P(y_i \mid i \in I) = P(x_i y_j \mid i \in I)$.*

The proof is completely similar to the proof of Corollary 1 to Lemma 1.

Theorem 4. *Let x, y and z be integers. Then:*
1) $xy = yx$ (the *commutativity of the product*);
2) $xyz = x(yz) = (xy)z$ (the *associativity of the product*).

The proof is similar to the proof of Theorem 2.

The element $\hat{1}$ is called the *unity element* in $\mathbb{Z}$. For every integer x, we have the equality $\hat{1}x = x\hat{1} = x$ (see Corollary 1 to Lemma 3 (1.3.5)).

Lemma 6. *Let $m \in \mathbb{N}$ and $x \in \mathbb{Z}$. Then, $\hat{m}x = \sum(x_i \mid i \in I)$ for every simple collection $(x_i \in \mathbb{Z} \mid i \in I)$ such that $x_i = x$ for every $i \in I$ and $I = m$.*

Proof. By Lemma 2 either $x = \hat{n}$ or $x = -\hat{n}$ for some $n \in \omega$. By Corollary 2 to Lemma 3 (1.3.5) $mn = \sum(n_i \mid i \in I)$, where $n_i \equiv n$ for every i. Therefore, in the first case, $\hat{m}x = \hat{m}\hat{n} = e(mn) = e\sum(n_i \mid i \in I) = \sum(\hat{n}_i \mid i \in I) = \sum(x_i \mid i \in I)$. In the second case, by Lemma 4, $\hat{m}x = \hat{m}(-\hat{n}) = -(\hat{m}\hat{n}) = -e(mn) = -e(\sum(n_i \mid i \in I)) = -\theta(\sum(n_i \mid i \in I), 0) = \theta(0, \sum(n_i \mid i \in I)) \equiv \sum(\theta(0, n_i) \mid i \in I) = \sum(-\hat{n}_i \mid i \in I) = \sum(x_i \mid i \in I)$. □

Lemma 7. *Let $m \in \omega$ and $x \equiv \theta(n, q) \in \mathbb{Z}$. Then, $\hat{m}x = \theta(mn, mq)$.*

Proof. Consider the simple collections $\nu \equiv (n_i \in \omega \mid i \in m)$ and $\varkappa \equiv (q_i \in \omega \mid i \in m)$ such that $n_i = n$ and $q_i = q$ for every i. By Lemma 6 $\hat{m}x = \sum(\theta(n_i, q_i) \mid i \in I) \equiv \theta(\sum \nu, \sum \varkappa)$. By Corollary 2 to Lemma 3 (1.3.5), $\sum \nu = mn$ and $\sum \varkappa = mq$. □

Lemma 8. *Let $m \in \omega$ and $x \in \mathbb{Z}$. Then, $(-\hat{m})x = -(\hat{m}x)$.*

Proof. By Lemma 2 either $x = \hat{n}$ or $x = -\hat{n}$ for some $n \in \omega$. In the first case by Lemma 4, $(-\hat{m})x = -(\hat{m}\hat{n}) = -(\hat{m}x)$. In the second case, $(-\hat{m}) = \hat{m}\hat{n} = -(-(\hat{m}\hat{n})) = -(\hat{m}(-\hat{n})) = -(\hat{m}x)$. □

The distributivity of the product with respect to the sum

Theorem 5. *Let $x, y, z \in \mathbb{Z}$. Then, $x(y + z) = xy + xz$ (the* distributivity of the product with respect to the sum*).*

Proof. By virtue of Lemma 2, we need to consider the following four cases: (1) $x, y+z \in \mathbb{Z}_+$; (2) $x \in \mathbb{Z}_-$ and $y + z \in \mathbb{Z}_+$; (3) $x \in \mathbb{Z}_+$ and $y + z \in \mathbb{Z}_-$; and (4) $x, y + z \in \mathbb{Z}_-$. Thus, either $x = \hat{l}$ or $x = -\hat{l}$ and either $y+z = \hat{k}$ or $y+z = -\hat{k}$ for some $l, k \in \omega$. Let $y \equiv \theta(m, p)$ and $z \equiv \theta(n, q)$.

In the first case, $y + z = (m + n, p + q) = \hat{k} \equiv (k, 0)$ implies $m + n = p + q + k$. Therefore, $x(y+z) = \hat{l}\hat{k} = \theta(lk, 0)$ in virtue of Lemma 7. But $lm+ln = lp+lq+lk$ implies $\theta(lk, 0) = \theta(lm + ln, lp + lq) = \theta(lm, lp) + \theta(ln, lq) = \hat{l}y + \hat{l}z = xy + xz$.

In the second case by Lemma 8 and Lemma 7, $x(y + z) = (-\hat{l})(y + z) = -(\hat{l}(y + z)) = -(\hat{l}\hat{k}) = -\theta(lm + ln, lp + lq) = \theta(lp, lq, lm + ln) = \theta(lp, lm) + \theta(lq, ln) = -\theta(lm, lp) - \theta(ln, lq) = -(\hat{l}y) - (\hat{l}z) = (-\hat{l})y + (-\hat{l})z = xy + xz$.

In the third case, $y + z = (m + n, p + q) = -\hat{k} = (0, k)$ implies $m + n + k = p + q$. Therefore, $x(y+z) = -(\hat{l}\hat{k}) = \theta(0, lk)$. But $lm+ln+lk = lp+lq$ implies $\theta(0, lk) = \theta(lm + ln, lp + lq) = \hat{l}y + \hat{l}z = xy + xz$.

Finally, in the fourth case, $x(y+z) = \hat{l}\hat{k} = \theta(lk, 0) = \theta(lp+lq, lm+ln) = \theta(lp, lm) + \theta(lq, ln) = -\theta(lm, lp) - \theta(ln, lq) = -(\hat{l}y) - (\hat{l}z) = (-\hat{l})y + (-\hat{l})z = xy + xz$. □

Now, we shall deduce the property of general distributivity of the product with respect to the sum using the property of binary distributivity from Theorem 5 and the properties of general commutativity and associativity of the sum and the product. This deduction does not depend on specific properties of the set $\mathbb{Z}$ and has quite general character.

Lemma 9. *Let $x \in \mathbb{Z}$ and $(y_j \in \mathbb{Z} \mid j \in J)$ be a simple finite collection. Then, $x \sum(y_j \mid j \in J) = \sum(xy_j \mid j \in J)$.*

Proof. Consider the set X of all natural numbers n such that for every collection $\varkappa \equiv (y_j \mid j \in J)$ with $\operatorname{card} J = n+1$, we have the property $x \sum \varkappa = \sum(xy_j \mid j \in J)$. By Corollary 1 to Proposition 1, $0 \in X$.

Let $n \in X$. Take any $\varkappa$ such that $\operatorname{card} J = (n + 1) + 1$. Fix some element $j_0 \in J$ and consider the sets $J_0 \equiv \{j_0\}$ and $J_1 \equiv J \setminus J_0$. It is clear that $\operatorname{card} J_1 = n+1$. Since $(J_k \mid k \in 2)$ is a partition of the set J, we get in virtue of Theorem 1 and Lemma 1 $\sum \varkappa = \sum(\sum(y_j \mid j \in J_k) \mid k \in 2) = \sum(y_j \mid j \in J_0) + \sum(y_j \mid j \in J_1) = y_{j_0} + \sum(y_j \mid j \in J_1)$. Therefore, by Theorem 5 $x \sum \varkappa = xy_{j_0} + \sum(xy_j \mid j \in J_1) = \sum(xy_j \mid j \in J_0) + \sum(xy_j \mid j \in J_1) = \sum(\sum(xy_j \mid j \in J_k) \mid k \in 2) = \sum(xy_j \mid j \in J)$. This means that $n + 1 \in X$. By the principle of natural induction from 1.2.6, we infer that $X = \omega$. □

Theorem 6. *Let $(I_m \mid m \in M)$ be a collection of finite sets and $(\varkappa_m \mid m \in M)$ be a simple collection of simple collections $\varkappa_m \equiv (x_{mi} \in \mathbb{Z} \mid i \in I_m)$ indexed by non-empty finite sets M and I_m. Consider the finite set $U \equiv \prod(I_m \mid m \in M)$. Then, $P(\sum(x_{mi} \mid i \in I_m) \mid m \in M) = \sum(P(x_{mu(m)} \mid m \in M) \mid u \in U)$ (the* general distributivity of the product with respect to the sum*).*

Proof. We shall denote the left and right parts of this equality by L and R, respectively.

Consider the set X of all natural numbers n such that $L = R$ for every collection $\pi \equiv (I_m \mid m \in M)$ with $\operatorname{card} M = n + 1$ and every collection $\sigma \equiv (\varkappa_m \mid m \in M)$ of simple collections $\varkappa_m \equiv (x_{mi} \mid i \in I_m)$ indexed by non-empty finite sets I_m.

Let $M = \{m\}$. Then, $L = \sum(x_{mi} \mid i \in I_m)$ and $U = \operatorname{Map}(\{m\}, I_m)$. Define a bijection v from I_m onto U setting $v(i) \equiv \{\langle m, i\rangle\}$ for every $i \in I_m$. Using assertion 1 of Theorem 1 and Corollary 1 to Proposition 1, we get $R = \sum(x_{mu(m)} \mid u \in U) = \sum(x_{mv(i)(m)} \mid i \in I_m) = \sum(x_{mi} \mid i \in I_m) = L$ because $v(i)(m) = i$. This means that $0 \in X$.

Let $n \in X$. Take any π and σ such that $\operatorname{card} M = (n+1)+1$. Fix some element $m_0 \in M$ and consider the sets $M_0 \equiv \{m_0\}$ and $M_1 \equiv M \setminus M_0$. It is clear that $\operatorname{card} M_1 = n+1$. Consider also the collections $\pi_0 \equiv (I_m \mid m \in M_0)$, $\pi_1 \equiv (I_m \mid m \in M_1)$, $\sigma_0 \equiv (\varkappa_m \mid m \in M_0)$, and $\sigma_1 \equiv (\varkappa_m \mid m \in M_1)$. Then, for $U_1 \equiv \prod \pi_1$, we have $P(\sum \varkappa_m \mid m \in M_1) = \sum(P(x_{mu(m)} \mid m \in M_1) \mid u \in U_1)$.

Using Theorem 3 and Lemma 9, we get $L = P(\sum \varkappa_m \mid m \in M) = P(P(\sum \varkappa_m \mid m \in M_k) \mid k \in 2) = P(\sum \varkappa_m \mid m \in M_0)P(\sum \varkappa_m \mid m \in M_1) = (\sum \varkappa_{m_0}) \sum(P(x_{mu(m)} \mid m \in M_1) \mid$

$u \in U_1) = \sum(x_{m_0 i} \sum(P(x_{mu(m)} \mid m \in M_1) \mid u \in U_1) \mid i \in I_{m_0}) = \sum(\sum(x_{m_0 i} P(x_{mu(m)} \mid m \in M_1) \mid u \in U_1) \mid i \in I_{m_0})$.

Consider the set $U_0 \equiv \prod \pi_0 = \mathrm{Map}(\{m_0\}, I_{m_0})$ and define a bijection w from U_0 onto I_{m_0} setting $w(u_0) \equiv u_0(m_0)$. Using assertion 1 of Theorem 1 and Corollary 1 to Proposition 1, we get $L = \sum(\sum(x_{m_0 w(u_0)} P(x_{mu_1(m)} \mid m \in M_1) \mid u_1 \in U_1) \mid u_0 \in U_0) = \sum(\sum(x_{m_0 u_0(m_0)} P(x_{mu_1(m)} \mid m \in M_1) \mid u_1 \in U_1) \mid u_0 \in U_0) = \sum(\sum(P(x_{mu_0(m)} \mid m \in M_0)$ $P(x_{mu_1(m)} \mid m \in M_1) \mid u_1 \in U_1) \mid u_0 \in U_0)$.

Consider a projection α from $U_0 \times U_1$ onto U_1 such that $\alpha(u_0, u_1) = u_1$. The mapping $\alpha|\{u_0\} \times U_1$ is a bijective mapping from $\{u_0\} \times U_1$ onto U_1 for every $u_0 \in U_0$. Therefore, by the property of general commutativity of the sum, we get $L = \sum(\sum(P(x_{mu_0(m)} \mid m \in M_0) P(x_{m\alpha(p)(m)} \mid m \in M_1) \mid p \in \{u_0\} \times U_1) \mid u_0 \in U_0)$. Since the collection $(\{u_0\} \times U_1 \mid u_0 \in U_0)$ is a partition of the set $U_0 \times U_1$, we can apply assertion 2 of Theorem 1. As a result, $L = \sum(P(x_{mu_0(m)} \mid m \in M_0) P(x_{m\alpha(p)(m)} \mid m \in M_1) \mid p \in U_0 \times U_1) = \sum(P(x_{mu_0(m)} \mid m \in M_0) P(x_{mu_1(m)} \mid m \in M_1) \mid (u_0, u_1) \in U_0 \times U_1)$.

Consider the bijection β from U onto $U_0 \times U_1$ such that $\beta(u) = (r_0(u), r_1(u))$, where $r_0(u) \equiv u|M_0$ and $r_1(u) \equiv u|M_1$. Denote temporarily the element $P(x_{mu_0(m)} \mid m \in M_0) P(x_{mu_1(m)} \mid m \in M_1)$ by $z_{(u_0,u_1)}$. Then, $L = \sum(z_{(u_0,u_1)} \mid (u_0, u_1) \in U_0 \times U_1) = \sum(z_{\beta(u)} \mid u \in U)$ by virtue of assertion 1 of Theorem 1. Since $r_0(u)(m) = u(m)$ for every $m \in M_0$ and $r_1(u)(m) = u(m)$ for every $m \in M_1$, we get $z_{\beta(u)} = P(x_{mu(m)} \mid m \in M_0) P(x_{mu(m)} \mid m \in M_1)$. As a result, $L = \sum(P(x_{mu(m)} \mid m \in M_0) P(x_{mu(m)} \mid m \in M_1) \mid u \in U)$. Since the collection $(M_k \mid k \in 2)$ is a partition of the set M, we get in virtue of Corollary 1 to Proposition 1 and Theorem 3 $L = \sum(P(P(x_{mu(m)} \mid m \in M_k) \mid k \in 2) \mid u \in U) = \sum(P(x_{mu(m)} \mid m \in M) \mid u \in U) = R$.

This means that $n+1 \in X$. By the principle of natural induction from 1.2.6, we infer that $X = \omega$. □

Corollary 1. *Let* $(y_j \in \mathbb{Z} \mid j \in J)$ *and* $(z_k \in \mathbb{Z} \mid k \in K)$ *be finite simple collections. Then,* $\sum(y_j \mid j \in J) \sum(z_k \mid k \in K) = \sum(y_j z_k \mid (j, k) \in J \times K)$.

Proof. Consider the collection $\pi \equiv (I_m \mid m \in 2)$ such that $I_0 \equiv J$ and $I_1 \equiv K$ and the simple collections $\varkappa_0 \equiv (x_{0i} \mid i \in I_0)$ and $\varkappa_1 \equiv (x_{i1} \mid i \in I_1)$ such that $x_{0i} \equiv y_i$ and $x_{1i} \equiv z_i$. Then, $L \equiv \sum(y_j \mid j \in J) \sum(z_k \mid k \in K) = P(\sum(x_{mi} \mid i \in I_m) \mid m \in 2) = \sum(P(x_{mu(m)} \mid m \in 2) \mid u \in U)$, where $U \equiv \prod \pi$. Define a bijection $r\colon J \times K \to U$ setting $r(j, k)(0) \equiv j$ and $r(j, k)(1) \equiv k$. Then, by Theorem 1 $L = \sum(P(x_{mr(j,k)(m)} \mid m \in 2) \mid (j, k) \in J \times K)$. Using Corollary 1 to Proposition 1, we get $P(x_{mr(j,k)(m)} \mid m \in 2) = x_{0j} x_{1k} = y_j z_k$. As a result, $L = \sum(y_j z_k \mid (j, k) \in J \times K)$. □

Rising to a degree for integers

Let $m, n \in \omega$. The integer $e(m^n) \in \mathbb{Z}_+$ is called the *degree of the number* $\hat{m}$ *with the exponent* $\hat{n}$ and is denoted by $\hat{m}^{\hat{n}}$.

Let $x \in \mathbb{Z}$. Define the *degree* $x^{\hat{n}}$ *of the number* x *with the exponent* $\hat{n}$ setting $x^{\hat{n}} \equiv \hat{m}^{\hat{n}}$ if $x = \hat{m}$, $x^{\hat{n}} \equiv \hat{m}^{\hat{n}}$ if $x = -\hat{m}$ and the number n is even, and $x^{\hat{n}} \equiv -\hat{m}^{\hat{n}}$ if $x = -\hat{m}$ and the number n is odd.

Lemma 10. *Let* $n \in \mathbb{N}$ *and* $x \in \mathbb{Z}$. *Then,* $x^{\hat{n}} = P(x_i \mid i \in I)$ *for every simple collection* $(x_i \in \mathbb{Z} \mid i \in I)$ *such that* $x_i \equiv x$ *for every* $i \in I$ *and* $\operatorname{card} I = n$.

Proof. By Lemma 2, either $x = \hat{m}$ or $x = -\hat{m}$ for some $m \in \omega$. By Lemma 7 (1.3.5), $m^n = P(m_i \mid i \in I)$, where $m_i \equiv m$ for every i. Therefore, in the first case $x^{\hat{n}} = e(m^n) = e(P(m_i \mid i \in I)) \equiv P(\hat{m}_i \mid i \in I) = P(x_i \mid i \in I)$. In the second case, if n is even, we get again $x^{\hat{n}} = \hat{m}^{\hat{n}} = P(\hat{m}_i \mid i \in I) = P(|x_i| \mid i \in I) \equiv P(x_i \mid i \in I)$. If n is odd, we get $x^{\hat{n}} = -\hat{m}^{\hat{n}} = -P(|x_i| \mid i \in I) \equiv P(x_i \mid i \in I)$. □

Proposition 2. *Let* $(x_i \in \mathbb{Z} \mid i \in I)$ *and* $(y_j \in \mathbb{Z}_+ \mid j \in J)$ *be simple finite collections,* $x \in \mathbb{Z}$ *and* $y, z \in \mathbb{Z}_+$. *Then:*
1) $x^{\hat{0}} = \hat{1}$, $x^{\hat{1}} = x$, *and* $\hat{1}^y = \hat{1}$;
2) $\hat{0}^y = \hat{0}$ *for* $y \neq \hat{0}$;
3) $x^{\Sigma(y_j \mid j \in J)} = P(x^{y_j} \mid j \in J)$;
4) $(P(x_i \mid i \in I))^y = P(x_i^y \mid i \in I)$;
5) $x^{yz} = (x^y)^z$.

Proof. Let $y \equiv \hat{m}$, $z \equiv \hat{n}$, and $y_j \equiv \hat{m}_j$.

1. If $x = \hat{l}$, then by Lemma 9 (1.3.5) $x^{\hat{0}} \equiv e(l^0) = \hat{1}$, $x^{\hat{1}} \equiv e(l^1) = x$, and $\hat{1}^y \equiv e(1^m) = \hat{1}$. If $x = -\hat{l}$, then $x^{\hat{0}} = e(l^0) = \hat{1}$ and $x^{\hat{1}} = -e(l^1) = x$.

2. Since $m \neq 0$, we get by Lemma 9 $\hat{0}^y \equiv e(0^m) = \hat{0}$.

3. If $x = \hat{l}$, then by Corollary 1 to Lemma 7 (1.3.5) $x^{\Sigma(y_j \mid j \in J)} \equiv e(l^{\Sigma(m_j \mid j \in J)}) = e(P(l^{m_j} \mid j \in J)) \equiv P(e(l^{m_j}) \mid j \in J) = P(x^{y_j} \mid j \in J)$. If $x = -\hat{l}$, then the arguments are completely the same.

4. Denote $(x_i \mid i \in I)$ by $\varkappa$. If $y = \hat{0}$, then 4) follows from 1). Now assume that $m \neq 0$. Consider the sets $I_0 \equiv \{i \in I \mid x_i = \hat{0}\}$, $I_1 \equiv \{i \in I \mid \exists l_i \in \mathbb{N}\ (x_i = -\hat{l}_i)\}$, and $I_2 \equiv \{i \in I \mid \exists l_i \in \mathbb{N}\ (x_i = \hat{l}_i)\}$. If $I_0 \neq \varnothing$, then $P\varkappa = \hat{0}$ implies $(P\varkappa)^y = \hat{0}$ by virtue of 2. Also, $x_i^y = \hat{0}$ for every $i \in I_0$ implies $P(x_i^y \mid i \in I) = \hat{0}$. Therefore, in this case we get again 4).

Further, assume that $I_0 = \varnothing$. Consider the set $K \equiv \{1, 2\}$ And the collections $\varkappa_k \equiv (x_i \mid i \in I_k)$. Since $[I_k \mid k \in K]$ is a partition of the set I, we get by Theorem 3 and Corollary 1 to Proposition 1 $P\varkappa = P(P(x_i \mid i \in I_k) \mid k \in K) = P\varkappa_1 P\varkappa_2$. At first assume that m is even. If $\operatorname{card} I_1$ is even, then $P\varkappa_1 \equiv P(\hat{l}_i \mid i \in I_1)$. Therefore, $P\varkappa = P\varkappa_1 P\varkappa_2 = P(\hat{l}_i \mid i \in I_1)P(\hat{l}_i \mid i \in I_2) = P(\hat{l}_i \mid i \in I) \equiv e(P(l_i \mid i \in I))$ implies by Lemma 8 (1.3.5) $(P\varkappa)^y \equiv (e(P(l_i \mid i \in I)))^y = e((P(l_i \mid i \in I))^m) = e(P(l_i^m \mid i \in I)) = P(e(l_i^m) \mid i \in I) = P(\hat{l}_i^{\hat{m}} \mid i \in I) = P(x_i^y \mid i \in I)$.

If $\operatorname{card} I_1$ is odd, then $P\varkappa_1 \equiv -P(\hat{l}_i \mid i \in I_1)$. Therefore, $P\varkappa = P\varkappa_1 P\varkappa_2 = -P(\hat{l}_i \mid i \in I_1)P(\hat{l}_i \mid i \in I_2) = -P(\hat{l}_i \mid i \in I) \equiv -e(P(l_i \mid i \in I))$ implies $(P\varkappa_1)^y \equiv (e(P(l_i \mid i \in I)))^y = P(\hat{l}_i^{\hat{m}} \mid i \in I) = P(x_i^y \mid i \in I)$.

Finally, assume that m is odd. Then, $x_i^y = -e(l_i^m)$ for every $i \in I_1$ and $x_i^y = e(l_i^m)$ for every $i \in I_2$. If $\operatorname{card} I_1$ is even, then as above $P\varkappa = e(P(l_i \mid i \in I))$ implies $(P\varkappa)^y \equiv (e(P(l_i \mid i \in I)))^y \equiv e(P(l_i^m \mid i \in I)) = P(e(l_i^m) \mid i \in I) = P(e(l_i^m) \mid i \in I_1)P(e(l_i^m) \mid i \in I_2) = P(-e(l_i^m) \mid i \in I_1)P(x_i^y \mid i \in I_2) = P(x_i^y \mid i \in I_1)P(x_i^y \mid i \in I_2) = P(x_i^y \mid i \in I)$.

If $\operatorname{card} I_1$ is odd, then as above $P\varkappa = -e(P(l_i \mid i \in I))$ implies $(P\varkappa)^y \equiv -(e(P(l_i \mid i \in I)))^y \equiv -e(P(l_i^m \mid i \in I)) = -P(e(l_i^m) \mid i \in I) = (-P(e(l_i^m) \mid i \in I_1))P(e(l_i^m) \mid i \in I_2) = P(-e(l_i^m) \mid i \in I_1)P(x_i^y \mid i \in I_2) = P(x_i^y \mid i \in I_1)P(x_i^y \mid i \in I_2) = P(x_i^y \mid i \in I)$.

5. If $x = \hat{l}$, then by Corollary 3 to Lemma 7 (1.3.5) $x^{yz} \equiv e(l^{mn}) = e((l^m)^n) = (e(l^m))^z = (x^y)^z$.

Now, assume that $x = -\hat{l}$. If m and n are even, then again $x^{yz} \equiv e(l^{mn}) = (e(l^m))^z = (x^y)^z$. If m is even and n is odd, then $x^{yz} \equiv e(l^{mn}) = e((l^m)^n) = (e(l^m))^z = ((-\hat{l})^y)^z = (x^y)^z$. If m is odd and n is even, then $x^{yz} \equiv e(l^{mn}) = e((l^m)^n) = (-e(l^m))^z = ((-\hat{l})^y)^z = (x^y)^z$. Finally, if m and n are even, then $x^{yz} \equiv -e(l^{mn}) = -e((l^m)^n) = (-e(l^m))^z = ((-\hat{l})^y)^z = (x^y)^z$. □

Order properties of $\mathbb{Z}$

Consider on $\mathbb{Z}$ the binary relation $\vartheta \equiv \{(x, y) \in \mathbb{Z}\times\mathbb{Z} \mid (\exists m, n \in \omega\,(x = \hat{m} \wedge y = \hat{n} \wedge m \leqslant n)) \vee (\exists m, n \in \omega\,(x = -\hat{m} \wedge y = \hat{n})) \vee (\exists m, n \in \omega\,(x = -\hat{m} \wedge y = -\hat{n} \wedge m \geqslant n))\}$. Using Lemma 2, we can easily check that ϑ is a linear order. Along with $(x, y) \in \vartheta$ we shall write also $x \leqslant y$. Numbers from $\mathbb{Z}_+$ [$\mathbb{Z}_+ \setminus \{\hat{0}\}$] are called *positive* [*strictly positive*], and numbers from $\mathbb{Z}_-$ [$\mathbb{Z}_- \setminus \{\hat{0}\}$] are called *negative* [*strictly negative*].

Lemma 11. *Let $m, n \in \omega$. Then:*
1) *$\hat{m} \leqslant \hat{n}$ iff $m \leqslant n$;*
2) *$-\hat{m} \leqslant -\hat{n}$ iff $m \geqslant n$.*

Proof. We check only the second assertion.

Let $-\hat{m} \leqslant -\hat{n}$. Suppose that $m < n$. Then, by definition $-\hat{m} \geqslant -\hat{n}$, where $-\hat{m} = -\hat{n}$ and $m = n$. It follows from this contradiction that $m \geqslant n$. □

Lemma 12. *Let $x, y \in \mathbb{Z}$. Then, $x \leqslant y$ iff $y - x \geqslant \hat{0}$.*

Proof. Let $x = -\hat{m}$ and $y = -\hat{n}$. If $x \leqslant y$, then by Lemma 11 $m \geqslant n$. Therefore, by Theorem 5 (1.3.6), $m = n + k$ for some $k \geqslant 0$. By definition $\hat{k} \geqslant \hat{0}$. As a result, $y - x = \theta(0, n) + \theta(m, 0) = \theta(m, n) = \theta(k, 0) = \hat{k} \geqslant \hat{0}$. Conversely, let $y - x \geqslant \hat{0}$. Supposing that $y - x = -\hat{l}$ for some $l > 0$, we get by definition $y - x = -\hat{l} < \hat{0}$. Thus, $y - x = \hat{l}$ for some $l \in \omega$. From $y - x = \theta(m, n) = \theta(l, 0)$, we infer that $m = n + l \geqslant n$. By definition, this implies $x \leqslant y$.

In the case when $x = \hat{m}$ and $y = \hat{n}$, the arguments are the same.

Let $x = -\hat{m}$ and $y = \hat{n}$. Then, $x \leqslant y$ and $y - x = \theta(m + n, 0) = e(m + n) \geqslant \hat{0}$.

Finally, let $x = \hat{m}$ and $y = -\hat{n}$. Then, $y \leqslant x$. If $x \leqslant y$, then $x = y$ implies $\hat{m} = -\hat{n}$, where $m + n = 0$, and so $m = n = 0$. As a result, $y - x = -\hat{0} - \hat{0} = \hat{0}$. Conversely, if $y - x \geqslant \hat{0}$, then simultaneously $y - x = -\hat{n} - \hat{m} = -e(n + m) \leqslant \hat{0}$. Therefore, $y - x = \hat{0}$ and $x = y$. □

Proposition 3. *Let $(x_i \in \mathbb{Z} \mid\in I)$ and $(y_i \in \mathbb{Z} \mid i \in I)$ be simple finite collections, $x, y, z \in \mathbb{Z}$, and $r, s \in \mathbb{Z}_+$. Then:*
1) *if $x_i \leqslant y_i$ for every $i \in I$, then $\sum(x_i \mid i \in I) \leqslant \sum(y_i \mid i \in I)$; if besides $x_i < y_i$ at least for one index, then $\sum(x_i \mid i \in I) < \sum(y_i \mid i \in I)$;*
2) *if $\hat{0} \leqslant x_i \leqslant y_i$ for every $i \in I$, then $P(x_i \mid i \in I) \leqslant P(y_i \mid i \in I)$; if besides $x_i < y_i$ at least for one index and $y_i > \hat{0}$ for every $i \in I$, then $P(x_i \mid i \in I) < P(y_i \mid i \in I)$;*
3) *if $x < y$, then $xz < yz$ for $z > \hat{0}$ and $xz > yz$ for $z < \hat{0}$;*
4) *if $\hat{0} \leqslant x < y$ and $r > \hat{0}$, then $x^r < y^r$;*
5) *if $x > \hat{1}$ and $r < s$, then $x^r < x^s$.*

Proof. 1. By Lemma 12 $x_i \leqslant y_i$ implies $y_i - x_i \geqslant \hat{0}$. By virtue of Lemma 2, either $y_i - x_i = -\hat{m}_i$ or $y_i - x_i = \hat{m}_i$. In the first case, $y_i - x_i \leqslant \hat{0}$ implies $y_i - x_i = \hat{0}$. Thus, in both cases, $y_i - x_i = \hat{m}_i$ for some $m_i \in \omega$. Therefore, $sum(y_i - x_i \mid i \in I) = \sum(\hat{m}_i \mid i \in I) \equiv e(\sum m_i \mid i \in I) \geqslant \hat{0}$. But by Theorem 1, $\sum(y_i - x_i \mid i \in I) = \sum(y_i \mid i \in I) + \sum(-x_i \mid i \in I) = \sum(y_i \mid i \in I) - \sum(x_i \mid i \in I)$. Therefore, by Lemma 12 $\sum(x_i \mid i \in I) \leqslant \sum(y_i \mid i \in I)$.

If besides $x_j < y_j$, then $m_j > 0$. Therefore, by Proposition 2 (1.3.6), $\sum(m_i \mid i \in I) > 0$. Consequently, $\sum(y_i \mid i \in I) - \sum(x_i \mid i \in I) = \sum(\hat{m}_i \mid i \in I) \equiv e\sum(m_i \mid i \in I) > \hat{0}$. By virtue of Lemma 12, this implies the necessary strict inequality.

2. From the condition $\hat{0} \leqslant x_i \leqslant y_i$, we infer as above that $x_i = \hat{m}_i$ and $y_i = \hat{n}_i$ for some $m_i, n_i \in \omega$. By Lemma 11 $m_i \leqslant n_i$. Therefore, by Lemma 11 (1.3.5), $P(m_i \mid i \in I) \leqslant P(n_i \mid n \in I)$. Finally, $P(x_i \mid i \in I) \equiv eP(m_i \mid i \in I) \leqslant eP(n_i \mid i \in I) \equiv P(y_i \mid i \in I)$.

If besides $x_j < y_j$ and $y_i > \hat{0}$ for every i, then $m_j < n_j$ and $n_i > 0$ for every i. Therefore, by Proposition 2 (1.3.6), $P(m_i \mid i \in I) < P(n_i \mid i \in I)$. Since the mapping e is injective, we conclude that $P(x_i \mid i \in I) < P(y_i \mid i \in I)$.

3. If $x < y$, then $y - x > \hat{0}$ implies as above that $y - x = \hat{m}$ for some $m > 0$. Similarly, if $z > \hat{0}$, then $z = \hat{n}$ for some $m > 0$. Therefore, by Corollary 1 to Proposition 2 (1.3.6) $mn > 0$. Thus, by Theorem 5, $yz - xz = (y - x)z = e(mn) > \hat{0}$, where by Lemma 12 $xz < yz$.

If $z < \hat{0}$, then $z = -\hat{n}$ for some $n > 0$. Therefore, by Lemma 4 $yz - xz = (y - x)z = \hat{m}(-\hat{n}) = -e(mn) < \hat{0}$, where $yz < xz$.

4. As above, $\hat{0} \leqslant x < y$ implies $x = \hat{m}$ and $y = \hat{n}$ for some $m, n \in \omega$. By Lemma 11, $m < n$. Besides, $r = \hat{k}$ for some $k > 0$. Therefore, by Corollary 3 to Proposition 2 (1.3.6), $m^k < n^k$. As a result, $x^r \equiv e(m^k) < e(n^k) \equiv y^k$.

5. As above, $x = \hat{m}$, $r = \hat{k}$, and $x = \hat{l}$ for some $m, k, l \in \omega$. By Lemma 11 $m > 1$ and $k < l$. Therefore, by Corollary 5 to Proposition 2 (1.3.6) $m^k < m^l$. As a result, $x^r < x^s$. □

Corollary 1. *Let $x, y, z \in \mathbb{Z}$. Then, $x = y$ iff $x + z = y + z$. When $z \neq \hat{0}$, then $x = y$ iff $xz = yz$.*

Proof. From $x \leqslant y$ and $y \leqslant x$, we infer $x + z \leqslant y + z \leqslant x + z$ and $xz \leqslant yz \leqslant xz$.

Let $x + z = y + z$. Then, the proven property implies $x = x + (z - z) = (x + z) - z = (u + z) - z = y$.

Finally, let $z \neq \hat{0}$ and $xz = yz$. Suppose that $x \neq y$. Then, by assertion 3 of Proposition 3, $xz \neq yz$. It follows from this contradiction that $x = y$. □

Corollary 2 (the *Archimedes principle*). *Let $x, y \in \mathbb{Z}_+$ and $x > \hat{0}$. Then, there is a number $n \in \mathbb{N}$ such that $\hat{n}x > y$.*

Proof. Since $x \geqslant \hat{1}$ and $y + \hat{1} > y$, we infer that $(y + 1)x > yx \geqslant y$. □

Corollary 3. *Let $x, y \in \mathbb{Z}$ and $xy = \hat{0}$. Then, either $x = \hat{0}$ or $y = \hat{0}$.*

Proof. Suppose that the conclusion is not valid. Then, there are the four opportunities. If $x > \hat{0}$ and $y > \hat{0}$, then by assertion 2 of Proposition 3, $xy > \hat{0}$. If $x > \hat{0}$ and $y < \hat{0}$, then by assertion 3 of this proposition, $(-\hat{1})y > (-\hat{1})\hat{0} = \hat{0}$, and so $(-\hat{1})xy > \hat{0}$ implies $xy < \hat{0}$. If $x < \hat{0}$ and $y > \hat{0}$, then by the same reason $xy < \hat{0}$. Finally, if $x < \hat{0}$ and $y < \hat{0}$, then $(-\hat{1})x > \hat{0}$ and $(-\hat{1})y > \hat{0}$ imply $xy > \hat{0}$. In all the four opportunities, we come to contradiction. □

Proposition 4. *Let $(z_i \in \mathbb{Z} \mid i \in I)$ be a simple finite collection, $x, y \in \mathbb{Z}$, and $z \in \mathbb{Z}_+$. Then:*

1) $|x| = |-x|$, $x \leqslant |x|$, *and* $-x \leqslant |x|$;
2) $|P(z_i \mid i \in I)| = P(|z_i| \mid i \in I)$; *in particular,* $|xy| = |x|\,|y|$;
3) $|x^z| = |x|^z$;
4) *if* $y > \hat{0}$, *then* $|x| \leqslant y$ *is equivalent to* $-y \leqslant x \leqslant y$, *and* $|x| < y$ *is equivalent to* $-y < x < y$;
5) $|\sum(z_i \mid i \in I)| \leqslant \sum(|z_i| \mid i \in I)$; *in particular,* $|x + y| \leqslant |x| + |y|$;
6) $||x| - |y|| \leqslant |x - y|$.

Proof. 1. It is clear that $x \leqslant |x|$. If $x \in \mathbb{Z}_+$, then $-x \in \mathbb{Z}_-$. This implies $|x| = x$ and $|-x| = -(-x) = x$. If $x \in \mathbb{Z}_-$, then $x = -\hat{m}$ implies $-x = \hat{m} \in \mathbb{Z}_+$. Therefore, $|x| = -x$ and $|-x| = -x$. In both cases, $|x| = |-x|$. Therefore, $-x \leqslant |-x| = |x|$.

2. By definition, $P(z_i \mid i \in I)$ is either $P(|z_i| \mid i \in I)$ or $-P(|z_i| \mid i \in I)$. Applying 1, we get the necessary equality.

3. Let $z = \hat{n}$. By Lemma 2 either $x = \hat{m}$ or $x = -\hat{m}$, where $|x| = \hat{m}$. By definition x^y is either $\hat{m}^{\hat{n}}$ or $-\hat{m}^{\hat{n}}$. In both cases, by 1, we get $|x^z| = \hat{m}^{\hat{n}} = |x|^z$.

4. By assertion 1, $|x| \leqslant y$ implies $x \leqslant |x| \leqslant y$ and $-x \leqslant |x| \leqslant y$, i. e. $x \geqslant -y$. Conversely, if $-y \leqslant x \leqslant y$, then $-y \leqslant -x \leqslant y$, where $|x| \leqslant y$. For strict inequalities, the argument is the same.

5. We always have $z_i \leqslant |z_i|$. By assertion 1 of Proposition 3, $\sum(z_i \mid i \in I) \leqslant \sum(|z_i| \mid i \in I)$. Similarly, $-z_i \leqslant |z_i|$ implies $-\sum(z_i \mid i \in I) = \sum(-z_i \mid i \in I) \leqslant \sum(|z_i| \mid i \in I)$. Now, applying 4) we get the necessary inequality.

6. Using 5 and 1, we get $|x| = |y + (x - y)| \leqslant |y| + |x - y|$ and $|y| = |x + (y - x)| \leqslant |x| + |y - x| = |x| + |x - y|$. Therefore, $-|x - y| \leqslant |x| - |y| \leqslant |x - y|$. Applying 4, we get 6. □

Further in the book, we shall identify positive integers $\hat{m} \in \mathbb{Z}_+$ with the corresponding natural numbers $m \in \omega$.

1.4.2 Rational numbers

Define of the set $\mathbb{Z} \times \mathbb{Z} \setminus \{0\}$ a binary relation θ setting $((m, p), (n, q)) \in \theta$ iff $mq = np$. We assert that θ is an equivalence relation. In fact, θ is obviously symmetric and reflexive. Let $(l, p)\theta(m, q)$ and $(m, q)\theta(n, r)$, i. e. $lq = mp$ and $mr = nq$. Then, $(lr)q = (lq)r = (mp)r = p(nq) = (np)q$ and $q \neq 0$ imply by Corollary 1 to Proposition 3 (1.4.1) that $lr = np$. This means that $(l, p)\theta(n, r)$. Thus, θ is transitive.

Consider the factor-set $\mathbb{Q} \equiv (\mathbb{Z} \times (\mathbb{Z} \setminus \{0\}))/\theta$ consisting of equivalence classes $x \equiv \theta(m, p)$ of all pairs $(m, p) \in \mathbb{Z} \times (\mathbb{Z} \setminus \{0\})$ (see 1.1.14). Elements of the set $\mathbb{Q}$ are called *rational numbers* or *rational fractions*, and the set $\mathbb{Q}$ is called the *set of all rational numbers*.

Consider the factor-mapping $f\colon \mathbb{Z} \times (\mathbb{Z} \setminus \{0\}) \to \mathbb{Q}$ from 1.1.14. By Lemma 1 (1.1.14) f is surjective. According to 1.4.1, the set $\mathbb{Z}$ is countable. Therefore, by Lemma 3 (1.3.9), the set $\mathbb{Z} \times \mathbb{Z}$ is countable. Consequently, by Lemma 1 and 2 (1.3.9), the set $\mathbb{Q}$ is countable.

Associate with every integer $m \in \mathbb{Z}$ the rational number $\hat{m} \equiv \theta(m, 1) \in \mathbb{Q}$, and consider the mapping e from $\mathbb{Z}$ into $\mathbb{Q}$ such that $em \equiv \hat{m}$. This mapping is injective. Consider the set $\mathbb{Q}_1 \equiv \{\hat{m} \mid m \in \mathbb{Z}\}$. Since $\omega = \operatorname{card} \mathbb{Z} = \operatorname{card} \mathbb{Q}_1 \leqslant \operatorname{card} \mathbb{Q} \leqslant \omega$, we infer that $\operatorname{card} \mathbb{Q} = \omega$, i. e. the set $\mathbb{Q}$ is denumerable.

Along with $\theta(m, p)$, we shall write also m/p.

Product of rational numbers

Let $(x_i \in \mathbb{Q} \mid i \in I)$ be a simple collection of rational numbers $x_i \equiv m_i/p_i$ indexed by a finite set I. The rational number $P(m_i \mid i \in I)/P(p_i \mid i \in I)$ is called the *product of the simple collection* $(x_i \in \mathbb{Q} \mid i \in I)$ and is denoted by $P(x_i \mid i \in I)$. If $I = n + 1$ for $n \in \omega \setminus 2$, then along with $P(x_i \mid i \in n + 1)$ we shall use the notation $x_0 \ldots x_n$.

It is clear that $e(P(m_i \in \mathbb{Z} \mid i \in I)) = P(em_i \in \mathbb{Q} \mid i \in I)$.

Let $x, x', x'', x''', \ldots$ be rational numbers. Then, (x, x'), (x, x', x''), (x, x', x'', x'''), ... are the corresponding simple collection (see 1.1.11).

The rational numbers $P(x, x')$, $P(x, x', x'')$, $P(x, x', x'', x''')$,... will be called the *product of the simple sequential pair* (x, x'), *triplet* (x, x', x''), *quadruplet* (x, x', x'', x'''),... and will be denoted also by xx', $xx'x''$, $xx'x''x'''$,....

Theorem 1. *Let* $(x_i \in \mathbb{Q} \mid i \in I)$ *be a simple collection indexed by a finite set* I*. Then:*
1) *if* K *is a finite set and* u *is a bijective mapping from* K *onto* I*, then* $P(x_i \mid i \in I) = P(x_{u(k)} \mid k \in K)$ (the *general commutativity of the product*);
2) *if a collection* $(I_m \subset I \mid m \in M)$ *is a partition of the set* I *indexed by a finite non-empty set* M*, then* $P(x_i \mid i \in I) = P(P(x_i \mid i \in I_m) \mid m \in M)$ (the *general associativity of the product*).

Proof. We shall denote the left parts of these equalities by L. Let $x_i \equiv \theta(m_i, p_i)$.

1. Using Proposition 1 (1.4.1), we get $L \equiv \theta(P(m_i \mid i \in I), P(p_i \mid i \in I)) = \theta(P(m_{u(k)} \mid k \in K), P(p_{u(k)} \mid k \in K)) \equiv P(\theta(m_{u(k)}, p_{u(k)}) \mid k \in K) \equiv P(x_{u(k)} \mid k \in K)$.

2. Analogously, using Theorem 3 (1.4.1), we get $L \equiv \theta(P(m_i \mid i \in I), P(p_i \mid i \in I)) = \theta(P(P(m_i \mid i \in I_m) \mid m \in M), P(P(p_i \mid i \in I_m) \mid m \in M)) \equiv P(\theta(P(m_i \mid i \in I_m), P(p_i \mid i \in I_m)) \mid m \in M) \equiv P(P(\theta(m_i, p_i) \mid i \in I_m) \mid m \in M) \equiv P(P(x_i \mid i \in I_m) \mid m \in M)$. □

Lemma 1.
1) *Let* $(x_i \in \mathbb{Q} \mid i \in \{p\})$ *be a simple collection indexed by a set* $\{p\}$*. Then,* $P(x_i \mid i \in \{p\}) = x_p$.
2) *Let* $(x_i \in \mathbb{Q} \mid i \in \{p, q\})$ *be a simple collection indexed by a set* $\{p, q\}$ *with different elements* $p \neq q$*. Then,* $P(x_i \mid i \in \{p, q\}) = x_p x_q$.

The proof is analogous to the proof of Lemma 1 (1.4.1).

Corollary 1. *Let* $(x_i \in \mathbb{Q} \mid i \in I)$ *and* $(y_i \in \mathbb{Q} \mid i \in I)$ *be simple collections indexed by a finite non-empty set* I*. Then,* $P(x_i \mid i \in I)P(y_i \mid i \in I) = P(x_i y_i \mid i \in I)$.

The proof is completely similar to the proof of Corollary 1 to Lemma 1 (1.4.1).

Theorem 2. *Let* x*,* y *and* z *be rational numbers. Then:*
1) $xy = yx$ (the *commutativity of the product*);
2) $xyz = x(yz) = (xy)z$ (the *associativity of the product*).

The proof is analogous to the proof of Theorem 2 (1.4.1).

The element $\hat{1}$ is called the *unity element* in $\mathbb{Q}$. For every rational number x, we have the equality $\hat{1}x = x\hat{1} = x$.

The element p/m is called the *inverse element to the element* $x \equiv m/p \neq \hat{0}$ and is denoted by $1/x$ or by x^{-1}. It is clear that $(x^{-1})^{-1}$. The unity and inverse elements are connected by the equality $xx^{-1} = x^{-1}x = \hat{1}$. Further, along with xy^{-1} we shall write also x/y; this number is called the *quotient of the numbers* x *and* y.

Consider the set $\mathbb{Q}_1^{-1} \equiv \{1/m \mid m \in \mathbb{Z} \setminus \{0\}\}$.

Lemma 2.
1) $\mathbb{Q}_1 \cap \mathbb{Q}_1^{-1} = \{-\hat{1}, \hat{1}\}$.
2) *For every* $x \in \mathbb{Q}$*, there exist* $y \in \mathbb{Q}_1$ *and* $z \in \mathbb{Q}_1^{-1}$ *such that* $x = yz$.

Proof. 1. Let $x \in \mathbb{Q}_1 \cap \mathbb{Q}_1^{-1}$. By definition $x = m/1$ and $x = 1/n$ for some $m \in \mathbb{Z}$ and $n \in \mathbb{Z} \setminus \{0\}$. Thus, $(m, 1)\theta(1, n)$ implies $mn = 1$. Therefore, either $m, n > 0$ or $m, n < 0$. At first, suppose that $m, n > 0$ and at least one of them is greater than 1. Then, by virtue of assertion 2 of Proposition 3 (1.4.1) we get $mn > 1 \cdot 1 = 1$. Now, suppose that $m, n < 0$ and at least one of them is smaller than -1. Then, by the same reason, $mn = |m|\,|n| > 1$. It follows from these contradictions that either $m = n = 1$ or $m = n = -1$.

2. By definition, $x \equiv m/p$ for some $m \in \mathbb{Z}$ and $p \in \mathbb{Z} \setminus \{0\}$. Consider the rational numbers $y \equiv m/1$ and $z \equiv 1/p$. Then, $yz \equiv m1/1p = x$. □

Let $(x_i \in \mathbb{Q} \mid i \in I)$ be a simple collection of rational numbers $x_i \equiv m_i/p_i$ indexed by a finite set I. To define a sum of this collection, we need to prove some preliminary assertion.

Lemma 3. *Let $x_i = m_i/p_i = n_i/q_i$ for every $i \in I$. Then, $\sum(m_i P(p_j \mid j \in I \setminus \{i\}) \mid i \in I)/P(p_i \mid i \in I) = \sum(n_i P(q_j \mid j \in I \setminus \{i\}) \mid i \in I)/P(q_i \mid i \in I)$.*

Proof. Consider the sets $I_i \equiv I \setminus \{i\}$. Since every pair $[\{i\}, I_i]$ is a partition of the set I, we infer by Corollary 1 to Proposition 1 (1.4.1) and by Theorem 3 (1.4.1) that $P(p_i \mid i \in I) = p_i P(p_j \mid j \in I)$ and $P(q_i \mid i \in I) = q_i P(q_j \mid j \in I_i)$. Therefore, using the equalities $m_i q_i = n_i p_i$ and Lemma 9 (1.4.1), we get $\sum(m_i P(p_j \mid j \in I_i) \mid i \in I) P(q_i \mid i \in I) = \sum(m_i P(q_i \mid i \in I) P(p_j \mid j \in I_i) \mid i \in I) = \sum(m_i q_i P(q_j \mid j \in I_i) P(p_j \mid j \in I_i) \mid i \in I) = \sum(n_i p_i P(p_j \mid j \in I_i) P(q_j \mid j \in I_i) \mid i \in I) = \sum(n_i P(p_i \mid i \in I) P(q_j \mid j \in I_i) \mid i \in I) = \sum(n_i P(q_j \mid j \in I_i) \mid i \in I) P(p_i \mid i \in I)$. This gives the necessary equality. □

Sum of rational numbers

The rational number $\sum(m_i P(p_j \mid j \in I \setminus \{i\}) \mid i \in I)/P(p_i \mid i \in I)$ is called the *sum of the simple collection* $(x_i \in \mathbb{Q} \mid i \in I)$ and is denoted by $\sum(x_i \mid i \in I)$. It follows from Lemma 3 that this definition is correct. If $I = n + 1$ for $n \in \omega \setminus 2$, then along with $\sum(x_i \mid i \in n + 1)$ we shall use the notation $x_0 + \cdots + x_n$.

It is clear that $e(\sum(m_i \in \mathbb{Z} \mid i \in I)) = \sum(em_i \in \mathbb{Q} \mid i \in I)$.

Let $x, x', x'', x''', \ldots$ be rational numbers. The rational numbers $\sum(x, x')$, $\sum(x, x', x'')$, $\sum(x, x', x'', x''')$, … will be called the *sums of the simple sequential pair* (x, x'), *triplet* (x, x', x''), *quadruplet* (x, x', x'', x'''), … and will be denoted also by $x + x'$, $x + x' + x''$, $x + x' + x'' + x'''$, …

Theorem 3. *Let $(x_i \in \mathbb{Q} \mid i \in I)$ be a simple collection indexed by a finite set I. Then:*

1) *if K is a finite set and u is a bijective mapping from K onto I, then $\sum(x_i \mid i \in I) = \sum(x_{u(k)} \mid k \in K)$ (the general associativity of the sum);*
2) *if a collection $[I_m \subset I \mid m \in M]$ is a partition of the set I indexed by a finite non-empty set M, then $\sum(x_i \mid i \in I) = \sum(\sum(x_i \mid i \in I_m) \mid m \in M)$ (the general associativity of the sum).*

Proof. We shall denote the left and right parts of these equalities by L and R, respectively. Let $x_i \equiv m_i/p_i$.

1. Using assertion 1 of Theorem 1 (1.4.1) and assertion 1 of Theorem 1, we get $L \equiv \theta(\sum(m_i P(p_j \mid j \in I \setminus \{i\}) \mid i \in I), P(p_i \mid i \in I)) = \theta(\sum(m_{u(k)} P(p_j \mid j \in I \setminus \{u(k) \mid k \in K\}), P(p_{u(k)} \mid k \in K)) \equiv R$.

2. Using assertion 2 of Theorem 1 (1.4.1), we get $L = \theta(\sum(\sum(m_i P(p_j \mid j \in I \setminus \{i\}) \mid i \in I_m) \mid m \in M), P(p_i \mid i \in I))$. On the other hand, $R_m \equiv \sum(x_i \mid i \in I_m) \equiv \theta(\sum(m_i P(p_j \mid j \in I_m \setminus \{i\}) \mid\in I_m), P(p_i \mid i \in I_m)) \equiv \theta(a_m, x_m)$, where $a_m \equiv \sum(m_i P(p_j \mid j \in I_m \setminus \{i\}) \mid i \in I_m)$ and $x_m \equiv P(p_i \mid i \in I_m)$. Therefore, $R = \sum(R_m \mid m \in M) \equiv \theta(\sum(a_m P(x_n \mid n \in M \setminus \{m\}) \mid m \in M), P(x_m \mid m \in M))$. By Theorem 3 (1.4.1), $P(x_m \mid m \in M) = P(p_i \mid i \in I) \equiv y$.

Using Lemma 9 (1.4.1), we get $b \equiv \sum(a_m P(x_n \mid n \in M \setminus \{m\}) \mid m \in M) = \sum(\sum(m_i P(p_j \mid j \in I_m \setminus \{i\}) \mid i \in I_m) P(x_n \mid n \in M \setminus \{m\}) \mid m \in M) = \sum(\sum(m_i P(p_j \mid j \in I_m \setminus \{i\}) P(x_n \mid n \in M \setminus \{m\}) \mid i \in I_m) \mid m \in M)$. Consider the sets $M_m \equiv M \setminus \{m\}$ and $J_m \equiv \bigcup(I_n \mid n \in M_m)$. Since the collection $(I_n \mid n \in M_m)$ is a partition of the set J_m, we get by virtue of Theorem 3 (1.4.1), $P(x_n \mid n \in M_m) = P(P(p_i \mid i \in I_n) \mid n \in M_m) = P(p_i \mid i \in J_m)$. Since for every $i \in I_m$ the pair $(I_m \setminus \{i\}, J_m)$ is a partition of the set $I \setminus \{i\}$, we get in the same manner $P(p_j \mid j \in I_m \setminus \{i\}) P(x_n \mid n \in M_m) = P(p_j \mid j \in I_m \setminus \{i\}) P(p_i \mid i \in J_m) = P(p_j \mid I \setminus \{i\})$. As a result, $b = \sum(\sum(m_i P(p_j \mid I \setminus \{i\}) \mid i \in I_m) \mid m \in M)$.

Thus, $R = \theta(b, y) = L$. □

Lemma 4.
1) *Let* $(x_i \in \mathbb{Q} \mid i \in \{p\})$ *be a simple collection indexed by a set* $\{p\}$. *Then,* $\sum(x_i \mid i \in \{p\}) = x_p$.
2) *Let* $(x_i \in \mathbb{Q} \mid i \in \{p, q\})$ *be a simple collection indexed by a set* $\{p, q\}$ *with different elements* $p \neq q$. *Then,* $\sum(x_i \mid i \in \{p, q\}) = x_p + x_q$.

The proof is analogous to the proof of Lemma 1 (1.4.1).

Corollary 1. *Let* $(x_i \in \mathbb{Q} \mid i \in I)$ *and* $(y_i \in \mathbb{Q} \mid i \in I)$ *be simple collections indexed by a finite non-empty set I. Then,* $\sum(x_i \mid i \in I) + \sum(y_i \mid i \in I) = \sum(x_i + y_i \mid i \in I)$.

The proof is completely the same as the proof of Corollary 1 to Lemma 1 (1.4.1).

Theorem 4. *Let x, y and z be rational numbers. Then:*
1) $x + y = y + x$ (the *commutativity of the sum);*
2) $x + y + z = x + (y + z) = (x + y) + z$ (the *associativity of the sum*).

The proof is analogous to the proof of Theorem 2 (1.4.1).

The element $\hat{0}$ is called the *zero element* in $\mathbb{Q}$. For every rational number x, we have the equality $\hat{0} + x = x + \hat{0} = x$.

The element $(-m)/p$ is called the *opposite element to the element* $x \equiv m/p$ and is denoted by $-x$. It is clear that $-(-x) = x$. The zero and opposite elements are connected by the equality $x + (-x) = -x + x = \hat{0}$. Further, along with $x + (-y)$, we shall write also $x - y$; this number is called the *difference of the numbers x and y*. If $x \in e[\mathbb{Z}_+]$, then $-x \in e[\mathbb{Z}_-]$; if $x \in e[\mathbb{Z}_-]$, then $-x \in e[\mathbb{Z}_+]$.

Consider the sets $\mathbb{Q}_+ \equiv \{m/p \mid m \in \mathbb{Z}_+ \wedge p \in \mathbb{Z}_+ \setminus \{0\}\}$ and $\mathbb{Q}_- \equiv \{-m/p \mid m \in \mathbb{Z}_+ \wedge p \in \mathbb{Z}_+ \setminus \{0\}\}$.

Lemma 5.
1) $\mathbb{Q} = \mathbb{Q}_+ \cup \mathbb{Q}_-$ *and* $\mathbb{Q}_+ \cap \mathbb{Q}_- = \{\hat{0}\}$.
2) *For every* $x \in \mathbb{Q}$, *there exists* $y \in \mathbb{Q}_+$ *and* $z \in \mathbb{Q}_-$ *such that* $x = y + z$.

Proof. 1. Let $x \equiv m/p \in \mathbb{Q}$. If $m \geqslant 0$ and $p > 0$, then $x \in \mathbb{Q}_+$. If $m \geqslant 0$ and $p < 0$, then $m|p| = (-m)p$ implies $x = -m/|p| \in \mathbb{Q}_-$. If $m \leqslant 0$ and $p > 0$, then $|m|p = m(-p)$ implies $x = -|m|/p \in \mathbb{Q}_-$. Finally, if $m \leqslant 0$ and $p < 0$, then $m|p| = |m|p$ implies $x = |m|/|p| \in \mathbb{Q}_+$.

If $x \in \mathbb{Q}_+ \cap \mathbb{Q}_-$, then $x = m/p$ and $x = -n/q = (-n)/q$ for some $m, n \geqslant 0$ and $p, q > 0$. Therefore, $mq = (-n)p = -np$ implies by virtue of Lemma 2 (1.4.1), $mq = -np = 0$. Hence, $m = n = 0$, and so $x = \hat{0}$.

2. The assertion follows from 1. □

Lemma 6. *Let* $x, y \in \mathbb{Q}$. *Then,* $(-x)y = -xy$, $x(-y) = -xy$, *and* $(-x)(-y) = xy$.

Proof. By definition for $x \equiv m/p$ and $y \equiv n/q$, we have $(-x)y \equiv ((-m)/p)(n/q) = (-mn)/pq = -mn/pq = -xy$. The other equalities follow from the first one. □

Lemma 7. *Let* $m \in \mathbb{N}$ *and* $x \in \mathbb{Q}$. *Then,* $\hat{m}x = \sum(x_i \mid i \in I)$ *for every simple collection* $(x_i \in \mathbb{Q} \mid i \in I)$ *such that* $x_i = x$ *for every* $i \in I$ *and* $\operatorname{card} I = m$.

Proof. By Lemma 5 either $x = n/q$ or $x = -n/q$ for some $n \geqslant 0$ and $q > 0$. By Lemma 6 (1.4.1) $mn = \sum(n_i \mid i \in I)$, where $n_i \equiv n$ for every i. Consider $q_i \equiv q$.

In the first case, $\hat{m}x = mn/q = \sum(n_i \mid i \in I)/q$ and $\sum(x_i \mid i \in I) \equiv \sum(n_i/q_i \mid i \in I) \equiv \sum(n_i P(q_j \mid j \in I \setminus \{i\}) \mid i \in I)/P(q_i \mid i \in I)$. Check that these fractions are equal. Using Lemma 9 (1.4.1) and Theorem 3 (1.4.1), we get $\sum(n_i \mid i \in I)P(q_i \mid i \in I) = \sum(n_i P(q_j \mid j \in I) \mid i \in I) = \sum(n_i q_i P(q_j \mid j \in I \setminus \{i\}) \mid i \in I) = (\sum(n_i P(q_j \mid j \in I \setminus \{i\}) \mid i \in I))q$. Thus, really $\hat{m}x = \sum(x_i \mid i \in I)$.

In the second case, the argument is the same. □

Theorem 5. *Let* $x, y, z \in \mathbb{Q}$. *Then,* $x(y + z) = xy + xz$ (the *distributivity of the product with respect to the sum*).

Proof. Let $x \equiv l/p$, $y \equiv m/q$, and $z \equiv n/r$. Then, $y + z = (mr + nq)/qr$, $xy = ln/pq$, and $xz = ln/pr$. Consequently, $x(y + z) = l(mr + nq)/pqr = (lmr + lnq)/pqr$ and $xy + xz = (lmpr + lnpq)/pqpr = (lmr + lnq)/pqr$. These equalities give the necessary equality. □

Lemma 8. *Let $x \in \mathbb{Q}$ and $(y_j \in \mathbb{Q} \mid j \in J)$ be a simple finite collection. Then, $x \sum(y_j \mid j \in J) = \sum(xy_j \mid j \in J)$.*

The proof is analogous to the proof of Lemma 9 (1.4.1).

Theorem 6. *Let $(I_m \mid m \in M)$ be a collection of finite sets and $(\varkappa_m \mid m \in M)$ be a simple collection of simple collections $\varkappa_m \equiv (x_{mi} \in \mathbb{Q} \mid i \in I_m)$ indexed by a non-empty finite sets M and I_m. Consider the finite set $U \equiv \prod(I_m \mid m \in M)$. Then, $P(\sum(x_{mi} \mid i \in I_m) \mid m \in M) = \sum(P(x_{mu(m)} \mid m \in M) \mid u \in U)$ (the* general distributivity of the product with respect to the sum).

The proof is completely the same as the proof of Theorem 6 (1.4.1).

Corollary 1. *Let $(y_j \in \mathbb{Q} \mid j \in J)$ and $(z_k \in \mathbb{Q} \mid k \in K)$ be finite collections. Then, $\sum(y_j \mid j \in J) \sum(z_k \mid k \in K) = \sum(y_j z_k \mid (j, k) \in J \times K)$.*

The proof is completely the same as the proof of Corollary 1 to Theorem 6 (1.4.1).

Rising to an integer degree

Let $x \equiv m/p \in \mathbb{Q}$ and $y \equiv \hat{n} \in \mathbb{Q}_1$. Define the *degree* x^y *of the number* x *with the exponent* y setting $x^y \equiv m^n/p^n$ if $n \in \mathbb{Z}_+$ and $x^y \equiv (x^{-1})^{-y}$ if $n \in \mathbb{Z}_- \setminus \{0\}$ and $x \neq \hat{0}$.

It is clear that $e(m^n) = (em)^{en}$ for every $m \in \mathbb{Z}$ and $n \in \mathbb{Z}_+$.

Lemma 9. *Let $n \in \mathbb{N}$ and $x \in \mathbb{Q}$. Then, $x^{\hat{n}} = P(x_i \mid i \in I)$ for every simple collection $(x_i \in \mathbb{Q} \mid i \in I)$ such that $x_i \equiv x$ for every $i \in I$ and* card *$I = n$.*

Proof. Let $x \equiv m/p$. Consider the collections $\mu \equiv (m_i \in \mathbb{Z} \mid i \in I)$ and $\pi \equiv (p_i \in \mathbb{Z}\setminus\{0\} \mid i \in I)$ such that $m_i \equiv m$ and $p_i \equiv p$. Then, by Lemma 10 (1.4.1), $x^{\hat{n}} \equiv m^p/p^n = P\mu/P\pi \equiv P(x_i \mid i \in I)$. □

Proposition 1. *Let $(x_i \in \mathbb{Q} \setminus \{0\} \mid i \in I)$ and $(y_j \in \mathbb{Q}_1 \mid j \in J)$ be simple finite collections, $x \in \mathbb{Q} \setminus \hat{0}$, and $y, z \in \mathbb{Q}_1$. Then:*
1) $x^{\hat{0}} = \hat{1}, \hat{0}^{\hat{0}} = \hat{1}, x^{\hat{1}} = x$, *and* $\hat{1}^y = \hat{1}$;
2) $\hat{0}^y = \hat{0}$ *for* $y \in e[\mathbb{N}]$;
3) $x^{\sum(y_j \mid j \in J)} = P(x^{y_j} \mid j \in J)$;
4) $(P(x_i \mid i \in I))^y = P(x_i^y \mid i \in I)$;
5) $x^{yz} = (x^y)^z$.

Proof. Let $x \equiv l/p$, $y \equiv \hat{m}$, $z \equiv \hat{n}$, $x_i \equiv l_i/p_i$, and $y_j = \hat{m}_j$.

1. By assertion 1 of Proposition 2 (1.4.1), $x^{\hat{0}} \equiv l^0/p^0 = 1/1 \equiv \hat{1}$. Similarly, $x^{\hat{1}} \equiv l^1/p^1 = l/p = x$. If $m \in \mathbb{Z}_+$, then by definition $\hat{1}^y \equiv 1^m/1^m = 1/1 \equiv \hat{1}$. Finally, $\hat{0}^{\hat{0}} \equiv 0^0/1^0 = 1/1 \equiv \hat{1}$.

2. Since $m \in \mathbb{Z}_+ \setminus \{0\}$, we get by assertion 2 of Proposition 2 (1.4.1) $\hat{0}^y \equiv 0^m/1^m = 0/1 \equiv \hat{0}$.

3. Denote $(y_j \mid j \in J)$ by $\varkappa$ and $(m_j \mid j \in J)$ by μ. It was mentioned above that $\sum \varkappa = e \sum \mu$.

Consider the sets $J_1 \equiv \{j \in J \mid m_j < 0\}$ and $J_2 \equiv J \setminus J_1$ and the collections $\mu_1 \equiv (m_j \mid j \in J_1)$ and $\mu_2 \equiv (m_j \mid j \in J_2)$. Since $(J_k \mid k \in \{1, 2\})$ is a partition of the set J, we infer from assertion 2 of Theorem 1 (1.4.1) that $\sum \mu = \sum \mu_1 + \sum \mu_2$.

First, assume that $\sum \mu \geqslant 0$. Then, using assertion 3 of Proposition 2 (1.4.1), we get $L \equiv x^{\sum \mu - \sum \mu_1} \equiv l^{(\sum \mu - \sum \mu_1)}/p^{(\sum \mu - \sum \mu_1)} = l^{\sum \mu} l^{-\sum \mu_1}/p^{\sum \mu} p^{-\sum \mu_1} = (l^{\sum \mu}/p^{\sum \mu})(l^{-\sum \mu_1}/p^{-\sum \mu_1}) = x^{\sum \varkappa}(P(l^{-m_j} \mid j \in J_1)/P(p^{-m_j} \mid j \in J_1)) = x^{\sum \varkappa} P(l^{-m_j}/p^{-m_j} \mid j \in J_1) = x^{\sum \varkappa} P(x^{-y_j} \mid j \in J_1) \equiv M$.

On the other hand, $\sum \mu - \sum \mu_1 = \sum \mu_2$ implies $L = x^{\sum \mu_2} \equiv l^{\sum \mu_2}/p^{\sum \mu_2} = P(l^{m_j} \mid j \in J_2)/P(p^{m_j} \mid j \in J_2) = P(l^{m_j}/p^{m_j} \mid j \in J_2) = P(x^{y_j} \mid j \in J_2) \equiv N$. As a result, $M = N$. Multiplying both sides by $Q \equiv P(x^{y_j} \mid j \in J_1)$ and applying assertion 2 of Theorem 1, Corollary 1 to Lemma 1, and the equality $x^{-y_j} x^{y_j} = (l^{-m_j}/p^{-m_j})(p^{-m_j}/l^{-m_j}) = \hat{1}$, we get $x^{\sum \varkappa} = x^{\sum \varkappa} P(x^{-y_j} x^{y_j} \mid j \in J_1) = MQ = QN = P(x^{y_j} \mid j \in J)$.

Now, assume that $\sum \mu < 0$. Then, as above $L \equiv x^{-\sum \mu_1} = l^{-\sum \mu_1}/p^{-\sum \mu_1} = P(l^{-m_j}/p^{-m_j} \mid j \in J_1) = P(x^{-y_j} \mid j \in J_1) \equiv M$. On the other hand, $-\sum \mu_1 = \sum \mu_2 - \sum \mu$ implies $L = x^{\sum \mu_2 - \sum \mu} = l^{\sum \mu_2 - \sum \mu}/p^{\sum \mu_2 - \sum \mu} = l^{\sum \mu_2} l^{-\sum \mu}/p^{\sum \mu_2} p^{-\sum \mu} = (l^{\sum \mu_2}/p^{\sum \mu_2})(l^{-\sum \mu}/p^{-\sum \mu}) = (P(l^{m_j} \mid j \in J_2)/P(p^{m_j} \mid j \in J_2))x^{-e\sum \mu} = P(x^{y_j} \mid j \in J_2)x^{-e\sum \mu} \equiv N$. As a result, $M = N$. Multiplying both sides by $Q \equiv P(x^{y_j} \mid j \in J_1)$, we get $\hat{1} = QM = QN = P(x^{y_j} \mid j \in J)x^{-e\sum \mu}$. Now, multiplying both sides by $x^{\sum \varkappa}$, we get $x^{\sum \varkappa} = P(x^{y_j} \mid j \in J)$.

4. Denote $(x_i \mid i \in I)$ by π. If $m \geqslant 0$, then by assertion 4 of Proposition 2 (1.4.1) $(P\pi)^y = (P(l_i \mid i \in I)/P(p_i \mid i \in I))^y \equiv (P(l_i \mid i \in I))^m/(P(p_i \mid i \in I))^m = P(l_i^m \mid i \in I)/P(p_i^m \mid i \in I) \equiv P(l_i^m/p_i^m \mid i \in I) = P(x_i^y \mid i \in I)$. If $m < 0$, then $(P\pi)^y \equiv ((P\pi)^{-1})^{-y} = (P(p_i \mid i \in I)/P(l_i \mid i \in I))^{-y} \equiv (P(p_i \mid i \in I))^{-y}/(P(l_i \mid i \in I))^{-y} = P(p_i^{-y} \mid i \in I)/P(l_i^{-y} \mid i \in I) \equiv P(p_i^{-y}/l_i^{-y} \mid i \in I) = P((x_i^{-1})^{-y} \mid i \in I) = P(x_i^y \mid i \in I)$.

5. If $m \geqslant 0$ and $n \geqslant 0$, then by assertion 5 of Proposition 2 (1.4.1) $x^{yz} \equiv l^{mn}/p^{mn} = (l^m)^n/(p^m)^n = (l^m/p^m)^z = (x^y)^z$. If $m \geqslant 0$ and $n < 0$, then $x^{yz} \equiv p^{-mn}/l^{-mn} = (p^m)^{-n}/(l^m)^{-n} = ((l^m/p^m)^{-1})^{-z} = (x^y)^z$. If $m < 0$ and $n \geqslant 0$, then the argument is the same. Finally, if $m < 0$ and $n < 0$, then $x^{yz} \equiv l^{mn}/p^{mn} = l^{(-m)(-n)}/p^{(-m)(-n)} = ((p^{-m}/l^{-m})^{-1})^{-z} = ((x^{-1})^{-y})^z = (x^y)^z$. □

Order properties of $\mathbb{Q}$

Consider on $\mathbb{Q}$ the binary relation $\vartheta \equiv \{(x, y) \in \mathbb{Q} \times \mathbb{Q} \mid \exists m, n \in \mathbb{Z}\ \exists p, q \in \mathbb{Z} \setminus \{0\}\ (x = m/p \wedge y = n/q \wedge (mq - np)pq \leqslant 0)\}$.

Lemma 10. *The relation ϑ is a linear order in $\mathbb{Q}$.*

Proof. It is clear that this relation is reflexive. Let $(m/p, n/q) \in \vartheta$ and $(n/q, m/p) \in \vartheta$. Then, $(mq - np)pq \leqslant 0$ and $(np - mq)pq \leqslant 0$ imply by virtue of Lemma 8 (1.4.1) and assertion 3 of Proposition 3 (1.4.1) $(mq - np)pq \geqslant 0$, and so $(mq - np)pq = 0$. Since

$pq \neq 0$, we conclude by virtue of assertion 2 of Proposition 3 (1.4.1) that $mq - np = 0$, i. e. $m/p = n/q$. Thus, this relation is antisymmetric.

Now, check the reflexivity. Let $(l/p, m/q) \in \vartheta$ and $(m/q, n/r) \in \vartheta$, i. e. $(lq-mp)pq \leqslant 0$ and $(mr - nq)qr \leqslant 0$. Since p^2, q^2, and r^2 belong to $\mathbb{Z}_+ \setminus \{0\}$, we can multiply the first inequality by r^2 and the second one by p^2. Then, adding the obtained inequalities, we get $0 \geqslant lqpqr^2 - mppqr^2 + mrqrp^2 - nqqrp^2 = (lr-np)prq^2$. It follows that $(lr-np)pr \leqslant 0$, i. e. $(l/p, n/r) \in \vartheta$. Thus, ϑ is an order relation.

Take any $x \equiv m/p$ and $y \equiv n/q$. Then, $x = mpq^2/p^2q^2$ and $y = np^2q/p^2q^2$. Since the order in $\mathbb{Z}$ is linear, we infer that either $mpq^2 = np^2q$, or $mpq^2 > np^2q$, or $mpq^2 < np^2q$. In the first case, $x = y$. In the second case, $(mq - np)pq > 0$ implies $y\vartheta x$. in the third case, $(np - mq)pq > 0$ implies $x\vartheta y$. □

Further, along with $(x, y) \in \vartheta$, we shall write also $x \leqslant y$.

Corollary 1. $\mathbb{Q}_+ = \{z \in \mathbb{Q} \mid x \geqslant \hat{0}\}$ *and* $\mathbb{Q}_- = \{x \in \mathbb{Q} \mid x \leqslant \hat{0}\}$.

Proof. Let $x \equiv m/p$. If $x \in \mathbb{Q}_+$, then by definition $m \geqslant 0$ and $p > 0$. Therefore, $(m1 - 0p)p1 \geqslant 0$ implies $x \geqslant \hat{0}$. Conversely, if $x \geqslant \hat{0}$, then $mp \geqslant 0$. If $m \geqslant 0$ and $p > 0$, then $x \in \mathbb{Q}_+$. If $m \leqslant 0$ and $p < 0$, then $(-m)p = m(-p)$ implies $x = (-m)/(-p) \in \mathbb{Q}_+$. The second equality is checked in a similar way. □

Numbers from $\mathbb{Q}_+$ [$\mathbb{Q}_+ \setminus \{\hat{0}\}$] are called *positive* [*strictly positive*], and numbers from $\mathbb{Q}_-$ [$\mathbb{Q}_- \setminus \{\hat{0}\}$] are called *negative* [*strictly negative*].

Lemma 11. *Let* $m, n \in \mathbb{Z}$. *Then,* $\hat{m} \leqslant \hat{n}$ *iff* $m \leqslant n$.

Proof. By Lemma 12 (1.4.1) $\hat{m} \leqslant \hat{n}$ is equivalent to $m \leqslant n$. □

Lemma 12. *Let* $x, y \in \mathbb{Q}$. *Then,* $x \leqslant y$ *iff* $y - x \geqslant \hat{0}$.

Proof. Let $x \equiv m/p$ and $y \equiv n/q$. Then, $(mq - np)pq \leqslant 0$ is equivalent to $(0qp - (np - mq)1)1qp \leqslant 0$. But the first inequality means that $x \leqslant y$, and the second one means that $0/1 \leqslant (np - mq)/qp = y - x$. □

Proposition 2. *Let* $(x_i \in \mathbb{Q} \mid i \in I)$ *and* $(y_i \in \mathbb{Q} \mid i \in I)$ *be simple collections,* $x, y, z \in \mathbb{Q}$, *and* $r, s \in \mathbb{Q}_1$. *Then:*

1) *if* $x_i \leqslant y_i$ *for every* $i \in I$, *then* $\sum(x_i \mid i \in I) \leqslant \sum(y_i \mid i \in I)$, *if besides* $x_i < y_i$ *at least for one index, then* $\sum(x_i \mid i \in I) < \sum(y_i \mid i \in I)$;
2) *if* $\hat{0} \leqslant x_i \leqslant y_i$ *for every* $i \in I$, *then* $P(x_i \mid i \in I) \leqslant P(y_i \mid i \in I)$, *if besides* $x_i < y_i$ *at least for one index and* $y_i > \hat{0}$ *for every* $i \in I$, *then* $P(x_i \mid i \in I) < P(y_i \mid i \in I)$;
3) *if* $x < y$, *then* $xz < yz$ *for* $z > \hat{0}$ *and* $xz > yz$ *for* $z < \hat{0}$;
4) *if* $\hat{0} \leqslant x < y$, *then* $x^r < y^r$ *for* $r > \hat{0}$ *and* $x^r > y^r$ *for* $r < \hat{0}$ *and* $x > \hat{0}$;
5) *if* $r < s$, *then* $x^r < x^s$ *for* $x > \hat{1}$ *and* $x^r > x^s$ *for* $\hat{0} < x < \hat{1}$.

Proof. Denote $(x_i \mid i \in I)$ by π and $(y_i \mid i \in I)$ by $\varkappa$. Let $x_i \equiv m_i/p_i$ and $y_i \equiv n_i/q_i$.

1. By definition $\sum \pi \equiv \sum(m_i P(p_j \mid j \in I \setminus \{i\}) \mid i \in I)/P(p_i \mid i \in I) = \sum(m_i P(q_i \mid i \in I)P(p_j \mid j \in I \setminus \{i\}) \mid i \in I)/P(p_i \mid i \in I)P(q_i \mid i \in I)$ and $\sum \varkappa = \sum(n_i P(p_i \mid i \in I)P(q_j \mid j \in I \setminus \{i\}) \mid i \in I)/P(p_i \mid i \in I)P(q_i \mid i \in I)$. By Corollary 1 to Lemma 1, the denominator of these fraction is equal to the number $P(p_i q_i \mid i \in I) \equiv Z$. Denote $P(p_i q_i \mid i \in I \setminus \{i\})$ by Z_i. Then, $Z = p_i q_i Z_i$ for every i because $(\{i\}, I \setminus \{i\})$ is a partition of the set I. By the same reasons $P(p_i \mid i \in I)P(q_j \mid j \in I \setminus \{i\}) = p_i P(p_j \mid j \in I \setminus \{i\})P(q_j \mid j \in I \setminus \{i\}) = p_i Z_i$ and $P(q_i \mid i \in I)P(p_j \mid j \in I \setminus \{i\}) = q_i Z_i$.

As a result, $\sum \pi = \sum(m_i q_i Z_i \mid i \in I)/Z$ and $\sum \varkappa = \sum(n_i p_i Z_i \mid i \in I)/Z$. Denote the numerator of the first fraction by X and the numerator of the second one by Y. Since $x_i \leqslant y_i$, we have $(m_i q_i - n_i p_i)p_i q_i \leqslant 0$. Therefore, $(XZ - YZ)Z^2 = (\sum((m_i q_i - n_i p_i)ZZ_i \mid i \in I))Z^2 = (\sum(((m_i q_i - n_i p_i)p_i q_i)Z_i^2 \mid i \in I))Z^2 \leqslant 0$. This means that $\sum \pi = X/Z \leqslant Y/Z = \sum \varkappa$.

Let in addition $x_j < y_j$ for some $j \in I$, i. e. $(m_j q_j - n_j p_j)p_j q_j < 0$. Then, using assertions 1 and 3 of Proposition 3 (1.4.1), we infer that $(XZ - YZ)Z^2 < 0$. This means that $\sum \pi < \sum \varkappa$.

2. By definition, $P\pi \equiv P(m_i \mid i \in I)/P(p_i \mid i \in I) \equiv A/R$ and $P\varkappa \equiv P(n_i \mid i \in I)/P(q_i \mid i \in I) \equiv B/S$. Since $\hat{0} \leqslant x_i \leqslant y_i$, we infer by Lemma 12 (1.4.1) that $m_i p_i \geqslant 0$ and $0 \leqslant m_i q_i p_i q_i \leqslant n_i p_i p_i q_i$. Therefore, using assertion 2 of Proposition 3 (1.4.1), we get $(AS - BR)RS = P(m_i q_i p_i q_i \mid i \in I) - P(n_i p_i p_i q_i \mid i \in I) \leqslant 0$. This means that $P\pi = A/R \leqslant B/S = P\varkappa$.

Let in addition $y_i > \hat{0}$ for every $i \in I$ and $x_j < y_j$ for some $j \in I$. Then, $n_i q_i > 0$ for every i and $0 \leqslant m_j q_j p_j q_j < n_j p_j p_j q_j$ imply by virtue of assertion 2 of Proposition 3 (1.4.1) that $ASRS < BRRS$, where $(AS - BR)RS < 0$. This means that $P\pi < P\varkappa$.

3. Let $x \equiv l/p$, $y \equiv m/q$, and $z \equiv n/r$. By condition $(lq - mp)pq < 0$. If $z > \hat{0}$, then $(0r - n1)1r < 0$, i. e. $nr > 0$. Multiplying the first inequality by $nr > 0$ and $r^2 > 0$ and using assertion 3 of Proposition 3 (1.4.1), we get $(lnqr - mnpr)prqr < 0$. This means that $xz = ln/pr < mn/qr = yz$. If $z < \hat{0}$, then $(n1 - 0r)r1 < 0$, i. e. $nr < 0$. Therefore, in this case $(lnqr - mnpr)prqr > 0$, where $xz > yz$.

4. Let $r \equiv u/1$. By conditions $lp \geqslant 0$ and $(lq-mp)pq < 0$, where $0 \leqslant lqpq < mppq$. If $r > \hat{0}$, then $u > 0$. Using assertion 4 of Proposition 3 (1.4.1), we get $(lqpq)^u < (mppq)^u$. Using assertion 4 of Proposition 2 (1.4.1), we get $(l^u q^u - m^u p^u)p^u q^u < 0$. This means $x^r \equiv l^u/p^u < m^u/q^u \equiv y^r$. If $r < \hat{0}$ and $x > \hat{0}$, then $u < 0$, $lp > 0$, and $mq > 0$. Multiplying $lqpq < mppq$ by $lp > 0$ and $mq > 0$, we get $(qllm)(p^2q^2) < (pmlm)(p^2q^2)$. By virtue of assertion 3 of Proposition 3 (1.4.1), we deduce that $qllm < pmlm$. As above, this implies $(qllm)^{-u} < (pmlm)^{-u}$, and so $(q^{-u}l^{-u} - p^{-u}m^{-u})l^{-u}m^{-u} < 0$. The obtained inequality means that $y^r \equiv q^{-u}/m^{-u} < p^{-u}/l^{-u} \equiv x^r$.

5. Let $s \equiv v/1$. From $r < s$, it follows by Lemma 11 that $u < v$. Since $x > \hat{0}$, we can presuppose that $l > 0$ and $p > 0$.

At first, assume that $u \geqslant 0$. Let $x > \hat{1}$. Then, $(1p-l1)1p < 0$, i. e. $p^2 < lp$. This implies $p < l$. From $u < v$, we infer that $v = u+w$ for $w \equiv v-u > 0$. Then, by virtue of assertion 4 of Proposition 3 (1.4.1) $p < l$ implies $p^w < l^w$. Multiplying this inequality by $l^u > 0$ and $p^u > 0$, we get $l^u p^v = (l^u p^u)p^w < (l^u p^u)l^w = l^v p^u$. Consequently, $(l^u p^v - l^v p^u)p^u p^v < 0$. This means that $x^r \equiv l^u/p^u < l^v/p^v \equiv x^s$.

Now, let $x < \hat{1}$. Then, $lp < p^2$ implies $l < p$. Therefore, $l^w < p^w$. Acting as above, we get $l^v p^u = (l^u p^u)l^w < (l^u p^u)p^w = l^u p^v$. Consequently, $(l^v p^u - l^u p^v)p^u p^v < 0$. This means that $x^s \equiv l^v/p^v < l^u/p^u \equiv x^r$.

Now, assume that $u < 0 \leqslant v$. Let $x > \hat{1}$. Then, $p < l$ implies $0 < p^{-u} < l^{-u}$ and $0 < p^v < l^v$. Multiplying these inequalities, we get $p^{-u}p^v < l^v l^{-u}$. Consequently, $(p^{-u}p^v - l^v l^{-u})l^{-u}p^v < 0$. This means that $x^r \equiv p^{-u}/l^{-u} < l^v/p^v \equiv x^s$.

Now, let $x < \hat{1}$. Then, $l < p$ implies $0 < l^{-u} < p^{-u}$ and $0 < l^v < p^v$. Multiplying these inequalities, we get $l^v l^{-u} < p^{-u}p^v$. Consequently, $(l^v l^{-u} - p^{-u}p^v)p^v l^{-u} < 0$. This means that $x^s \equiv l^v/p^v < p^{-u}/l^{-u} \equiv x^r$.

Finally, assume that $u < v \leqslant 0$, i. e. $0 \leqslant -v < -u$. Then, $-u = -v + w$ for $w \equiv -u-(-v) > 0$. Let $x > \hat{1}$. Then, $p < l$ implies $p^w < l^w$. Multiplying this inequality by p^{-v} and l^{-v}, we get $p^{-u}l^{-v} = (p^{-v}l^{-v})p^w < (p^{-v}l^{-v})l^w = p^{-v}l^{-u}$. Consequently, $(p^{-u}l^{-v} - p^{-v}l^{-u})l^{-u}l^{-v} < 0$. This means that $x^r \equiv p^{-u}/l^{-u} < p^{-v}/l^{-v} \equiv x^s$.

Now, let $x < \hat{1}$. Then, $l < p$ implies $l^w < p^w$. Acting as above, we get $p^{-v}l^{-u} = (p^{-v}l^{-v})l^w < (p^{-v}l^{-v})p^w = p^{-u}l^{-v}$. Consequently, $(p^{-v}l^{-u} - p^{-u}l^{-v})l^{-v}l^{-u} < 0$. This means that $x^s \equiv p^{-v}/l^{-v} < p^{-u}/l^{-u} \equiv x^r$. □

Corollary 1. *Let $x, y, z \in \mathbb{Q}$. Then, $x = y$ iff $x + z = y + z$. When $z \neq \hat{0}$, then $x = y$ iff $xz = yz$.*

The proof is the same as the proof of Corollary 1 to Proposition 3 (1.4.1).

Corollary 2 (the *Archimedes principle*). *Let $x, y \in \mathbb{Q}_+$ and $x > \hat{0}$. Then, there is a number $n \in \mathbb{N}$ such that $\hat{n}x > y$.*

Proof. Let $x \equiv l/p$ and $y \equiv m/q$. Since $x > \hat{0}$ and $y \geqslant \hat{0}$, we infer by assertion 2 of Proposition 2 that $lqpq > 0$ and $mppq \geqslant 0$. Using Corollary 2 to Proposition 3 (1.4.1), we find a number n such that $nlqpq > mppq$. Consequently, $((nl)q - mp)pq > 0$ implies $nx > y$. □

Corollary 3. *Let $x, y \in \mathbb{Q}$ and $xy = \hat{0}$. Then, either $x = \hat{0}$ or $y = \hat{0}$.*

The proof is the same as the proof of Corollary 3 to Proposition 3 (1.4.1).

Corollary 4. *Let $x, y \in \mathbb{Q}$ and $x < y$. Then, there is $z \in \mathbb{Q}$ such that $x < z < y$.*

Proof. Take $z \equiv x + (y - x)/2 = (x + y)/2 = y - (y - x)/2$. □

Corollary 5. *Let $x, y \in \mathbb{Q}$ and $\hat{0} < x < y$. Then, $\hat{0} < y^{-1} < x^{-1}$.*

Proof. From $-1 < 0$, we infer by Lemma 11 that $e(-1) < e0$. Therefore, by virtue of assertion 4 of Proposition 2 $x^{-1} = x^{e(-1)} > y^{e(-1)} = y^{-1}$. Let $y \equiv n/q$. Since $y > \hat{0}$, we infer that $nq > 0$. Consequently, $y^{-1} = q/n > \hat{0}$. □

Modulus of a rational number

For every rational number, x we can define correctly its *modulus* $|x| \in \mathbb{Q}_+$ setting $|x| \equiv x$ if $x \in \mathbb{Q}_+$ and $|x| \equiv -x$ if $x \in \mathbb{Q}_-$. It is clear that $|x| = x \vee (-x)$.

Lemma 13. *Let* $x \equiv m/p \in \mathbb{Q}$. *Then,* $|x| = |m|/|p|$.

Proof. If $x \in \mathbb{Q}_+$, then $m \geqslant 0$ and $p > 0$. This implies $|x| \equiv x = m/p = |m|/|p|$. If $x \in \mathbb{Q}_-$, then by Corollary 1 to Lemma 10 $x \leqslant \hat{0}$. Therefore, $mp \leqslant 0$. If $m \geqslant 0$ and $p < 0$, then $|m| \equiv m$ and $|p| \equiv -p$ imply $|x| \equiv -x \equiv (-m)/p = (-|m|)/(-|p|) = |m|/|p|$. If $m \leqslant 0$ and $p > 0$, then $|m| \equiv -m$ and $|p| \equiv p$ imply $|x| = -x \equiv (-m)/p = |m|/|p|$. □

Proposition 3. *Let* $(z_i \in \mathbb{Q} \mid i \in I)$ *be a simple finite collection,* $x, y \in \mathbb{Q}$, *and* $z \in \mathbb{Q}_1$. *Then:*

1) $|x| = |-x|$, $x \leqslant |x|$, *and* $-x \leqslant |x|$;
2) $|P(z_i \mid i \in I)| = P(|z_i| \mid i \in I)$; *in particular,* $|xy| = |x|\,|y|$;
3) $|x^z| = |x|^z$ *if* $x \neq \hat{0}$;
4) *if* $y > \hat{0}$, *then* $|x| \leqslant y$ *is equivalent to* $-y \leqslant x \leqslant y$, *and* $|x| < y$ *is equivalent to* $-y < x < y$;
5) $|\sum(z_i \mid i \in I)| \leqslant \sum(|z_i| \mid i \in I)$; *in particular,* $|x + y| \leqslant |x| + |y|$;
6) $||x| - |y|| \leqslant |x - y|$.

Proof. 1. It is clear that $x \leqslant |x|$. If $x \in \mathbb{Q}_+$, then $-x \in \mathbb{Q}_-$. This implies $|x| \equiv x$ and $|-x| \equiv -(-x) = x$. If $x \in \mathbb{Q}_-$, then $x = -y$ for some $y \in \mathbb{Q}_+$. Therefore, $-x = -(-y) = y$. This implies $|x| \equiv -x = y$ and $|-x| \equiv y$. In both cases, $|x| = |-x|$. Therefore, $-x \leqslant |-x| = |x|$.

2. Let $z_i \equiv n_i/r_i$. Denote $(z_i \mid i \in I)$ by π, $(|z_i| \mid i \in I)$ by $\varkappa$, $(n_i \mid i \in I)$ by μ, $(|n_i| \mid i \in I)$ by ν, $(r_i \mid i \in I)$ by ρ, and $(|r_i| \mid i \in I)$ by σ. By Lemma 13 $|z_i| = |n_i|/|r_i|$. Therefore, using assertion 2 of Proposition 4 (1.4.1), we get $|P\pi| = |P\mu/P\rho| = |P\mu|/|P\rho| = P\nu/P\sigma \equiv P(|n_i|/|r_i| \mid i \in I) = P\varkappa$.

3. Let $x \equiv l/p$ and $z \equiv \hat{n}$. At first, assume that $n \geqslant 0$. Then, $x^z \equiv l^n/p^n$. By Lemma 13 and assertion 3 of Proposition 4 (1.4.1) we infer that $|x^z| = |l^n|/|p^n| = |l|^n/|p|^n = (|l|/|p|)^z = |x|^z$. Now, assume that $n \leqslant 0$, i. e. $n = -k$ for some $k \geqslant 0$. Then, $x^z \equiv (x^{-1})^{-z} = p^k/l^k$ implies $|x^z| = |p^k|/|l^k| = |p|^k/|l|^k = (|p|/|l|)^{-z} = ((|l|/|p|)^{-1})^{-z} = (|x|^{-1})^{-z} \equiv |x|^z$.

The other assertions are checked as the corresponding assertions of Proposition 4 (1.4.1). □

Corollary 1. *Let* $x, y \in \mathbb{Q}$ *and* $y \neq \hat{0}$. *Then,* $|x/y| = |x|/|y|$.

Proof. By definition $y^{-1} = y^{-e(1)}$. Therefore, using assertions 2 and 3 of this proposition, we get $|x/y| = |xy^{-e(1)}| = |x|\,|y^{-e(1)}| = |x|\,|y|^{-e(1)} = |x|/|y|$. □

Lemma 14. *Let $x, y \in \mathbb{Q}_+$ and $r \in e[\mathbb{N}]$. Then, $(x+y)^r \geqslant x^r + y^r$.*

Proof. Consider the set $N \subset \omega$ of all natural numbers n such that $(x+y)^{e(n+1)} \geqslant x^{e(n+1)} + y^{e(n+1)}$. It is clear that $0 \in N$. Assume that $n \in N$. Then, $(x+y)^{e(n+2)} = (x+y)^{e(n+1)}(x+y) \geqslant (x^{e(n+1)} + y^{e(n+1)})(x+y) \geqslant x^{e(n+2)} + y^{e(n+2)}$ means that $n+1 \in N$. By Theorem 1 (1.2.6), $N = \omega$. □

Further in the book, we shall identify rational numbers $\hat{m} \in \mathbb{Q}_1 \equiv e[\mathbb{Z}]$ with the corresponding integers $m \in \mathbb{Z}$.

1.4.3 Real and extended real numbers

A sequence $\alpha \equiv (a_n \in \mathbb{Q} \mid n \in \omega)$ is called *bounded* if there is a number $b \in \mathbb{Q}$ such that $|a_n| \leqslant b$ for every n (see also 1.1.15). A sequence α is called *inner convergent* ($\equiv$ *fundamental*), a *Cauchy sequence* if for every $\varepsilon \in \mathbb{Q}_+ \setminus \{0\}$, there is a natural number n such that $|a_p - a_q| < \varepsilon$ for all $p, q \geqslant n$. A sequence α is called *null* ($\equiv$ *negligible*) if for every such an ε, there is a natural number n such that $|a_p| < \varepsilon$ for all $p \geqslant n$. The sets of all inner convergent and all null sequences α will be denoted by $\mathcal{R}$ and $\mathcal{N}$, respectively.

For a sequence $\alpha \equiv (a_n \in \mathbb{Q} \mid n \in \omega)$ and a number $\varepsilon \in \mathbb{Q}_+ \setminus \{0\}$, we shall consider the sets $I(\alpha, \varepsilon) \equiv \{n \in \omega \mid \forall p, q \in \omega\, (p, q \geqslant n \Rightarrow |a_p - a_q| < \varepsilon)\}$ and $N(\alpha, \varepsilon) \equiv \{n \in \omega \mid \forall p \in \omega\, (p \geqslant n \Rightarrow |a_p| < \varepsilon)\}$. If α is fixed, then we shall denote these sets simply by $I(\varepsilon)$ and $N(\varepsilon)$, respectively. If a sequence α is inner convergent, then $I(\alpha, \varepsilon) \neq \varnothing$ for every ε. If a sequence α is null, then $N(\alpha, \varepsilon) \neq \varnothing$ for every ε.

Lemma 1. *Every null sequence α is inner convergent, and every inner convergent sequence α is bounded.*

Proof. If α is null, then for every $n \in N(\varepsilon/2)$ and every $p, q \geqslant n$, we have $|a_p - a_q| \leqslant |a_p| + |a_q| < \varepsilon/2 + \varepsilon/2 = \varepsilon$. It follows that α is inner convergent.

If α is inner convergent, then for every $n \in I(1)$ and every $p, q \geqslant n$ we have $|a_p - a_q| < 1$. In particular, $|a_{n+k} - a_n| < 1$ for every $k \in \omega$. Consider the number $b \equiv \mathrm{gr}(|a_0|, \ldots, |a_n|, |a_n| + 1)$. Then, $|a_p| \leqslant b$ for every $p \in \omega$. Thus, α is bounded. □

Lemma 2. *Let $(\alpha_i \mid i \in I)$ and $(\beta_i \mid i \in I)$ be finite simple collections of null sequences $\alpha_i \equiv (a_{in} \in \mathbb{Q} \mid n \in \omega)$ and bounded sequences $\beta_i \equiv (b_{in} \in \mathbb{Q} \mid n \in \omega)$, respectively. Then, the sequence $(\sum(a_{in} b_{in} \mid i \in I) \mid n \in \omega)$ is null.*

Proof. We may assume that $c \equiv \operatorname{card} I \in \mathbb{N}$. Fix any $\varepsilon \in \mathbb{Q}_+ \setminus \{0\}$. Let $|b_{in}| \leqslant b_i$. Consider the number $b \equiv \operatorname{gr}(b_i \mid i \in I)$. Take any numbers $n_i \in N(\alpha_i, \varepsilon/(c_i b))$, where $c_i \equiv c$ for every $i \in I$. Using assertion 1 of Proposition 2 (1.4.2) and Lemma 7 (1.4.2), we get $|\sum(a_{ip}b_{ip} \mid i \in I)| \leqslant \sum(|a_{ip}||b_{ip}| \mid i \in I) \leqslant c(\varepsilon/(cb))b = \varepsilon$ for every $p \geqslant \operatorname{gr}(n_i \mid i \in I)$. □

Lemma 3. *Let* $(\alpha_i \mid i \in I)$ *and* $(\beta_i \mid i \in I)$ *be finite simple collections of inner convergent sequences* $\alpha_i \equiv (a_{in} \in \mathbb{Q} \mid n \in \omega)$ *and null sequences* $\beta_i \equiv (b_{in} \in \mathbb{Q} \mid i \in I)$*, respectively. Then, the sequence* $(\sum(a_{in} + b_{in} \mid i \in I) \mid n \in \omega)$ *is inner convergent.*

Proof. We may assume that $c \equiv \operatorname{card} I \in \mathbb{N}$. Denote $\sum(a_{in}+b_{in} \mid i \in I)$ by c_n. Fix any $\varepsilon \in \mathbb{Q}_+ \setminus \{0\}$; take any numbers $m_i \in I(\alpha_i, \varepsilon/3c)$ and $n_i \in N(\beta_i, \varepsilon/3c)$. Consider the numbers $m \equiv \operatorname{gr}(m_i \mid i \in I)$, $n \equiv \operatorname{gr}(n_i \mid i \in I)$, and $l \equiv \operatorname{gr}(m, n)$. Then by Proposition 3 (1.4.2), Lemma 7 (1.4.2), and Corollary 1 to Lemma 4 (1.4.2), we have $|c_p - c_q| \leqslant \sum(|(a_{ip} + b_{ip}) - (a_{iq} + b_{iq})| \mid i \in I) \leqslant \sum(|a_{ip} - a_{iq}| + |b_{ip}| + |b_{iq}| \mid i \in I) = \sum(|a_{ip} - a_{iq}| \mid i \in I) + \sum(|b_{ip}| \mid i \in I) + \sum(|b_{iq}| \mid i \in I) < 3(c(\varepsilon/3c)) = \varepsilon$. □

With every rational number a, we shall associate the constant sequence $\alpha_a \equiv (a_n \in \mathbb{Q} \mid n \in \omega)$ such that $a_n \equiv a$ for every n. It is clear that α_a is inner convergent and bounded.

Define on the set $\mathcal{R}$ a binary relation θ setting $((a_n \mid n \in \omega), (b_n \mid n \in \omega)) \in \theta$ iff $(a_n - b_n \mid n \in \omega) \in \mathcal{N}$. We assert that θ is an equivalence relation. In fact, θ is obviously reflexive and symmetric. Let $(\alpha, \beta) \in \theta$ and $(\beta, \gamma) \in \theta$ for some $\alpha \equiv (a_n)$, $\beta \equiv (b_n)$, and $\gamma \equiv (c_n)$. Then, $(a_n - b_n) \in \mathcal{N}$ and $(b_n - c_n) \in \mathcal{N}$. Since $a_n - c_n = (a_n - b_n) + (b_n - c_n)$, we infer by Lemma 2 that $(a_n - c_n) \in \mathcal{N}$, i. e. $(\alpha, \gamma) \in \theta$. Thus, θ is transitive.

Consider the factor-set $\mathbb{R} \equiv \mathcal{R}/\theta$ consisting of equivalence classes $x \equiv \theta\alpha \equiv \bar{\alpha}$ of all sequences $\alpha \in \mathcal{R}$ (see 1.1.14). Elements of the set $\mathbb{R}$ are called *real numbers*; and the set $\mathbb{R}$ is called the *set of all real numbers*.

Associate with every rational number $a \in \mathbb{Q}$ the real number $\hat{a} \equiv \theta\alpha_a \equiv \bar{\alpha}_a \in \mathbb{R}$, and consider the mapping e from $\mathbb{Q}$ into $\mathbb{R}$ such that $ea \equiv \hat{a}$. This mapping is injective.

Let $(x_i \in \mathbb{R} \mid i \in I)$ be a simple collection of real numbers $x_i \equiv \theta(a_{in} \in \mathbb{Q} \mid n \in \omega)$ indexed by a finite set I. It easily deduced from Lemma 2 that we can introduce the following definitions.

Sum and product of real numbers

The real number $\theta(\sum(a_{in} \mid i \in I) \mid n \in \omega)$ is called the *sum of the simple collection* $(x_i \in \mathbb{R} \mid i \in I)$ and is denoted by $\sum(x_i \mid i \in I)$. If $I = n+1$ for $n \in \omega \setminus 2$, then along with $\sum(x_i \mid i \in n + 1)$, we shall use the notation $x_0 + \ldots + x_n$.

It is clear that $e(\sum(a_i \in \mathbb{Q} \mid i \in I)) = \sum(ea_i \in \mathbb{R} \mid i \in I)$.

Let $x, x', x'', x''', \ldots$ be real numbers. Then, (x, x'), (x, x', x''), (x, x', x'', x'''), ... are corresponding simple collections (see 1.1.11).

The real numbers $\sum(x, x')$, $\sum(x, x', x'')$, $\sum(x, x', x'', x''')$,... will be called the *sums of the simple sequential pair* (x, x'), *triplet* (x, x', x''), *quadruplet* (x, x', x'', x'''),... and will be denoted also by $x + x'$, $x + x' + x''$, $x + x' + x'' + x'''$, ...

In the similar manner, the real number $\theta(P(a_{in} \mid i \in I) \mid n \in \omega)$ is called the *product of the simple collection* $(x_i \in \mathbb{R} \mid i \in I)$ and is denoted by $P(x_i \mid i \in I)$. If $I = n+1$ for $n \in \omega \setminus 2$, then along with $P(x_i \mid i \in n + 1)$, we shall use the notation $x_0 \ldots x_n$.

It is clear that $e(P(a_i \in \mathbb{Q} \mid i \in I)) = P(ea_i \in \mathbb{R} \mid i \in I)$.

The real numbers $P(x, x')$, $P(x, x', x'')$, $P(x, x', x'', x''')$,... will be called the *products of the simple sequential pair* (x, x'), *triplet* (x, x', x''), *quadruplet* (x, x', x'', x'''),... and will be denoted also by xx', $xx'x''$, $xx'x''x'''$,...

Theorem 1. *Let* $(x_i \in \mathbb{R} \mid i \in I)$ *be a simple collection indexed by a finite set* I. *Then:*
1) *if* K *is a finite set and* u *is a bijective mapping from* K *onto* I, *then* $\sum(x_i \mid i \in I) = \sum(x_{u(k)} \mid k \in K)$ *and* $P(x_i \mid i \in I) = P(x_{u(k)} \mid k \in K)$ (the *general commutativity of the sum and the product*, respectively);
2) *if a collection* $[I_m \subset I \mid m \in M]$ *is a partition of the set* I *indexed by a finite non-empty set* M, *then* $\sum(x_i \mid i \in I) = \sum(\sum(x_i \mid i \in I_m) \mid m \in M)$ *and* $P(x_i \mid i \in I) = P(P(x_i \mid i \in I_m) \mid m \in M)$ (the *general associativity of the sum and the product*, respectively).

All the assertions are direct consequences of definitions and the corresponding assertions of Theorems 1 and 3 from 1.4.2.

The following assertion represents some special form of the general associativity.

Proposition 1. *Let* $[J_i \mid i \in I)$ *be a collection of finite non-empty sets indexed by a finite non-empty set* I *and* $[x_k \in \mathbb{R} \mid k \in K]$ *be a simple collection indexed by the set* $K \equiv \bigcup(\{i\} \times J_i \mid i \in I)$. *Then,* $\sum(x_k \mid k \in K) = \sum\left(\sum\left(x_{ij} \mid j \in J_i\right) \mid i \in I\right)$ *and* $P(x_k \mid k \in K) = P\left(P\left(x_{ij} \mid j \in J_i\right) \mid i \in I\right)$.

Proof. Consider the sets $K_i \equiv \{i\} \times J_i$. Then, $[K_i \mid i \in I)$ is a partition of the set K. Therefore, according to assertion 2 of Theorem 1, we get the equality $B \equiv \sum(x_k \mid k \in K) = \sum(\sum(x_k \mid k \in K_i) \mid i \in I)$. Consider the bijective mappings $u_i : J_i \rightarrowtail\!\!\!\!\twoheadrightarrow K_i$ such that $u_i(j) = (i, j)$ for every $j \in J_i$. Then, assertion 1 of Theorem 1 implies $\sum(x_k \mid k \in K_i) = \sum\left(x_{ij} \mid j \in J_i\right)$. Thus, $B = \sum\left(\sum\left(x_{ij} \mid j \in J_i\right) \mid i \in I\right)$.

For the products the arguments are the same. □

Corollary 1. *Let* I *and* J *be finite non-empty sets and* $([x_{ij} \in \mathbb{R} \mid j \in J] \mid i \in I]$ *be a collection of collections. Then:*
1) $\sum\left(\sum\left(x_{ij} \mid j \in J\right) \mid i \in I\right) = \sum\left(\sum\left(x_{ij} \mid i \in I\right) \mid j \in J\right) = \sum\left(x_{ij} \mid (i, j) \in I \times J\right)$;
2) $P\left(P\left(x_{ij} \mid j \in J\right) \mid i \in I\right) = P\left(P\left(x_{ij} \mid i \in I\right) \mid j \in J\right) = P\left(x_{ij} \mid (i, j) \in I \times J\right)$.

Proof. Consider the collections $(J_i \mid i \in I)$, where $J_i \equiv J$ and $(\iota_i \mid i \in I)$, where $\iota_i \equiv \{i\}$. Then, $\bigcup(\iota_i \mid i \in I) = I$ and $\bigcup(J_i \mid i \in I) = J$. Take also the sets $K_i \equiv \{i\} \times J_i$ and $K \equiv \bigcup(K_i \mid i \in I)$. Since the mapping $u : I \times I \to I$ such that $u(i, i') = i$ is surjective, assertion 1 of Proposition 1 (1.1.10) and assertion 5 of Corollary 2 to Theorem 1 (1.1.13) imply $K = \bigcup(\{i\} \times J_i \mid i \in I) = \bigcup(\iota_i \times J_{i'} \mid (i, i') \in I \times I) = \bigcup(\iota_i \mid i \in I) \times \bigcup(J_{i'} \mid i' \in I) = I \times J$. According to Proposition 1, we infer that $\sum\left(\sum\left(x_{ij} \mid j \in J\right) \mid i \in I\right) = \sum\left(x_k \mid k \in K\right) = \sum\left(x_{ij} \mid (i, j) \in I \times J\right)$.

Similarly, $\sum\left(\sum\left(x_{ij} \mid i \in I\right) \mid j \in J\right) = \sum\left(x_{ij} \mid (j, i) \in J \times I\right)$. Since the mapping $v : J \times I \to I \times J$ such that $v(i, j) = (j, i)$ is bijective, by assertion 1 of Theorem 1, we get $\sum\left(x_{ij} \mid (i, j) \in I \times J\right) = \sum\left(x_{ij} \mid (j, i) \in J \times I\right)$.

The second assertion is proven in the same way. □

Lemma 4.

1) *Let $(x_i \in \mathbb{R} \mid i \in \{p\})$ be a simple collection indexed by a set $\{p\}$. Then, $\sum(x_i \mid i \in \{p\}) = x_p$ and $P(x_i \mid i \in \{p\}) = x_p$.*
2) *Let $(x_i \in \mathbb{R} \mid i \in \{p, q\})$ be a simple collection indexed by a set $\{p, q\}$ with different elements $p \neq q$. Then, $\sum(x_i \mid i \in \{p, q\}) = x_p + x_q$ and $P(x_i \mid i \in \{p, q\}) = x_p x_q$.*

The proof is analogous to the proof of Lemma 1 (1.4.1).

Corollary 1. *Let $(x_i \in \mathbb{R} \mid i \in I)$ and $(y_i \in \mathbb{R} \mid i \in I)$ be simple collections indexed by a finite non-empty set I. Then, $\sum(x_i \mid i \in I) + \sum(y_i \mid i \in I) = \sum(x_i + y_i \mid i \in I)$ and $P(x_i \mid i \in I)P(y_i \mid i \in I) = P(x_i y_i \mid i \in I)$.*

The proof is completely similar to the proof of Corollary 1 to Lemma 1 (1.4.1).

Theorem 2. *Let x, y and z be real numbers. Then:*

1) $x + y = y + x$ *and* $xy = yx$ (the *commutativity of the sum and the product*, respectively);
2) $x + y + z = x + (y + z) = (x + y) + z$ *and* $xyz = x(yz) + (xy)z$ (the *associativity of the sum and the product*, respectively).

The proof is analogous to the proof of Theorem 2 (1.4.1).

Lemma 5. *Let $m \in \mathbb{N}$ and $x \in \mathbb{R}$. Then, $\hat{m}x = \sum(x_i \mid i \in I)$ for every simple collection $(x_i \in \mathbb{R} \mid i \in I)$ such that $x_i = x$ for every $i \in I$ and* card $I = m$.

The assertion is a direct consequence of definition and Lemma 7 (1.4.2).

Theorem 3. *Let $x, y, z \in \mathbb{R}$. Then, $x(y + z) = xy + xz$* (the *distributivity of the product with respect to the sum*).

The assertion is a direct consequence of definitions and Theorem 5 (1.4.2).

Lemma 6. *Let $x \in \mathbb{R}$ and $(y_j \in \mathbb{R} \mid j \in J)$ be a simple finite collection. Then, $x \sum(y_j \mid j \in J) = \sum(xy_j \mid j \in J)$.*

The assertion is a direct consequence of definitions and Lemma 8 (1.4.2). It can also be proven in the same manner as Lemma 9 (1.4.1).

Theorem 4. *Let $(I_m \mid m \in M)$ be a collection of finite sets and $(\varkappa_m \mid m \in M)$ be a simple collection of simple collections $\varkappa_m \equiv (x_{mi} \in \mathbb{R} \mid i \in I_m)$ indexed by non-empty finite sets M and I_m. Consider the finite set $U \equiv \prod(I_m \mid m \in M)$. Then, $P(\sum(x_{mi} \mid i \in I_m) \mid m \in M) = \sum(P(x_{mu(m)} \mid m \in M) \mid u \in U)$ (the general distributivity of the product with respect to the sum).*

The assertion is a direct consequence of definitions and Theorem 6 (1.4.2). It can also be proven in the same manner as Theorem 6 (1.4.1).

Corollary 1. *Let $(y_j \in \mathbb{R} \mid j \in J)$ and $(z_k \in \mathbb{R} \mid k \in K)$ be finite simple collections. Then, $\sum(y_j \mid j \in J) \sum(z_k \mid k \in K) = \sum(y_j z_k \mid (j, k) \in J \times K)$.*

The proof is completely the same as the proof of Corollary 1 to Theorem 6 (1.4.1). It can also be deduced directly from Corollary 1 to Theorem 6 (1.4.2).

The element $\hat{0}$ is called the *zero element* in $\mathbb{R}$. For every real number x, we have the equality $\hat{0} + x = x + \hat{0} = x$. By virtue of Lemmas 1 and 2, we also have the equality $\hat{0}x = x\hat{0} = \hat{0}$.

The number $\theta(-a_n \mid n \in \omega)$ is called the *opposite number to the number* $x \equiv \theta(a_n \in \mathbb{Q} \mid n \in \omega)$ and is denoted by $-x$. It is clear that $-(-x) = x$. The zero and opposite elements are connected by the equality $x + (-x) = -x + x = \hat{0}$. Further, along with $x + (-y)$ we shall write also $x - y$; this number is called the *difference of the numbers* x *and* y.

It is clear that $e(-a) = -ea$ for every $a \in \mathbb{Q}$.

The element $\hat{1}$ is called the *unity element* in $\mathbb{R}$. For every real number, we have the equality $\hat{1}x = x\hat{1} = x$.

Proposition 2. *Let $x \in \mathbb{R}$ and $x \neq \hat{0}$. Then, there is a unique number $y \in \mathbb{R}$ such that $xy = \hat{1}$.*

Proof. Let $x \equiv \theta\alpha$ for some $\alpha \equiv (a_n \in \mathbb{Q} \mid n \in \omega)$. Since $x \neq \hat{0}$, there exists a rational number $\varepsilon > 0$ such that for every natural number r, there is some natural number $s \geqslant r$ for which $|a_s| \geqslant \varepsilon$. Take any number $m \in I(\alpha, \varepsilon/2)$, and for it take a number $s > m$ for which $|a_s| \geqslant \varepsilon$. Then, for any $p > m$ we have $\varepsilon \leqslant |a_s| = |a_s - a_p + a_p| \leqslant \varepsilon/2 + |a_p|$. Hence, $|a_p| \geqslant \varepsilon/2$. We now define a sequence $\beta \equiv (b_n \mid n \in \omega)$, supposing $b_n \equiv 1$ for every $n < m$ and $b_n \equiv 1/a_n$ for every $n \geqslant m$. If $p, q \geqslant m$, then we have $|b_p - b_q| = |a_p - a_q|/|a_p|\,|a_q| < 4|a_p - a_q|/\varepsilon^2$.

Take any rational number $\delta > 0$, and for it take a number $l \in I(\alpha, \varepsilon^2\delta/4)$. Then, for every $p, q \geqslant \mathrm{gr}(m, l)$, we have $|b_p - b_q| < \delta$. This means that $y \equiv \bar{\beta} \in \mathbb{R}$. From $b_p a_p - 1 = 0$ for every $p \geqslant m$, we infer that $xy = \hat{1}$.

Let $xz = \hat{1}$ and $\gamma \equiv (c_n \mid n \in \omega) \in z$. Then, $x(y - z) = \hat{0}$ implies $\sigma \equiv (a_n(b_n - c_n) \mid n \in \omega) \in \mathcal{N}$. Take any rational number, $\zeta > 0$ and a number $k \in N(\sigma, \zeta\varepsilon/2)$. Then, for $p \geqslant \mathrm{gr}(m, k)$ we have $|b_p - c_p| = |a_p(b_p - c_p)|/|a_p| < \zeta$. This means that $(v_n - c_n \mid n \in \omega) \in \mathcal{N}$, i. e. $\beta\theta\gamma$. Thus, $y = z$. □

The number y from Proposition 2 is called the *inverse number to the number x* and is denoted by $1/x$ or by x^{-1}. It is clear that $(x^{-1})^{-1} = x$. The unity and inverse elements are connected by the equality $xx^{-1} = x^{-1}x = \hat{1}$. Further along with xy^{-1} we shall write also x/y; this number is called the *quotient of the numbers x and y*.

Corollary 1. *Let $x \in \mathbb{R}$, $x \neq \hat{0}$, $\alpha \equiv (a_n \mid n \in \omega) \in x$, $k \in \omega$, $a \in \mathbb{Q}$, and $|a_p| \geqslant a > 0$ in $\mathbb{Q}$ for every $p \geqslant k$. Let $\beta \equiv (b_n \in \mathbb{Q} \mid n \in \omega)$ be a sequence such that $b_p = 1/a_p$ for every $p \geqslant k$. Then, $\beta \in x^{-1}$.*

Proof. Take any rational number $\varepsilon > 0$ and a number $l \in I(\alpha, \varepsilon a^2)$. Then, for $p, q \geqslant \mathrm{gr}(k, l)$ we have $|b_p - b_q| = |a_q - a_p|/(|a_p|\,|a_q|) < \varepsilon$. This means that the sequence β is inner convergent. Thus, we can consider the real number $y \equiv \theta\beta$. Since $a_p b_p - 1 = 0$ for $p \geqslant k$, we infer that $xy \equiv \theta(a_n b_n \mid n \in \omega) = \theta\alpha_1 = \hat{1}$. By Proposition 2, $y = x^{-1}$. □

Rising to an integer degree

Let $x \equiv \bar{\alpha} \in \mathbb{R}$, $\alpha \equiv (a_n \mid n \in \omega)$, $y \equiv \hat{k}$, and $k \in \mathbb{Z}$. Define the *degree x^y of the number x with the exponent y* setting $x^y \equiv \theta(a^k \mid n \in \omega)$ if $k \in \mathbb{Z}_+$ and $x^y \equiv (x^{-1})^{-y}$ if $k \in \mathbb{Z}_- \setminus \{0\}$ and $x \neq \hat{0}$.

It is clear that $e(a^k) = (ea)^{ek}$ for every $a \in \mathbb{Q}$ and $k \in \mathbb{Z}$.

Lemma 7. *Let $k \in \mathbb{N}$ and $x \in \mathbb{R}$. Then, $x^{\hat{k}} = P(x_i \mid i \in I)$ for every simple collection $(x_i \in \mathbb{R} \mid i \in I)$ such that $x_i \equiv x$ for every $i \in I$ and* card $I = k$.

Proof. Let $x \equiv \theta(a_n \mid n \in \omega)$. Consider the numbers $a_{ni} \equiv a_n$ for $i \in I$. Then, $x_i \equiv x = \theta(a_{ni} \mid n \in \omega)$ in virtue of Lemma 9 (1.4.2) implies $x^{\hat{k}} \equiv \theta(a_n^k \mid n \in \omega) = \theta(P(a_{ni} \mid i \in I) \mid n \in \omega) \equiv P(x_i \mid i \in I)$. □

Proposition 3. *Let $(x_i \in \mathbb{R}\setminus\{\hat{0}\} \mid i \in I)$ and $(y_j \in e[\mathbb{Z}] \mid j \in J)$ be simple finite collections, $x \in \mathbb{R} \setminus \{\hat{0}\}$, and $y, z \in e[\mathbb{Z}]$. Then:*
1) $x^{\hat{0}} = \hat{1}$, $\hat{0}^{\hat{0}} = \hat{1}$, $x^{\hat{1}} = x$, *and* $\hat{1}^y = \hat{1}$;
2) $\hat{0}^y = \hat{0}$ *for* $y \in e[\mathbb{N}]$;
3) $x^{\Sigma(y_j \mid j \in J)} = P(x^{y_j} \mid j \in J)$;
4) $(P(x_i \mid i \in I))^y = P(x_i^y \mid i \in I)$;
5) $x^{yz} = (x^y)^z$.

All the assertions are direct consequences of definitions and the corresponding assertions of Proposition 1 (1.4.2).

Further in the book, we shall identify real numbers $\hat{a} \in e[\mathbb{Z}]$ with the corresponding rational numbers $a \in \mathbb{Q}$. Real numbers from $\mathbb{R} \setminus \mathbb{Q}$ are called *irrational*.

Basic order properties of the real line

Consider on $\mathbb{R}$ the binary relation ϑ such that $(x, y) \in \vartheta$ iff there are sequences $(a_n \mid n \in \omega) \in x$ and $(b_n \mid n \in \omega) \in y$ and a natural number m such that $a_p \leqslant b_p$ for every $p \geqslant m$.

Lemma 8. *The relation ϑ is a linear order in $\mathbb{R}$.*

Proof. It is clear that this relation is reflexive. Let $(x, y) \in \vartheta$ and $(y, x) \in \vartheta$. Then, there are $(a_n \mid n \in \omega)$, $(a'_n \mid n \in \omega) \in x$, $(b_n \mid n \in \omega)$, $(b'_n \mid n \in \omega) \in y$, and $m, m' \in \omega$ such that $a_p \leqslant b_p$ for $p \geqslant m$ and $b'_p \leqslant a'_p$ for $p \geqslant m'$. Take any $\varepsilon \in \mathbb{Q}_+ \setminus \{0\}$. By definition there are $k, l \in \omega$ such that $|a_p - a'_p| < \varepsilon$ for $p \geqslant k$ and $|b_p - b'_p| < \varepsilon$ for $p \geqslant l$. Therefore, for every $p \geqslant \mathrm{gr}(k, l, m, m')$, we have $a_p - b'_p \leqslant b_p - b'_p < \varepsilon$ and $a_p - b'_p \geqslant a_p - a'_p > -\varepsilon$, where $|a_p - b'_p| < \varepsilon$. This means that $x = y$, i. e. ϑ is antisymmetric.

Let $(x, y) \in \vartheta$ and $(y, z) \in \vartheta$. Then, there are $(a_n \mid n \in \omega) \in x$, $(b_n \mid n \in \omega)$, $(b'_n \mid n \in \omega) \in y$, $\gamma \equiv (c_n \mid n \in \omega) \in z$, and $m, m' \in \omega$ such that $a_p \leqslant b_p$ for $p \geqslant m$ and $b'_p \leqslant c_p$ for $p \geqslant m'$. Therefore, for every $p \geqslant \mathrm{gr}(m, m')$ we have $a_p \leqslant b_p = (b_p - b'_p) + b'_p \leqslant (b_p - b'_p) + c_p$. Consider the sequence $\gamma' \equiv (c'_n \mid n \in \omega)$ such that $c'_n \equiv c_n + (b_n - b'_n)$. By Lemma 3 γ' is inner convergent. Now, from $c'_n - c_n = b_n - b'_n$ we infer that $\gamma' \in z$. This means that $(x, z) \in \vartheta$. Thus, ϑ is transitive.

Let $x \equiv \vartheta\alpha$, $y \equiv \vartheta\beta$, $\alpha \equiv (a_n \mid n \in \omega)$, $\beta \equiv (b_n \mid n \in \omega)$, and $x \neq y$. Consider the sequence $\gamma \equiv (c_n \mid n \in \omega)$, where $c_n \equiv a_n - b_n$. Since $\gamma \notin \mathcal{N}$, there exists a rational number $\varepsilon > 0$ such that for every natural number r, there is some natural number $s \geqslant r$ for which $|c_s| \geqslant \varepsilon$. Take $m \in I(\gamma, \varepsilon/2)$ and $s \geqslant m$ such that $|c_s| \geqslant \varepsilon$. Then, for any $p \geqslant m$, we have $\varepsilon \leqslant |c_s| = |c_s - c_p + c_p| \leqslant \varepsilon/2 + |c_p|$. Hence, $|c_p| \geqslant \varepsilon/2$. If $c_{m+1} \geqslant 0$, then for every $p \geqslant m$ we have $a_p - b_p = (c_p - c_{m+1}) + |c_{m+1}| > -\varepsilon/2 + \varepsilon = \varepsilon/2$, i. e. $a_p \geqslant b_p$. In this case, we infer that $y\vartheta x$. If $c_{m+1} < 0$, then we have $a_p - b_p = (c_p - c_{m+1}) - |c_{m+1}| < \varepsilon/2 - \varepsilon = -\varepsilon/2$, i. e. $a_p \leqslant b_p$. In this case, we infer that $x\vartheta y$. This means that ϑ is a linear order. □

Further, along with $(x, y) \in \vartheta$ we shall write also $x \leqslant y$.

Lemma 9. *Let $x \equiv \theta(a_n \mid n \in \omega) \in \mathbb{R}$. Then, $x \neq 0$ iff there are $a \in \mathbb{Q}_+ \setminus \{0\}$ and $m \in \omega$ such that either $a_p > a$ or $a_p < -a$ for every $p \geqslant m$.*

Proof. Let $x \neq 0$. Then, there exists a rational number $\varepsilon > 0$ such that for every natural number r, there is some natural number $s \geqslant r$ for which $|a_s| \geqslant \varepsilon$. Take $k \in I(\varepsilon/2)$ and $s \geqslant k$ such that $|a_s| \geqslant \varepsilon$. Then, for any $p \geqslant k$, we have $\varepsilon \leqslant |a_s| = |a_s - a_p + a_p| \leqslant \varepsilon/2 + |a_p|$. Hence, $|a_p| \geqslant \varepsilon/2$. Take $l \in I(\varepsilon/4)$ and consider $m \equiv \mathrm{gr}(k, l)$. If $a_{m+1} \geqslant 0$, then for every

$p \geqslant m$, we have $a_p = (a_p - a_{m+1}) + |a_{m+1}| > -\varepsilon/4 + \varepsilon/2 = \varepsilon/4$. If $a_{m+1} < 0$, then for every $p \geqslant m$ we have $a_p = (a_p - a_{m+1}) - |a_{m+1}| < \varepsilon/4 - \varepsilon/2 = -\varepsilon/4$. Now, take $a \equiv \varepsilon/4$.

Conversely, let either $a_p > a$ or $a_p < -a$ for $p \geqslant m$. Then, $|a_p| > a$ for $p \geqslant m$. Thus, $x \neq 0$. □

Lemma 10. *Let $x \equiv \theta(a_n \mid n \in \omega)$ and $y \equiv \theta(b_n \mid n \in \omega)$ be real numbers. Then, $x < y$ iff there are $a \in \mathbb{Q}_+ \setminus \{0\}$ and $m \in \omega$ such that $a_p + a < b_p$ for every $p \geqslant m$.*

Proof. Consider the sequence $\gamma \equiv (c_n \mid n \in \omega)$, where $c_n \equiv b_n - a_n$. Since $\gamma \notin \mathcal{N}$, there exists a rational number $\varepsilon > 0$ such that for every natural number r, there is some natural number $s \geqslant r$ for which $|c_s| \geqslant 2\varepsilon$. Take $k \in I(\gamma, \varepsilon)$ and $s \geqslant k$ such that $|c_s| \geqslant 2\varepsilon$. Then, for any $p \geqslant k$, we have $2\varepsilon \leqslant |c_s| = |c_s - c_p + c_p| \leqslant \varepsilon + |c_p|$. Hence, $|c_p| \geqslant \varepsilon$. By condition $x < y$, there are $(a'_n \mid n \in \omega) \in x$ and $(b'_n \mid n \in \omega) \in y$, and $l \in \omega$ such that $a'_p \geqslant b'_p$ for $p \geqslant l$. Take some $u \in I((a_n - a'_n \mid n \in \omega), \varepsilon/2)$ and $v \in I((b_n - b'_n \mid n \in \omega), \varepsilon/2)$, and consider the number $m \equiv \mathrm{gr}(k, l, u, v)$. Then, for every $p \geqslant m$, we have $c_p = (b_p - b'_p) + b'_p + (a'_p - a_p) - a'_p > -\varepsilon/2 - \varepsilon/2 = -\varepsilon$ and simultaneously $|c_p| \geqslant \varepsilon$, where $c_p \geqslant \varepsilon > \varepsilon/2$. Take the number $a \equiv \varepsilon/2$. Then, $a_p + a < b_p$.

Conversely, let $a_p + a < b_p$ for some a and m and every $p \geqslant m$. Then, $x \leqslant y$. From $b_p - a_p > a > 0$, we infer that $x < y$. □

Corollary 1. *Let $x \in \mathbb{R}$ and $x > 0$. Then, $x^{-1} > 0$.*

Proof. Let $x \equiv \theta(a_n \mid n \in \omega)$. By Lemma 10, there are a and m such that $a < a_p$ for every $p \geqslant m$. Besides, by Lemma 1, $a_n < b$ for some $b \in \mathbb{Q}$. Therefore, by Corollary 5 to Proposition 2 (1.4.2) $a_p^{-1} > b^{-1} > 0$. Consider the sequence $\beta \equiv (b_n \in \mathbb{Q} \mid n \in \omega)$ such that $b_p \equiv 1$ for $p < m$ and $b_p \equiv a_p^{-1}$ for $p \geqslant m$. By Corollary 1 to Proposition 2 $\beta \in x^{-1}$. From the condition $0 + b^{-1} < b_p$ for $p \geqslant m$ by Lemma 10, we infer that $x^{-1} > 0$. □

Consider the sets $\mathbb{R}_+ \equiv \{x \vee 0 \mid x \in \mathbb{R}\} = \{x \in \mathbb{R} \mid x \geqslant 0\} \equiv \mathbb{R}_0$ and $\mathbb{R}_- \equiv \{x \wedge 0 \mid x \in \mathbb{R}\} = \{x \in \mathbb{R} \mid x \leqslant 0\}$ (see 1.1.15). Numbers from $\mathbb{R}_+$ [$\mathbb{R}_+ \setminus \{0\}$] are called *positive* [*strictly positive*], and numbers from $\mathbb{R}_-$ [$\mathbb{R}_- \setminus \{0\}$] are called *negative* [*strictly negative*] (*with respect to the neutral element* 0). It follows from Lemma 8 that $\mathbb{R} = \mathbb{R}_- \cup \mathbb{R}_+$ and $\mathbb{R}_- \cap \mathbb{R}_+ = \{0\}$.

Lemma 11. *Let $a, b \in \mathbb{Q}$. Then, $a < b$ in $\mathbb{Q}$ iff $ea < eb$ in $\mathbb{R}$.*

Proof. Let $a < b$. Then, $a + c < b$ for $c \equiv (b - a)/2 \in \mathbb{Q}$. Therefore, by Lemma 10 $ea < eb$. Conversely, let $ea < eb$ and suppose that $a > b$. Then, there are $(a_n \mid n \in \omega) \in ea$, $(b_n \mid n \in \omega) \in eb$, and $m \in \omega$ such that $a_p \leqslant b_p$ for $p \geqslant m$. Take $\varepsilon \equiv a - b$, $k \in N((a_n - a \mid n \in \omega), \varepsilon/2)$, and $l \in N((b_n - b \mid n \in \omega), \varepsilon/2)$. Then, for every $p \geqslant \mathrm{gr}(k, l, m)$, we have $a_p - b_p = (a_p - a) + a + (b - b_p) - b > -\varepsilon/2 - \varepsilon/2 + \varepsilon = 0$, where $a_p > b_p$. But this contradicts $a_p \leqslant b_p$. Thus, $a < b$. □

Lemma 12. *For every real number x, there exists a strictly positive rational number a such that $-a < x < a$.*

Proof. Let $x \equiv \theta(a_n \mid n \in \omega)$. By Lemma 1 $|a_n| \leqslant b < b+1$ for some rational number b. Therefore, $-(b+2)+1 < -b < a_n + 1 < b+2$. Consider $a \equiv b+2$. Then, by Lemma 10, we infer that $-a < x < a$. □

Lemma 13 (the *Archimedes principle*). *Let $x, y \in \mathbb{R}_+$ and $x > 0$. Then, there is a number $n \in \mathbb{N}$ such that $nx > y$.*

Proof. By Lemma 9, there is a rational number $a > 0$ such that $x > a > 0$. By Lemma 12, there is a rational number $b > 0$ such that $y < b$. Therefore, by Corollary 2 to Proposition 2 (1.4.2) $na > b$ for some n. As a result, we get $na > b > y$.

Let $(a_n \mid n \in \omega) \in x$. By Lemma 10, there are a strictly positive rational number c and a natural number m such that $a + c < a_p$ for every $p \geqslant m$. Then, the inequality $na + nc < na_p$ implies by Lemma 10 the inequality $na < nx$. Thus, $nx > y$. □

Corollary 1. *Let $x \in \mathbb{R}$ and $x > 0$. Then, there is a number $n \in \mathbb{N}$ such that $nx > 1$.*

Proposition 4. *Let $(x_i \in \mathbb{R} \mid i \in I)$ and $(y_i \in \mathbb{R} \mid i \in I)$ be simple finite collections, $x, y, z \in \mathbb{R}$, and $r, s \in \mathbb{Z}$. Then:*
1) *if $x_i \leqslant y_i$ for every $i \in I$, then $\sum(x_i \mid i \in I) \leqslant \sum(y_i \mid i \in I)$; if besides $x_i < y_i$ at least for one index, then $\sum(x_i \mid i \in I) < \sum(y_i \mid i \in I)$;*
2) *if $0 \leqslant x_i \leqslant y_i$ for every $i \in I$, then $P(x_i \mid i \in I) \leqslant P(y_i \mid i \in I)$; if besides $x_i < y_i$ at least for one index and $y_i > 0$ for every $i \in I$, then $P(x_i \mid i \in I) < P(y_i \mid i \in I)$;*
3) *if $x < y$, then $xz < yz$ for $z > 0$ and $xz > yz$ for $z < 0$;*
4) *if $0 \leqslant x < y$, then $x^r < y^r$ for $r > 0$ and $x^r > y^r$ for $r < 0$ and $x > 0$;*
5) *if $r < s$, then $x^r < x^s$ for $x > 1$ and $x^r > x^s$ for $0 < x < 1$.*

Proof. Denote $(x_i \mid i \in I)$ by π and $(y_i \mid i \in I)$ by $\varkappa$.

1. By definition, there are $(a_{in} \mid n \in \omega) \in x_i$, $(b_{in} \mid n \in \omega) \in y_i$, and $m_i \in \omega$ such that $a_{ip} \leqslant b_{ip}$ for every $p \geqslant m_i$. By assertion 1 of Proposition 2 (1.4.2), we get $a_p \equiv \sum(a_{ip} \mid i \in I) \leqslant \sum(b_{ip} \mid i \in I) \equiv b_p$ for every $p \geqslant m \equiv \mathrm{gr}(m_i \mid i \in I)$. This means that $\sum \pi \leqslant \sum \varkappa$. Let now $x_j < y_j$ for some $j \in I$. By Lemma 10, there are $a \in \mathbb{Q}_+ \setminus \{0\}$ and $n \in \omega$ such that $a_{jp} + a < b_{jp}$ for every $p \geqslant n$. Consider the collection $\rho \equiv (a'_{ip} \in \mathbb{Q} \mid i \in I)$ such that $a'_{jp} \equiv a_{jp} + a$ and $a'_{ip} \equiv a_{ip}$ for every $i \neq j$. Then, $\sum(a_{ip} \mid i \in I) + a = \sum \rho < \sum(b_{ip} \mid i \in I)$ for every $p \geqslant \mathrm{gr}(m, n)$.

By Lemma 10, we conclude that $\sum \pi < \sum \varkappa$.

2. We shall use the notations from 1. By assertion 2 of Proposition 2 (1.4.2), we get $a_p \equiv P(a_{ip} \mid i \in I) \leqslant P(b_{ip} \mid i \in I) \equiv b_p$ for every $p \geqslant m \equiv \mathrm{gr}\{m_i \mid i \in I\}$. This means that $P\pi \leqslant P\varkappa$. Now, let $x_j < y_j$ for some j and $y_i > 0$ for every i. Then, as above $a_{jp} + a < b_{jp}$ for every $p \geqslant n$. By the same reason, there are $b_i \in \mathbb{Q}_+$ and $n_i \in \omega$ such that

$0 < b_i < b_{ip}$ for every $p \geqslant n_i$. Therefore, in $\mathbb{Q}$, we have $0 < P(b_i \mid i \in I) < P(b_{ip} \mid i \in I)$ for every $p \geqslant \mathrm{gr}(n_i \mid i \in I)$. Consequently, by Lemma 10 $0 < P\varkappa$. If $x_k = 0$ for some k, then $P\pi = 0 < P\varkappa$. Therefore, we now assume that $x_i > 0$ for every i. Then, by Lemma 10, there are $l_i \in \omega$ such that $a_{ip} > 0$ for every i and every $p \geqslant l_i$. By assertion 2 of Proposition 2 (1.4.2), we get $b \equiv aP(a_{ip} \mid i \in I \setminus \{j\}) > 0$. Consider the number $l \equiv \mathrm{gr}(l_i \mid i \in I)$.

Since $a_p + b \leqslant P(a_{ip} \mid i \in I) + (b_{jp} - a_{jp})P(a_{ip} \mid i \in I \setminus \{j\}) = b_{jp}P(a_{ip} \mid i \in I \setminus \{j\}) \leqslant b_p$ for every $p \geqslant \mathrm{gr}(l, m, n)$, we conclude by Lemma 10 that $P\pi < P\varkappa$.

3. Let $(a_n \mid n \in \omega) \in x$, $(b_n \mid n \in \omega) \in y$, and $(c_n \mid n \in \omega) \in z$. By Lemma 10, there are $a \in \mathbb{Q}_+ \setminus \{0\}$ and $k \in \omega$ such that $a_p + a < b_p$ for $p \geqslant k$. At first, assume that $z > 0$. Then, by the same reason, there are b and l such that $0 < b < c_p$ for $p \geqslant l$. Using assertions 2 and 3 of Proposition 2 (1.4.2), we get $a_p c_p + ab \leqslant (a_p + a)c_p < b_p c_p$ for every $p \geqslant \mathrm{gr}(k, l)$. Since $ab > 0$, we infer by Lemma 10 that $xz < yz$.

Now, assume that $z < 0$. Then, there are c and m such that $c_p + c < 0$ for every $p \geqslant m$. Therefore, $b_p c_p + ac < b_p c_p + (b_p - a_p)(-c_p) = a_p c_p$ for every $p \geqslant \mathrm{gr}(k, m)$. By Lemma 10, this implies $yz < xz$.

4. Let $(a_n \mid n \in \omega) \in x$, $(b_n \mid n \in \omega) \in y$, and $r \equiv em$. As above $a_p + a < b_p$ for $p \geqslant k$. At first, assume that $r > 0$. By Lemma 11 $m > 0$. If $x = 0$, then by Proposition 3 $x^r = 0$. In this case, we can suppose that $a_n = 0$ for every $n \in \omega$. Therefore, by assertion 4 of Proposition 2 (1.4.2), $0 < a^m < b_p^m$ for $p \geqslant k$. By Lemma 10, we infer that $x^r = 0 < y^r$.

Now, assume that $x > 0$. Then, by Lemma 10, there are b and l such that $0 < b < a_p$ for $p \geqslant l$. Therefore, by Lemma 14 (1.4.2), $a_p^m + a^m \leqslant (a_p + a)^m < b_p^m$ for $p \geqslant \mathrm{gr}(k, l)$. Since $a^m > 0$, we infer by Lemma 10 that $x^r < y^r$.

Finally, assume that $r < 0$ and $x > 0$. Then, $m < 0$. By Corollary 1 to Lemma 10 $x^{-1} > 0$ and $y^{-1} > 0$. By Lemma 1, $|a_n| \leqslant u$ and $|b_n| \leqslant v$ for some rational numbers u and v and for every n. Consequently, $1/a_p - 1/b_p = (b_p - a_p)/a_p b_p > a/uv > 0$ for $p \geqslant \mathrm{gr}(k, l)$. By Corollary 1 to Proposition 2 and Lemma 10, we get $x^{-1} - y^{-1} > 0$. Adding this inequality to the inequality $y^{-1} > y^{-1}$, we get by assertion 1 that $x^{-1} > y^{-1} > 0$. Having $-r > 0$ and using the inequality prove above, we get $x^r \equiv (x^{-1})^{-r} > (y^{-1})^{-r} \equiv y^r$.

5. Let $(a_n \mid n \in \omega) \in x$, $r \equiv eu$, $s \equiv ev$, and $r < s$. By Lemma 11, $w \equiv v - u > 0$.

At first, consider the case $x > 1$. By Lemma 10, there are a rational number $a > 0$ and a natural number k such that $1 + a < a_p$ for every $p \geqslant k$. Let $u \geqslant 0$. By Lemma 14 (1.4.2), $a_p^u > (1+a)^u \geqslant 1 + a^u$ and $a_p^w > 1 + a^w$. Therefore, $a_p^v - a_p^u = a_p^u(a_p^w - 1) \geqslant (1 + a^u)((1 + a)^w - 1) \geqslant (1 + a^u)(1 + a^w - 1) = (1 + a^u)a^w > 0$ for $p \geqslant k$. By Lemma 10 $x^r \equiv \theta(a_n^u \mid n \in \omega) < \theta(a_n^v \mid n \in \omega) \equiv x^s$.

Let $u < 0 \leqslant v$. By Lemma 1 $|a_n| \leqslant b$ for some rational number b. Consider the sequence $\beta \equiv (b_n \mid n \in \omega)$ such that $b_p \equiv 1$ for $p < k$ and $b_p \equiv 1/a_p$ for $p \geqslant k$. By Corollary 1 to Proposition 2 $\beta \in x^{-1}$. Therefore, $x^r \equiv (x^{-1})^{-r} \equiv (\theta\beta)^{-r} \equiv \theta(b_n^{-u} \mid n \in \omega)$. Besides $x^s \equiv \theta(a_n^v \mid n \in \omega)$. Since $a_p^v - b_p^{-u} = a_p^v - a_p^u = a_p^u(a_p^w - 1) \geqslant ((1+a)^w - 1)/a_p^{-u} \geqslant (1 + a^w - 1)/b^{-u} = a^w/b^{-u} > 0$ for $p \geqslant k$, we infer by Lemma 10 that $x^r < x^s$.

Finally, let $v < 0$. Then, as above, $x^s = \theta(b_n^{-v} \mid n \in \omega)$. Since as above, $b_p^{-v} - b_p^{-u} = a_p^v - a_p^u \geqslant a^w/b^{-u} > 0$ for $p \geqslant k$, we infer by Lemma 10 that $x^r < x^s$.

Now, consider the case $0 < x < 1$. By Lemma 10, there are rational numbers c and d and a natural number l such that $0 < c < a_p < a_p + d < 1$ for every $p \geqslant l$. It is clear that $d < 1$. Let $u \geqslant 0$. By assertion 4 of Proposition 2 (1.4.2), $0 < 1 - d < 1$ implies $0 < (1-d)^w < 1^w = 1$. Therefore, $a_p^u - a_p^v = a_p^u(1 - a_p^w) > c^u(1 - (1-d)^w) > 0$ for $p \geqslant l$. By Lemma 10, this implies $x^s \equiv \theta(a_n^v \mid n \in \omega) < \theta(a_n^u \mid n \in \omega) \equiv x^r$.

Let $u < 0 \leqslant v$. Consider the number b and the sequence β as above. Since $b_p^{-u} - a_p^v = a_p^u - a_p^v = a_p^u(1-a_p^w) \geqslant (1-(1-d)^w)/a_p^{-u} \geqslant (1-(1-d)^w)/b^{-u} > 0$ for $p \geqslant l$, we infer by Lemma 10 that $x^s \equiv \theta(a_n^v \mid n \in \omega) < \theta(b_n^{-u} \mid n \in \omega) = x^r$.

Finally, let $v < 0$. Then, as above, $b_p^{-u} - b_p^{-v} \geqslant (1-(1-d)^w)/b^{-u} > 0$ for $p \geqslant l$ implies $x^s = \theta(b_n^{-v} \mid n \in \omega) < \theta(b_n^{-u} \mid n \in \omega) = x^r$. □

Corollary 1. *Let $x, y, z \in \mathbb{R}$. Then, $x = y$ iff $x + z = y + z$. When $z \neq 0$, then $x = y$ iff $xz = yz$.*

The proof is the same as the proof of Corollary 1 to Proposition 3 (1.4.1).

Corollary 2. *Let $x, y \in \mathbb{R}$ and $xy = 0$. Then, either $x = 0$ or $y = 0$.*

The proof is the same as the proof of Corollary 3 to Proposition 3 (1.4.1).

Corollary 3. *Let $x, y \in \mathbb{R}$ and $0 < x < y$. Then, $0 < y^{-1} < x^{-1}$.*

Proof. By Corollary 1 to Lemma 10 $x^{-1} > 0$ and $y^{-1} > 0$. Therefore, by assertion 2 of Proposition 4 $y^{-1} = y^{-1}x^{-1}x < y^{-1}x^{-1}y = x^{-1}$. □

Interval density of rational numbers in $\mathbb{R}$

The following lemma shows the *interval density* of rational numbers in the set of all real numbers.

Lemma 14. *Let $x, y \in \mathbb{R}$ and $x < y$. Then, there is a rational number r such that $x < r < y$.*

Proof. At first, assume that $y > 0$. By Corollary 1 to Lemma 13, there is a number $n \in \mathbb{N}$ such that $a \equiv y - x > 1/n$. Similarly, by Lemma 13, there is a number $k \in \mathbb{N}$ such that $k(1/n) \geqslant y$. Consider the number $m \equiv \mathrm{sm}(k \in \omega \mid k/n \geqslant y)$. Then, $(m-1)/n < y$. Adding the inequalities $y \leqslant m/n$ and $-a < -1/n$, we get $x = y - a < (m-1)/n$. Denote $(m-1)/n$ by r.

Now, assume that $y \leqslant 0$. Then, as above for $0 \leqslant -y < -x$, there is a rational number s such that $-y < s < -x$. Hence, $x < -s < y$. Denote $-s$ by r. □

Corollary 1. *Let $x, y \in \mathbb{R}$ and $x < y$. Then, there are rational numbers r and s such that $x < r < s < y$.*

Lemma 15. *Let $x, y \in \mathbb{R}$ and $x < y$. Then, there are sequences $\alpha \equiv (r_n \in \mathbb{Q} \mid n \in \omega) \uparrow$ and $\beta \equiv (s_n \in \mathbb{Q} \mid n \in \omega) \downarrow$ such that $x < r_p < r_{p+1} < s_{q+1} < s_q < y$ for every $p, q \in \omega$.*

Proof. By Corollary 1 to Lemma 14, there are r_0 and s_0 such that $x < r_0 < s_0 < y$. Consider the class $B \equiv \mathbb{Q} \times \mathbb{Q}$, the element $b_0 \equiv (r_0, s_0)$, and a choice mapping $p\colon \mathcal{P}(B) \setminus \{\varnothing\} \to B$ from the axiom of choice in 1.1.12. Define a mapping $V\colon B \times \omega \to B$ setting $V((r, s), n) \equiv p\{(r', s') \in B \mid r < r' < s' < r\}$. This definition is correct since the set $\{(r', s') \in B \mid r < r' < s' < r\}$ is non-empty by Corollary 1 to Lemma 14.

Now, by Theorem 1 (1.2.7), there is a unique mapping $u\colon \omega \to B$ such that $u(0) = b_0$ and $u(n+1) = V(u(n), n)$. Consider the projections pr_0 and pr_1 from B onto $\mathbb{Q}$ such that $(\mathrm{pr}_0(b), \mathrm{pr}_1(b)) = b$. Define sequences $\alpha \equiv (r_n \mid n \in \omega)$ and $\beta \equiv (s_n \mid n \in \omega)$ setting $r_n \equiv \mathrm{pr}_0(u(n))$ and $s_n \equiv \mathrm{pr}_1(u(n))$. Then, $(r_{n+1}, s_{n+1}) = u(n+1) = V(u(n), n) = V((r_n, s_n), n)$ implies $r_n < r_{n+1} < s_{n+1} < s_n$. Thus, $\alpha \uparrow$ and $\beta \downarrow$.

Take any $p, q \in \mathbb{N}$. If $p = q$, then $x < r_0 < r_p < s_q < s_0 < y$. If $p < q$, then $x < r_0 < r_p < r_q < s_q < s_0 < y$. If $p > q$, then $x < r_0 < r_p < s < s_q < s_0 < y$. □

Lemma 16. *Let $x, y \in \mathbb{R}$ and $x < y$. Then, there are sequences $\alpha \equiv (r_n \in \mathbb{Q} \mid n \in \omega) \downarrow$ and $\beta \equiv (s_n \in \mathbb{Q} \mid n \in \omega) \uparrow$ such that $x < r_{p+1} < r_p < s_q < s_{q+1} < y$ for every $p, q \in \omega$.*

The proof is similar to the proof of Lemma 15.

Proposition 5 (the *Bernoulli inequality*). *Let $x \in \mathbb{R}$, $n \in \mathbb{N}$, and $x \geqslant -1$. Then, $(1+x)^n \geqslant 1 + nx$.*

Proof. Consider the set $N \subset \omega$ of all natural numbers n such that $(1+x)^{n+1} \geqslant 1+(n+1)x$. It is clear that $0 \in N$.

Suppose that $n \in N$. Since $1 + x \geqslant 0$, we infer by virtue of assertion 2 of Proposition 4 that $(1+x)^{n+2} = (1+x)^{n+1}(1+x) \geqslant (1+(n+1)x)(1+x) \geqslant 1+(n+2)x$. This means that $n+1 \in N$. By the principe of natural induction (Theorem 1 (1.2.6)) $N = \omega$. □

Corollary 1. *Let $x, y \in \mathbb{R}$, $y \geqslant 0$, and $x > 1$. Then, there is a number $n \in \mathbb{N}$ such that $x^n > y$.*

Proof. If $y \leqslant 1$, then by assertion 4 of Proposition 4 $x^n > 1 \geqslant y$ for every n. Thus, further we can presuppose that $y > 1$. Consider $z \equiv x - 1 > 0$. Then, by Proposition 5, $x^n = (1+z)^n \geqslant 1 + nz$ for every n. By Lemma 13, there is n such that $y - 1 < nz$. For this n, we infer that $x^n > y$. □

Corollary 2. *Let $x, y \in \mathbb{R}$, $y > 0$, and $0 < x < 1$. Then, there is a number $n \in \mathbb{N}$ such that $x^n < y$.*

Proof. Consider $z \equiv 1/x > 1$. By Corollary 1 to Lemma 13, there is $m \in \mathbb{N}$ such that $y > 1/m$. By Corollary 1 to Proposition 5, there is n such that $z^n > m$. As a result, we get $x^n < 1/m < y$. □

Corollary 3. *Let $y, z \in \mathbb{R}_+ \setminus \{0\}$, $y \neq z$, and $n \in \mathbb{N}$. Then, $y^{n+1}/z^n \geqslant (n+1)y - nz$.*

Proof. By Proposition 5 $(y/z)^{n+1} = (1 + (y/z - 1))^{n+1} \geqslant 1 + (n+1)(y/z - 1)$. Then, using Proposition 3, Proposition 4, and Theorem 3, we get $y^{n+1}/z^n = z(y/z)^{n+1} \geqslant z(1 + (n + 1)(y/z - 1)) = z + (n+1)y - (n+1)z = (n+1)y - nz$. □

Corollary 4. *Let $x \in \mathbb{R}$, $n \in \mathbb{N}$, and $x > -1$. Then, $(1 + x)^{-n} \geqslant 1 - nx$.*

Proof. For $x = 0$ we have the equality. If $x > -1$ and $x \neq 0$, then applying Corollary 3 to $y \equiv 1$ and $z \equiv 1 + x$, we get $(1 + x)^{-n} = 1^{n+1}/(1 + x)^n \geqslant n + 1 - n(1 + x) = 1 - nx$. □

The following lemma shows the interval density of dyadic-rational numbers in $[0, 1] \subset \mathbb{R}$.

Lemma 17. *Let $x, y \in \mathbb{R}$ and $0 \leqslant x < y \leqslant 1$. Then, there are natural numbers k and n such that $x < k/2^n < y$.*

Proof. Consider the number $z \equiv y - x > 0$. By Corollary 2 to Proposition 5, there is n such that $1/2^n < z < y$. If $x < 1/2^n$, then we have the necessary inequality. Thus, further we can presuppose that $1/2^n \leqslant x$. Consider the non-empty set $K \equiv \{k \in \mathbb{N} \mid k/2^n \leqslant x\}$. By Lemma 13, there is m such that $m/2^n > x$. If $k \in K$, then $k/2^n \leqslant x < m/2^n$ implies $k < m$. Consequently, $K \subset m$. Thus, the set K is finite. Therefore, by Theorem 3 (1.2.6) K has the greatest element l. It is clear that $l/2^n \leqslant x < (l+1)/2^n = l/2^n + 1/2^n \leqslant x + 1/2^n = y - z + 1/2^n < y$. □

Modulus of a real number

For every real number x, we can define correctly its *modulus* $|x| \in \mathbb{R}_+$ setting $|x| \equiv x$ if $x \in \mathbb{R}_+$ and $|x| \equiv -x$ if $x \in \mathbb{R}_-$. It is clear that $|x| = x \vee (-x)$.

Lemma 18. *Let $\alpha \equiv (a_n \mid n \in \omega) \in x \in \mathbb{R}$. Then, $(|a_n| \mid n \in \omega) \in |x|$.*

Proof. Let $x = 0$. Then, α is null. Therefore, the sequence $\beta \equiv (|a_n| \mid n \in \omega)$ is also null. Thus, $\beta \in 0 = |x|$.

Now, let $x > 0$. Then, by Lemma 10, there are a and m such that $0 < a < a_p$ for $p \geqslant m$. Since $|a_p| - a_p = a_p - a_p = 0$ for $p \geqslant m$, we infer that $\beta \in x = |x|$.

Finally, set $x < 0$. Then, by the same reason, there are a and m such that $a_p + a < 0$ for $p \geqslant m$. Since $|a_p| - (-a_p) = -a_p + a_p = 0$ for $p \geqslant m$, we infer that $\beta \in -x = |x|$. □

Proposition 6. *Let $(z_i \in \mathbb{R} \mid i \in I)$ be a simple finite collection, $x, y \in \mathbb{R}$, and $z \in e[\mathbb{Z}]$. Then:*
1) $|x| = |-x|$, $x \leqslant |x|$, *and* $-x \leqslant |x|$;
2) $|P(z_i \mid i \in I)| = P(|z_i| \mid i \in I)$; *in particular,* $|xy| = |x|\,|y|$;
3) $|x^z| = |x|^z$ *if* $x \neq 0$;
4) *if* $y > 0$, *then* $|x| \leqslant y$ *is equivalent to* $-y \leqslant x \leqslant y$, *and* $|x| < y$ *is equivalent to* $-y < x < y$;
5) $|\sum(z_i \mid i \in I)| \leqslant \sum(|z_i| \mid i \in I)$; *in particular,* $|x + y| \leqslant |x| + |y|$;
6) $||x| - |y|| \leqslant |x - y|$.

Proof. 1. It is clear that $x \leqslant |x|$. If $x \geqslant 0$, then $-x \leqslant 0$ implies $|x| \equiv x$ and $|-x| \equiv -(-x) = x = |x|$. If $x < 0$, then $-x > 0$ implies $|x| = |-(-x)| = |-x|$. Therefore, $-x \leqslant |-x| = |x|$.

2. Let $(c_{in} \mid n \in \omega) \in z_i$. By Lemma 18, $(|c_{in}| \mid n \in \omega) \in |z_i|$. Since $u \equiv P(z_i \mid i \in I) = \theta(P(c_{in} \mid i \in I) \mid n \in \omega)$, we infer by Lemma 18 and assertion 2 of Proposition 3 (1.4.2) that $|u| = \theta(|P(c_{in} \mid i \in I)| \mid n \in \omega) = \theta(P(|c_{in}| \mid i \in I) \mid n \in \omega) = P(|z_i| \mid i \in I)$.

3. Let $\alpha \equiv (a_n \mid n \in \omega) \in x$ and $z \equiv ek$. Let $k \geqslant 0$. By Lemma 18, $\mu \equiv (|a_n| \mid n \in \omega) \in |x|$. Applying assertion 3 of Proposition 3 (1.4.2), we get $|x^z| = |\theta(a_n^k \mid n \in \omega)| = \theta(|a_n^k| \mid n \in \omega) = \theta(|a_n|^k \mid n \in \omega) = (\theta\mu)^z = |x|^z$.

Now, let $k < 0$. By Lemma 9, there are a rational number $a > 0$ and a natural number m such that $|a_p| > a$ for $p \geqslant m$. Consider the sequence $\beta \equiv (b_n \mid n \in \omega)$ such that $b_p \equiv 1$ for $p < m$ and $b_p \equiv 1/a_p$ for $p \geqslant m$. By Corollary 1 to Proposition 2 $x^{-1} = \theta\beta$. Therefore, by Lemma 18, $|x^{-1}| = \theta(|b_n| \mid n \in \omega)$. Since $|b_p| = 1/|a_p|$ for $p \geqslant m$, we infer as above that $|x|^{-1} = |x^{-1}|$. Finally, using for $-z$, the property proven above, we get $|x^z| \equiv |(x^{-1})^{-z}| = |x^{-1}|^{-z} = (|x|^{-1})^{-z} \equiv |x|^z$.

The other assertions are checked as the corresponding assertions of Proposition 4 (1.4.1). □

Corollary 1. *Let $x, y \in \mathbb{R}$ and $y \neq 0$. Then, $|x/y| = |x|/|y|$.*

The proof is completely the same as the proof of Corollary 1 to Proposition 3 (1.4.2).

Extended real numbers

Now, we shall consider some important extension of the set of all real numbers.

A sequence $\alpha \equiv (a_n \in \mathbb{Q} \mid n \in \omega)$ will be called *uniformly upper* [*lower*] *unbounded* if for every $b \in \mathbb{Q}$, there is $n \in \omega$ such that $a_p > b$ [$a_p < b$] for all $p \geqslant n$. The sets of all uniformly upper unbounded and all uniformly lower unbounded sequences α will be denoted by $\mathcal{R}^u$ and $\mathcal{R}^l$, respectively.

Consider the set $\bar{\mathcal{R}} \equiv \mathcal{R} \cup \mathcal{R}^l \cup \mathcal{R}^u$, in which all component sets are mutually disjoint. Define on the set $\bar{\mathcal{R}}$ a binary relation $\bar{\theta} \equiv \theta \cup (\mathcal{R}^l \times \mathcal{R}^l) \cup (\mathcal{R}^u \times \mathcal{R}^u)$. In the set $\bar{\theta}$, all component sets are also mutually disjoint. It is clear that $\bar{\theta}$ is an equivalence relation.

Consider the factor-set $\overline{\mathbb{R}} \equiv \bar{\mathcal{R}}/\bar{\theta}$ consisting of equivalence classes $x \equiv \bar{\theta}\alpha \equiv \bar{\alpha}$ of all sequences $\alpha \in \bar{\mathcal{R}}$. Elements of the set $\overline{\mathbb{R}}$ are called *extended real numbers*, and the set $\overline{\mathbb{R}}$ is called the *set of all extended real numbers*. It is clear that $\mathbb{R} \subset \overline{\mathbb{R}}$.

The extended real number $\infty \equiv \bar{\theta}(n \in \mathbb{Q} \mid n \in \omega) = \mathcal{R}^u$ is called the *upper* (≡ *right, plus*) *infinity* or simply the *infinity*. The extended real number $-\infty \equiv \bar{\theta}(-n \in \mathbb{Q} \mid n \in \omega) = \mathcal{R}^l$ is called the *lower* (≡ *left, minus*) *infinity*. Note that many writers use the symbol $+\infty$ for what we write as ∞. The $+$ sign is a more nuisance and so we omit it.

Consider the sets $\overline{\mathbb{R}}_+ \equiv \mathbb{R}_+ \cup \{\infty\}$ and $\overline{\mathbb{R}}_- \equiv \mathbb{R}_- \cup \{-\infty\}$. It is clear that $\overline{\mathbb{R}} = \overline{\mathbb{R}}_- \cup \overline{\mathbb{R}}_+$ and $\overline{\mathbb{R}}_- \cap \overline{\mathbb{R}}_+ = \{0\}$.

Let $\varkappa \equiv (x_i \in \overline{\mathbb{R}} \mid i \in I)$ be a simple finite collection. The number $\sum(x_i \mid i \in I) \equiv \sum \varkappa$, called the *sum of the collection* $\varkappa$, is defined in the following cases:

1) if $x_i \in \mathbb{R}$ for all $i \in I$, then $\sum \varkappa$ is the common sum of $\varkappa$ in $\mathbb{R}$;
2) if $x_i \in \mathbb{R} \cup \{\infty\}$ for all $i \in I$ and $x_i = \infty$ at least for one index, then $\sum \varkappa \equiv \infty$;
3) if $x_i \in \mathbb{R} \cup \{-\infty\}$ for all $i \in I$ and $x_i = -\infty$ at least for one index, then $\sum \varkappa \equiv -\infty$.

In all other cases, the sum $\sum \varkappa$ is undefined.

The notations $x_0 + \cdots + x_n$, $x + x'$, $x + x' + x''$, $x + x' + x'' + x'''$, ... are defined in the manner as it was made above.

In partial cases, we have $\infty + \infty = \infty$, $-\infty + (-\infty) = -\infty$, and $\infty + x = x + \infty = \infty$ and $-\infty + x = x + (-\infty) = -\infty$ for every $x \in \mathbb{R}$. Besides, $0 + x = x + 0 = x$ for every $x \in \overline{\mathbb{R}}$.

The number $-x$, called the *opposite number to the number* $x \in \overline{\mathbb{R}}$, is defined in the following cases:

1) if $x \in \mathbb{R}$, then $-x$ is the common opposite number in $\mathbb{R}$;
2) $-(\infty) \equiv -\infty$;
3) $-(-\infty) \equiv \infty$.

Consider on $\overline{\mathbb{R}}$ the binary relation $\bar{\vartheta} \equiv \vartheta \cup (\{-\infty\} \times \mathbb{R}) \cup \{(-\infty, \infty)\} \cup (\mathbb{R} \times \{\infty\})$. It is clear that $\bar{\vartheta}$ is a linear order. Since all component sets are mutually disjoint, we infer that $\bar{\vartheta}|\mathbb{R} \times \mathbb{R} = \vartheta$, i. e. $\bar{\vartheta}$ is an extension of ϑ. Further, along with $(x, y) \in \bar{\vartheta}$ we shall write also $x \leqslant y$.

It is clear that $\overline{\mathbb{R}}_+ = \{x \in \overline{\mathbb{R}} \mid x \geqslant 0\}$ and $\overline{\mathbb{R}}_- = \{x \in \overline{\mathbb{R}} \mid x \leqslant 0\}$. Numbers from $\overline{\mathbb{R}}_+$ [$\overline{\mathbb{R}}_+ \setminus \{0\}$] are called *positive* [*strictly positive*], and numbers from $\overline{\mathbb{R}}_-$ [$\overline{\mathbb{R}}_- \setminus \{0\}$] are called *negative* [*strictly negative*].

For every extended real number x, we can define its *modulus* $|x| \in \overline{\mathbb{R}}_+$ setting $|x| \equiv x$ if $x \in \overline{\mathbb{R}}_+$ and $|x| \equiv -x$ if $x \in \overline{\mathbb{R}}_-$.

Let $\varkappa \equiv (x_i \in \overline{\mathbb{R}} \mid i \in I)$ be a simple finite collection. Consider the set $I_\varkappa^- \equiv \{i \in I \mid x_i < 0\}$. The number $P(x_i \mid i \in I) \equiv P\varkappa$, called the *product of the collection* $\varkappa$, is defined in the following cases:

1) if $x_i \in \mathbb{R}$ for all $i \in I$, then $P\varkappa$ is the common product of $\varkappa$ in $\mathbb{R}$;
2) if $x_i = 0$ at least for one index, then $P\varkappa \equiv 0$;
3) if $x_i > 0$ for all $i \in I$ and $x_i = \infty$ at least for one index, then $P\varkappa \equiv \infty$;

4) if $x_i \neq 0$ for all $i \in I$, then $P\varkappa \equiv P(|x_i| \mid i \in I)$ if card $I_\varkappa^-$ is even and $P\varkappa \equiv -P(|x_i| \mid i \in I)$ if card $I_\varkappa^-$ is odd.

The notations $x_0 \dots x_n$, xx', $xx'x''$, $xx'x''x'''$,... are defined in the same manner as it was made above.

In partial cases, we have $\infty x = x\infty = \infty$ and $(-\infty)x = x(-\infty) = -\infty$ for $x > 0$, and $\infty x = x\infty = -\infty$ and $(-\infty)x = x(-\infty) = \infty$ for $x < 0$. Besides, $1x = x1 = x$ and $0x = x0 = 0$ for every $x \in \overline{\mathbb{R}}$. The latter property is not so "natural" as others, but it is very convenient for measure theory (see Chapter 3 of the book and also [*Hewitt and Stromberg*, 1965, ch.II, 6.1]).

For such defined sums and products, we have the common properties of commutativity, associativity, and distributivity; however, with some reservations.

Further in the book, initial intervals $]\leftarrow, x[$ and $]\leftarrow, x]$ and final intervals $]x, \rightarrow[$ and $[x, \rightarrow[$ in the ordered set $(\mathbb{R}, \leqslant)$ (see 1.1.15) will be denoted also by $]-\infty, x[$, $]-\infty, x]$, $]x, \infty[$, and $[x, \infty[$, respectively.

If we consider the ordered set $(\overline{\mathbb{R}}, \leqslant)$, then we have the following useful equalities: $[-\infty, \infty] = \overline{\mathbb{R}}$, $[-\infty, \infty[= \{-\infty\} \cup \mathbb{R}$, $]-\infty, \infty[= \mathbb{R}$, and $]-\infty, \infty] = \mathbb{R} \cup \{\infty\}$.

1.4.4 The Cantor completeness of the real line

Sequences of real numbers

In this subsection, we shall consider simple sequences $s \equiv (x_n \in \mathbb{R} \mid n \in N)$ of real numbers indexed by infinite subsets N of the set of all natural numbers ω (see 1.2.6). According to 1.2.6, such sequences are called *infinite*.

As in 1.1.15 for a preordered set $(M, \leqslant)$ closed final intervals $[m, \rightarrow[\equiv \{p \in M \mid p \geqslant m\}$ with the beginnings $m \in M$ will be denoted also by M_m.

According to 1.1.15, a subset N is *cofinal to the set* ω iff for every $m \in \omega$, there is $n \in N_m$.

Lemma 1. *Let $N \subset \omega$. Then, N is infinite iff N is cofinal to ω.*

Proof. Let N is infinite. Suppose that there is $m \in \omega$ such that $N \cap \omega_m = \varnothing$. Then, $N \subset m$. Since the set m is finite, the set N is also finite. It follows from this contradiction that N is cofinal.

Conversely, let N is cofinal. Suppose that it is finite. Then, by Theorem 3 (1.2.6), it has the greatest element m. Therefore, $N \cap \omega_{m+1} = \varnothing$, but this contradicts the cofinality of N. Thus, N is infinite. □

According to 1.2.6, a sequence $t \equiv (y_n \in \mathbb{R} \mid n \in N)$ is called a *subsequence of a sequence* $s \equiv (x_m \in \mathbb{R} \mid m \in M)$ if there exists a sequence $(m_n \in M \mid n \in N)$ such that:

1) for every number $m \in M$, there exists a number $n \in N$ such that $k \in N_n$ implies $m_k \in M_m$;
2) $y_n = x_{m_n}$ for every $n \in N$.

According to 1.1.15, a sequence $s \equiv (x_n \in \mathbb{R} \mid n \in N)$ is called *bounded above* [*below*] if there is a number $b \in \mathbb{R}$ such that $x_n \leqslant b$ [$x_n \geqslant b$] for every $n \in N$. A sequence s is called *bounded* if it is bounded above and below simultaneously or equivalently if there is a number $b \in \mathbb{R}$ such that $|x_n| \leqslant b$ for every $n \in N$.

A sequence s is called *inner convergent* ($\equiv$ *fundamental*, a *Cauchy sequence*) if for every real number $\varepsilon > 0$, there is a number $n \in N$ such that $|x_p - x_q| < \varepsilon$ for all $p, q \in N$ such that $p, q \geqslant n$.

A sequence s is called *convergent to a number* $x \in \mathbb{R}$ and the number x is called a *limit of the sequence* s if for every real number $\varepsilon > 0$, there is a number $n \in N$ such that $|x_p - x| < \varepsilon$ for all $p \in N$ such that $p \geqslant n$.

For a sequence s, a number x, and a number $\varepsilon \in \mathbb{R}_+ \setminus \{0\}$, we shall consider the sets $I(s, \varepsilon) \equiv \{n \in N \mid \forall p, q \in N\ (p, q \geqslant n \Rightarrow |x_p - x_q| < \varepsilon)\}$ and $C(s, x, \varepsilon) \equiv \{n \in N \mid \forall p \in N\ (p \geqslant n \Rightarrow |x_p - x| < \varepsilon)\}$. If s is fixed, then we shall denote these sets simply by $I(\varepsilon)$ and $C(x, \varepsilon)$. If a sequence s is inner convergent, then $I(s, \varepsilon) \neq \varnothing$ for every ε. If a sequence s is convergent to a number x, then $C(s, x, \varepsilon) \neq \varnothing$ for every ε.

Lemma 2. *Let $s \equiv (s_n \in \mathbb{R} \mid n \in N)$ be an infinite sequence and $x, y \in \mathbb{R}$. Then:*
1) *if s is inner convergent, then s is bounded;*
2) *if s is convergent to x, then s is inner convergent;*
3) *if s is convergent to x and to y, then $x = y$.*

Proof. 1. It is checked completely in the same manner as the corresponding assertion in Lemma 1 (1.4.3).

2. Take any $\varepsilon > 0$ and some $m \in C(x, \varepsilon/2)$. Then, $|x_p - x_q| \leqslant |x_p - x| + |x - x_q| < \varepsilon/2 + \varepsilon/2 = \varepsilon$ for $p, q \geqslant m$.

3. Suppose that $x \neq y$. Then, for $\varepsilon \equiv |x - y| > 0$, there are $m \in C(x, \varepsilon/2)$ and $n \in C(y, \varepsilon/2)$. Therefore, $|x - y| \leqslant |x - x_p| + |x_p - y| < \varepsilon/2 + \varepsilon/2 = \varepsilon$ for $p \geqslant \mathrm{gr}(m, n)$, but this is impossible. Thus, $x = y$. □

Thus, a sequence s can have a unique limit. To denote the property that x is a limit of s, we shall write $x = \lim s$ or $x = \lim(x_n \mid n \in N)$.

Lemma 3. *Let $s \equiv (x_n \mid n \in N)$ be an infinite sequence and $x \in \mathbb{R}$. Then, the following conclusions are equivalent:*
1) $x = \lim s$;
2) $0 = \lim(x_n - x \mid n \in N)$;
3) $0 = \lim(|x_n - x| \mid n \in N)$.

Proof. The assertion follows from the equalities $|x_p - x| = |0 - (x_p - x)| = |0 - |x_p - x||$. □

Lemma 4. *Let $x \in \mathbb{R}$ and $\alpha \equiv (a_n \in \mathbb{Q} \mid n \in \omega) \in x$. Then, $x = \lim(a_n \mid n \in \omega)$.*

Proof. Take any $\varepsilon \in \mathbb{R}_+ \setminus \{0\}$. By Lemma 14 (1.4.3), there is $e \in \mathbb{Q}$ such that $0 < e < \varepsilon$. Take some $m \in I(\alpha, e)$. Then, for every $p, q \geqslant m$, we have $a_p - e < a_q < a_p + e$ in $\mathbb{Q}$. Using the definition of the order in $\mathbb{R}$, we get $\hat{a}_p - \varepsilon < \hat{a}_p - \hat{e} \leqslant x \equiv \theta\alpha \leqslant \hat{a}_p + e < \hat{a}_p + \varepsilon$ for every $p \geqslant m$. This implies $|x - \hat{a}_p| < \varepsilon$ for $p \geqslant m$. □

Lemma 5. *Let $t \equiv (x_{n_k} \mid k \in K)$ be a subsequence of an infinite sequence $s \equiv (x_n \in \mathbb{R} \mid n \in N)$, $x \in \mathbb{R}$, and $x = \lim s$. Then, the sequence t is infinite and $x = \lim t$.*

Proof. Suppose that the set K is finite. Then, by assertion 1 of Theorem 3 (1.2.6), there is $p \equiv \mathrm{gr}(n_k \mid k \in K)$. Since for every $n \in N$, there is $l \in K$ such that $n_l \geqslant n$, we infer that $n \leqslant n_l \leqslant p$ for every $n \in N$. But this is impossible because N is cofinal to ω. Thus, K is infinite.

Take any $\varepsilon > 0$ and some $m \in C(s, x, \varepsilon)$. Then, $|x_p - x| < \varepsilon$ if $m \leqslant p \in N$. By definition of m, there is $l \in K$ such that $n_k \geqslant m$ if $l \leqslant k \in K$. Therefore, $|x_{n_k} - x| < \varepsilon$ if $l \leqslant k \in K$. □

Lemma 6. *Let $s \equiv (x_n \in \mathbb{R} \mid n \in N)$ be an inner convergent sequence, $t \equiv (x_{n_k} \mid k \in K)$ be a subsequence of s, and $x = \lim t$. Then, $x = \lim s$.*

Proof. Take any $\varepsilon > 0$ and some $m \in I(s, \varepsilon/2)$ and $n \in C(t, x, \varepsilon/2)$. By Lemma 1, there is $l \in N$ such that $l > \mathrm{gr}(m, n)$. By definition of a subsequence for l, there is $k \in K$ such that $n_p \geqslant l$ if $k \leqslant p \in K$. Take any $q \in N$ such that $q \geqslant \mathrm{gr}(l, k)$. By Lemma 1 for q, there is $p \in K$ such that $p \geqslant q$. Therefore, $k \leqslant p \in K$ implies $n_p \geqslant l$. Consequently, $|x_q - x| \leqslant |x_q - x_{n_p}| + |x_{n_p} - x| < \varepsilon/2 + \varepsilon/2 = \varepsilon$. This means that $x = \lim s$. □

Lemma 7. *Let $x \in \mathbb{R}$ and N be an infinite subset in ω. Then:*
1) $\lim (x^n \mid n \in N) = 0$ *for every* $x \in [0, 1[$;
2) $\lim (x^n \mid n \in N) = \infty$ *for every* $x \in]1, \infty[$.

Proof. 1. For $x > 0$, this assertion follows from Corollary 2 to Proposition 5 (1.4.3).

2. It follows from Corollary 1 to Proposition 5 (1.4.3). □

The Cantor completeness of $\mathbb{R}$

The following theorem may be understood as the theorem about the *Cantor completeness of the real line* $(\mathbb{R}, \leqslant)$

Theorem 1. *Let an infinite sequence* $s \equiv (x_n \in \mathbb{R} \mid n \in N)$ *be inner convergent. Then, there exists* $x \in \mathbb{R}$ *such that* $x = \lim s$.

Proof. By Theorem 1 (1.3.7), there is an isotone bijection $\nu\colon \omega \to N$. Consider the sequence $t \equiv (y_n \mid n \in \omega)$ such that $y_n \equiv x_{\nu(n)}$. If t is finally constant in the sense of 1.2.7, then the assertion is evidently fulfilled. If t is finally non-constant, then by Proposition 1 (1.2.7), there exists a subsequence $(y_{n_k} \mid k \in \omega)$ such that $k \leqslant n_k < n_{k+1}$ and $y_{n_k} \neq y_{n_{k+1}}$.

Consider the sequences $u \equiv (z_k \mid k \in \omega)$ and $r \equiv (d_k \mid k \in \omega)$ such that $z_k \equiv y_{n_k}$ and $d_k \equiv |z_k - z_{k+1}|$. It is clear that the sequence t is inner convergent. For any $\varepsilon > 0$ and any $n \in I(t, \varepsilon)$ by virtue of $n_p \geqslant p$ and $n_q \geqslant q$, we infer that $|z_p - z_q| < \varepsilon$ for every $p, q \geqslant n$. In particular, $d_p < \varepsilon$ for every $p \geqslant n$.

By Lemma 4 for z_p, there is $a \in \mathbb{Q}$ such that $|a - z_p| < d_p$. Thus, the set $Q_p \equiv \{a \in \mathbb{Q} \mid |a - z_p| < d_p\}$ is non-empty. Take some choice mapping $c\colon \mathcal{P}(\mathbb{Q}) \setminus \{\varnothing\} \to \mathbb{Q}$ from the axiom of choice in 1.1.12. Define a sequence $\alpha \equiv (a_p \in \mathbb{Q} \mid p \in \omega)$ setting $a_p \equiv c(Q_p)$. Take any rational number $e > 0$ and some number $n \in I(r, \varepsilon/3)$. Then, for every $p, q \geqslant n$, we have $|a_p - a_q| \leqslant |a_p - z_p| + |z_p - z_q| + |z_q - a_q| < d_p + e/3 + d_q < \varepsilon$. Thus, the sequence α is inner convergent. Therefore, we can take the real number $x \equiv \theta\alpha$.

Now, take any $\varepsilon > 0$ and some $n \in I(t, \varepsilon/2)$ and $m \in C(\alpha, x, \varepsilon/2)$ (the latter set is non-empty by virtue of Lemma 4). Then, for every $p \geqslant \mathrm{gr}(m, n)$, we have $|z_p - x| \leqslant |z_p - a_p| + |a_p - x| < d_p + \varepsilon/2 < \varepsilon/2 + \varepsilon/2 = \varepsilon$. Consequently, $x = \lim u$. By Lemma 6 $x = \lim t$. Since ν is an isotone bijection, we conclude that $x = \lim s$. □

Theorem 1 and conclusion 2 of Lemma 2 give us the well-known *Cauchy criterion* of convergence of sequences.

According to 1.1.15, a sequence $s \equiv (x_n \in \mathbb{R} \mid n \in N)$ is called *increasing* [*strictly increasing, decreasing, strictly decreasing*] if $m, n \in N$ and $m < n$ imply $x_m \leqslant x_n$ [$x_m < x_n$, $x_m \geqslant x_n$, $x_m > x_n$].

Proposition 1. *Let an infinite sequence* $s \equiv (x_n \in \mathbb{R} \mid n \in N)$ *be strictly increasing* [*decreasing*] *and bounded above* [*below*]. *Then, s is inner convergent.*

Proof. As in the proof of Theorem 1, take the isotone bijection $\nu\colon \omega \to N$ and the sequence $t \equiv (y_n \mid n \in \omega)$ such that $y_n \equiv x_{\nu(n)}$. Then, $y_n < y_{n+1} \leqslant y$ for some real number y.

Suppose that the sequence t is not inner convergent, i. e. there exists a real number $\varepsilon > 0$ such that for every natural number k, there are some natural numbers l and m such that $l, m \geqslant k$, $l < m$, and $y_m - y_l \geqslant \varepsilon$. Consider in the set $B \equiv \omega \times \omega$ the non-empty subsets $P_k \equiv \{(l, m) \in B \mid (l, m \geqslant k) \wedge (l < m) \wedge (y_m - y_l \geqslant \varepsilon)\}$. Take some choice mapping $c\colon \mathcal{P}(B) \setminus \{\varnothing\} \to B$ from the axiom of choice in 1.1.12. Consider the projections pr_0 and pr_1 from B onto ω such that $(\mathrm{pr}_0(b), \mathrm{pr}_1(b)) = b$ for every $b \in B$. Define sequences $(l_k \in \omega \mid k \in \omega)$ and $(m_k \in \omega \mid k \in \omega)$ setting $l_k \equiv \mathrm{pr}_0(c(P_k))$ and $m_k \equiv \mathrm{pr}_1(c(P_k))$. By definition, $(l_k, m_k) \in P_k$.

Take $b_0 \equiv (l_0, m_0)$ and define a mapping $V\colon B \times \omega \to B$ setting $V((p,q),n) \equiv (l_k, m_k)$, where $k \equiv \mathrm{gr}(p,q)$. By Theorem 1 (1.2.7), there is a mapping $u\colon \omega \to B$ such that $u(0) \equiv b_0$ and $u(n+1) = V(u(n), n)$ for $n \in \omega$. Define sequences $(p_n \in \omega \mid n \in \omega)$ and $(q_n \in \omega \mid n \in \omega)$ setting $p_n \equiv \mathrm{pr}_0(u(n))$ and $q_n \equiv \mathrm{pr}_1(u(n))$. Then, $(p_{n+1}, q_{n+1}) = u(n+1) = V(u(n), n) = V((p_n, q_n), n) \in P_k$, where $k \equiv \mathrm{gr}(p_n, q_n)$. Therefore, p_{n+1}, $q_{n+1} \geqslant k$, $p_{n+1} < q_{n+1}$, and $y_{q_{n+1}} - y_{p_{n+1}} \geqslant \varepsilon$. Hence, $p_n < q_n < p_{n+1} < q_{n+1}$. Consequently, $y_{p_n} < y_{q_n} < y_{p_{n+1}} < y_{q_{n+1}}$ and $y_{q_{n+1}} - y_{q_n} > y_{q_{n+1}} - y_{p_{n+1}} \geqslant \varepsilon$.

For $\varepsilon > 0$ and $y - y_{q_0} \geqslant 0$ by Lemma 13 (1.4.3), there is $n \in \mathbb{N}$ such that $n\varepsilon > y - y_{q_0}$. As a result, $y_{q_n} = (y_{q_n} - y_{q_{n-1}}) + (y_{q_{n-1}} - y_{q_{n-2}}) + \cdots + (y_{q_1} - y_{q_0}) + y_{q_0} > n\varepsilon + y_{q_0} > y > y_{q_n}$. It follows from the obtained contradiction that the sequence t is inner convergent. Thus, the sequence s is also inner convergent. □

Proposition 2. *Let an infinite sequence $s \equiv (x_n \in \mathbb{R} \mid n \in N)$ be increasing [decreasing] and bounded above [below]. Then, there exists $x \in \mathbb{R}$ such that $x = \lim s$.*

Proof. As in the proof of Theorem 1, take the isotone bijection $v\colon \omega \to N$ and the sequence $t \equiv (y_n \mid n \in \omega)$ such that $y_n \equiv x_{v(n)}$. Then, $y_n \leqslant y_{n+1} \leqslant y$ for some real number y. If t is finally constant in the sense of 1.2.7, then the assertion is evidently fulfilled. If t is finally non-constant, then by Proposition 1 (1.2.7), there exists a subsequence $(y_{n_k} \mid k \in \omega)$ such that $k \leqslant n_k < n_{k+1}$ and $y_{n_k} \neq y_{n_{k+1}}$. Since t is increasing, we infer that $y_{n_k} < y_{n_{k+1}}$.

Consider the strictly increasing sequence $u \equiv (z_k \mid k \in \omega)$ such that $z_k \equiv y_{n_k}$. By Proposition 1, there is x such that $x = \lim u$. Take any $\varepsilon > 0$ and some $m \in C(u, x, \varepsilon)$. Then, for every $p \geqslant m$, we have $|x - z_p| < \varepsilon$. Therefore, for every $p \geqslant n_m \geqslant m$, we infer that $x - y_p \leqslant x - y_{n_m} = x - z_m < \varepsilon$ and $x - y_p \geqslant x - y_{n_p} = x - z_p > -\varepsilon$ since $p \leqslant n_p$. As a result, $|x - y_p| < \varepsilon$ for $p \geqslant n_m$. Consequently, $x = \lim t$. Since v is an isotone bijection, we conclude that $x = \lim s$. □

Lemma 8. *Let an infinite sequence $s \equiv (x_n \in \mathbb{R} \mid n \in N)$ be increasing [decreasing] and $x = \lim s$. Then, $x_n \leqslant x$ [$x_n \geqslant x$] for every $n \in N$.*

Proof. Suppose that there is $n \in N$ such that $x_n > x$. Then, for $\varepsilon \equiv x_n - x$, there is $m \in N$ such that $|x_p - x| < \varepsilon$ if $m \leqslant p \in N$. Take $p \equiv \mathrm{gr}(m, n)$. Then, $x_m \leqslant x_p$ and $x_n \leqslant x_p$ imply $\varepsilon \equiv x_n - x \leqslant x_p - x < \varepsilon$. It follows from this contradiction that $x_n \leqslant x$ for every n. □

Theorem 2. *Let a sequence $s \equiv (x_m \in \mathbb{R} \mid m \in M)$ be increasing, a sequence $t \equiv (y_n \in \mathbb{R} \mid n \in N)$ be decreasing, and $x_m \leqslant y_n$ for every $m \in M$ and $n \in N$. Then, there exist $x, y \in \mathbb{R}$ such that $x_m \leqslant x \leqslant y \leqslant y_n$ for every $m \in M$ and $n \in N$.*

Proof. By Proposition 2, there exist $x = \lim s$ and $y = \lim t$. By Lemma 8 $x_m \leqslant x$ and $y_n \geqslant y$. Suppose that $x > y$. Take $\varepsilon \equiv x - y$ and some numbers $m \in C(s, x, \varepsilon/2)$ and $n \in C(t, y, \varepsilon/2)$. By Lemma 1, there are $p \in M$ and $q \in N$ such that $p, q \geqslant \mathrm{gr}(m, n)$. Therefore, $0 \leqslant x - x_p < \varepsilon/2$ and $0 \leqslant y_q - y < \varepsilon/2$ imply $y_q - x_p = (y_q - y) + (y - x) + (x - x_p) < \varepsilon/2 - \varepsilon + \varepsilon/2 = 0$, i. e. $y_q < x_p$. But this contradicts the condition. Thus, $x \leqslant y$. □

Corollary 1. *Let* $(I_n \subset \mathbb{R} \mid n \in N)$ *be an infinite sequence of closed intervals such that* $I_n \subset I_m$ *if* $n \geqslant m$. *Then,* $\bigcap(I_n \mid n \in N) \neq \varnothing$.

Theorem 3. *The set* $\mathbb{R}$ *is uncountable.*

Proof. Since $\omega \subset \mathbb{R}$, the latter set is at least denumerable. Suppose that there exists a bijective mapping $f\colon\ \omega \to \mathbb{R}$. Consider the set $B \equiv \{I \subset \mathbb{R} \mid \exists\, a, b \in \mathbb{R}\ (a < b \wedge I = [a, b])\}$. For every interval $I \equiv [a, b] \in B$, consider the intervals $I' \equiv [a, a + (b - a)/3]$, $I'' \equiv [a + (b - a)/3, a + 2(b - a)/3]$, and $I''' \equiv [a + 2(b - a)/3, b]$. It is clear that $I = I' \cup I'' \cup I'''$.

Take the interval $I_0 \equiv [f(0) + 1, f(0) + 2]$. Define the mapping $V\colon\ B \times \omega \to B$ by the following conditions: (1) if $f(n + 1) \notin I$, then $V(I, n) \equiv I$; (2) if $f(n + 1) \in I$ and $f(n + 1) \notin I'$, then $V(I, n) \equiv I'$; (3) if $f(n + 1) \in I'$ and $f(n + 1) \notin I''$, then $V(I, n) \equiv I''$; (4) if $f(n+1) \in I' \cap I''$ and $f(n+1) \notin I'''$, then $V(I, n) \equiv I'''$. It follows from this definition that $f(n + 1) \notin V(I, n) \subset I$.

By Theorem 1 (1.2.7), there is a mapping $u\colon\ \omega \to B$ such that $u(0) = I_0$ and $u(n + 1) = V(u(n), n)$. Denote $u(n)$ by I_n. Then, $I_{n+1} = V(I_n, n)$ implies $f(n + 1) \notin I_{n+1} \subset I_n$. By Corollary 1 to Theorem 2, there is $x \in \bigcap(I_n \mid n \in \omega)$. But $x = f(m)$ for some m. Therefore, $x \notin I_m$ and simultaneously $x \in I_m$. Thus, $\mathbb{R}$ is not denumerable. □

The cardinal number card $\mathbb{R}$ is called the *power of continuum* and is denoted by $\mathfrak{c}$.

1.4.5 The Dedekind completeness and order properties of the extended real line

The Dedekind completeness of $\mathbb{R}$

The greatest lower bound $\inf \sigma$ and the smallest upper bound $\sup \sigma$ of a collection $\sigma \equiv (x_i \in \mathbb{R} \mid i \in I)$ indexed by a non-empty set I (see 1.1.15) can be characterized in $\mathbb{R}$ by the following *property of countability*.

Lemma 1. *For a collection* $\sigma \equiv (x_i \in \mathbb{R} \mid i \in I)$ *and a number* $y \in \mathbb{R}$ *the following conditions are equivalent:*

1) $y = \sup \sigma$ $[y = \inf \sigma]$;
2) y *is an upper [lower] bound of* σ *and for every* $n \in \mathbb{N}$, *there is* $i \in I$ *such that* $y - 1/n < x_i \leqslant y$ $[y + 1/n > x_i \geqslant y]$.

Proof. (1) ⊢ (2). Suppose that $y - 1/n \geqslant x_i$ for every i. Then, $y - 1/n \geqslant y$, but this is impossible. Thus, $y - 1/n < x_i$ for some i.

(2) ⊢ (1). Let b be any upper bound of σ. Suppose that $y > b$. Then, by Corollary 1 to Lemma 13 (1.4.3) $y - b > 1/n$ for some $n \in \mathbb{N}$. By condition for n, there is i such that $y-1/n < x_i$. As a result, $x_i > y-1/n > b \geqslant x_i$, but this is impossible. Thus, $y \leqslant b$, where $y = \sup \sigma$. □

Lemma 2. *Let $x \in \mathbb{R}$. Then, $x = \sup(x - 1/n \mid n \in \mathbb{N}) = \inf(x + 1/n \mid n \in \mathbb{N})$.*

Proof. By assertion 1 of Proposition 4 (1.4.3), $0 \geqslant -1/n$ and $x \geqslant x$ imply $x \geqslant x - 1/n$. Let $b \geqslant x - 1/n$ for every n. Suppose that $b < x$. By Corollary 1 to Lemma 13 (1.4.3) $m(x - b) > 1$ for some $m \geqslant 1$. Consequently, $b < x - 1/m$, but this is impossible. Thus, $b \geqslant x$. The first equality is checked. The second one is checked analogously. □

Let $(A, \leqslant)$ be some ordered class (see 1.1.15). A pair (R, S) of non-empty subsets of the set A is called a *Dedekind cut in A* if $r \leqslant s$ for every $r \in R$ and $s \in S$ and $R \cup S = A$.

Lemma 3. *Let $x \in \mathbb{R}$. Then, there exists a Dedekind cut (R, S) in $\mathbb{Q}$ such that $x = \sup(r \mid r \in R) = \inf(s \mid s \in S)$.*

Proof. Consider the sets $R \equiv \{r \in \mathbb{Q} \mid r \leqslant x\}$ and $S \equiv \{s \in \mathbb{Q} \mid x \leqslant s\}$. By Lemma 14 (1.4.3), there are rational numbers r and s such that $x-1 < r < x < s < x+1$. Therefore, the pair (R, S) is a Dedekind cut in $\mathbb{Q}$.

Let a real number y is an upper bound of the set R. Suppose that $y < x$. By the same lemma there is a rational number t such that $y < t < x$. By definition $t \in R$ and so $t \leqslant y$. It follows from this contradiction that $x \leqslant y$. Consequently, $x = \sup(r \mid r \in R)$.

In the same manner, it is checked that $x = \inf(s \mid s \in S)$. □

Corollary 1. *Let $x \in \mathbb{R}$. Then, there exist sequences $(r_n \in \mathbb{Q} \mid n \in \omega) \uparrow$ and $(s_n \in \mathbb{Q} \mid n \in \omega) \downarrow$ such that $x = \sup(r_n \mid n \in \omega) = \inf(s_n \mid n \in \omega)$.*

Proof. Consider the identical collections $\mathrm{id}_R \equiv (r_r \mid r \in R)$ and $\mathrm{id}_S \equiv (s_s \mid s \in S)$ (see 1.1.9) for the Dedekind cut (R, S) from Lemma 3. By Lemma 1, the sets $R_{n-1} \equiv \{r \in R \mid x-1/n < r \leqslant x\}$ and $S_{n-1} \equiv \{s \in S \mid x + 1/n > s \geqslant x\}$ are non-empty. Take some choice mapping $p\colon \mathcal{P}(\mathbb{Q}) \setminus \{\varnothing\} \to \mathbb{Q}$ from the axiom of choice in 1.1.12. Define sequences $u \equiv (x_n \in \mathbb{Q} \mid n \in \omega)$ and $v \equiv (y_n \in \mathbb{Q} \mid n \in \omega)$ setting $x_n \equiv p(R_n)$ and $y_n \equiv p(S_n)$. By definition, $x - 1/n < x_{n-1} \leqslant x$ and $x + 1/n > y_{n-1} \geqslant x$. Therefore, by Lemma 1 $x = \sup u = \inf v$. Consider the rational numbers $r_n \equiv \sup(x_k \mid k \in n + 1)$ and $s_n \equiv \inf(y_k \mid k \in n + 1)$. It is easy check that $x = \sup(r_n \mid n \in \omega) = \inf(s_n \mid n \in \omega)$. □

The following theorem shows the Dedekind completeness (see 1.1.15) of the real line $(\mathbb{R}, \leqslant)$.

Theorem 1. *Let a collection $\sigma \equiv (x_i \in \mathbb{R} \mid i \in I)$ be bounded above [below]. Then, there is a number $y \in \mathbb{R}$ such that $y = \sup \sigma$ [$y = \inf \sigma$].*

Proof. Let b be an upper bound of σ. Take some $i \in I$ and $n \in \omega$. Since $b - x_i \geqslant 0$, there is a number $m \in \mathbb{N}$ such that $b - x < m2^{-n}$. Therefore, $b \leqslant x + m2^{-n}$, i. e. $x + m2^{-n}$ is an upper bound of σ. Consider the number $p(n) \equiv \mathrm{sm}(m \in \mathbb{N} \mid \forall i \in I\ (x_i \leqslant x + m2^{-n}))$. Then, for the interval $I_n \equiv [x + (p(n) - 1)2^{-n}, x + p(n)2^{-n}]$, there is an index i such that $x_i \in I_n$. Compare the intervals I_n and I_{n+1}. Since the number $x + 2p(n)2^{-(n+1)} = x + p(n)2^{-n}$ is an upper bound and the number $x + (2p(n) - 2)2^{-(n+1)} = x + (p(n) - 1)2^{-n}$ is not an upper bound, we infer that either $p(n + 1) = 2p(n)$ or $p(n+1) = 2p(n) - 1$. In any case, $I_{n+1} \subset I_n$. By Corollary 1 to Theorem 2 (1.4.4), $I \equiv \bigcap(I_n \mid n \in \omega) \neq \varnothing$.

Suppose that there are $a, b \in I$ such that $a < b$. Then, $[a, b] \subset I_n$ implies $b - a \leqslant 2^{-n}$ for every n. But by virtue of Corollary 2 to Proposition 5 (1.4.3), $2^{-m} < b - a$ for some m. It follows from this contradiction that $I = \{y\}$ for some y.

Suppose that $y < x_j$ for some j. Then, $2^{-n} < x_j - y$ for some n. Since $y \in I_n$, we infer that $x_j > y + 2^{-n} \geqslant x + p(n)2^{-n}$. But this contradicts the definition of the number $p(n)$. Therefore, y is an upper bound of σ. Take any upper bound d of σ and suppose that $y > d$. Then, $y - d > 2^{-n}$ for some n. Since $c \in I_n$, we infer that $d < y - 2^{-n} \leqslant x + (p(n) - 1)2^{-n}$, i. e. the latter number is an upper bound of σ. But this contradicts the definition of the number $p(n)$. Thus, $y \leqslant d$. Thus, $y = \sup \sigma$. □

Corollary 1. *Let a set $X \subset \mathbb{R}$ be bounded above [below]. Then, there is a number $y \in \mathbb{R}$ such that $y = \sup(x \mid x \in X)$ [$y = \inf(x \mid x \in X)$].*

Corollary 2. *Let (R, S) be a Dedekind cut in $\mathbb{R}$. Then, there is $x \in \mathbb{R}$ such that $x = \sup(r \mid r \in R) = \inf(s \mid s \in S)$.*

Proof. By Corollary 1, there are $x = \sup R$ and $y = \inf S$. Then, $x \leqslant s$ for all $s \in S$ implies $x \leqslant y$. Suppose that $x < y$. By definition of a Dedekind cut, $z \equiv (x + y)/2 \in R \cup S$. If $z \in R$, then $x < z \leqslant x$. If $z \in S$, then $y > z \geqslant y$. It follows from these contradictions that $x = y$. □

Corollary 3. *Let (R, S) be a Dedekind cut in $\mathbb{Q}$. Then, there is $x \in \mathbb{R}$ such that $x = \sup(r \mid r \in R) = \inf(s \mid s \in S)$.*

Proof. By Corollary 1, there are $x = \sup(r \mid r \in R)$ and $y = \inf(s \mid s \in S)$. Then, $x \leqslant y$. Suppose that $x < y$. By Lemma 14 (1.4.3) $x < t < y$ for some rational number t. But $t \in R \cup S$. If $t \in R$, then $x < t \leqslant x$. If $t \in S$, then $y > t \geqslant y$. It follows from these contradictions that $x = y$. □

The order completeness of $\overline{\mathbb{R}}$

Now, we shall generalize Theorem 1 for the ordered set $(\overline{\mathbb{R}}, \leqslant)$, i. e. we shall prove that the ordered set $(\overline{\mathbb{R}}, \leqslant)$ is order complete (see 1.1.15 and 1.4.3).

Theorem 2. *Let $\sigma \equiv (x_i \in \overline{\mathbb{R}} \mid i \in I)$ be a collection indexed by a non-empty set I. Then, there are numbers $y, z \in \overline{\mathbb{R}}$ such that $y = \sup \sigma$ and $z = \inf \sigma$.*

Proof. At first, assume that there is $a \in \mathbb{R}$ such that $x_i \leqslant a$ for every $i \in I$. If $x_i = -\infty$ for every $i \in I$, then $-\infty = \sup \sigma$. If there is $j \in I$ such that $x_j > -\infty$, then the set $J \equiv \{j \in I \mid x_j > -\infty\}$ is non-empty. Consider the subcollection $\tau \equiv (x_j \in \mathbb{R} \mid j \in J)$. Since τ is bounded above, there is by Theorem 1 $y \in \mathbb{R}$ such that $y = \sup \tau$. If $i \in I \setminus J$, then $x_j = -\infty < y$. Thus, y is an upper bound of σ. If $b \in \overline{\mathbb{R}}$ and $b \geqslant x_i$ for every $i \in I$, then $b \geqslant y$. Consequently, $y = \sup \sigma$.

Now, assume that for every $a \in \mathbb{R}$, there is $i \in I$ such that $x_i > a$. By definition of the order in $\overline{\mathbb{R}}$ from 1.4.3 $y \equiv \infty$ is an upper bound of σ. If $b \in \overline{\mathbb{R}}$ and $b \geqslant x_i$ for every $i \in I$, then $b > a$ for every $a \in \mathbb{R}$. Thus, $b = \infty \geqslant y$. Consequently, $y = \sup \sigma$.

For the infimum, the arguments are the same. □

Corollary 1. *Let $X \subset \overline{\mathbb{R}}$. Then, there are numbers $y, z \in \overline{\mathbb{R}}$ such that $y = \sup(x \mid x \in X)$ and $z = \inf(x \mid x \in X)$.*

Corollary 2. *Let (R, S) be a Dedekind cut in $\overline{\mathbb{R}}$. Then, there is $x \in \overline{\mathbb{R}}$ such that $x = \sup(r \mid r \in R) = \inf(s \mid s \in S)$.*

Proof. At first, assume that $R \neq \{-\infty\}$ and $S \neq \{\infty\}$. Then, (R_1, S_1), where $R_1 \equiv R \setminus \{-\infty\}$, $S_1 \equiv S \setminus \{\infty\}$ is a Dedekind cut in $\mathbb{R}$. By Corollary 2 to Theorem 1, there is $x \in \mathbb{R}$ such that $x = \sup R = \inf S$.

Now, consider the case $S = \{\infty\}$. By Corollary 1, there are $x = \sup R$ and $y = \inf S$. Then, $x \leqslant s$ for all $s \in S$ implies $x \leqslant y$. Suppose that $x < y = \infty$. By definition of a Dedekind cut, $z \equiv x + 1 \in R \cup S$. If $z \in R$, then $x < z \leqslant x$. It follows from this contradiction that $z \in S$. Therefore, $z = \infty$ and $x = z - 1 = \infty = y$.

The case $R = \{-\infty\}$ is considered in a similar way. □

Some further order properties of $\mathbb{R}$

Consider some order properties of $\mathbb{R}$.

By Theorem 3 (1.2.6) every simple collection $\sigma \equiv (x_i \in \mathbb{R} \mid i \in I)$ indexed by a finite non-empty set I has the largest element $\operatorname{gr} \sigma$ and the smallest element $\operatorname{sm} \sigma$. Therefore, according to 1.1.15, we have $\sup \sigma = \operatorname{gr} \sigma$ and $\inf \sigma = \operatorname{sm} \sigma$.

In particular, if $I = n + 1$, then $x_0 \vee \ldots \vee x_n = x_0 \veebar \ldots \veebar x_n$ and $x_0 \wedge \ldots \wedge x_n = x_0 \barwedge \ldots \barwedge x_n$ (see 1.2.6).

Also, if $x, x', x'', x''', \dots \in \mathbb{R}$, then $x \vee x' = x \veebar x'$, $x \vee x' \vee x'' = x \veebar x' \veebar x''$, $x \vee x' \vee x'' \vee x''' = x \veebar x' \veebar x'' \veebar x''', \dots$ and $x \wedge x' = x \barwedge x'$, $x \wedge x' \wedge x'' = x \barwedge x' \barwedge x''$, $x \wedge x' \wedge x'' \wedge x''' = x \barwedge x' \barwedge x'' \barwedge x''', \dots$

Lemma 4. *Let $\sigma \equiv (z_i \in \mathbb{R} \mid i \in I)$ be a simple collection indexed by a non-empty set I and $z, a \in \mathbb{R}$. Then:*
1) *if $z = \sup \sigma$, then $a + z = \sup(a + z_i \mid i \in I)$;*
2) *if $z = \inf \sigma$, then $a + z = \inf(a + z_i \mid i \in I)$;*
3) *if $z = \sup \sigma$, then $az = \sup(az_i \mid i \in I)$ when $a \geqslant 0$ and $az = \inf(az_i \mid i \in I)$ when $a \leqslant 0$;*
4) *if $z = \inf \sigma$, then $az = \inf(az_i \mid i \in I)$ when $a \geqslant 0$ and $az = \sup(az_i \mid i \in I)$ when $a \leqslant 0$;*
5) *if $z_i > 0$ for every $i \in I$ and $z = \sup \sigma$, then $z^{-1} = \inf(z_i^{-1} \mid i \in I)$;*
6) *if $z > 0$ and $z = \inf \sigma$, then $z^{-1} = \sup(z_i^{-1} \mid i \in I)$.*

Proof. 1. It is clear that $a + z \geqslant z + z_i$ for every i. Let $b \geqslant a + z_i$ for every i. Then, $b - a \geqslant z_i$ implies $b - a \geqslant z$, where $b \geqslant a + z$.

Assertion 2 is checked in a similar way.

3. Let $a > 0$. By assertion 3 of Proposition 4 (1.4.3) $az \geqslant az_i$ for every i. Let $b \geqslant az_i$ for every i. Then, $b/a \geqslant z_i$ implies $b/a \geqslant z$, where $b \geqslant az$.

Let $a < 0$. By the same reason $az \leqslant az_i$ for every i. Let $b \leqslant az_i$ for every i. Then, $b/a \leqslant z_i$ implies $b/a \leqslant z$, where $b \leqslant az$.

If $a = 0$, then the equalities are obvious.

Assertion 4 is checked in a similar way.

5. Since $0 < z_i \leqslant z$ for every i, we infer by Corollary 3 to Proposition 4 (1.4.3) that $0 < z^{-1} \leqslant z_i^{-1}$. Let $b \leqslant z_i^{-1}$ for every i. If $b \leqslant 0$, then $b \leqslant z^{-1}$. If $b > 0$, then by the same reason $b^{-1} > (z_i^{-1})^{-1} = z_i$ implies $b^{-1} \geqslant z$. Consequently, $b = (b^{-1})^{-1} \leqslant z^{-1}$.

6. Since $0 < z \leqslant z_i$ for every i, we infer as in 5) that $0 < z_i^{-1} \leqslant z^{-1}$. Let $b \geqslant z_i^{-1}$ for every i. Then, $0 < b^{-1} \leqslant z_i$ implies $b^{-1} \leqslant z$. Consequently, $b \geqslant z^{-1}$. □

Corollary 1. *Let $x, y, z \in \mathbb{R}$. Then:*
1) $a + x \veebar y = (a + x) \veebar (a + y)$;
2) $a + x \barwedge y = (a + x) \barwedge (a + y)$;
3) $a(x \veebar y) = ax \veebar ay$ *when* $a \geqslant 0$ *and* $a(x \veebar y) = ax \barwedge ay$ *when* $a \leqslant 0$;
4) $a(x \barwedge y) = ax \barwedge ay$ *when* $a \geqslant 0$ *and* $a(x \barwedge y) = ax \veebar ay$ *when* $a \leqslant 0$;
5) *if* $x, y > 0$, *then* $(x \veebar y)^{-1} = x^{-1} \barwedge y^{-1}$;
6) *if* $x, y > 0$, *then* $(x \barwedge y)^{-1} = x^{-1} \veebar y^{-1}$.

Lemma 5. *Let $x, y \in \mathbb{R}$. Then:*
1) $|x| = x \veebar 0 + (-x) \veebar 0 = x \veebar 0 - x \barwedge 0$;
2) $x \veebar y = (x - y) \veebar 0 + y = (y - x) \veebar 0 + x$;

3) $x \barwedge y = x - (x - y) \veebar 0 = y - (y - x) \veebar 0$;
4) $x \veebar y + x \barwedge y = x + y$;
5) $x \veebar y - x \barwedge y = |x - y|$;
6) $|x| = x \veebar (-x)$.

Proof. 1. If $x \geqslant 0$, then $|x| \equiv x \veebar 0 + (-x) \veebar 0$. If $x \leqslant 0$, then $|x| \equiv -x = x \veebar 0 + (-x) \veebar 0$. The second equality follows from the first one and Corollary 1 to Lemma 4.

Assertions 2 and 3 follow from Corollary 1 to Lemma 4.

Assertion 4 follows from assertions 2 and 3.

5. Subtracting the second equality in 3) from the first equality in (2) and applying (1), we get $x \veebar y - x \barwedge y = (x - y) \veebar 0 + (y - x) \veebar 0 = |x - y|$.

6. It is clear. □

Corollary 1. *Let $x, y \in \mathbb{R}$. Then:*
1) $x \veebar y = (x + y)/2 + |x - y|/2$;
2) $x \barwedge y = (x + y)/2 - |x - y|/2$.

Lemma 6. *Let $a, b, x \in \mathbb{R}$ and $a \geqslant b$. Then:*
1) $x \veebar a \geqslant x \veebar b$;
2) $x \barwedge a \geqslant x \barwedge b$.

Proof. 1. Since $x \veebar a \geqslant x$ and $x \veebar a \geqslant a$, we have $x \veebar a \geqslant x$ and $x \veebar a \geqslant b$, where $x \veebar a \geqslant x \vee b = x \veebar b$.

2. Analogously, $x \barwedge b \leqslant x$ and $x \barwedge b \leqslant b \leqslant a$ imply $x \barwedge b \leqslant x \wedge a = x \barwedge a$. □

Now, in addition to Proposition 1 and Theorem 1 from 1.1.15 we can prove for $\mathbb{R}$ the following properties of distributivity.

Proposition 1. *Let $\sigma \equiv (x_i \in \mathbb{R} \mid i \in I)$ be a simple collection indexed by a non-empty set I and $x, e \in \mathbb{R}$. Then:*
1) *if $e = \sup \sigma$, then $x \wedge e = \sup(x \wedge x_i \mid i \in I)$;*
2) *if $e = \inf \sigma$, then $x \vee e = \inf(x \vee x_i \mid i \in I)$.*

Proof. 1. Since $e \geqslant x_i$, we infer by Lemma 6 that $x \wedge e \geqslant x \wedge x_i$. Take any $u \in \mathbb{R}$ such that $u \geqslant x \wedge x_i$ for every i. By assertion 4 of Lemma 5 $x \wedge x_i = x + x_i - x \vee x_i$ and $x \wedge e = x + e - x \vee e$. Thus, $u \geqslant x + x_i - x \vee x_i$ implies $u - x \geqslant x_i - x \vee x_i$, where $x \vee x_i + u - x \geqslant x_i$. By Corollary 1 to Theorem 1 (1.1.15) $x \vee e = \sup(x \vee x_i \mid i \in I)$. Consequently, by assertions 2 and 3 of Lemma 4 $x \vee e + u - x = \sup(x \vee x_i + u - x) \geqslant x \vee x_i + u - x \geqslant x_i$ for every i. This implies $x \vee e + u - x \geqslant e$, where $u \geqslant x + e - x \vee e = x \wedge e$. Thus, $x \wedge e = \sup(x \wedge x_i \mid i \in I)$.

2. The proof is a simple modification of the proof of assertion 1. □

Corollary 1. *Let $x, y, z, e \in \mathbb{R}$. Then:*
1) $x \barwedge (y \veebar z) = (x \barwedge y) \veebar (x \barwedge z)$ (the *distributivity of the smallest element with respect to the greatest element*);
2) $x \veebar (y \barwedge z) = (x \veebar y) \barwedge (x \veebar z)$ (the *distributivity of the greatest element with respect to the smallest element*).

Proof. 1. Consider the set $I \equiv 2$ and a simple collection $\sigma \equiv (x_i \mid i \in I)$ such that $x_0 \equiv y$ and $x_1 \equiv z$. Then, by Corollary 1 to Proposition 1 (1.1.15) $e \equiv y \veebar z = y \vee z = x_0 \vee x_1 = \sup(x_i \mid i \in I)$. Therefore, by Proposition 1 $x \barwedge e = x \wedge e = \sup(x \wedge x_i \mid i \in I) = \sup(x \barwedge x_i \mid i \in I) = (x \barwedge x_0) \vee (x \barwedge x_1) = (x \barwedge y) \veebar (x \barwedge z)$.

2. The proof is a simple modification of the proof of assertion 1. □

Corollary 2. *Let $\sigma \equiv (y_j \in \mathbb{R} \mid j \in J)$ and $\tau \equiv (z_k \in \mathbb{R} \mid k \in K)$ be simple collections indexed by non-empty sets J and K and $f, g \in \mathbb{R}$. Then:*
1) *if $f = \sup \sigma$ and $g = \sup \tau$, then $f \wedge g = \sup(y_j \wedge z_k \mid (j, k) \in J \times K)$;*
2) *if $f = \inf \sigma$ and $g = \inf \tau$, then $f \vee g = \inf(y_j \vee z_k \mid (j, k) \in J \times K)$.*

Proof. 1. By Proposition 1 $y_j \wedge g = \sup(y_j \wedge z_k \mid k \in K)$ and $f \wedge g = \sup(y_j \wedge g \mid j \in J)$. Since $f \wedge g \geqslant y_j \wedge z_k$ for every j and k, by Theorem 1, there is a number u such that $f \wedge g \geqslant u = \sup(y_j \wedge z_k \mid (j, k) \in J \times K)$. It is clear that $u \geqslant y_j \wedge g$ and so $u \geqslant f \wedge g$. As a result, $f \wedge g = u$.

Assertion 2 is checked in a similar way. □

Theorem 3. *Let $(I_m \mid m \in M)$ be a total multivalued collection of sets indexed by a non-empty set M and $U \equiv \prod(I_m \mid m \in M)$. Let $(\varkappa_m \mid m \in M)$ be a collection of collections of numbers $\varkappa_m \equiv (x_{mi} \in \mathbb{R} \mid i \in I_m)$ and $(e_m \in \mathbb{R} \mid m \in M)$ and $(g_u \in \mathbb{R} \mid u \in U)$ be collections of numbers. Then:*
1) *if $e_m = \sup(x_{mi} \mid i \in I_m)$ and $g_u = \inf(x_{mu(m)} \mid m \in M)$, then there exists $x \in \mathbb{R}$ such that $x = \inf(e_m \mid m \in M)$ and $x = \sup(g_u \mid u \in U)$ (the* general distributivity of the infimum with respect to the supremum*);*
2) *if $e_m = \inf(x_{mi} \mid i \in I_m)$ and $g_u = \sup(x_{mu(m)} \mid m \in M)$, then there exists $x \in \mathbb{R}$ such that $x = \sup(e_m \mid m \in M)$ and $x = \inf(g_u \mid u \in U)$ (the* general distributivity of the supremum with respect to the infimum*).*

Proof. 1. If $e_m = \sup(x_{mi} \mid i \in I_m)$ and $g_u = \inf(x_{mu(m)} \mid m \in M)$, then for every m and u the condition $u(m) \in I_m$ implies the inequality $e \geqslant x_{mu(m)}$. Therefore, $e_m \geqslant g_u$. By Theorem 1, there exists unique numbers x and y such that $x = \inf(e_m \mid m \in M)$, $y = \sup(g_u \mid u \in U)$, and $x \geqslant y$. We need to prove that $x = y$.

By Lemma 1, for every $n \in \mathbb{N}$ and every $m \in M$, there is $i \in I_m$ such that $e_m - 1/n < x_{mi}$. Therefore, we can consider the non-empty sets $J_{nm} \equiv \{i \in I_m \mid e_m - 1/n < x_{mi}\}$. By the axiom of choice from 1.1.12 for the non-empty set $I \equiv \bigcup(I_m \mid m \in M)$, there exists a choice mapping $p\colon \mathcal{P}(I) \setminus \{\varnothing\} \to I$. Define some mapping u from M into I setting $u(m) \equiv p(J_{nm}) \in J_{nm} \subset I_m$. Then, $u \in U$ and $x - 1/n \leqslant e_m - 1/n < x_{mu(m)}$ for every $m \in M$. Consequently, $x - 1/n \leqslant g_u \leqslant y$. By Lemma 2 $x \leqslant y$, where $x = y$.

2. The proof is a simple modification of the proof of assertion 1. □

Corollary 1. *Let I be a finite set, $(J_i \mid i \in I)$ be a collection of non-empty sets, $U \equiv \prod(J_i \mid i \in I)$. Let $(\sigma_i \mid i \in I)$ be a collection of collections of numbers $\sigma_i \equiv \left(x_{ij_i} \in \mathbb{R} \mid j_i \in J_i\right)$, and $(z_i \in \mathbb{R} \mid i \in I)$ be a collection of numbers. Then:*

1) *if $z_i = \sup \sigma_i$, then $\inf(z_i \mid i \in I) = \sup(\inf(x_{iu(i)} \mid i \in I) \mid u \in U)$;*
2) *if $z_i = \inf \sigma_i$, then $\sup(z_i \mid i \in I) = \inf(\sup(x_{iu(i)} \mid i \in I) \mid u \in U)$.*

Theorem 4 (The *Birkhoff identity*). *Let $x, y, z \in \mathbb{R}$. Then, $|x \veebar z - y \veebar z| + |x \barwedge z - y \barwedge z| = |x - y|$.*

Proof. Applying the formula $|p - q| = p \veebar q - p \barwedge q$ from Lemma 5 to the case that $p = x \veebar z$ and $q = y \veebar z$ as well as to the case that $p = x \barwedge z$ and $q = y \barwedge z$, we obtain $L \equiv |x \veebar z - y \veebar z| + |x \barwedge z - y \barwedge z| = (x \veebar z) \veebar (y \veebar z) - (x \veebar z) \barwedge (y \veebar z) + (x \barwedge z) \veebar (y \barwedge z) - (x \barwedge z) \barwedge (y \barwedge z)$. Using Corollary 1 to Proposition 1 and Corollary 1 to Theorem 1 (1.1.15), we obtain $L = (x \veebar z) \veebar z - (x \barwedge y) \veebar z + (x \veebar y) \barwedge z - (x \barwedge y) \barwedge z = ((x \veebar y) \veebar z + (x \veebar y) \barwedge z) - ((x \barwedge y) \veebar z + (x \barwedge y) \barwedge z)$. Finally, using equalities 4) and 5) from Lemma 5, we obtain $L = (x \veebar y + z) - (x \barwedge y + z) = x \veebar y - x \barwedge y = |x - y|$. □

Corollary 1 (the *Birkhoff inequalities*). *Let $x, y, z \in \mathbb{R}$. Then, $|x \veebar z - y \veebar z| \leqslant |x - y|$ and $|x \barwedge z - y \barwedge z| \leqslant |x - y|$.*

1.4.6 Natural roots of positive real numbers. Raising to a rational degree

Consider the set $B \equiv \omega$ and take the number $b_0 \equiv 1$. Define for the set B a productive mapping V from $B \times \omega$ into B setting $V(m, n) \equiv m(n + 1)$. Then, by Theorem 1 (1.2.7), there is a unique mapping u from ω into B such that $u(0) = b_0$ and $u(n + 1) = V(u(n), n) = u(n)(n + 1)$. The mapping u is called the *factorial function*. The number $u(n)$ is called the *factorial of the number n* and is denoted by $n!$. It follows from this recursive formula that $0! = 1$, $1! = 1$, $2! = 1! \cdot 2 = 1 \cdot 2$, $3! = 2! \cdot 3 = 1 \cdot 2 \cdot 3, \ldots$

For numbers $k, m \in \omega$, the number $m!/(k!(m-k)!)$ is called the *number of combinations of m things k at a time* and is denoted by C_m^k or by $\binom{m}{k}$. It follows from the definition that $C_m^0 = 1$ and $C_m^k = C_m^{m-k}$.

Theorem 1 (the *Newton binomial theorem*). *Let $a, b \in \mathbb{R}$ and $m \in \mathbb{N}$. Then, $(a + b)^m = \sum(C_m^k a^{m-k} b^k \mid k \in m + 1) = a^m + C_m^1 a^{m-1} b^1 + \ldots + C_m^k a^{m-k} b^k + \ldots + C_m^{m-1} a^1 b^{m-1} + b^m$.*

Proof. Consider the set X of all numbers $n \in \omega$ such that for $m \equiv n + 1$, we have the equality written above. If $n = 0$, then for $m = 1$ by Lemma 4 (1.4.3), we have the equality $\sum(C_1^k a^{1-k} b^k \mid k \in 2) = C_1^0 a^{1-0} b^0 + C_1^1 a^{1-1} b^1 = (a+b)^1$. Thus, $0 \in X$.

Suppose that $n \in X$. Take $m \equiv n+1$. Then, $(a+b)^{m+1} = (a+b)^m(a+b) = \sum(C_m^k a^{m+1-k} b^k \mid k \in m+1) + \sum(C_m^k a^{m-k} b^{k+1} \mid k \in m+1) = a^{m+1} + \sum(C_m^k a^{m+1-k} b^k \mid k \in (m+1) \setminus 1) + \sum(C_m^{k-1} a^{m+1-k} b^k \mid k \in (m+2) \setminus 1) = a^{m+1} + \sum((C_m^k + C_m^{k-1}) a^{m+1-k} b^k \mid k \in (m+1) \setminus 1) + b^{m+1}$. Since $C_m^k + C_m^{k-1} = m!/k!(m-k)! + m!/(k-1)!(m+1-k)! = (m+1)!/k!(m+1-k)! = C_{m+1}^k$, we conclude that $(a+b)^{m+1} = a^{m+1} + \sum(C_{m+1}^k a^{m+1-k} b^k \mid k \in (m+1) \setminus 1) + b^{m+1} = \sum(C_{m+1}^k a^{m+1-k} b^k \mid k \in m+2)$. This means that $n+1 \in X$. By Theorem 1 (1.2.6), $X = \omega$. □

Theorem 2. *Let $x \in \mathbb{R}_+$ and $m \in \mathbb{N}$. Then, there is a unique number $a \in \mathbb{R}_+$ such that $a^m = x$.*

Proof. If $x = 0$, then by Proposition 3 (1.4.3) $a = 0$. Therefore, we shall assume that $x > 0$. Consider the sets $R \equiv \mathbb{R}_- \cup \{r \in \mathbb{R}_+ \mid r^m \leqslant x\}$ and $S \equiv \{s \in \mathbb{R}_+ \mid s^m \geqslant x\}$. By Corollary 1 to Lemma 13 (1.4.3), there is $n \in \mathbb{N}$ such that $x > 1/n \geqslant (1/n)^m$. Thus, $R \neq \varnothing$. If $x \leqslant 1$, then $1^m \geqslant x$; if $x > 1$, then by Proposition 4 (1.4.3) $x^m \geqslant x$. Thus, $S \neq \varnothing$. Besides, $\mathbb{R} = R \cup S$. Take any $r \in R$ and $s \in S$ and suppose that $r > s$. Then, by Proposition 4 (1.4.3), $x \geqslant r^m > s^m \geqslant x$, but this is impossible. It follows from this contradiction that $r \leqslant s$. Thus, (R, S) is a Dedekind cut in $\mathbb{R}$ in the sense of 1.4.3. By Corollary 2 to Theorem 1 (1.4.5), there is a such that $a = \sup(r \mid r \in R) = \inf(s \mid s \in S)$.

By Lemma 1 (1.4.5) for every $n \in \mathbb{N}$, there are $r \in R$ and $s \in S$ such that $a - 1/n < r \leqslant a \leqslant s < a + 1/n$. Therefore, $(a - 1/n)^m < r^m \leqslant x$ and $(a + 1/n)^m > s^m \geqslant x$. Since $(1/n)^k \leqslant 1/n$ for every $k \in \mathbb{N}$, we get by virtue of Theorem 1 the inequality $x < (a + 1/n)^m \leqslant a^m + \sum(C_m^k a^{m-k} \mid k \in (m+1) \setminus 1)/n$. Denote $\sum(C_m^k a^{m-k} \mid k \in (m+1) \setminus 1)$ by z. Then, again by Theorem 1 $x > (a - 1/n)^m = a^m + \sum(C_m^k a^{m-k}(-1/n)^k \mid k \in (m+1) \setminus 1) \geqslant a^m - \sum(C_m^k a^{m-k}(1/n)^k \mid k \in (m+1) \setminus 1) \geqslant a^m - z/n$. Thus, $a^m - z/n < x < a^m + z/n$ for every $n \in \mathbb{N}$.

Suppose that $x > a^m$. Then, by Lemma 13 (1.4.3), $n(x - a^m) > z$ for some n, i. e. $x > a^m + z/n$. But this contradicts the proven inequality. Now, suppose that $x < a^m$. Then, in the same way, $n(a^m - x) > z$ for some n, i. e. $x < a^m - z/n$. It follows from these contradictions that $x = a^m$.

Now, suppose that $x = b^m$ for some b. If $a < b$, then by assertion 4 of Proposition 4 (1.4.3) $x = a^m < b^m = x$. If $a > b$, then by the same reason, $x = a^m > b^m = x$. It follows from these contradictions that $a = b$. □

The number a from Theorem 2 is called the *root of the number $x \in \mathbb{R}_+$ with the natural exponent m* and is denoted by $\sqrt[m]{x}$. It is also called the *degree of the number $x \in \mathbb{R}_+$ with the exponent* $1/m$ and in this case is denoted by $x^{1/m}$.

Let $r \equiv m/p$ be a rational number with a denominator $p \in \mathbb{N}$ and a numerator $m \in \mathbb{Z}$. Define the degree x^r *of the number $x \in \mathbb{R}_+$ with the rational exponent r* setting $x^r \equiv (x^{1/p})^m$ if either $x > 0$ or $m \geqslant 0$.

Proposition 1. *Let* $\pi \equiv (x_i \in \mathbb{R}_+ \setminus \{0\} \mid i \in I)$ *and* $\varkappa \equiv (y_j \in \mathbb{Q} \mid j \in J)$ *be simple finite collections,* $x \in \mathbb{R}_+ \setminus \{0\}$, *and* $y, z \in \mathbb{Q}$. *Then:*
1) $1^y = 1$;
2) $0^y = 0$ *for* $y > 0$;
3) $x^{\sum(y_j \mid j \in J)} = P(x^{y_j} \mid j \in J)$;
4) $(P(x_i \mid i \in I))^y = P(x_i^y \mid i \in I)$;
5) $x^{yz} = (x^y)^z$.

Proof. Let $y \equiv m/p$, $z \equiv n/q$, and $y_j \equiv m_j/p_j$ for some $m, n, m_j \in \mathbb{Z}$ and $p, q, p_j \in \mathbb{N}$.

1. Since $1^p = 1$, we infer that $1^{1/p} = 1$. Therefore, $1^y \equiv (1^{1/p})^m = 1^m = 1$.

2. Since $0^p = 0$, we infer that $0^{1/p} = 0$. From $y > 0$, we infer that $m > 0$. Therefore, $0^y \equiv (0^{1/p})^m = 0^m = 0$.

3. Denote $P(p_j \mid j \in J)$ by v, $P(p_i \mid i \in J \setminus \{j\})$ by v_j, and $\sum(m_j v_j \mid j \in J)$ by u. Then, by definition from 1.4.2 $\sum \varkappa = u/v$. Therefore, $L \equiv x^{\sum \varkappa} = (x^{1/v})^u$. By virtue of assertion 5 of Proposition 3 (1.4.3), we have $L^v = ((x^{1/v})^v)^u = x^u$.

On the other hand, $x^{y_j} \equiv (x^{1/p_j})^{m_j}$ implies $R \equiv P(x^{y_j} \mid j \in J) = P((x^{1/p_j})^{m_j} \mid j \in J)$. By virtue of assertions 4 and 5 of Proposition 3 (1.4.3), we have $R^v = P(((x^{1/p_j})^v)^{m_j} \mid j \in J)$. Since by Theorem 1 (1.4.2) $v = p_j v_j$, we infer that $(x^{1/p_j})^v = x^{v_j}$. Consequently, by assertion 3 of Proposition 3 (1.4.3), $R^v = P(x^{m_j v_j} \mid j \in J) = x^u$.

As a result, we get the equality $L^v = R^v$. By virtue of Theorem 2, we conclude that $L = R$.

4. Since $L \equiv (P\pi)^y \equiv ((P\pi)^{1/p})^m$, we infer that $L^p = (P\pi)^m = P(x_i^m \mid i \in I)$. On the other hand, $R \equiv P(x_i^y \mid i \in I) = P((x_i^{1/p})^m \mid i \in I)$ implies $R^p = P(((x_i^{1/p})^m)^p \mid i \in I) = P(x_i^m \mid i \in I)$. As a result, we get equality $L^p = R^p$, which implies $L = R$.

5. From $yz \equiv mn/pq$ we infer that $L \equiv x^{yz} \equiv (x^{1/pq})^{mn}$. Therefore, $L^{pq} = x^{mn}$. On the other hand, for $R \equiv (x^y)^z \equiv (((x^{1/p})^m)^{1/q})^n$, we have $R^q = (x^{1/p})^{mn}$, and as a result, $R^{pq} = (R^q)^p = x^{mn}$. From the equality $L^{pq} = R^{pq}$, we conclude that $L = R$. □

Proposition 2. *Let* $x, y \in \mathbb{R}_+ \setminus \{0\}$ *and* $r, s \in \mathbb{Q}$. *Then:*
1) *if* $x < y$, *then* $x^r < y^r$ *for* $r > 0$ *and* $x^r > y^r$ *for* $r < 0$;
2) *if* $r < s$, *then* $x^r < x^s$ *for* $x > 1$ *and* $x^r > x^s$ *for* $x < 1$.

Proof. Let $r \equiv m/p$ and $s \equiv n/q$ for $m, n \in \mathbb{Z}$ and $p, q \in \mathbb{N}$.

1. By definition $L \equiv x^r \equiv (x^{1/p})^m$ and $R \equiv y^r \equiv (y^{1/p})^m$. Therefore, $L^p = x^m$ and $R^p = y^m$.

If $r > 0$, then $m > 0$. Consequently, by virtue of assertion 4 of Proposition 4 (1.4.3) $x^m < y^m$, where $L^p < R^p$. Suppose that $L \geqslant R$. Then, by the same reason, $L^p \geqslant R^p$. Since this inequality contradicts the previous one, we conclude that $L < R$.

For $r < 0$, the arguments are the same.

2. By definition, $L \equiv x^r \equiv (x^{1/p})^m$ and $R \equiv x^s \equiv (x^{1/q})^n$. Therefore, $L^{pq} = x^{qm}$ and $R^{pq} = x^{pn}$.

By definition from 1.4.2 $r < s$ implies $mq \neq np$ and $(mq-np)pq \leqslant 0$. Suppose that $mq - np > 0$. Then, by assertion 3 of Proposition 3 (1.4.1), $(mq - np)pq > 0$, but this contradicts the previous inequality. Thus, $mq < np$.

If $x > 1$, then by assertion 5 of Proposition 4 (1.4.3) $x^{qm} < x^{pn}$, where $L^{pq} < R^{pq}$. As above, this implies $L < R$.

If $x < 1$, then using the similar arguments we deduce that $L > R$. □

Now, we shall prove that rising to a natural degree and taking a natural root possess properties of "acceleration" and "deceleration", respectively.

Lemma 1. *Let $x, y \in \mathbb{R}$, $m \in \mathbb{N}$, and $0 < y < x$. Then, $(x+y)^m - x^m > x^m - (x-y)^m$.*

Proof. By Theorem 1 $(x+y)^m - x^m = \sum(C_m^k x^{m-k}y^k \mid k \in (m+1)\setminus 1) > \sum((-1)^{k+1}C_m^k x^{m-k}y^k \mid k \in (m+1)\setminus 1) = x^m - (x-y)^m$. □

Corollary 1. *Let $x, y \in \mathbb{R}$, $m \in \mathbb{N}$, and $0 < y < x$. Then, $\sqrt[m]{x+y} - \sqrt[m]{x} < \sqrt[m]{x} - \sqrt[m]{x-y}$.*

Proof. Denote $\sqrt[m]{x}$ by a and $\sqrt[m]{x+y} - \sqrt[m]{x}$ by b. Then, $b < \sqrt[m]{2x} - a \leqslant \sqrt[m]{2^m x} - a = a$. Suppose that $b \geqslant a - \sqrt[m]{x-y}$, i. e. $0 < a - b \leqslant \sqrt[m]{x-y}$. By Lemma 1, $y = x + y - x = (a+b)^m - a^m > a^m - (a-b)^m \geqslant x-(x-y) = y$, but this is impossible. Now, the necessary inequality follows from this contradiction. □

Proposition 3. *Let $x \in \mathbb{R}_+ \setminus \{0\}$. Then, $\lim(\sqrt[m]{x} \mid m \in \mathbb{N}) = 1$.*

Proof. At first, assume that $x > 1$. Then, by virtue of Proposition 2 $\sqrt[m]{x} > 1$. Take $\varepsilon > 0$ and suppose that for every n, there is $m \geqslant n$ such that $\sqrt[m]{x} - 1 \geqslant \varepsilon$. By Lemma 13 (1.4.3), there is $n \in \mathbb{N}$ such that $n\varepsilon > x$, where $m\varepsilon \geqslant n\varepsilon > x$. At the same time, by Theorem 1 for this m, we have $x \geqslant (1+\varepsilon)^m = 1 + m\varepsilon + \dots + m\varepsilon^{m-1} + \varepsilon^m \geqslant m\varepsilon$. It follows from this contradiction that there is n such that $m \geqslant n$ implies $0 < \sqrt[m]{x} - 1 < \varepsilon$.

Now, assume that $x < 1$. Take $\varepsilon > 0$. Then, for $y \equiv 1/x > 1$, there is n such that $\sqrt[m]{y} - 1 < \varepsilon$ for every $m \geqslant n$. Using assertion 5 of Proposition 1, we get $\varepsilon > \sqrt[m]{y} - 1 = (x^{-1})^{1/m} - 1 = (x^{1/m})^{-1} - 1 = 1/\sqrt[m]{x} - 1 = (1 - \sqrt[m]{x})/\sqrt[m]{x}$. This implies $1 > \sqrt[m]{x} > 1/(1+\varepsilon)$ for every $m \geqslant n$. Consequently, $0 < 1 - \sqrt[m]{x} < 1 - 1/(1+\varepsilon) = \varepsilon/(1+\varepsilon) < \varepsilon$. □

In conclusion, we shall prove that $\mathbb{R} \neq \mathbb{Q}$.

Lemma 2. *Let $r \in \mathbb{Q}$. Then, $r^2 \neq 2$.*

Proof. Suppose that $r^2 = 2$. Let $r = m/p$ for some $m \in \mathbb{Z}$ and $p \in \mathbb{N}$. We can assume that m and p have no common divisor. From $m^2/p^2 = 2$, we infer that $m^2 = 2p^2$, i. e. m^2 is even. Suppose that $m = 2k+1$ for some $k \in \mathbb{Z}$. Then, $m^2 = 2k^2 + 2k + 1 = 2(2k^2 + k) + 1$ means that m^2 is odd. It follows from this contradiction that $m = 2k$. But then,

$2p^2 = m^2 = 4k^2$ implies $p^2 = 2k^2$, i. e. p^2 is even. As above, this implies that $p = 2l$ for some $l \in \mathbb{N}$. Thus, m and p have the common divisor 2. It follows from this contradiction that $r^2 \neq 2$. □

Corollary 1. $\sqrt{2} \in \mathbb{R} \setminus \mathbb{Q}$.

Proof. The assertion follows from Theorem 2 and Lemma 2. □

1.4.7 Convergence of nets in the extended real line

Limits of nets in $\overline{\mathbb{R}}$

In this subsection, we shall consider nets $s \equiv (x_n \in \overline{\mathbb{R}} \mid n \in N)$ of extended real numbers indexed by the principal set N of an upward directed preordered infinite set $(N, \leqslant)$ (see 1.1.15). According to 1.2.6, such nets are called *infinite.*

As in 1.4.4, for a preordered set $(N, \leqslant)$ closed final intervals $[n, \rightarrow[\equiv \{p \in N \mid p \geqslant n\}$ with the beginnings $n \in N$ will be denoted also by N_n.

A net s is called *convergent to a number* $x \in \mathbb{R}$ and the number x is called a *limit of the net* s if for every number $\varepsilon > 0$, there is an index $n \in N$ such that $|x_p - x| < \varepsilon$ for all $p \in N$ such that $p \geqslant n$.

Lemma 1. *Let* $s \equiv (x_n \in \overline{\mathbb{R}} \mid n \in N)$ *be an infinite net,* $x, y \in \mathbb{R}$*, and* s *convergent to* x *and to* y*. Then,* $x = y$*.*

Proof. Suppose that $x \neq y$. Then, for $\varepsilon \equiv |x - y| > 0$, there are $m, n \in N$ such that $|x_p - x| < \varepsilon/2$ and $|x_q - y| < \varepsilon/2$ for all $p, q \in N$ such that $p \geqslant m$ and $q \geqslant n$. Take some $k \in N$ such that $k \geqslant m$ and $k \geqslant n$. It follows from these inequalities that $x_r \in \mathbb{R}$ for all $r \in N$ such that $r \geqslant k$. Therefore, $|x - y| \leqslant |x - x_r| + |x_r - y| < \varepsilon$, but this is impossible. Thus, $x = y$. □

Thus, the net s can have a unique real limit. To denote the property that $x \in \mathbb{R}$ is a limit of s, we shall write $x = \lim s$ or $x = \lim(x_n \mid n \in N)$.

Lemma 2. *Let* $s \equiv (x_n \mid n \in N)$ *be a net and* $x \in \mathbb{R}$*. Then, the following conclusions are equivalent:*

1) $x = \lim s$;
2) $0 = \lim(x_n - x \mid n \in N)$;
3) $0 = \lim(|x_n - x| \mid n \in N)$.

Proof. The assertion follows from the equalities $|x_p - x| = |0 - (x_p - x)| = |0 - |x_p - x||$. □

The net s is called *convergent to the number* $x \equiv \infty$ $[x \equiv -\infty]$ and the number x is called the *limit of the net* s if for every real number $\delta > 0$ $[\delta < 0]$ there is an index $n \in N$ such that $x_p > \delta$ $[x_p < \delta]$ for all $p \in N$ such that $p \geqslant n$. This property is denoted by $x = \lim s$ or by $x = \lim(x_n \mid n \in N)$.

As above, for a net $s \equiv (x_n \in \overline{\mathbb{R}} \mid n \in N)$, a number $x \in \mathbb{R}$, and a number $\varepsilon \in \mathbb{R}_+ \setminus \{0\}$, we shall consider the set $C(s, x, \varepsilon) \equiv \{n \in N \mid \forall p \in N\,(p \geqslant n \Rightarrow |x_p - x| < \varepsilon)\}$; and besides for the net s, a number $x = \infty$ $[x = -\infty]$, and a real number $\delta > 0$ $[\delta < 0]$ we shall consider the set $C(s, x, \delta) \equiv \{n \in N \mid \forall p \in N\,(p \geqslant n \Rightarrow x_p > \delta\ [x_p < \delta])\}$.

Lemma 3. *Let $s \equiv (x_n \in \overline{\mathbb{R}} \mid n \in N)$ be a net and $x \in \overline{\mathbb{R}}$. Then, for the ordered set $\overline{\mathbb{R}}$, the following conclusions are equivalent:*
1) $x = \lim s$;
2) $x = \text{o-}\lim s$ (see 1.1.15).

Proof. (1) ⊢ (2). For every $n \in N$, consider the set $N_n \equiv \{p \in N \mid p \geqslant n\}$. Since the ordered set $\overline{\mathbb{R}}$ is order complete in the sense of 1.1.15, there are nets $\underline{s} \equiv (y_n \in \overline{\mathbb{R}} \mid n \in N) \uparrow$ and $\overline{s} \equiv (z_n \in \overline{\mathbb{R}} \mid n \in N) \downarrow$ such that $y_n = \inf(x_p \mid p \in N_n)$ and $z_n = \sup(x_p \mid p \in N_n)$. If $l, m \in N$, then for $q \in N_l \cap N_m$, we have $y_l \leqslant y_q \leqslant x_q \leqslant z_q \leqslant z_m$. Therefore, there are $y, z \in \overline{\mathbb{R}}$ such that $y = \sup \underline{s}$, $z = \inf \overline{s}$, and $y \leqslant z$.

At first, assume that $x \in \mathbb{R}$. Then, by definition, for any real $\varepsilon > 0$, there is $k \in N$ such that $|x_p - x| < \varepsilon$ for every $p \in N_k$. It follows from this inequality that $x_p \in \mathbb{R}$ for all $p \in N_k$. Consequently, $x - \varepsilon \leqslant y_k \leqslant x_p \leqslant z_k \leqslant x + \varepsilon$, where $y_k, z_k \in \mathbb{R}$. Then, $y_l \leqslant z_k \in \mathbb{R}$ and $z_m \geqslant y_k \in \mathbb{R}$ for all $l, m \in N$. Therefore, $y \leqslant z$ and $z \geqslant y_k$ imply $y, z \in \mathbb{R}$. Besides, $x - \varepsilon \leqslant y_k \leqslant y$ and $z \leqslant z_k \leqslant x + \varepsilon$. Now, using Lemma 2 (1.4.5) we infer that $x \leqslant y \leqslant z \leqslant x$, where $y = z = x$. Thus, $\underline{s} \uparrow x$ and $\overline{s} \downarrow x$. By definition from 1.1.15 $x = \text{o-}\lim s$.

Now, assume that $x = \infty$. Then, by definition, for every real $\delta > 0$, there is $k \in N$ such that $x_p > \delta$ for every $p \in N_k$. Therefore, $y \geqslant y_k \geqslant \delta$ implies $y = \infty = x$. Thus, $\underline{s} \uparrow x$. Therefore, the net $\overline{t} \equiv (v_n \in \overline{\mathbb{R}} \mid n \in N) \downarrow x$ such that $v_n \equiv \infty$ for every n. Since $y_n \leqslant x_n \leqslant v_n$, we conclude that $x = \text{o-}\lim s$.

Finally, assume that $x = -\infty$. Then, for every real $\delta < 0$, there is $k \in N$ such that $x_p < \delta$ for every $p \in N_k$. Therefore, $z \leqslant z_k \leqslant \delta$ implies $z = -\infty = x$. Thus, $\overline{s} \downarrow x$. Consider the net $\underline{t} \equiv (u_n \in \overline{\mathbb{R}} \mid n \in N) \uparrow x$ such that $u_n \equiv -\infty$ for every n. Since $u_n \leqslant x_n \leqslant z_n$, we conclude that $x = \text{o-}\lim s$.

2) ⊢ 1). Let there be nets $(y_n \in \overline{\mathbb{R}} \mid n \in N) \uparrow x$ and $(z_n \in \overline{\mathbb{R}} \mid n \in N) \downarrow x$ such that $y_n \leqslant x_n \leqslant z_n$ for every $n \in N$.

At first, assume that $x \in \mathbb{R}$. Then, for every real $\varepsilon > 0$, there are $l, m \in N$ such that $x \geqslant y_p > x - \varepsilon$ and $x \leqslant z_q < x + \varepsilon$ for all $p \in N_l$ and $q \in N_m$. Take $k \in N_l \cap N_m$. Then, $x - \varepsilon < y_r \leqslant x_r \leqslant z_r < x + \varepsilon$ implies $x_r \in \mathbb{R}$ and $-\varepsilon < x_r - x < \varepsilon$ for all $r \in N_k$. This means that $x = \lim s$.

Now, assume that $x = \infty$. Take any real $\delta > 0$ and suppose that $y_n \leqslant \delta$ for every $n \in N$. Then, $\infty = x = \sup(y_n \mid n \in N) \leqslant \delta$. It follows from this contradiction that there is $n \in N$ such that $y_n > \delta$. Therefore, $x_p \geqslant y_p \geqslant y_n > \delta$ for all $p \in N_n$. This means that $x = \lim s$.

Finally, assume that $x = -\infty$. Take any real $\delta < 0$ and suppose that $z_n \geqslant \delta$ for every $n \in N$. Then, $-\infty = x = \inf(z_n \mid n \in N) \geqslant \delta$. It follows from this contradiction that there is $n \in N$ such that $z_n < \delta$. Therefore, $x_p \leqslant z_p \leqslant z_n < \delta$ for all $p \in N_n$. Thus, $x = \lim s$. □

Properties of the limits

Proposition 1. *Let $X \subset \overline{\mathbb{R}}$, $s \equiv (x_n \in X \mid n \in N)$ and $t \equiv (y_n \in X \mid n \in N)$ be infinite nets, $x, y \in X$, $x = \lim s$, and $y = \lim t$. Then:*

1) $x + y = \lim(x_n + y_n \mid n \in N)$ *for* $X = \mathbb{R} \cup \{\infty\}$ *or* $X = \mathbb{R} \cup \{-\infty\}$;
2) $xy = \lim(x_n y_n \mid n \in N)$ *for* $X = \mathbb{R}$ *or* $X = \overline{\mathbb{R}} \setminus \{0\}$;
3) $1/x = \lim(1/x_n \mid n \in N)$ *for* $X = \mathbb{R} \setminus \{0\}$;
4) $x \veebar y = \lim(x_n \veebar y_n \mid n \in N)$ *for* $X = \overline{\mathbb{R}}$;
5) $x \barwedge y = \lim(x_n \barwedge y_n \mid n \in N)$ *for* $X = \overline{\mathbb{R}}$.

Proof. Denote the set $\{l \in N \mid l \geqslant k\}$ for $k \in N$ by N_k.

1. We shall consider only the case $X = \mathbb{R} \cup \{\infty\}$. The other case is considered in a similar way.

At first, assume that $x, y \in \mathbb{R}$. Take any $\varepsilon > 0$ and some $m \in C(s, x, \varepsilon/2)$, $n \in C(t, y, \varepsilon/2)$, and $k \in N_m \cap N_n$. Then, $|x - x_p| < \varepsilon/2$ and $|y - y_p| < \varepsilon/2$ imply x_p, $y_p \in \mathbb{R}$ for every $p \in N_k$. Therefore, by assertion 5 of Proposition 6 (1.4.3) $|x + y - (x_p + y_p)| \leqslant |x - x_p| + |y - y_p| < \varepsilon$ for every $p \in N_k$. This means that $x + y = \lim(x_n + y_n \mid n \in N)$.

Now, assume that $x \in \mathbb{R}$ and $y = \infty$. Fix any $\delta > 0$ and take some $m \in C(s, x, 1)$, $n \in C(t, y, (\delta - x + 1) \veebar 1)$, and $k \in N_m \cap N_n$. Then, $|x - x_p| < 1$ and $y_p > (\delta - x + 1) \veebar 1$ for every $p \in N_k$ imply $x_p + y_p > (x - 1) + (\delta - x + 1) = \delta$. This means that $x + y = \infty = \lim(x_n + y_n \mid n \in N)$.

If $x = \infty$ and $y \in \mathbb{R}$, then the arguments are the same. Finally, assume that $x = y = \infty$. Fix any $\delta > 0$ and take some $m \in C(s, x, \delta)$, $n \in C(t, y, \delta)$, and $k \in N_m \cap N_n$. Then, $x_p > \delta$ and $y_p > \delta$ for every $p \in N_k$ imply $x_p + y_p > \delta$. This means that $x + y = \infty = \lim(x_n + y_n \mid n \in N)$.

2. At first, consider the case $X = \mathbb{R}$. Take some $l \in C(t, y, 1)$. Then, $|y_p| \leqslant |y - 1| \veebar |y + 1| \equiv b$ for every $p \in N_l$. Take any $\varepsilon > 0$ and some $m \in C(s, x, \varepsilon/2(|x| + 1))$, $n \in C(t, y, \varepsilon/2b)$, and $k \in N_l \cap N_m \cap N_n$. Then, by Proposition 6 (1.4.3), $|xy - x_p y_p| \leqslant |xy - xy_p + xy_p - x_p y_p| \leqslant |x|\,|y - y_p| + |y_p|\,|x - x_p| < |x|\varepsilon/2(|x| + 1) + b\varepsilon/2b \leqslant \varepsilon$ for every $p \in N_k$. This means that $xy = \lim(x_n y_n \mid n \in N)$.

Now, consider the case $X = \overline{\mathbb{R}} \setminus \{0\}$. At first, assume that $x, y \in \mathbb{R} \setminus \{0\}$. Then, as above, $|y_p| \leqslant b$ for every $p \in N_l$. Take any $\varepsilon > 0$ and some m, n and k as above. Then, $x_p, y_p \in \mathbb{R}$ for every $p \in N_k$. Therefore, by the same arguments as above, we infer that $|xy - x_p y_p| < \varepsilon$ for every $p \in N_k$. This means that $xy = \lim(x_n y_n \mid n \in N)$.

Now, assume that $-\infty < x < 0$ and $y = \infty$. Take any $\delta < 0$ and some $m \in C(s, x, -x/2)$, $n \in C(t, y, 2\delta/x)$, and $k \in N_m \cap N_n$. Then, $|x - x_p| < -x/2$ and $y_p > 2\delta/x$ imply $x_p y_p \leqslant (x/2)(2\delta/x) = \delta$ for every $p \in N_k$. This means that $xy = -\infty = \lim(x_n y_n \mid n \in N)$.

Finally, assume that $x = -\infty$ and $y = \infty$. Take any $\delta < 0$ and some $m \in C(s, x, -\sqrt{-\delta})$, $n \in C(t, y, \sqrt{-\delta})$, and $k \in N_m \cap N_n$ (see Theorem 2 (1.4.6)). Then, $x_p < -\sqrt{-\delta}$ and $y_p > \sqrt{-\delta}$ imply $x_p y_p < \delta$ for every $p \in N_k$. This means that $xy = -\infty = \lim(x_n y_n \mid n \in N)$.

All the other opportunities are considered in a similar way.

3. Take any $\varepsilon > 0$ and some $m \in C(s, x, |x|/2)$. Then, $|x - x_p| < |x|/2$ implies $x - |x|/2 < x_p < x + |x|/2$ for every $p \in N_m$. If $x > 0$, then $x/2 < x_p$; if $x < 0$, then $x_p < x/2$. In both cases, $|x_p| > |x|/2$. Take some $n \in C(s, x, |x|^2\varepsilon/2)$ and $k \in N_m \cap N_n$. Then, $|1/x - 1/x_p| \leqslant |x - x_p|/|x|\,|x_p| < \varepsilon$ for every $p \in N_k$. This means that $1/x = \lim(1/x_n \mid n \in N)$.

4. At first, assume that $x, y \in \mathbb{R}$. Take any $\varepsilon > 0$ and some $m \in C(s, x, \varepsilon/2)$, $n \in C(t, y, \varepsilon/2)$, and $k \in N_m \cap N_n$. Then, $x_p, y_p \in \mathbb{R}$ for every $p \in N_k$. Therefore, by Corollary 1 to Theorem 4 (1.4.5), $|x \veebar y - x_p \veebar y_p| \leqslant |x \veebar y - x \veebar y_p| + |x \veebar y_p - x_p \veebar y_p| \leqslant |y - y_p| + |x - x_p| < \varepsilon$ for every $p \in N_k$. This means that $x \veebar y = \lim(x_n \veebar y_n \mid n \in N)$.

Now, assume that $x \in \mathbb{R}$ and $y = -\infty$. Take any real $\varepsilon > 0$ and some $m \in C(s, x, \varepsilon)$, $n \in C(t, y, -|x| - \varepsilon)$, and $k \in N_m \cap N_n$. Then, $|x - x_p| < \varepsilon$ and $y_p < -|x| - \varepsilon$ imply $y_p < x - \varepsilon < x_p$ for every $p \in N_k$. Thus, $x_p \veebar y_p = x_p$ and $x \veebar y = x \in \mathbb{R}$ imply $|x \veebar y - x_p \veebar y_p| = |x - x_p| < \varepsilon$ for every $p \in N_k$. This means that $x \veebar y = \lim(x_n \veebar y_n \mid n \in N)$.

Finally, assume that $x = \infty$ and $y = -\infty$. Take any $\delta > 0$ and some $m \in C(s, x, \delta)$, $n \in C(t, y, -\delta)$, and $k \in N_m \cap N_n$. Then, $x_p > \delta$ and $y_p < -\delta$ for every $p \in N_k$. Since $x \veebar y = \infty$ and $x_p \veebar y_p = x_p > \delta$, we infer that $x \veebar y = \lim(x_n \veebar y_n \mid n \in N)$.

All the other opportunities are considered in a similar way.

Assertion 5 is checked analogously to assertion 4. □

Corollary 1. *Let X be $\mathbb{R}$ or $\overline{\mathbb{R}} \setminus \{0\}$, $(x_n \in X \mid n \in N)$ be an infinite net, $x, y \in X$, and $x = \lim(x_n \mid n \in N)$. Then, $yx = \lim(yx_n \mid n \in N)$.*

Proof. Consider the constant net $t \equiv (y_n \mid n \in N)$ such that $y_n \equiv y$. Then, we get a partial case of assertion 2 of Proposition 1. □

Corollary 2. *Let $(x_n \in \overline{\mathbb{R}} \mid n \in N)$ be an infinite net, $x \in \overline{\mathbb{R}}$, and $x = \lim(x_n \mid n \in N)$. Then, $|x| = \lim(|x_n| \mid n \in N)$.*

Proof. By Corollary 1, we have $-x = \lim(-x_n \mid n \in N)$. Therefore, by assertion 4 of Proposition 1 we obtain $|x| = x \veebar (-x) = \lim(x_n \veebar (-x_n) \mid n \in N) = \lim(|x_n| \mid n \in N)$. □

Corollary 3. *Let $s \equiv (x_n \in \mathbb{R} \mid n \in N)$ and $t \equiv (y_n \in \overline{\mathbb{R}} \mid n \in N)$ be infinite nets, $x, y \in \overline{\mathbb{R}}$, $x = \lim s$, $y = \lim t$, and $x_n \leqslant y_n$ for every $n \in N$. Then, $x \leqslant y$.*

Proof. By condition $y_n = x_n \veebar y_n$. Therefore, by assertion 4 of Proposition 1 we get $y = \lim t = \lim(x_n \veebar y_n \mid n \in N) = x \veebar y$. Hence, $x \leqslant y$. □

Proposition 2. *Let $X \subset \overline{\mathbb{R}}$, $\sigma \equiv (s_i \mid i \in I)$ be a finite collection of infinite nets $s_i \equiv (x_{in} \in X \mid n \in N)$, $\varkappa \equiv (x_i \in X \mid i \in I)$ be a finite collection, and $x_i = \lim(x_{in} \mid n \in N)$ for every $i \in I$. Then:*
1) $\sum(x_i \mid i \in I) = \lim(\sum(x_{in} \mid i \in I) \mid n \in N)$ *for* $X = \mathbb{R} \cup \{\infty\}$ *or* $X = \mathbb{R} \cup \{-\infty\}$;
2) $P(x_i \mid i \in I) = \lim(P(x_{in} \mid i \in I) \mid n \in N)$ *for* $X = \mathbb{R}$ *or* $X = \overline{\mathbb{R}} \setminus \{0\}$;
3) $\mathrm{gr}(x_i \mid i \in I) = \lim(\mathrm{gr}(x_{in} \mid i \in I) \mid n \in N)$ *for* $X = \overline{\mathbb{R}}$;
4) $\mathrm{sm}(x_i \mid i \in I) = \lim(\mathrm{sm}(x_{in} \mid i \in I) \mid n \in N)$ *for* $X = \overline{\mathbb{R}}$.

Proof. We shall prove only assertion 2. All the other assertions are proven completely in the same manner.

2. Consider the set E consisting of all numbers $e \in \omega$ such that the assertion 2 is valid for all collections σ and $\varkappa$ with the condition $\mathrm{card}\, I = e + 2$.

Let $\mathrm{card}\, I = 2$. Then, $I = \{j, k\}$ for some different element j and k. It can be checked that $P\varkappa = x_j x_k$ and $P(x_{in} \mid i \in I) = x_{jn} x_{kn}$. Therefore, by virtue of assertion 2 of Proposition 1 $P\varkappa = x_j x_k = \lim(x_{jn} x_{kn} \mid n \in N) = \lim(P(x_{in} \mid i \in I) \mid n \in N)$. This means that $0 \in X$.

Suppose that $e \in E$. Take arbitrary collections σ and $\varkappa$ such that there is a bijective mapping u from $e + 3$ onto I. Consider the element $k \equiv e(e + 2) \in I$ and the sets $I_0 \equiv u[e + 2]$ and $I_1 \equiv \{k\}$. Then, $(I_m \mid m \in 2)$ is a partition of I and $\mathrm{card}\, I_0 = e + 2$.

Consider the collections $\sigma_m \equiv (s_i \mid i \in I_m)$ and $\varkappa_m \equiv (x_i \mid i \in I_m)$. By the condition, $P\varkappa_1 = x_k = \lim(x_{kn} \mid n \in N) = \lim(P(x_{in} \mid i \in I_1) \mid n \in N)$. By our supposition, $P\varkappa_0 = \lim(P(x_{in} \mid i \in I_0) \mid n \in N)$. It can be checked that $P\varkappa = P\varkappa_0 P\varkappa_1$ and $P(x_{in} \mid i \in I) = P(x_{in} \mid i \in I_0)P(x_{in} \mid i \in I_1)$ for every $n \in N$. Now, applying assertion 2 of Proposition 1, we get $P\varkappa = P\varkappa_0 P\varkappa_1 = \lim(P(x_{in} \mid i \in I_0)P(x_{in} \mid i \in I_1) \mid n \in N) = \lim(P(x_{in} \mid i \in I) \mid n \in N)$.

This means that $e + 1 \in E$. Consequently, by Theorem 1 (1.2.6) we get $E = \omega$. □

Corollary 1. *Let $(x_n \in \mathbb{R} \mid n \in N)$ be an infinite net, $x \in \mathbb{R}$, $k \in \omega$, and $x = \lim(x_n \mid n \in N)$. Then, $x^k = \lim(x_n^k \mid n \in N)$.*

Proof. If $k = 0$, then $x^0 = x_n^0 = 1$ implies the necessary equality. If $k \geqslant 1$, then the assertion follows from Lemma 7 (1.4.3) and Proposition 2. □

Lemma 4. *Let $s \equiv (x_n \in \mathbb{R} \setminus \{0\} \mid n \in N)$ be an infinite net and $\lim s = \infty$ or $\lim s = -\infty$. Then, $\lim (1/x_n \mid n \in N) = 0$.*

Proof. First let $\lim s = \infty$. Take any $\varepsilon > 0$ and some $m \in C(s, \infty, 1/\varepsilon) \equiv \{n \in N \mid \forall p \in N\ (p \geqslant n \Rightarrow x_p > 1/\varepsilon)\}$. Then, $x_p > 1/\varepsilon$ for every $p \in N_m$. Hence, by Corollary 3 to

Proposition 4 (1.4.3) $|1/x_p - 0| = 1/x_p < \varepsilon$ for every $p \in N_m$. This means that $\lim(1/x_n \mid n \in N) = 0$.

Now, let $\lim s = -\infty$. Take any $\varepsilon > 0$ and some $m \in C(s, -\infty, -1/\varepsilon) \equiv \{n \in N \mid \forall p \in N\,(p \geqslant n \Rightarrow x_p < -1/\varepsilon)\}$. Then, $x_p < -1/\varepsilon$ for every $p \in N_m$. Hence, by assertion 3 of Proposition 4 (1.4.3) and Corollary 3 to Proposition 4 (1.4.3) $|1/x_p - 0| = -1/x_p = 1/(-x_p) < \varepsilon$ for every $p \in N_m$. This means that $\lim(1/x_n \mid n \in N) = 0$. □

According to 1.1.15, a net $t \equiv (y_n \in \overline{\mathbb{R}} \mid n \in N)$ is called a *subnet of a net* $s \equiv (x_m \in \overline{\mathbb{R}} \mid m \in M)$ if there exists a collection $(m_n \in M \mid n \in N)$ such that:

1) for every index $m \in M$, there exists an index $n \in N$ such that $k \in N_n$ implies $m_k \in M_m$;
2) $y_n = x_{m_n}$ for every $n \in N$.

In this case, we shall say that t is a *subnet of s with respect to a thinning collection* $(m_n \in M \mid n \in N)$.

Lemma 5. *Let $(M, \leqslant)$ and $(N, \leqslant)$ be upward-directed preordered sets, a net $t \equiv (y_n \in \overline{\mathbb{R}} \mid n \in N)$ be a subnet of a net $s \equiv (x_m \in \overline{\mathbb{R}} \mid m \in M)$, $x \in \overline{\mathbb{R}}$, and $x = \lim s$. Then, $x = \lim t$.*

Proof. At first, assume that $x \in \mathbb{R}$. Take any $\varepsilon > 0$ and some $m \in C(s, x, \varepsilon)$. By definition for m, there is $n \in N$ such that $p \in N_n$ implies $m_p \in M_m$. Therefore, $|x - y_p| = |x - x_{m_p}| < \varepsilon$ for every $p \in N_n$. This means that $x = \lim t$.

Now, assume that $x = \infty$. Take any $\delta > 0$ and some $m \in C(s, x, \delta)$. Then, take $n \in N$ as above. Since $y_p = x_{m_p} > \delta$ for every $p \in N_n$, we infer that $x = \infty = \lim t$.

The case $x = -\infty$ is considered in a similar way. □

Lemma 6. *Let $r \equiv (x_n \in \overline{\mathbb{R}} \mid n \in N)$, $s \equiv (y_n \in \overline{\mathbb{R}} \mid n \in N)$, and $t \equiv (z_n \in \overline{\mathbb{R}} \mid n \in N)$ be nets, $x \in \overline{\mathbb{R}}$, $x = \lim s$, $x = \lim t$, and $y_n \leqslant x_n \leqslant z_n$ for every $n \in N$. Then, $x = \lim r$.*

Proof. At first, assume that $x \in \mathbb{R}$. Take any $\varepsilon > 0$ and some $m \in C(s, x, \varepsilon)$, $n \in C(t, x, \varepsilon)$, and $k \in N_m \cap N_n$. Then, $|x - y_p| < \varepsilon$ and $|x - z_p| < \varepsilon$ for every $p \in N_k$. Therefore, y_p, $z_p \in \mathbb{R}$ for every $p \in N_k$. By the condition, the same is valid for x_p. Using assertion 4 of Proposition 6 (1.4.3), we get $-\varepsilon < y_p - x \leqslant x_p - x \leqslant z_p - x < \varepsilon$, where $|x - x_p| < \varepsilon$. This means that $x = \lim r$.

Now, assume that $x = \infty$. Therefore, any $\delta > 0$ and some $n \in C(t, x, \delta)$. Then, $x_p \geqslant y_p > \delta$ for every $p \in N_n$. Thus, $x = \infty = \lim r$.

The case $x = -\infty$ is checked analogously. □

Lemma 7. *Let $s \equiv (x_n \in \overline{\mathbb{R}} \mid n \in N)$ be an increasing [a decreasing] net and $x \in \overline{\mathbb{R}}$. Then, the following conclusions are equivalent:*

1) $x = \sup s$ $[x = \inf s]$;
2) $x = \lim s$.

Proof. (1) ⊢ (2). Let $x \in \mathbb{R}$. Take any $\varepsilon > 0$. Then, there is $n \in N$ such that $x - \varepsilon < x_n$. Therefore, for every $p \in N_n$ we have $x - \varepsilon < x_n \leqslant x_p \leqslant x < x + \varepsilon$, where $|x_p - x| < \varepsilon$.

Let $x = -\infty$. Then, $x_n = -\infty$ for every $n \in N$. Thus, $x = -\infty = \lim s$.

Finally, let $x = \infty$. Take any $\delta > 0$. Then, there is $n \in N$ such that $x_n > \delta$. Therefore, $x_p \geqslant x_n > \delta$ for every $p \in N_n$. Thus, $x = \infty = \lim s$.

2) ⊢ 1). By Lemma 3, there are nets $\underline{s} \equiv (y_n \in \overline{\mathbb{R}} \mid n \in N) \uparrow x$ and $\overline{s} \equiv (z_n \in \overline{\mathbb{R}} \mid n \in N) \downarrow x$ such that $y_n \leqslant x_n \leqslant z_n$ for every $n \in N$. Take any $q \in N$. If $q \leqslant n$, then $x_n \leqslant z_n \leqslant z_q$; if $q \geqslant n$, then $x_n \leqslant x_q \leqslant z_q$. Since $x = \inf \overline{s}$, we infer that $x_n \leqslant x$ for every $n \in N$. Let $b \in \overline{\mathbb{R}}$ and $b \geqslant x_n$ for every $n \in N$. Take any $p \in N$. If $p \leqslant n$, then $b \geqslant x_n \geqslant x_p \geqslant y_p$; if $p \geqslant n$, then $b \geqslant x_p \geqslant y_p$. Since $x = \sup \underline{s}$, we infer that $b \geqslant x$. This means that $x = \sup s$. □

Proposition 3. *Let a net* $(x_n \in \mathbb{R} \mid n \in N)$ *be increasing and bounded above [decreasing and bounded below]. Then, it has the limit.*

Proof. By Theorem 1 (1.4.5) on the Dedekind completeness of the real line the collection $(x_n \mid n \in N)$ have the supremum [infimum] x. By virtue of Lemma 7, $x = \lim (x_n \mid n \in N)$. □

The exponential function

Define the sequence $(a_n \mid n \in \mathbb{N})$ of functions (mappings) $a_n : \mathbb{R} \to \mathbb{R}$ setting $a_n(x) \equiv (1 + x/n)^n$ for every $n \in \mathbb{N}$ and $x \in \mathbb{R}$. Note that $a_n(0) = 1$ for every $n \in \mathbb{N}$.

Lemma 8. *Let* $x \in \mathbb{R}$, $n \in \mathbb{N}$, $n > -x$. *Then,* $a_{n+1}(x) \geqslant a_n(x)$.

Proof. If $x = 0$, then $a_{n+1}(x) = 1 = a_n(x)$.

For $x \neq 0$, put $z \equiv 1 + x/n$ and $y \equiv 1 + x/(n+1) \neq z$. It follows from $x/n > -1$ that $z > 0$, $y > 0$. Applying Corollary 3 to Proposition 5 (1.4.3), we obtain $y^{n+1}/z^n \geqslant (n+1)y - nz = n + 1 + x - n - x = 1$. Hence, $a_{n+1}(x) = y^{n+1} \geqslant z^n = a_n(x)$. □

Thus, for every $x \in \mathbb{R}$, the sequence $(a_n(x) \mid n \in \mathbb{N}_{-x})$, where $\mathbb{N}_{-x} \equiv \{n \in \mathbb{N} \mid n > -x\}$, is increasing. Prove that this sequence is bounded above.

Lemma 9. *Let* $x \in \mathbb{R}$, $n, m \in \mathbb{N}$, $n > -x$, $m > x$. *Then,* $a_n(x) \leqslant (1 - x/m)^{-m}$.

Proof. Since $x > -n \geqslant -nm$, it follows from Corollary 4 to Proposition 5 (1.4.3) that $(1 + x/(mn))^{-n} \geqslant 1 - x/m$. Then, by assertion 4 of Proposition 4 (1.4.3), we get $(1 + x/(mn))^{mn} \leqslant (1 - x/m)^{-m}$. Using Lemma 8, we conclude that $a_n(x) \leqslant a_{nm}(x) = (1 + x/(mn))^{mn} \leqslant (1 - x/m)^{-m}$. □

Thus, for every $x \in \mathbb{R}$, the sequence $(a_n(x) \mid n \in \mathbb{N}_{-x})$ is bounded above by every number $(1 - x/m)^{-m}$ for $m > x$. Such $m \in \mathbb{N}$ exists by virtue of the Archimedes principle (Lemma 13 (1.4.3)). By Proposition 3 for every $x \in \mathbb{R}$, there is the limit $\lim(a_n(x) \mid n \in \mathbb{N})$.

This allows us to define the *exponential function* $\exp : \mathbb{R} \to \mathbb{R}$ setting $\exp x \equiv \lim((1 + x/n)^n \mid n \in \mathbb{N})$ for every $x \in \mathbb{R}$. It is obvious that $\exp 0 = 1$.

Lemma 10. *Let* $x \in \mathbb{R}$. *Then,* $1+x \leqslant a_n(x)$ *for every* $n > -x$. *If, besides,* $x < 1$, *then* $a_n(x) \leqslant 1/(1-x)$.

Proof. Since $n > -x$, we have $x/n > -1$. Then, Proposition 5 (1.4.3) guarantees that $a_n(x) \equiv (1 + x/n)^n \geqslant 1 + x$.

Corollary 4 to Proposition 5 (1.4.3) implies that $1/a_n(x) = (1+x/n)^{-n} \geqslant 1-x$. Hence, by Corollary 3 to Proposition 4 (1.4.3), we get $a_n(x) \leqslant 1/(1-x)$ for $x < 1$. □

Corollary 1. *Let* $x \in \mathbb{R}$. *Then,* $\exp x \geqslant 1 + x$. *If, besides,* $x < 1$, *then* $\exp x \leqslant 1/(1-x)$.

Proof. By Lemma 10 $a_n(x) \geqslant 1 + x$ for every $n > -x$. According to Corollary 3 to Proposition 1, this implies $\exp x \geqslant 1 + x$.

The second inequality for $x < 1$ is proven in the same way. □

Corollary 2. *Let* $x \in \mathbb{R}$. *Then,* $\exp x > 1$ *for every* $x > 0$ *and* $\exp x < 1$ *for every* $x < 0$.

Proof. For $x > 0$ by Corollary 1, we get $\exp x \geqslant 1 + x > 1$.

For $x < 0$ by Corollary 1, we obtain $\exp x \leqslant 1/(1-x) < 1$. □

Lemma 11. *Let* $(x_n \in \mathbb{R} \mid n \in \mathbb{N})$ *be a sequence such that* $\lim(x_n \mid n \in \mathbb{N}) = 0$. *Then,* $\lim(a_n(x_n) \mid n \in \mathbb{N}) = 1$.

Proof. Since $\lim(x_n \mid n \in \mathbb{N}) = 0$, by the definition, there exists $m \in \mathbb{N}$ such that $-1/2 < x_p < 1/2$ for all $p \geqslant m$. Then, $\mathbb{N}_{-x_p} \equiv \{n \in \mathbb{N} \mid n > -x_p\} \subset \{n \in \mathbb{N} \mid n > 1/2\} = \mathbb{N}$ for every $p \geqslant m$. Therefore, by virtue of Lemma 10, we have $1 + x_n \leqslant a_n(x_n) \leqslant 1/(1-x_n)$ for all $n \geqslant m$. By Proposition 1, we get $\lim(1 + x_n \mid n \in \mathbb{N}) = 0$ and $\lim 1/(1-x_n) \mid n \in \mathbb{N} = 1$. According to Lemma 6, this implies $\lim(a_n(x_n) \mid n \in \mathbb{N}) = 1$. □

Theorem 1. *Let* $x, y \in \mathbb{R}$. *Then,* $\exp(x + y) = \exp x \cdot \exp y$.

Proof. For every $n > -(x+y)$, we have the equality $(1+x/n)(1+y/n) = (1+(x+y)/n)(1+z_n/n)$, where $z_n = xy/(n + x + y)$. This provides $a_n(x)a_n(y) = a_n(x + y)a_n(z_n)$. Applying the definition of the exponential function and Lemma 11 we get $\exp x \cdot \exp y = \exp(x + y) \cdot 1$. □

Corollary 1. *Let* $x, y \in \mathbb{R}$. *Then:*
1) $\exp(-x) = 1/\exp x$;

2) $\exp x > 0$;
3) $\exp(x - y) = \exp x / \exp y$.

Proof. By Theorem 1 we have $\exp(-x)\exp x = \exp 0 = 1$.

2. Assertion 1 implies $\exp x \neq 0$ for every $x \in \mathbb{R}$. Then, using Theorem 1, we get $\exp x = \exp(x/2)\exp(x/2) > 0$.

3. Assertion 1 and Theorem 1 imply $\exp(x-y) = \exp x \cdot \exp(-y) = \exp x / \exp y$. □

Theorem 2. *Let* $(x_n \in \mathbb{R} \mid n \in N)$ *be a net,* $x \in \overline{\mathbb{R}}$, *and* $x = \lim (x_n \mid n \in N)$. *Then:*
1) *if* $x \in \mathbb{R}$, *then* $\lim (\exp x_n \mid n \in N) = \exp x$;
2) *if* $x = \infty$, *then* $\lim (\exp x_n \mid n \in N) = \infty$;
3) *if* $x = -\infty$, *then* $\lim (\exp x_n \mid n \in N) = 0$.

Proof. 1. By the definition of limit, there exists $m \in N$ such that $x_n - x < 1$ for all $n \geqslant m$. It follows from Corollary 1 to Lemma 10 that $x_n - x = 1 + x_n - x - 1 \leqslant \exp(x_n - x) - 1 \leqslant 1/(1 - (x_n - x)) - 1 = (x_n - x)/(1 - (x_n - x))$ for all $n \geqslant m$. Corollary 1 to Theorem 1 implies that $(\exp x_n - \exp x)/\exp x = \exp(x_n - x) - 1$. By Lemma 2, $\lim (x_n - x \mid n \in N) = 0$. Therefore, using Lemma 6, we get $\lim ((\exp x_n - \exp x)/\exp x \mid n \in N) = 0$. Hence, $\lim (\exp x_n - \exp x \mid n \in N) = 0$. Finally, again, by Lemma 2, $\lim (\exp x_n \mid n \in N) = x$.

2. By Corollary 1 to Lemma 10, we have $\exp x_n \geqslant 1 + x_n \equiv y_n$ for every $n \in N$. By Proposition 1, $\lim (y_n \mid n \in N) = 1 + \lim (x_n \mid n \in N) = \infty$. Taking $z_n \equiv \infty$ for every $n \in N$ we have $y_n \leqslant \exp x_n < z_n$. Therefore, Lemma 6 implies $\lim (\exp x_n \mid n \in N) = \infty$.

3. By Corollary 1 to Theorem 1, $\exp x_n = 1/\exp(-x_n)$. By Corollary 1 to Proposition 1 $\lim (-x_n \mid n \in N) = -x = \infty$. It follows from (2) that $\lim (\exp(-x_n) \mid n \in N) = \infty$. Finally, by Lemma 4 $\lim (\exp x_n \mid n \in N) = \lim (1/\exp(-x_n) \mid n \in N) = 0$. □

1.4.8 Netful and sequential series in the extended real line

Let $(x_i \in X \mid i \in I)$ be a collection of extended real numbers from the set $X \equiv \mathbb{R} \cup \{\infty\}$ or the set $X \equiv \mathbb{R} \cup \{-\infty\}$ [from the set $X \equiv \mathbb{R}$ or the set $X \equiv \overline{\mathbb{R}} \setminus \{0\}$, respectively] indexed by a non-empty set I.

Consider the ensemble $\mathcal{P}^f(I)$ of all finite non-empty subsets J of the set I. Endowed $\mathcal{P}^f(I)$ with the order by inclusion $J \leqslant K \equiv J \subset K$. With respect to this order $\mathcal{P}^f(I)$ is upwards directed. For every finite subset J of I, we can consider the extended real number $s_J \equiv \sum(x_j \mid j \in J)$ [$p_J \equiv P(x_j \mid j \in J)$] (see 1.4.3). It is called a *partial sum* [*product*] *of the collection* $(x_i \mid i \in I)$. The collection $(s_J \in X \mid J \in \mathcal{P}^f(I))$ [$(p_J \in X \mid J \in \mathcal{P}^f(I))$] is a net in $\overline{\mathbb{R}}$. It is called the *additive* [*multiplicative*] *netful series of the collection* $(x_i \in X \mid i \in I)$ and is denoted by $S^a_{\text{net}}(x_i \mid i \in I)$ [$S^m_{\text{net}}(x_i \mid i \in I)$].

If there is an extended real number, $s \in \overline{\mathbb{R}}$ [$p \in \overline{\mathbb{R}}$] such that $s = \lim S^a_{\text{net}}(x_i \mid i \in I)$ [$p = \lim S^m_{\text{net}}(x_i \mid i \in I)$], then s [p] is called the *netful sum* [*product*] *of the collection* $(x_i \mid i \in I)$ and is denoted by $\sum_{\text{net}}(x_i \mid i \in I)$ [$P_{\text{net}}(x_i \mid i \in I)$].

If $x_i \in \mathbb{R}$ for every $i \in I$ and $s \in \mathbb{R}$ [$p \in \mathbb{R}$], then the netful series $S^a_{\text{net}}(x_i \mid i \in I)$ [$S^m_{\text{net}}(x_i \mid i \in I)$] is called *convergent* (*in* $\mathbb{R}$), and the collection $(x_i \mid i \in I)$ is called *well-summarized* [*well-multiplied*]. Along with the word "well", the words "commutatively", "unconditionally", and "unorderedly" are used.

Let now the considered collection $(x_i \in X \mid i \in I)$ be a sequence $(x_i \in X \mid i \in N)$ for some infinite set $N \subset \omega$. In this particular case, by virtue of Theorem 1 (1.3.7), there is a unique isotone (see 1.1.15) bijection u from $\mathbb{N}$ into N. Therefore, we can consider for every number $n \in \mathbb{N}$ the finite set $N(n) \equiv u[n] \in \mathcal{P}^f(N)$, consisting of the first n elements of the set N, and the corresponding partial sum $s_n \equiv s_{N(n)} \equiv \sum(x_i \mid i \in N(n))$ [product $p_n \equiv p_{N(n)} \equiv P(x_i \mid i \in N(n))$]. The sequence $(s_n \in X \mid n \in \mathbb{N})$ [$(p_n \mid n \in \mathbb{N})$] is called the *additive* [*multiplicative*] (*sequential*) *series of the sequence* $(x_i \in X \mid i \in N)$ and is denoted by $S^a(x_i \mid i \in N)$ [$S^m(x_i \mid i \in N)$].

If there is an extended real number $s \in \overline{\mathbb{R}}$ [$p \in \overline{\mathbb{R}}$] such that $s = \lim S^a(x_i \mid i \in N)$ [$p = \lim S^m(x_i \mid i \in N)$], then s [p] is called the (*sequential*) *sum* [*product*] *of the sequence* $(x_i \mid i \in N)$ and is denoted by $\sum(x_i \mid i \in N)$ [$P(x_i \mid i \in N)$].

If $x_i \in \mathbb{R}$ for every $i \in N$ and $s \in \mathbb{R}$ [$p \in \mathbb{R}$], then the series $S^a(x_i \mid i \in N)$ [$S^m(x_i \mid i \in N)$] is called *convergent* (*in* $\mathbb{R}$), and the sequence $(x_i \mid i \in N)$ is called (*sequentially*) *summarized* [*multiplied*]. Along with the word "sequentially" the words "conditionally" and "orderedly" are used.

Lemma 1. *Let X be the set $\mathbb{R} \cup \{\infty\}$ or the set $\mathbb{R} \cup \{-\infty\}$ [the set $\mathbb{R}$ or the set $\overline{\mathbb{R}} \setminus \{0\}$] and $(x_i \in X \mid i \in N)$ be an infinite sequence. Then, the net $S^a(x_i \mid i \in N) \equiv (s_n \in X \mid n \in \mathbb{N})$ [$S^m(x_i \mid i \in N) \equiv (p_n \in X \mid n \in \mathbb{N})$] is a subnet of the net $S^a_{\text{net}}(x_i \mid i \in N) \equiv (s_J \in X \mid J \in \mathcal{P}^f(N))$ [$S^m_{\text{net}}(x_i \mid i \in N) \equiv (p_J \in X \mid J \in \mathcal{P}^f(N))$].*

Proof. Consider the corresponding collection $(N(n) \in \mathcal{P}^f(N) \mid n \in \mathbb{N})$. By definition, $s_n = s_{N(n)}$. Take any $J \in \mathcal{P}^f(N)$. Since $J \subset N \subset \omega$, by Theorem 3 (1.2.6), the collection $(j \in \omega \mid j \in J)$ has the greatest element $j_0 \in J$. Since u is bijective, $j_0 = u(q)$ for some $q \in \mathbb{N}$. If $j \in J$, then $j = u(p)$ for some $p \in \mathbb{N}$. From the inequality $u(p) = j \leqslant j_0 = u(q)$, we infer that $p \leqslant q$, where $p \in q + 1$. Denote $q + 1$ by n. We proven that $J \subset u[n] \equiv N(n)$. If $k \in \mathbb{N}_n$, then $n \subset k$ implies $N(n) \subset N(k)$, i. e. $N(k) \geqslant N(n) \geqslant J$ in the ordered set $(\mathcal{P}^f(N), \leqslant)$. □

Corollary 1. *In the conditions of Lemma 1, let $s \in \overline{\mathbb{R}}$ [$p \in \overline{\mathbb{R}}$] and $s = \sum_{\text{net}}(x_i \mid i \in N)$ [$p = P_{\text{net}}(x_i \mid i \in N)$]. Then, $s = \sum(x_i \mid i \in N)$ [$p = P(x_i \mid i \in N)$].*

Proof. By conditions $s = \lim S^a_{\text{net}}(x_i \mid i \in N)$. Therefore, by Lemma 1 and Lemma 5 (1.4.7) $s = \lim S^a(x_i \mid i \in N)$. □

It follows from this corollary that if a sequence $(x_i \in \mathbb{R} \mid i \in N)$ is well-summarized [well-multiplied], then it is sequentially summarized [multiplied]. But the converse assertions are not true. It follows from the following lemma.

Lemma 2. *Suppose $N \equiv \mathbb{N}$ and $x \equiv (x_i \mid i \in \mathbb{N})$, such that $x_i = \frac{(-1)^i}{i}$ for every $i \in \mathbb{N}$; then, the sequence x is sequentially summarized but is not well-summarized.*

Proof. 1. Prove that the sequence $s \equiv (s_n \mid n \in \mathbb{N})$, where $s_n = \sum(x_i \mid i \in n)$, is inner convergent. Take $\varepsilon > 0$, $n \in \mathbb{N}$ such that $n\varepsilon > 1$ from Lemma 13 (1.4.3), $p, k \in \mathbb{N}$, $p \geqslant n$, and $q = p + k$. Then, $|s_q - s_p| = \frac{1}{p}$ if $k = 1$,

$$\begin{aligned}|s_q - s_p| &= \frac{1}{p} - \frac{1}{p+1} + \frac{1}{p+2} - \ldots - \frac{1}{p+k-2} + \frac{1}{p+k-1} = \\ &= \frac{1}{p} - \frac{1}{(p+1)(p+2)} - \ldots - \frac{1}{(p+k-2)(p+k-1)} < \frac{1}{p}\end{aligned}$$

if k is odd and $k \neq 1$, and

$$\begin{aligned}|s_q - s_p| &= \frac{1}{p} - \frac{1}{p+1} + \frac{1}{p+2} - \ldots - \frac{1}{p+k-3} + \frac{1}{p+k-2} - \frac{1}{p+k-1} = \\ &= \frac{1}{p} - \frac{1}{(p+1)(p+2)} - \ldots - \frac{1}{(p+k-2)(p+k-1)} - \frac{1}{p+k-1} < \frac{1}{p}\end{aligned}$$

if k is even. In all the cases, $|s_q - s_p| \leqslant \frac{1}{n} < \varepsilon$; hence, s is inner convergent. By Theorem 1 (1.4.4), the sequence s is convergent and, therefore, the sequence x is sequentially summarized.

2. For every $k \in \mathbb{N}$, consider the subsets $I_k = u[2k+1] \cup v[k+1]$ and $J_k = u[2^k+1] \cup v[k+1]$, where u and v are mappings from $\mathbb{N}$ to $\mathbb{N}$ such that $u(n) = 2n$, $v(n) = 2n - 1$ for all $n \in \mathbb{N}$. We claim that for the subnets of partial sums $t_{I_k} \equiv \sum(x_i \mid i \in I_k)$ and $t_{J_k} \equiv \sum(x_i \mid i \in J_k)$, we have the inequalities $t_{I_k} < \frac{1}{2} < t_{J_k}$ for every $k \geqslant 10$. Indeed, $t_{I_k} = \sum\left(\frac{1}{2n} \mid n \in 2k+1\right) - \sum\left(\frac{1}{2n-1} \mid n \in k+1\right) < \sum\left(\frac{1}{2n} \mid n \in 2k+1\right) - \sum\left(\frac{1}{2n} \mid n \in k+1\right) = \frac{1}{2(k+1)} + \ldots + \frac{1}{2 \cdot 2k} < \frac{k}{2(k+1)} < \frac{1}{2}$ and

$$\begin{aligned}t_{J_k} &= \sum\left(\frac{1}{2n} \mid n \in 2^k+1\right) - \sum\left(\frac{1}{2n-1} \mid n \in k+1\right) > \\ &> \sum\left(\frac{1}{2n} \mid n \in 2^k+1\right) - \sum\left(\frac{1}{2n-2} \mid n \in (k+1) \setminus 2\right) - 1 = \\ &= \frac{1}{2}\left(\frac{1}{k} + \frac{1}{k+1} + \ldots + \frac{1}{2^k}\right) - 1 > \\ &> \frac{1}{2}\left(\frac{2^k - 2^{k-1}}{2^k} + \frac{2^{k-1} - 2^{k-2}}{2^{k-1}} + \ldots + \frac{2^{k-5} - 2^{k-6}}{2^{k-6}}\right) - 1 = \\ &= \frac{1}{2} \cdot 6 \cdot \frac{1}{2} - 1 = \frac{1}{2}.\end{aligned}$$

Then, it follows from Lemma 5 (1.4.7) that there is no $t \in \mathbb{R}$ such that $t = \lim S^a_{\text{net}}(x_i \mid i \in I)$ and the sequence x is not well-summarized. □

Proposition 1. *Let $X = [0, \infty]$ [respectively, $X = [1, \infty]$], $\sigma \equiv (x_i \in X \mid i \in N)$ be an infinite sequence, $m \equiv \mathrm{sm}(n \in \omega \mid n \in N)+1$, and $\tau \equiv (t_n \in X \mid n \in \mathbb{N}_m)$ be a sequence such that $t_n = \sum(x_i \mid i \in N \cap n)$ [$t_n = P(x_i \mid i \in N \cap n)$] for every $n \in \mathbb{N}_m$. Then, there is $s \in X$ [$p \in X$] such that $s = \sup \tau$ [$p = \sup \tau$]. Moreover, $s = \sum_{\mathrm{net}} \sigma$ and $s = \sum \sigma$ [$p = P_{\mathrm{net}}\sigma$ and $p = P\sigma$].*

Proof. The sequence τ is increasing. By Theorem 2 (1.4.5), there is $s \in \overline{\mathbb{R}}$ such that $s = \sup \tau$. It is clear that $s \in X$.

Take any $J \in \mathcal{P}^f(N)$. By Theorem 3 (1.2.6), the collection $(j \in \omega \mid j \in J)$ has the greatest element $j_0 \in J$. Take $n \equiv j_0+1$. Then, $n \geqslant m$. From $J \subset N \cap n$, we infer that $s_J \equiv \sum(x_i \mid i \in J) \leqslant t_n \leqslant s$. Thus, s is an upper bound of the collection $\eta \equiv (s_J \mid J \in \mathcal{P}^f(N))$. If $b \in \overline{\mathbb{R}}$ and b is another upper bound of η, then b is an upper bound of τ. Therefore, $b \geqslant s$. Consequently, $s = \sup \eta$.

At first, assume that s is a real number. Then, s_J is also a real number for every J. Take any $\varepsilon > 0$. Then, by Lemma 1 (1.4.5), there is $J \in \mathcal{P}^f(N)$ such that $s - \varepsilon < s_J$. Since the collection η is increasing, we get $s-\varepsilon < s_J \leqslant s_K \leqslant s < s+\varepsilon$ for every $K \in \mathcal{P}^f(N)$ such that $K \geqslant J$. Thus, $|s - s| < \varepsilon$ means that $s = \lim \eta \equiv \sum_{\mathrm{net}} \sigma$.

Now, assume that $s = \infty$. Take any $\delta > 0$. Then, $t_n > \delta$ for some $n \in \mathbb{N}_m$. Denote $N \cap n$ by J. If $K \in \mathcal{P}^f(N)$ and $K \geqslant J$, then $s_K \geqslant s_J = t_n > \delta$. This means that $s = \infty = \lim \eta \equiv \sum_{\mathrm{net}} \sigma$. By Corollary 1 to Lemma 1 $s = \sum \sigma$. □

Consider now some important example of an additive series. Let x be a real number such that $0 < x \neq 1$. The sequence $(x^i \mid i \in \omega)$ is called the *infinite geometric progression with the base* x. The corresponding additive (sequential) series $S^a(x^i \mid i \in \omega)$ of this progression consists of the partial sums $s_n = \sum(x^i \mid i \in n)$. We shall consider this series in the ordered set $(\overline{\mathbb{R}}, \leqslant)$.

Lemma 3 (on the sum of the infinite geometric progression). *Let $x \in \mathbb{R}$ and $0 < x \neq 1$. Then, $s_n = (1 - x^n)/(1 - x) < s_{n+1}$. If $x < 1$, then $s_n < \sum(x^i \mid i \in \omega) = 1/(1 - x)$. If $x > 1$, then $n \leqslant s_n < \sum(x^i \mid i \in \omega) = \infty$.*

Proof. It is easy to check that $(1 - x)s_n = 1 - x^n$. Therefore, $s_n = (1 - x^n)/(1 - x)$.

If $x < 1$, then $0 < x^{n+1} < x^n$ and by Lemma 7 (1.4.4) $\lim(x^n \mid n \in \omega) = 0$. By virtue of Proposition 1 (1.4.7), we get $\lim(1-x^n \mid n \in \omega) = 1$ and $\sum(x^i \mid i \in \omega) \equiv \lim(s_n \mid n \in \mathbb{N}) = 1/(1 - x)$. Besides, by virtue of Proposition 4 (1.4.3), $s_n < s_{n+1} < 1/(1 - x)$.

Now, let $x > 1$. If $n \in \mathbb{N}$, then by Proposition 5 (1.4.3) $x^{n+1} > x^n \geqslant 1 + n(x - 1)$, where $1 - x^{n+1} < 1 - x^n \leqslant n(1 - x)$ and, respectively $s_{n+1} > s_n \geqslant n$. Take any $\delta > 0$. By Lemma 13 (1.4.3) $\delta < m$ for some $m \in \mathbb{N}$. Therefore, $s_p \geqslant s_m \geqslant m > \delta$ for every $p \in \mathbb{N}$ such that $p \geqslant m$. This means that $\sum(x^i \mid i \in \omega) = \lim(s_n \mid n \in \mathbb{N}) = \infty$ (see 1.4.4). □

Lemma 4. *Let $x \in \mathbb{R}$, $0 < x < 1$, and $(y_n \in \mathbb{R} \mid n \in \omega)$ be a sequence such that $|y_n - y_{n+1}| \leqslant x^n$ for every n. Then:*

1) $|y_p - y_q| < 2x^n/(1-x)$ *for all* $p, q \geqslant n$;
2) *the sequence* $(y_n \mid n \in \omega)$ *is inner convergent.*

Proof. 1. If $p > n$, then by virtue of Lemma 3, $|y_n - y_p| \leqslant \sum(|y_i - y_{i+1}| \mid i \in p \setminus n) \leqslant \sum(x^i \mid i \in p \setminus n) = x^n \sum(x^k \mid k \in p - n) < x^n/(1-x)$. If $p = n$, then $|y_n - y_p| = 0 < x^n/(1-x)$. Consequently, for all $p, q \geqslant n$ we have $|y_p - y_q| \leqslant |y_p - y_n| + |y_n - y_q| < 2x^n/(1-x)$.

2. Take any real $\varepsilon > 0$. Then, by Corollary 2 to Proposition 5 (1.4.3), $x^n < \varepsilon(1-x)/2$ for some $n \in \mathbb{N}$. Using assertion 1, we infer that $|y_p - y_q| < \varepsilon$ for all $p, q \geqslant n$. □

In conclusion, we shall prove the properties of general commutativity and associativity for netful sums and products.

Theorem 1.
1) *Let X be the set $\mathbb{R} \cup \{\infty\}$ or the set $\mathbb{R} \cup \{-\infty\}$ [the set $\mathbb{R}$ or the set $\overline{\mathbb{R}} \setminus \{0\}$], I and K be non-empty sets, $\chi \equiv (x_i \in X \mid i \in I)$ be a collection, u be a bijective mapping from the set K into the set I, and $s \in X$ [$p \in X$]. Then, $s = \sum_{\text{net}}(x_i \mid i \in I)$ iff $s = \sum_{\text{net}}(x_{u(k)} \mid k \in K)$ [$p = P_{\text{net}}(x_i \mid i \in I)$ iff $p = P_{\text{net}}(x_{u(k)} \mid k \in K)$] (the* general commutativity of the netful sum and the netful product, *respectively).*
2) *Let X be the set $\mathbb{R}$, or the set $[0, \infty]$, or the set $[-\infty, 0]$ [the set $\mathbb{R}$, or the set $[1, \infty]$, or the set $[0, 1]$], I and M be non-empty sets, $\chi \equiv (x_i \in X \mid i \in I)$ and $\alpha \equiv (a_m \in X \mid m \in M)$ be collections, a total collection $(I_m \subset I \mid m \in M)$ be a partition of the set I, and $a_m = \sum_{\text{net}}(x_i \mid i \in I_m)$ [$a_m = P_{\text{net}}(x_i \mid i \in I_m)$] for every $m \in M$. Then, $s = \sum_{\text{net}}(x_i \mid i \in I)$ implies $s = \sum_{\text{net}}(a_m \mid m \in M)$ [$p = P_{\text{net}}(x_i \mid i \in I)$ implies $p = P_{\text{net}}(a_m \mid m \in M)$]. Moreover, if X is the set $[0, \infty]$ or the set $[-\infty, 0]$ [the set $[1, \infty]$ or the set $[0, 1]$], then $s = \sum_{\text{net}} \chi$ iff $s = \sum_{\text{net}} \alpha$ [$p = P_{\text{net}}\chi$ iff $p = P_{\text{net}}\alpha$] (the* general associativity of the netful sum and the netful product, *respectively*).

Proof. 1. We shall consider only the case $X = \mathbb{R} \cup \{\infty\}$. Denote $x_{u(k)}$ by y_k, $(y_k \mid k \in K)$ by ψ, $\mathcal{P}^f(I)$ by $\mathcal{M}$, and $\mathcal{P}^f(K)$ by $\mathcal{N}$. Consider nets $\eta \equiv (s_J \mid J \in \mathcal{M})$ and $\vartheta \equiv (t_L \mid L \in \mathcal{N})$ such that $s_J \equiv \sum(x_i \mid i \in J)$ and $t_L \equiv \sum(y_k \mid k \in L)$. Let $s = \sum_{\text{net}} \chi$.

At first, assume that s is a real number. Take any real $\varepsilon > 0$ and some $J \in C(\eta, s, \varepsilon)$. Then, $|s - s_P| < \varepsilon$ for every $P \in \mathcal{M}_J$. Consequently, $s_P \in \mathbb{R}$ for such indices P. Consider the set $L \equiv u^{-1}[J] \in \mathcal{N}$.

Take any set $R \in \mathcal{N}_L$ and consider the set $P \equiv u[R]$. Then, $L \subset R \subset K$ implies $J \subset P \subset I$, i.e. $P \in \mathcal{M}_J$. From $s_P \equiv \sum(x_i \mid i \in P) \in \mathbb{R}$, we infer that $x_i \in \mathbb{R}$ for every $i \in P$. Therefore, by virtue of assertion 1 of Theorem 1 (1.4.3) $s_P = \sum(x_{u(k)} \mid k \in R) = \sum(y_k \mid k \in R) \equiv t_R$. As a result, we get $|s - t_R| = |s - s_P| < \varepsilon$. This means that $s = \lim \vartheta = \sum_{\text{net}} \psi = \sum_{\text{net}}(x_{u(k)} \mid k \in K)$.

Now, assume that $s = \infty$. Take any real $\delta > 0$ and some $J \in C(\eta, s, \delta)$. Then, $s_P > \delta$ for every $P \in \mathcal{M}_J$. Consider the set $L \equiv u^{-1}[J]$.

Take any set $R \in \mathcal{N}_L$ and consider the set $P \equiv u[R] \in \mathcal{M}_J$. If $x_i < \infty$ for every $i \in P$, then as above $s_P = \sum(x_{u(k)} \mid k \in R) = \sum(y_k \mid k \in R) \equiv t_R$. If $x_p = \infty$ for some $p \in P$, then $y_r = s_{u(r)} = x_p = \infty$ for $r \equiv u^{-1}(p) \in R$. Therefore, by definition of sums in $\overline{\mathbb{R}}$ from 1.4.3, we have $s_P = \infty = t_R$. In both cases, as a result, we get $t_R = s_P > \delta$. This means that $s = \infty = \lim \vartheta = \sum_{\text{net}} \psi = \sum_{\text{net}}(x_{u(k)} \mid k \in K)$.

Conversely, let $s = \sum_{\text{net}}(x_{u(k)} \mid k \in K) = \sum_{\text{net}} \psi$. Then, we can apply the proven property to the bijection $v \equiv u^{-1}$ from I into K. As a result, $y_{v(i)} = x_{u(v(i))} = x_i$ yields $s = \sum_{\text{net}}(y_{v(i)} \mid i \in I) = \sum_{\text{net}} \chi$.

2. We shall use some notations from 1). Denote the collection $(x_i \mid i \in I_m)$ by χ_m. Consider nets $\eta_m \equiv (s_J \mid J \in \mathcal{P}^f(I_m))$ and $\theta \equiv (S_N \mid N \in \mathcal{P}^f(M))$ such that $s_J \equiv \sum(x_i \mid i \in J)$ and $S_N \equiv \sum(a_m \mid m \in N)$.

At first, consider the case $X = [0, \infty]$. Assume that $a_m = \infty$ for some $m \in M$. Then, $\sum_{\text{net}} \alpha = \infty$. Take any real $\delta > 0$. Then, there is $J \in \mathcal{P}^f(I_m)$ such that $s_J > \delta$. If $P \in \mathcal{M}_J$, then $s_P \equiv \sum(x_i \mid i \in P) \geqslant s_J > \delta$. This means that $\infty = \lim \eta = \sum_{\text{net}} \chi$. As a result, $\sum_{\text{net}} \chi = \sum_{\text{net}} \alpha$.

Now, assume that $a_m < \infty$ for every $m \in M$. Then, $x_i < \infty$ for every $i \in I$. Let $s = \sum_{\text{net}} \chi$. At first, assume that s is a real number. Take any real $\varepsilon > 0$ and some $J \in C(\eta, s, \varepsilon/2)$. Then, $|s - s_P| < \varepsilon/2$ for every $P \in \mathcal{M}_J$. Consider the finite sets $J_m \equiv J \cap I_m$ and the non-empty set $N \equiv \{m \in M \mid J_m \neq \varnothing\}$. Then, $(J_n \mid n \in N)$ is a partition of J. Define a mapping $e\colon N \to \mathcal{P}(J)$ setting $e(n) \equiv J_n$. By Lemma 6 (1.3.3) the set $\mathcal{P}(J)$ is finite. Thus, by Lemma 7 (1.3.3), the set $e[N]$ is finite. Since e is injective, we infer that the set N is also finite.

Take any set $U \in \mathcal{P}^f(M)$ such that $U \geqslant N$. Consider the number $c \equiv \operatorname{card} U$. For every $u \in U$, take some $K_u \in \mathcal{P}^f(I_u)$ such that $|a_u - s_L| < \varepsilon/2c$ for every $L \in \mathcal{P}^f(I_u)$ such that $L \geqslant K_u$. Define a collection $\lambda \equiv (L_u \mid u \in U)$, setting $L_n \equiv K_n \cup J_n$ for every $n \in N$ and $L_u \equiv K_u$ for every $u \in U \setminus N$. This collection is a partition of the set $L \equiv \bigcup \lambda$. By Lemma 3 (1.3.3), the set L is finite. Besides, $L \geqslant J$. Therefore, $|s - s_L| < \varepsilon/2$. By virtue of assertion 2 of Theorem 1 (1.4.3), we have $s_L \equiv \sum(x_i \mid i \in L) = \sum(\sum(x_i \mid i \in L_u) \mid u \in U) = \sum(s_{L_u} \mid u \in U)$. Since $L_u \geqslant K_u$, we have $|a_u - s_{L_u}| < \varepsilon/2c$. As a result, $|s - S_U| \leqslant |s - s_L| + |\sum(s_{L_u} \mid u \in U) - \sum(a_u \mid u \in U)| < \varepsilon/2 + \sum(|s_{L_u} - a_u| \mid u \in U) < \varepsilon/2 + (\varepsilon/2c)c = \varepsilon$. This means that $s = \lim \theta = \sum_{\text{net}} \alpha$.

Now, assume that $s = \infty$. Take any real $\delta > 0$ and some $J \in C(\eta, s, \delta)$. Then, $s_J > \delta$. Consider as above the sets I_m and N and the partition $(J_n \mid n \in N)$ of the set J. Take any set $U \in \mathcal{P}^f(M)$ such that $U \geqslant N$. Since the net η_n is increasing, we infer that $a_n \geqslant s_{J_n}$. By virtue of assertion 2 of Theorem 1 (1.4.3), we have $s_j = \sum(s_{J_n} \mid n \in N)$. Therefore, $S_U \geqslant S_N = \sum(a_n \mid n \in N) = \sum(a_n - s_J \mid n \in N) + \sum(s_{J_n} \mid n \in N) \geqslant s_J > \delta$. This means that $s = \infty = \lim \theta = \sum_{\text{net}} \alpha$.

Conversely, let $s = \sum_{\text{net}} \alpha$. At first, assume that s is a real number. Take any real $\varepsilon > 0$ and some $N \in C(\alpha, s, \varepsilon/2)$. Then, $|s - S_U| < \varepsilon/2$ for every $U \in \mathcal{P}^f(M)$ such that $U \geqslant N$. Consider the number $c \equiv \operatorname{card} N$. For every $n \in N$ take some $K_n \in \mathcal{P}^f(I_n)$ such that $|a_n - s_L| < \varepsilon/2c$ for every $L \in \mathcal{P}^f(I_n)$ such that $L \geqslant K_n$. Then, the collection $\varkappa \equiv (K_n \mid n \in N)$ is a partition of the finite set $K \equiv \bigcup \varkappa$.

Take any $P \in \mathcal{M}_K$. Consider the finite sets $P_m \equiv P \cap I_m$ and the non-empty set $U \equiv \{m \in M \mid P_m \neq \varnothing\}$. Then, $(P_u \mid u \in U)$ is a partition of P. As above, we check that the set U is finite. From $K \subset P$, we infer that $s_K \leqslant s_P$. As above, we have $s_K = \sum(s_{K_n} \mid n \in N)$ and $s_P = \sum(s_{P_u} \mid u \in U)$. Since $a_u = \lim \eta_u$ and the net η_u is increasing, we infer by Lemma 7 (1.4.7) that $\infty > a_u \geqslant s_{P_u}$. By the same reason, $s = \lim \theta$ implies $s \geqslant S_U$. Applying assertion 1 of Proposition 4 (1.4.3), we get $s_P \leqslant \sum(a_u \mid u \in U) \equiv S_U \leqslant s$. Therefore, $0 \leqslant s - s_P \leqslant s - s_K = (s - S_N) + (S_N - s_K) < \varepsilon/2 + \sum(a_n - s_{K_n} \mid n \in N) < \varepsilon/2 + (\varepsilon/2c)c = \varepsilon$. This means that $s = \lim \eta = \sum_{\text{net}} \chi$.

Now, assume that $s = \infty$. Take any real $\delta > 0$ and some $N \in C(\alpha, s.2\delta)$. Then, $S_U > 2\delta$ for every $U \in \mathcal{P}^f(M)$ such that $U \geqslant N$. Consider the number $c \equiv \operatorname{card} N$. For every $n \in N$, take some $K_n \in \mathcal{P}^f(I_n)$ such that $|a_n - s_L| < \delta/c$ for every $L \in \mathcal{P}^f(I_n)$ such that $L \geqslant K_n$. Then, the collection $\varkappa \equiv (K_n \mid n \in N)$ is a partition of the finite set $K \equiv \bigcup \varkappa$.

Take any $P \in \mathcal{M}_K$. Then, $s_P \geqslant s_K = \sum(s_{K_n} \mid n \in N) = \sum(s_{K_n} - a_n \mid n \in N) + \sum(a_n \mid n \in N) \geqslant c(-\delta/c) + S_N > -\delta + 2\delta = \delta$. This means that $s = \infty = \lim \eta = \sum_{\text{net}} \chi$.

In the case $X = [-\infty, 0]$, the arguments are completely the same.

Finally, if $X = \mathbb{R}$, then we can prove only the first implication by slight modification of the previous arguments in the case when $a_m < \infty$ for every $m \in M$ and $s \in \mathbb{R}$. □

1.4.9 The order equivalence of intervals of the real line

Define a mapping u from $\mathbb{R}$ into $]-1, 1[$ setting $u(x) \equiv x/(1 + |x|)$ and a mapping v from $]-1, 1[$ into $\mathbb{R}$ setting $v(y) \equiv y/(1 - |y|)$.

Lemma 1. *The mappings u and v are bijective and isotone, $v = u^{-1}$ and $u = v^{-1}$.*

Proof. Let $0 < x' < x''$. Then, $1 + x' < 1 + x''$ implies $1/(1 + x') > 1/(1 + x'')$, where $u(x') = 1 - 1/(1 + x') < 1 - 1/(1 + x'') = u(x'')$. If $x' < x'' < 0$, then $u(x') = -1 + 1/(1 - x') < -1 + 1/(1 - x'') = u(x'')$. If $x' < 0 \leqslant x''$, then $u(x') < 0 \leqslant u(x'')$. Finally, if $x' \leqslant 0 < x''$, then $u(x') \leqslant 0 < u(x'')$. This means that u is strictly monotone.

Let $ux' < ux''$ and suppose that $x' \geqslant x''$. Then, $ux' \geqslant ux''$. This contradiction shows that $x' < x''$. Thus, u is isotone in the sense of 1.1.15.

The similar arguments prove that v is also isotone. Take any point $y \in]-1, 1[$ and the corresponding point $x \equiv v(y)$. Then, it is easy to check that $u(x) = y$. Thus, u is surjective. By Lemma 1 (1.1.15), u is bijective. Analogously, take any point $x \in \mathbb{R}$ and the corresponding point $y \equiv u(x)$. Then, $v(y) = x$. Thus, v is surjective, and by the same reason as above, v is bijective.

Besides, the equalities $v(u(x)) = x$ and $u(v(y)) = y$ for every $x \in \mathbb{R}$ and $y \in]-1, 1[$ show that $v = u^{-1}$ and $u = v^{-1}$. □

Define a mapping f from $]-1, 1[$ into $]a, b[$ setting $f(x) \equiv (b - a)x/2 + (a + b)/2$ and a mapping g from $]a, b[$ into $]-1, 1[$ setting $g(x) \equiv 2x/(b - a) - (a + b)/(b - a)$.

Lemma 2. *The mappings f and g are bijective and isotone, $g = f^{-1}$ and $f = g^{-1}$.*

Proof. Using Proposition 4 (1.4.3), we can easily check that $x' < x''$ implies $f(x') < f(x'')$. Thus, f is strictly monotone. As in the proof of Lemma 2, it is checked that f is isotone. Since $g(f(x)) = x$ and $f(g(y)) = y$ for every $x \in]-1, 1[$ and $y \in]a, b[$, we infer that f and g are bijective and mutually inverse. □

Corollary 1. *The mapping $f \circ u$ from $\mathbb{R}$ into $]a, b[$ and the mapping $v \circ g$ from $]a, b[$ into $\mathbb{R}$ are bijective and isotone, $v \circ g = (f \circ u)^{-1}$, and $f \circ u = (v \circ g)^{-1}$.*

It follows from the proven properties that the ordered set $(\mathbb{R}, \leqslant)$ and all its ordered open intervals are order equivalent. According to 1.4.4, card$]a, b[=$ card$\,\mathbb{R} = \mathfrak{c}$.

For every number, $a \in \mathbb{R}$ consider the mapping t_a from $\mathbb{R}$ into $\mathbb{R}$ such that $t_a(x) = x + a$. It is called the *translation on* $\mathbb{R}$.

Lemma 3. *The mapping t_a is bijective and isotone.*

Proof. The assertion follows from Proposition 4 (1.4.3). □

Corollary 1. *The mapping $t_a \circ v$ from $]-1, 1[$ into $\mathbb{R}$ is bijective and isotone and maps the intervals $]-1, 0[$, $]-1, 0]$, $[0, 1[$, and $]0, 1[$ onto the intervals $]\leftarrow, a[$, $]\leftarrow, a]$, $[a, \rightarrow[$, and $]a, \rightarrow[$, respectively.*

Lemma 4. *Let $x \in \mathbb{R}$, $(x_n \in \mathbb{R} \mid n \in N)$ be an infinite sequence, and $x = \lim(x_n \mid n \in N)$. Then:*
1) $u(x) = \lim(u(x_n) \mid n \in N)$ *and* $t_a(x) = \lim(t_a(x_n) \mid n \in N)$;
2) *if* $x \in]-1, 1[$ *and* $x_n \in]-1, 1[$ *for every* $n \in N$, *then* $v(x) = \lim(v(x_n) \mid n \in N)$ *and* $f(x) = \lim(f(x_n) \mid n \in N)$;
3) *if* $x \in]a, b[$ *and* $x_n \in]a, b[$ *for every* $n \in N$, *then* $g(x) = \lim(g(x_n) \mid n \in N)$.

Proof. All the assertions follow from Proposition 1 (1.4.7) and Corollary 1 to it. □

Remark. Define a mapping $\bar{u}$ from $\overline{\mathbb{R}}$ into $[-1, 1]$, a mapping $\bar{f}$ from $[-1, 1]$ into $[a, b]$, and a mapping $\bar{t}_a$ from $\overline{\mathbb{R}}$ into $\overline{\mathbb{R}}$, extending the corresponding mappings u, f, and t_a in the following way: $\bar{u}(-\infty) \equiv -1$, $\bar{u}(\infty) \equiv 1$, $\bar{f}(-1) \equiv a$, $\bar{f}(1) \equiv b$, $\bar{t}_a(-\infty) \equiv -\infty$, and $\bar{t}_a(\infty) \equiv \infty$. Then, these mappings are bijective and isotone. Thus, the corresponding ordered sets are order equivalent.

A Characterization of all natural models of Neumann – Bernays – Gödel and Zermelo – Fraenkel set theories

Introduction

The crises that arose in the naive set theory at the beginning of the 20th century brought to the origin of some strict axiomatic theories. The most widely used are the *theory of sets in Zermelo – Fraenkel's axiomatics* (ZF) [*Zermelo*, 1908; *Fraenkel*, 1922] and the *theory of classes and sets in Neumann – Bernays – Gödel's axiomatics* (NBG) [*Neumann*, 1929; *Bernays*, 1976; *Gödel*, 1940].

D. Mirimanov [1917], using transfinite induction, constructed the *cumulative collection* (≡ *hierarchy*) of sets V_α for all ordinal numbers α having the following properties:

1) $V_0 = \varnothing$;
2) $V_{\alpha+1} = V_\alpha \cup \mathcal{P}(V_\alpha)$, where $\mathcal{P}(V_\alpha)$ denotes the set of all subsets of the set V_α;
3) $V_\alpha = \bigcup(V_\beta \mid \beta \in \alpha)$ for every limit ordinal number α.

It turns out that *cumulative sets* V_α themselves and the collection $(V_\alpha \mid \alpha \in \mathbf{On})$ as a whole have many remarkable properties. In particular, J. von Neumann proved [1929] that the regularity axiom in ZF is equivalent to the property $\forall x \exists \alpha$ (α is an ordinal number $\wedge x \in V_\alpha$) and the class $\bigcup(V_\alpha \mid \alpha \in \mathbf{On})$ is an abstract (≡class) standard model for the ZF theory in ZF. Models of the ZF and NBG theories of the form $(V_\alpha, =, \in)$ are called *natural*.

After the introduction of the concept of a (*strongly*) *inaccessible cardinal number* in [*Zermelo*, 1930] and [*Sierpiński and Tarski*, 1930], E. Zermelo [1930] (not strictly) and J. Shepherdson [1951, 1952, 1953] (strictly) proved that *a set U is a supertransitive standard model for the NBG theory iff it has the form $V_{\varkappa+1}$ for a certain inaccessible cardinal number $\varkappa$*. Thus, the natural model of the NBG theory was described.

The Zermelo – Shepherdson theorem admits the following equivalent reformulation: *a set U is a supertransitive standard model for the ZF theory with the strong substitution property* ($\forall x \forall f(x \in U \wedge f \in U^x \Rightarrow \operatorname{rng} f \in U)$) *iff it has the form $V_\varkappa$ for a certain inaccessible cardinal number $\varkappa$*.

Starting from the requirements of category theory, instead of the metaconcept of a supertransitive standard model set with the strong substitution property for the ZF theory C. Ehresmann [1957], P. Dedecker [1959], J. Sonner [1962], and A. Grothendieck [*Gabriel*, 1962] introduced an equivalent set-theoretic concept of a *universal set U* (see [*MacLane*, 1971, I.6] and [*Forster*, 1995; *Holmes*, 1998]), which is defined by the following properties:

https://doi.org/10.1515/9783110550948-002

1) $x \in U \Rightarrow x \subset U$;
2) $x \in U \Rightarrow \mathcal{P}(x), \cup x \in U$;
3) $x, x \in U \Rightarrow x \cup x, \{x, y\}, \langle x, y \rangle, x \times y \in U$;
4) $x \in U \wedge (f \in U^x) \Rightarrow \operatorname{rng} f \in U$ (*strong substitution property*);
5) $\omega \in U$, where $\omega \equiv \{0, 1, 2, \dots\}$ is the set of all finite ordinal numbers.

To deal with categories in the set-theoretic framework, they suggested to strengthen the ZF theory by adding the *universality axiom* AU: *each set is an element of a certain universal set.* The equivalent form of the Zermelo – Shepherdson theorem states that the universality axiom AU is equivalent to the *inaccessibility axiom* AI: *for every ordinal number there exists an inaccessible cardinal number strictly greater than it.*

For axiomatic construction of inaccessible cardinal numbers, in [*Tarski*, 1938] (see also [*Kuratowski and Mostowski*, 1967, IX, § 1, § 5]) A. Tarski introduced the concept of a *Tarski set* U, which is defined by the following properties:
1) $x \in U \Rightarrow x \subset U$ (the *transitivity property*);
2) $x \in U \Rightarrow \mathcal{P}(x) \in U$ (the *exponentiality property*);
3) $((x \subset U) \wedge \forall f(f \in U^x \Rightarrow \operatorname{rng} f \neq U)) \Rightarrow x \in U$ (the *Tarski property*).

In [*Tarski*, 1938], it was also proven that the set $V_\varkappa$ ($\equiv$*inaccessible cumulative set*) is a Tarski set for each cardinal number $\varkappa$. In this paper, A. Tarski also proved that the inaccessibility axiom AI is equivalent to the *Tarski axiom* AT: *every set is an element of a certain Tarski set.* In connection with the Tarski theorem, the following problem remained open: *to what extent is the axiomatic concept of Tarski set is wider than the constructive concept of inaccessible cumulative set?*

In this appendix, we give an answer to this question: *the concepts of an inaccessible cumulative set and of an uncountable Tarski set are equivalent.*

The equivalence of the concepts of an inaccessible cumulative set and an uncountable Tarski set was proven using the concept of a universal set. More precisely, it was proven that *every uncountable Tarski set is universal.*

As a result, we obtain the following theorem on the characterization of natural models for the NBG set theory: *the following properties are equivalent for a set U:*
1) *U is an inaccessible cumulative set, i. e., $U = V_\varkappa$ for a certain inaccessible cardinal number $\varkappa$;*
2) *$\mathcal{P}(U)$ is a supertransitive standard model for the NBG theory;*
3) *U is a supertransitive standard model with the strong substitution property for the ZF theory;*
4) *U is a universal set;*
5) *U is an uncountable Tarski set.*

The Zermelo – Shepherdson theorem yields a canonical form of supertransitive standard models for the NBG theory and an (equivalent) canonical form of standard models with the strong substitution property for the ZF theory. However, R. Montague and L. Vaught [1959] proved that for any inaccessible cardinal number $\varkappa$, there exists an cardinal number $\theta < \varkappa$ such that it is not inaccessible and the cumulative set V_θ is a supertransitive standard model for the ZF theory. Therefore, the problem on the canonical forms of supertransitive standard models for the ZF theory turned out to be more complicated.

Since the concept of model in the ZF theory cannot be defined by a finite set of formulas, in this appendix, using the formula scheme and its relativization to the set V_θ, we introduce the concept of a *(strongly) scheme-inaccessible cardinal number* θ and prove a scheme analogue of the Zermelo – Shepherdson theorem.

To prove this theorem, we introduce the concept of a *scheme-universal set*, which is a scheme analogue of the concept of a universal set. Moreover, here we introduce the concept of a *scheme Tarski set*, which is a scheme analogue of the concept of a Tarski set.

As a result, we prove the theorem on the characterization of natural models for the ZF theory: *the following properties are equivalent for a set U:*

1) *U is a scheme-inaccessible cumulative set, i. e., $U = V_\theta$ for a certain scheme-inaccessible cardinal number θ;*
2) *U is a supertransitively standard model for the ZF theory;*
3) *U is a scheme-universal set;*
4) *U is a scheme Tarski set.*

In this appendix, the problems mentioned above are solved for the ZF set theory (with the axiom of choice). For the NBG set theory, all things are equally true. For the reader's convenience, we present all the necessary facts that are not sufficiently reflected in the literature or related to the mathematical folklore, with complete proofs.

The exposition of the material is based on papers [*Bunina and Zakharov*, 2003; 2005; 2006; 2007].

A.1 First-order theories

A.1.1 The language of first-order theories

The proposed theory is a *first-order theory*. We will give definition of a first-order theory basing on [*Mendelson*, 1997].

The *special symbols* of every first-order theory T are the following:

parentheses (,);

connectives $\Rightarrow$ ("implies") and $\neg$ ("not");

quantifier $\forall$ (for all);

a countable set of variables v_i, ($i \geqslant 0$) (in our case variables are denoted by letters $x, X, y, Y, z, Z, u, U, v, V, w, W$, and also these letters with primes);

a non-empty countable set of predicate letters P_i^n $(n \geqslant 1, i \geqslant 0)$;
a countable set of functional letters F_i^n $(n \geqslant 1, i \geqslant 0)$;
and, finally, a countable set of constants a_i $(i \geqslant 0)$.

General symbols are symbols that are not special, but are often used in mathematics. The special and general symbols compose the *initial alphabet*.

A *symbol-string* is defined by induction in the following way: (1) every symbol α of the initial alphabet, except the blank-symbol, is a *symbol-string*; (2) if σ and ρ are symbol-strings, then $\sigma\rho$ and $\rho\sigma$ are *symbol-strings*.

A *designating* (≡ *shortening*) *symbol-string* σ *for a symbol-string* ρ is introduced in the form of the symbol-string $\sigma \equiv \rho$ or $\rho \equiv \sigma$ (σ *is a designation for* ρ).

If a symbol-string ρ is a part of a symbol-string σ, staying in one of the three following positions: $\ldots\rho$, $\rho\ldots$, $\ldots\rho\ldots$, then ρ is an *occurrence in* σ (≡ ρ *occurs in* σ).

A *text* is defined by induction in the following way: (1) every symbol-string σ is a text; (2) if Φ and Ψ are texts, then Φ_Ψ and Ψ_Φ are texts.

If a text Φ is a part of a text Σ, staying in one of the three following positions: $\ldots\_\Phi$, $\Phi\_\ldots$, $\ldots\_\Phi\_\ldots$, then Φ is an *occurrence in* Σ (≡ Φ *occurs in* Σ).

Some symbol-strings constructed from the mentioned above special symbols are called *terms* and *formulas* of the first-order theory T.

Terms are defined in the following way:

1) a variable is a *term*;
2) a constant symbol is a *term*;
3) if F_i^n is a n-placed functional letter, $t_0, \ldots, t_{n-1}$ are terms, then $F_i^n(t_0, \ldots, t_{n-1})$ is a *term*;
4) a symbol-string is a *term* if and only if it follows from the rules 1 – 3.

If P_i^n is some n-placed predicate letter, $t_0, \ldots, t_{n-1}$ are terms, then the symbol-string $P_i^n(t_0, \ldots, t_{n-1})$ is called an *elementary formula*.

Formulas of a first-order theory T are defined in the following way:

1) every elementary formula is a *formula*;
2) if φ and ψ are formulas, v is a variable then every symbol-string $(\neg\varphi)$, $(\varphi \Rightarrow \psi)$, and $\forall v(\varphi)$ is a *formula*;
3) a symbol-string is a *formula* if and only if it follows from the rules 1 and 2.

Let us introduce the following abbreviations:
$(\varphi \wedge \psi)$ for $\neg(\varphi \Rightarrow \neg\psi)$;
$(\varphi \vee \psi)$ for $(\neg\varphi) \Rightarrow \psi$;
$(\varphi \equiv \psi)$ for $(\varphi \Rightarrow \psi) \wedge (\psi \Rightarrow \varphi)$;
$\exists v\, \varphi$ is an abbreviation for $(\neg(\forall v\, (\neg\varphi)))$.

Introduce a notion of *free* and *connected* occurrence of a variable in a formula. An occurrence of a variable v in a given formula is called *connected*, if v is either a variable of an occurring in this formula quantifier $\forall v$ or is under the action of occurring in this formula quantifier $\forall v$; otherwise an occurrence of a variable in a given formula is called *free*. Thus, one variable can have free and connected occurrences in the same formula. A variable is called a *free* (*connected*) *variable* in a given formula, if there exist free (connected) occurrences of this variable in this formula, i. e. a variable can in the same time be free and connected in one formula.

A *sentence* is a formula with no free variables.

If ζ is a term or a formula, θ is a term, v is a variable then $\zeta(v \parallel \theta)$ denotes a symbol-string, obtained by replacing every free occurrence of the variable v in the symbol-string ζ by the symbol-string θ.

The substitution $v \parallel \theta$ in ζ is called *admissible*, if for every free occurrence of a variable w in the symbol-string θ every free occurrence v in ζ is not a free occurrence in some formula ψ, occurring in some formulas $\forall w\, \psi(w)$ and $\exists w\, \psi(w)$, occurring in the symbol-string ζ.

In the sequel, if the substitution $v \parallel \theta$ in ζ is admissible, then together with $\zeta(v \parallel \theta)$ we will write $\zeta(\theta)$.

If ζ is a term or a formula, θ is a term, v is a variable such that the substitution $v \parallel \theta$ in ζ is admissible, then the substitution $\zeta(v \parallel \theta)$ is a term or a formula respectively.

Every free occurrence of some variable u (except v) in a symbol-string ζ and every free occurrence of some variable w in a symbol-string θ are free occurrences of these variables in a symbol-string $\zeta(v \parallel \theta)$.

A.1.2 Deducibility in a first-order theory

A symbol-string γ, equipped with some rule, is called a *formula scheme of a theory T*, if:
1) this rule marks some letters (in particular, free and connected variables), occurring in γ;
2) this rule determines the necessary substitution of these marked letters in γ by some terms (in particular, variables);
3) after every such a substitution in γ some propositional formula φ of the theory T is obtained.

Every such a propositional formula φ is called a *propositional formula obtained by the application of the formula scheme γ*.

A text Γ consisting of symbol-strings separated by the blank-symbols is called an *axiom text*, if every symbol-string γ occurring in Γ is either a formula or a formula scheme of the theory T. If γ is a formula, then γ is called an *explicit axiom of the theory T*. If γ is a formula scheme, then it is called an *axiom scheme of the theory T*. Every formula, obtained by the application of the axiom scheme γ, is called an *implicit axiom of the theory T*.

Axioms and axiom schemes of every first-order theory are divided in two classes: *logical* and *proper* (or *mathematical*).

Logical axiom schemes of any first-order theory are cited below:

LAS1. $\varphi \Rightarrow (\psi \Rightarrow \varphi)$;
LAS2. $(\varphi \Rightarrow (\psi \Rightarrow \chi)) \Rightarrow ((\varphi \Rightarrow \psi) \Rightarrow (\varphi \Rightarrow \chi))$;
LAS3. $(\varphi \wedge \psi) \Rightarrow \varphi$;
LAS4. $(\varphi \wedge \psi) \Rightarrow \psi$;
LAS5. $\varphi \Rightarrow (\psi \Rightarrow (\varphi \wedge \psi))$;
LAS6. $\varphi \Rightarrow (\varphi \vee \psi)$;
LAS7. $\psi \Rightarrow (\varphi \vee \psi)$;
LAS8. $(\varphi \Rightarrow \chi) \Rightarrow ((\psi \Rightarrow \chi) \Rightarrow ((\varphi \vee \psi) \Rightarrow \chi))$;
LAS9. $(\varphi \Rightarrow \psi) \Rightarrow ((\varphi \Rightarrow \neg\psi) \Rightarrow \neg\varphi)$;
LAS10. $(\neg(\neg\varphi)) \Rightarrow \varphi$;
LAS11. $(\forall v \varphi) \Rightarrow \varphi(v \parallel \theta)$, if v is a variable, θ is a term such that a substitution $v \parallel \theta$ in φ is admissible.
LAS12. $\varphi(v \parallel \theta) \Rightarrow (\exists v\, \varphi)$ in the same conditions as in LAS11;
LAS13. $(\forall v(\psi \Rightarrow \varphi(v))) \Rightarrow (\psi \Rightarrow (\forall v \varphi))$, if ψ does not contain a free variable v;
LAS14. $(\forall v(\varphi(v) \Rightarrow \psi)) \Rightarrow ((\exists v\, \varphi) \Rightarrow \psi)$ in the same condition as in LAS13.

Proper axioms and *axiom schemes* can not be formulated in general case because they depend on a theory. The first-order theory which does not contain any proper axioms is called the *first order predicate calculus*.

The rules of deduction in the first-order theory are the following:
- *the rule of implication* ($\equiv$ *modus ponens* (MP)): from φ and $\varphi \Rightarrow \psi$ it follows that ψ;
- *the rule of generalization* (Gen): from φ it follows that $\forall v\, \varphi$.

Let Φ be a totality of formulas and ψ be a formula of the theory T. A sequence $f \equiv (\varphi_i \mid i \in n+1) \equiv (\varphi_0, \ldots, \varphi_n)$ of formulas of the theory T is called a *deduction of the formula ψ from the totality* Φ, if $\varphi_n = \psi$ and for any $0 < i \leqslant n$ one of following conditions is fulfilled:
1) φ_i belongs to Φ;
2) there exist $0 \leqslant k < j < i$ such that φ_j is $(\varphi_k \Rightarrow \varphi_i)$, i. e. φ_i is obtained from φ_k and $\varphi_k \Rightarrow \varphi_i$ by the rule of implication MP;
3) there exists $0 \leqslant j < i$ such that φ_i is $\forall x\, \varphi_j$, where x is not a free variable of every formula from Φ, i. e. φ_i is obtained from φ_j by the rule of generalization Gen with the given *structural requirement.*

Denote this deduction either by $f \equiv (\varphi_0, \ldots, \varphi_n) : \Phi \vdash \psi$, or by $(\varphi_0, \ldots, \varphi_n) : \Phi \vdash \psi$, or by $f : \Phi \vdash \psi$.

A totality Φ_a is called a *totality of axioms of the theory T* if Φ_a consists of all explicit proper axioms of the theory T, all implicit proper axioms of the theory T, and all implicit logical axioms of the predicate calculus. If there exists a deduction

$f : \Phi_a \vdash \psi$, then the formula ψ is called *deducible in the axiomatic theory* (T, Φ_a) and the deduction f is called a *proof of the formula* ψ.

A totality of formulas Φ is called *contradictory* ($\equiv$ *non-consistent*) if every formula of the theory T is deducible from it. In the opposite case, Φ is called *non-contradictory* ($\equiv$ *consistent*).

An axiomatic theory (T, Φ_a) is called *contradictory* [*non-contradictory*] if the totality of its axioms Φ_a is contradictory [non-contradictory]. The proposition expressing the consistency of the theory (T, Φ_a) will be denoted by $cons(T, \Phi_a)$ or simply $cons(T)$.

Lemma 1. *A totality of formulas Φ is contradictory if and only if the formulas φ and $\neg\varphi$ for some sentence φ are deducible from Φ.*

Proof. If the totality Φ is contradictory, then every sentence of the theory T is deducible from it, in particular, φ and $\neg\varphi$ for arbitrary sentence φ are deducible. Suppose now that sentences φ and $\neg\varphi$ are deducible from the totality Φ, and σ is an arbitrary formula. Show that the formula σ can be deduced from φ and $\neg\varphi$. This is a deduction: 1. $\varphi \Rightarrow (\neg\sigma \Rightarrow \varphi)$ (LAS1); 2. $\neg\varphi \Rightarrow (\neg\sigma \Rightarrow \neg\varphi)$ (LAS1); 3. φ; 4. $\neg\varphi$; 5. $\neg\sigma \Rightarrow \varphi$ (MP, 1 and 3); 6. $(\neg\sigma \Rightarrow \varphi) \Rightarrow (\neg\sigma \Rightarrow \neg\varphi) \Rightarrow \neg(\neg\sigma))$ (LAS9); 7. $(\neg\sigma \Rightarrow \neg\varphi) \Rightarrow \neg(\neg\sigma)$ (MP, 5 and 6); 8. $\neg\sigma \Rightarrow \neg\varphi$ (MP, 2 and 4); 9. $\neg(\neg\sigma)$ (MP, 7 and 8); 10. $(\neg(\neg\sigma)) \Rightarrow \sigma$ (LAS10); 11. σ (MP, 9, and 10). □

A.1.3 An interpretation of a first-order theory in a set theory

Consistency of first-order theories is often proven by the method of interpretations, going back to A. Tarski (see [*Mendelson*, 1997, 2.2]).

A first-order theory S is called a *set theory*, if the binary predicate symbol $\in$ belongs to the set of its predicates symbols. This symbol denotes the *belonging ratio* ($\in (x, y)$ is read as "*x belongs to y*", "*x is an element of y*", and so on.)

Let some object D be selected by means of the set theory S. We will call this selected object D of the set theory S *equipped*, if in S for all $n \geqslant 1$ the notions of n-finite sequence $(x_i \in D \mid i \in n)$ of elements of the object D, n-placed relation $R \subset D^n$, and n-placed operation $O : D^n \to D$ and also a notion of an infinite sequence $x_0, \dots, x_q, \dots$ of elements of the object D are defined.

Let S be some fixed set theory with some fixed equipped object D.

An *interpretation of a first-order theory T in the set theory S with the equipped object D* is a pair M, consisting of the object D and some correspondence I, assigning to every predicate letter P_i^n some n-placed relation $I(P_i^n)$ in D, every functional letter F_i^n some n-placed operation $I(F_i^n)$ in D, and every constant symbol a_i some element $I(a_i)$ of D.

Let s be an infinite sequence $x_0, \dots, x_q, \dots$ of elements of the object D.

Define the *value of a term t of the theory T on the sequence s under the interpretation M of the theory T in the set theory S* (in notation $t_M[s]$) by induction in the following way:

- if $t \equiv v_i$, then $t_M[s] \equiv x_i$;
- if $t \equiv a_i$, then $t_M[s] \equiv I(a_i)$;
- if $t \equiv F(t_0, \ldots, t_{n-1})$, where F is a n-placed functional symbol and $t_0, \ldots, t_{n-1}$ are terms, then $t_M[s] \equiv I(F)(t_{0M}[s], \ldots, t_{n-1M}[s])$.

Define the *translation (satisfaction) of a formula φ on the sequence s under the interpretation M of the theory T in the set theory S* (in notation $M \vDash \varphi[s]$) by induction in the following way:

- if $\varphi \equiv (P(t_0, \ldots, t_{n-1}))$, where P is an n-placed predicate symbol and $t_0, \ldots, t_{n-1}$ are terms, then $M \vDash \varphi[s] \equiv ((t_{0M}[s], \ldots, t_{n-1M}[s]) \in I(P))$;
- if $\varphi \equiv (\neg\theta)$, then $M \vDash \varphi[s] \equiv (\neg M \vDash \theta[s])$;
- if $\varphi \equiv (\theta_1 \Rightarrow \theta_2)$, then $M \vDash \varphi[s] \equiv (M \vDash \theta_1[s] \Rightarrow M \vDash \theta_2[s])$;
- if $\varphi \equiv (\forall v_i \theta)$, then $M \vDash \varphi[s] \equiv (\forall x(x \in D \Rightarrow M \vDash \theta[x_0, \ldots, x_{i-1}, x, x_{i+1}, \ldots, x_q, \ldots]))$.

Using the abbreviations cited above, we have also the following:

- if $\varphi \equiv (\theta_1 \wedge \theta_2)$, then $M \vDash \varphi[s] \equiv (M \vDash \theta_1[s] \wedge M \vDash \theta_2[s])$;
- if $\varphi \equiv (\theta_1 \vee \theta_2)$, then $M \vDash \varphi[s] \equiv (M \vDash \theta_1 \vee M \vDash \theta_2[s])$;
- if $\varphi \equiv (\exists v_i \theta)$, then $M \vDash \varphi[s] \equiv (\exists x(x \in D \wedge M \vDash \theta[x_0, \ldots, x_{i-1}, x, x_{i+1}, \ldots, x_q, \ldots]))$;
- if $\varphi \equiv (\theta_1 \Leftrightarrow \theta_2)$, then $M \vDash \varphi[s] \equiv (M \vDash \theta_1[s] \Leftrightarrow M \vDash \theta_2[s])$.

If, in the theory S, the symbol-string $\sigma(s) \equiv ((t_{0M}[s], \ldots, t_{n-1M}[s]) \in I(P))$ is a formula of the theory S, then this definition implies that $M \vDash \varphi[s]$ is always a formula of the theory S.

Further in this section, we will consider the set theory S for which all symbol-strings $\sigma(s)$ for every sequence s from D are formulas of the theory S. All concrete set theories considered later in this paper will possess this property.

An interpretation M is called a *model of the axiomatic theory (T, Φ_a) in the axiomatic set theory (S, Ξ_a) with the selected equipped object D* if for every sequence s from D the translation $M \vDash \varphi[s]$ of every axiom φ of the theory T is a deducible formula in the theory (S, Ξ_a).

Define now the *translation of the deduction $f \equiv (\varphi_0, \ldots, \varphi_n) : \Phi \vdash \psi$ of the formula ψ from the totality Φ of formulas of the theory T on the sequence s under the interpretation M of the theory T in the set theory S* in the form of the sequence $g \equiv (M \vDash \varphi_0[s], \ldots, M \vDash \varphi_n[s])$, which is a *$D$-bounded deduction of the formula $M \vDash \psi[s]$ from the totality $M \vDash \Phi[s] \equiv \{M \vDash \varphi[s] \mid \varphi \in \Phi\}$* in such a sense that the rule of generalization $Gen \equiv \frac{\sigma}{\forall x \sigma}$ is used in the following *D-bounded form*

$$Gen_D \equiv \frac{\sigma}{\forall x(x \in D \Rightarrow \sigma)},$$

where x and σ are a variable and a formula of the theory T, respectively.

Lemma 1. *A sequence* $g \equiv (M \vDash \varphi_0[s], \dots, M \vDash \varphi_n[s])$ *can be canonically extended to some sequence* $g_{ext} \equiv (M \vDash \varphi_0[s], \dots, M \vDash \varphi_n[s])_{ext}$ *so that* $g_{ext} : M \vDash \Phi[s] \vdash M \vDash \psi[s]$*, i. e.* g_{ext} *is a usual deduction of the formula* $M \vDash \psi[s]$ *from the totality* $M \vDash \Phi[s]$*.*

Proof. We will look through all j from 1 to n. Let φ_j be $\forall y\, \varphi_i$ for some $i < j$, where y is not a free variable of every formula φ of the totality Φ. The parameters of every formula $M \vDash \varphi[s]$ from $M \vDash \Phi[s]$ are only some members of the sequence s. Then $M \vDash \varphi_j[s]$ is $\forall x(x \in D \Rightarrow M \vDash \varphi_i[s])$, where x differs from all the parameters of the totality $M \vDash \Phi[s]$. We insert in g, right after the formula $M \vDash \varphi_i[s]$, the explicit axiom

$$\xi \equiv (M \vDash \varphi_i(s) \Rightarrow (x \in D \Rightarrow M \vDash \varphi_i[s])),$$

obtained from logical axiom scheme LAS1. Then applying the rule MP to the two previous formulas $M \vDash \varphi_i[s]$ and ξ, we can insert in g, after ξ, the formula $\chi \equiv (x \in D \Rightarrow M \vDash \varphi_i[s])$. Then the formula $M \vDash \varphi_j[s] \equiv \forall x\, \chi$ is obtained as the usual application of the rule Gen to the formula χ. □

This lemma implies that the translation of a deduction $f \equiv (\varphi_0, \dots, \varphi_n) : \Phi \vdash \psi$ leads to the deduction $g_{ext} \equiv (M \vDash \varphi_0[s], \dots, M \vDash \varphi_n[s])_{ext} : M \vDash \Phi(s) \vdash M \vDash \psi[s]$. This procedure is always used without any special mentioning.

Lemma 2. *Let every formula from the translation* $M \vDash \Phi[s]$ *of a totality* Φ *of formulas of the theory* T *is deducible in the axiomatic theory* (S, Ξ_a) *from the totality* Ξ_a *of axioms of the theory* S*. Besides, let* $f \equiv (\varphi_0, \dots, \varphi_n) : \Phi \vdash \psi$*. Then the deduction* $g_{ext} : M \vDash \Phi(s) \vdash M \vDash \psi[s]$ *can be extended to some deduction* $h : \Xi_a \vdash M \vDash \psi[s]$ *in the theory* (S, Ξ_a)*.*

Proof. Consider the deduction $g_{ext} : M \vDash \Phi[s] \vdash M \vDash \psi[s]$. Let $g_{ext} = (\xi_0, \dots, \xi_m)$. Every formula ξ_i of the theory S in this deduction either is one of the formulas of the totality $M \vDash \Phi[s]$ or follows from the previous formulas of this sequence as a result of application of one of the rules of deduction. At first, we consider only such ξ_i that do not follow from previous formulas of the deduction. These formulas belong to the totality $M \vDash \Phi[s]$. By the lemma condition each of these formulas, ξ_i is deduced in the theory (S, Ξ_a) from the totality Ξ_a of axioms of this theory, i. e. for each ξ_i there exists a deduction $g_i \equiv (\eta_i^0, \dots, \eta_i^{k_i}) : (\Xi_a)_i \vDash \xi_i$, where $(\Xi_a)_i$ is some finite subtotality of the totality Ξ_a. Change in the finite subtotality $(\Xi_a)_0, \dots, (\Xi_a)_m$ all free variables in such a way that they became different from those variables which were touched by the application of rule of generalization in the deduction g_{ext}. For $i \in m+1$ such that ξ_i is a corollary of the previous formulas we put $k_i \equiv 0$, $\eta_i^0 \equiv \xi_i$. Then $h \equiv (\eta_0^0, \dots, \eta_0^{k_0}, \eta_1^0, \dots, \eta_1^{k_1}, \dots, \eta_m^0, \dots, \eta_m^{k_n})$ is a deduction of the formula $M \vDash \psi[s]$ from the totality Ξ_a in the theory (S, Ξ_a). □

Lemma 3. *Let for every sequence s from D, every formula from the translation $M \vDash \Phi_a[s]$ of axioms of the theory (T, Φ_a) is deduced in the theory (S, Ξ_a) from the totality Ξ_a of axioms of the theory (S, Ξ_a), i. e. M is a model of the theory (T, Φ_a) in the theory (S, Ξ_a). Under this condition, if the theory (S, Ξ_a) is consistent, then the theory (T, Φ_a) is also consistent.*

Proof. Suppose that the theory T is contradictory, i. e. there exist some formula ψ of the theory T and some deduction $f \equiv (\varphi_0, \ldots, \varphi_n) : \Phi_a \vdash \psi \wedge \neg\psi$. Consider on an arbitrary sequence s from D its translation $g \equiv (M \vDash \varphi_0[s], \ldots, M \vDash \varphi_n[s])$ and its canonical extension $g_{ext} : M \vDash \Phi_a[s] \vdash (M \vDash \psi[s] \wedge \neg M \vDash \psi[s])$ from Lemma 1. Then, according to Lemma 2, there exists a deduction $h : \Xi_a \vdash (M \vDash \psi[s] \wedge \neg M \vDash \psi[s])$. However by virtue of consistency of the theory S such a deduction is impossible. So the theory T is consistent. □

A.2 Some elements of the Zermelo – Fraenkel set theory

A.2.1 The proper axioms and axiom schemes of the ZF set theory

At the beginning we will cite the list of proper axioms and axiom schemes of the theory ZF (the Zermelo – Fraenkel set theory with the choice axiom) (see [*Zermelo*, 1908; *Fraenkel*, 1922; *Kolmogorov and Dragalin*, 1982; *Jech*, 1971]).

This theory is a first-order theory with two binary predicate symbols of *belonging* $\in$ (we write $A \in B$) and *equality* $=$ (we write $A = B$).

The predicate of equality $=$ satisfies the following axiom and axiom scheme:

- $\forall x(x = x)$ (*reflexivity of equality*);
- $(x = y) \Rightarrow (\varphi(x, x) \Rightarrow \varphi(x, y))$ (*replacement of equals*), where x and y are variables, $\varphi(x, x)$ is an arbitrary formula, $\varphi(x, y)$ is constructed from $\varphi(x, x)$ by changing some (not necessarily all) free occurrences of x by occurrences of y with such a condition that y is free for such occurrences of x that are changed.

Objects of the given theory are called *sets*.

As above it is useful to consider the totality $\mathbf{C}$ of all sets A, satisfying a given formula $\varphi(x)$. The totality $\mathbf{C}$ is called the *class* (*ZF*), *defined by the formula* φ. The totality $\mathbf{C}(\vec{u})$ of all sets A, satisfying a formula $\varphi(x, \vec{u})$, is called the *class* (*ZF*) *defined by the formula* φ *through the parameter* $\vec{u}$. Along with these words we will use the notations

$$A \in \mathbf{C} \equiv \varphi(A), \quad A \in \mathbf{C}(\vec{u}) \equiv \varphi(A, \vec{u})$$

and

$$\mathbf{C} \equiv \{x \mid \varphi(x)\}, \quad \mathbf{C}(\vec{u}) \equiv \{x \mid \varphi(x, \vec{u})\}.$$

If $\mathbf{C} \equiv \{x \mid \varphi(x)\}$ and φ contain only one free variable x, then the class $\mathbf{C}$ is called *well-defined* (≡ *completely determined*) *by the formula* φ.

Every set A can be considered as the class $\{x \mid x \in A\}$.

A class $\mathbf{C} \equiv \{x \mid \varphi(x)\}$ is called the *subclass of a class* $\mathbf{D} \equiv \{x \mid \psi(x)\}$ (denoted by $\mathbf{C} \subset \mathbf{D}$) if $\forall x(\varphi(x) \Rightarrow \psi(x))$. Classes $\mathbf{C}$ and $\mathbf{D}$ are called *equal* if $(\mathbf{C} \subset \mathbf{D}) \wedge (\mathbf{D} \subset \mathbf{C})$.

Further, we will use the notation $\{x \in A \mid \varphi(x)\} \equiv \{x \mid x \in A \wedge \varphi(x)\}$.

If a class $\mathbf{C}$ is not equal to any set, then $\mathbf{C}$ is called a *proper class*. Not every class is a set: the class $\{x \mid x \notin x\}$ is a proper class.

The *universal class* is the class of all sets $\mathbf{V} \equiv \{x \mid x = x\}$.

For classes $\mathbf{C} \equiv \{x \mid \varphi(x)\}$ and $\mathbf{D} \equiv \{x \mid \psi(x)\}$ define the *binary union* $\mathbf{C} \cup \mathbf{D}$ and the *binary intersection* $\mathbf{C} \cap \mathbf{D}$ as the classes

$$\mathbf{C} \cup \mathbf{D} \equiv \{x \mid \varphi(x) \vee \psi(x)\} \text{ and } \mathbf{C} \cap \mathbf{D} \equiv \{x \mid \varphi(x) \wedge \psi(x)\}.$$

A1. (The *extensionality axiom.*) $\forall X \forall Y(\forall u(u \in X \Leftrightarrow u \in Y) \Rightarrow X = Y)$.

This axiom postulates that if two sets consist of the same elements, then they are equal.

For sets A and B, define the *unordered pair* $\{A, B\}$ as the class $\{A, B\} \equiv \{z \mid z = A \vee z = B\}$.

A2. (The *pair axiom.*) $\forall u \forall v \exists x \forall z(z \in x \Leftrightarrow z = u \vee z = v)$.

Axioms A2 and A1 imply that the unordered pair of sets is a set.

For sets A and B, define

– the *solitary set* $\{A\} \equiv \{A, A\}$;

– the *ordered pair* $\langle A, B\rangle \equiv \{\{A\}, \{A, B\}\}$;

From the previous assertions, we infer that $\{A\}$ and $\langle A, B\rangle$ are sets.

Lemma 1. *$\langle A, B\rangle = \langle A', B'\rangle$ if and only if $A = A'$ and $B = B'$.*

AS3. (The *separation axiom scheme.*) $\forall X \exists Y \forall u(u \in Y \Leftrightarrow u \in X \wedge \varphi(u, \vec{p}))$, where the formula $\varphi(u, \vec{p})$ does not contain Y as a free variable.

This axiom scheme postulates that the class $\{u \mid u \in X \wedge \varphi(u, \vec{p})\}$ is a set. This set is unique by A1. Suppose that there exist some sets Y and Y' such that $\forall u(u \in Y \Leftrightarrow u \in X \wedge \varphi(u, p))$ and $\forall u(u \in Y' \Leftrightarrow u \in X \wedge \varphi(u, p))$. Then, by LAS11, $u \in Y \Leftrightarrow u \in X \wedge \varphi(u, \vec{p})$ and $u \in Y' \Leftrightarrow u \in X \wedge \varphi(u, \vec{p})$, where, by LAS3 and LAS4, $u \in Y \Rightarrow u \in X \wedge \varphi(u, \vec{p})$, $u \in Y' \Rightarrow u \in X \wedge \varphi(u, \vec{p})$, $\varphi(u, \vec{p}) \wedge u \in X \Rightarrow u \in Y$, and $u \in X \wedge \varphi(u, \vec{p}) \Rightarrow u \in Y'$. Consequently, $u \in Y \Leftrightarrow u \in Y'$, and by the rule of generalization (Gen), $\forall u(u \in Y \Leftrightarrow u \in Y')$, where now, by A1, $Y = Y'$.

Consider the class $\mathbf{C} = \{u \mid \varphi(u, \vec{p})\}$. Then axiom scheme AS3 can be expressed in the following form: $\forall X \exists Y(Y = \mathbf{C}(\vec{p}) \cap X)$.

For classes $\mathbf{A}$ and $\mathbf{B}$, define the *difference* $A \backslash B$ as the class $A \backslash B \equiv \{x \in A \mid x \notin B\}$.

If A is a set, then, by AS3, the difference $A \backslash \mathbf{B}$ is a set.

Since $A \cap B = \{x \in A \mid x \in B\} \subset A$, we infer, by AS3 that for any sets A and B, the binary intersection $A \cap B$ is a set.

For a class $\mathbf{C} \equiv \{x \mid \varphi(x)\}$, define the *union* $\cup\mathbf{C}$ as the class $\cup\mathbf{C} \equiv \{z \mid \exists x(\varphi(x) \wedge z \in x)\}$.

A4. (The *union axiom.*)

$$\forall X \exists Y \forall u(u \in Y \Leftrightarrow \exists z(u \in z \wedge z \in X)) \wedge z \in X)).$$

We can deduce from A4 and AS3 that for every set A its union $\cup A$ is a set.

We have the equality $A \cup B = \cup\{A, B\}$. Therefore, for every sets A and B, their binary union $A \cup B$ is a set.

We will call the *full ensemble* of a class $\mathbf{C}$ the class $\mathcal{P}(\mathbf{C}) \equiv \{u \mid u \subseteq \mathbf{C}\}$.

A5. (The *power set* ($\equiv$ *full ensemble*) *axiom.*) $\forall X \exists Y \forall u(u \in Y \Rightarrow u \subset X)$.

If A is a set, then, according to A5 and A1, $\mathcal{P}(A)$ is a set.

For classes $\mathbf{A}$ and $\mathbf{B}$, define the (*coordinate*) *product*

$$\mathbf{A} * \mathbf{B} \equiv \{x \mid \exists u \exists v(u \in \mathbf{A} \wedge v \in \mathbf{B} \wedge x = \langle u, v\rangle\}.$$

The fact that $A * B$ is a set for sets A and B follows from AS3, because $A * B \subseteq \mathcal{P}(\mathcal{P}(A \cup B))$.

A class (in particular, a set) $\mathbf{C}$ is called a *correspondence* if

$$\forall u(u \in \mathbf{C} \Rightarrow \exists x \exists y(u = \langle x, y\rangle))).$$

For a correspondence $\mathbf{C}$ consider the classes:

$$\operatorname{dom} \mathbf{C} \equiv \{u \mid \exists v(\langle u, v\rangle \in \mathbf{C})\};$$
$$\operatorname{rng} \mathbf{C} \equiv \{v \mid \exists u(\langle u, v\rangle \in \mathbf{C}\}.$$

If $\mathbf{C}$ is a set, then $\operatorname{dom} \mathbf{C} \subset \cup \cup \mathbf{C}$, by A4 and AS3, implies that $\operatorname{dom} \mathbf{C}$ also is a set.

A correspondence $\mathbf{F}$ is called a *function* ($\equiv$ a *mapping*) if

$$\forall x \forall y \forall y'(\langle x, y\rangle \in \mathbf{F} \wedge \langle x, y'\rangle \in \mathbf{F} \Rightarrow y = y').$$

The formula expressing for a class $\mathbf{F}$ the property to be a mapping will be denoted by $func(\mathbf{F})$. For the expression $\langle x, y\rangle \in \mathbf{F}$, we also use the notations: $y = \mathbf{F}(x)$, $\mathbf{F} : x \mapsto y$, and others.

A correspondence $\mathbf{C}$ is called a *correspondence from a class* $\mathbf{A}$ *into a class* $\mathbf{B}$ if $\operatorname{dom} \mathbf{C} \subset \mathbf{A}$ and $\operatorname{rng} \mathbf{C} \subset \mathbf{B}$ (it is denoted by $\mathbf{C} : \mathbf{A} \longrightarrow\!\!\!\!\!< \mathbf{B}$). A function $\mathbf{F}$ is called a *function from a class* $\mathbf{A}$ *into a class* $\mathbf{B}$ if $\operatorname{dom} \mathbf{F} = \mathbf{A}$ and $\operatorname{rng} \mathbf{F} \subset \mathbf{B}$ (it is denoted by $\mathbf{F} : \mathbf{A} \to \mathbf{B}$).

The formula expressing the property of the class $\mathbf{F}$ to be a function from the class $\mathbf{A}$ into the class $\mathbf{B}$ will be denoted by $\mathbf{F} \leftrightharpoons \mathbf{A} \to \mathbf{B}$. The formulas $(\mathbf{F} \leftrightharpoons \mathbf{A} \to \mathbf{B}) \wedge \forall x, y \in \mathbf{A}(\mathbf{F}(x) = \mathbf{F}(y) \Rightarrow x = y)$ and $(\mathbf{F} \leftrightharpoons \mathbf{A} \to \mathbf{B}) \wedge \operatorname{rng} \mathbf{F} = \mathbf{B}$ will be denoted by $\mathbf{F} \leftrightharpoons \mathbf{A} \rightarrowtail \mathbf{B}$ and $\mathbf{F} \leftrightharpoons \mathbf{A} \twoheadrightarrow \mathbf{B}$, respectively. The conjunction of these formulas will be denoted by $\mathbf{F} \leftrightharpoons \mathbf{A} \rightarrowtail\!\!\!\!\twoheadrightarrow \mathbf{B}$. These formulas define the *injectivity*, the *surjectivity*, and the *bijectivity* of the function $\mathbf{F} : \mathbf{A} \to \mathbf{B}$, respectively.

The class $\{f \mid func(f) \wedge \operatorname{dom} f = A \wedge \operatorname{rng} f \subseteq B\}$ of all functions from a class $\mathbf{A}$ into a class $\mathbf{B}$ which are sets is denoted by $\mathbf{B}^{\mathbf{A}}$ or by $\operatorname{Map}(\mathbf{A}, \mathbf{B})$. Since $B^A \subset \mathcal{P}(A * B)$, we infer that the class B^A is a set for any sets A and B.

The *restriction of the function* $\mathbf{F}$ *on the class* $\mathbf{A}$ is defined as the class

$$\mathbf{F}|\mathbf{A} \equiv \{x \mid \exists u \exists v(x = \langle u, v\rangle \wedge x \in \mathbf{F} \wedge u \in \mathbf{A}\}.$$

The *image* and the *inverse image* of the *class* $\mathbf{A}$ *with respect to the function* $\mathbf{F}$ are defined as the classes $\mathbf{F}[\mathbf{A}] \equiv \{v \mid \exists u \in \mathbf{A}(v = \mathbf{F}(u))\}$ and $\mathbf{F}^{-1}[\mathbf{A}] \equiv \{u \mid \mathbf{F}(u) \in \mathbf{A}\}$.

A correspondence $\mathbf{C}$ from a class $\mathbf{A}$ into a class $\mathbf{B}$ is called also a (*multivalued*) *collection of subclasses* $\mathbf{B}_a \equiv \mathbf{C}\langle a\rangle \equiv \{y \mid y \in \mathbf{B} \wedge \langle a, y\rangle \in \mathbf{C}\}$ of the *class* $\mathbf{B}$, *indexed by the class* $\mathbf{A}$. In this case, the correspondence $\mathbf{C}$ and the class $\operatorname{rng} \mathbf{C}$ are denoted also by $(\mathbf{B}_a \subset \mathbf{B} \mid a \in \mathbf{A})$ and $\cup(\mathbf{B}_a \subset \mathbf{B} \mid a \in \mathbf{A})$ respectively. The class $\cup(\mathbf{B}_a \subset \mathbf{B} \mid a \in \mathbf{A})$ is called also the *union of the collection* $(\mathbf{B}_a \subset \mathbf{B} \mid a \in \mathbf{A})$. The class $\{y \mid \forall x \in \mathbf{A}(y \in \mathbf{B}_x)\}$ is called the *intersection of the collection* $(\mathbf{B}_a \subset \mathbf{B} \mid a \in \mathbf{A})$ and is denoted by $\cap(\mathbf{B}_a \subset \mathbf{B} \mid a \in \mathbf{A})$. With every class $\mathbf{A}$ it is associated in the canonical way the *collection* $(a \in \mathbf{V} \mid a \in \mathbf{A})$ *of element sets of the class* $\mathbf{A}$. For this collection, the equality $\cup\mathbf{A} = \cup(a \subset \mathbf{V} \mid a \in \mathbf{A})$ is valid.

A function $\mathbf{F}$ from a class $\mathbf{A}$ into the class $\mathbf{B}$ is called also the *simple collection of elements* $b_a \equiv \mathbf{F}(a)$ *of the class* $\mathbf{B}$ *indexed by the class* $\mathbf{A}$. In this case, the function $\mathbf{F}$ and the class $\operatorname{rng} \mathbf{F}$ are also denoted by $(b_a \in \mathbf{B} \mid a \in \mathbf{A})$ and $\{b_a \in \mathbf{B} \mid a \in \mathbf{A}\}$, respectively. The collection $(b_a \in \mathbf{V} \mid a \in \mathbf{A})$ is also denoted by $(b_a \mid a \in \mathbf{A})$. With every class $\mathbf{A}$ it is associated in the canonical way, the *simple collection* $(a \in \mathbf{A} \mid a \in \mathbf{A})$ *of elements of the class* $\mathbf{A}$. It is clear that $\{a \in \mathbf{A} \mid a \in \mathbf{A}\} = \mathbf{A}$.

AS6. (The *replacement axiom scheme*.)

$$\forall x \forall y \forall y'(\varphi(x, y, \vec{p}) \wedge \varphi(x, y', \vec{p}) \Rightarrow y = y') \Rightarrow \forall X \exists Y \forall x \in X \forall y(\varphi(x, y, \vec{p}) \Rightarrow y \in Y),$$

where the formula $\varphi(x, y, \vec{p})$ does not contain Y as a free variable.

To explain the essence of this axiom scheme, consider the class

$$\mathbf{F} \equiv \{u \mid \exists x \exists y(u = \langle x, y\rangle \wedge \varphi(x, y, \vec{p}))\}.$$

The premise in AS6 states that $\mathbf{F}$ is a function. Therefore, scheme AS6 can be expressed in the following way: $func(\mathbf{F}) \Rightarrow \forall X \exists Y(\mathbf{F}[X] \subseteq Y)$. In other words, if $\mathbf{F}$ is a function, then for every set X the class $\mathbf{F}[X]$ is a set.

If A is a set, then by AS6 we infer that the class $\operatorname{rng} \mathbf{F} \equiv \{b_a \in \mathbf{B} \mid a \in \mathbf{A}\}$ is a set. Then $\mathbf{F} \subset A \times \operatorname{rng} \mathbf{F}$ implies that the class $\mathbf{F} \equiv (b_a \in \mathbf{B} \mid a \in A)$ also is a set. Therefore, if A is a set we use the notations $F : A \to \mathbf{B}$ and $F \equiv (b_a \in \mathbf{B} \mid a \in A)$.

A7. (The *empty set axiom.*) $\exists x \forall z(\neg(z \in x))$.

Axiom A1 implies that the set containing no elements is unique. It is denoted by $\varnothing$.

A8. (The *infinity axiom.*) $\exists Y(\varnothing \in Y \wedge \forall u(u \in Y \Rightarrow u \cup \{u\} \in Y))$.

According to this axiom, there exists a set I, containing $\varnothing, \{\varnothing\}, \{\varnothing, \{\varnothing\}\}$, and so on.

Note that any set X with property $u \in X \Rightarrow u \cup \{u\} \in X$ is called *inductive.*

A9. (The *regularity axiom.*) $\forall X(X \neq \varnothing \Rightarrow \exists x(x \in X \wedge x \cap X = \varnothing))$.

A function $f : \mathcal{P}(A)\backslash\{\varnothing\} \to A$ is called a *choice function for the set* A, if $f(X) \in X$ for every $X \in \mathcal{P}(A)\backslash\{\varnothing\}$.

The last axiom postulates the existence of a choice function for every non-empty set.

A10≡AC. (The *choice axiom.*)

$$\forall X(X \neq \varnothing \Rightarrow \exists z((z \leftrightharpoons \mathcal{P}(X)\backslash\{\varnothing\} \to X) \wedge \forall Y(Y \in \mathcal{P}(X)\backslash\{\varnothing\} \Rightarrow z(Y) \in Y))).$$

The described first-order theory is called the *Zermelo – Fraenkel axiomatic set theory (ZF) (with the choice axiom).*

A.2.2 Ordinals and cardinals in the ZF set theory

Let $\mathbf{A}$, $\mathbf{A}'$, $\mathbf{A}''$, ... be classes, where the prime symbol (′) is used only for the sake of uniformity of notations.

The collection $[\mathbf{A}_i \subset \mathbf{V} \mid i \in 2]$ such that $\mathbf{A}_0 \equiv \mathbf{A}$ and $\mathbf{A}_1 \equiv \mathbf{A}'$ will be called the *(multivalued) sequential pair of the classes* $\mathbf{A}$ *and* $\mathbf{A}'$ and will be denoted by $[\mathbf{A}, \mathbf{A}']$. The collection $[\mathbf{A}_i \subset \mathbf{V} \mid i \in 3]$ such that $\mathbf{A}_0 \equiv \mathbf{A}$, $\mathbf{A}_1 \equiv \mathbf{A}'$, and $\mathbf{A}_2 \equiv \mathbf{A}''$ will be called the *(multivalued) sequential triplet of the classes* $\mathbf{A}$, $\mathbf{A}'$, *and* $\mathbf{A}''$ and will be denoted by $[\mathbf{A}, \mathbf{A}', \mathbf{A}'']$, and so on.

Let now $a, a', a'', \ldots$ be sets.

The simple collection $(a_i \in \mathbf{V} \mid i \in 2)$ such that $a_0 \equiv a$ and $a_1 \equiv a'$ will be called the *simple sequential pair of the sets* a *and* a' and will be denoted by (a, a'). The simple collection $(a_i \in \mathbf{V} \mid i \in 3)$ such that $a_0 \equiv a$, $a_1 \equiv a'$, and $a_2 \equiv a''$, will be called the *simple sequential triplet of the sets* a, a', *and* a'' and will be denoted by (a, a', a''), and so on.

If $\mathbf{A}$, $\mathbf{A}'$, $\mathbf{B}$, $\mathbf{B}'$ are classes and $[\mathbf{A}, \mathbf{A}'] = [\mathbf{B}, \mathbf{B}']$, then $\mathbf{A} = \mathbf{B}$ and $\mathbf{A}' = \mathbf{B}'$. If a, a', b, b' are sets and $(a, a') = (b, b')$, then $a = b$ and $a' = b'$. The similar properties are valid also for every finite collections.

Let some collection $u \equiv [\mathbf{A}_i \subset \mathbf{V} \mid i \in I]$ be indexed by a class $I \neq \varnothing$. The class $\prod[\mathbf{A}_i \mid i \in I] \equiv \{z \in \mathbf{V} \mid (z : I \to \mathbf{V}) \wedge (\forall x\,(x \in I \Rightarrow z(x) \in \mathbf{A}_x))\}$ will be called the *product of the collection* u. In the particular case, if $\mathbf{A}$, $\mathbf{A}'$, $\mathbf{A}''$, ... are classes, then the classes $\prod(\mathbf{A}, \mathbf{A}')$, $\prod(\mathbf{A}, \mathbf{A}', \mathbf{A}'')$, ...will be called the *product of the pair* $(\mathbf{A}, \mathbf{A}')$, *the triplet* $(\mathbf{A}, \mathbf{A}', \mathbf{A}'')$, ... and will be denoted by $\mathbf{A} \times \mathbf{A}'$, $\mathbf{A} \times \mathbf{A}' \times \mathbf{A}''$, ...

One can check that $\mathbf{A} \times \mathbf{A}' = \{x \in \mathbf{V} \mid (\exists y \exists y'(y \in \mathbf{A} \wedge y' \in \mathbf{A}' \wedge x = (y, y')))\}$. It is easily seen from this equality that the product $\mathbf{A} \times \mathbf{A}'$ is similar to the coordinate product $\mathbf{A} * \mathbf{A}'$, but in contrast to the latter one it is a partial case of the general product $\prod(\mathbf{A}_i \subset \mathbf{V} \mid i \in I)$.

If $\mathbf{A} = \mathbf{A}' = \mathbf{A}'' = \dots$, then $\mathbf{A} \times \mathbf{A} = \mathbf{A}^2 \equiv \mathrm{Map}(2, A)$, $\mathbf{A} \times \mathbf{A} \times \mathbf{A} = \mathbf{A}^3 \equiv \mathrm{Map}(3, A)$, ... At the same time, $\mathbf{A} * \mathbf{A} \neq \mathbf{A}^2$ and between the classes $\mathbf{A} * \mathbf{A}$ and $\mathbf{A}^2$ there exists only a bijective mapping of the canonical form $\langle a, a' \rangle \mapsto (a, a')$. Namely, this stipulates the necessity of introducing the non-coordinate product $\mathbf{A} \times \mathbf{A}'$, $\mathbf{A} \times \mathbf{A}' \times \mathbf{A}''$, ...

If $n \in \omega$, then a subclass $\mathbf{R}$ of the class $\mathbf{A}^n \equiv \mathrm{Map}(n, A)$ is called an *n-placed correspondence on the class* $\mathbf{A}$. A mapping $O : \mathbf{A}^n \to \mathbf{A}$ is called an *n-placed operation on the class* $\mathbf{A}$. Note that $O \subset \mathbf{A}^n * \mathbf{A} \neq \mathbf{A}^{n+1}$. Therefore, an n-placed operation O can not be considered as an $(n+1)$-placed correspondence.

It can be checked that $\mathbf{A} \times \mathbf{A}' = \{x \mid \exists y \exists y'(y \in \mathbf{A} \wedge y' \in \mathbf{A}' \wedge x = (y, y'))\}$ and $\mathbf{A}^2 \neq \mathbf{A} * \mathbf{A}$.

A class $\mathbf{P}$ is called *ordered by a binary relation* $\leqslant \subset \mathbf{P}^2 = \mathbf{P} \times \mathbf{P}$ *on* $\mathbf{P}$, if:

1) $\forall p \in \mathbf{P}(p \leqslant p)$;
2) $\forall p, q \in \mathbf{P}(p \leqslant q \wedge q \leqslant p \Rightarrow p = q)$;
3) $\forall p, q, r \in \mathbf{P}(p \leqslant q \wedge q \leqslant r \Rightarrow p \leqslant r)$.

If, in addition,

4) $\forall p, q \in \mathbf{P}(p \leqslant q \vee q \leqslant p)$,
   then the relation $\leqslant$ is called the *linear order* on the class $\mathbf{P}$.
   An ordered class $\mathbf{P}$ is called *well-ordered* if
5) $\forall Q(\varnothing \neq Q \subseteq \mathbf{P} \Rightarrow \exists x \in Q(\forall y \in Q(x \leqslant y)))$, i. e. every non-empty subset of the class $\mathbf{P}$ has the smallest element.

Let $\mathbf{P}$ and $\mathbf{Q}$ be ordered classes. A mapping $\mathbf{F} : \mathbf{P} \to \mathbf{Q}$ is called *monotone* ($\equiv$ *increasing, order preserving*) if $p \leqslant p'$ implies $\mathbf{F}(p) \leqslant \mathbf{F}(p')$. The mapping $\mathbf{F}$ is called *strictly monotone* ($\equiv$ *strictly increasing*) if $p < p'$ implies $\mathbf{F}(p) < \mathbf{F}(p')$. The mapping $\mathbf{F}$ is called *isotone* if it is monotone and $\mathbf{F}(p) \leqslant \mathbf{F}(p')$ implies $p \leqslant p'$. It can be checked that: (1) if $\mathbf{F}$ is isotone, then $\mathbf{F}$ is injective and strictly monotone; (2) if $\mathbf{F}$ is isotone and surjective, then $\mathbf{F}$ is bijective and the inverse mapping $\mathbf{F}^{-1} : \mathbf{Q} \to \mathbf{P}$ is also isotone.

Ordered classes $(\mathbf{P}, \leqslant)$ and $(\mathbf{Q}, \leqslant)$ are called *order equivalent* (in notation $(\mathbf{P}, \leqslant) \approx (\mathbf{Q}, \leqslant)$) if there exists some isotone bijective mapping $\mathbf{F} : \mathbf{P} \rightarrowtail\!\!\!\!\twoheadrightarrow \mathbf{Q}$.

If a class $\mathbf{P}$ is ordered by a relation $\leqslant$ and $\mathbf{A}$ is a non-empty subclass of the class $\mathbf{P}$, then an element $p \in \mathbf{P}$ is called *the smallest upper bound* or *the supremum of the subclass* $\mathbf{A}$ if $\forall x \in \mathbf{A}(x \leqslant p) \wedge \forall y \in \mathbf{P}((\forall x' \in \mathbf{A}(x' \leqslant y)) \Rightarrow p \leqslant y)$. This formula is denoted by $p = \sup \mathbf{A}$.

A class $\mathbf{S}$ is called *transitive* if $\forall x(x \in \mathbf{S} \Rightarrow x \subseteq \mathbf{S})$. The class $\mathbf{S}$ is called *quasitransitive* if $\forall x \forall y(x \in \mathbf{S} \wedge y \subset x \Rightarrow x \in \mathbf{S})$. A transitive and quasitransitive class is said to be *supertransitive*.

A class [a set] $\mathbf{S}$ is called an *ordinal* [an *ordinal number*] if $\mathbf{S}$ is transitive and well-ordered by the relation $\in \cup =$ on $\mathbf{S}$. For the property of a class $\mathbf{S}$ to be an ordinal, we will denote by $On(\mathbf{S})$.

In the form of formula

$$\begin{aligned} On(\mathbf{S}) \equiv {} & \forall x(x \in \mathbf{S} \Rightarrow x \subseteq \mathbf{S}) \wedge \\ & \wedge \forall x, y, z(x \in \mathbf{S} \wedge y \in \mathbf{S} \wedge z \in \mathbf{S} \wedge x \in y \wedge y \in z \Rightarrow x \in z) \wedge \\ & \wedge \forall x, y(x \in \mathbf{S} \wedge y \in \mathbf{S} \Rightarrow x \in y \vee x = y \vee y \in x) \wedge \\ & \wedge \forall T(\varnothing \neq T \subseteq \mathbf{S} \Rightarrow \exists x(x \in T \wedge \forall y(y \in T \Rightarrow x \in y))). \end{aligned}$$

Ordinal numbers are usually denoted by Greek letters α, β, γ, and so on. The class of all ordinal numbers is denoted by $\mathbf{On}$. The natural ordering of the class of ordinal numbers is the relation $\alpha \leqslant \beta \equiv \alpha = \beta \vee \alpha \in \beta$. The class $\mathbf{On}$ is transitive and linearly ordered by the relation $\in \cup =$.

There are some simple assertions about ordinal numbers:
1) if α is an ordinal number, A is a set, and $A \in \alpha$, then A is an ordinal number;
2) $\alpha = \{\beta \mid \beta \in \alpha\}$ for every ordinal number α;
3) $\alpha + 1 \equiv \alpha \cup \{\alpha\}$ is the smallest of all ordinal numbers that are greater than α;
4) every non-empty set of ordinal numbers has the smallest element.

Therefore, the ordered class $\mathbf{On}$ is well-ordered. Thus, $\mathbf{On}$ is an ordinal.

Lemma 1. *Let* $\mathbf{A}$ *be a non-empty subclass of the class* $\mathbf{On}$. *Then* $\mathbf{A}$ *has the smallest element.*

Proof. By the condition there exists some ordinal number $\alpha \in \mathbf{A}$. Consider the class $\mathbf{B} \equiv \{x \mid x \in \mathbf{A} \wedge x \in \alpha+1\}$. By axiom scheme of separation AS3 this class is a set. Since $\alpha \in \mathbf{B} \subset \mathbf{On}$ and the class $\mathbf{On}$ is well-ordered, the set $\mathbf{B}$ has the smallest element β. Take an arbitrary element $\gamma \in \mathbf{A}$. If $\gamma \leqslant \alpha$, then $\gamma \in \mathbf{B}$ and therefore $\gamma \geqslant \beta$. If $\gamma > \alpha$, then $\gamma > \beta$. Thus, β is the smallest element of the class $\mathbf{A}$. □

Lemma 2. *If* A *is a non-empty set of ordinal numbers, then*
1) *the class* $\cup A$ *is an ordinal number;*
2) $\cup A = \sup A$ *in the ordered class* $\mathbf{On}$.

Proof. (1) The set $\cup A$ is transitive. In fact, if $x \in y \in \cup A$, then $y \in \alpha \in A$ for some ordinal α. By virtue of transitivity of α we get $x \in \alpha$, where $x \in \cup A$. It is clear that the set $\cup A$ is well-ordered by the relation $\in \cup =$. (2) Show that $p \equiv \cup A$ satisfies the formula $p = \sup A$. First p is an ordinal number. Second, assume that there is $x \in A$ such that $p < x$, i. e. $p \in x$. Since $p \in x$ and $x \in A$, we infer that $p \in \cup A \equiv p$, but it is impossible. Therefore, $\forall x \in A(x \leqslant p)$. Third, let $\exists y \in \mathbf{On}((\forall x' \in A(x' \leqslant y)) \wedge y \in p)$. Since $y \in p$, we infer that $\exists x \in A(y \in x)$, but it contradicts $\forall x' \in A(x' \leqslant y)$. So $p = \sup A$. □

Corollary 1. *The class* **On** *is a proper class.*

An ordinal number α is called *successive*, if $\alpha = \beta + 1$ for some ordinal number β. In the opposite case, α is called *limit*. This unique number β, we will denote by $\alpha - 1$. The formula expressing for an ordinal number α, the property of being successive [limit] will be denoted by $Son(\alpha)$ [$Lon(\alpha)$].

Lemma 3. *An ordinal number α is limit if and only if $\alpha = \sup \alpha$.*

The smallest (in the class **On**) not zero limit ordinal is denoted by ω. The existence of such an ordinal follows from A7, AS6, and AS3. Ordinals which are smaller than ω are called *natural numbers*.

Remark. The *principle of natural induction* in ZF is quite similar to the principle of natural induction in NBG (see Theorem 1 (1.2.6)). Moreover, all assertions about sets in NBG hold also in ZF and have the same proofs. Therefore, in what follows, we directly refer to such assertions from Chapter 1 when it is needed.

Collections $(\mathbf{B}_n \subset \mathbf{B} \mid n \in N \subset \omega)$ and $(b_n \in \mathbf{B} \mid n \in N \subset \omega)$, where N is an arbitrary subset of ω, are called *sequences*. If $N \subset n \in \omega$, then these collections are called *finite*; in the opposite case, they are called *infinite*.

Theorem 1 (the principle of transfinite induction). *Let* **C** *be a class of ordinal numbers such that:*
1) $\varnothing \in \mathbf{C}$;
2) $\alpha \in \mathbf{C} \Rightarrow \alpha + 1 \in \mathbf{C}$;
3) $Lon(\alpha) \wedge \alpha \subset \mathbf{C}) \Rightarrow S \in \mathbf{C}$.

Then $\mathbf{C} = \mathbf{On}$.

Proof. Let it be false. Consider the subclass $\mathbf{D} \equiv \mathbf{On} \backslash \mathbf{C}$. The class **D** is not empty, and therefore, according to Lemma 1, has the smallest element γ. Now, $\gamma \neq \varnothing$, because $\varnothing \in \mathbf{C}$. Thus, γ is either the successive ordinal number, or a limit one. Suppose that $\gamma = \gamma' + 1$. Since $\gamma' \in \gamma$, then $\gamma' \notin \mathbf{D}$ and so $\gamma' \in \mathbf{C}$. Then, by condition 2 of the theorem, $\gamma = \gamma' + 1 \in \mathbf{C}$. Suppose that γ is a limit ordinal number. Then, $\gamma \subset \mathbf{C}$, and by condition 3 of the theorem, $\gamma \in \mathbf{C}$. In both the cases, we arrive at a contradiction with the fact that $\gamma \notin \mathbf{C}$. Therefore, $\mathbf{C} = \mathbf{On}$. □

Theorem 2 (the construction by transfinite induction). *For every function* $\mathbf{G} : \mathbf{V} \to \mathbf{V}$ *there exists a unique function* $\mathbf{F} : \mathbf{On} \to \mathbf{V}$, *such that for every* $\alpha \in \mathbf{On}$ *the equality* $\mathbf{F}(\alpha) = \mathbf{G}(\mathbf{F}|\alpha)$ *is fulfilled.*

Proof. Consider the class $\mathbf{C} \equiv \{f \mid func(f) \wedge On(\mathrm{dom} f) \wedge \forall x \in \mathrm{dom} f(f(x) = \mathbf{G}(f|x))\}$. Take any function $f, g \in \mathbf{C}$ and consider the numbers $\alpha \equiv \mathrm{dom} f$ and $\beta \equiv \mathrm{dom}\, g$. Let $\alpha \subset \beta$. If $\alpha = 9$, then $f = \varnothing \subset g$. If $\alpha \neq 0$, then consider the set $P \equiv \{x \in \alpha \mid f(x) \neq g(x)\}$. Suppose that the set P is not empty. Then, P contains the smallest element π. Since $f(0) = \mathbf{G}(f|\varnothing) = \mathbf{G}(\varnothing) = \mathbf{G}(g|\varnothing) = g(0)$, we infer that $\pi \neq 0$. By definition, for every $x \in \pi$, we have $f(x) = g(x)$, where $f|\pi = g|\pi$. This implies $f(\pi) = \mathbf{G}(f|\pi) = \mathbf{G}(g|\pi) = g(\pi)$. But it follows from $\pi \in P$ that $f(\pi) \neq g(\pi)$. From this contradiction, we infer that the set P is empty. Therefore, $f \subset g$. Thus, we proved that $\alpha \subset \beta$ implies $f \subset g$. It follows from this property that $\alpha = \beta$ implies $f = g$. Consider the correspondence $\mathbf{E} \equiv \{z \mid \exists \alpha \in \mathbf{On} \exists f \in \mathbf{C}(\alpha = \mathrm{dom} f \wedge z = \langle \alpha, f \rangle)\}$. From the proof above, we infer that $\mathbf{E}$ is a mapping from the class $\mathbf{D} \equiv \mathrm{dom}\, \mathbf{E}$ into the class $\mathbf{V}$. We will consider this mapping in the form of simple collection $\mathbf{E} \equiv (f_\alpha \in \mathbf{C} \mid \alpha \in \mathbf{D})$. As was proven above, $\alpha \subset \beta$ implies $f_\alpha \subset f_\beta$. Prove by transfinite induction that $\mathbf{D} = \mathbf{On}$. Since $\varnothing \in \mathbf{C}$ and $\langle 0, \varnothing \rangle \in \mathbf{E}$, we infer that $0 \in \mathbf{D}$. Let $\alpha \in \mathbf{D}$. Using the function f_α, define a function $g : \alpha + 1 \to \mathbf{V}$ setting $g|\alpha \equiv f_\alpha$ and $g(\alpha) \equiv \mathbf{G}(f_\alpha)$. From $\alpha \cap \{\alpha\} = \varnothing$, $\alpha + 1 = \alpha \cup \{\alpha\}$, $g = f_\alpha \cup \{\langle \alpha, \mathbf{G}(f_\alpha) \rangle\}$, and axiom of union A4, it follows that this definition is correct. Let $x \in \alpha + 1$. If $x \in \alpha$, then $g(x) = f_\alpha(x) = \mathbf{G}(f_\alpha|x)$. By virtue of transitivity, $x \subset \alpha = \mathrm{dom} f_\alpha$. Therefore, $f_\alpha|x = g|x$ implies $g(x) = \mathbf{G}(g|x)$. If $x \in \{\alpha\}$, then $g(x) \equiv \mathbf{G}(f_\alpha) = \mathbf{G}(g|x)$. Consequently $g \in \mathbf{C}$ and $\langle \alpha + 1, g \rangle \in \mathbf{E}$ implies $\alpha + 1 \in \mathbf{D}$. Let α be a limit ordinal number and $\alpha \subset \mathbf{D}$. By Lemmas 3 and 2, $\alpha = \sup \alpha = \cup \alpha$. Let $x \in \alpha$. Then, $x \in y$ for some $y \in \alpha \subset \mathbf{D}$. Let $x \in z \in \alpha$. If $y = z$, then $f_y(x) = f_z(x)$. If $y \in z$, then $y \subset z$ by virtue of the embedding $f_y \subset f_z$ implies $f_y(x) = f_z(x)$. If $z \in y$, then, in a similar way, $f_z(x) = f_y(x)$. Define a function $g : \alpha \to \mathbf{V}$, setting $g(x) \equiv f_y(x)$ for any $y \in \alpha$ such that $x \in y$. It is clear that $g|y = f_y$. From $\alpha = \{y \mid y \in \alpha\}$, $g = \cup [f_y \mid y \in g]$, axiom scheme of replacement AS6 and axiom of union A4, it follows that this definition is correct. Check that $g \in \mathbf{C}$. Let $x \in \mathrm{dom}\, g = \alpha$. Then, $x \in y \in \alpha$ implies $g(x) = f_y(x) = \mathbf{G}(f_y|x) = \mathbf{G}(g|x)$ since $x \subset y \subset \alpha$. Now, from $g \in \mathbf{C}$ and $\alpha = \mathrm{dom}\, g$, it follows that $\alpha \in \mathbf{D}$. By virtue of the principle of transfinite induction, we conclude that $\mathbf{D} = \mathbf{On}$. Let $x \in \mathbf{On}$. Then, $x \in x + 1 \equiv \alpha$. Let $x \in \beta$ and $x \in \gamma$ for some $\beta, \gamma \in \mathbf{On}$. Since $\alpha \subset \beta$, $\alpha \subset \gamma$, $f_\alpha \subset f_\beta$, and $f_\alpha \subset f_\gamma$, we infer that $f_\beta(x) = f_\alpha(x) = f_\gamma(x)$. It follows that we can define correctly a function $\mathbf{F} : \mathbf{On} \to \mathbf{V}$ setting $\mathbf{F}(x) \equiv f_\beta(x)$ for every $\beta \in \mathbf{On}$ such that $x \in \beta$. It is clear that $\mathbf{F}|\beta = f_\beta \in \mathbf{V}$ for every $\beta \in \mathbf{On}$. If $x \in \mathbf{On}$, then $x \in x + 1 \equiv \alpha$ and $x \subset \alpha$ implies $\mathbf{F}(x) = f_\alpha(x) = \mathbf{G}(f_\alpha|x) = \mathbf{G}(\mathbf{F}|x)$. It remains to show that the function $\mathbf{F}$ is unique. Assume that there is a function $\mathbf{F}' : \mathbf{On} \to \mathbf{V}$ such that $\mathbf{F}'(\alpha) = \mathbf{G}(\mathbf{F}'|\alpha)$ for any $\alpha \in \mathbf{On}$. Note that by axiom scheme of replacement AS6, $\mathbf{F}'|\alpha \in \mathbf{V}$ for any $\alpha \in \mathbf{On}$. Consider the class $\mathbf{A} \equiv \{\alpha \in \mathbf{On} \mid \mathbf{F}(\alpha) = \mathbf{F}'(\alpha)\}$. Since $\mathbf{F}(0) = \mathbf{G}(\mathbf{F}|\varnothing) = \mathbf{G}(\varnothing) = \mathbf{G}(\mathbf{F}'|\varnothing) = \mathbf{F}'(0)$, we infer that $0 \in \mathbf{A}$, i.e. this class is non-empty. Assume that the class $\mathbf{B} \equiv \mathbf{On} \backslash \mathbf{A}$ is non-empty.

Then, it contains the smallest element β. If $\alpha \in \beta$, then $\alpha \in \mathbf{A}$ implies $\mathbf{F}(\alpha) = \mathbf{F}'(\alpha)$. Thus, $\mathbf{F}|\beta = \mathbf{F}'|\beta$. Hence, we get $\mathbf{F}(\beta) = \mathbf{G}(\mathbf{F}|\beta) = \mathbf{G}(\mathbf{F}'|\beta) = \mathbf{F}'(\beta)$, but this contradicts the inequality $\mathbf{F}(\beta) \neq \mathbf{F}'(\beta)$. It follows from the obtained contradiction that $\mathbf{A} = \mathbf{On}$. □

In ZF, there exists the following *principle of* $\in$*-induction.*

Lemma 4. *If a class* $\mathbf{C}$ *satisfies the condition* $\forall x \subseteq \mathbf{C} \Rightarrow x \in \mathbf{C}$, *then* $\mathbf{C} = \mathbf{V}$.

Proof. Suppose that $\mathbf{C} \neq \mathbf{D}$, i. e. $\mathbf{D} \equiv \mathbf{V} \backslash \mathbf{C} \neq \varnothing$. Then, there exists $P \in \mathbf{D}$. If $P \cap \mathbf{D} = \varnothing$, then put $X \equiv P$. Let $P \cap \mathbf{D} \neq \varnothing$. Consider the set N containing all $n \in \omega$ satisfying the condition that there exists a unique sequence $u \equiv u(n) \equiv (R_k \mid k \in n+1)$ of sets R_k such that $R_0 = P$ and $R_{k+1} = \cup R_k$ for every $k \in n$. Since the sequence $R_k \mid k \in 1$ such that $R_0 \equiv X$ satisfies this property, we infer that $0 \in N$. Let $n \in N$, i. e. for n there exists a unique sequence $u \equiv (R_k \mid k \in n+1)$. Define the sequence $v \equiv (S_k \mid k \in n+2)$, putting $S_k \equiv R_k$ for every $k \in n+1$ and $S_{n+1} \equiv \cup R_n = \cup S_n$, i. e. $v = u \cup \{\langle n+1, \cup R_n \rangle\}$. It is clear that the sequence v possesses all necessary properties. Check its uniqueness. Suppose that there exists a sequence $w \equiv (T_k \mid k \in n+2)$ such that $T_0 = P$ and $\forall k \in n+1 (T_{k+1} = \cup T_k)$. Consider the set M' consisting of all $m \in n+2$ such that $S_m = T_m$. Let $M'' \equiv \omega \backslash (n+2)$ and $M \equiv M' \cup M''$. Since $S_0 = P = T_0$, we infer that $0 \in M' \subset M$. Let $m \in M'$. If $m = n+1$, then $m+1 = n+2 \in M'' \subset M$. If $m < n+1$, then $m+1 \in n+2$ and $S_{m+1} = \cup S_m = \cup T_m = T_{m+1}$ implies $m+1 \in M' \subset M$. If $m \in M'$, then $m+1 \in M'' \subset M$. Therefore, $m \in M$ implies $m+1 \in M$. By the principle of transfinite induction (Theorem 1), $M = \omega$. Hence, $M' = n+2$ and $v = w$, i. e. the sequence v is unique. Therefore, $n+1 \in N$. By the principle of natural induction (see Theorem 1 (1.2.6) and Remark before Theorem 1), $N = \omega$. Thus, for every $n \in \omega$ there exists a unique sequence $u(n)$. By virtue of the uniqueness we can denote it by $(R_k^n \mid k \in n+1)$. Consider the following formula of the ZF set theory: $\varphi(x, y) \equiv (x \in \omega \Rightarrow y = R_x^x) \wedge (x \notin \omega \Rightarrow y = \varnothing)$. By axiom scheme of replacement AS6, for ω there exists a set Y such that $\forall x \in \omega (\forall y (\varphi(x, y) \Rightarrow y \in Y))$. If $n \in \omega$, then $\varphi(n, R_n^n)$ implies $R_n^n \in Y$. Therefore, we can define in the set $\omega \times Y$ an infinite sequence $u \equiv (R_n \in Y \mid n \in \omega)$ setting $u \equiv \{z \in \omega \times Y \mid \exists x \in \omega (z = \langle x, R_x^x \rangle)\}$. The property of uniqueness mentioned above implies that $u(m) = u(n)|(m+1)$ for all $m \leqslant n$. Thus, $u|(n+1) = u(n)$. Hence, the sequence u satisfies the following properties: $R_0 = P$ and $R_{k+1} = \cup R_k$ for every $k \in \omega$. Having the set u, we can take the set $A \equiv \operatorname{rng} u \equiv \{R_n \mid n \in \omega\}$ and the set $Q \equiv \cup A = \{y \mid \exists x \in \omega (y \in R_x)\} = \cup \{R_n | n \in \omega\}$. It is clear that $R_n \subset Q$ for every $n \in \omega$, and therefore, $P = R_0 \subset Q$. Since $P \cap \mathbf{D} \neq \varnothing$, we infer that $R \equiv Q \cap \mathbf{D} \neq \varnothing$. By regularity axiom A9, there exists $X \in R$ such that $X \cap R = \varnothing$. Check that $X \cap \mathbf{D} = \varnothing$. Suppose that there exists $x \in X \cap \mathbf{D}$. Since $X \in Q$, we infer that $X \in R_n$ for some $n \in \omega$. Therefore, $x \in X \in R_n$ implies $x \in \cup R_n = R_{n+1} \subset Q$. Thus, $x \in R$. As a result, we have $x \in X \cap R = \varnothing$, but it is impossible. It follows from this contradiction that $X \in \mathbf{D}$ and $X \cap \mathbf{D} = \varnothing$. Thus, in both cases, $X \subset \mathbf{C}$. By condition, $X \in \mathbf{C}$, but it is impossible because $X \in \mathbf{D}$. From this contradiction, we infer that $\mathbf{C} = \mathbf{V}$. □

Sets A and B are called *equivalent* ($A \sim B$) if there exists a bijective function $u : A \rightarrowtail\!\!\!\twoheadrightarrow B$.

An ordinal number α is called *cardinal* if for every ordinal number β the conditions $\beta \leqslant \alpha$ and $\beta \sim \alpha$ imply $\beta = \alpha$. The property of an ordinal number α to be a cardinal we will denote by $Cn(\alpha)$. The class of all cardinal numbers will be denoted by **Cn**. The class **Cn** with the order induced from the class **On** is well-ordered.

Lemma 5. *For every set A there exists an ordinal number α such that $A \sim \alpha$.*

Now, for a set A consider the class $\{\beta \mid \beta \in \mathbf{On} \wedge \beta \sim A\}$. By Lemma 5 this class is not empty and therefore it contains the smallest element α. It is clear that α is a cardinal number. Besides, this class contains only one cardinal number α. This number α is called the *power of the set A* (it is denoted by $|A|$ or $\operatorname{card} A$). A set of the power ω is called *denumerable*. Sets of the power $n \in \omega$ are called *finite*. A set is called *countable* if it is finite or denumerable. A set is called *infinite* if it is not finite.

Note that if $\varkappa$ is an infinite cardinal number then $\varkappa$ is a limit ordinal number.

If $\varkappa = \alpha + 1$, then $\varkappa = \operatorname{card} \varkappa = \operatorname{card}(\alpha + 1) = \operatorname{card} \alpha \leqslant \alpha < \varkappa$, but it is impossible.

Let α be an ordinal. A *confinality* of α is the ordinal number $cf(\alpha)$, which is equal to the smallest ordinal number β for which there exists a function f from β into α such that $\cup \operatorname{rng} f = \alpha$.

A cardinal $\varkappa$ is called *regular* if $cf(\varkappa) = \varkappa$, i. e. for every ordinal number β for which there exists a function $f : \beta \to \varkappa$ such that $\cup \operatorname{rng} f = \varkappa$ it is valid $\varkappa \leqslant \beta$.

A cardinal $\varkappa > \omega$ is called (*strongly*) *inaccessible* if $\varkappa$ is regular and $\operatorname{card} \mathcal{P}(\lambda) < \varkappa$ for all ordinal numbers $\lambda < \varkappa$. The property of a cardinal number $\varkappa$ to be inaccessible will be denoted by $Icn(\varkappa)$. The class of all inaccessible cardinal numbers will be denoted by s **In**.

A.3 Cumulative sets in the ZF set theory

A.3.1 Construction of cumulative sets

Now, we will apply the construction by transfinite induction in the following situation. Consider the class

$$\begin{aligned}\mathbf{G} \equiv \{Z \mid \exists X \exists Y (Z = \langle X, Y \rangle) \wedge ((X = \varnothing \Rightarrow Y = \varnothing) \vee \\ \vee (X \neq \varnothing \Rightarrow (\neg func(X) \Rightarrow Y = \varnothing) \vee \\ \vee (func(X) \Rightarrow (\neg On(\operatorname{dom} X) \Rightarrow Y = \varnothing) \vee \\ \vee (On(\operatorname{dom} X) \Rightarrow (Son(\operatorname{dom} X) \Rightarrow Y = X(\operatorname{dom} X - 1) \cup \mathcal{P}(X(\operatorname{dom} X - 1))) \vee \\ \vee (Lon(\operatorname{dom} X) \Rightarrow Y = \cup \operatorname{rng} X))))))\}.\end{aligned}$$

If we express the definition of the class $\mathbf{G}$ less formally, then $\mathbf{G}$ consists of all pairs $\langle X, Y\rangle$ for which there are the following five disjunctive cases:

1) if $X = \varnothing$, then $Y = \varnothing$;
2) if $X \neq \varnothing$ and X is not a function, then $Y = \varnothing$;
3) if $X \neq \varnothing$, X is a function, and $\operatorname{dom} X$ is not an ordinal number, then $Y = \varnothing$;
4) if $X \neq \varnothing$, X is a function, $\operatorname{dom} X$ is an ordinal number, and $\operatorname{dom} X = \alpha + 1$, then $Y = X(\alpha) \cup \mathcal{P}(X(\alpha))$;
5) if $X \neq \varnothing$, X is a function, and $\operatorname{dom} X$ is a limit ordinal number, then $Y = \cup \operatorname{rng} X$.

By definition, $\mathbf{G}$ is a correspondence. Since any set X possesses one of these properties, we have $\operatorname{dom} \mathbf{G} = \mathbf{V}$. Since in each of these five cases the set Y is defined by a set X in a unique way, using the property of an ordered pair from Lemma 1 (A.2.1), we infer that $\mathbf{G}$ is a function from $\mathbf{V}$ into $\mathbf{V}$.

According to Theorem 2 (A.2.2), for the function $\mathbf{G}$ there exists a function $\mathbf{F} : \mathbf{On} \to \mathbf{V}$ for which for any $\alpha \in \mathbf{On}$, we have $\mathbf{F}(\alpha) = \mathbf{G}(\mathbf{F}|\alpha)$.

From case 1, for the function $\mathbf{G}$, it follows that $\mathbf{F}(\varnothing) = \mathbf{G}(\mathbf{F}|\varnothing) = \mathbf{G}(\varnothing) = \varnothing$.

From case 4, it follows that if β is a successive ordinal number and $\beta = \alpha + 1$, then $\mathbf{F}(\beta) = \mathbf{G}(\mathbf{F}|\beta) = (\mathbf{F}|\beta)(\alpha) \cup \mathcal{P}((\mathbf{F}|\beta)(\alpha)) = \mathbf{F}(\alpha) \cup \mathcal{P}(\mathbf{F}(\alpha))$.

Finally, from case 5, it follows that if α is a limit ordinal number, then $\mathbf{F}(\alpha) = \mathbf{G}(\mathbf{F}|\alpha) = \cup \operatorname{rng}(\mathbf{F} \mid \alpha) = \cup\{\mathbf{F}(\beta) \mid \beta \in \alpha\}$.

Denote $\mathbf{F}(\alpha)$ by V_α. Thus, we obtained the collection $\{V_\alpha \subset \mathbf{V} \mid \alpha \in \mathbf{On}\}$ satisfying the following conditions:

1) $V_0 = \varnothing$;
2) $V_\alpha = \cup\{V_\beta \mid \beta \in \alpha\}$, if α is a limit ordinal number;
3) $V_{\alpha+1} = V_\alpha \cup \mathcal{P}(V_\alpha)$.

This collection is called the *Mirimanov – Neumann collection*, and its elements V_α are called *cumulative (Mirimanov – Neumann) sets.*

A.3.2 Properties of cumulative sets

Prove now some lemmas about the sets V_α, which we will need later.

Lemma 1. *If α and β are ordinal numbers, then*

1) $\alpha < \beta \Leftrightarrow V_\alpha \in V_\beta$;
2) $\alpha = \beta \Leftrightarrow V_\alpha = V_\beta$;
3) $\alpha \subset V_\alpha$ *and* $\alpha \in V_{\alpha+1}$.

Proof. (1) and (2). By means of transfinite induction we will prove that for any ordinal number β $(\alpha \in \beta \Rightarrow V_\alpha \in V_\beta)$. If $\beta = \varnothing$, then it is clear because $\forall\alpha\neg(\alpha \in \beta)$. If for some ordinal number β $(\alpha \in \beta \Rightarrow V_\alpha \in V_\beta)$, then consider the ordinal number $\beta + 1$. From

$\alpha \in \beta + 1$, we infer that $\alpha \in \beta \vee \alpha = \beta$. If $\alpha \in \beta$, then, by the inductive assumption, $V_\alpha \in V_\beta$, and since $V_{\beta+1} = V_\beta \cup \mathcal{P}(V_\beta)$, we infer that $V_\alpha \in V_{\beta+1}$. If $\alpha = \beta$, then $V_\alpha = V_\beta \in V_{\beta+1}$, because $V_\beta \in \mathcal{P}(V_\beta)$. Therefore, for $\beta + 1$, the property $\alpha \in \beta + 1 \Rightarrow V_\alpha \in V_{\beta+1}$ is fulfilled. Suppose now that β is a limit ordinal number and $\forall\gamma \in \beta\forall\alpha(\alpha \in \gamma \Rightarrow V_\alpha \in V_\gamma)$. Let α belongs to β. Since β is a limit ordinal number, it follows that $\alpha + 1 \in \beta$. From $V_\beta = \cup(V_\gamma \mid \gamma \in \beta)$, it follows that $V_{\alpha+1} \subset V_\beta$. In this case, $V_\alpha \in V_{\alpha+1}$, implies $V_\alpha \in V_\beta$. It is clear that $\alpha = \beta \Rightarrow V_\alpha = V_\beta$. If $V_\alpha = V_\beta$, then either $\alpha < \beta$ or $\alpha = \beta$ or $\beta < \alpha$. If $\alpha < \beta$, then $V_\alpha \in V_\beta$; if $\beta < \alpha$, then $V_\beta \in V_\alpha$. Therefore, $\beta = \alpha$. If $V_\alpha \in V_\beta$, then $\alpha < \beta$, because for $\alpha = \beta$ it is fulfilled $V_\alpha = V_\beta$, and for $\beta < \alpha$ it is fulfilled $V_\beta \in V_\alpha$. (3) Consider the class $\mathbf{C} \equiv \{x \mid x \in \mathbf{On} \wedge x \subset V_x\}$. Since $0 \subset \varnothing = V_0$, we infer that $0 \in \mathbf{C}$. If $\alpha \in \mathbf{C}$, then $\alpha \subset V_\alpha$ implies $\alpha + 1 \equiv \alpha \cup \{\alpha\} \subset V_\alpha \subset V_{\alpha+1}$. Let α be a limit ordinal number and $\alpha \subset \mathbf{C}$. By construction, $V_\alpha = \cup(V_\beta \mid \beta \in \alpha)$. If $x \in \alpha$, then $x \in \mathbf{C}$ means that $x \subset V_x$. Therefore, $x \in \mathcal{P}(V_x) \subset V_{x+1}$. Since α is a limit ordinal number then $x + 1 \in \alpha$ implies $x \in V_\alpha$. Thus, $\alpha \subset V_\alpha$ and so $\alpha \in \mathbf{C}$. By Theorem 1 (A.2.2), $\mathbf{C} = \mathbf{On}$. The lemma is proven. □

Lemma 2. *For every ordinal number α the condition $z \subset x \in V_\alpha$ implies $z \in V_\alpha$.*

Proof. We will prove this assertion by transfinite induction. Let $\mathbf{C} = \{\alpha \mid \alpha \in \mathbf{On} \wedge \forall x\forall z(z \subset x \in V_\alpha \Rightarrow z \in V_\alpha)\}$. Show that $\mathbf{C} = \mathbf{On}$. If $\alpha = \varnothing$, then it is clear that $\alpha \in \mathbf{C}$. Suppose that $\alpha \in \mathbf{C}$. Prove that in this case $\alpha + 1 \in \mathbf{C}$. Let $z \subset x \in V_{\alpha+1}$. Since $V_{\alpha+1} = V_\alpha \cup \mathcal{P}(V_\alpha)$, we infer that $x \in V_\alpha$ or $x \subset V_\alpha$. If $z \subset x \in V_\alpha$, then $z \in V_\alpha$ by the inductive assumption and therefore $z \in V_{\alpha+1}$. If $x \subset V_\alpha$ and $z \subset x$, then $z \subset V_\alpha$, and therefore $z \in V_{\alpha+1}$. Thus, $\alpha + 1 \in \mathbf{C}$. If α is a limit ordinal number and $\forall\beta \in \alpha(\beta \in \mathbf{C})$, then from $z \subset x \in V_\alpha$ we infer that $\exists\beta \in \alpha(z \subset x \in V_\beta)$, and, by inductive assumption, we conclude that $\exists\beta \in \alpha(z \in V_\beta)$. From $V_\alpha = (V_\beta \mid \beta \in \alpha)$ it follows now that $z \in V_\alpha$. Therefore, by transfinite induction, $\mathbf{C} = \mathbf{On}$. □

This lemma shows that any cumulative set V_α is quasitransitive.

Lemma 3. *For any ordinal number α, we have $\forall x(x \in V_\alpha \Rightarrow x \subset V_\alpha)$.*

Proof. This lemma also will be proven by transfinite induction. For $\alpha = \varnothing$ the given formula is valid, because $\forall x\neg(x \in V_\varnothing)$. Let for some ordinal number α it is valid $\forall x(x \in V_\alpha \Rightarrow x \subset V_\alpha)$. Consider the ordinal number $\alpha+1$. If $x \in V_{\alpha+1}$, then $x \in V_\alpha \vee x \in \mathcal{P}(V_\alpha)$, or more presicely $x \in V_\alpha \vee x \subset V_\alpha$. In the case $x \in V_\alpha$, by the inductive assumption, $x \subset V_\alpha$, and $V_\alpha \subset V_{\alpha+1}$ implies $x \subset V_{\alpha+1}$. If $x \subset V_\alpha$, then we infer from $V_\alpha \subset V_{\alpha+1}$ that $x \subset V_{\alpha+1}$. Now, let α be a limit ordinal number and $\forall\beta \in \alpha\forall x(x \in V_\beta \Rightarrow x \subset V_\beta)$. Then, from $x \in V_\alpha$ we infer that $\exists\beta \in \alpha(x \in V_\beta)$. By the inductive assumption $\exists\beta \in \alpha(x \subset V_\beta)$, and therefore $x \subset V_\alpha$. The lemma is proved. □

This lemma shows that any cumulative set is transitive. Thus, any cumulative set is supertransitive.

Corollary 1. *If α and β are ordinal numbers and $\alpha \leqslant \beta$, then $V_\alpha \subset V_\beta$.*

Corollary 2. *For every ordinal number α the inclusion $V_\alpha \subset \mathcal{P}(V_\alpha)$ and the equality $V_{\alpha+1} = \mathcal{P}(V_\alpha)$ are valid.*

Proof. If $x \in V_\alpha$, then by the given lemma $x \subset V_\alpha$, i. e. $x \in \mathcal{P}(V_\alpha)$. Thus, $V_\alpha \subset \mathcal{P}(V_\alpha)$. Therefore. $V_{\alpha+1} = V_\alpha \cup \mathcal{P}(V_\alpha) = \mathcal{P}(V_\alpha)$. □

Corollary 3. *If α and β are ordinal numbers and $\alpha < \beta$, then $|V_\alpha| < |V_\beta|$.*

Proof. By the previous two corollaries, $V_\alpha \subset \mathcal{P}(V_\alpha) = V_{\alpha+1} \subset V_\beta$. Using Cantor's theorem, we infer that $|V_\alpha| < |\mathcal{P}(V_\alpha)| = |V_{\alpha+1}| \leqslant |V_\beta|$. □

Lemma 4. *For every ordinal number α if $x \in V_{\alpha+1}$, then $x \subset V_\alpha$.*

Proof. Suppose that $x \in V_{\alpha+1}$. It means that $x \in V_\alpha \vee x \subset V_\alpha$. If $x \subset V_\alpha$, then everything is proven. If $x \in V_\alpha$, then by the previous lemma $x \subset V_\alpha$. □

Lemma 5. *For every ordinal number α $\forall x \forall y(x \in V_\alpha \wedge y \in V_\alpha \Rightarrow x \cup y \in V_\alpha)$.*

Proof. We will again use the principle of transfinite induction. If $\alpha = \varnothing$, then the conclusion of lemma is valid, because $\forall x \neg(x \in V_\varnothing)$. Let now $\alpha = \beta + 1$ for some ordinal number β. Then, from the formula $x \in V_\alpha \wedge y \in V_\alpha$ by Lemma 4 we infer that $x \subset V_\beta \wedge y \subset V_\beta$. Therefore, $x \cup y \subset V_\beta$, where $x \cup y \in V_{\beta+1}$, i. e. $x \cup y \in V_\alpha$. Now, suppose that α is a limit ordinal number and $\forall \beta \in \alpha \forall x \forall y(x \in V_\beta \wedge y \in V_\beta \Rightarrow x \cup y \in V_\beta)$. Then, $x, y \in V_\alpha$ implies $\exists \beta \in \alpha(x, y \in V_\beta)$. Therefore, by the inductive assumption, $\exists \beta \in \alpha(x \cup y \in V_\beta)$, and so $x \cup y \in V_\alpha$. □

Lemma 6. *For every limit ordinal number α the condition $x \in V_\alpha$ implies $\mathcal{P}(x) \in V_\alpha$.*

Proof. Suppose that α is some limit ordinal number and $x \in V_\alpha$. Then, there exists $\beta \in \alpha$ such that $x \in V_\beta$. Show that in this case $\mathcal{P}(x) \subset V_\beta$. By Lemma 2, from $x \in V_\beta$ and $z \subset x$ we infer $z \in V_\beta$, where $\forall z(z \in \mathcal{P}(x) \Rightarrow z \in V_\beta)$, and it means that $\mathcal{P}(x) \subset V_\beta$. If $\mathcal{P}(x) \subset V_\beta$, then $\mathcal{P}(x) \in V_{\beta+1} \subset V_\alpha$. □

Corollary 1. *For every limit ordinal number α the condition $x, y \in V_\alpha$ implies $\{x\}, \{x, y\}, \langle x, y\rangle \in V_\alpha$.*

Proof. By Lemma 6, $\mathcal{P}(x) \in V_\alpha$. By Lemma 2, $\{x\} \subset \mathcal{P}(x)$ implies $\{x\} \in V_\alpha$. Now, by Lemma 5, $\{x, y\} \in V_\alpha$. It follows from the proved properties that $\langle x, y\rangle \in V_\alpha$. □

Corollary 2. *For every limit ordinal number α the conditions $X, Y \in V_\alpha$ implies $X * Y \in V_\alpha$.*

Proof. Let $x \in X$ and $y \in Y$. Then, $\{x\} \subset X \cup Y$ and $\{y\} \subset X \cup Y$ imply $\{x, y\} \subset X \cup Y$. By Lemma 5, $X \cup Y \in V_\alpha$. Since $\{x\} \in \mathcal{P}(X \cup Y)$ and $\{x, y\} \in \mathcal{P}(X \cup Y)$, we infer that $\langle x, y\rangle \equiv \{\{x\}, \{x, y\}\} \subset \mathcal{P}(X \cup Y)$. Hence, $\langle x, y\rangle \in \mathcal{P}(\mathcal{P}(X \cup Y))$. Therefore, $X * Y \subset \mathcal{P}(\mathcal{P}(X \cup Y))$. By Lemmas 5, 6, and 2, $X * Y \in V_\alpha$. □

Lemma 7. *If $\alpha \geqslant \omega$, then $\omega \subset V_\alpha$. If $\alpha > \omega$, then $\omega \in V_\alpha$.*

Proof. By Lemma 1, $\omega \subset V_\omega \subset V_\alpha$. If $\alpha > \omega$, then $\omega \subset V_\omega \in V_{\omega+1}$, by Lemma 2, implies $\omega \in V_{\omega+1} \subset V_\alpha$. □

Let λ be an ordinal number. Consider a collection $K(\lambda) \equiv (M_\beta \mid \beta \in \lambda + 1)$ of the sets $M_\beta \equiv \{x \mid x \leftrightharpoons \mathcal{P}(|V_\beta|) \rightarrowtail\!\!\!\!\twoheadrightarrow |\mathcal{P}(|V_\beta|)|\}$ of all corresponding bijective mappings for all $\beta \in \lambda + 1$ and the set $M(\lambda) \equiv \cup(M_\beta \mid \beta \in \lambda + 1)$. By the choice axiom there exists a choice function $ch(\lambda) : \mathcal{P}(M(\lambda))\backslash\{\varnothing\} \to M(\lambda)$ such that $ch(\lambda)(P) \in P$ for every $P \in \mathcal{P}(M(\lambda))\backslash\{\varnothing\}$. Since $M_\beta \subset M(\lambda)$ for $\beta \in \lambda+1$, we infer that $c_\beta(\lambda) \equiv ch(\lambda)(M_\beta) \in M_\beta$, i. e. $c_\beta(\lambda)$ is a bijection from $\mathcal{P}(|V_\beta|)$ onto $|\mathcal{P}(|V_\beta|)|$.

Theorem 1 (the Zakharov theorem on initial synchronization of powers of cumulative sets). *Let λ be an ordinal number. Then, for every ordinal number $\alpha \leqslant \lambda$ there exists a unique collection $u(\alpha) \equiv u(\lambda)(\alpha) \equiv (f_\beta \mid \beta \in \alpha+1)$ of bijective functions $f_\beta : V_\beta \rightarrowtail\!\!\!\!\twoheadrightarrow |V_\beta|$ such that:*
1) $f_0 \equiv \varnothing$;
2) *if $\gamma < \beta \in \alpha + 1$, then $f_\gamma = f_\beta|V_\gamma$;*
3) *if $\beta \in \alpha + 1$ and $\beta = \gamma + 1$, then $f_\beta|V_\gamma = f_\gamma$ and $f_\beta(x) = c_\gamma(\lambda)(f_\gamma[x])$ for every $x \in V_\beta\backslash V_\gamma = \mathcal{P}(V_\gamma)\backslash V_\gamma$;*
4) *if $\beta \in \alpha + 1$ and β is a limit ordinal number then $f_\beta = \cup(f_\gamma \mid \gamma \in \beta)$.*

It follows from the uniqueness property that $u(\alpha)|\delta + 1 = u(\delta)$ for every $\delta \leqslant \alpha$, i. e. these collections continue each other.

Proof. In the beginning check the uniqueness of the collection $u \equiv u(\alpha)$. Let for α there exist a collection $v \equiv (g_\beta \mid \beta \in \alpha + 1)$ of bijective functions $g_\beta : V_\beta \rightarrowtail\!\!\!\!\twoheadrightarrow |V_\beta|$, possessing properties 1–4. Consider the set $D' \equiv \{\beta \in \alpha + 1 \mid f_\beta = g_\beta\}$, the class $\mathbf{D}'' \equiv \mathbf{On}\backslash(\alpha + 1)$, and the class $\mathbf{D} \equiv \mathbf{D}' \cup \mathbf{D}''$. It is clear that $0 \in D' \subset \mathbf{D}$. Let $\beta \in \mathbf{D}$. If $\beta \geqslant \alpha$, then $\beta + 1 \in \mathbf{D}'' \subset \mathbf{D}$. Let $\beta < \alpha$. Then, $\beta \in D'$ and $\beta + 1 \in \alpha + 1$. Therefore, by property 3, $f_{\beta+1}(x) = f_\beta(x) = g_\beta(x) = g_{\beta+1}(x)$ for every $x \in V_\beta$ and $f_{\beta+1}(x) = c_\beta(\lambda)(f_\beta[x]) = c_\beta(\lambda)(g_\beta[x]) = g_{\beta+1}(x)$ for every $x \in V_{\beta+1}\backslash V_\beta$, i. e. $f_{\beta+1} = g_{\beta+1}$. So $\beta + 1 \in D' \subset \mathbf{D}$. Thus, $\beta \in \mathbf{D}$ implies $\beta + 1 \in \mathbf{D}$. Let β be a limit ordinal number and $\beta \subset \mathbf{D}$. If $\beta \cap \mathbf{D}'' \neq \varnothing$, then there exists $\gamma \in \beta$ such that $\gamma \geqslant \alpha + 1$. Therefore, $\beta > \gamma \geqslant \alpha + 1$ implies $\beta \in \mathbf{D}'' \subset \mathbf{D}$. Let $\beta \cap \mathbf{D}'' = \varnothing$, i. e. $\beta \subset D'$. Then, for every $\gamma \in \beta$ it is valid $f_\gamma = g_\gamma$. Since $\beta \subset \alpha + 1$, then $\beta \leqslant \alpha + 1$. If $\beta = \alpha + 1$, we infer that $\beta \in \mathbf{D}'' \subset \mathbf{D}$. Let $\beta \in \alpha + 1$. If $x \in V_\beta = \cup(V_\gamma \mid \gamma \in \beta)$, then $x \in V_\gamma$ for some $\gamma \in \beta$. Therefore, by property 2, $f_\beta(x) = f_\gamma(x) = g_\gamma(x) = g_\beta(x)$ for every $x \in V_\beta$, i. e. $f_\beta = g_\beta$. So $\beta \in D' \subset \mathbf{D}$. Thus, the properties

$Lon(\beta)$ and $\beta \subset \mathbf{D}$ imply $\beta \in \mathbf{D}$. By the principle of transfinite induction, $\mathbf{D} = \mathbf{On}$. Consequently, $D' = \alpha + 1$. Therefore, $u = v$. Now, we will write c_γ instead of $c_\gamma(\lambda)$. Consider a set C', consisting of all ordinal numbers $\alpha \leqslant \lambda$, for which there exists a collection $u(\alpha)$ with properties 1–4. Consider also the classes $\mathbf{C}'' \equiv \mathbf{On}\backslash(\lambda + 1)$ and $\mathbf{C} \equiv C' \cup \mathbf{C}''$. Since $V_0 = \varnothing$ and $|V_0| = 0$, we infer that the collection $u(0) \equiv (f_\beta \mid \beta \in 1)$ with the bijective function $f_0 = \varnothing : V_0 \rightarrowtail\!\!\!\twoheadrightarrow |V_0|$ possesses all properties 1–4, and therefore $0 \in \mathbf{C}$. Let $\alpha \in \mathbf{C}$. If $\alpha \geqslant \lambda$, then $\alpha + 1 \in \mathbf{C}'' \subset \mathbf{C}$. Let now $\alpha < \lambda$. Then, $\alpha + 1 \in \lambda + 1$ means that we can use the function c_α. Since $\alpha \in C'$, we infer that for α there exists a unique collection $u \equiv (f_\beta \mid \beta \in \alpha + 1)$. Define a collection $v \equiv (g_\beta \mid \beta \in \alpha + 2)$ of bijective functions $g_\beta : V_\beta \rightarrowtail\!\!\!\twoheadrightarrow |V_\beta|$, setting $g_\beta \equiv f_\beta$ for every $x \in V_\alpha$ and $g_{\alpha+1}(x) \equiv c_\alpha(f_\alpha[x])$ for every $x \in V_{\alpha+1}\backslash V_\alpha = \mathcal{P}(V_\alpha)\backslash V_\alpha$. Check that v possesses properties 1–4. Let $\beta \in \alpha + 2$. If $\beta \in \alpha + 1$, then properties 1–4 are evidently true. Let $\beta = \alpha + 1$. Then, $g_\beta(x) = g_{\alpha+1}(x) = f_\alpha(x) = g_\alpha(x)$ for every $x \in V_\alpha$ and $g_\beta(x) = g_{\alpha+1}(x) = c_\alpha(f_\alpha[x]) = c_\alpha(g_\alpha[x])$ for every $x \in V_\beta\backslash V_\alpha$. Besides, $g_\beta|V_\alpha = f_\alpha = g_\alpha$. Therefore, $\gamma < \beta$ implies $g_\beta|V_\gamma = g_\alpha|V_\gamma = f_\alpha|V_\gamma = f_\gamma = g_\gamma$. So $\alpha + 1 \in C' \subset \mathbf{C}$. Let α be a limit ordinal number and $\alpha \subset \mathbf{C}$. If $\alpha \cap \mathbf{C}'' \neq \varnothing$, then there exists $\beta \in \alpha$ such that $\beta \geqslant \alpha + 1$. Consequently, $\alpha > \beta \leqslant \lambda + 1$ implies $\alpha \in \mathbf{C}'' \subset \mathbf{C}$. Let $\alpha \cap \mathbf{C}'' = \varnothing$, i. e. $\alpha \subset C'$. Then, for every $\beta \in \alpha$ there exists a unique collection $u_\beta \equiv (f_\gamma^\beta \mid \gamma \in \beta + 1)$ of bijective functions $f_\gamma^\beta : V_\gamma \rightarrowtail\!\!\!\twoheadrightarrow |V_\gamma|$ with properties 1–4. Since $\alpha \subset \lambda + 1$, it follows that $\alpha \leqslant \lambda + 1$. If $\alpha = \lambda + 1$, then $\alpha \in \mathbf{C}'' \subset \mathbf{C}$. Let now $\alpha \in \lambda + 1$. For every $\delta \leqslant \beta \in \alpha$ consider the collection $w \equiv u_\beta|\delta + 1 \equiv (f_\gamma^\beta \mid \gamma \in \delta + 1)$. The collection w possesses properties 1–4. By the uniqueness, which was proved above, $w = u_\delta$. Therefore, $u_\delta = u_\beta|\delta + 1$, i. e. $f_\gamma^\delta = f_\gamma^\beta$ for every $\gamma \in \delta + 1$. In particular, $f_\delta^\delta = f_\delta^\beta$ for every $\delta \leqslant \beta$. Define a collection $v \equiv (g_\beta \mid \beta \in \alpha + 1)$ of functions g_β, setting $g_\beta \equiv f_\beta^\beta$ for every $\beta \in \alpha$ and $g_\alpha(x) \equiv f_\gamma^\beta(x)$ for every $x \in V_\alpha = \cup(V_\gamma \mid \gamma \in \alpha)$ and every $\gamma \leqslant \beta \in \alpha$ such that $x \in V_\gamma$. It is clear that $g_\beta \leftrightharpoons V_\beta \rightarrowtail\!\!\!\twoheadrightarrow |V_\beta|$ for every $\beta \in \alpha$. Check that $g \leftrightharpoons V_\alpha \to |V_\alpha|$. By Corollary 1 to Lemma 3, $V_\gamma \subset V_\alpha$. Consequently, $|V_\gamma| \subset |V_\alpha|$. Therefore, for every $x \in V_\alpha$ it is valid $g_\alpha(x) \equiv f_\gamma^\beta(x) \in |V_\gamma| \subset \varkappa \equiv \cup(|V_\gamma| \subset |V_\alpha| \mid \gamma \in \alpha) \subset |V_\alpha|$. Let $x, y \in V_\alpha$ and $g_\alpha(x) = g_\alpha(y)$. Then, $x \in V_\gamma$ and $y \in V_\delta$ for some $\gamma, \delta \in \alpha$. Consider the number β, which is greatest of the numbers γ and δ. By definition, $f_\beta^\beta(x) = g_\alpha(x) = g_\alpha(y) = f_\beta^\beta(y)$. From injectivity of this function we infer that $x = y$. Therefore, the function g_α is surjective. Let $z \in \varkappa$. Then, $z \in |V_\gamma|$ for some $\gamma \in \alpha$. Since the function $f_\gamma^\gamma : V_\gamma \rightarrowtail\!\!\!\twoheadrightarrow |V_\gamma|$ is injective, we infer that $z = f_\gamma^\gamma(x)$ for some $x \in V_\gamma \subset V_\alpha$. Consequently $z = g_\alpha(x)$. Thus, g_α is a bijective function from V_α onto $\varkappa$, i. e. $V_\alpha \sim \varkappa$. By Corollary 3 to Lemma 3 $|V_\gamma| \in |V_\alpha|$. Therefore, there exists a set $A \equiv \{x \in |V_\alpha| \mid \exists y \in \alpha(x = |V_y|)\} = \{|V_\gamma| \mid \gamma \in \alpha\}$ of all ordinal numbers $|V_\gamma|$. Since α is a limit ordinal number, we infer that $A \neq \varnothing$. Therefore, by Lemma 2 (A.2.2), the set $\cup A = \sup A$ is an ordinal number. If $z \in \cup A = \{z \mid \exists x \in A(z \in x)\}$, then $z \in |V_\gamma| \subset \varkappa$ for some $\gamma \in \alpha$. Conversely, if $z \in \varkappa$, then $z \in |V_\gamma| \in A$ for some $\gamma \in \alpha$. Therefore, $z \in \cup A$. Consequently, $\varkappa = \cup A$, i. e. $\varkappa$ is an ordinal number. Prove that $\varkappa$ is a cardinal number. Let β be an ordinal number, $\beta \leqslant \varkappa$, and $\beta \sim \varkappa$. Suppose that $\beta < \varkappa$. Then, $\beta \in \varkappa$ implies $\beta \in |V_\gamma|$ for some

$\gamma \in \alpha$. Consequently, $\beta < |V_\gamma| = \operatorname{card}|V_\gamma| \leqslant |\varkappa| = |\beta|$. Since β is an ordinal number, we infer that $|\beta| \leqslant \beta$. As a result, we come to the inequality $\beta < \beta$, which is impossible. It follows from this contradiction that $\beta = \varkappa$. It means that $\varkappa$ is a cardinal number. Since $\varkappa$ is a cardinal number and $\varkappa \sim V_\alpha$, we infer that $\varkappa = |V_\alpha|$. Therefore, $g_\alpha \coloneqq V_\alpha \rightarrowtail\!\!\!\twoheadrightarrow |V_\alpha|$. Check that the collection v possesses properties 1–4. By definition of this collection, $g_0 \equiv f_0^0 = \varnothing$. Let $\gamma < \beta \in \alpha + 1$. If $\beta \in \alpha$, then the equality $f_\gamma^\gamma = f_\gamma^\beta$, which was proved above, implies $g_\beta | V_\gamma = f_\beta^\beta | V_\gamma = f_\gamma^\beta = f_\gamma^\gamma \equiv g_\gamma$. If $\beta = \alpha$, then, by construction, $g_\beta | V_\gamma = g_\alpha | V_\gamma = f_\gamma^\gamma \equiv g_\gamma$. Thus, for v property 2 is valid. Let $\beta \in \alpha + 1$, $\beta = \gamma + 1$ and $x \in V_\beta = \mathcal{P}(V_\gamma)$. If $\beta \in \alpha$, then the equality $f_\gamma^\gamma = f_\gamma^\beta$, which was proved above, implies $g_\beta(x) = f_\beta^\beta(x) = f_\gamma^\beta(x) = f_\gamma^\gamma(x) = g_\gamma(x)$ for every $x \in V_\gamma$ and $g_\beta(x) = f_\beta^\beta(x) = c_\gamma(f_\gamma^\beta[x]) = c_\gamma(f_\gamma^\gamma[x]) = c_\gamma(g_\gamma[x])$ for every $x \in V_\beta \backslash V_\gamma$. Therefore, for v property 3 fulfilled. Property 4 follows from property 2. From the properties which were already checked we infer that $\alpha \in C' \subset \mathbf{C}$. By the principle of transfinite induction, $\mathbf{C} = \mathbf{On}$, and therefore $C' = \lambda + 1$. □

Note that, since the functions $c_\gamma(\lambda)$ depend on the number λ, we can not componate the (continuing each other) collections $u(\lambda)(\alpha)$ into one global collection indexed by all ordinal numbers.

Corollary 1. *For every limit ordinal number α the equalities $|V_\alpha| = \cup(|V_\beta| \mid \beta \in \alpha) = \cup\{|V_\beta| \mid \beta \in \alpha\} = \sup\{|V_\beta| \mid \beta \in \alpha\}$ are valid.*

Proof. Consider the number $\lambda \equiv \alpha$. By Theorem 1, there exists the corresponding collection $u(\alpha) \equiv (f_\beta \mid \beta \in \alpha + 1)$. Since α is a limit ordinal number and $\alpha \in \alpha + 1$, it follows that by property 4, $f_\alpha = \cup(f_\beta \mid \beta \in \alpha)$. Therefore, $|V_\alpha| = \operatorname{rng} f_\alpha = \cup(\operatorname{rng} f_\beta \mid \beta \in \alpha) = \cup(|V_\beta| \mid \beta \in \alpha) = \cup\{|V_\beta| \mid \beta \in \alpha\} = \sup\{|V_\beta| \mid \beta \in \alpha\}$, where the latter equality follows from Lemma 2 (A.2.2). □

A.3.3 Properties of inaccessible cumulative sets

The sets $V_\varkappa$ for inaccessible cardinal numbers $\varkappa$ will be called *inaccessible cumulative sets*. They have a number of specific properties. We present these properties with complete proofs. Note that their proofs are practically absent in the corresponding literature and are not obvious.

Lemma 1. *For every inaccessible cardinal number $\varkappa$ and every ordinal number $\alpha \in \varkappa$ the property $|V_\alpha| < \varkappa$ is valid.*

Proof. Consider the set $C' \equiv \{x \in \varkappa \mid |V_x| < \varkappa\}$ and the classes $\mathbf{C}'' \equiv \mathbf{On} \backslash \varkappa$ and $\mathbf{C} \equiv C' \cup \mathbf{C}''$. Since $V_0 = \varnothing$, we have $|V_0| = 0 < \varkappa$. Therefore, $0 \in \mathbf{C}$. Let $\alpha \in \mathbf{C}$. If $\alpha \geqslant \varkappa$, then $\alpha + 1 \in \mathbf{C}'' \subset \mathbf{C}$. Let $\alpha < \varkappa$. Then, $\alpha \in C'$. If $\alpha + 1 = \varkappa$, then $\alpha + 1 \in \mathbf{C}'' \subset \mathbf{C}$. Let $\alpha + 1 < \varkappa$. Since $V_\alpha \sim |V_\alpha|$, we have $\mathcal{P}(V_\alpha) \sim \mathcal{P}(|V_\alpha|)$. Therefore, $|\mathcal{P}(V_\alpha)| = |\mathcal{P}(|V_\alpha|)|$. By Corollary 2 to Lemma 3 (A.3.2), $|V_{\alpha+1}| = |\mathcal{P}(V_\alpha)| = |\mathcal{P}(|V_\alpha|)|$. Since $|V_\alpha| < \varkappa$ and the cardinal number $\varkappa$ is inaccessible, we obtain $|\mathcal{P}(|V_\alpha|)| < \varkappa$. Hence, $|V_{\alpha+1}| < \varkappa$. Thus, $\alpha + 1 \in C' \subset \mathbf{C}$. Let α be a limit ordinal number and $\alpha \subset \mathbf{C}$. If $\alpha \cap \mathbf{C}'' \neq \varnothing$, then there exists $\beta \in \alpha$ such that $\beta \geqslant \varkappa$. Therefore, $\alpha > \beta \geqslant \varkappa$ implies $\alpha \in \mathbf{C}'' \subset \mathbf{C}$. Let $\alpha \cap \mathbf{C}'' = \varnothing$, i. e. $\alpha \subset C' \subset \varkappa$. If $\alpha = \varkappa$, then $\alpha \in \mathbf{C}'' \subset \mathbf{C}$. Let $\alpha < \varkappa$. By virtue of $\alpha \subset C'$, for every $\beta \in \alpha$, we have $|V_\beta| < \varkappa$. Therefore, $\sup\{|V_\beta| \mid \beta \in \alpha\} \leqslant \varkappa$. Using the property $|V_\beta| \in \varkappa$, we can correctly define a function $f : \alpha \to \varkappa$, setting $f(\beta) \equiv |V_\beta|$. It is clear that $\operatorname{rng} f = \{|V_\beta| \mid \beta \in \alpha\}$. By Corollary 1 to Theorem 1 (A.3.2), $\cup \operatorname{rng} f = \cup\{|V_\beta| \mid \beta \in \alpha\} = \sup\{|V_\beta| \mid \beta \in \alpha\} = |V_\alpha|$. By virtue of the inequality, which was proved above, we infer that $|V_\alpha| \leqslant \varkappa$. Suppose that $|V_\alpha| = \varkappa$. Then, $\varkappa = \cup \operatorname{rng} f$ by virtue of regularity of the number $\varkappa$ implies $\varkappa \leqslant \alpha$, but this contradicts the initial inequality $\alpha < \varkappa$. Therefore, $|V_\alpha| < \varkappa$. Consequently, $\alpha \in C' \subset \mathbf{C}$. By the principle of transfinite induction, $\mathbf{C} = \mathbf{On}$. Thus, $C' = \varkappa$. □

Lemma 2. *If $\varkappa$ is an inaccessible cardinal, then $\varkappa = |V_\varkappa|$.*

Proof. By Lemma 1 (A.3.2), $\varkappa \subset V_\varkappa$. Therefore, $\varkappa = |\varkappa| \leqslant |V_\varkappa|$. By Corollary 1 to Theorem 1 (A.3.2), $|V_\varkappa| = \sup(|V_\beta| \mid \beta \in \varkappa)$. Since by Lemma 1 $|V_\beta| < \varkappa$, we have $|V_\varkappa| \leqslant \varkappa$. As a result, we have $\varkappa = |V_\varkappa|$. □

Lemma 3. *If $\varkappa$ is an inaccessible cardinal number, α is an ordinal number such that $\alpha < \varkappa$, and f is a correspondence from V_α into $V_\varkappa$ such that* $\operatorname{dom} f = V_\alpha$ *and $f\langle x\rangle \in V_\varkappa$ for every $x \in V_\alpha$, then $\operatorname{rng} f \in V_\varkappa$.*

Proof. Since $\varkappa$ is a limit ordinal number, we have $V_\varkappa = \cup(V_\delta \mid \delta \in \varkappa)$. For $x \in V_\alpha$ there exists $\delta \in \varkappa$ such that $f\langle x\rangle \in V_\delta$. Therefore, a non-empty set $\{y \in \varkappa \mid f\langle x\rangle \in V_y\}$ contains the smallest element z. By virtue of the uniqueness of the element z we can correctly define a function $g : V_\alpha \to \varkappa$, setting $g(x) \equiv z$. Consider the ordinal number $\beta \equiv |V_\alpha|$ and take some bijective mapping $h : \beta \rightarrowtail\!\!\!\!\twoheadrightarrow V_\alpha$. Consider the mapping $\varphi \equiv g \circ h : \beta \to \varkappa$ and the ordinal number $\gamma \equiv \cup \operatorname{rng} \varphi = \sup \operatorname{rng} \varphi \leqslant \varkappa$. Suppose that $\gamma = \varkappa$. Since the cardinal $\varkappa$ is regular, the supposition $\cup \operatorname{rng} \varphi = \varkappa$ implies $\varkappa \leqslant \beta \equiv |V_\alpha|$. But, by Lemma 3 $|V_\alpha| < \varkappa$. It follows from this contradiction that $\gamma < \varkappa$. Since h is bijective, $\operatorname{rng} \varphi = \operatorname{rng} g$. Therefore, $\gamma = \sup \operatorname{rng} g$. If $x \in V_\alpha$, then $f\langle x\rangle \in V_z = V_{g(x)}$. From $g(x) \leqslant \gamma$, we infer, by Lemma 1 (A.3.2) that $V_{g(x)} \subset V_\gamma$. Consequently, $f\langle x\rangle \in V_\gamma$ by Lemma 3 (A.3.2) implies $f\langle x\rangle \subset V_\gamma$. Therefore, $\operatorname{rng} f \subset V_\gamma$. By Lemma 1 (A.3.2) $\operatorname{rng} f \in V_{\gamma+1} \subset V_\varkappa$. □

Lemma 4. *If $\varkappa$ is an inaccessible cardinal number, $A \in V_\varkappa$, and f is a correspondence from A into $V_\varkappa$ such that $f\langle x\rangle \in V_\varkappa$ for every $x \in A$, then $\operatorname{rng} f \in V_\varkappa$.*

Proof. Since $\varkappa$ is a limit ordinal number, $V_\varkappa = \cup(V_\alpha \mid \alpha \in \varkappa)$. Therefore, $A \in V_\alpha$ for some $\alpha \in \varkappa$. By Lemma 3 (A.3.2), $A \subset V_\alpha$. Define a correspondence g from V_α into $V_\varkappa$, setting $g|A \equiv f$ and $g\langle x\rangle \equiv \varnothing \subset V_\varkappa$ for every $x \in V_\alpha \backslash A$. Then, $\operatorname{dom} g = V_\alpha$ and $\operatorname{rng} g = \operatorname{rng} f$. If $x \in A$, then $g\langle x\rangle = f\langle x\rangle \in V_\varkappa$, and if $x \in V_\alpha \backslash A$, then $g\langle x\rangle = \varnothing \in V_\varkappa$. Therefore, by Lemma 3, we obtain $\operatorname{rng} f = \operatorname{rng} g \in V_\varkappa$. □

Corollary 1. *If $\varkappa$ is an inaccessible cardinal number and $(B_a \mid a \in A)$ is a collection of sets such that $A \in V_\varkappa$ and $B_a \in V_\varkappa$ for every $a \in A$, then $\bigcup(B_a \mid a \in A) \in V_\varkappa$.*

Corollary 2. *If $\varkappa$ is an inaccessible cardinal number and $A \in V_\varkappa$, then $\cup A \in V_\varkappa$.*

The following assertion is due to A. Tarski [1938] (see also [*Kolmogorov and Dragalin*, 1982, IX, § 1, Theorem 6]). Here, we present another proof of this assertion.

Lemma 5. *If $\varkappa$ is an inaccessible cardinal number, $A \subset V_\varkappa$, and $|A| < |V_\varkappa|$, then $A \in V_\varkappa$.*

Proof. By Lemma 2 $|A| \in |V_\varkappa| = \varkappa \subseteq V_\varkappa$. Consider the bijection $b : |A| \rightarrowtail\!\!\!\twoheadrightarrow A \subset V_\varkappa$. Lemma 4 implies $A = \operatorname{rng} b \in V_\varkappa$. □

Lemma 6. *If $\varkappa$ is an inaccessible cardinal number, ε is an ordinal number, and $\varepsilon \in V_\varkappa$, then $\varepsilon \in \varkappa$.*

Proof. Since $V_\varkappa = \bigcup(V_\alpha \mid \alpha \in \varkappa)$, it follows that $\varepsilon \in V_\alpha$ for some $\alpha \in \varkappa$. By Lemma 3 (A.3.2) $\varepsilon \subset V_\alpha$. By Lemma 1 $|\varepsilon| \leqslant |V_\alpha| < \varkappa$. Suppose that $\varepsilon \geqslant \varkappa$. Then, $\varkappa \subset \varepsilon$ implies $\varkappa = |\varkappa| \leqslant |\varepsilon|$, what contradicts the previous inequality. Hence, $\varepsilon < \varkappa$. □

Consider the class $\Pi \equiv \{x \mid \exists \alpha \in \mathbf{On}(x \in V_\alpha)\} \equiv \cup(V_\alpha \mid \alpha \in \mathbf{On})$.

Lemma 7 (the von Neumann identity). $\Pi = \mathbf{V}$.

Proof. Show that Π satisfies the principle of $\in$-induction. Introduce the function $ran : \Pi \to \mathbf{On}$, setting $ran(x) \equiv$ the smallest ordinal α such that $x \in V_{\alpha+1}$. It follows from Lemma 1 (A.3.2) that all ordinal numbers are contained in Π. Check that $x \subset \Pi$ implies $x \in \Pi$ for every set x. If $x = \varnothing$, then, by Lemma 1 (A.3.2), $x = 0 \in V_1 \subset \Pi$. Let $x \neq \varnothing$. Consider the following formula of the ZF set theory: $\varphi(y, z) \equiv (y \in \Pi \Rightarrow z = ran(y)) \wedge (y \notin \Pi \Rightarrow z = \varnothing)$. By axiom scheme of replacement AS6, for the set x there exists a set B such that $\forall y \in x \forall z(\varphi(y, z) \Rightarrow z \in B)$. If $y \in x$, then $\varphi(y, ran(y))$ implies $ran(y) \in B$. Therefore, $A \equiv \{z \in B \mid \exists y \in x(z = ran(y))\} \subset B$. By axiom scheme of separation AS3, A is a set. By Lemma 2 (A.2.2), $\alpha \equiv \cup A = \sup A$ is an ordinal number. If $y \in x$, then $z \equiv ran(y) \in A$ implies $z \leqslant \alpha$. Therefore, by Lemma 1 (A.3.2), $y \in V_{z+1} \subset V_{\alpha+1}$. Thus, $x \subset V_{\alpha+1}$, where $x \in \mathcal{P}(V_{\alpha+1}) \subset V_{\alpha+2} \subset \Pi$. By the principle of $\in$-induction, we now infer that $\Pi = \mathbf{V}$. □

A.4 Universal sets and their connection with inaccessible cumulative sets

A.4.1 Universal sets and their properties

A set U is called *universal* in the ZF set theory (see [*MacLane*, 1971, I, 6] and [*Forster*, 1995; *Holmes*, 1998]) if it has the following properties:
1) $x \in U \Rightarrow x \subset U$ (the *transitivity property*);
2) $x \in U \Rightarrow \mathcal{P}(x), \cup x \in U$;
3) $x \in U \wedge y \in U \Rightarrow x \cup y, \{x, y\}, \langle x, y \rangle, x * y \in U$;
4) $x \in U \wedge (f \in U^x) \Rightarrow \operatorname{rng} f \in U$ (the *strong substitution property*);
5) $\omega \in U$.

Clearly, not all of these properties are independent.

The property that a set U is universal will be denoted by $U\bowtie$. Denote by $\mathbf{U}$ the class (possibly, empty) of all universal sets. It immediately follows from the definition of a universal set that the intersection $\cap\mathbf{A} \equiv \{x \mid \forall U \in \mathbf{A}(x \in U)\}$ of any non-empty subclass $\mathbf{A}$ of the class of universal sets is a universal set. Let us deduce several properties of universal sets from these conditions.

Lemma 1. *If a set U is universal, then $x \in U \wedge y \subset x \Rightarrow y \in U$.*

Proof. If $x \in U$, then (2) implies $\mathcal{P}(x) \in U$ and (1) implies $\mathcal{P}(x) \subset U$. Since $y \in \mathcal{P}(x)$, we have $y \in U$. □

This lemma shows that a universal set is quasitransitive. This fact and the transitivity property imply that a universal set is supertransitive.

Lemma 2. *If a set U is universal, then $\varnothing \in U$.*

Proof. This follows directly from properties 1 and 5. □

Lemma 3. *Let $(A_i \mid i \in I)$ be a collection such that $I \in U$ and $A_i \in U$ for every $i \in I$. Then, $\cup(A_i \mid i \in I) \in U$.*

Proof. Consider the function $f : I \to U$ such that $f(i) \equiv A_i$. Then, (4) implies $\operatorname{rng} f \in U$ and (2) implies $\cup(A_i \mid i \in I) = \cup \operatorname{rng} f \in U$. □

Lemma 4. *If U is a universal set, then $x \in U \Rightarrow |x| \in U$.*

Proof. Consider the class $\mathbf{C} \equiv \{\alpha \in \mathbf{On} \mid \alpha \notin U)\}$. Since the class $\mathbf{On}$ is not a set, the class $\mathbf{C}$ is non-empty. Denote its minimal element by $\varkappa$. Suppose that there exists an element $x \in U$ such that $\alpha \equiv |x| \notin U$. Then, there exists a bijection $f : \alpha \to x$. It follows from $\alpha \in \mathbf{C}$ that $\varkappa \leqslant \alpha$. Since $\varkappa \subset \alpha$, we can consider the mapping $g \equiv f|\varkappa$. The mapping g is a bijection from $\varkappa$ onto $y \equiv \operatorname{rng} g \subset x$. The inclusion $y \subset x$ implies $y \in U$. Hence, $h \equiv g^{-1}$ is a function from $y \in U$ onto $\varkappa \notin U$. Since $\varkappa$ is a minimal element of then class $\mathbf{C}$, we have $\forall \beta \in \varkappa(\beta \in U)$. Therefore, $h(z) \in \varkappa$ provides that $h(z) \in U$ for every $z \in y$. Then (4) implies $\varkappa \in U$ what contradicts the definition of $\varkappa$. This contradiction yields $\forall x \in U(|x| \in U)$. □

Let us prove that in a universal set there exists a $\in$-induction principle similar to the $\in$-induction principle in the ZF set theory (see Lemma 4 (A.2.2)).

Lemma 5. *Let U be an universal set, $C \subset U$, and $\forall x \in U(x \subset C \Rightarrow x \in C)$. Then, $C = U$.*

Proof. Suppose that $C \neq U$, i. e. $D \equiv U \backslash C \neq \varnothing$. Then, there is $P \in D$. It is clear that $P \in U$. If $P \cap D = \varnothing$, then put $X \equiv P$. Let $P \cap D \neq \varnothing$. Consider the set N consisting of all $n \in \omega$ such that there is a unique sequence $u \equiv u(n) \equiv (R_k \in U \mid k \in n+1)$ of the sets $R_k \in U$ such that $R_0 = P$ and $R_{k+1} = \cup R_k$ for every $k \in n$. Since the sequence $(R_k \mid k \in 1)$ such that $R_0 \equiv P$ possesses this property, it follows that $0 \in N$. Let $n \in N$, i. e. for n there is a unique $u \equiv (R_k \in U \mid k \in n+1)$. Define the sequence $v \equiv (S_k \in U \mid k \in n+2)$ setting $S_k \equiv R_k \in U$ for every $k \in n+1$ and $S_{n+1} \equiv \cup R_n = \cup S_n$, i. e. $v = u \cup \{\langle n+1, \cup R_n \rangle\}$. Since U is an universal set, $R_n \in U$ implies $S_{n+1} \in U$. Clearly, v has the necessary properties. Check its uniqueness.

Suppose there is a sequence $w \equiv (T_k \in U \mid k \in n+2)$ such that $T_0 = P$ and $\forall k \in n+1(T_{k+1} = \cup T_k)$. Consider the set M' consisting of all $m \in n+2$ such that $S_m = T_m$. Put $M'' \equiv \omega \backslash (n+2)$ and $M \equiv M' \cup M''$. It follows from $S_0 = P = T_0$ that $0 \in M' \subset M$.

Let $m \in M'$. If $m = n+1$, then $m+1 = n+2 \in M'' \subset M$. If $m < n+1$, then $m+1 \in n+2$ and $S_{m+1} = \cup S_m = \cup T_m = T_{m+1}$ imply $m+1 \in M' \subset M$. If $m \in M''$, then $m+1 \in M'' \subset M$. Hence, $m \in M$ guarantees that $m+1 \in M$. By the principle of natural induction (Theorem 1 (1.2.6)), $M = \omega$. Thus, $M' = n+2$, and therefore, $v = w$, i. e. the sequence v is unique. Hence, $n+1 \in N$. By the principle of natural induction, $N = \omega$. Therefore, for every $n \in \omega$ there is a unique sequence $u(n)$. Since it is unique, it will be denoted by $(R_k^n \mid k \in n+1)$.

Consider the following formula of the ZF set theory: $\varphi(x, y) \equiv (x \in \omega \Rightarrow y = R_x^x) \wedge (x \notin \omega \Rightarrow y = \varnothing)$. By the replacement axiom scheme (AS6) there exists a set Y such that $\forall x \in \omega(\forall y(\varphi(x, y) \Rightarrow y \in Y))$. If $n \in \omega$, then $\varphi(n, R_n^n)$ implies $R_n^n \in Y$. Therefore, we can define in the set $\omega \times Y$ the infinite sequence $u \equiv (R_n \in Y \mid n \in \omega)$ setting $u \equiv \{z \in \omega \times Y \mid \exists x \in \omega(z = \langle x, R_x^x \rangle)\}$. The property of uniqueness mentioned above implies that $u(m) = u(n)|(m+1)$ for all $m \leqslant n$. Thus, $u|(n+1) = u(n)$. Hence, the sequence u has the following properties: $R_0 = P$ and $R_{k+1} = \cup R_k$ for every $k \in \omega$. Having the sequence u we can take the set $A \equiv \operatorname{rng} u \equiv \{R_n \mid n \in \omega\} \in U$ and the set

$Q \equiv \cup A = \{y \mid \exists x \in \omega(y \in R_x)\} = \cup(R_n \mid n \in \omega) \in U$. It is clear that $R_n \subset Q$ for every $n \in \omega$, and therefore, $P = R_0 \subset Q$.

Since $P \cap D \neq \varnothing$, we infer that $R \equiv Q \cap D \neq \varnothing$. By regularity axiom A9, there exists $X \in R$ such that $X \cap R = \varnothing$. Clearly, $X \in U$ and $X \subset U$. Check that $X \cap D = \varnothing$. Suppose that there exists $x \in X \cap D$. Since $X \in Q$, we infer that $X \in R_n$ for some $n \in \omega$. Therefore, $x \in X \in R_n$ implies $x \in \cup R_n = R_{n+1} \subset Q$. Thus, $x \in R$. As a result, we obtain $x \in X \cap R = \varnothing$, which is impossible. It follows from this contradiction that $X \in D$ and $X \cap D = \varnothing$.

Thus, $X \in U$ and $X \subset C$ in both the cases. Therefore, $X \in C$. This contradicts $X \in D$. Hence, $C = U$. □

For a universal set the following analogue of the von Neumann identity from Lemma 7 (A.3.3) holds.

Lemma 6. *Let U be a universal set. Then, $V_\alpha \in U$ for every $\alpha \in \mathbf{On} \cap U$ and $U = \cup(V_\alpha \subset U \mid \alpha \in \mathbf{On} \cap U)$.*

Proof. Consider the sets $A \equiv \mathbf{On} \cap U$ and $C' \equiv \{\alpha \in A \mid V_\alpha \in U\}$ and classes $\mathbf{C}'' \equiv \mathbf{On} \backslash U$ and $\mathbf{C} \equiv C' \cup \mathbf{C}''$. By Lemma 7 (A.3.2) $0 = V_0 = \varnothing \in U$. Hence, $0 \in \mathbf{C}$. Let $\alpha \in \mathbf{C}$. Suppose that $\alpha + 1 \in A$. Since $\alpha \in \alpha + 1 \in U$, property 1 implies $\alpha \in U$, and therefore, $\alpha \in A \cap \mathbf{C} = C'$. Then, it follows from $V_\alpha \in U$ and properties 2 and 3 that $V_{\alpha+1} = V_\alpha \cup \mathcal{P}(V_\alpha) \in U$, where $\alpha + 1 \in C' \subset \mathbf{C}$. In the case $\alpha + 1 \notin A$, we immediately get $\alpha + 1 \in \mathbf{C}'' \subset \mathbf{C}$.

Let α be a limit ordinal number and $\alpha \subset \mathbf{C}$. Suppposse that $\alpha \in A$. If $\beta \in \alpha$, then $\beta \in \alpha \in U$ implies $\beta \in A \cap \mathbf{C} = C'$. Then, the condition $V_\beta \in U$ and Lemma 1 (A.3.3) provide the equality $V_\alpha = \cup(V_\beta \mid \beta \in \alpha) \in U$. Hence, $\alpha \in C' \subset \mathbf{C}$. In the case $\alpha \notin A$ we immediately get $\alpha \in \mathbf{C}'' \subset \mathbf{C}$.

By the principle of transfinite induction (Theorem 1 (A.2.2)), $\mathbf{C} = \mathbf{On}$, and therefore, $C' = A$.

By the above, we have $V_\alpha \subset U$ for every $\alpha \in A$, where $P \equiv \cup(V_\alpha \mid \alpha \in A |\subset) U$. Show that P satisfies the $\in$-induction principle from Lemma 5. Define the function $r : P \to A$ setting $r(p) \equiv \mathrm{sm}\{\alpha \in A | p \in V_\alpha\}$ for every $p \in P \subset U$.

Let $x \in U$ and $x \subset P$. If $x = \varnothing$, then $x \in P$. In what follows, we assume that $x \neq \varnothing$. If $y \in x \subset P$, then $y \in V_\alpha$ for some $\alpha \in A$. Hence, it follows from $r(y) \leqslant \alpha \in U$ and Lemma 6 (A.3.2) that $r(y) \in A$. Therefore, we can consider the function $s \equiv r|x$ from x to A. By property 4, $R \equiv \mathrm{rng}\, s \in U$ and by property 2 $\rho \equiv \cup R \in U$. Since $\varnothing \neq R \subset \mathbf{On}$, ρ is an ordinal number by virtue of Lemma 2 (A.2.2). Therefore, $\rho \in A$.

If $y \in x$, then $s(y) \subset \rho$ implies $y \in V_{s(y)} \subset V_\rho$ due to Lemma 1 (A.3.2). According to Lemma 2 (A.3.2), $x \subset V_\rho \in V_{\rho+1}$ implies $x \in V_{\rho+1}$. Property 3 and $\rho + 1 = \rho \cup \{\rho\} \in U$ provide $\rho + 1 \in A$. Therefore, $x \in P$.

Now, it follows from Lemma 5 that $P = U$. □

A.4.2 Description of the class of all universal sets

The following theorem is deduced from the Zermelo – Shepherdson theorem (see [*Zermelo*, 1930] (incomplete proof) and [*Shepherdson*, 1951; 1952; 1953] (complete proof)) on the canonical form of standard supertransitive model sets for the NBG set theory in the ZF set theory (see A.6 below). Here, we give another proof.

Theorem 1.
1) *Let U be an arbitrary universal set. Then,* $\varkappa \equiv \sup(\mathbf{On} \cap U) = \cup(\mathbf{On} \cap U) \subset U$ *is an inaccessible cardinal number and* $U = V_\varkappa$.
2) *The correspondence* $\mathbf{q} : U \mapsto \varkappa$ *such that* $U = V_\varkappa$ *is is an isotone injective mapping from the class* $\mathbf{U}$ *of all universal sets into the class* $\mathbf{In}$ *of all inaccessible cardinal numbers.*

Proof. 1. Since $A \equiv \mathbf{On} \cap U$ contains the element ω by virtue of property 5 from the definition of a universal set, A is non-empty. Hence, Lemma 2 (A.2.2) implies that $\varkappa$ is an ordinal number.

Suppose that $\varkappa$ is not a cardinal number. Then, there are an ordinal number $\alpha < \varkappa$ and a bijective function $f : \alpha \to \varkappa$. Since $\alpha \in \varkappa \subset U$, we get $\alpha \in U$. If $\beta \in \alpha$, then $f(\beta) \in U$. Then, by property 4 $\varkappa = \operatorname{rng} f \in U$ and by property 3 $\{\varkappa\} \in U$ and $\varkappa^+ \equiv \varkappa \cup \{\varkappa\} \in U$. It follows from $\varkappa^+ \in \mathbf{On}$ that $\varkappa^+ \in A$, i. e. $\varkappa^+ \leqslant \varkappa$, which is impossible. Having this contradiction, we infer that $\varkappa$ is a cardinal number.

Suppose now that $\varkappa$ is not regular. Then, $\alpha \equiv cf(\varkappa) < \varkappa$. By definition, there is a function $f : \alpha \to \varkappa$ such that $\cup \operatorname{rng} f = \varkappa$. As above, we get $\alpha \in U$ and $f(\beta) \in U$ for all $\beta \in \alpha$, where $\operatorname{rng} f \in U$ by virtue of property 4. Since $\cup \operatorname{rng} f \in U$, property 2 guarantees that $\varkappa \in U$. Similarly to the previous indentation, we arrive at a contradiction. Hence, $\varkappa$ is a regular cardinal.

Let λ is a cardinal number such that $\lambda < \varkappa$. Since $\lambda \in \varkappa \subset U$, property 2 implies $\mathcal{P}(\lambda) \in U$. By Lemma 4 (A.4.1), $|\mathcal{P}(\lambda)| \in U$. Therefore, $|\mathcal{P}(\lambda)| \leqslant \varkappa$. If we suppose that $\varkappa = |\mathcal{P}(\lambda)| \in U$, then, as above, we arrive at a contradiction. Hence, $|\mathcal{P}(\lambda)| < \varkappa$.

Besides, $\omega \in U$ implies $\omega + 1 = \omega \cup \{\omega\} \in U$. Then, it follows from $\omega \in \omega + 1 \in A$ that $\omega \in \cup A = \varkappa$.

Prove now that $U = V_\varkappa$. As was shown above, $\varkappa$ is a limit ordinal number, where $V_\varkappa = \cup\{V_\beta \mid \beta \in \varkappa\}$. Lemma 6 (A.4.1) provides that $U = \cup\{V_\alpha \mid \alpha \in A\}$. If $\alpha \in A$, then $\alpha \leqslant \varkappa$ implies $V_\alpha \subset V_\varkappa$. Hence, $U \subset V_\varkappa$. If $\beta \in \varkappa = \cup A$, then $\beta \in \alpha \in A$ for some α. By property 1, $\beta \in A$. Therefore, $V_\varkappa \subset U$. Thus, $U = V_\varkappa$.

2. Lemma 1 (A.3.2) implies that $\varkappa$ is unique. Therefore, we can define the mapping $\mathbf{q} : \mathbf{U} \to \mathbf{In}$ such that $\mathbf{q}(U) = \varkappa$, where $U = V_\varkappa$. Lemma 1 (A.3.2) also guarantees that q is isotone. □

Corollary 1. *If U is a universal set, then $|U|$ is an inaccessible cardinal number,* $|U| = \sup(\mathbf{On} \cap U)$, *and* $U = V_{|U|}$.

Proof. By Theorem 1 $U = V_\varkappa$ for the inaccessible cardinal number $\varkappa \equiv \sup(\mathbf{On} \cap U)$. By Lemma 2 (A.3.3) $\varkappa = |V_\varkappa| = |U|$. □

Corollary 2. *If U is a universal set, then $|U| = \sup(|V_\alpha| \mid \alpha \in \mathbf{On} \cap U)$.*

Proof. By Theorem 1, $U = V_\varkappa$ for the inaccessible cardinal number $\varkappa \equiv \sup A$, where $A \equiv \mathbf{On} \cap U$. Since $\varkappa$ is a limit ordinal number, Corollary 1 to Theorem 1 (A.3.2) implies $|V_\alpha| = \sup(|V_\alpha| \mid \alpha \in \varkappa)$. If $\alpha \in \varkappa$, then $\alpha \in a$ for some $a \in A$. By the transitivity property we get $\alpha \in A$. Conversely if $\alpha \in A$, then $\alpha \leqslant \varkappa$. Suppose that $\alpha = \varkappa$. Then, $\varkappa \in U$. However, in the proof of Theorem 1 we obtain that the condition $\varkappa \in U$ leads to a contradiction. Hence, $\alpha \in \varkappa$. □

Theorem 2. *For any set U the following conclusions are equivalent:*
1) *U is a inaccessible cumulative set;*
2) *U is a universal set.*

Proof. (1) $\vdash$ (2). Let $U = V_\varkappa$ for some inaccessible cardinal number $\varkappa$. Show that the set U possesses each of five properties from the definition of a universal set.

The property $x \in U \Rightarrow x \subset U$ follows from Lemma 3 (A.3.2). The property $x \in U \Rightarrow \mathcal{P}(x) \in U$ follows from Lemma 6 (A.3.2). The property $x \in U \wedge y \in U \Rightarrow x \cup y \in U$ follows from Lemma 5 (A.3.2). Corollaries 1 and 2 to Lemma 6 (A.3.2) implies the property $x \in U \wedge y \in U \Rightarrow \{x, y\}, \langle x, y\rangle, x \times y \in U$. The property $\omega \in U$ follows from Lemma 7 (A.3.2). Lemma 4 (A.3.3) and its Corollaries implies the properties $x \in U \Rightarrow \cup x \in U$ and $x \in U \wedge (f \in U^x) \Rightarrow \operatorname{rng} f \in U$.

Thus, the set U is universal.

(2) $\vdash$ (1). This deduction follows directly from Theorem 1. □

Corollary 1. *The mapping $\mathbf{q} : \mathbf{U} \rightarrowtail \mathbf{In}$ from Theorem 1 is an isotone bijection.*

Thus, the cardinalities of universal sets exhaust all inaccessible cardinal numbers.

The last theorem allows us to make the following conclusions on the structure of the class $\mathbf{U} \equiv \{U \mid U\bowtie\}$ of all universal sets.

The relation $\in \cup =$ is an order relation on the class $\mathbf{U}$. It will be denoted by $\leqslant$, i. e. $U \leqslant V$ if $U \in V$ or $U = V$. Lemma 3 (A.3.2) guarantees that the class $\mathbf{U}$ is transitive. Hence, $U \in V$ implies $U \subset V$, and therefore, $U \leqslant V$ implies $U \subset V$. Prove that these relations are equivalent.

Proposition 1. *Let U and V be universal sets. Then, the relation $U \leqslant V$ is equivalent to the relation $U \subset V$.*

Proof. We only need to verify that $U \subset V$ implies $U \leqslant V$. By Theorem 1 $U = V_\pi$ and $V = V_\varkappa$ for some inaccessible cardinal numbers π è $\varkappa$. If $\pi = \varkappa$, then $U = V_\pi = V_\varkappa = V$.

If $\pi < \varkappa$, then by Lemma 1 (A.3.2) $U = V_\pi \in V_\varkappa = V$. Finally, if $\pi > \varkappa$, then by the same lemma $V = V_\varkappa \in V_\pi = U \subset V$, which is impossible. Hence, $U \leqslant V$. □

Theorem 3. *Let the class* **U** *of all universal sets is non-empty in the ZF set theory. Then, it is well-ordered with respect to the order* $\subset$. *Moreover, any non-empty subclass of the class* **U** *has a minimal element.*

Proof. Let $\varnothing \neq \mathbf{A} \subset \mathbf{U}$, i. e. $\forall U \in \mathbf{A}(U\bowtie)$. The isotone injection $\mathbf{q} : \mathbf{U} \rightarrowtail \mathbf{On}$ from Theorem 1 maps a class **A** to some subclass $\mathbf{B} \equiv q[\mathbf{A}] \equiv \{x \mid x \in \mathbf{On} \wedge \exists U \in \mathbf{A}(x = \mathbf{q}(U))\}$ of the class **On**. By Lemma 1 (A.2.2) it has a minimal element π, which is an inaccessible cardinal. Since $\pi \in \mathbf{B}$, we have $\pi = \mathbf{q}(U)$ for some $U \in \mathbf{A}$, i. e. $U = V_\pi$. Since q is injective and strictly monotone, U is a minimal element in the class **A**. □

A.4.3 Enumeration of the class of all universal sets in the ZF+AU set theory and the structural form of the universality axiom

Now, we shall consider the ZF set theory with some additional axioms. The first of them is the universality axiom, which means that for any set X there exists a universal set containing X.

AU. (The *universality axiom.*) $\forall X \exists U(U\bowtie \wedge X \in U)$.

Consider the following class in the ZF+AU set theory:

$$\begin{aligned}\mathbf{G} \equiv \{Z \mid \exists X \exists Y(Z = \langle X, Y\rangle \wedge ((X = \varnothing \Rightarrow Y = \cap\{U \mid U\bowtie\}) \vee \\ (X \neq \varnothing \Rightarrow (\neg func(X) \Rightarrow Y = \varnothing) \vee \\ (func(X) \Rightarrow (\neg On(\operatorname{dom} X) \Rightarrow Y = \varnothing) \vee \\ \vee (On(\operatorname{dom} X) \Rightarrow (Son(\operatorname{dom} X) \Rightarrow Y = \cap\{U \mid U\bowtie \wedge X(\operatorname{dom} X - 1) \in U\}) \vee \\ (Lon(\operatorname{dom} X) \Rightarrow Y = \cap\{U \mid U\bowtie \wedge \cap \operatorname{rng} X \subset U\})))))))\}.\end{aligned}$$

If we reformulate the definition of the class G less formally, then we can say that G consists of all pairs $\langle X, Y\rangle$ such that the following five mutually exclusive possibilities take place:

i) if $X = \varnothing$, then Y is the intersection of all universal sets (the existence of a non-empty intersection follows from the universality axiom);
ii) if $X \neq \varnothing$ and X is not a function, then $Y = \varnothing$;
iii) if $X \neq \varnothing$, X is a function, and $\operatorname{dom} X$ is not ordinal number, then $Y = \varnothing$;
iv) if $X \neq \varnothing$, X is a function, $\operatorname{dom} X$ is an ordinal number, and $\operatorname{dom} X = \alpha + 1$, then Y is the intersection of all universal sets U such that $X(\alpha) \in U$ (the existence of this non-empty intersection follows from the universality axiom);

v) if $X \neq \varnothing$, X is a function, and $\operatorname{dom} X$ is a limit ordinal number, then Y is the intersection of all universal sets U such that $\cup \operatorname{rng} X \subset U$ (the existence of this non-empty intersection follows from the universality axiom).

As in A.3.1 we can check that the class $\mathbf{G}$ is a function from $\mathbf{V}$ to $\mathbf{V}$.

By Theorem 2 (A.2.2) there exists a function $\mathbf{F} : \mathbf{On} \to \mathbf{V}$ such that $\mathbf{F}(\alpha) = \mathbf{G}(\mathbf{F}|\alpha)$. For every $\alpha \in \mathbf{On}$.

Case i) for the function $\mathbf{G}$ implies that $\mathbf{F}(\varnothing) = \mathbf{G}(\mathbf{F}|\varnothing) = \mathbf{G}(\varnothing) = \cap\{U \mid U\bowtie\}$.

It follows from case iv) that if β is a subsequent ordinal number and $\beta = \alpha+1$, then $\mathbf{F}(\beta) = \mathbf{G}(\mathbf{F}|\beta) = \cap\{U \mid U\bowtie \wedge \mathbf{F}(\alpha) \in U\}$.

Finally, case v) implies that if α is a limit ordinal number, then $\mathbf{F}(\alpha) = \mathbf{G}(\mathbf{F}|\alpha) = \cap\{U \mid U\bowtie \wedge \cup(\mathbf{F}(\beta) \mid \beta \in \alpha) \subset U\}$.

Denote $\mathbf{F}(\alpha)$ by U_α. We have obtained the collection $(U_\alpha \in \mathbf{U} \mid \alpha \in \mathbf{On})$ possessing the following properties:

1) $U_0 = \cap(U \mid U\bowtie)$;
2) $U_{\alpha+1} = \cap\{U \mid U\bowtie \wedge U_\alpha \in U\}$;
3) $U_\alpha = \cap\{U \mid U\bowtie \wedge \cup(U_\beta \mid \beta \in \alpha) \subset U\}$ for any limit ordinal number α.

Let us establish several properties of this collection.

Lemma 1. *In the ZF+AU set theory, the collection $(U_\alpha \in \mathbf{U} \mid \alpha \in \mathbf{On})$ has the following properties:*

1) *$\alpha \in \beta \Leftrightarrow U_\alpha \in U_\beta$ (strict increasing);*
2) *U_0 is the minimal universal set (initiality);*
3) *if V is a universal set and $U_0 \subset V \in U_\alpha$, then $V = U_\beta$ for some $\beta \in \alpha$ (incompressibility);*
4) *if V is a universal set, then $V = U_\alpha$ for some α (surjectivity);*
5) *$\alpha \subset U_\alpha$ (absorptivity).*

Proof. 1. Prove by transfinite induction that $(\alpha \in \beta \Rightarrow U_\alpha \in U_\beta)$ for every ordinal number β.

If $\beta = 0$, then it follows from $\forall\alpha\neg(\alpha \in \beta)$.

Let $\forall\alpha(\alpha \in \beta \Rightarrow U_\alpha \in U_\beta)$ for some ordinal number β. Consider the ordinal number $\beta + 1$. It follows from $\alpha \in \beta + 1$ that $\alpha \in \beta \vee \alpha = \beta$. If $\alpha \in \beta$, then by the inductive assumption $U_\alpha \in U_\beta$ and $U_\alpha \in U_{\beta+1}$ because $U_\beta \subset U_{\beta+1}$. If $\alpha = \beta$, then $U_\alpha = U_\beta \in U_{\beta+1}$. Hence, for $\beta + 1$ we get $\alpha \in \beta + 1 \Rightarrow U_\alpha \in U_{\beta+1}$.

Now, suppose that β is a limit ordinal number and $\forall\gamma \in \beta\forall\alpha(\alpha \in \gamma \Rightarrow U_\alpha \in U_\gamma)$. Let $\alpha \in \beta$. Since β is a limit ordinal number, we have $\alpha+1 \in \beta$. Since $\cup(U_\gamma \mid \gamma \in \beta) \subset U_\beta$, we obtain $U_{\alpha+1} \subset U_\beta$. In this case, $U_\alpha \in U_{\alpha+1}$ implies $U_\alpha \in U_\beta$.

Clearly, $\alpha = \beta \Rightarrow U_\alpha = U_\beta$. Since $U_\alpha = U_\beta$ for $\alpha = \beta$ and $U_\beta \in U_\alpha$ for $\beta \in \alpha$, $U_\alpha \in U_\beta$ implies $\alpha \in \beta$.

2. This property holds by construction.

3. Let V be an arbitrary universal set. If $V = U_0$, then the property is proved. Suppose that $V \neq U_0$.

Consider the class $\mathbf{A} \equiv \{\alpha \in \mathbf{On} \mid U_\alpha \in V\}$. By construction $U_0 \subset V$. By Proposition 1 (A.4.2) $U_0 \in V$. Therefore, $0 \in \mathbf{A}$.

Consider the class $\mathbf{F} \equiv \{z \mid \exists x \in \mathbf{A} \exists y \in \mathbf{In}(z = \langle x, y\rangle \wedge y = \mathbf{q}(U_x))\}$, where $\mathbf{q}$ is the mapping from Theorem 1 (A.4.2). It is clear that $\mathbf{F}$ is a mapping from $\mathbf{A}$ into $\mathbf{In}$. If $\alpha \in \mathbf{A}$, then $\mathbf{F}(\alpha) = \mathbf{q}(U_\alpha) \in \mathbf{q}(V)$. Therefore, $\operatorname{rng}\mathbf{F} \subset \mathbf{q}(V)$. This implies that $B \equiv \operatorname{rng}\mathbf{F}$ is a set. Let $\alpha, \beta \in \mathbf{A}$ and $\alpha \neq \beta$. If $\alpha \in \beta$, then property 1 proven above implies $U_\alpha \in U_\beta$. Since q is isotone, we get $\mathbf{F}(\alpha) < \mathbf{F}(\beta)$. If $\beta \in \alpha$, then similarly, $\mathbf{F}(\beta) < \mathbf{F}(\alpha)$. This means that the mapping $\mathbf{F} : \mathbf{A} \to B$ is bijective and the inverse mapping $\mathbf{F}^{-1} : B \rightarrowtail\!\!\twoheadrightarrow \mathbf{A}$ is defined. Since B is a set, the replacement axiom scheme AS6 implies that $\mathbf{A} = \operatorname{rng}\mathbf{F}^{-1}$ is a set. Therefore, in what follows, instead of $\mathbf{A}$, we will write A.

Consider the non-empty class $\mathbf{C} \equiv \mathbf{On}\backslash A$ and its minimal element β. Clearly, $U_\beta \notin V$. Then, by Proposition 1 (A.4.2) $V \subset U_\beta$. If $V = U_\beta$, then the property is proven.

Suppose $V \in U_\beta$ and $\beta = \gamma + 1$. Then, $\gamma \in A$ implies $U_\gamma \in V$, and therefore, property 2 of the collection $(U_\alpha \in \mathbf{U} \mid \alpha \in \mathbf{On})$ provides $U_\beta \subset V$. Thus, in this case, $V = U_\beta$.

Suppose that β is limit. If $\gamma \in \beta$, then $\gamma \in A$ implies $U_\gamma \in V$ and, according to the transitivity property, we have $U_\gamma \subset V$. Hence, $\cup(U_\gamma \mid \gamma \in \beta) \subset V$. In view of property 3 of the collection $(U_\alpha \in \mathbf{U} \mid \alpha \in \mathbf{On})$, this entails $U_\beta \subset V$. Thus, in this case, we also have $V = U_\beta$.

4. This property follows from properties 1 and 3 proven above.

5. Using property 1 and induction prove that $\alpha \subset U_\alpha$ for every α. It is clear that $\alpha = 0 = \varnothing \subset U_0$.

Let $\alpha \subset U_\alpha$. Since $\alpha + 1 \equiv \alpha \cup \{\alpha\}$, it follows from $\alpha \subset U_\alpha \in U_{\alpha+1}$ that $\alpha \in U_{\alpha+1}$. Therefore, $\{\alpha\} \in U_{\alpha+1}$. The transitivity properties $\alpha \subset U_{\alpha+1}$ and $\{\alpha\} \subset U_{\alpha+1}$ imply $\alpha + 1 \subset U_{\alpha+1}$.

Let α be a limit ordinal number and $\beta \subset U_\beta$ for every $\beta \in \alpha$. Lemmas 2 and 3 (A.2.2) $\alpha = \sup\alpha = \cup\alpha = \cup(\beta \mid \beta \in \alpha)$. Since $\beta \subset U_\beta \subset U_\alpha$, we get $\alpha = \cup\alpha \subset U_\alpha$. □

This lemma implies that the collection $(U_\alpha \in \mathbf{U} \mid \alpha \in \mathbf{On})$ is a natural enumeration of the class of all universal sets in the ZF+AU theory. The following lemma shows that this enumeration is unique.

Lemma 2. *In the ZF+AU set theory, the collection* $(U_\alpha \in \mathbf{U} \mid \alpha \in \mathbf{On})$ *with properties 1 – 3 from Lemma 1 is unique.*

Proof. Assume that there exists a collection $(W_\alpha \in \mathbf{U} \mid \alpha \in \mathbf{On})$ possessing the properties 1 – 3 from Lemma 1. Consider the classes $\mathbf{A} \equiv \{\alpha \in \mathbf{On} \mid U_\alpha = W_\alpha\}$ and $\mathbf{B} = \mathbf{On}\backslash\mathbf{A}$.

Since $U_0 = W_0$, we get $0 \in \mathbf{A}$. Suppose that $\mathbf{B} \neq \varnothing$. Then, by Theorem 3 (A.4.2) there is $\beta = \operatorname{sm} \mathbf{B}$. If $U_\beta \in W_\beta$, then the condition $W_0 = U_0 \subset U_\beta \in W_\beta$ implies $U_\beta = W_\gamma$ for some $\gamma \in \beta$ by virtue of property 3.

It follows from $\gamma \in \mathbf{A}$ that $W_\gamma = U_\gamma$, and therefore, $U_\beta = U_\gamma$ and $\gamma \in \beta$. This contradicts property 1. If $W_\beta \in U_\beta$, then we arrive to a contradiction in a similar way. Hence, $U_\beta = W_\beta$ by virtue of Theorem 3 (A.4.2) and Proposition 1 (A.4.2). However, this contradicts the definition of the class $\mathbf{B}$. Thus, we arrive to a contradiction. Therefore, $\mathbf{B} = \varnothing$ and $\mathbf{A} = \mathbf{On}$. □

Below, we shall also consider the following inaccessibility axiom, which means that for any ordinal number α there is an inaccessible cardinal number greater than α.

AI. (The *inaccessibility axiom.*) $\forall \alpha (On(\alpha) \Rightarrow \exists \varkappa (Icn(\varkappa) \wedge \alpha \in \varkappa))$

The following theorem yields the structural form of the universality axiom.

Theorem 1. *In the ZF set theory, the following conclusions are equivalent:*
1) *AU;*
2) *there is collection $(U_\alpha \in \mathbf{U} \mid \alpha \in \mathbf{On})$ of universal sets having properties 1–5 from Lemma 1;*
3) *AI.*

Proof. (1) ⊢ (2). This deduction was proven in Lemma 1.

(2) ⊢ (3). Take an arbitrary order number α. By Corollary 1 to Theorem 2 (1.3.2) (see also Remark before Theorem 1 (A.2.2)), there is a cardinal number β such that $\alpha < \beta$. By property 5, $\beta \subset U_\beta$. Consider the cardinal number $\varkappa \equiv |U_{\beta+1}|$. The universality implies $\beta \in U_{\beta+1}$, and therefore, $\beta \subset U_{\beta+1}$. Hence, $\beta = |\beta| \leqslant |U_{\beta+1}| \equiv \varkappa$. Suppose that $\beta = \varkappa$. Then, $\varkappa \in U_{\beta+1}$ implies $\mathcal{P}(\varkappa) \in U_{\beta+1}$, and therefore, $\mathcal{P}(\varkappa) \subset U_{\beta+1}$. Applying the Cantor theorem on the cardinality of the set of all subsets (Theorem 2 (1.3.2)), we get $\varkappa = |\varkappa| < |\mathcal{P}(\varkappa)| \leqslant |U_{\beta+1}| \equiv \varkappa$. It follows from this contradiction that $\beta < \varkappa$. Therefore, $\alpha < \varkappa$. By Corollary 1 to Theorem 1 (A.4.2) $\varkappa$ is an inaccessible cardinal number.

(3) ⊢ (1). Lemma 7 (A.3.3) guarantees that $X \in V_\alpha$ for some ordinal number α. By condition 3, $\alpha < \varkappa$ for some inaccessible cardinal number $\varkappa$. By Lemma 1 (A.3.2) $V_\alpha \in V_\varkappa$. By virtue of Theorem 2 (A.4.2) the set $V_\varkappa$ is universal. By Corollary 1 to Lemma 3 (A.3.2) $V_\alpha \subset V_\varkappa$. Thus, $X \in V_\varkappa$. □

Note that the equivalence of the universality and inaccessibility axioms was proven in [*Da Costa and Caroli*, 1967] by another method. Theorem 1 shows the structure of the class of all universal sets in the ZF+AU set theory. The amount of all universal sets and is the same as that of ordinal numbers in the ZF set theory.

A.4.4 Enumeration of the class of all inaccessible cardinals in the ZF+AI theory and the structural form of the inaccessibility axiom

Now, let us enumerate all inaccessible cardinal numbers in the ZF+AI set theory. For this purpose, consider the class

$$\begin{aligned}\mathbf{G} \equiv \{Z \mid \exists X \exists Y (Z = \langle X, Y\rangle \wedge ((X = \varnothing \Rightarrow Y = \mathrm{sm}\{\varkappa \mid Icn(\varkappa)\}) \vee \\ (X \neq \varnothing \Rightarrow (\neg func(X) \Rightarrow Y = \varnothing) \vee \\ (func(X) \Rightarrow (\neg On(\mathrm{dom}\, X) \Rightarrow Y = \varnothing) \vee \\ \vee\ (On(\mathrm{dom}\, X) \Rightarrow (\mathrm{rng}\, X \not\subset \mathbf{On} \Rightarrow Y = \varnothing) \vee \\ \vee\ (\mathrm{rng}\, X \subset \mathbf{On} \Rightarrow (Son(\mathrm{dom}\, X) \Rightarrow Y = \mathrm{sm}\{\varkappa \mid Icn(\varkappa) \wedge X(\mathrm{dom}\, X - 1) \in \varkappa\}) \vee \\ (Lon(\mathrm{dom}\, X) \Rightarrow Y = \mathrm{sm}\{\varkappa \mid Icn(\varkappa) \wedge \cup \mathrm{rng}\, X \subset U\}))))))))\}.\end{aligned}$$

If we reformulate the definition of the class G less formally, then we can say that G consists of all pairs $\langle X, Y\rangle$ such that the following six mutually exclusive possibilities take place:

(i) if $X = \varnothing$, then Y is is a minimal inaccessible cardinal number (its existence follows from the inaccessibility axiom);

(ii) if $X \neq \varnothing$ and X is not a function, then $Y = \varnothing$;

(iii) if $X \neq \varnothing$, X is a function, and $\mathrm{dom}\, X$ is not ordinal number, then $Y = \varnothing$;

(iv) if $X \neq \varnothing$, X is a function, $\mathrm{dom}\, X$ is an ordinal number, and $\mathrm{rng}\, X \not\subset \mathbf{On}$, then $Y = \varnothing$;

(v) if $X \neq \varnothing$, X is a function, $\mathrm{dom}\, X$ is an ordinal number, $\mathrm{rng}\, X \subset \mathbf{On}$, and $\mathrm{dom}\, X = \alpha + 1$, then Y is minimal among all inaccessible cardinals $\varkappa$ such that $X(\alpha) \in \varkappa$ (its existence follows from the inaccessibility axiom);

(vi) if $X \neq \varnothing$, X is a function, $\mathrm{rng}\, X \subset \mathbf{On}$, and $\mathrm{dom}\, X$ is a limit ordinal number, then Y is is minimal among all inaccessible cardinals $\varkappa$ such that $\cup \mathrm{rng}\, X \subset \varkappa$ (its existence follows from Lemmas 1 and 2 (A.2.2), the inaccessibility axiom, and the transitivity of $\varkappa$).

As in A.3.1 we can check that the class $\mathbf{G}$ is a function from $\mathbf{V}$ to $\mathbf{V}$.

By Theorem 2 (A.2.2) there exists a function $\mathbf{F} : \mathbf{On} \to \mathbf{V}$ such that $\mathbf{F}(\alpha) = \mathbf{G}(\mathbf{F}|\alpha)$. for every $\alpha \in \mathbf{On}$.

Case i) for the function $\mathbf{G}$ implies that $\mathbf{F}(\varnothing) = \mathbf{G}(\mathbf{F}|\varnothing) = \mathbf{G}(\varnothing) = sm\{\varkappa \mid Icn(\varkappa)\}$.

It follows from case v) that if β is a subsequent ordinal number and $\beta = \alpha + 1$, then $\mathbf{F}(\beta) = \mathbf{G}(\mathbf{F}|\beta) = \mathrm{sm}\{\varkappa \mid Icn(\varkappa) \wedge \mathbf{F}(\alpha) \in \varkappa\}$.

Finally, case vi) implies that if α is a limit ordinal number, then $\mathbf{F}(\alpha) = \mathbf{G}(\mathbf{F}|\alpha) = \mathrm{sm}\{\varkappa \mid Icn(\varkappa) \wedge \cup\, (\mathbf{F}(\beta) \mid \beta \in \alpha) \subset \varkappa\}$.

Denote $\mathbf{F}(\alpha)$ by q_α. We have obtained the collection $(q_\alpha \in \mathbf{In} \mid \alpha \in \mathbf{On})$ of inaccessible cardinal numbers possessing the following properties:

1) $q_0 = \mathrm{sm}\{\varkappa \mid Icn(\varkappa)\}$;
2) $q_{\alpha+1} = \mathrm{sm}\{\varkappa \mid Icn(\varkappa) \wedge q_\alpha \in \varkappa\}$;
3) $q_\alpha = \mathrm{sm}\{\varkappa \mid Icn(\varkappa) \wedge \cup(q_\beta \mid \beta \in \alpha) \subset \varkappa\}$ for any limit ordinal number α.

Let us establish several properties of this collection.

Lemma 1. *In the ZF+AI set theory, the collection* $(q_\alpha \in \mathbf{In} \mid \alpha \in \mathbf{On})$ *has the following properties:*
1) $\alpha \in \beta \Leftrightarrow q_\alpha \in q_\beta$ *(strict increasing);*
2) q_0 *is a minimal inaccessible cardinal number (initiality);*
3) *if* p *is an inaccessible cardinal and* $q_0 \subset p \in q_\alpha$, *then* $p = q_\beta$ *for some* $\beta \in \alpha$ *(incompressibility);*
4) *if* p *is an inaccessible cardinal, then* $p = q_\alpha$ *for some* α *(surjectivity);*
5) $\alpha \subset q_\alpha$ *(absorptivity).*

The proof is analogous to the proof of Lemma 1 (A.4.3). However, it can be obtained from Lemma 1 (A.4.3) by using the isotone bijection $\mathbf{q} : \mathbf{U} \rightarrowtail\!\!\!\!\twoheadrightarrow \mathbf{In}$ from the Corollary 1 to Theorem 2 (A.4.2).

Lemma 2. *In the ZF+AI set theory, the collection* $(q_\alpha \in \mathbf{In} \mid \alpha \in \mathbf{On})$ *with properties 1 – 3 from Lemma 1 is unique.*

The proof is analogous to the proof of Lemma 2 (A.4.3).

The following theorem yields the structural form of the inaccessibility axiom.

Theorem 1. *In the ZF set theory, the following conclusions are equivalent:*
1) *the inaccessibility axiom AI;*
2) *there is collection* $(q_\alpha \in \mathbf{In} \mid \alpha \in \mathbf{On})$ *of inaccessible cardinal numbers having properties 1 – 5 from Lemma 1.*

Proof. (1) $\vdash$ (2). This deduction was proven in Lemma 1.

(2) $\vdash$ (1). Consider an arbitrary order number α. It follows from properties 5 and 1 that $\alpha \subset q_\alpha \in q_{\alpha+1} \equiv \beta$ and $\alpha \neq \beta$. The transitivity implies $\alpha \subset q_\alpha \subset \beta$, and therefore, the non-empty set $\beta \backslash \alpha$ has a minimal element y. Check that $\alpha = y$.

If $x \in y$, then $x \in y \in \beta$ implies $x \in \beta$. Since $x < y$, we get $x \in \alpha$. This means that $y \subset \alpha$. Conversely, let $x \in \alpha$. It follows from $y \notin \alpha$ that $y \neq x$.

If $y \in x$, then $y \in x \in \alpha$ implies $y \in \alpha$. Therefore, $y \in \alpha \cap (\beta \backslash \alpha) = \varnothing$. This contradiction provides $x \in y$. As a result, we obtain $\alpha \subset y$, and therefore, $\alpha = y \in \beta$. □

Theorem 1 shows the structure of the class of all inaccessible cardinal numbers in the ZF+AI set theory. The amount of all inaccessible cardinal numbers is the same as that of ordinal numbers in the ZF set theory.

Now, let us connect the collections $(V_\alpha \in \mathbf{V} \mid \alpha \in \mathbf{On})$, $(U_\alpha \in \mathbf{U} \mid \alpha \in \mathbf{On})$, and $(q_\alpha \in \mathbf{In} \mid \alpha \in \mathbf{On})$ with each other.

Theorem 2. *In the equivalent ZF+AU and ZF+AI set theories, the equality* $V_{q_\alpha} = U_\alpha$ *holds for every ordinal number* α.

Proof. Since V_{q_0} is a universal set by virtue of Theorem 2 (A.4.2), we get $U_0 \subset V_{q_0}$. Let U be an arbitrary universal set. By Theorem 1 (A.4.2), $U = V_\varkappa$ for some inaccessible cardinal number $\varkappa$. Lemma 1 provides $\varkappa = q_\alpha$ for some α. Since $q_0 \subset q_\alpha$, we see that $V_{q_0} \subset V_{q_\alpha} = V_\varkappa = U$. Therefore, $V_{q_0} \subset \cap\{U \mid U\bowtie\} = U_0$. As a result, we have $V_{q_0} = U_0$.

Consider the non-empty class $\mathbf{A} \equiv \{\alpha \in \mathbf{On} \mid V_{q_\alpha} = U_\alpha\}$ and the class $\mathbf{B} \equiv \mathbf{On}\backslash\mathbf{A}$. Suppose that $\mathbf{B} \neq \varnothing$. Then, there is a number $\beta \equiv \mathrm{sm}\,\mathbf{B} > 0$. Consider universal sets V_{q_β} and U_β.

Suppose that $V_{q_\beta} \in U_\beta$. Then, according to Lemma 1 (A.4.3), the condition $U_0 = V_{q_0} \subset V_{q_\beta} \in U_\beta$ implies $V_{q_\beta} = U_\gamma$ for some $\gamma \in \beta$. It follows from $\gamma < \beta$ that $\gamma \in \mathbf{A}$, and therefore, $V_{q_\gamma} = U_\gamma$. As a result, we obtain the equality $V_{q_\beta} = V_{q_\gamma}$. Applying Lemma 1 (A.3.2), we conclude that $q_\beta = q_\gamma$ and $\gamma \in \beta$, but this contradicts Lemma 1.

On the other hand, suppose that $U_\beta \in V_{q_\beta}$. Since U_β is a universal set, by Theorem 1 (A.4.2) we get $U_\beta = V_\varkappa$ for some inaccessible cardinal number $\varkappa$. Then, it follows from $V_{q_0} = U_0 \subset U_\beta = V_\varkappa \in V_{q_\beta}$ that $q_0 \subset \varkappa \in q_\beta$ in view of Lemma 1 (A.3.2). By virtue of Lemma 1 this implies $\varkappa = q_\gamma$ for some $\gamma \in \beta$. Since $\gamma \in \mathbf{A}$, we get $V_{q_\gamma} = U_\gamma$. As a result, we obtain the equality $U_\beta = V_\varkappa = V_{q_\gamma} = U_\gamma$, where $\gamma \in \beta$, but this contradicts Lemma 1 (A.4.3).

Using Theorem 3 (A.4.2) and Proposition 1 (A.4.2), we conclude that $V_{q_\beta} = U_\beta$. But this contradicts the definition of the class **B**. Hence, $\mathbf{B} = \varnothing$ and $\mathbf{A} = \mathbf{On}$. □

Corollary 1. *In the equivalent ZF+AU and ZF+AI set theories, the equality* $|U_\alpha| = q_\alpha$ *holds for every ordinal number* α.

Proof. By Theorem 2 and Lemma 2 (A.3.3), we get $|U_\alpha| = |V_{q_\alpha}| = q_\alpha$. □

A.5 Weak forms of the universality and inaccessibility axioms

A.5.1 The ω-universality and ω-inaccessibility axioms

Along with universality axiom AU, the following weaker ω-universality axiom is considered in the ZF set theory.

AU(ω). (The *ω-universality axiom.*) $\exists X(\forall U \in X(U\bowtie) \wedge X \neq \varnothing \wedge \forall U \in X \exists V \in X (U \in V))$.

The explanation of such a name of this axiom is given by the following theorem and proposition, which is proven using Theorem 1 (A.4.2).

Theorem 1. *In the ZF set theory, the following conclusions are equivalent:*

1) *AU(ω);*

2) *for every $n \in \omega$ there exists a finite set of universal sets with the cardinality $n + 1$;*
3) *for every $n \in \omega$, there exists a finite sequence $u \equiv (U_k \mid k \in n + 1)$ of universal sets such that $U_k \in U_l$ for any $k \in l \in n + 1$, i. e. the sequence u is strictly increasing;*
4) *there exists a universal set U^* and for every $n \in \omega$ there is a unique finite strictly increasing sequence $u(n) \equiv (U_k^n \mid k \in n+1)$ of universal sets such that $U_0^n = U^*$ and if V is a universal set and $U_0^n \leqslant V \leqslant U_n^n$, then $V = U_k^n$ for some $k \in n+1$ (the incompressibility property);*
5) *there exists a denumerable set of universal sets;*
6) *there exists an infinite sequence $u \equiv (U_n \mid n \in \omega)$ of universal sets such that $U_k < U_l$ for some $k \in l \in \omega$, i. e. the sequence u is strictly increasing;*
7) *there exists an infinite strictly increasing sequence $u \equiv (U_n \mid n \in \omega)$ of universal sets such that if $n \in \omega$, V is a universal set, and $U_0 \leqslant V \leqslant U_n$, then $V = U_k$ for some $k \in n + 1$ (the incompressibility property);*
8) *there exists an infinite set of universal sets.*

Proof. (1) $\vdash$ (4). Consider a non-empty set W whose existence is ensured by axiom AU(ω). Consider also the non-empty class $\mathbf{W} \equiv \{x \mid x \bowtie \ \wedge \exists y \in W(x \leqslant y)\}$. If $x \in \mathbf{W}$, then $x \leqslant y$ for some $y \in W$. Axiom AU(ω) guarantees that for $y \in W$ there is $z \in W$ such that $y < z$. Hence, $x < z \in \mathbf{W}$. Thus, for the class $\mathbf{W}$, all the properties are listed in formula AU(ω).

Since $\varnothing \neq \mathbf{W} \subset \mathbf{U}$, by Theorem 3 (A.4.2) there is a minimal element U^* in $\mathbf{W}$. Since $U^* \leqslant y$ for every $y \in W$, we get $W^* \in \mathbf{W}$. The class $\mathbf{W}$ has the following property: if $z \in \mathbf{U}$ and $z \leqslant y$ for some $y \in \mathbf{W}$, then $z \in \mathbf{W}$.

Consider the set N consisting of all $n \in \omega$ such that there is a unique finite strictly increasing sequence $u = u(n) \equiv (U_k \in \mathbf{W} \mid k \in n + 1)$ such that $U_0 = U^*$ and if V is a universal set and $U_0 \leqslant V \leqslant U_n$, then $V = U_k$ for some $k \in n + 1$.

Since the sequence $(U_k \in \mathbf{W} \mid k \in (1)$ such that $U_0 \equiv U^*$ has all the properties listed above, we see that $0 \in N$. Let $n \in N$. By the property of the class $\mathbf{W}$ for $U_n \in \mathbf{W}$ there is $z \in \mathbf{W}$ such that $U_n < z$. Hence, the class $\mathbf{J} \equiv \{x \in \mathbf{W} \mid U_n < x\}$ is non-empty. Therefore, by Theorem 3 (A.4.2) it has a minimal element A.

Now, we can define the sequence $v \equiv (P_k \in \mathbf{W} \mid k \in n+2)$ setting $P_k \equiv U_k$ for every $k \in n+1$ and $P_{n+1} \equiv A$, i. e. $v = u \cup \{\langle n+1, A\rangle\}$. It is clear that, $P_0 = U^*$ and $P_k < P_l$ for all $k \in l \in n+2$. Let $V \in \mathbf{U}$ and $P_0 \leqslant V < P_{n+1}$. Then, $V \in \mathbf{W}$ and $U_0 \leqslant V < A$. If $V = U_n$, then $V = P_n$. If $V < U_n$, then $U_0 \leqslant V < U_n$ implies $V = U_k = P_k$ for some $k \in n$. Finally, if $V > U_n$, then $V \in \mathbf{J}$. This means that $A \leqslant V$, but it contradicts the property $V < A$, and therefore, this case is impossible. In the previous two cases, we have $V = P_k$ for some $k \in n+1$. This means that the sequence v possesses necessary properties. Check its uniqueness.

Suppose that there exists a strictly increasing sequence $w \equiv (V_k \in \mathbf{W} \mid k \in n + 2)$ such that $V_0 = U^*$, and if $V \in \mathbf{U}$ and $V_0 \leqslant V < V_{n+1}$, then $V = V_k$ for some $k \in n + 1$. Since the sequence $w|n + 1 \equiv (V_k \in \mathbf{W} \mid k \in n + 1)$ has all the properties listed above for n, the uniqueness of the sequence u implies $u = w|(n+1)$, i. e. $V_k = U_k \equiv P_k$ for all

$k \in n+1$. If $V_{n+1} < P_{n+1}$, then $P_0 = V_0 \leqslant V_{n+1} < P_{n+1}$ implies by above $V_{n+1} = P_k = V_k$ for some $k \in n+1$, which is impossible. If $P_{n+1} < V_{n+1}$, then $V_0 = P_0 \leqslant P_{n+1} < V_{n+1}$ similarly implies $P_{n+1} = V_k = P_k$ for some $k \in n+1$, which is also impossible. Hence, $V_{n+1} = P_{n+1}$. Thus, the uniqueness of the sequence v is proven, where $n+1 \in N$. By the principle of natural induction, $N = \omega$. Therefore, for every $n \in \omega$, there exists the indicated unique sequence $u(n)$. Its uniqueness allows us to denote it by $(U_k^n \mid k \in n+1)$.

4. $\vdash$ (7). Consider the following formula of the ZF set theory: $\varphi(x, y) \equiv (x \in \omega \Rightarrow y = U_x^x) \wedge (x \notin \omega \Rightarrow y = \varnothing)$. By the replacement axiom scheme AS6 for ω there is a set Y such that $\forall x \in \omega(\forall y(\varphi(x, y) \Rightarrow y \in Y))$. If $n \in \omega$, then $\varphi(n, U_n^n)$ implies $U_n^n \in Y$. Therefore, we can define the infinite sequence $u \equiv (U_n \in Y \mid n \in \omega)$ setting $u \equiv \{z \in \omega \times Y \mid \exists x \in \omega(z = \langle x, U_x^x \rangle)\}$. The property of uniqueness mentioned above guarantees that $u(m) = u(n)|m+1$ for all $m \leqslant n$. Hence, $u|n+1 = u(n)$. It is clear that the sequence u possesses necessary properties. (6) $\vdash$ (1). Consider the following formula of the ZF set theory: $\varphi(x, y) \equiv (x \in \omega \Rightarrow y = U_x) \wedge (x \notin \omega \Rightarrow y = \varnothing)$. By the replacement axiom scheme AS6 for ω, there is a set Y such that $\forall x \in \omega(\forall y(\varphi(x, y) \Rightarrow y \in Y))$. If $n \in \omega$, then $\varphi(n, U_n)$ implies $U_n \in Y$. By the separation axiom scheme AS3, the class $X \equiv \{U_n \mid n \in \omega\} \equiv \{y \mid \exists x \in \omega(y = U_x)\} = \{y \mid y \in Y \wedge \exists x \in \omega(y = U_x)\}$ is a set. Since the sequence u strictly increases, the set X satisfies axiom AU(ω).

The deductions (7) $\vdash$ (6) $\vdash$ (5) $\vdash$ (2) are obvious.

The deductions (4) $\vdash$ (3) $\vdash$ (2) are also obvious.

(2) $\vdash$ (3) and (2) $\vdash$ (6). Consider the non-empty class $\mathbf{A}$ of all finite sets consisting of universal sets. Then, the class $\mathbf{W} \equiv \cup\mathbf{A}$ is also non-empty, and therefore, it has a minimal element U^* in view of Theorem 3 (A.4.2).

Consider the set N consisting of all $n \in \omega$ such that there is a unique finite strictly increasing sequence $u = u(n) \equiv (U_k \in \mathbf{W} \mid k \in n+1)$ such that $U_0 = U^*$ and if $V \in \mathbf{W}$ and $U_0 \leqslant V \leqslant U_n$, then $V = U_k$ for some $k \in n+1$ (the *property of* **W***-incompressibility*).

Since the sequence $(U_k \in \mathbf{W} \mid k \in (1)$ such that $U_0 \equiv U^*$ has all the properties listed above, we see that $0 \in N$. Let $n \in N$, i. e. the sequence $u \equiv (U_k \in \mathbf{W} \mid k \in n+1)$ is constructed for n. Consider the finite set $A \equiv \{U_k \in \mathbf{W} \mid k \in n+1\}$ of the cardinality $n+1$. By conclusion 2 for $n+2$, there is a finite set $B \in \mathbf{A}$ of the cardinality $n+2$. Take a minimal element a and a maximal element b in B. By definition, $a \geqslant U^*$. Suppose that $b \leqslant U_n$. Then, for every $c \in B$ the inequality $U_0 = U^* \leqslant a \leqslant c \leqslant b \leqslant U_n$ is valid. If $c < U_n$, then $c \in \mathbf{W}$ implies $c = U_k$ for some $k \in n$ by virtue of property of $\mathbf{W}$-incompressibility, i. e. $c \in A$. If $c = U_n$, then we again get $c \in A$. As a result, we obtain the inclusion $B \subset A$, which is impossible. This contradiction implies $U_n < b$. Since $b \in \mathbf{W}$, the class $\mathbf{J} \equiv \{x \in \mathbf{W} \mid U_n < x\}$ is non-empty. Therefore, it has a minimal element Λ.

Therefore, we can define the sequence $v \equiv (P_k \in \mathbf{W} \mid k \in n+2)$ setting $P_k \equiv U_k$ for every $k \in n+1$ and $P_{n+1} \equiv \Lambda$, i. e. $v = u \cup \{\langle n+1, \Lambda \rangle\}$. Further, almost in the same way as in the proof of deduction (1) $\vdash$ (4) with the replacement of $\mathbf{U}$ by $\mathbf{W}$, we conclude that the sequence v possesses all necessary properties and is unique. Hence, $n+1 \in N$. By the principle of natural induction, $N = \omega$. Therefore, for every $n \in \omega$, there exists the indicated unique sequence $u(n)$. Its uniqueness allows us to denote it by $(U_k^n \mid k \in n+1)$. This completes the deduction (2) $\vdash$ (3).

Further, almost in the same way as in the proof of deduction (4) $\vdash$ (7) starting from sequences $(U_k^n \mid k \in n+1)$, we construct infinite strictly increasing sequence $u \equiv (U_n \mid n \in \omega)$ of universal sets. This yields the deduction (2) $\vdash$ (6).

Thus, we obtain deductions (1) $\vdash$ (4) $\vdash$ (7) $\vdash$ (6) $\vdash$ (1) and (6) $\vdash$ (5) $\vdash$ (2) $\vdash$ (6) and the equivalence 2) $\sim$ (3). This provides the equivalence of conclusions 1 – 7.

(8) $\vdash$ (6). Let W be an infinite set of universal sets. By Theorem 3 (A.4.2) in W, there is a minimal element U^*.

Consider the set N consisting of all $n \in \omega$ such that there is a unique finite strictly increasing sequence $u = u(n) \equiv (U_k \in W \mid k \in n+1)$ such that $U_0 = U^*$ and if $V \in W$ and $U_0 \leqslant V \leqslant U_n$, then $V = U_k$ for some $k \in n+1$ (the *property of W-incompressibility*).

Since the sequence $(U_k \in W \mid k \in 1)$ such that $U_0 \equiv U^*$ has all the properties listed above, we see that $0 \in N$. Let $n \in N$. Consider the set $J \equiv W \backslash \{U_k \mid k \in n+1\}$. It is non-empty because the set W is infinite; hence, J has a minimal element Λ. Clearly, $\Lambda \neq U_n$ and $\Lambda \geqslant U^* = U_0$. Suppose that $\Lambda < U_n$. Then, $U_0 \leqslant \Lambda < U_n$ implies $\Lambda = U_k$ for some $k \in n$, which is impossible. Therefore, $U_n < \Lambda$.

Now, we can define the sequence $v \equiv (P_k \in W \mid k \in n+2)$ setting $P_k \equiv U_k$ for every $k \in n+1$ and $P_{n+1} \equiv \Lambda$, i. e. $v = u \cup \{\langle n+1, \Lambda\rangle\}$. It is clear that, $P_0 = U^*$ and $P_k < P_l$ for all $k \in l \in n+2$. Let $V \in W$ and $P_0 \leqslant V < P_{n+1}$. Then, $U_0 \leqslant V < \Lambda$. If $V = U_n$, then $V = P_n$. If $V < U_n$, then $U_0 \leqslant V < U_n$ implies $V = U_k = P_k$ for some $k \in n$. Finally, if $V > U_n$, then $V > U_k$ for all $k \in n+1$, and therefore, $V \in J$. This means that $\Lambda \leqslant V$, but it contradicts the property $V < \Lambda$, and therefore, this case is impossible. In the previous two cases, we have $V = P_k$ for some $k \in n+1$. This means that the sequence v possesses necessary properties. Check its uniqueness.

Suppose that there exists a strictly increasing sequence $w \equiv (V_k \in W \mid k \in n+2)$ such that $V_0 = U^*$, and if $V \in W$ and $V_0 \leqslant V < V_{n+1}$, then $V = V_k$ for some $k \in n+1$. Since the sequence $w|n+1 \equiv (V_k \in W \mid k \in n+1)$ has all the properties listed above for n, the uniqueness of the sequence u implies $u = w|(n+1)$, i. e. $V_k = U_k \equiv P_k$ for all $k \in n+1$. If $V_{n+1} < P_{n+1}$, then $P_0 = V_0 \leqslant V_{n+1} < P_{n+1}$ implies by above $V_{n+1} = P_k = V_k$ for some $k \in n+1$, which is impossible. If $P_{n+1} < V_{n+1}$, then $V_0 = P_0 \leqslant P_{n+1} < V_{n+1}$ similarly implies $P_{n+1} = V_k = P_k$ for some $k \in n+1$, which is also impossible. Hence, $V_{n+1} = P_{n+1}$. Thus, the uniqueness of the sequence v is proven, where $n+1 \in N$. By the principle of natural induction (see Theorem 1 (1.2.6) and Remark before Theorem 1 (A.2.2)), $N = \omega$. Therefore, for every $n \in \omega$, there exists the indicated unique sequence $u(n)$. Its uniqueness allows us to denote it by $(U_k^n \mid k \in n+1)$.

Further, as in the proof of deduction (4) $\vdash$ (7) starting from sequences $(U_k^n \mid k \in n+1)$, we construct infinite strictly increasing sequence $u \equiv (U_n \mid n \in \omega)$ of universal sets.

(6) $\vdash$ (8). As in the proof of the deduction (6) $\vdash$ (1) consider for the sequence u the set $X \equiv \{U_n \mid n \in \omega\}$. Suppse that the set X is finite. Then X has a maximal element V. This contradicts strict increasing of u. □

The fact that the ω-universality axiom is weaker than the universality axiom is established in the following proposition.

Proposition 1. *In the ZF set theory, the ω-universality axiom is deduced from the universality axiom.*

Proof. Show that assertion 2 of Theorem 1 is deduced from AU. For this purpose, prove by induction that for every $n \in \omega$ there exists the finite set of universal sets with the cardinality $n + 1$.

For $n = 0$, this means that there is at least one universal set. This assertion obviously holds.

Suppose that for some $n \in \omega$ there is a set of cardinality $n+1$ consisting of universal sets. Denote this set by A. The universality axiom provides the existence of a universal set U such that $A \in U$, and therefore, $A \subset U$. If $V \in A$, then $V \neq U$, since otherwise, $U \in U$, which is impossible. Consider the set $B \equiv A \cup \{U\}$. Clearly, this set is of cardinality $n + 2$. □

Along with inaccessibility axiom AI, in the ZF set theory, the following weaker ω-inaccessibility axiom is considered.

AI(ω). (The *ω-inaccessibility axiom.*) $\exists X(\forall x \in X(Icn(x)) \wedge X \neq \varnothing \wedge \forall x \in X \exists y \in X(x \in y))$.

The explanation of such a name of this axiom is given by the following theorem and proposition.

Theorem 2. *In the ZF set theory, the following conclusions are equivalent:*

1) *$AI(\omega)$;*
2) *for every $n \in \omega$, there exists a finite set of inaccessible cardinals with the cardinality $n + 1$;*
3) *for every $n \in \omega$, there exists a finite sequence $u \equiv (\iota_k \mid k \in n + 1)$ of inaccessible cardinals such that $\iota_k < \iota_l$ for any $k \in l \in n + 1$, i. e. the sequence u strictly increase;*
4) *there exists a inaccessible cardinal $\varkappa^*$, and for every $n \in \omega$, there is a unique finite strictly increasing sequence $u(n) \equiv (\iota_k^n \mid k \in n+1)$ of inaccessible cardinals such that $\iota_0^n = \varkappa^*$ and if $\varkappa$ is an inaccessible cardinal and $\iota_0^n \leqslant \varkappa \leqslant \iota_n^n$, then $\varkappa = \iota_k^n$ for some $k \in n + 1$ (the incompressibility property);*
5) *there exists a denumerable set of inaccessible cardinals;*
6) *there exists an infinite sequence $u \equiv (\iota_n \mid n \in \omega)$ of inaccessible cardinals such that $\iota_k < \iota_l$ for some $k \in l \in \omega$, i. e. the sequence u is strictly increasing;*
7) *there exists an infinite strictly increasing sequence $u \equiv (\iota_n \mid n \in \omega)$ of inaccessible cardinals such that if $n \in \omega$, $\varkappa$ is an inaccessible cardinal, and $\iota_0 \leqslant \varkappa \leqslant \iota_n$, then $\varkappa = \iota_k$ for some $k \in n + 1$ (the incompressibility property);*
8) *there exists an infinite set of inaccessible cardinals.*

The proof of this theorem is completely analogous to the proof of Theorem 1. However, it can be also obtained from Theorem 1 by using the isotone bijection $\mathbf{q} : \mathbf{U} \rightarrowtail\!\!\!\!\twoheadrightarrow \mathbf{In}$ from the Corollary 1 to Theorem 2 (A.4.2). Moreover, this result is repeated as Theorem 1 (B.4.1) with the complete proof.

The following proposition is an ω-analogue of Theorem 1 (A.4.3).

Proposition 2. *In the ZF set theory, the following axioms are equivalent:*
1) *the ω-universality axiom $AU(\omega)$;*
2) *the ω-inaccessibility axiom $AI(\omega)$.*

Proof. To prove the equivalence, it is sufficient to apply the isotone bijection $\mathbf{q} : \mathbf{U} \rightarrowtail\!\!\!\!\twoheadrightarrow \mathbf{In}$ from Corollary 1 to Theorem 2 (A.4.2). □

A.5.2 Comparison of various forms of the universality and inaccessibility axioms

Along with axioms AU and AU(ω) the following axiom is considered in ZF.

ATU(ω). (The *axiom of transitive ω-universality.*) *There exists a set Y such that:*
a) $Y \neq \varnothing$;
b) $\forall U \in Y(U\bowtie)$;
c) $\forall U \forall V(U \bowtie \wedge U \in V \wedge V \in Y \Rightarrow U \in Y)$ *(the transitivity property with respect to universal sets);*
d) $\forall V \in Y \exists W \in Y(V \in W)$ *(the unboundedness property).*

Lemma 1. *In the ZF set theory, the following axioms are equivalent:*
1) *$AU(\omega)$;*
2) *$ATU(\omega)$.*

Proof. (1) ⊢ (2). Denote by D a set whose existence is ensured by AU(ω). Consider the set $E \equiv \{U \in \cup D \mid U\bowtie\}$. It satisfies conditions (a) and (b).

If $U \in D$, then AU(ω) implies $\exists V \in D(U \in V)$. Consequently, $D \subset E$. Show that the set E satisfies condition (c). Indeed, if $U\bowtie$ and $U \in V \in E$, then $U \in V \in W \in D$ for some $W \in D$. By virtue of transitivity of the set W we obtain $U \in W \in D$, i. e. $U \in E$.

If $V \in E$, then by definition $V \in W \in D \subset E$ for some W. Hence, E satisfies condition (d).

(2) ⊢ (1). This deduction is obvious. □

An analogous lemma holds for inaccessible cardinals with the replacements of AU(ω) by AI(ω) and ATU(ω) by ATI(ω) (the *axiom of transitive ω-inaccessibility*).

Lemma 2. *Let E be a non-empty set of universal sets with the transitivity property with respect to universal sets, i. e. E satisfies conditions* a) – c) *from Lemma* 1. *Then E contains a minimal universal set* $\mathfrak{a}_0 \equiv U_0 \equiv \cap \mathbf{U}$.

Proof. Let $V \in E$. By Proposition 1 (A.5.1) $V = \mathfrak{a}$ or $\mathfrak{a} \in V$. In the first case, $\mathfrak{a} \in E$. In the second case, $\mathfrak{a} \in V \in E$ implies $\mathfrak{a} \in E$ in view of condition (c). □

An analogous lemma holds for inaccessible cardinals with the replacement $\mathfrak{a} \equiv U_0$ by $\mathfrak{q} \equiv q_0 \equiv \mathrm{sm}\,\mathbf{In}$.

Along with axioms AU and AU(ω) consider one more weaker 1-universality axiom asserting the existence of at least one universal set.

AU(1)≡AUS. (The 1-*universality axiom* or the *axiom of universal set*.) $\exists U(U\bowtie)$

In the ZF+AU(1) the class **U** of all universal sets is non-empty, and therefore, contains a minimal element $\mathfrak{a}_0 \equiv U_0 \equiv \cap\mathbf{U}$.

Similarly, along with axioms AI and AI(ω) consider one more weaker 1-inaccessibility axiom asserting the existence of at least one inaccessible cardinal number.

AI(1)≡AIC. (The 1-*inaccessibility axiom* or the *axiom of inaccessible cardinal.*) $\exists\varkappa(Icn(\varkappa))$.

In the ZF+AI(1), the class **In** of all inaccessible cardinal numbers is non-empty, and therefore, contains a minimal element $\mathfrak{q} \equiv q_0 \equiv \mathrm{sm}\,\mathbf{In}$.

The following proposition is a 1-analog of Theorem 1 (A.4.2) and Proposition 2 (A.5.1).

Proposition 1. *In the ZF set theory, the following axioms are equivalent:*
1) *AU*(1);
2) *AI*(1).

Proof. To prove the equivalence, it is sufficient to apply the isotone bijection $\mathbf{q} : \mathbf{U} \to \mathbf{In}$ from Corollary 1 to Theorem 2 (A.4.2). □

The following relations between these axioms hold:

$$AU \vdash AU(\omega) \vdash AU(1) \text{ and } AI \vdash AI(\omega) \vdash AI(1).$$

Let us show that these axioms are really different.

Statement 1.
1) *If the theory ZF + AU(1) is consistent, then the theory ZF + AU(1) + ¬AU(ω) is consistent.*
2) *If the theory ZF + AU(1) is consistent, then axiom AU(ω) is not deducible in ZF + AU(1).*

Proof. 1. Let U_0 be a minimal universal set whose existence is ensured by axiom AU(1). Consider the classes $\mathbf{W} \equiv \{W \mid W\bowtie \wedge U_0 \in W\}$ and $\mathbf{D} \equiv \{X \mid \forall W(W\bowtie \wedge U_0 \in W \Rightarrow X \in W)\}$.

The following two cases are possible. If the class **W** is non-empty, the in contains a minimal element U_1. Clearly, $\mathbf{D} \subset U_1$. If $X \in U_1$ and $W \in \mathbf{W}$, then $X \in U_1 \subset W$ implies $X \in W$. Consequently, $X \in \mathbf{D}$. Hence, $\mathbf{D} = U_1$. If the class **W** is empty, then $\mathbf{D} = \mathbf{V}$.

By Lemma 1 axiom AU(ω) is equivalent to axiom ATU(ω). Therefore, we consider the equivalent theory $T \equiv ZF + AUS + \neg ATU(\omega)$. Consider a class standard interpretation $\mathbf{M} \equiv (\mathbf{D}, I)$ of the theory T in the set theory $S \equiv ZF + AUS$ such that the correspondence I (see A.1.3) assigning to predicate symbols = and $\in$ in T 2-placed relation $\mathbf{E} \equiv \{z \mid \exists x \exists y(x \in \mathbf{D} \wedge y \in \mathbf{D} \wedge z = (x, y) \wedge x = y)\}$ and $\mathbf{B} \equiv \{z \mid \exists x \exists y(x \in \mathbf{D} \wedge y \in \mathbf{D} \wedge z = (x, y) \wedge x \in y)\}$ on **D**.

If $\mathbf{D} = U_1$, then Proposition 1 (A.5.1) guarantees that the interpretation $M \equiv \mathbf{M} = (U_1, I)$ is a model of the ZF set theory in the theory S. If $\mathbf{D} = \mathbf{V}$, then it is clear that the interpretation **M** is a class model of ZF in the theory S.

Check that axiom AUS holds in **M**. The axiom can be written as follows:

$$\begin{aligned} AUS \equiv \exists X(\forall x(x \in X \Rightarrow x \subset X \wedge \mathcal{P}(x) \in X \wedge \\ \wedge \cup x \in X) \wedge \forall x \forall y(x \in X \wedge y \in X \Rightarrow \{x, y\} \in X) \wedge \\ \wedge \forall x \forall f(x \in X \wedge f \leftrightarrows x \to X \Rightarrow \operatorname{rng} f \in X) \wedge \omega \in X). \end{aligned}$$

Consider the first case. Let s be some sequence $x_0, \dots, x_q, \dots$ of elements of the domain U_1. Taking into account the equivalences $(u \in v)^t \Leftrightarrow u^t \in v^t$, $(v \subset w)^t \Leftrightarrow v^t \subset w^t$, and $(u = v)^t \Leftrightarrow u^t = v^t$ proved in Proposition 1 (A.6.1) below and the notation from its proof, we obtain

$$\begin{aligned} \widetilde{AUS^t} = \exists X \in U_1(\forall x \in U_1(x \in X \Rightarrow x \subset X \wedge \mathcal{P}(x)^\tau \in X \wedge \\ \wedge (\cup x)^\tau \in X) \wedge \forall x \in U_1 \forall y \in U_1(x \in X \wedge y \in X \Rightarrow \{x, y\}^\sigma \in X) \wedge \\ \wedge \forall x \in U_1 \forall f \in U_1(x \in X \wedge (f \leftrightarrows x \to X)^\rho \Rightarrow (\operatorname{rng} f)^\rho \in X) \wedge \omega^\pi \in X). \end{aligned}$$

In the proof of Proposition 1 (A.6.1), it was established that $\mathcal{P}(x)^\tau = \mathcal{P}(x)$, $(\cup x)^\tau = \cup x$, $\{x, y\}^\sigma = \{x, y\}$, $(f \leftrightarrows x \to X)^\rho \Leftrightarrow (f \leftrightarrows x \to X)$, and $(\operatorname{rng} f)^\rho = \operatorname{rng} f$. In a similar way, we can prove that $\omega^\pi = \omega$. Therefore, $\widetilde{AUS^t} \Leftrightarrow \exists X \in U_1 \chi(X)$, where the formula

$$\begin{aligned} \chi(X) \equiv \forall x \in U_1(x \in X \Rightarrow x \subset X \wedge \mathcal{P}(x) \in X \wedge \cup x \in X) \wedge \\ \wedge \forall x \in U_1 \forall y \in U_1(x \in X \wedge y \in X \Rightarrow \{x, y\} \in X) \wedge \\ \wedge \forall x \in U_1 \forall f \in U_1(x \in X \wedge f \leftrightarrows x \to X \Rightarrow \operatorname{rng} f \in X) \wedge \omega \in X \end{aligned}$$

is obtained by deleting indices τ, σ, and ρ in the conjunctive kernel of formula $\widetilde{AUS^t}$. Since U_0 is a universal set, the formula $\chi(U_0)$ is valid for it. This means that the formula $\chi(U_0)$ is deduced from axiom AUS in the theory S. Consequently, the formula $\exists X \in U_1 \chi(X)$ is deduced, and therefore, formula AUS^t is also deduced.

In the second case, it is clear that formula AUS translates (see A.1.3) on the sequence s of elements $x_0, \dots, x_q, \dots$ of the domain **V** into formula AUS again. Hence, axiom AUS holds in **M**.

It remains to be checked that the formula $\neg ATU(\omega)$ is fulfilled. By Lemma 2, we can insert the formula $U_0 \in Y$ into the conjunctive kernel of axiom $ATU(\omega)$. Consider, therefore, the formulas

$$\varphi \equiv ATU(\omega) \equiv \exists Y(\forall U(U \in Y \Rightarrow U\bowtie) \wedge U_0 \in Y \wedge$$
$$\wedge \forall U \forall V(U \bowtie \ \wedge U \in V \wedge V \in Y \Rightarrow U \in Y) \wedge \forall V(V \in Y \Rightarrow \exists W(W \in Y \wedge V \in W)))$$

and $\varphi^t \equiv \mathbf{M} \models \varphi[s]$.

Let us consider the first case. Taking into account the elucidations made after rewriting of axiom AUS, we obtain

$$\varphi^t \Leftrightarrow \widetilde{\varphi^t} = \exists Y \in U_1(\forall U \in U_1(U \in Y \Rightarrow (U\bowtie)^\sigma) \wedge U_0^\tau \in Y \wedge \forall U \in U_1 \forall V \in U_1$$
$$((U\bowtie)^\rho \wedge U \in V \wedge V \in Y \Rightarrow U \in Y) \wedge \forall V \in U_1(V \in Y \Rightarrow$$
$$\Rightarrow \exists W \in U_1(W \in Y \wedge V \in W))).$$

Considering the translation of the previous axiom, we have proven that $(U\bowtie)^\sigma \Leftrightarrow \chi(U)$ and $(U\bowtie)^\rho \Leftrightarrow \chi(U)$.

Since the set U_0 can be determined by the formula $\exists! Z(Z \bowtie \ \wedge \forall U(U\bowtie \Rightarrow Z \subset U))$, the set U_0^τ is determined by the formula $\exists! Z \in U_1((Z\bowtie)^* \ \wedge \forall U \in U_1((U\bowtie)^{**} \Rightarrow Z \subset U))$.

As above, $(Z\bowtie)^* \Leftrightarrow \chi(Z)$ è $(U\bowtie)^{**} \Leftrightarrow \chi(U)$. Hence, U_0^τ defines from the formula $\exists! Z \in U_1(\chi(Z) \wedge \forall U \in U_1(\chi(U) \Rightarrow Z \subset U))$. Then, it is clear that $U_0^\tau = U_0$. Thus,

$$\varphi^t \Leftrightarrow \exists Y \in U_1(\forall U \in U_1(U \in Y \Rightarrow \chi(U)) \wedge U_0 \in Y \wedge \forall U \in U_1 \forall V \in U_1(\chi(U) \wedge$$
$$\wedge U \in V \wedge V \in Y \Rightarrow U \in Y) \wedge \forall V \in U_1(V \in Y \Rightarrow \exists W \in U_1(W \in Y \wedge V \in W))).$$

Suppose that the condition φ^t is fixed and consider the set $E \in U_1 = \mathbf{D}$, which existence follows from this condition. By condition, $U_0 \in E$. Therefore, φ^t implies that for $U_0 \in U_1$ there is $W \in U_1$ such that $W \in E$ and $U_0 \in W$. Deduce that the set W is universal.

Since $W \in E$, we get $\chi(W)$. Let $x \in W$. It follows from $W \in U_1$ that $x \in U_1$ by virtue of the transitivity of U_1. Hence, $\chi(W)$ implies $x \subset W$, $\mathcal{P}(x) \in W$, and $\cup x \in W$. Similarly, if $x, y \in W$, then $x, y \in U_1$ and $\chi(W)$ implies $\{x, y\} \in W$. Finally, suppose $x \in W$ and $f \leftrightharpoons x \to W$. Then, $x \in U_1$ and $W \in U_1$ imply $f \subset x * W \in U_1$. Lemma 1 (A.4.1) provides $f \in U_1$, and therefore, it follows from $\chi(W)$ that $\operatorname{rng} f \in W$. The properties $y \subset x \wedge x \in W \Rightarrow y \in W$ and $x, y \in W \Rightarrow (\langle x, y\rangle \in W \wedge x \cup y \in W)$ are easily derived from the properties proven above. Since $x * y \subset \mathcal{P}(\mathcal{P}(x \cup y))$ we obtain $x, y \in W \Rightarrow x * y \in W$. Finally, it follows directly from $\chi(W)$ that $\omega \in W$. Thus, W is universal.

Moreover, $U_0 \in W$. Consequently, $W \in \mathbf{W}$, and therefore, $U_1 \subset W$. In view of Proposition 1 (A.4.2), we conclude that $W \notin U_1$. On the other hand, we have deduced from φ^t that $W \in U_1$.

Thus, in the theory S, we have deduced from the formula φ^t the formulas $\eta \equiv W \in U_1$ and $\neg\eta = W \notin U_1$. By the deduction theorem, we derive the formulas $(\varphi^t \Rightarrow \chi)$ and $(\varphi^t \Rightarrow \neg\chi)$ in the theory S.

Applying now the implicit logical axiom $(\varphi^t \Rightarrow \chi) \Rightarrow ((\varphi^t \Rightarrow \neg\chi) \Rightarrow \neg\varphi^t)$ (LAS9), we sequentially deduce the formulas $(\varphi^t \Rightarrow \neg\chi) \Rightarrow \neg\varphi^t$ and $\neg\varphi^t$. Thus, we have deduced the formula $\neg\varphi^t$ from the condition $\mathbf{W} \neq \varnothing$. By the deduction theorem, the formula $\mathbf{W} \neq \varnothing \Rightarrow \neg\varphi^t$ is deduced in the theory S.

In the second case, it is clear that the formula φ translates on the sequence s of elements $x_0, \ldots, x_q, \ldots$ of the domain $\mathbf{V}$ into the formula φ again. i. e. $\varphi^t = \varphi$.

Suppose that the condition $\varphi^t = \varphi$ is fixed and consider the set $E \in \mathbf{V} = \mathbf{D}$, which existence follows from this condition. By condition, $U_0 \in E$. Then φ^t guarantees that for U_0 there is a universal set $W \in E$ such that $U_0 \in W$, where $\mathbf{W} \neq \varnothing$. By the deduction theorem, the formula $\varphi^t \Rightarrow \mathbf{W} = \varnothing$ is deduced in the theory S. Applying logical formula $(\varphi^t \Rightarrow \neg(\mathbf{W} = \varnothing)) \Rightarrow (\mathbf{W} = \varnothing \Rightarrow \neg\varphi^t)$, we obtain $\mathbf{W} = \varnothing \Rightarrow \neg\varphi^t$. Thus, we have deduced the formula $\mathbf{W} = \varnothing \Rightarrow \neg\varphi^t$ from the condition $\mathbf{W} = \varnothing$. Therefore, the formula $\neg\varphi^t$ is deduced from $\mathbf{W} = \varnothing$. By the deduction theorem, the formula $\mathbf{W} = \varnothing \Rightarrow \neg\varphi^t$ is deduced in the theory S. Denote the formula $\mathbf{W} = \varnothing$ by ξ.

Applying now the logical formula $(\xi \Rightarrow \neg\varphi^t) \Rightarrow ((\neg\xi \Rightarrow \neg\varphi^t) \Rightarrow ((\xi \vee \neg\xi) \Rightarrow \neg\varphi^t))$, we sequentially deduce in the theory S the formulas $(\neg\xi \Rightarrow \neg\varphi^t) \Rightarrow (\xi \vee \neg\xi \Rightarrow \neg\varphi^t)$ and $\xi \vee \neg\xi \Rightarrow \neg\varphi^t$. Since in the first-order theory for any formula ξ the formula $\xi \vee \neg\xi$ is deduced, we obtain $\neg\varphi^t$ in S.

The last formula equals to the formula $\mathbf{M} \models (\neg\varphi)[s]$. This means that $\mathbf{M}$ is a model of T in S.

2. We will proceed in the naive propositional logic with the implication symbol $\supset$.

Denote the totalities of the axioms of the theories T and S by Φ_a and Ξ_a, respectively.

Consider the propositions $A \equiv cons(S) \supset \neg(\Xi_a \vdash AU(\omega))$ and $B \equiv cons(S) \wedge (\Xi_a \vdash AU(\omega))$. Then, $\neg A = cons(S) \wedge \neg\neg(\Xi_a \vdash AU(\omega))$. Using the axiom $\neg\neg C \supset C$, we get $\neg A \supset B$.

Clearly, $B \supset (\Phi_a \vdash AU(\omega))$ and $\Phi_a \vdash \neg AU(\omega)$. Therefore, the proposition $B \supset (\Phi_a \vdash AU(\omega)) \wedge (\Phi_a \vdash \neg AU(\omega))$, i. e. the proposition $B \supset \neg cons(T)$ is true. By the deduction rule, $\neg A \supset \neg cons(T)$.

According to (1), the proposition $cons(S) \supset cons(T)$ is deduced. Hence, $B \supset cons(T)$ is true. By the deduction rule, $\neg A \supset cons(T)$.

Thus, the proposition $(\neg A \supset cons(T)) \wedge (\neg A \supset \neg cons(T))$ is deduced. Applying the tautology ("reductio ad absurdum") $(\neg A \supset C) \wedge (\neg A \supset \neg C) \supset A$ (see, e. g., [*Kolmogorov and Dragalin*, 1982, I, § 7]), we deduce the proposition A. □

Corollary 1. *If the theory ZF+AU(1) is consistent, then axiom AU is not deducible in ZF+AU(1).*

Remark. In fact, we have prove that the existence of a second universal set U_1, i. e. of a set U_1 such that $U_0 \in U_1$ and $U_1 = \cap\{U \mid U\bowtie \wedge U_0 \in U\}$ is not deducible in ZF+AU(1).

Analogous assertions hold for inaccessible cardinals with the replacement of AU(1), AU(ω), and AU by AI(1), AI(ω), and AI, respectively.

Statement 2.
1) *If the theory ZF+AU(ω) is consistent, then the theory ZF+AU(ω) + ¬AU is consistent.*
2) *If the theory ZF+AU(ω) is consistent, then axiom AU is not deducible in ZF+AU(ω).*

Proof. 1. Let D be a set whose existence follows from Axiom AU(ω). Consider the classes $\mathbf{W} \equiv \{W \mid W\bowtie \wedge D \in W\}$ and $\mathbf{D} \equiv \{X \mid \forall W(W\bowtie \wedge D \in W \Rightarrow X \in W)\}$.

The following two cases are possible. If the class **W** is non-empty, the it contains a minimal element U^*. Clearly, $\mathbf{D} = U^*$. If the class **W** is empty, then $\mathbf{D} = \mathbf{V}$.

Consider a class standard interpretation $\mathbf{M} \equiv (\mathbf{D}, I)$ of the theory $T \equiv ZF + AU(\omega) + \neg AU$ in the set theory $S \equiv ZF + AU(\omega)$ with the same correspondence I as in the proof of Statement 1. According to that proof, **M** is a class model of the theory ZF in the theory S.

Check that that axiom AU(ω) of the theory T holds in **M**. This axiom has the form

$$AU(\omega) \equiv \exists X(\forall U(U \in X \Rightarrow U\bowtie) \wedge X \neq \varnothing \wedge \forall V(V \in X \Rightarrow \exists W(W \in X \wedge V \in W))).$$

Consider the first case. In the same way as in the proof of Statement 1, we establish that

$$AU(\omega)^t \Leftrightarrow \exists X \in U^*(\forall U \in U^*(U \in X \Rightarrow \chi(U)) \wedge X \neq \varnothing \wedge \\ \wedge \forall V \in U^*(V \in X \Rightarrow \exists W \in U^*(W \in X \wedge V \in W))).$$

Consider the set $D \neq \varnothing$. If $U \in D$, then U is a universal set, and therefore, the formula $\chi(U)$ holds for it. Let $V \in D$. It follows from AU(ω) that there is $W \in D$ such that $V \in W$. By transitivity of U^* we derive $W \in U^*$ from $W \in D \in U^*$. This means that the formula AU$(\omega)^t$ is deduced from the formula AU$(\omega)^t$.

In the second case, it is clear that the formula AU(ω) translates into the formula AU(ω) again, and therefore, axiom AU(ω) holds in **M**.

It remains to verify the fulfilment of the formula ¬AU.

Consider the formula $\varphi \equiv AU \equiv \forall X \exists V(V \bowtie \wedge X \in V)$. Consider the first case. Then, $\varphi^t \Leftrightarrow \forall X \in U^* \exists V \in U^*(\chi(V) \wedge X \in V)$.

Suppose that the condition φ^t is fixed. Since $D \in U^*$, this condition guarantees that there is a set $W \in U^*$ such that $\chi(W)$ and $D \in W$. As in the proof of Statement 1, we deduce from $W \in U^*$ and $\chi(W)$ that the set W is universal. Moreover, $D \in W$. Consequently, $W \in \mathbf{W}$. This implies that $W \in \mathbf{W}$, and therefore, $U^* \subset W$. In view of Proposition 1 (A.4.2), we conclude that $W \notin U^*$. On the other hand, we have deduced from φ^t that $W \in U^*$.

Thus, as in the proof of Statement 1, we conclude that the formula $\mathbf{W} \neq \varnothing \Rightarrow \neg\varphi^t$ is deduced in the theory S.

In the second case, it is clear that φ translates into φ again, i. e. $\varphi^t = \varphi$.

Suppose that the condition $\varphi^t = \varphi$ is fixed. By this condition for the set D there is a universal set W such that $D \in W$. This implies $W \in \mathbf{W}$, and therefore, $\mathbf{W} \neq \varnothing$. By the deduction theorem, in S the formula $\varphi^t \Rightarrow \mathbf{W} \neq \varnothing$ is deduced. As in the proof of Statement 1, we deduce from this formula that $\mathbf{W} = \varnothing \Rightarrow \neg\varphi^t$.

As in the proof of Statement 1, we deduce from the formulas $\xi \Rightarrow \neg\varphi^t$ and $\neg\xi \Rightarrow \varphi^t$ the formula $\neg\varphi^t$ equal to the formula $\mathbf{M} \vDash (\neg\varphi)[s]$. This means that $\mathbf{M}$ is a model of T in S.

2. The proof is the same as the proof of assertion 2 of Statement 1, □

Thus, in fact, axiom AU(ω) is strictly weaker than axiom AU and axiom AU(1) is strictly weaker than axiom AU(ω). The quite similar relation holds for axioms AI, AI(ω), and AI(1).

Note that axiom AI(1) is not deducible in ZF. Moreover, by methods formalized in the ZF set theory it is not possible to show that that axiom AI(1) is consistent with ZF (see [*Jech*, 2003, Theorem 12.12]). The similar assertions are valid for axioms AI(ω) and AI and for the universality axioms equivalent to them.

A.6 Characterization of all supertransitive standard models of the ZF and NBG set theories in the ZF set theory

A.6.1 Supertransitive standard model sets with the strong substitution property for the ZF set theory

Let U be a set in the ZF set theory. Consider on U binary *relation of equality* $E \equiv \{z \in U * U \mid \exists x, y \in U(z = (x, y) \wedge x = y)\}$ and *relation of membership* $B \equiv \{z \in U * U \mid \exists x, y \in U(z = (x, y) \wedge x \in y)\}$. An interpretation $M \equiv (U, I)$ (see A.1.3) of the ZF or NBG theories such that the correspondence I assigning to predicate symbols $=$ and $\in$ binary relations E and B on the set U is called *standard*.

A set U is called *standard model for the theory ZF* [*for the theory NBG*] if the standard interpretation $M \equiv (U, I)$ is a model of ZF [of NBG, respectively].

If $\forall x \forall f(x \in U \wedge f \in U^x \Rightarrow \operatorname{rng} f \in U)$, then we say that the *set U has the strong substitution property*.

As usual for a formula $\varphi(x, y, \dots)$, we denote by $\varphi^U(x, y, \dots)$ the *relativization of the formula φ to the set U*, i. e. the formula obtained by replacement in φ all quantifier prefixes $\forall t$ and $\exists t$ by quantifier prefixes $\forall t \in U$ and $\exists t \in U$, respectively.

Proposition 1. *In the ZF set theory, the following conclusions are equivalent:*
1) *U is a supertransitive standard model for ZF and U has the strong substitution property;*
2) *U is universal.*

Proof. (1) $\vdash$ (2). Consider an arbitrary sequence $s \equiv x_0, \dots, x_q, \dots$ of elements of the set U and translations of some axioms and axiom schemes of the theory ZF with respect to the standard interpretation $M \equiv (U, I)$ on the sequence s (see A.1.3).

Instead of $\theta_M[s]$ and $M \models \varphi[s]$, we shall write θ^t and φ^t for terms θ and formulas φ, respectively.

To simplify the further presentation, we first consider the translations of certain simple formulas. Let u and v be some sets.

The formula $u \in v$ translates into the formula $(u \in v)^t = (\langle u^t, v^t\rangle \in B)$. Denote the last formula by γ. By definition, this formula is equivalent to the formula $(\exists x \exists y(x \in U \wedge y \in U \wedge \langle u^t, v^t\rangle = \langle x, y\rangle \wedge x \in y))$. Using the property of an ordered pair, we conclude that $u^t = x$ and $v^t = y$. Therefore, it is deduced from γ that $\delta \equiv (u^t \in v^t)$. By the deduction theorem, $\gamma \Rightarrow \delta$. Conversely, consider the formula δ. It is proven in ZF that for sets u^t and v^t, there is a set z such that $z = \langle u^t, v^t\rangle$. By virtue of logical axiom scheme LAS3 (A.1.2) we deduce from the formula δ the formula $(z = \langle u^t, v^t\rangle \Rightarrow u^t \in U \wedge v^t \in U \wedge z = \langle u^t, v^t\rangle \wedge u^t \in v^t)$. Since the formula $z = \langle u^t, v^t\rangle$ is deduced from the axioms, the formula $(u^t \in U \wedge v^t \in U \wedge z = \langle u^t, v^t\rangle \wedge u^t \in v^t)$ is also deduced. By LAS13, we deduce the formula $\exists x \exists y(x \in U \wedge y \in U \wedge z = \langle x, y\rangle \wedge x \in y)$ equivalent to the formula $z \in B$, and therefore, to the formula γ. By the deduction theorem, $\delta \Rightarrow \gamma$. Thus, the first equivalence $(u \in v)^t \Leftrightarrow u^t \in v^t$ holds.

The formula $v \subset w$ translates into the formula $(v \subset w)^t$. Denote the last formula by ε. The first equivalence proven above is equivalent to the formula $\varepsilon' \equiv \forall u \in U(u \in v^t \Rightarrow u \in w^t)$. According to LAS11, from the formula ε', we deduce the formula $\varepsilon'' \equiv (x \in U \Rightarrow (x \in v^t \Rightarrow x \in w^t))$. If $x \in v^t$, then $v^t \in U$ and transitivity of U imply $x \in U$. Then, the formula ε'' implies $x \in v^t \Rightarrow x \in w^t$. Consequently, by the deduction theorem we deduce $(\varepsilon \Rightarrow (x \in v^t \Rightarrow x \in w^t))$. By the rule of generalization (Gen), the formula $\forall x(\varepsilon \Rightarrow (x \in v^t \Rightarrow x \in w^t))$ is deduced. By LAS12, we deduce the formula $(\varepsilon \Rightarrow \forall x(x \in v^t \Rightarrow x \in w^t))$, i. e. the formula $(\varepsilon \Rightarrow v^t \subset w^t)$.

Conversely, let the formula $v^t \subset w^t$ be given. Using the logical axioms, we sequentially deduce from it the formulas $(u \in v^t \Rightarrow u \in w^t)$ and $(u \in U \Rightarrow (u \in v^t \Rightarrow u \in w^t))$. By (Gen) we deduce the formula ε'. Hence, by the deduction theorem, we get the formula $(v^t \subset w^t \Rightarrow \varepsilon)$. Thus, the second equivalence $(v \subset w)^t \Leftrightarrow v^t \subset w^t$ holds.

We obtain the third equivalence $(u = v)^t \Leftrightarrow u^t = v^t$ in exactly the same way as the first equivalence.

In what follows, we will write not literal transformations of axioms but their equivalent variants obtained using the mentioned equivalences.

The extensionality axiom A1 translates into the formula $A1^t \Leftrightarrow A1^U = \forall X \in U \forall Y \in U(\forall u \in U(u \in X \Leftrightarrow u \in Y) \Rightarrow X = Y)$.

The pair axiom A2 translates into the formula $A2^t \Leftrightarrow A2^U = \forall u \in U \forall v \in U \exists x \in U \forall z \in U(z \in x \Leftrightarrow z = u \vee z = v)$.

The union axiom A4 translates into the formula $A4^t \Leftrightarrow A4^U = \forall X \in U \exists Y \in U \forall u \in U(u \in X \Leftrightarrow \exists z \in U(u \in z \wedge z \in X))$.

The power set axiom A5 translates into the formula $A5^t \Leftrightarrow A5^U = \forall X \in U \exists Y \in U \forall u \in U(u \subset X \Leftrightarrow u \in Y)$.

The replacement axiom scheme AS6 translates into the formula scheme

$$AS6^t \Leftrightarrow \forall x \in U \forall y \in U \forall y' \in U(\varphi^\tau(x, y) \wedge \varphi^\tau(x, y') \Rightarrow y = y') \Rightarrow$$

$$\Rightarrow \forall X \in U \exists Y \in U \forall x \in U(x \in X \Rightarrow \forall y \in U(\varphi^\sigma(x, y) \Rightarrow y \in Y)),$$

where φ^τ and φ^σ are denotations of the formulas $M \vDash \varphi[s^\tau]$ and $M \vDash \varphi[s^\sigma]$ and s^τ and s^σ denote the corresponding changes of the sequence s under translation of the quantifier overformulas indicated above. Denote the last formula scheme by $\alpha \Rightarrow \beta$.

The empty set axiom A7 translates into the formula $A7^t \Leftrightarrow A7^U = \exists x \in U \forall z \in U(z \notin x)$.

The infinity axiom A8 translates into the formula $A8^t \Leftrightarrow A8^\tau \equiv \exists Y \in U(\varnothing^t \in Y \wedge \forall y \in U(y \in Y \Rightarrow (y \cup \{y\})^\tau \in Y))$, where

- the set $\varnothing^t$ is determined by the formula $A7^U$;
- the set $Z_1 \equiv Z_1(y) \equiv (y \cup \{y\})^\tau$ is determined by the formula $\exists Z_1 \in U \forall u \in U(u \in Z_1 \Leftrightarrow \exists z \in U(u \in z \wedge z \in \{y, \{y\}\}^\sigma))$;
- the set $Z_2 \equiv Z_2(y) \equiv \{y, \{y\}\}^\sigma$ is determined by the formula $\exists Z_2 \in U \forall u \in U(u \in Z_2 \Leftrightarrow u = y \vee u = \{y\}^\rho)$;
- the set $Z_3 \equiv Z_3(y) \equiv \{y\}^\rho$ is determined by the formula $\exists Z_3 \in U \forall u \in U(u \in Z_3 \Leftrightarrow u = y)$.

Since M is a model of the ZF theory, all the translations written above are deducible formulas in the ZF theory.

Therefore, the formula $A7^U$ asserts the existence of some $x \in U$ denoted by $\varnothing^t$. If $z \in U$, then $A7^U$ implies $z \notin x$. Now, suppose that $z \notin U$ and $z \in x$. Then, by virtue of transitivity of U we obtain $z \in U$, but it contradicts the condition. Hence, $z \notin x$. Thus, we deduce $z \notin x$. By (Gen) the formula $\forall z(z \notin x)$ meant $x = \varnothing$ is deduced. Thus, $\varnothing^t = \varnothing$ and $\varnothing \in U$.

Check now that if $y \in U$, then $Z_3 = \{y\}$. Let $u \in Z_3$. Since $Z_3 \in U$ and U is transitive, we get $u \in U$. If $u \in U$, then the formula for Z_3 presented above implies $u = y$, where

$u \in \{y\}$. Therefore, $Z_3 \subset \{y\}$. Conversely, suppose $u \in \{y\}$. Then, $u = y$. Since $y \in U$, we get $u \in U$, where, by the same formula, we obtain $u \in Z_3$. Consequently, $\{y\} \subset Z_3$, which implies the required equality. This equality eliminates the index ρ in the formula for Z_2.

Using this equality, show that $Z_2 = \{y, \{y\}\}$. Let $u \in Z_2$. Then, as above, $u \in U$. Therefore, the formula for Z_2 presented above implies $u = y$ or $u = \{y\}$, where $u \in \{y, \{y\}\}$. Consequently, $Z_2 \subset \{y, \{y\}\}$. Conversely, suppose $u \in \{y, \{y\}\}$. Then, $u = y \in U$ or $u = \{y\} = Z_3 \in U$, i. e. $u \in U$ in both cases. Hence, by the same formula, we get $u \in Z_2$, where $\{y, \{y\}\} \subset Z_2$. This implies the required equality. This equality eliminates the index σ in the formula for Z_1.

Finally, we verify that if $y \in U$, then $Z_1 = y \cup \{y\}$. Let $u \in Z_1$. Since $Z_1 \in U$ and U is transitive, we get $u \in U$. It follows from the formula for Z_1 that there exists $z \in U$ such that $u \in z$ and $z \in \{y, \{y\}\}$. Therefore, $u \in \cup\{y, \{y\}\} \equiv Z$, i. e. $Z_1 \subset Z$. Conversely, suppose $u \in Z$. Then, there exists $z \in \{y, \{y\}\}$ such that $u \in z$. It follows from $z = y \in U$ or $z = \{y\} = Z_3 \in U$ that $z \in U$. Then, the formula presented above implies $u \in Z_1$. Hence, $Z \subset Z_1$, which implies the required equality. This equality eliminates the index τ in the formula for $A8^\tau$.

All said above implies $A8^\tau = \exists Y \in U(\varnothing \in Y \wedge \forall y \in U(y \in Y \Rightarrow y \cup \{y\} \in Y))$. If $y \in Y$, then it follows from $Y \in U$ and transitivity of U that $y \in U$. Then, $y \cup \{y\} \in Y$ deduced from this formula. By the deduction theorem, we deduce $y \in Y \Rightarrow y \cup \{y\} \in Y$. By the generalization rule we deduce $\forall y \in Y(y \cup \{y\} \in Y)$. Thus, we deduce from $A8^t$ the formula $\exists Y \in U(\varnothing \in Y \wedge \forall y \in Y(y \cup \{y\} \in Y))$ almost coinciding with the infinity axiom and asserting the existence of an inductive set $Y \in U$.

Using the obtained translations, let us prove that the set U is universal.

Consider the formula $A2^U$. According to it, for any $u, v \in U$ there is a corresponding set $x \in U$. If $z \in x$, then by transitivity of U we get $z \in U$. Therefore, the formula $z = u \vee z = v$ is deduced from it. If $z = u \vee z = v$, then $z \in U$, and therefore, it is deduced from $A2^U$ that $z \in x$. Since $A2^U$ is deducible in ZF, by the deduction theorem and the generalization rule the formula $\forall z(z \in x \Leftrightarrow z = u \vee z = v)$ is deduced. This formula means that $x = \{u, v\}$. Hence, $\{u, v\} \in U$. By the deduction theorem, we deduce the formula $u, v \in U \Rightarrow \{u, v\} \in U$. This implies $\{u\} \in U$ and $\langle u, v\rangle \in U$.

Consider the formula $A4^U$. According to it, for any $X \in U$ there is a corresponding set $Y \in U$. As above, transitivity of U implies $Y = \cup X$. Consequently, $\cup X \in U$, and by the deduction theorem, we deduce the formula $X \in U \Rightarrow \cup X \in U$. This implies that it follows from $X, Y \in U$ that $X \cup Y \equiv \cup\{X, Y\} \in U$.

Consider the formula $A5^U$. According to it, for any $X \in U$, there is a corresponding set $Y \in U$. Clearly, $Y \subset \mathcal{P}(X)$. Let $y \in \mathcal{P}(X)$. Then, $y \subset X \in U$ implies $y \in U$ in view of quasitransitivity of U. Hence, $Y = \mathcal{P}(X)$. Therefore, $\mathcal{P}(X) \in U$, and by the deduction theorem, we deduce $X \in U \Rightarrow \mathcal{P}(X) \in U$.

If $X, Y \in U$, then $X * Y \subset \mathcal{P}(\mathcal{P}(X \cup Y)) \in U$ implies $X * Y \in U$ in view of quasitransitivity of U.

Consider the inductive set $Y \in U$, whose existence was proven above. Since ω is the smallest among all inductive sets, we get $\omega \subset Y$. By the quasitransitivity property, this implies $\omega \in U$.

It is obvious that property 4 from the definition of a universal set holds.

Thus, we have proven that (1) $\vdash$ (2).

(2) $\vdash$ (1). Let U be a universal set. According to A.4.1, it is supertransitive. Consider the standard interpretation $M \equiv (U, I)$ of the theory ZF. We have translated above some axioms and axiom schemes of ZF under the interpretation M on the sequence s. Prove that they are deducible in ZF.

Consider the formula $A1^U$. Let $X, Y \in U$ and $\chi \equiv \forall u \in U(u \in X \Leftrightarrow u \in Y)$. Take an arbitrary set u. If $u \in X$, then by transitivity of U, we obtain $u \in U$, and therefore, the formula $u \in Y$ is deduced. Similarly, we deduce $u \in Y$ from $u \in X$. Then, by the deduction theorem, the formula $u \in X \Leftrightarrow u \in Y$ is deduced, and by the generalisation rule (Gen), the formula $\forall u(u \in X \Leftrightarrow u \in Y)$ is deduced. According to the extensionality axiom A1, the equality $X = Y$ is deduced. By the deduction theorem, in ZF the formula $\chi \Rightarrow X = Y$ is deduced. Further, by logical tools, we deduce $A1^t$.

Consider the formula $A2^U$. Let $u, v \in U$. By the property of a universal set $\{u, v\} \in U$. It follows from the pair axiom A2 that $\forall z \in U(z \in \{u, v\} \Leftrightarrow z = u \vee z = v)$. Then, by LAS13, we deduce $\exists x \in U \forall z \in U(z \in x \Leftrightarrow z = u \vee z = v)$. Further, by logical tools we deduce $A2^t$.

The separation axiom scheme AS3 translates into the formula scheme $AS3^t \Leftrightarrow \forall X \in U \exists Y \in U \forall u \in U(u \in Y \Leftrightarrow u \in X \wedge \varphi^\tau(u))$, where Y is not a free variable in $\varphi(u)$ and φ^τ denotes the formula $M \models \varphi[s^\tau]$, where s^τ denote the corresponding changes of the sequence s under translation of the quantifier overformulas $\forall x(\dots)$, $\exists Y(\dots)$, and $\forall u(\dots)$ indicated above. According to AS3 for $X \in U$ there is Y such that $\forall u \in U(u \in Y \Leftrightarrow u \in X \wedge \varphi^\tau(u))$. Since $Y \subset X \in U$, by Lemma 1 (A.4.1), we get $Y \in U$. Therefore, $AS3^t$ is deduced in ZF.

Similar to the deducibility of $A2^t$, we verify the deducibility of $A4^t$ and $A5^t$.

Let us verify the deducibility of $AS6^t$. Suppose that the formula α holds. Consider the set $X \in U$. According to the separation axiom scheme AS3, the set $F \equiv \{z \in U \mid \exists x, y \in U(z = \langle x, y\rangle \wedge \varphi^\sigma(x, y))\}$ exists. Clearly, $F \subset U * U$. It follows from transitivity of U that $X \subset U$. Therefore, there is a set $Z \equiv F[X] \subset U$. Consider the set $G \equiv \{z \in U \mid \exists x, y \in U(z = \langle x, y\rangle \wedge \varphi^\sigma(x, y) \wedge x \in X)\} = F|X \subset X * Z$. Let $x \in X \subset U$. If $x \notin \operatorname{dom} G$, then $G\langle x\rangle = \varnothing \in U$. Let $x \in \operatorname{dom} G$, i. e. $G\langle x\rangle \neq \varnothing$. If $y, y' \in G\langle x\rangle \subset U$, then the formula $\varphi^\sigma(x, y) \wedge \varphi^\sigma(x, y')$ or, more precisely, the formula $\varphi^\sigma(x, y, X, Y) \wedge \varphi^\sigma(x, y', X, Y)$ holds (since X and Y can be free variables of the formula φ^σ). Since $\varphi^\tau(x, y) = \varphi^\sigma(x, y, X \parallel X_M[s], Y \parallel Y_M[s])$ and, similarly, for y', by virtue of LAS11 we obtain $\varphi^\tau(x, y) \wedge \varphi^\tau(x, y')$. Hence, the formula α implies $y = y'$. Therefore, $G\langle x\rangle = \{y\} \in U$. Thus, $G\langle x\rangle \in U$ for every $x \in X$. By Lemma 3 (A.4.1), we get $Y_0 \equiv \operatorname{rng} G = \cup(G\langle x\rangle \mid x \in X) \in U$.

If $x \in X \subset U$, $y \in U$, and $\varphi^\sigma(x, y)$, then $\langle x, y\rangle \in G$ implies $y \in Y_0$. This means that the formula β deduced from the formula α. By the deduction theorem, the formula $\alpha \Rightarrow \beta$ is deduced, and therefore, the scheme $AS6^t$ is deduced.

According to Lemma 2 (A.4.1), $\varnothing \in U$. Then, we deduce $A7^t$ from this and A7.

Consider the formula $A8^\tau$ and the set $\omega \in U$. It follows from the above that $\varnothing^t = \varnothing \in \omega$. Let $y \in U$ and $y \in \omega$. Then, as above, we check that $Z_3 = \{y\}$, $Z_2 = \{y, \{y\}\}$ and $Z_1 = y \cup \{y\} \in \omega$. By the deduction theorem, we deduce $(y \in \omega \Rightarrow Z_1 \in \omega)$. Further, by logical tools we deduce $(\varnothing^t \in \omega \wedge \forall y \in U(y \in \omega \Rightarrow (y \cup \{y\})^\tau \in \omega))$, and therefore, the formula $A8^t$.

The regularity axiom translates into the formula $A9^t \Leftrightarrow A9^\tau \equiv \forall X \in U(X \neq \varnothing^t \Rightarrow \exists x \in U(x \in X \wedge (x \cap X)^\tau = \varnothing^t))$, where

- the set $\varnothing^t$ is determined by $A7^U$ and, as was proven above, it coincides with the empty set $\varnothing$,
- the set $Z \equiv (x \cap X)^\tau$ is determined by the formula $\exists Z \in U \forall u \in U(u \in Z \Leftrightarrow u \in x \wedge u \in X)$.

Check now that if $X \in U$ and $x \in U$, then $Z = x \cap X$. Let $u \in Z$. Since $Z \in U$ and U is transitive, we get $u \in U$. Therefore, it follows from the formula for Z that $u \in x \wedge u \in X$, i. e. $u \in x \cap X$. Hence, $Z \subset x \cap X$. Conversely, suppose $u \in x \cap X$, i. e. $u \in x \wedge u \in X$. Then, by virtue of transitivity we get $u \in U$ and the mentioned formula implies $u \in Z$. Thus, $x \cap X \subset Z$, which implies the required equality. This equality eliminates the index τ in the formula $A9^\tau$.

Let $X \in U$ and $X \neq \varnothing^t = \varnothing$. By the regularity axiom, there is $x \in X$ such that $x \cap X = \varnothing$. By virtue of transitivity we get $x \in U$. Further, by logical tools, we deduce $A9^t$.

Finally, the choice axiom A10 translates into the formula

$$
\begin{aligned}
&A10^t \Leftrightarrow A10^\tau \equiv \\
&\equiv \forall X \in U(X \neq \varnothing^t \Rightarrow \exists z \in U((z \leftrightarrows \mathcal{P}(X)\backslash\{\varnothing\} \to X)^\tau \wedge \forall Y \in U(Y \in (\mathcal{P}(X)\backslash\{\varnothing\})^\sigma \Rightarrow \\
&\qquad \Rightarrow \forall x \in U(x \in X \wedge \langle Y, x\rangle^\sigma \in z \Rightarrow x \in Y))))),
\end{aligned}
$$

where

- the set $Z_1 \equiv Z_1(X) \equiv (\mathcal{P}(X)\backslash\{\varnothing\})^\sigma$ is determined by the formula $\exists Z_1 \in U \forall u \in U(u \in Z_1 \Leftrightarrow u \in \mathcal{P}(X)^\rho \wedge u \notin \{\varnothing\}^\rho)$,
- the set $Z_2 \equiv \langle Y, x\rangle^\sigma$ is determined by the formula $\exists Z_2 \in U \forall u \in U(u \in Z_2 \Leftrightarrow (u = \{Y\}^\sigma \vee u = \{Y, x\}^\sigma))$,
- the set $Z_3 \equiv \{Y, x\}^\sigma$ is determined by the formula $\exists Z_3 \in U \forall u \in U(u \in Z_3 \Leftrightarrow (u = Y \vee u = x))$,
- the set $Z_4 \equiv \{Y\}^\sigma$ is determined by the formula $\exists Z_4 \in U \forall u \in U(u \in Z_4 \Leftrightarrow u = y)$,

and $\varphi^\tau \equiv (z \leftrightarrows \mathcal{P}(X)\backslash\{\varnothing\} \to X)^\tau$ denotes the formula $M \models \varphi[s^\tau]$, where s^τ denote the corresponding changes of the sequence s under translation of the quantifier overformulas $\forall X(\ldots)$ and $\exists z(\ldots)$ indicated above.

Fix the conditions $X \in U$ and $X \neq \varnothing^t = \varnothing \in U$. As was shown above, this implies $\mathcal{P}(X)^\rho = \mathcal{P}(X)$ and $\{\varnothing\}^\rho = \{\varnothing\}$. This equality eliminates the index ρ in the formula for Z_1.

Check that $Z_1 = \mathcal{P}(X)\backslash\{\varnothing\} \equiv Z$. Let $u \in Z_1$. Since $Z_1 \in U$ and U is transitive, we get $u \in U$. Then, the formula for Z_1 implies $u \in Z$. Hence, $Z_1 \subset Z$. Conversely, suppose $u \in Z$. Since $\mathcal{P}(X) \in U$ and U is transitive, we get $\mathcal{P}(X) \subset U$. This implies $u \in U$. Consequently, the mentioned formula implies $u \in Z_1$. Therefore, $Z \subset Z_1$, which implies the required equality. This guarantees that Z_1 is replaced by Z in the formula A10$^\tau$.

Consider the formula $\varphi \equiv (z \leftrightharpoons Z \to X)$. It is the conjunction of the following three formulas: $\varphi_1 \equiv (z \subset Z * X)$, $\varphi_2 \equiv (\operatorname{dom} z = Z)$, and $\varphi_3 \equiv (\forall x(x \in Z \Rightarrow \forall y(y \in X \Rightarrow \forall y'(y' \in X \Rightarrow (\langle x, y\rangle \in z \wedge \langle x, y'\rangle \in z \Rightarrow y = y')))))$.

Then, $\varphi^\tau = \varphi_1^\tau \wedge \varphi_2^\tau \wedge \varphi_3^\tau$. Since $\varphi_1 = (\forall u(u \in z \Rightarrow \exists x \exists y(x \in Z \wedge y \in X \wedge u = \langle x, y\rangle)))$, we obtain $\varphi_1^\tau \Leftrightarrow (\forall u \in U(u \in z \Rightarrow \exists x \in U \exists y \in U(x \in Z \wedge y \in X \wedge u = \langle x, y\rangle^\sigma)))$. Similarly, it follows from $\varphi_2 = (\forall x(x \in Z \Rightarrow \exists y(y \in X \wedge \langle x, y\rangle \in z)))$ that $\varphi_2^\tau \Leftrightarrow (\forall x \in U(x \in Z \Rightarrow \exists y \in U(y \in X \wedge \langle x, y\rangle^\sigma \in z)))$.

Finally, $\varphi_3^\tau \Leftrightarrow (\forall x \in U(x \in Z \Rightarrow \forall y \in U(y \in X \Rightarrow \forall y' \in U(y' \in X \Rightarrow (\langle x, y\rangle^\sigma \in z \wedge \langle x, y'\rangle^\sigma \in z \Rightarrow y = y')))))$.

By the transitivity property for x, y, and y' in the formulas φ_1^τ, φ_2^τ è φ_3^τ, we have $x, y, y' \in U$. Therefore, as was shown above, the equalities $\langle x, y\rangle^\sigma = \langle x, y\rangle$ and $\langle x, y'\rangle^\sigma = \langle x, y'\rangle$ hold in these formulas. This implies that the formulas φ_1^τ, φ_2^τ, and φ_3^τ differ from the formulas φ_1, φ_2, and φ_3, respectively, only by bounded quantifier prefixes $\forall \cdots \in U$ and $\exists \cdots \in U$.

For X by the choice axiom A10 there is z such that $\chi \equiv (z \leftrightharpoons Z \to X) \wedge \forall Y(Y \in Z \Rightarrow \forall x(x \in X \wedge \langle Y, x\rangle \in z \Rightarrow x \in Y))$.

Hence, the formula $\varphi = \varphi_1 \wedge \varphi_2 \wedge \varphi_3$ is deduced, and therefore, the formulas φ_1, φ_2, and φ_3 are also deduced.

Let $u \in U$ and $u \in z$. Then, it is deduced from the formula φ_1 that there are $x \in Z$ and $y \in X$ such that $u = \langle x, y\rangle$. Since $x \in Z \in U$, $y \in X \in U$, and U is transitive, we get $x, y \in U$. This means that for the given conditions $u \in U$ and $u \in z$, the formula $\exists x \in U \exists y \in U(x \in Z \wedge y \in X \wedge u = \langle x, y\rangle^\sigma)$ is deduced. Applying the deduction theorem and the deduction rules twice, we deduce the formula φ_1^τ.

Let $x \in U$ and $x \in Z$. Then, we deduce from the formula φ_2 that for x there is $y \in X$ such that $\langle x, y\rangle \in z$. It follows from $y \in X \in U$ that $y \in U$. This means that for the given conditions $x \in U$ and $x \in Z$ the formula $\exists y \in U(y \in X \wedge \langle x, y\rangle^\sigma \in z)$ is deduced. Therefore, as above, we deduce the formula φ_2^τ.

Let $x \in U$, $x \in Z$, $y \in U$, $y \in X$, $y' \in U$, $y' \in X$, $\langle x, y\rangle \in z$, and $\langle x, y'\rangle \in z$. Then, it is deduced from φ_3 that $y = y'$. Applying alternately the deduction theorem and the deduction rules several times, we deduce the formula φ_3^τ.

Thus, the formula φ^τ is deduced.

Check that $Z_4 = \{Y\}$ under the conditions $X \in U$, $Y \in U$, and $Y \in Z$. Let $u \in \{Y\}$, i. e. $u = Y \in U$. Then, the formula for Z_4 implies $u \in Z_4$. Conversely, if $u \in Z_4 \in U$, then $u \in U$, and therefore, $u = Y \in \{Y\}$. This yields the necessary equality.

Check that $Z_3 = \{Y, x\}$ under the conditions $X \in U$, $x \in X$, $Y \in U$, and $Y \in Z$. Let $u \in \{Y, x\}$. Then, $u = Y \in U$ or $u = x \in X \in U$ implies $u \in U$, and therefore, $u \in Z_3$. Conversely, if $u \in Z_3 \in U$, then $u \in U$ and the formula for Z_3 imply $u = Y \vee u = x$, i. e. $u \in \{Y, x\}$. This yields the necessary equality.

Finally, check that $Z_2 = \langle Y, x\rangle$ under the indicated conditions. Let $u \in \langle Y, x\rangle$, i. e. $u = \{Y\}$ or $u = \{Y, x\}$. The previous equalities eliminate the index σ in the formula for Z_2. Since $Y \in U$ and $x \in X \in U$, we see that $x \in U$ and universality of U imply $u = \{Y\} \in U$ or $u = \{Y, x\} \in U$. Hence, $u \in U$ implies $u \in Z_2$. Conversely, if $u \in Z_2 \in U$, then $u \in U$ and the formula for Z_2 imply $u = \{Y\}$ or $u = \{Y, x\}$, i. e. $u = \langle Y, x\rangle$. This yields the necessary equality.

Since $Z \in U$ and $X \in U$, we get $Z * X \in U$. By Lemma 1 (A.4.1), it follows from $z \subset Z * X$ that $z \in U$.

Thus, it can be deduced from axiom A10 that there exists the object $z \in U$ satisfying the formula χ, implying the formula $\xi \equiv (\varphi^\tau \wedge \forall Y \in U(Y \in Z \Rightarrow \forall x \in U(x \in X \wedge \langle Y, x\rangle \in z \Rightarrow x \in Y))$. Consequently, we deduce the formula $\exists z \in U\ \xi$ from the fixed conditions. Applying the deduction theorem and the generalization rule several times, we, as a result, deduce the formula $A10^t$.

Thus, M is a supertransitive standard model of the ZF set theory. □

This proposition implies that for supertransitive standard model sets all the assertions presented in A.4 for universal sets hold.

Theorem 1. *In the ZF set theory, the following conclusions are equivalent for a set U:*
1) $U = V_\varkappa$ *for the inaccessible cardinal number* $\varkappa = |U| = \sup(\mathbf{On} \cap U)$;
2) *U is a supertransitive standard model for ZF and U has the strong substitution property.*

Proof. (1) $\vdash$ (2). By Theorem 2 (A.4.2), the set $U = V_\varkappa$ is universal. By Proposition 1, it satisfies (2). (2) $\vdash$ (1). By Proposition 1, U is universal. By Theorem 1 (A.4.2), $U = V_\varkappa$ and $\varkappa = \sup(\mathbf{On} \cap U)$. By Corollary 1 to Theorem 1 (A.4.2), $\varkappa = |U|$. □

This theorem gives the canonical form of supertransitive standard model sets with the strong substitution property. It is equivalent to the Zermelo – Shepherdson theorem [*Zermelo*, 1930; *Shepherdson*, 1951; 1952; 1953] on the canonical forms of supertransitive standard model sets for the NBG theory in the ZF set theory (see A.6.2 below).

Unfortunately, this theorem does not yield the description of all natural models and all supertransitive standard models of the ZF set theory. This description will be given in A.8.3.

A.6.2 Supertransitive standard model of the NBG set theory in the ZF set theory

The NBG set theory is a first-order theory (without equalities) with a single binary predicate symbol of belonging $\in$ (we write $A \in B$). The objects of the NBG theory are called *classes.* All proper axioms and axiom schemes of NBG and all corresponding definitions and comments are given in subsections 1.1.5 – 1.1.12. Here we only list these axioms and axiom schemes in more formal way as it was done for the axioms of ZF in A.2. In this list, the formula $\exists X(A \in X)$ meaning that A is a set is denoted by $S(A)$.

A1. (The *extensionality axiom.*) $\forall y \forall z((y = z) \Rightarrow \forall X(y \in X \Leftrightarrow z \in X))$.

AS2. (The *full comprehension axiom scheme.*) Let $\varphi(x)$ be a predicative formula such that the substitution $\varphi(x \parallel y)$ is admissible and such that Y is not a free variable of φ. Then, $\exists Y \forall y(y \in Y \Leftrightarrow (S(y) \wedge \varphi(y)))$.

A3. (The *axiom of the full ensemble.*)

$$\forall X(S(X) \Rightarrow \exists Y(S(Y) \wedge \forall Z(Z \subset X \Leftrightarrow Z \in Y))).$$

Axiom A3 is equivalent to the conjunction of the following two axioms.

A3′. (The *axiom of subset.*) $\forall X \forall Y(S(X) \wedge Y \subset X \Rightarrow S(Y))$.

A3″. (The *power set axiom.*) $\forall X(S(X) \Rightarrow S(\mathcal{P}(X)))$.

A4. (The *axiom of binary union.*) $\forall X \forall Y(S(X) \wedge S(Y) \Rightarrow S(X \cup Y))$.

A5. (The *axiom of general union.*)

$$\forall X \forall Y \forall Z(S(X) \wedge (Z \subset X * Y) \wedge \forall x(x \in X \Rightarrow S(Z\langle x \rangle)) \Rightarrow S(\operatorname{rng} Z)).$$

Axiom A5 is equivalent to the conjunction of the following two axioms.

A5′. (The *axiom of values.*)

$$\forall X \forall Y \forall Z(S(X) \wedge (Z \leftrightharpoons X \to Y) \Rightarrow S(\operatorname{rng} Z)).$$

A5″. (The *axiom of the union.*) $\forall X(S(X) \Rightarrow S(\cup X))$.

A6. (The *axiom of regularity.*) $\forall X(X \neq \varnothing \Rightarrow \exists x(x \in X \wedge x \cap X = \varnothing))$.

A7. (The *infinity axiom.*) $\exists X(S(X) \wedge \varnothing \in X \wedge \forall x(x \in X \Rightarrow x \cup \{x\} \in X))$.

A8. (The *axiom of choice.*)

$$\forall X(S(X) \wedge X \neq \varnothing \Rightarrow \exists z((z \leftrightharpoons \mathcal{P}(X) \backslash \{\varnothing\} \to X) \wedge \forall Y(Y \in \mathcal{P}(X) \backslash \{\varnothing\} \Rightarrow z(Y) \in Y))).$$

Theorem 1. *In the ZF set theory, the following conclusions are equivalent for a set P:*
1) *P is a supertransitive standard model set for the NBG set theory;*
2) $P = \mathcal{P}(U)$ *for some universal set U.*

Proof. (1) ⊢ (2). Consider an arbitrary sequence $s \equiv x_0, \ldots, x_q, \ldots$ of elements of the set P and translations of the axioms and the axiom schemes of the NBG theory on the sequence s with respect to the standard interpretation $M \equiv (P, I)$.

We shall write θ^t and φ^t instead of $\theta_M[s]$ and $M \models \varphi[s]$ for terms θ and formulas φ, respectively.

To simplify the further presentation we first consider translations of some simple formulas. Let u and v be some classes.

Exactly in the same way as in the proof of Proposition 1 (A.6.1) we verify that the equivalences $(u \in_{NBG} v)^t \Leftrightarrow u^t \in_{ZF} v^t$ and $(u \subset_{NBG} v)^t \Leftrightarrow u^t \subset_{ZF} v^t$ hold. This implies that the equivalence $(u =_{NBG} v)^t \Leftrightarrow (u^t \subset_{ZF} v^t) \wedge (v^t \subset_{ZF} u^t)$ holds.

The formula $X \subset Y \wedge Y \subset X$ of the ZF theory, which is equivalent to the formula $\forall a(a \in X \Leftrightarrow a \in Y)$, will be temporary denoted by $X \overset{*}{=} Y$.

Suppose that $u^t \overset{*}{=} v^t$. By the extensionality axiom A1 the formula $u^t =_{ZF} v^t$ is deduced in ZF. By the deduction theorem, the formula $u^t \overset{*}{=} v^t \Rightarrow u^t =_{ZF} v^t$ is deduced in ZF. Conversely, suppose $u^t =_{ZF} v^t$. Take $a \in u^t$. By the scheme of replacement of equals we deduce the formula $a \in v^t$ from the last equality. Then, by the deduction theorem in ZF we obtain $a \in u^t \Rightarrow a \in v^t$, and by the generalization rule, we get $u^t \subset v^t$. Similarly, the formula $v^t \subset U^t$ is deduced. Consequently, we deduce the formula $u^t \overset{*}{=} v^t$. By the deduction theorem, in ZF we deduce $u^t =_{ZF} v^t \Rightarrow u^t \overset{*}{=} v^t$. Thus, the equivalence $u^t \overset{*}{=} v^t \Leftrightarrow u^t =_{ZF} v^t$ holds. Hence, the equivalence $(u =_{NBG} v)^t \Leftrightarrow (u^t =_{ZF} v^t)$ is also holds.

In what follows, we will write not literal transformations of axioms but their equivalent variants obtained using the mentioned equivalences.

The extensionality axiom A1 translates into the formula $A1^t \Leftrightarrow A1^P = \forall y \in P \forall z \in P(y = z \Rightarrow \forall X \in P(y \in X \Leftrightarrow z \in X))$.

The full comprehension axiom scheme AS2 translates into the formula scheme $AS2^t \Leftrightarrow \exists Y \in P \forall y \in P(y \in Y \Leftrightarrow \exists X \in P(y \in X) \wedge \varphi^\tau(y))$, where Y is not a free variable in $\varphi(y)$ and φ^τ denotes the formula $M \vDash \varphi[s^\tau]$, where s^τ denotes the corresponding changes of the sequence s under translation of the quantifier overformulas $\exists Y(\dots)$, $\forall y(\dots)$, and $\exists X(\dots)$ indicated above.

The axiom of subset A3′ translates into the formula $(A3')^t \Leftrightarrow (A3')^P = \forall X \in P \forall Y \in P(\exists E \in P(X \in E) \wedge Y \subset X \Rightarrow \exists F \in P(Y \in F))$.

The power set axiom A3″ translates into the formula $(A3'')^t \Leftrightarrow (A3'')^\tau = \forall X \in P(\exists E \in P(X \in E) \Rightarrow \exists F \in P(\mathcal{P}(X)^\tau \in F))$, where the set $Z \equiv \mathcal{P}(X)^\tau$ is determined by the formula $\exists Z \in P \forall z \in P(z \in Z \Leftrightarrow (\exists G \in P(z \in G) \wedge z \subset X))$.

The axiom of binary union A4 translates into the formula $A4^t \Leftrightarrow A4^\tau = \forall X \in P \forall Y \in P(\exists E \in P(X \in E) \wedge \exists F \in P(Y \in F) \Rightarrow \exists G \in P((X \cup Y)^\tau \in G))$, where the set $Z \equiv (X \cup Y)^\tau$ is determined by the formula $\exists Z \in P \forall z \in P(z \in Z \Leftrightarrow (\exists H \in P(z \in H) \wedge (z \in X \vee z \in Y)))$.

The axiom of general union A5 translates into the formula

$$A5^t \Leftrightarrow A5^\tau = \forall X \in P \forall Y \in P \forall Z \in P(\exists E \in P(X \in E) \wedge (Z \subset (X * Y)^\tau) \wedge \\ \wedge \forall x \in P(x \in X \Rightarrow \exists F \in P(Z\langle x\rangle^\sigma \in F)) \Rightarrow \exists G \in P((\operatorname{rng} Z)^\tau \in G)),$$

where

- the class $Z_1 \equiv (X * Y)^\tau$ is determined by the formula $\exists Z_1 \in P \forall z \in P(z \in Z_1 \Leftrightarrow (\exists H \in P(z \in H) \wedge \exists x \in P \exists y \in P(x \in X \wedge y \in Y \wedge z = \langle x, y\rangle^*)))$;

- the class $Z_2 \equiv Z_2(x) \equiv Z\langle x\rangle^\sigma$ is determined by the formula $\exists Z_2 \in P \forall y \in P(y \in Z_2 \Leftrightarrow (\exists K \in P(y \in K) \wedge y \in Y \wedge \langle x, y\rangle^* \in Z))$;
- the class $Z_3 \equiv (\operatorname{rng} Z)^\tau$ is determined by the formula $\exists Z_3 \in P \forall y \in P(y \in Z_3 \Leftrightarrow (\exists L \in P(y \in L) \wedge y \in Y \wedge \exists x \in P(x \in X \wedge \langle x, y\rangle^* \in Z))$;
- the class $Z_4 \equiv \langle x, y\rangle^*$ is determined by the formula $\exists Z_4 \in P \forall z \in P(z \in Z_4 \Leftrightarrow \exists M \in P(z \in M) \wedge (z = \{x\}^* \vee z = \{x, y\}^*))$;
- the class $Z_5 \equiv \{x, y\}^*$ is determined by the formula $\exists Z_5 \in P \forall z \in P(z \in Z_5 \Leftrightarrow \exists N \in P(z \in N) \wedge (z = x \vee z = y))$;
- the class $Z_6 \equiv \{x\}^*$ is determined by the formula $\exists Z_6 \in P \forall z \in P(z \in Z_6 \Leftrightarrow \exists Q \in P(z \in Q) \wedge z = x)$.

The regularity axiom A6 translates into the formula $A6^t \Leftrightarrow A6^\tau \equiv \forall X \in P(X \neq \varnothing^t \Rightarrow \exists x \in P(x \in X \wedge (x \cap X)^\tau = \varnothing^t))$, where
- the class $Z_1 \equiv \varnothing^t$ is determined by the formula $\exists Z_1 \in P \forall z \in P(z \in Z_1 \Leftrightarrow (\exists E \in P(z \in E) \wedge z \neq z))$;
- the class $Z_2 \equiv (x \cap X)^\tau$ is determined by the formula $\exists Z_2 \in P \forall z \in P(z \in Z_2 \Leftrightarrow (\exists F \in P(z \in F) \wedge z \in x \wedge z \in X))$.

The infinity axiom A7 translates into the formula $A7^t \Leftrightarrow A7^\tau \equiv \exists X \in P(\exists E \in P(X \in E) \wedge \varnothing^t \in X \wedge \forall x \in P(x \in X \Rightarrow (x \cup \{x\})^\tau \in X))$, where
- the class $Z_1 \equiv \varnothing^t$ is determined by the formula presented above;
- the class $Z_2 \equiv Z_2(x) \equiv (x \cup \{x\})^\tau$ is determined by the formula $\exists Z_2 \in P \forall z \in P(z \in Z_2 \Leftrightarrow (\exists F \in P(z \in F) \wedge (z \in x \vee z \in \{x\}^\sigma)))$;
- the class $Z_3 \equiv Z_3(x) \equiv \{x\}^\sigma$ is determined by the formula $\exists Z_3 \in P \forall z \in P(z \in Z_3 \Leftrightarrow (\exists G \in P(z \in G) \wedge z = x))$.

Since M is a model of the NBG theory, all the translations written above are deducible formulas in the ZF theory.

Using the obtained translations, we prove that $P = \mathcal{P}(U)$ for some set U.

Consider the formula $\varphi(x) \equiv (x = x)$ in NBG theory. Then, AS2 defines in NBG the implicit axiom of the form $\exists Y \forall y(y \in Y \Leftrightarrow \exists X(y \in X) \wedge y = y)$. According to the translation obtained above, this implicit axiom translates into the formula equivalent to the formula $\Phi \equiv \exists Y \in P \forall y \in P(y \in Y \Leftrightarrow \exists X \in P(y \in X) \wedge y = y)$. Since this formula is deducible in ZF, it defines in ZF some element $U \in P$.

Consider an arbitrary element $X \in P$. Let $y \in X$. Since P is transitive, we get $y \in P$, where for y the formula $\exists X \in P(y \in X \wedge y = y)$ is deduced. By the formula Φ, we have $y \in U$. Therefore, $X \subset U$, i. e. $X \in \mathcal{P}(U)$. Thus, we have derived the embedding $P \subset \mathcal{P}(U)$.

Conversely, if $X \in \mathcal{P}(U)$, then quasitransitivity of P implies $X \in P$. Hence, $P = \mathcal{P}(U)$.

Prove that the set U is universal.

Let $y \in x \in U \in P$. By the transitivity of P we get $x \in P = \mathcal{P}(U)$. Therefore, $y \in x \subset U$ implies $y \in U$. Thus, the set U is transitive.

Let $y \subset x \in U \in P$. Then, $x \in P = \mathcal{P}(U)$ and $y \subset x \subset U$ imply $y \in P$. By virtue of $A3'^{P}$ we conclude that $y \in F$ for some $F \in P$. Hence, $y \in F \subset U$ implies $y \in U$. Therefore, the set U quasitransitive.

Check that in $A3''^{\tau}$ for $X \in E \in P$ the equality $\mathcal{P}(X)^{\tau} = \mathcal{P}(X)$ holds. Let $z \in \mathcal{P}(X)$. Then, $z \subset X \in P$ and the quasitransitivity of P imply $z \in P$. Further, $z \subset X \in E \in P = \mathcal{P}(U)$ implies $z \subset X \in E \subset U$. In view of quasitransitivity of U, it follows from $z \subset X \in U$ that $z \in U \in P$. Then, the formula for $Z \equiv \mathcal{P}(X)^{\tau}$ presented above implies $z \in \mathcal{P}(X)^{\tau}$. Therefore, $\mathcal{P}(X) \subset \mathcal{P}(X)^{\tau}$. The mentioned formula provides also the inverse embedding.

Let $X \in U \in P$. Then, $X \in P$ by virtue of $A3''^{\tau}$ implies $\mathcal{P}(X) = \mathcal{P}(X)^{\tau} \in F$ for some $F \in P$. Hence, $\mathcal{P}(X) \in F \subset U$ implies $\mathcal{P}(X) \in U$.

Check that in $A4^{\tau}$ for $X \in E \in P$ and $Y\imath F \in P$ the equality $(X \cup Y)^{\tau} = X \cup Y$ holds. Let $z \in X \cup Y$. Then, $z \in X$ or $z \in Y$. By virtue of transitivity of P we get $X \in P$ and $Y \in P$, where $z \in P$. Besides, $z \in X \in E \subset U$ or $z \in Y \in F \subset U$ implies $z \in U \in P$ in view of the transitivity of U. Then, the formula for $Z \equiv (X \cup Y)^{\tau}$ presented above implies $z \in (X \cup Y)^{\tau}$. Therefore, $X \cup Y \subset (X \cup Y)^{\tau}$. The mentioned formula provides also the inverse embedding.

Let $X, Y \in U$. Then, $X, Y \in P$ implies $X \cup Y = (X \cup Y)^{\tau} \in G$ for some $G \in P$ by virtue of $A4^{\tau}$. Hence, $X \cup Y \in G \subset U$ implies $X \cup Y \in U$.

Let $X \in U \in P$. By the above, we have $\mathcal{P}(X) \in U$. Then, $\{X\} \subset \mathcal{P}(X) \in U$ and the quasitransitivity of the set U imply $\{X\} \in U$.

Let $X, Y \in U$. By the above, we have $\{X, Y\} = \{X\} \cup \{Y\} \in U$, where $\langle X, Y\rangle \in U$.

If $X, Y \in U$, then it follows from $X * Y \subset \mathcal{P}(\mathcal{P}(X \cup Y)) \in U$ and the quasitransitivity of the set U that $X * Y \in U$.

To prove other universality properties, we need some simplification of the formula $A5^{\tau}$ obtained by translation of axiom A5.

Let $z \in \{x\}$. Then, $z = x \in X \in P$ implies $z \in P$, and therefore, $z \in Z_6$. Conversely, if $z \in Z_6 \in P$, then $z \in P$ and the formula for Z_6 imply $z = x \in \{x\}$. Therefore, $Z_6 = \{x\}$.

Let $z \in \{x, y\}$. Then, it follows from $z = x \in X \in P$ or $z = y \in Y \in P$ that $z \in Z_5$. Conversely, if $z \in Z_5$, then $z = x$ or $z = y$ implies $z \in \{x, y\}$. Therefore, $Z_5 = \{x, y\}$.

These equalities eliminate the asterisk in the formula for Z_4. Let $z \in \langle x, y\rangle$. Then, $z = \{x\}$ or $z = \{x, y\}$. Since $x \in X \in P = \mathcal{P}(U)$, we get $x \in U$. Similarly, $y \in U$. By the above, this implies $\{x\} \in U$ or $\{x, y\} \in U$. Hence, $z \in U \in P$ implies $z \in Z_4$. Conversely, if $z \in Z_4 \in P$, then it follows from $z \in P$ and the formula for Z_4 that $z = \{x\}$ or $z = \{x, y\}$, i. e. $z \in \langle x, y\rangle$. Consequently, $Z_4 = \langle x, y\rangle$.

This equality eliminates the asterisk in the formulas for Z_3, Z_2, and Z_1.

Using this conclusion, verify that $Z_1 = X * Y$. Let $z \in Z_1 \in P$. Since P is transitive, we get $z \in P$. Therefore, the formula for Z_1 guarantees that $z = \langle x, y\rangle$ for some $x \in X$ and $y \in Y$. Hence, $z \in X * Y$. Conversely, suppose $z \in X * Y$. Then, $z = \langle x, y\rangle$ for some $x \in X \in P$ and $y \in Y \in P$. Since P is transitive, we get $x, y \in P$. By the above, $x \in X \subset U$

and $y \in Y \subset U$ imply $z = \langle x, y\rangle \in U \in P$ and $z \in P$. Hence, the formula for Z_1 implies $z \in Z_1$. This yields the necessary equality.

Consequently, $Z \subset X * Y$.

Using this conclusion, verify that $Z_3 = \operatorname{rng} Z$. Let $y \in Z_3 \in P$. Since P is transitive, we get $y \in P$. Therefore, the formula for Z_3 implies $y \in \operatorname{rng} Z$. Conversely, suppose $y \in \operatorname{rng} Z \subset Y \in P$. The there is $x \in X \in P$ such that $\langle x, y\rangle \in Z$. Since P is transitive, we get $x, y \in P$. Then, by the formula for Z_3 we get $y \in Z_3$. This yields the necessary equality.

Finally, check that $Z_2 = Z\langle x\rangle$.

Let $y \in Z_2 \in P$. Since P is transitive, we get $y \in P$. Therefore, the formula for Z_2 implies $y \in Y$ and $\langle x, y\rangle \in Z$, where $y \in Z\langle x\rangle$. Conversely, suppose $y \in Z\langle x\rangle \subset Y \in P$. Then, $\langle x, y\rangle \in Z$. Since P is transitive, we get $y \in P$. Then, by the formula for Z_2 we get $y \in Z_2$. This yields the necessary equality.

We can conclude now that the indices τ and σ disappears in the formula $A5^\tau$.

Using this conclusion, prove that $X \in U$ implies $\cup X \in U$. Consider in ZF the sets $Y \equiv \cup X$ and $Z \equiv \{z \in X * Y \mid \exists x \in X \exists y \in y(z = \langle x, y\rangle \wedge y \in x)\}$. If $y \in x \in X \in U$, then we get $y \in U$ in view of the transitivity of U. Therefore, $Y \subset U$ implies $Y \in P$. Let $z \in Z$, i. e. $z = \langle x, y\rangle$ for some $x \in X$ and $y \in Y$ such that $y \in x$. Then, $y \in x \in U$ implies $y \in U$. By the above, we have $z = \langle x, y\rangle \in U$. Consequently, $Z \subset U$, i. e. $Z \in P$.

Check that for every $x \in P$ such that $x \in X$, we have $Z\langle x\rangle = x$. If $y \in Z\langle x\rangle$, then $\langle x, y\rangle \in Z$ implies $\langle x, y\rangle = \langle x', y'\rangle$ for some $x' \in X$ and $y' \in Y$ such that $y' \in x'$. Hence, $y = y' \in x' = x$. Conversely, if $y \in x \in X$, then $y \in Y$ and $\langle x, y\rangle \in Z$ provide $y \in Z\langle x\rangle$.

This implies that $Z\langle x\rangle = x \in U \in P$ for every $x \in P$ such that $x \in X \in U \in P$. Since the formula $A5^\tau$ is deducible in ZF, this formula guarantees that $Y = \operatorname{rng} Z \in G$ for some $G \in P$. Hence, $Y \in G \subset U$.

Check that $X \in U$ and $f \in U^X$ imply $\operatorname{rng} f \in U$. If $x \in X \in U$ and $y \in U$, then $x \in U$, by above, implies $\langle x, y\rangle \in U$. Consequently, $f \subset X * U \subset U$ provides $f \in P$. Moreover, by above again, $f(x) \in U$ provides $f\langle x\rangle = \{f(x)\} \in U \in P$ for every $x \in X$. Applying the formula $A5^\tau$, we infer that $\operatorname{rng} f \in G$ for some $G \in P$. Hence, $\operatorname{rng} f \in U$.

Simplify now the formula $A7^\tau$. Check that $Z_1 = \varnothing_{ZF}$. Let $z \in P$. Suppose that $z \in Z_1$. Then, by the formula for Z_1 we obtain $z \neq z$. But, according to the equality axiom, $z = z$. This contradiction implies $z \notin Z_1$. Suppose now $z \notin P$. Since $Z_1 \subset P$, we get $z_1 \notin Z_1$. Thus, for every z, we have $z \notin Z_1$. According to the empty set axiom A7 of the ZF theory, we conclude that $Z_1 = \varnothing_{ZF}$.

When we simplified the formula $A5^\tau$, we established that the formula for $Z_3 \equiv \{x\}^\sigma$ implies $Z_3 = \{x\}$.

Let $x \in X \in P$. By the above, we have that $x \in U$ implies $\{x\} \in U \in P$. By the transitivity of P we get $\{x\} \in P$. When we simplified the formula $A4^\tau$, we established that these properties provides the equality $Z_2 = x \cup \{x\}$.

Thus, the formula $A7^\tau$ take the form $\exists X \in P(\exists E \in P(X \in E) \wedge \varnothing_{ZF} \in X \wedge \forall x \in P(x \in X \Rightarrow x \cup \{x\} \in X))$. Let $x \in X$, where $X \in E \in P$. Since P is transitive, we get $x \in P$. Then, the formula $x \cup \{x\} \in X$ is deduced from $A7^\tau$. By the deduction theorem, we deduce $(x \in X \Rightarrow x \cup \{x\} \in X)$ and the generalization rule we deduce $\forall x \in X(x \cup \{x\} \in X)$.

Thus, it is deduced from $A7^\tau$ that $\exists X \in P(\exists E \in P(X \in E) \wedge \varnothing_{ZF} \in X \wedge \forall x \in X(x \cup \{x\} \in X))$, almost coinciding with the infinity axiom A8 in ZF and asserting the existence of an inductive set $X \in E \in P$. Since ω is the smallest among all inductive sets, we get $\omega \subset X \in U$. By the quasitransitivity property, this implies $\omega \in U$.

Thus, we have proven that (1) $\vdash$ (2).

(2) $\vdash$ (1). Let $P = \mathcal{P}(U)$ for some universal set U. Consider the standard interpretation $M \equiv (P, I)$ of the NBG theory. We have translated above some axioms and axiom schemes of NBG under the interpretation M on the sequence s. Prove that they are deducible in ZF.

Check that P is supertransitive. Let $x \in y \in P$. Then, $x \in y \subset U$ implies $x \in U$. Since U is transitive, we get $x \subset U$, and therefore, $x \in P$. Hence, P is transitive. Let $x \subset y \in P$. Then, $x \subset y \subset U$ implies $x \in P$. Hence, P is quasitransitive.

Let $y, z \in P$, $y = z$, and $X \in P$. Consider the formula $\varphi(y) \equiv (y \in X)$. By the scheme of replacement of equals in ZF, we deduce the formula $\varphi(z) = (z \in X)$ for the formula $y = z$. By the deduction theorem, we deduce $y \in X \Rightarrow z \in X$. Similarly, the formula $z \in X \Rightarrow y \in X$ is deduced. Thus, we deduce the formula $y \in X \Leftrightarrow z \in X$, and therefore, the formula $X \in P \Rightarrow (y \in X \Leftrightarrow z \in X)$. By the generalization rule, the formula $\psi \equiv \forall X \in P(y \in X \Leftrightarrow z \in X)$ is deduced. Further, by the deduction theorem we get $y = z \Rightarrow \psi$ and by logical tools we deduce the formula $A1^t$.

According to AS3 in ZF, for the formula $\varphi^\tau(y)$ and the set U there is a set Y such that $\forall y(y \in Y \Leftrightarrow y \in U \wedge \varphi^\tau(y))$. Let $y \in Y$. Then, $y \in U \wedge \varphi^\tau(y)$. Since $U \in P$, we obtain $\exists X \in P(y \in X) \wedge \varphi^\tau(y)$. By the deduction theorem, the formula $y \in Y \Rightarrow \exists X \in P(y \in X) \wedge \varphi^\tau(y)$ is deduced. Conversely, let $\exists X \in P(y \in X) \wedge \varphi^\tau(y)$. Then, $y \in X \subset U$ implies $y \in U$. Hence, $y \in U \wedge \varphi^\tau(y)$ implies $y \in Y$. By the deduction theorem, the formula $\exists X \in P(y \in X) \wedge \varphi^\tau(y) \Rightarrow y \in Y$ is deduced. Thus, we deduce the formula $y \in Y \Leftrightarrow \exists X \in P(y \in X) \wedge \varphi^\tau(y)$. It deduced from it that $\forall y \in P(y \in Y \Leftrightarrow \exists X \in P(y \in X) \wedge \varphi^\tau(y))$. Since $Y \subset U \in P$ and P is quasitransitive, we obtain $Y \in P$. Consequently, $AS2^t$ is deduced in ZF.

Let $X, Y \in P, X \in E \in P$, and $Y \subset X$. Then, $Y \subset X \in E \subset U$ and the quasitransitivity of the universal set U imply $Y \in U \in P$. This means that $A3'^t$ is deducible in ZF.

We have established above that for $X \in E \in P$ the equality $\mathcal{P}(X)^\tau = \mathcal{P}(X)$ holds. By axiom A5 in ZF, $\mathcal{P}(X)$ exists. Since U is universal, we see that $X \in E \subset U$ implies $\mathcal{P}(X) \in U \in P$. This means that $A3''^t$ is deducible in ZF.

Let $X, Y \in P$, $X \in E \in P$, and $Y \in F \in P$. We have established above that in these conditions, the equality $(X \cup Y)^\tau = X \cup Y$ holds. It follows from $X \in U$ and $Y \in U$ that $X \cup Y \in U \in P$ by virtue of the universality of U. This means that $A4^t$ is deducible in ZF.

Let $X, Y, Z \in P$ and $X \in E \in P$. We have derived above that in these conditions the equality $(X * Y)^\tau = X * Y$ holds, and if $Z \subset X * Y$, then the equalities $Z\langle x\rangle^\sigma = Z\langle x\rangle$ and $(\operatorname{rng} Z)^\tau = \operatorname{rng} Z$ hold. If $x \in X$ and $Z\langle x\rangle \in F \in P$, then $X \in U$, $x \in U$, and $Z\langle x\rangle \in U$. Since U is universal, we get $U * U \subset U$. Consider the set $f \equiv \{s \in U * U \mid \forall x \in X(s = \langle x, Z\langle x\rangle\rangle) \wedge \forall x(x \notin X \Rightarrow s = \langle x, \varnothing\rangle)\}$. It is clear that f is a function from X in U such that $f(x) = Z\langle x\rangle$. Since U is universal, we infer that $S \equiv \operatorname{rng} f \in U$, and therefore, $T \equiv \cup S \in U$. If $t \in T$, then $t \in s \in S$ implies $t \in Z\langle x\rangle$ for some $x \in X$. Hence, $t \in \operatorname{rng} Z$. Conversely, if $t \in \operatorname{rng} Z$, then $\langle x, t\rangle \in Z$ for some $x \in \operatorname{dom} Z \subset X$. Therefore, $t \in Z\langle x\rangle = f(x) \in S$, and therefore, $t \in T$. Thus, $\operatorname{rng} Z = T \in U \in P$. This means that $\mathrm{A5}^t$ is deducible in ZF.

We have derived above that $\varnothing^t = \varnothing_{ZF}$. Let $X \in P$ and $X \neq \varnothing_{ZF}$. Verify that $Z_2 \equiv (x \cap X)^\tau$ for $x \in X$ in the formula $\mathrm{A6}^\tau$ coincides with $x \cap X$. Suppose that $z \in x \cap X$; then, $z \in X \in P$ implies $z \in P$ by virtue of the transitivity of P. By the formula for Z_2 we obtain $z \in Z_2$. Conversely, suppose $z \in Z_2 \in P$; then, $z \in P$ implies $z \in x \cap X$ in view of the formula for Z_2. Hence, $Z_2 = x \cap X$.

By regularity axiom A9 in ZF there is $x \in X$ such that $x \cap X = \varnothing_{ZF}$. It follows from $x \in X \in P$ that $x \in P$. This means that $\mathrm{A6}^t$ is deducible in ZF.

We have established above that in the formula $A7^\tau$, we have $\varnothing^t = \varnothing_{ZF}$ and if $x \in X \in E \in P$, then $(x \cup \{x\})^\tau = x \cup \{x\}$. Since U is universal, we get $\omega \in U$, where $\omega \in P$. Since ω is an inductive set, we have $\varnothing_{ZF} \in X$ and $x \in X \Rightarrow x \cup \{x\} \in X$. Further, by logical tools, we deduce the formula $\mathrm{A7}^t$.

The axiom of choice A8 in NBG translates into the formula

$$A8^t \Leftrightarrow A8^\tau \equiv \forall X \in P(\exists E \in P(X \in E) \wedge X \neq \varnothing^t \Rightarrow \exists z \in P((z \leftrightharpoons \mathcal{P}(X)\backslash\{\varnothing\} \to X)^\tau \wedge$$
$$\wedge \forall Y \in P(Y \in (\mathcal{P}(X)\backslash\{\varnothing\})^\sigma \Rightarrow \forall x \in P(x \in X \wedge \langle Y, x\rangle^\sigma \in z \Rightarrow x \in Y),$$

where

- the set $Z_1 \equiv Z_1(X) \equiv (\mathcal{P}(X)\backslash\{\varnothing\})^\sigma$ is determined by the formula $\exists Z_1 \in P \forall u \in P(u \in Z_1 \Leftrightarrow u \in \mathcal{P}(X)^\rho \wedge u \notin \{\varnothing\}^\rho)$;
- the set $Z_2 \equiv \langle Y, x\rangle^\sigma$ is determined by the formula $\exists Z_2 \in P \forall u \in P(u \in Z_2 \Leftrightarrow \exists F \in P(u \in F) \wedge (u = \{Y\}^\sigma \vee u = \{Y, x\}^\sigma))$;
- the set $Z_3 \equiv \{Y, x\}^\sigma$ is determined by the formula $\exists Z_3 \in P \forall u \in P(u \in Z_3 \Leftrightarrow \exists G \in P(u \in G) \wedge (u = Y \vee u = x))$;
- the set $Z_4 \equiv \{Y\}^\sigma$ is determined by the formula $\exists Z_4 \in P \forall u \in P(u \in Z_4 \Leftrightarrow \exists H \in P(u \in H) \wedge u = Y)$;
- the set $Z_5 \equiv \{\varnothing\}^\rho$ is determined by the formula $\exists Z_5 \in P \forall z \in P(z \in Z_5 \Leftrightarrow (\exists K \in P(z \in K) \wedge z = \varnothing^t))$;

and $\varphi^\tau \equiv (z \leftrightharpoons \mathcal{P}(X)\backslash\{\varnothing\} \to X)^\tau$ denotes the formula $M \models \varphi[s^\tau]$, where s^τ denotes the corresponding changes of the sequence s under translation of the quantifier overformulas $\forall X(\dots)$ and $\exists z(\dots)$ indicated above.

We have established above the equality $\varnothing^t = \varnothing_{ZF}$. Since $\varnothing_{ZF} \in \omega \in U \in P$, as was shown above, these conditions implies $Z_5 = \{\varnothing_{ZF}\}$.

Fix the conditions $X \in P$, $X \in E \in P$, and $X \neq \varnothing^t = \varnothing_{ZF}$. As was shown before, this implies $\mathcal{P}(X)^\rho = \mathcal{P}(X)$.

Check that $Z_1 = \mathcal{P}(X)\backslash\{\varnothing_{ZF}\} \equiv Z$. Let $u \in Z_1 \in P$. Since $X \in E \subset U$ and U is universal, we get $\mathcal{P}(X) \in U$. The quasitransitivity of U implies now $Z \in U$. Since $u \in P$, the formula for Z_1 provides that $u \in Z$. Hence, $Z_1 \subset Z$. Conversely, let $u \in Z \in U \in P$. Since

P is transitive, we get $u \in P$. The formula for Z_1 implies now $u \in Z$. Hence, $Z \subset Z_1$, which yields the necessary equality. This leads to the replacement of Z_1 by Z in the formula A8$^\tau$.

Consider the formula $\varphi \equiv (z \leftrightharpoons Z \to X)$. It is the conjunction of three formulas: $\varphi_1 \equiv (z \subset Z * X)$, $\varphi_2 \equiv (\text{dom}\, z = Z)$, and $\varphi_3 \equiv (\forall x(x \in Z \Rightarrow \forall y(y \in X \Rightarrow \forall y'(y' \in X \Rightarrow ((\langle x,y\rangle \in z \wedge \langle x,y'\rangle \in z \Rightarrow y = y')))))$. Therefore, $\varphi^\tau = \varphi_1^\tau \wedge \varphi_2^\tau \wedge \varphi_3^\tau$. Since $\varphi_1 = (\forall u(u \in z \Rightarrow \exists x \exists y(x \in Z \wedge y \in X \wedge u = \langle x,y\rangle)))$, we obtain $\varphi_1^\tau \Leftrightarrow (\forall u \in P(u \in z \Rightarrow \exists x \in P \exists y \in P(x \in Z \wedge y \in X \wedge u = \langle x,y\rangle^\sigma)))$. Similarly, $\varphi_2 = (\forall x(x \in Z \Rightarrow \exists y(y \in X \wedge \langle x,y\rangle \in z)))$ implies $\varphi_2^\tau \Leftrightarrow (\forall x \in P(x \in Z \Rightarrow \exists y \in P(y \in X \wedge \langle x,y\rangle^\sigma \in z)))$.

Finally, $\varphi_3^\tau \Leftrightarrow (\forall x \in P(x \in Z \Rightarrow \forall y \in P(y \in X \Rightarrow \forall y' \in P(y' \in X \Rightarrow ((\langle x,y\rangle^\sigma \in z \wedge \langle x,y'\rangle^\sigma \in z \Rightarrow y = y')))))$. This guarantees that φ_1^τ, φ_2^τ, and φ_3^τ differ from the formulas φ_1, φ_2, and φ_3, respectively, only by bounded quantifier prefixes $\forall \cdots \in P$ and $\exists \cdots \in P$.

By the axiom of choice in ZF for X there is z such that $\chi \equiv (z \leftrightharpoons Z \to X) \wedge \forall Y(Y \in Z \Rightarrow \forall x(x \in X \wedge \langle Y,x\rangle \in z \Rightarrow x \in Y))$. Hence, the formula $\varphi = \varphi_1 \wedge \varphi_2 \wedge \varphi_3$ is deduced, and therefore, the formulas φ_1, φ_2, and φ_3 are deduced.

Let $u \in P$ and $u \in z$. Then, we derive from the formula φ_1 that there are $x \in Z$ and $y \in X$ such that $u = \langle x,y\rangle$. Since $x \in Z \in U \in P$ and $y \in X \in P$, by the transitivity property we get $x, y \in P$. This means that under the conditions $u \in P$ and $u \in z$ the formula $\exists x \in P \exists y \in P(x \in Z \wedge y \in X \wedge u = \langle x,y\rangle)$ is deduced. Applying the deduction theorem and the deduction rules twice, we deduce the formula φ_1^τ.

Let $x \in P$ and $x \in Z$. Then, we derive from the formula φ_2 that for x there is $y \in X$ such that $\langle x,y\rangle \in z$. It follows from $y \in X \in P$ that $y \in P$. This means that under the conditions $x \in P$ and $x \in Z$ the formula $\exists y \in P(y \in X \wedge \langle x,y\rangle \in z)$ is deduced. Therefore, as above, we deduce the formula φ_2^τ.

Let $x \in P$, $x \in Z$, $y \in P$, $y \in X$, $y' \in P$, $y' \in X$, $\langle x,y\rangle \in z$, and $\langle x,y'\rangle \in z$. Then we deduce from the formula φ_3 the formula $y = y'$. Applying alternately the deduction theorem and the deduction rules several times, we deduce the formula φ_3^τ.

Thus, the formula φ^τ is deduced.

Check that $Z_4 = \{Y\}$ under the conditions $X \in E \in P$, $Y \in P$, and $Y \in Z$. Let $u \in \{Y\}$, i. e. $u = Y \in P$. Since $u = Y \in Z \in U \in P$, the transitivity implies $u \in U \in P$. Then, the formula for Z_4 implies $u \in Z_4$. Conversely, if $u \in Z_4 \in P$, then $u \in P$ and the formula for Z_4 imply $u = Y \in \{Y\}$. This yields the necessary equality.

Check that $Z_3 = \{Y,x\}$ under the conditions $X \in E \in P$, $x \in X$, $Y \in P$, and $Y \in Z$. Let $u \in \{Y,x\}$. Then, $u = Y \in Z \in U \in P$ or $u = x \in X \in E \in P$ implies $u \in P$, and therefore, $u \in Z_3$. Conversely, if $u \in Z_3 \in P$, then $u \in P$ and the formula for Z_3 imply $u = Y \vee u = x$, i. e. $u \in \{Y,x\}$. This yields the necessary equality.

Finally check that $Z_2 = \langle Y,x\rangle$ under the previous condition. Let $u \in \langle Y,x\rangle$, i. e. $u = \{Y\}$ or $u = \{Y,x\}$. The previous equalities eliminate the index σ in the formula for Z_2. Since $Y \in Z \in U$, we get $Y \in U$. Moreover, $x \in X \in E \in P$ implies $x \in X \in P$, i. e. $x \in X \subset U$. It follows now from the universality of U that $u = \{Y\} \in U$ or $u = \{Y,x\} \in U$. Hence, $u \in U \in P$ and $u \in P$ provides $u \in Z_2$. Conversely, if $u \in Z_2 \in P$, then $u \in P$ and the formula for Z_2 imply $u = \{Y\}$ or $u = \{Y,x\}$, i. e. $u \in \langle Y,x\rangle$. This yields the necessary equality.

Since $Z \in U \in P$ and $X \in E \in P$, i. e. $X \in E \subset U$, we obtain $z \subset Z * X \in U$ by virtue of the universality of U. By Lemma 1 (A.4.1), we get $z \in U \in P$, and therefore, $z \in P$.

Thus, we see that it is deduced under the fixed conditions from the axiom of choice in ZF that there exists the object $z \in P$ satisfying the formula χ implying the formula $\xi \equiv (\varphi^\tau \wedge \forall Y \in P(Y \in Z \Rightarrow \forall x \in P(x \in X \wedge \langle Y, x \rangle \in z \Rightarrow x \in Y)))$. As a result, we conclude that for the fixed conditions the formula $\exists z \in P\xi$ is deduced. Applying alternately the deduction theorem and the generalization rule several times, we deduce the formula $A8^t$.

Thus, M is a model of the NBG set theory. □

Now, we can prove the *Zermelo – Shepherdson theorem* [*Zermelo*, 1930; *Shepherdson*, 1951; 1952; 1953].

Theorem 2. *In the ZF set theory, the following conclusions are equivalent for a set P:*
1) *P is a supertransitive standard model set for the NBG set theory;*
2) *$P = V_{\varkappa+1} = \mathcal{P}(V_\varkappa)$ for some inaccessible cardinal number $\varkappa$.*

Proof. (1) $\vdash$ (2). By Theorem 1 $P = \mathcal{P}(U)$ for some universal set U. By Theorem 1 (A.4.2) $U = V_\varkappa$ for some inaccessible cardinal number $\varkappa$. By Corollary 2 to Lemma 3 (A.3.2) $P = \mathcal{P}(V_\varkappa) = V_{\varkappa+1}$.

(2) $\vdash$ (1). By Corollary 2 to Lemma 3 (A.3.2) $V_{\varkappa+1} = \mathcal{P}(V_\varkappa)$. By Theorem 2 (A.4.2) the set $V_\varkappa$ is universal. Now, the assertion follows from Theorem 1. □

A.7 Characterization of all natural models of the NBG set theory

A.7.1 Tarski sets and their properties

A set U in ZF is called a *Tarski set* if it has the following properties (see [*Tarski*, 1938] and [*Kuratowski and Mostowski*, 1967, IX, § 5]):
1) $x \in U \Rightarrow x \subset U$ (the *transitivity property*);
2) $x \in U \Rightarrow \mathcal{P}(x) \in U$ (the *exponentiality property*);
3) $((x \subset U) \wedge \forall f(f \in U^x \Rightarrow \operatorname{rng} f \neq U)) \Rightarrow x \in U$ (the *Tarski property*).

A. Tarski added to the ZF theory the following axiom.

AT. (The *Tarski axiom.*) *Every set is an element of a certain Tarski set.*

In [*Tarski*, 1938], it was proven that AT is equivalent to inaccessibility axiom AI (see also [*Kuratowski and Mostowski*, 1967, IX, § 1, Theorem 6 and § 5, Theorem 1]).

Lemma 1. *For any sets U and X, the following conclusions are equivalent:*
3) $(x \subset U) \wedge \forall f(f \in U^x \Rightarrow \operatorname{rng} f \neq U) \Rightarrow x \in U;$
3′) $(x \subset U) \wedge (|x| < |U|) \Rightarrow x \in U.$

Proof. (3′) ⊢ (3). Let $x \subset U$ and $\forall f(f \in U^x \Rightarrow \operatorname{rng} f \neq U)$. Clearly, $|x| \leqslant |U|$. Suppose that $|x| = |U|$. Then, there is a bijection $f : x \rightarrowtail\!\!\!\!\twoheadrightarrow U$. This contradicts the condition. Hence, $|x| < |U|$. By property 3′ we get $x \in U$.

(3) ⊢ (3′). Let $f \in U^x$. Then, $|\operatorname{rng} f| \leqslant |x| < |U|$ implies $\operatorname{rng} f \neq U$. By property 3 $x \in U$. □

Let us derive other properties of Tarski sets from these properties.

Lemma 2. *If U is a Tarski set and* $x \in U$, *then* $|x| \in |U|$.

Proof. By properties 1 and 2, $x \in U$ implies $\mathcal{P}(x) \in U$ and $\mathcal{P}(x) \subset U$. By the Cantor theorem (Theorem 2 (1.3.2)), $|x| < |\mathcal{P}(x)| \leqslant |U|$. □

Lemma 3. *If U is a Tarski set, then* $x \in U \land y \subset x \Rightarrow y \in U$.

Proof. If $x \in U$, then by property 2, we get $\mathcal{P}(x) \in U$ and by property 1, we get $\mathcal{P}(x) \subset U$. It follows from $y \in \mathcal{P}(x)$ that $y \in U$. □

Lemma 4. *If U is a Tarski set, then* $x \in U \land (f \in U^x) \Rightarrow \operatorname{rng} f \in U$.

Proof. If $x \in U$, then by Lemma 2, we get $|x| < |U|$. Since $f \in U^x$, we have $\operatorname{rng} f \subset U$ and $|\operatorname{rng} f| \leqslant |x| < |U|$, where, by property 3′, we obtain $\operatorname{rng} f \in U$. □

Lemma 5. *If U is a Tarski set, then* $|U| \subset U$.

Proof. Consider the class $\mathbf{C} \equiv \{x \mid x \in \mathbf{On} \land x \notin U\}$. This class is non-empty, since otherwise the class **On** is a set. Therefore, it has a minimal element $\varkappa$. Since $\forall \alpha \in \varkappa(\alpha \in U)$, we get $\varkappa \subset U$. Consequently, $|\varkappa| \leqslant |U|$. Suppose that $|\varkappa| < |U|$. Then, by Lemma 1 $\varkappa \in U$, which is false. Hence, $|U| = |\varkappa| \leqslant \varkappa$, i. e. $|U| \subset \varkappa \subset U$. □

Lemma 6. *If U is a Tarski set, then* $|U| \notin U$.

Proof. Suppose that $\varkappa \equiv |U| \in U$. Then, by property 2, we infer $\mathcal{P}(\varkappa) \in U$. Lemma 2 implies $\alpha \equiv |\mathcal{P}(\varkappa)| \in |U|$. On account of Lemma 5, we conclude that $\alpha \in U$ and $\alpha \subset U$ by virtue of property 1. By the Cantor theorem (Theorem 2 (1.3.2)) $\alpha > |U|$. But since $\alpha \subset U$, we get $\alpha \leqslant |U|$. This contradiction proves that $|U| \notin U$. □

Lemma 7. *If U is a Tarski set, then* $|\mathcal{P}(\alpha)| \in |U|$ *holds for every ordinal number* $\alpha \in |U|$.

Proof. Since $\alpha \in |U|$ and $|U| \subset U$ by virtue of Lemma 5, we get $\alpha \in U$. Then, property 2 implies $\mathcal{P}(\alpha) \in U$. By Lemma 2 we obtain $|\mathcal{P}(\alpha)| \in |U|$. □

Lemma 8. *If U is a non-empty Tarski set, then* $\varnothing \in U$ *and* $|U| \geqslant 5$.

Proof. Since $\forall x(\varnothing \subset x)$, we get $x_0 \equiv \varnothing \subset U$. Since $|\varnothing| = 0 < |U|$, Lemma 1 implies $x_0 \in U$. By property 2, $x_1 \equiv \{x_0\} = \mathcal{P}(\varnothing) \in U$. It follows from $x_0 \neq x_1$ that $|U| \geqslant 2$. These properties provide $x_2 \equiv \{x_1\} \subset U$ and $|x_2| = 1 < |U|$. Hence, property 3′ implies $x_2 \in U$. Consequently, by $x_3 \equiv \{x_2\} \subset U$ and $|x_3| = 1 < |U|$ implies $x_3 \in U$. Similarly, $x_4 \equiv \{x_3\} \in U$. Since all x_i are different for $i \in 5$, we obtain $|U| \geqslant 5$. □

Lemma 9. *If U is a Tarski set, then* $x, y \in U \Rightarrow \{x\}, \{x, y\}, \langle x, y\rangle \in U$.

Proof. By Lemma 8, we have $|U| \geqslant 5$. Therefore, $\{x\}, \{x, y\} \subset U$ and $|\{x\}| = 1 \leqslant |\{x, y\}| \leqslant 2 < |U|$ imply $\{x\}, \{x, y\} \in U$ in view of property 3′. Hence, $\langle x, y\rangle \equiv \{\{x\}, \{x, y\}\} \in U$. □

Lemma 10. *If U is a Tarski set, then* $|x| \geqslant |U| \Rightarrow x \notin U$.

Proof. Assume the converse, i. e. there is x such that $|x| \geqslant |U| \wedge x \in U$. By property 2, we get $y \equiv \mathcal{P}(x) \in U$. Since $|\mathcal{P}(x)| > |x|$, we get $|y| > |U|$. Property 1 implies $y \subset U$. But in this case $|y| \leqslant |U|$. This contradiction proves that $x \notin U$. □

Lemma 11. *If U is a Tarski set, then* $x, y \in U \Rightarrow x \cup y \in U$.

Proof. Since $\forall z(z \in x \vee z \in y \Rightarrow z \in U)$, we get $x \cup y \subset U$. It follows from $x, y \in U$ and Lemma 10 that $\alpha \equiv |x| < |U| \equiv \varkappa$ and $\beta \equiv |y| < \varkappa$. We need to prove that $|x \cup y| < \varkappa$. First, consider the case where $\alpha \leqslant 2$ and $\beta \leqslant 2$. Then, it is easily seen that $|x \cup y| \leqslant 4 < |U|$ by virtue of Lemma 8. Therefore, $x \cup y \in U$ in view of property 3′. Further, suppose that $\alpha \geqslant \beta > 2$. Consider the sets $P \equiv \{0\} \times x$, $Q \equiv \{1\} \times y$, $S \equiv x \cup y$, and $T \equiv P \cup Q$. Define the mapping $u : T \to S$ setting $u(0, a) \equiv a$ for every $(0, a) \in P$ and $u(1, b) \equiv b$ for every $(1, b) \in Q$. Since u is surjective, we get $|S| \leqslant |T|$.

It is clear that there exist bijections $g : P \rightarrowtail\!\!\!\!\twoheadrightarrow \alpha$ and $h : Q \rightarrowtail\!\!\!\!\twoheadrightarrow \beta \subset \alpha$. Define the function $f : T \to \mathcal{P}(\alpha)$ setting $f(p) \equiv \{g(p)\}$ for every $p \in P$ and $f(q) \equiv \alpha\backslash\{h(q)\}$ for every $q \in Q$. Since $P \cap Q = \varnothing$, this function is well defined. The function f is injective. Indeed, f is injective on P and on Q. Let $p \in P$, $q \in Q$, and $f(p) = f(q)$. Then, $\{g(p)\} = \alpha\backslash\{h(q)\}$ implies $\alpha = \{g(p)\} \cup \{h(q)\} = \{g(p), h(q)\} \leqslant 2$, but it contradicts our assumption. Consequently, $f(p) \neq f(q)$. The injectivity of f implies $|S| \leqslant |T| \leqslant |\mathcal{P}(\alpha)| < \varkappa$ by virtue of Lemma 7. Then, by property 3′ we get $S \in U$. □

Corollary 1. *If U is a Tarski set, then* $\omega \subset U$.

Corollary 2. *If U is a Tarski set, then* $|U| \geqslant \omega$.

Corollary 3. *If U is a Tarski set, then* $x, y \in U \Rightarrow x * y \in U$.

Proof. Lemma 11 and property 2 imply $B \equiv \mathcal{P}(\mathcal{P}(x \cup y)) \in U$. By Lemma 3, it follows from $A \equiv x * y \subset B$ that $A \in U$. □

Lemma 12. *If U is a Tarski set, then $\alpha < |U| \Rightarrow |\alpha * \alpha| < |U|$ holds for every ordinal number α.*

Proof. First, consider the case where $|\alpha| \leqslant 2$. Then, by Lemma 8 we get $|\alpha * \alpha| \leqslant |2 * 2| = 4 < |U|$. Further, suppose that $|\alpha| > 2$. Since $\alpha < |U| \equiv \varkappa$, Lemma 7 implies $|\mathcal{P}(\alpha)| < \varkappa$. The set $X \equiv \alpha * \alpha$ consists of ordered pairs $\langle\beta, \gamma\rangle$ such that $\beta, \gamma \in \alpha$. Divide the set X into three disjoint subsets $X_1 \equiv \{\langle\beta, \gamma\rangle \mid \beta < \gamma < \alpha\}$, $X_2 \equiv \{\langle\beta, \beta\rangle \mid \beta < \alpha\}$, and $X_3 \equiv \{\langle\beta, \gamma\rangle \mid \gamma < \beta < \alpha\}$. Obviously, $X_1 \cup X_2 \cup X_3 = X$. Define the function $f : X \to \mathcal{P}(\alpha)$ in the following way: if $x_1 = \langle\beta, \gamma\rangle \in X_1$, then $f(x_1) \equiv \{\beta, \gamma\} \in \mathcal{P}(\alpha)$; if $x_2 \equiv \langle\beta, \beta\rangle \in X_2$, then $f(x_2) \equiv \{\beta\} \in \mathcal{P}(\alpha)$; if $x_3 = \langle\beta, \gamma\rangle \in X_3$, then $f(x_3) \equiv \alpha \backslash \{\beta, \gamma\} \in \mathcal{P}(\alpha)$. The function f is injective on X_1, X_2, and X_3. If $f(x_1) = f(x_2)$, then $\{\beta, \gamma\} = \{\beta\}$ implies $\gamma = \beta < \gamma$, which is impossible. If $f(x_1) = f(x_3)$, then $\{\beta, \gamma\} = \alpha \backslash \{\beta, \gamma\}$, which is impossible in view of $\alpha \neq \varnothing$. Finally, if $f(x_2) = f(x_3)$, then $\{\beta\} = \alpha \backslash \{\beta, \gamma\}$ implies $\alpha = \{\beta\} \cup \{\beta, \gamma\} = \{\beta, \gamma\}$, and therefore, $|\alpha| \leqslant 2$, but it contradicts our assumption. This contradictory guarantees that f is injective. Consequently, $|X| \leqslant |\mathcal{P}(\alpha)| < \varkappa$. □

The following theorem and its Corollary 1 were proved by A. Tarski [1938] (see also [*Kuratowski and Mostowski*, 1967, IX, § 5]). We give here another proof.

Theorem 1. *If U is a Tarski set, then $\varkappa \equiv |U|$ is a regular cardinal number.*

Proof. Suppose that the cardinal number $\varkappa$ is not regular; then, $\alpha' \equiv cf(\varkappa) < \varkappa$ and Lemma 5 implies $\alpha' \in U$. By definition, there is a function $\varphi : \alpha' \to \varkappa$ such that $\cup \operatorname{rng} \varphi = \varkappa$. Denote $\operatorname{rng} \varphi$ by A and consider the cardinal number $\alpha \equiv |A| \leqslant \alpha' < \varkappa$. By Lemma 5, $A \subset U$ and $\alpha \in U$. Define the function $g : A \to \mathcal{P}(\varkappa)$ in the following way. Consider an arbitrary ordinal number $\beta \in A$ and the set $A_\beta \equiv \{\gamma \in A \mid \gamma < \beta\}$. Put $\beta' \equiv \sup A_\beta = \cup A_\beta$ for $A_\beta \neq \varnothing$ (see Lemma 2 (A.2.2)) and $\beta' \equiv 0$ for $A_\beta = \varnothing$. Consider the set $C_\beta \equiv \{\gamma \in \varkappa \mid \beta' \leqslant \gamma < \beta\}$ and put $g(\beta) \equiv C_\beta$. Show that $g(\beta_1) \cap g(\beta_2) = \varnothing$ for $\beta_1 \neq \beta_2$. Indeed, let $\beta_1 < \beta_2$. Then, $\beta_1 \in A_{\beta_2}$, and therefore, $\beta_1 \leqslant \beta_2{}'$. If $x \in g(\beta_1) \wedge x \in g(\beta_2)$, then $x \in C_{\beta_1} \wedge x \in C_{\beta_2}$, where $x \in \varkappa \wedge x < \beta_1 \wedge \beta_2{}' \leqslant x$, which is impossible. This contradiction implies that $g(\beta_1) \cap g(\beta_2) = \varnothing$. Check that $B \equiv \cup(g(\beta) \mid \beta \in A) = \varkappa$. It follows from the definitions of the sets $g(\beta)$ that $B \subset \varkappa$. Suppose now that $x \in \varkappa$. Since $\cup A = \varkappa$, there is $\beta \in A$ such that $x \in \beta$. Therefore, the set $D \equiv \{\gamma \in A \mid x \in \gamma\}$ is non-empty, and therefore, it has a minimal element λ. By the definition of D we get $\forall \gamma \in A(\gamma < \lambda \Rightarrow \gamma \leqslant x)$, where $x \geqslant \lambda'$. Hence, $\lambda' \leqslant x < \lambda$, i. e. $x \in g(\lambda)$. Consequently, $B = \varkappa$.

Since U is a Tarski set and $\varkappa$ is its power, there is a bijection $f : \varkappa \rightarrowtail\!\!\!\!\twoheadrightarrow U$. Since $\varkappa = \cup(g(\beta) \mid \beta \in A)$ and the sets $g(\beta)$ are pairwise disjoint, we conclude that $U = \cup(f[g(\beta)] \mid \beta \in A)$. Denote the sets $f[g(\beta)]$ by U_β. Fix $\beta \in A$. It follows from $C_\beta \subset \beta$ that $|U_\beta| = |C_\beta| \leqslant |\beta|$.

Consider (possibly, empty) the set $F_\beta \equiv \{q \in U_\beta \mid |q| = \alpha\}$. By above, $|F_\beta| \leqslant |U_\beta| \leqslant |\beta| \leqslant \beta$. Hence, $|\cup F_\beta| = |\cup(q \mid q \in F_\beta)| \leqslant |\cup(q * \{q\} \mid q \in F_\beta)| \equiv \sum(|q| \mid q \in F_\beta) =$

$\alpha|F_\beta| \leqslant \alpha|\beta| = \sum(\alpha_q \mid q \in \beta) \equiv |\cup(\alpha * \{q\} \mid q \in \beta)|$, where $\alpha_q \equiv \alpha$ for every $q \in \beta$. Since $\cup(\alpha * \{q\} \mid q \in \beta) \subset \alpha * \beta \subset \max(\alpha, \beta) * \max(\alpha, \beta)$ and $\max(\alpha, \beta) < \varkappa$, by Lemma 12, we get $|\cup F_\beta| < \varkappa$. Consequently, $\cup F_\beta \in U$ and $\mathcal{P}(\cup F_\beta) \in U$.

It follows from $F_\beta \subset U_\beta \subset U$ and the transitivity of U that $\cup F_\beta \subset U$. Therefore, by the inequality proven above, we conclude that $V_\beta \equiv U \setminus \cup F_\beta \neq \varnothing$ for every $\beta \in A$. Suppose that $\mathcal{P}(\cup F_\beta) \in \cup F_\beta$. Since $\cup F_\beta \in \mathcal{P}(\cup F_\beta)$, we obtain an infinite decreasing sequence $\mathcal{P}(\cup F_\beta) \ni \cup F_\beta \ni \mathcal{P}(\cup F_\beta) \ni \cup F_\beta \ni \ldots$. This contradicts the regularity axiom. Hence, $\mathcal{P}(\cup F_\beta) \in V_\beta$. Define the function $h : A \to U$ setting $h(\beta) \equiv \mathcal{P}(\cup F_\beta)$. Consider the function $h' \equiv h \circ \varphi : \alpha' \to U$. By Lemma 4, we get $M \equiv \operatorname{rng} h = \operatorname{rng} h' \in U$.

Evidently, $|M| \leqslant \alpha$. The transitivity of U implies $\alpha \subset U$. If α is an infinite number, then for the set $M' \equiv M \cup \alpha \subset U$ the inequalities $\alpha \leqslant |M'| \leqslant |(M * \{0\}) \cup (\alpha * \{1\})| \equiv |M| + \alpha = \alpha$ hold, where we get the equality $|M'| = \alpha$. By Lemma 11, it follows from $\alpha \in U$ and $M \in U$ that $M' \in U$. If α is a finite (i. e. natural) number, then the set $U \setminus M$ is infinite by virtue of Corollary 2 to Lemma 11. Consequently, there is an injective mapping $v : \omega \to U \setminus M$. Consider the natural number $n \equiv \alpha - |M|$ and the finite set $N \equiv v[n] \subset U \setminus M$. In this case, the equality $|M'| = \alpha$ hold for the set $M' \equiv M \cup N$. By Corollary 1 to Lemma 11, we get $n \in \omega \subset U$. Therefore, by Lemma 4, we get $N = \operatorname{rng}(u|n) \in U$. It follows from $M \in U$ and $N \in U$ that $M' \in U$ by virtue of Lemma 11.

Since we have proven that $U = \cup(U_\beta \mid \beta \in A)$, we have $M' \in U_\beta$ for some $\beta \in A$. Besides, $|M'| = \alpha$, where $M' \in F_\beta$. If $x \in M' \in F_\beta$, then $x \in \cup F_\beta$, i. e. $M' \subset \cup F_\beta$. It follows from $h(\beta) \in V_\beta = U \setminus \cup F_\beta$ that $h(\beta) \notin M'$. However, by definition, $h(\beta) \in M \subset M'$. This contradiction implies that the cardinal $\varkappa$ is regular. □

Corollary 1. *If U is a Tarski set and $\varkappa \equiv |U| > \omega$, then $\varkappa$ is an inaccessible cardinal number.*

Proof. By Theorem 1 the cardinal number $\varkappa$ is regular. By Lemma 7 for every $\alpha < \varkappa$ we get $|\mathcal{P}(\alpha)| \in \varkappa$. By condition, $\varkappa > \omega$. Consequently, $\varkappa$ is an inaccessible cardinal number. □

Theorem 2. *If U is a Tarski set, then $x \in U \Rightarrow \cup x \in U$.*

Proof. Consider the numbers $\alpha \equiv |x|$ and $\varkappa \equiv |U|$ and some bijection $u : \alpha \rightarrowtail\!\!\!\rightarrow x$. By Lemma 2 $\alpha \in \varkappa$. Since U is transitive, we get $\cup x \subset U$. Hence, $|\cup x| \leqslant \varkappa$.

Suppose that $|\cup x| = \varkappa$; then, there is a bijection $f : \cup x \rightarrowtail\!\!\!\rightarrow \varkappa$. Fix an element $a \in \alpha$. Then, $u(a) \in x$ implies $u(a) \subset \cup x$. Therefore, we can consider the injective mapping $g_a \equiv f|u(a) : u(a) \rightarrowtail \varkappa$. Consider the number $\beta_a \equiv |u(a)|$, a bijection $v_a : \beta_a \rightarrowtail\!\!\!\rightarrow u(a)$, and the injective mapping $h_a \equiv g_a \circ v_a : \beta \rightarrowtail \varkappa$. Suppose that $\cup \operatorname{rng} h_a = \varkappa$. Then, by Theorem 1, we get $\beta_a \geqslant \varkappa$. However, $u(a) \in x \subset U$ implies $\beta_a \equiv |u(a)| < \varkappa$ by virtue of Lemma 2. This contradiction provides $\sup \operatorname{rng} h_a = \cup \operatorname{rng} h_a < \varkappa$.

Thus, we can define the function $\eta : \alpha \to \varkappa$ setting $\eta(a) \equiv \sup \operatorname{rng} h_a$. Consider the set $Z \equiv \operatorname{rng} \eta \subset \varkappa$. By the above, $z \leqslant \varkappa$ for every $z \in Z$. Let π be an order number such that $z \leqslant \pi$ for every $z \in Z$. Take any element $q \in \varkappa$ and consider the element $p \equiv f^{-1}(q) \in \cup x$. Then, $p \in y \in x$ for some $y \in x$. Consider the element $a \equiv u^{-1}(y) \in \alpha$. Since $p \in y = u(a)$, we get $q = f(p) \in f[u(a)] = \operatorname{rng} q_a = \operatorname{rng} h_a$. Hence, $q \leqslant \sup \operatorname{rng} h_a \equiv \eta(a) \leqslant \pi$. This means that $\sup \varkappa \leqslant \pi$. Since $\varkappa$ is a limit number, Lemma 3 (A.2.2) guarantees that $\varkappa = \sup \varkappa \leqslant \pi$. This implies $\varkappa = \sup Z = \cup Z = \cup \operatorname{rng} \eta$. Now, we have $\alpha \geqslant \varkappa$ by virtue of Theorem 1. This contradicts the inequality $\alpha < \varkappa$.

Hence, $|\cup x| < \varkappa$. By property 3′, we conclude that $\cup x \in U$. □

A.7.2 Galactic sets and their connection with Tarski sets

Let x be a set. Any finite sequence $(x_i \mid i \in n+1)$ such that $x_0 = x$ and $x_{i+1} \in x_i$ for every $i \in n$ will be called a *chain of subelements of the set x (of the length n)*.

A set U is said to be *dominant* if for any set x the following conclusions are equivalent:

1) $x \in U$;
2) all elements of any chains of subelements of the set x are of cardinality less than $|U|$.

Lemma 1. *Any dominant set is transitive.*

Proof. Let a set U be dominant, $x \in U$, and $y \in x$. Any chain of subelements of the set y is a subchain of some chain of subelements of the set x, and therefore, all its elements are of cardinality less than $|U|$. Hence, $y \in U$. □

Lemma 2. *Any dominant set U has property 3′ of a Tarski set, i. e. $x \subset U \wedge |x| < |U| \Rightarrow x \in U$.*

Proof. Let $x \subset U$ and $|x| < |U|$. Consider an arbitrary chain of subelements $(x_0, x_1, \dots, x_n)$ of the set x. Since $x_1 \in x_0 = x \subset U$ and U is transitive by virtue of Lemma 1, $x_1 \in U$ implies by induction that $x_i \in U$ for every $i \in n + 1 \setminus 1$. Hence, $|x_i| < |U|$ for every $i \in n + 1 \setminus 1$. Moreover, by condition, $|x_0| = |x| < |U|$. Consequently, $x \in U$. □

Proposition 1. *Any Tarski set is dominant.*

Proof. Let $(x_i \mid i \in n + 1)$ be a chain of subelements of the set x and $x \in U$. Since U is transitive, using induction, we infer that $x_i \in U$ for every $i \in n + 1$. By Lemma 2 (A.7.1) $|x_i| < |U|$.

Denote by $\mathbf{C}$ the class consisting of sets satisfying property 2 from the definition of dominant sets. Show that $\mathbf{C} \subset U$. Consider the class $\mathbf{D} \equiv \{x \mid (x \in U \wedge x \in \mathbf{C}) \vee x \notin \mathbf{C}\}$ and prove that it satisfies the $\in$-induction principle (Lemma 4 (A.2.2)). Take some $y \subset \mathbf{D}$. Then, for every $z \in y$ we get $(z \in U \wedge z \in \mathbf{C}) \vee z \notin \mathbf{C}$. If $z \notin \mathbf{C}$ for some $z \in y$, then

$y \notin \mathbf{C}$, and therefore, $y \in \mathbf{D}$. Suppose now that $\forall z \in y(z \in U \wedge z \in \mathbf{C})$. Consider $\alpha \equiv |y|$ and $\varkappa \equiv |U|$. If $\alpha \geqslant \varkappa$, then $y \notin \mathbf{C}$, where $y \in \mathbf{D}$. Let $\alpha < \varkappa$. In this case, $y \in U$ by virtue of Lemma 1 (A.7.1). Show that $y \in \mathbf{C}$. Indeed, consider an arbitrary chain $(y_i \mid i \in n+1)$ of subelements of the set y. Then, the sequence $(y_i \mid i \in (n+1)\backslash 1)$ is a chain of subelements of the set $y_1 \in y = y_0$. Since, by assumption, $y_1 \in \mathbf{C}$, all elements of the sequence $(y_i \mid i \in (n+1)\backslash 1)$ are of cardinality less than $\varkappa$. Thus, $y \in \mathbf{C}$. Hence, $y \in U \wedge y \in \mathbf{C}$, and therefore, $y \in \mathbf{D}$. Then, we conclude that the class $\mathbf{D}$ satisfies the $\in$-induction principle, which yields $\mathbf{D} = \mathbf{V}$. Therefore, $\forall x((x \in U \wedge x \in \mathbf{C}) \vee x \notin \mathbf{C})$, i. e. $\mathbf{C} \subset U$. □

Lemma 3. *For every cardinal number α, there exists no more than one Tarski set of cardinality α.*

Proof. Suppose that there exist two Tarski sets U_1 and U_2 of the same cardinality α. Let $x \in U_1$. Then, by Proposition 1, we get $|x| < \alpha = |U_1|$ and any chain of subelements of the set x consists of sets of cardinality less than $\alpha = |U_1|$. This implies that $|x| < |U_2|$ and any chain of subelements of the set x consists of sets of cardinality less than $|U_2|$. Consequently, the same Proposition provides $x \in U_2$, where $U_1 \subset U_2$. Similarly, we obtain $U_2 \subset U_1$. Thus, $U_1 = U_2$. □

A set U is said to be *exponential* if $\forall x \in U(\mathcal{P}(x) \in U)$. A dominant and exponential set will be called *galactic*.

Theorem 1 (the Bunina theorem on galactic sets). *The following conclusions are equivalent for a set U:*
1) *U is a Tarski set;*
2) *U is a galactic set.*

Proof. (1) $\vdash$ (2). This deduction follows from the exponentiality property of Tarski sets and Proposition 1.

(2) $\vdash$ (1). This deduction follows from Lemmas 1 and 2 and Lemma 1 (A.7.1). □

Let us show that under the assumption of the continuum hypothesis, there exists a dominant non-exponential set.

Lemma 4. *If $|2^\omega| = \omega_1$, then there is a dominant set of cardinality ω_1.*

Proof. If $|2^\omega| = \omega_1$, then to prove the existence of a dominant set of cardinality ω_1 it is sufficient to prove that for a set X consisting of sets such that all their chains of subelements consist of countable sets only, we have $|X| = \omega_1$.

Since $\omega_1 \subset X$, we get $|X| \geqslant \omega_1$.

Check now that $|X| \leqslant \omega_1$, i. e. there is a injection from the set X into the set of infinite sequences of zeros and units.

Any set $x \in X$ can be represented as a tree whose root is the set x itself, branches are chains of subelements, and leaves are last elements of these chains, i. e. sets containing no sets (empty sets). All branches of such a tree are of finite length, and, moreover, the number of these branches is countable. The number of stores of the tree is also countable, and on each store, there are countably many sets (nodes or leaves of the tree). Clearly, certain trees correspond to the same set in X (these are the trees obtained from each other by a renumbering of vertices), but only one set in X corresponds to each three X. We will consider not trees themselves but their "isomorphism classes".

Let us enumerate leaves of the tree in a certain way (this can be done, since the set of leaves is countable). we put into correspondence to every such "numbered" tree the function $f \in \omega^{\omega\times\omega}$ in the following way: $f(n, m)$ is the maximal natural k such that n-th and m-th leaves end branches of a certain set from the k-th store (if $n = m$ set $f(n, m) \equiv n$). We can always define such a number k, since, first, any two leaves end the branches of the initial set x, i. e. $k \geqslant 1$; second, $k \leqslant \min(m, n)$. Such a function $f \in \omega^{\omega\times\omega}$ determines uniquely the isomorphism class of a tree. Therefore, $|X| \leqslant |\omega^{\omega\times\omega}|$.

Show that $|\omega^{\omega\times\omega}| = |2^{\omega}| = \omega_1$. Since $|\omega\times\omega| = \omega$, we get $|\omega^{\omega\times\omega} = |\omega^{\omega}|$. The set ω^{ω} is the set of infinite sequences of natural numbers. Since $|\omega^{\omega}| \geqslant |2^{\omega}|$, it remains to check that $|\omega^{\omega}| \leqslant |2^{\omega}|$, i. e. to construct an injective mapping from the set of infinite sequences of natural numbers into the set of infinite sequences consisting of zeros and units. We do this as follows. Let $N \equiv (n_i \in \omega \mid i \in \omega)$ be an infinite sequences of natural numbers. Put into the correspondence to this sequence the sequence $M \equiv \{m_j \in 2 \mid j \in \omega\}$ of zeros and units such that for every $i \in \omega$ put $m_j \equiv 0$ for $j = \sum(n_k \mid k \in i) + i$ and $m_j \equiv 1$ for all other j. For example, the sequence $1, 2, 3, 4, 5, \ldots$ maps into the sequence $1, 0, 1, 1, 0, 1, 1, 1, 0, 1, 1, 1, 1, 0, 1, 1, 1, 1, 1, 0 \ldots$. This mapping is injective, where $|\omega^{\omega}| = |2^{\omega}|$, and therefore, $|X| \leqslant |2^{\omega}|$. Since $|X| \geqslant |2^{\omega}|$, the Cantor theorem (Theorem 2 (1.3.2)) implies $|X| = |2^{\omega}|$. Since, by assumption, $|2^{\omega}| = \omega_1$, we get $|X| = \omega_1$. Thus, the set X is dominant. □

The fact that such a set X is not exponential is obvious, since ω_1 is not an inaccessible cardinal number.

A.7.3 Characterization of Tarski sets. Characterization of all natural models of the NBG theory

Proposition 1. *Let U is a Tarski set and $|U| = \omega$. Then, $U = V_{\omega}$.*

Proof. By Lemma 1 (A.3.2) $\omega \subset V_{\omega}$. Hence, $\omega \leqslant |V_{\omega}|$. Since $V_{\omega} = \cup(V_n \mid n \in \omega)$ and $|V_n| < \omega$, we have $|V_{\omega}| \leqslant \omega$, and therefore, $|V_{\omega}| = \omega$. Prove that V_{ω} is a Tarski set. By Lemma 3 (A.3.2) the set V_{ω} is transitive, by Lemma 6 (A.3.2), it is exponential.

Check that V_ω has property $3'$. Consider some set $x \subset V_\omega$ such that $|x| < |V_\omega| = \omega$. If $y \in x \subset V_\omega = \cup\{V_n \mid n \in \omega\}$, then $N(y) \equiv \{n \in \omega \mid y \in V_n\} \neq \varnothing$. Consequently, the set $N(y)$ has a minimal element $n(y) \in \omega$. Since the set x is finite, the set $M \equiv \{m \in \omega \mid \exists y \in x(m = n(y))\}$ has a maximal element n. Therefore, $x \subset V_n$ implies $x \in V_{n+1} \subset V_\omega$. Thus, V_ω is a Tarski set. Since by Lemma 3 (A.7.2) a Tarski set of cardinality ω is unique, we get $U = V_\omega$. □

Theorem 1. *Let U be a Tarski set and $\varkappa \equiv |U| > \omega$. Then,*
1) *U is a universal set;*
2) *$U = V_\varkappa$ for the inaccessible cardinal number $\varkappa = \sup(\mathbf{On} \cap U)$.*

Proof. 1. Show that the set U has all the properties of a universal set.

Property 1 follows from property 1 of a Tarski set. The property $x \in U \Rightarrow \mathcal{P}(x) \in U$ follows from property 2 of a Tarski set. The property $x \in U \Rightarrow \cup x \in U$ follows from Theorem 2 (A.7.1). The property $x, y \in U \Rightarrow x \cup y \in U$ is a consequence of Lemma 11 (A.7.1). The properties $x, y \in U \Rightarrow \{x, y\}, \langle x, y\rangle \in U$ follows from Lemma 9 (A.7.1). Corollary 3 to Lemma 11 (A.7.1) implies the property $x, y \in U \Rightarrow x * y \in U$. Property 4 follows from Lemma 4 (A.7.1).

Since, by condition, $|U| > \omega$, Lemma 5 (A.7.1) implies $\omega \in U$.

Thus, the set U is universal.

2. Theorem 1 (A.4.2) guarantees that $U = V_\varkappa$ for the inaccessible cardinal number $\varkappa = \sup(\mathbf{On} \cap U)$. □

Now, we can prove the main *theorem on the characterization of natural models of the NBG set theory.*

Theorem 2. *In the ZF set theory, the following conclusions are equivalent for a set U:*
1) *U is an uncountable Tarski set;*
2) *U is a universal set;*
3) *U is an inaccessible cumulative set, i. e. $U = V_\varkappa$ for a certain inaccessible cardinal number $\varkappa$;*
4) *U is a supertransitive standard model set for the ZF set theory and U has the strong substitution property;*
5) *$\mathcal{P}(U)$ is a supertransitive standard model set for the NBG set theory;*
6) *U is an uncountable galactic set.*

Proof. The deduction (1) $\vdash$ (3) follows from Theorem 1.

The deduction (3) $\vdash$ (1) follows from Lemmas 3 and 6 (A.3.2) and Lemma 5 (A.3.3).

The equivalence of (2) and (3) follows from Theorem 2 (A.4.2).

The equivalence of (2) and (4) follows from Proposition 1 (A.6.1).

The equivalence of (1) and (6) follows from Theorem 1 (A.7.2).

The equivalence of (2) and (5) follows from Theorem 1 (A.6.2). □

The deduction (3) ⊢ (1) was proven by A. Tarski [1938]. The equivalence of (3) and (4) was, in fact, proven by Zermelo [1930] and Shepherdson [1951, 1952, 1953] (see also [*Kanamori*, 2003, Theorem 1.3]). All other assertions of this Theorem belong to E. I. Bunina and V. K. Zakharov [2006].

Consider one more axiom in ZF.

AG. (The *galacticity axiom*). Every set is an element of a certain galactic set.

Corollary 1. *In the ZF set theory, axioms AT, AU, AI, and AG are equivalent.*

The equivalence of axioms AT and AI in this corollary was proven by A. Tarski [1938] (see also [*Kuratowski and Mostowski*, 1967, IX, § 5, Theorem 1]). Here another proof using the theorem on characterization of natural models is given.

A.8 Characterization of all natural models of the ZF set theory in the ZF set theory

A.8.1 Scheme-inaccessible cardinal numbers and scheme-inaccessible cumulative sets

If all free variables of a formula φ are among the variables $x_0, \ldots, x_{m-1}, p_0, \ldots, p_{n-1}$, then this situation will be denoted in the form $\varphi(\vec{x}, \vec{p})$. In this case, the variables $p_0, \ldots, p_{n-1}$ will be called *parameters*. Instead of $x_0 \in A \wedge \ldots \wedge x_{m-1} \in A$, $\forall x_0 \in A \ldots \forall x_{m-1} \in A$, and $\exists x_0 \in A \ldots \exists x_{m-1} \in A$ we shall write $\vec{x} \in A$, $\forall \vec{x} \in A$, and $\exists \vec{x} \in A$, respectively.

For every transitive set A every formula $\varphi(x, y, \vec{p})$ of the ZF theory defines the correspondence $[\varphi(x, y, \vec{p})|A] \equiv \{z \in A * A \mid \exists x, y \in A(z = \langle x, y\rangle \wedge \varphi^A(x, y, \vec{p}))\} \subset A * A$ depending on the parameter $\vec{p}$ (see A.6.1).

An ordinal number $\varkappa$ is said to be *scheme-regular* if

$$\forall \vec{p} \in V_\varkappa \forall \alpha(\alpha \in \varkappa \wedge ([\varphi(x, y, \vec{p})|V_\varkappa] \leftrightharpoons \alpha \to \varkappa \Rightarrow \cup \operatorname{rng}[\varphi(x, y, \vec{p})|V_\varkappa] \in \varkappa),$$

where φ is a metavariable denoting an arbitrary formula of ZF.

An ordinal number $\varkappa > \omega$ is said to be (*strongly*) *scheme-inaccessible* if

1) $\forall \alpha(\alpha \in \varkappa \Rightarrow |\mathcal{P}(\alpha)| \in \varkappa)$;
2) $\varkappa$ is scheme-regular.

Lemma 1. *Let an ordinal number $\varkappa$ satisfy the quasiexponentiality condition $\forall \alpha(\alpha \in \varkappa \Rightarrow |\mathcal{P}(\alpha)| \subset \varkappa)$. Then, $\varkappa$ is a cardinal number.*

Proof. Let α be an ordinal number such that $\alpha \leqslant \varkappa$ and $\alpha \sim \varkappa$. Then, $|\alpha| = |\varkappa|$. Suppose that $\alpha < \varkappa$. By the condition, $|\mathcal{P}(\alpha)| \subset \varkappa$. Applying the Cantor theorem

(Theorem 2 (1.3.2)), we obtain $|\alpha| < |\mathcal{P}(\alpha)| \leqslant |\varkappa|$. It contradicts the preceding equality. Hence, $\alpha = \varkappa$. □

Corollary 1. *An scheme-inaccessible ordinal number $\varkappa$ is a cardinal number.*

Proof. If $\alpha \in \varkappa$, then property 1 implies $|\mathcal{P}(\alpha)| \in \varkappa$. By virtue of transitivity, we get $|\mathcal{P}(\alpha)| \subset \varkappa$, where $\varkappa$ satisfies the conditions of Lemma 1, and therefore, $\varkappa$ is a cardinal number. □

The sets $V_\varkappa$ for scheme-inaccessible cardinal numbers $\varkappa$ will be called *scheme-inaccessible cumulative sets.*

Lemma 2. *Let $\varkappa$ be an scheme-inaccessible cardinal number and α is an ordinal number such that $\alpha \in \varkappa$. Then, $|V_\alpha| < \varkappa$.*

Proof. Consider the set $C' \equiv \{x \in \varkappa \mid |V_x| < \varkappa\}$ and the classes $\mathbf{C}'' \equiv \mathbf{On} \backslash \varkappa$ and $\mathbf{C} \equiv C' \cup \mathbf{C}''$. Since $V_0 = \varnothing$, we get $|V_0| = 0 < \varkappa$, where $0 \in \mathbf{C}$.

Let $\alpha \in \mathbf{C}$. If $\alpha \geqslant \varkappa$, then $\alpha + 1 \in \mathbf{C}'' \subset \mathbf{C}$. If $\alpha < \varkappa$, then $\alpha \in C'$. If $\alpha + 1 = \varkappa$, then $\alpha + 1 \in \mathbf{C}'' \subset \mathbf{C}$. Let $\alpha + 1 < \varkappa$. Since $V_\alpha \sim |V_\alpha|$, we get $\mathcal{P}(V_\alpha) \sim \mathcal{P}(|V_\alpha|)$, where$|\mathcal{P}(V_\alpha)| = |\mathcal{P}(|V_\alpha|)|$. By Corollary 2 to Lemma 3 (A.3.2), we have $|V_{\alpha+1}| = |\mathcal{P}(V_\alpha)| = |\mathcal{P}(|V_\alpha|)|$. Since $|V_\alpha| < \varkappa$ and the ordinal number $\varkappa$ is scheme-inaccessible, we obtain $|\mathcal{P}(|V_\alpha|)| < \varkappa$. Consequently, $|V_{\alpha+1}| < \varkappa$, and therefore, $\alpha + 1 \in C' \subset \mathbf{C}$.

Let α be a limit ordinal number and $\alpha \subset \mathbf{C}$. If $\alpha \cap \mathbf{C}'' \neq \varnothing$, then there is $\beta \in \alpha$ such that $\beta \geqslant \varkappa$. Therefore, $\alpha > \beta \geqslant \varkappa$ implies $\alpha \in \mathbf{C}'' \subset \mathbf{C}$. Let $\alpha \cap \mathbf{C}'' = \varnothing$, i. e. $\alpha \subset C' \subset \varkappa$. If $\alpha = \varkappa$, ôî $\alpha \in \mathbf{C}'' \subset \mathbf{C}$. Suppose $\alpha < \varkappa$; then, for every $\beta \in \alpha$ the inequality $|V_\beta| < \varkappa$ holds since $\alpha \subset C'$. Consequently, $\sup\{|V_\beta| \mid \beta \in \alpha\} \leqslant \varkappa$.

Consider the formula $\varphi(x, y) \equiv (x \in \alpha \Rightarrow y = |V_x|) \wedge (x \notin \alpha \Rightarrow y = \varnothing)$.

Show that under our conditions $x \in \alpha \in V_\varkappa$ and $y \in V_\varkappa$ the equivalence $(y = |V_x|)^{V_\varkappa} \Leftrightarrow y = |V_x|$ holds.

The formula $(y = |V_x|)^{V_\varkappa}$ is rewritten as $(Cn(y))^{V_\varkappa} \wedge \exists f \in V_\varkappa(f \leftrightharpoons y \rightarrowtail\!\!\!\rightarrow V_x)^{V_\varkappa}$. The formula $Cn(y)^{V_\varkappa}$ can be rewritten as $On(y)^{V_\varkappa} \wedge \forall \alpha \in V_\varkappa(On(\alpha)^{V_\varkappa} \wedge (\alpha \subset y)^{V_\varkappa} \wedge \exists h \in V_\varkappa(h \leftrightharpoons \alpha \rightarrowtail\!\!\!\rightarrow y)^{V_\varkappa} \Rightarrow \alpha = y)$. Consider the formula $On(y)^{V_\varkappa}$ under the condition $y \in V_\varkappa$. This formula has the form

$$\begin{aligned} On(y)^{V_\varkappa} \equiv{}& \forall x \in V_\varkappa(x \in y \Rightarrow (x \subset y)^{V_\varkappa}) \wedge \\ &\wedge \forall x, x', x'' \in V_\varkappa(x \in y \wedge x' \in y \wedge x'' \in y \wedge x \in x' \wedge x' \in x'' \Rightarrow x \in x'') \wedge \\ &\wedge \forall x, x' \in V_\varkappa(x \in y \wedge x' \in y \Rightarrow x \in x' \vee x = x' \vee x' \in x) \wedge \\ &\wedge \forall T \in V_\varkappa((\varnothing \neq T \subset y)^{V_\varkappa} \Rightarrow \exists x \in V_\varkappa(x \in T \wedge \forall x' \in V_\varkappa(x' \in T \Rightarrow x \in x'))). \end{aligned}$$

Note that under the condition $y \in V_\varkappa$ the formula $(x \subseteq y)^{V_\varkappa} \equiv \forall z \in V_\varkappa(z \in x \Rightarrow z \in y)$ is equivalent to the formula $x \subseteq y$ by virtue of the supertransitivity of $V_\varkappa$. Similarly,

$(\varnothing \neq T \subseteq y)^{V_\varkappa} \Leftrightarrow \varnothing \neq T \subseteq y$. The formula $\forall x \in V_\varkappa(x \in y \Rightarrow x \subseteq y)$ is equivalent to the formula $\forall x(x \in y \Rightarrow x \subseteq y)$ since $x \in y$ implies $x \in V_\varkappa$. The formula $\forall x, x', x'' \in V_\varkappa(x \in y \wedge x' \in y \wedge x'' \in y \wedge x \in x' \wedge x' \in x'' \Rightarrow x \in x'')$ is equivalent to the formula $\forall x, x', x''(x \in y \wedge x' \in y \wedge x'' \in y \wedge x \in x' \wedge x' \in x'' \Rightarrow x \in x'')$ since $x, x', x'' \in y$ implies $x, x', x'' \in V_\varkappa$. The formula $\forall x, x' \in V_\varkappa(x \in y \wedge x' \in y \Rightarrow x \in x' \vee x = x' \vee x' \in x)$ is equivalent to the formula $\forall x, x'(x \in y \wedge x' \in y \Rightarrow x \in x' \vee x = x' \vee x' \in x)$ since $x, x' \in y$ implies $x, x' \in V_\varkappa$. Finally, the formula $\forall T \in V_\varkappa(\varnothing \neq T \subseteq y \Rightarrow \exists x \in V_\varkappa(x \in T \wedge \forall x' \in V_\varkappa(x' \in T \Rightarrow x \in x')))$ is equivalent to the formula $\forall T(\varnothing \neq T \subset y \Rightarrow \exists x(x \in T \wedge \forall x'(x' \in T \Rightarrow x \in x')))$ since $T \subset y$ implies $T \in V_\varkappa$ and $x, x' \in T$ implies $x, x' \in V_\varkappa$. Thus, $On(y)^{V_\varkappa} \Leftrightarrow On(y)$.

This guarantees that the formula $Cn(y)^{V_\varkappa}$ can be rewritten as $On(y) \wedge \forall \alpha \in V_\varkappa(On(\alpha) \wedge (\alpha \subset y)^{V_\varkappa} \wedge \exists h \in V_\varkappa(h \leftrightharpoons \alpha \rightarrowtail\!\!\!\!\twoheadrightarrow y)^{V_\varkappa} \Rightarrow \alpha = y)$. Since $y \in V_\varkappa$, we obtain $(\alpha \subset y)^{V_\varkappa} \Leftrightarrow \alpha \subset y$, and therefore, $\alpha \in V_\varkappa$.

The formula $\exists h \in V_\varkappa(h \leftrightharpoons \alpha \rightarrowtail\!\!\!\!\twoheadrightarrow y)^{V_\varkappa}$ is written as

$$\begin{aligned}\exists h \in V_\varkappa(\forall x \in V_\varkappa(x \in h \Leftrightarrow \exists z \in V_\varkappa \exists z' \in V_\varkappa(z \in \alpha \wedge \exists z' \in y \wedge x = \langle z, z'\rangle)) \wedge \\ \wedge \forall z \in V_\varkappa(z \in \alpha \Rightarrow \exists z' \in V_\varkappa(z' \in y \wedge \langle z, z'\rangle \in h)) \wedge \\ \wedge \forall z' \in V_\varkappa(z' \in y \Rightarrow \exists z \in V_\varkappa(z \in \alpha \wedge \langle z, z'\rangle \in h)) \wedge \\ \wedge \forall z, z', z'' \in V_\varkappa(z \in \alpha \wedge z', z'' \in y \wedge \langle z, z'\rangle \in h \wedge \langle z, z''\rangle \in h \Rightarrow z' = z'') \wedge \\ \wedge \forall z, z', z'' \in V_\varkappa(z, z' \in \alpha \wedge z'' \in y \wedge \langle z, z''\rangle \in h \wedge \langle z', z''\rangle \in h \Rightarrow z = z')).\end{aligned}$$

The formula $\forall x \in V_\varkappa(x \in h \Leftrightarrow \exists z, z' \in V_\varkappa(z \in \alpha \wedge z' \in y \wedge x = \langle z, z'\rangle))$ is equivalent to the formula $\forall x(x \in h \Leftrightarrow \exists z \in \alpha \exists z' \in y(x = \langle z, z'\rangle))$ since $z \in \alpha$ implies $z \in V_\varkappa$, $z' \in y$ implies $z' \in V_\varkappa$, and $x = \langle z, z'\rangle$ implies $x \in V_\varkappa$.

The formula $\forall z \in V_\varkappa(z \in \alpha \Rightarrow \exists z' \in V_\varkappa(z' \in y \wedge \langle z, z'\rangle \in h))$ is equivalent $\forall z \in \alpha \exists z' \in y(\langle z, z'\rangle \in h)$ since $z \in \alpha$ implies $z \in V_\varkappa$ and $z' \in y$ implies $z' \in V_\varkappa$. Similarly, the formula $\forall z' \in V_\varkappa(z' \in y \Rightarrow \exists z \in V_\varkappa(z \in \alpha \wedge \langle z, z'\rangle \in h))$ is equivalent $\forall z' \in y \exists z \in \alpha(\langle z, z'\rangle \in h)$.

The formula $\forall z, z', z'' \in V_\varkappa(z \in \alpha \wedge z', z'' \in y \wedge \langle z, z'\rangle \in h \wedge \langle z, z''\rangle \in h \Rightarrow z' = z'')$ is equivalent $\forall z \in \alpha \forall z', z'' \in y(\langle z, z''\rangle \in h \wedge \langle z, z''\rangle \in h \Rightarrow z' = z'')$ since $z \in \alpha$ and $z', z'' \in y$ imply $z, z', z'' \in V_\varkappa$. Similarly, the formula $\forall z, z', z'' \in V_\varkappa(z, z' \in \alpha \wedge z'' \in y \wedge \langle z, z''\rangle \in h \wedge \langle z', z''\rangle \in h \Rightarrow z = z')$.

Thus, the formula $Cn(y)^{V_\varkappa}$ is equivalent to the formula $On(y) \wedge \forall \alpha(On(\alpha) \wedge \alpha \subset y \wedge \exists h \in V_\varkappa(h \leftrightharpoons \alpha \rightarrowtail\!\!\!\!\twoheadrightarrow y) \Rightarrow \alpha = y)$. Since $h \leftrightharpoons \alpha \rightarrowtail\!\!\!\!\twoheadrightarrow y$, we get $h \subset \alpha * y$. By Corollary 2 to Lemma 6 (A.3.2), it follows from $\alpha, y \in V_\varkappa$ that $\alpha * y \in V_\varkappa$, and, therefore, $h \in V_\varkappa$. Hence, we obtain $Cn(y)^{V_\varkappa} \Leftrightarrow Cn(y)$.

We know that for $x < \varkappa$, we have $V_x \in V_\varkappa$. Therefore, as above, it can be shown that the formula $\exists f \in V_\varkappa(f \leftrightharpoons y \rightarrowtail\!\!\!\!\twoheadrightarrow V_x)^{V_\varkappa}$ is equivalent to the formula $\exists f(f \leftrightharpoons y \rightarrowtail\!\!\!\!\twoheadrightarrow V_x)$.

Now, we can conclude that $(y = |V_x|)^{V_\varkappa} \Leftrightarrow (y = |V_x|)$.

Then, $[\varphi|V_\varkappa] = \{z \mid \exists x \in V_\varkappa \exists y \in V_\varkappa(z = \langle x, y\rangle \wedge (x \in \alpha \Rightarrow y = |V_x|) \wedge (x \notin \alpha \Rightarrow y = \varnothing) \wedge \alpha \in V_\varkappa)\}$. If $y \in \operatorname{rng}[\varphi|V_\varkappa]$, then $\exists x(\langle x, y\rangle \in [\varphi|V_\varkappa])$, i.e. $y \in V_\varkappa \wedge \exists x(x \in V_\varkappa \wedge ((x \in \alpha \wedge y = |V_x|) \vee (x \notin \alpha \wedge y = \varnothing)))$. Thus, $y = \varnothing$ or $y = V_x$ for some $x \in \alpha$. Conversely, if $y = V_x$ for some $x \in \alpha$, then $y \in \operatorname{rng}[\varphi|V_\varkappa]$. Hence, $\operatorname{rng}[\varphi|V_\varkappa] = \{|V_\beta| \mid \beta \in \alpha\}$. By Corollary 1 to Theorem 1 (A.3.2) $\cup \operatorname{rng} f = \cup\{|V_\beta| \mid \beta \in \alpha\} = \sup\{|V_\beta| \mid \beta \in \alpha\} = |V_\alpha|$. By the inequality proven above, we obtain $|V_\alpha| \leqslant \varkappa$.
Suppose that $|V_\alpha| = \varkappa$; then, $\varkappa = \cup \operatorname{rng}[\varphi|V_\varkappa]$ implies $\varkappa \leqslant \alpha$ since $\varkappa$ is scheme-regular. This contradicts the initial inequality $\alpha < \varkappa$. Hence, $|V_\alpha| < \varkappa$. Therefore, $\alpha \in C' \subset \mathbf{C}$.

By the principle of transfinite induction, $\mathbf{C} = \mathbf{On}$, and therefore, $C' = \varkappa$. □

Lemma 3. *Let $\varkappa$ be a scheme-inaccessible cardinal number. Then, $\varkappa = |V_\varkappa|$.*

Proof. By Lemma 2, we get $\varkappa \subset V_\varkappa$. Therefore, $\varkappa = |\varkappa| \leqslant |V_\varkappa|$. By Corollary 1 to Theorem 1 (A.3.2) $|V_\varkappa| = \sup(|V_\beta| \mid \beta \in \varkappa)$. Since by Lemma 2 $|V_\beta| < \varkappa$, we obtain $|V_\varkappa| \leqslant \varkappa$. As a result, we obtain $\varkappa = |V_\varkappa|$. □

Lemma 4. *Let $\varkappa$ be a scheme-inaccessible cardinal number, α be an ordinal number such that $\alpha < \varkappa$, and $\varphi(x, y, \vec{p})$ be a formula. Then, $\forall \vec{p} \in V_\varkappa([\varphi(x, y, \vec{p})|V_\varkappa] \fallingdotseq V_\alpha \to V_\varkappa \Rightarrow \operatorname{rng}[\varphi(x, y, \vec{p})|V_\varkappa] \in V_\varkappa)$.*

Proof. Since $\varkappa$ is a limit ordinal number, we get $V_\varkappa = \cup(V_\delta \mid \delta \in \varkappa)$. For $x \in V_\alpha$ there is $\delta \in \varkappa$ such that $[\varphi|V_\varkappa](x) \in V_\delta$. Hence, the non-empty set $\{y \leqslant \delta \mid [\varphi|V_\varkappa](x) \in V_y\}$ has a minimal element z.

Consider some bijection $h : |V_\alpha| \to V_\alpha$.

Since, by condition, $v \in V_\varkappa$, the formula $\forall x \in V_\alpha \exists v[\varphi|V_\varkappa](x) = v \wedge \varphi^{V_\varkappa}(x, v, \vec{p}) \wedge \vec{p} \in V_\varkappa$ holds. Consider the formula $\psi(u, z) \equiv (u \in |V_\alpha| \Rightarrow z = sm\{y \leqslant \delta \mid [\varphi|V_\varkappa](h(u)) \in V_y\}) \wedge (u \notin |V_\alpha| \Rightarrow z = \varnothing)$. In this case, $[\psi|V_\varkappa] = \{v \mid \exists u \in V_\varkappa \exists z \in V_\varkappa(v = \langle u, z\rangle \wedge \psi^{V_\varkappa}(u, z))$. Consider the formula $\psi^{V_\varkappa}(u, z)$ in more detail. It is equivalent to the formula $((u \in |V_\alpha|)^{V_\varkappa} \Rightarrow (z = sm\{y \leqslant \delta \mid [\varphi|V_\varkappa](h(u)) \in V_y\})^{V_\varkappa}) \wedge ((u \notin |V_\alpha|)^{V_\varkappa} \Rightarrow (z = \varnothing)^{V_\varkappa})$ equivalent to the formula $(u \in |V_\alpha|^{V_\varkappa} \Rightarrow ((\forall y \leqslant \delta([\varphi|V_\varkappa](h(u)) \in V_y \Rightarrow z \subset y) \wedge ([\varphi|V_\varkappa](h(u)) \in V_z)^{V_\varkappa}) \wedge (u \notin |V_\alpha|^{V_\varkappa} \Rightarrow z = vrn)$. The last formula is equivalent to $(u \in |V_\alpha| \Rightarrow ([\varphi|V_\varkappa](h(u)) \in V_z)^{V_\varkappa} \wedge \forall y \in V_\varkappa(y \leqslant \delta \wedge ([\varphi|V_\varkappa](h(u)) \in V_y)^{V_\varkappa} \Rightarrow z \subset y) \wedge (u \notin |V_\alpha| \Rightarrow z = \varnothing)$.

The formula $([\varphi|V_\varkappa](h(u)) \in V_z)^{V_\varkappa} \wedge \forall y \in V_\varkappa(y \leqslant \delta \wedge ([\varphi|V_\varkappa](h(u)) \in V_y)^{V_\varkappa} \Rightarrow z \subset y)$ is equivalent to the formula $([\varphi|V_\varkappa](h(u)) \in V_z)^{V_\varkappa} \wedge \forall y \leqslant \delta(([\varphi|V_\varkappa](h(u)) \in V_y)^{V_\varkappa} \Rightarrow z \subset y)$ since $y \leqslant \delta$ implies $y \in V_\varkappa$.

Consider the formula $([\varphi|V_\varkappa](h(u)) \in V_z)^{V_\varkappa}$ in more detail. It is equivalent to the formula $\exists w(w \in V_z \wedge \langle h(u), w\rangle \in [\varphi|V_\varkappa]))^{V_\varkappa}$ equivalent to $\exists w \in V_\varkappa(w \in V_z \wedge (\langle h(u), w\rangle \in \{a \mid \exists b \in V_\varkappa \exists c \in V_\varkappa(a = \langle b, c\rangle \wedge \varphi^{V_\varkappa}(b, c, \vec{p}) \wedge \vec{p} \in V_\varkappa)\})^{V_\varkappa})$. This means that $\exists w \in V_\varkappa(w \in V_z \wedge (u \in V_\varkappa \wedge w \in V_\varkappa \wedge \varphi^{V_\varkappa}(h(u), w, \vec{p}) \wedge \vec{p} \in V_\varkappa)^{V_\varkappa})$, i.e. $\exists w(w \in V_z \wedge \varphi^{V_\varkappa}(h(u), w, \vec{p}) \wedge \vec{p} \in V_\varkappa)$.

Thus, the formula $[\psi|V_\varkappa]$ is equivalent to the formula

$$\{v \mid \exists u \in V_\varkappa \exists z \in V_\varkappa(v = \langle u, z\rangle \wedge (u \in |V_\alpha| \Rightarrow$$
$$\Rightarrow \exists w(w \in V_z \wedge \varphi^{V_\varkappa}(h(u), w, \vec{p}) \wedge \vec{p} \in V_\varkappa) \wedge$$
$$\wedge \forall y \in \varkappa(\exists w(w \in V_y \wedge \varphi^{V_\varkappa}(h(u), w, \vec{p}) \wedge \vec{p} \in V_\varkappa) \Rightarrow z \subset y)) \wedge (u \notin |V_\alpha| \Rightarrow z = \varnothing))\}.$$

This formula is equivalent to the formula $\{v \mid \exists u \in V_\varkappa \exists z \in V_\varkappa(v = \langle u, z\rangle \wedge (u \in |V_\alpha| \Rightarrow z = sm\{y \mid \exists w(w \in V_y \wedge \varphi^{V_\varkappa}(h(u), w, \vec{p}) \wedge \vec{p} \in V_\varkappa)\}) \wedge (u \notin |V_\alpha| \Rightarrow z = \varnothing))\}$. Now, we easily derive that $[\psi|V_\varkappa] \leftrightharpoons |V_\alpha| \to \varkappa$.

Consider the ordinal number $\gamma \equiv \cup \mathrm{rng}[\psi|V_\varkappa] = \sup \mathrm{rng}[\psi|V_\varkappa] \leqslant \varkappa$.

Suppose $\gamma = \varkappa$; but it is impossible since the cardinal $\varkappa$ is quasiregular. Therefore, $\gamma < \varkappa$.

Let $\mathrm{rng}[\varphi|V_\varkappa] \notin V_\varkappa$. Then, there is $t \in \mathrm{rng}[\varphi|V_\varkappa]$ such that $t \notin V_\gamma$. To this set, some $s \in \mathrm{dom}[\varphi|V_\varkappa]$ such that $\langle s, t\rangle \in [\varphi|V_\varkappa]$ and $s \in V_\alpha$ corresponds. Moreover, $h^{-1}(s) \in |V_\alpha|$.

Consider $\beta \equiv [\psi|V_\varkappa](h^{-1}(s))$. Since $h^{-1}(s) \in |V_\alpha|$, we get $\beta = sm\{y \mid \exists w(w \in V_y \wedge \varphi^{V_\varkappa}(s, w, \vec{p}) \wedge \vec{p} \in V_\varkappa)\}$ and since the formula φ is functional, we get $w = t$, i. e. $\beta = sm\{y \mid t \in V_y\}$. Since, by condition, $t \notin V_\gamma$, we get $\beta > \gamma$, what contradicts the condition $\mathrm{rng}[\psi|V_\varkappa] \subset \gamma$.

Thus, $\mathrm{rng}[\varphi|V_\varkappa] \subset V_\gamma \in V_\varkappa$. □

Corollary 1. *Let $\varkappa$ be a scheme-inaccessible cardinal number, $A \in V_\varkappa$, $\varphi(x, y, \vec{p})$ be a formula, and $[\varphi(x, y, \vec{p})|V_\varkappa] \leftrightharpoons A \to V_\varkappa$. Then, $\mathrm{rng}[\varphi(x, y, \vec{p})|V_\varkappa] \in V_\varkappa$.*

Proof. Since $\varkappa$ is a limit number, we get $V_\varkappa = \cup(V_\alpha \mid \alpha \in \varkappa)$, and therefore, $A \in V_\alpha$ for some $\alpha \in \varkappa$. By Lemma 3 (A.3.2) $A \subset V_\alpha$. Consider the formula $\psi(x, y, \vec{p}, A, \alpha) \equiv (x \in A \wedge \varphi(x, y, \vec{p}) \vee (x \in V_\alpha \backslash A \wedge y = \varnothing)$. It follows from $x \in V_\alpha\backslash A \subset V_\alpha \in V_\varkappa$ and the supertransitivity of V_α that $x, V_\alpha\backslash A \in V_\varkappa$. Hence, $(x \in V_\alpha\backslash A)^{V_\varkappa} \Leftrightarrow x \in V_\alpha\backslash A$. Therefore, $\psi^{V_\varkappa} \Leftrightarrow (x \in A \wedge \varphi^{V_\varkappa}) \vee (x \in V_\alpha\backslash A) \wedge y = \varnothing$. Since $g \equiv [\psi|V_\varkappa] \leftrightharpoons V_\alpha \to V_\varkappa$, we obtain $\mathrm{rng}\, g \in V_\varkappa$. It follows from $B \equiv \mathrm{rng}[\varphi|V_\varkappa] \subset \mathrm{rng}\, g \in V_\varkappa$ that $B \in V_\varkappa$. □

Lemma 5. *Let $\varkappa$ be a scheme-inaccessible cardinal number and $A \in V_\varkappa$. Then, $\cup A \in V_\varkappa$.*

Proof. Since $\varkappa$ is a limit ordinal number, we get $V_\varkappa = \cup(V_\delta \mid \delta \in \varkappa)$. For $A \in V_\varkappa$ there is $\delta \in \varkappa$ such that $A \in V_\delta$. Then, for every $a \in A$, we have $a \in V_\delta$. This implies that for every $a \in A$ the non-empty set $\{y \leqslant \delta \mid a \in V_y\}$ has a minimal element z_a.

Consider some bijection $h : |V_\delta| \to V_\delta$.

Consider the formula $\psi(u, z) \equiv (u \in |V_\delta| \wedge h(u) \in A \Rightarrow z = sm\{y \leqslant \delta \mid h(u) \in V_y\})$ $\wedge\ (u \notin |v_\delta| \vee h(u) \notin A \Rightarrow z = \varnothing)$. In this case, $[\psi|V_\varkappa] = \{v \mid \exists u \in V_\varkappa \exists z \in V_\varkappa(v = \langle u, z\rangle$ $\wedge\ \psi^{V_\varkappa}(u, z))$. Consider the formula $\psi^{V_\varkappa}(u, z)$ in more details. It is equivalent to the formula $((u \in |V_\delta|)^{V_\varkappa} \wedge (h(u) \in A)^{V_\varkappa} \Rightarrow (z = sm\{y \leqslant \delta \mid h(u) \in V_y\})^{V_\varkappa}) \wedge ((u \notin |V_\delta|)^{V_\varkappa} \vee$

$(h(u) \notin A)^{V_\varkappa} \Rightarrow (z = \varnothing)^{V_\varkappa})$. As in the proof of the previous lemma, we establish its equivalence to the formula $(u \in |V_\delta| \wedge h(u) \in A \Rightarrow ((\forall y \leqslant \delta(h(u) \in V_y) \Rightarrow z \subset y) \wedge (h(u) \in V_z))^{V_\varkappa}) \wedge (u \notin |V_\delta| \vee h(u) \notin A \Rightarrow z = \varnothing)$.

The formula $((\forall y \leqslant \delta(h(u) \in V_y) \Rightarrow z \subset y) \wedge (h(u) \in V_z))^{V_\varkappa}$ is equivalent to the formula $\forall y \in V_\varkappa(y \leqslant \delta \wedge h(u) \in V_y \Rightarrow z \subset y) \wedge (h(u) \in V_z)$ by virtue of $h, h(u), \delta, V_y, V_z \in V_\varkappa$. Since $y \leqslant \delta$ implies $y \in V_\varkappa$ in this case, this formula is equivalent to the formula $(\forall y \leqslant \delta(h(u) \in V_y) \Rightarrow x \subset y) \wedge (h(u) \in V_z)$, i.e. the formula $z = sm\{y \leqslant \delta \mid h(u) \in V_y\}$.

Thus, the formula $[\psi|V_\varkappa] = \{v \mid \exists u \in V_\varkappa \exists z \in V_\varkappa(v = \langle u, z\rangle \wedge ((u \in |V_\delta| \wedge h(u) \in A \Rightarrow z = sm\{v \leqslant \delta \mid h(u) \in V_y\}) \wedge (u \notin |V_\delta| \vee h(u) \notin A \Rightarrow z = \varnothing))$. We easily derive from this that $[\psi|V_\varkappa] \leftrightharpoons |V_\delta| \to \varkappa$.

Consider the ordinal number $\gamma \equiv \cup \operatorname{rng}[\psi|V_\varkappa] \in \varkappa$.

Suppose $\cup A \notin V_\varkappa$; then, there is $t \in \cup A$ such that $t \notin V_\gamma$. Since $t \in \cup A$, there exists $a \in A$ such that $t \in a$, where $a \notin V_\gamma$. If we consider $s \equiv [\psi|V_\varkappa](h^{-1}(a))$, the we obtain $s \leqslant \gamma \wedge a \in V_s$, where $a \in V_\gamma$. This contradiction yields $\cup A \in V_\varkappa$. □

Any transitive set A and any arbitrary formula $\sigma(x; \vec{u})$ of the ZF set theory define the *scheme set* $\langle\sigma(x;\vec{u})|A\rangle \equiv \{x \in A \mid \sigma^A(x;\vec{u})\}$ depending on the parameter $\vec{u}$.

Lemma 6. *Let $\varkappa$ be a scheme-inaccessible cardinal number and $\varphi(x, y, \vec{p})$ and $\sigma(x;\vec{u})$ be formulas. Then, $\forall \vec{p}, \vec{u} \in V_\varkappa \forall \varepsilon \in |V_\varkappa|([\varphi(x, y, \vec{p})|V_\varkappa] \leftrightharpoons \langle\sigma(x;\vec{u})|V_\varkappa\rangle \rightarrowtail\!\!\!\twoheadrightarrow \varepsilon \Rightarrow \langle\sigma(x;\vec{u})|V_\varkappa\rangle \in V_\varkappa)$.*

Proof. Denote $[\varphi|V_\varkappa]$, $\operatorname{rng}[\varphi|V_\varkappa]$, and $\langle\sigma|V_\varkappa\rangle$ for given $\vec{p}, \vec{u} \in V_\varkappa$ by f, R, and S, respectively. Consider the formula $\rho(y, \vec{p}, \vec{u}) \equiv \exists x(\sigma(x;\vec{u}) \wedge \varphi(x, y, \vec{p}))$. Then, $\rho^{V_\varkappa} = \exists x \in V_\varkappa(\sigma^{V_\varkappa}(x;\vec{u}) \wedge \varphi^{V_\varkappa}(x, y, \vec{p}))$ implies $\langle\rho(y, \vec{p}, \vec{u})|V_\varkappa\rangle \equiv \{y \in V_\varkappa \mid \exists x(x \in V_\varkappa \wedge \sigma^{V_\varkappa}(x;\vec{u}) \wedge \varphi^{V_\varkappa}(x, y, \vec{p}))\} = R$ for given $\vec{p}, \vec{u} \in V_\varkappa$. By Lemma 3, we have $R \subset \varepsilon \in |V_\varkappa| = \varkappa \subset V_\varkappa$. Hence, $R \in V_\varkappa$. Consider the formula $\psi(y, x, \vec{p}, \vec{u}) \equiv \sigma(x;\vec{u}) \wedge \varphi(x, y, \vec{p})$. Then, $\psi^{V_\varkappa} = \sigma^{V_\varkappa}(x;\vec{u}) \wedge \varphi^{V_\varkappa}(x, y, \vec{p})$ implies $g \equiv [\psi|V_\varkappa] \equiv \{t \in V_\varkappa * V_\varkappa \mid \exists y, x \in V_\varkappa(t = \langle y, x\rangle \wedge \sigma^{V_\varkappa}(x;\vec{u}) \wedge \varphi^{V_\varkappa}(x, y, \vec{p}))\} = f^{-1}$. Therefore, g is a bijective mapping from R onto S. Since $R \in V_\varkappa$, Corollary 1 to Lemma 4, we get $S = \operatorname{rng} g \in V_\varkappa$. □

A.8.2 Scheme-universal sets and their connection with scheme-inaccessible cumulative sets

A set U is said to be *scheme-universal* if it has the following properties:
1) $x \in U \Rightarrow x \subset U$ (the *transitivity property*);
2) $x \in U \Rightarrow \mathcal{P}(x), \cup x \in U$;
3) $x, y \in U \Rightarrow x \cup y, \{x, y\}, \langle x, y\rangle, x * y \in U$;
4) $\forall \vec{p} \in U \forall x(x \in U \wedge [\varphi(x, y, \vec{p})|U] \leftrightharpoons x \to U \Rightarrow \operatorname{rng}[\varphi(x, y, \vec{p})|U] \in U)$;
5) $\omega \in U$.

Similarly to Lemmas 1 and 2 (A.4.1), one can prove the following two lemmas.

Lemma 1. *If a set U is scheme-universal, then $x \in U \wedge y \subset x \Rightarrow y \in U$.*

This lemma shows that a scheme-universal set is quasitransitive. This and the transitivity property imply that a scheme-universal set is supertransitive.

Lemma 2. *If a set U is scheme-universal, then $\varnothing \in U$ and $1 \in U$.*

The following theorem is similar to Lemma 4 (A.4.1), but it has a completely different proof.

Theorem 1. *If U is a scheme-universal set, then $X \in U \Rightarrow |X| \in U$.*

Proof. If $X = \varnothing$, then $|X| = 0 \in U$. In what follows, we will assume that $X \neq \varnothing$. By the Zermelo principle (Theorem 1 (1.2.11)), we can assume that the set X is well-ordered. Take a minimal element m of the set X and consider the non-empty set $A \equiv \mathbf{On} \cap U$.

For every $x \in X$, denote by X_x the initial interval $]\leftarrow, x[\equiv \{t \in X \mid t < x\}$. By Lemma 1, it follows from $X_x \subset X \in U$ that $X_x \in U$. If for X_x there is a mapping f such that $\operatorname{dom} f = X_x$ and $\operatorname{rng} f \in A$, then $f \subset X_x * \operatorname{rng} f \in U$ implies $f \in U$ by virtue of Lemma 1.

Assume that for $x \in X$ there are isotone bijections f and g such that $\operatorname{dom} f = \operatorname{dom} g = X_x$ and $\operatorname{rng} f, \operatorname{rng} g \in \mathbf{On}$. If $x = m$, then $f = g = \varnothing$. If $x > m$, then consider the set $X' \equiv \{y \in X_x \mid f(y) \neq g(y)\}$. Suppose that $X' \neq \varnothing$; then X' has a minimal element n. Consider the set $X_n \subset X_x$. Suppose that $f(m) > 0$ and consider $z \in X_x$ such that $f(z) = 0$. Since f is isotone, it follows from $f(m) > f(z)$ that $m > z$, which is impossible. Consequently, $f(m) = 0 = g(m)$ implies $m < n$, i. e. $m \in X_n$. Clearly, $f|X_n = g|X_n$. Since an isotone bijection preserves any exact bounds, we obtain $f(n) = f(\sup X_n) = \sup f[X_n] = \sup g[X_n] = g(\sup X_n) = g(n)$. This contradicts the property $n \in X'$. This contradiction implies that $X' = \varnothing$, i. e. $f = g$.

Denote by $bij(f)$ and $isot(f)$ the formulas expressing the properties of the mapping f to be bijective and to be isotone, respectively. Consider the formula $\varphi(x, a; X) \equiv (X \neq \varnothing \wedge x \in X \wedge \exists f(func(f) \wedge \operatorname{dom}(f) = X_x \wedge \operatorname{rng}(f) = a \wedge bij(f) \wedge isot(f) \wedge On(a))$. By the above, we infer that for $x \in X$ only unique f can exists, and therefore, only unique a, i. e. the formula $\varphi(x, a; X)$ is functional. Consider the function $H \equiv [\varphi|U] \equiv \{z \in U * U \mid \exists x, a \in U(z = \langle x, a\rangle \wedge \varphi^U(x, a; X))\} \subset U * U$ depending on parameter $X \in U$.

Consider the formula $\varphi^U(x, a; X) = (X \neq \varnothing \wedge x \in X \wedge \exists f \in U(func^U(f) \wedge (\operatorname{dom}(f) = X_x)^U \wedge (\operatorname{rng}(f) = a)^U \wedge bij^U(f) \wedge isot^U(f)) \wedge On^U(a))$ for $x, a, X \in U$. By virtue of transitivity of U and the property of absoluteness of the corresponding subformulas of this formula (see Lemma 12.10 from [*Jech*, 2003]) we have $func^U(f) \Leftrightarrow func(f)$, $(\operatorname{dom}(f) = X_x)^U \Leftrightarrow (\operatorname{dom}(f) = X_x)$, $(\operatorname{rng}(f) = a)^U \Leftrightarrow (\operatorname{rng}(f) = a)$, $On^U(a) \Leftrightarrow On(a)$, $bij^U(f) \Leftrightarrow bij(f)$, and $isot^U(f) \Leftrightarrow isot(f)$. Therefore, $\varphi^U(x, a; X) \Leftrightarrow (X \neq \varnothing \wedge x \in X \wedge$

$\exists f \in U(func(f) \wedge \operatorname{dom} f = X_x \wedge \operatorname{rng}(f) = a \wedge bij(f) \wedge isot(f)) \wedge On(a))$ for $a, x, X \in U$. Consider the set $Z \equiv \operatorname{dom} H$. Since $Z \subset X \in U$ and U is quasitransitive, we get $Z \in U$. Hence, property 4 from the definition of a scheme-universal set implies $c \equiv \operatorname{rng} H \in U$. Thus, H is a function from Z into c. Since the set $e \equiv \varnothing \in U$ is an isotone bijection such that $\operatorname{dom} e = X_m = \varnothing \in U$ and $\operatorname{rng} e = \varnothing = 0 \in A$, we conclude that $H \neq \varnothing$.

Let $\alpha \in \beta \in c$. Then, $\beta = H(y)$ for some $\beta \in \mathbf{On}$ and $y \in Z$ such that $\varphi^U(y, \beta; X)$. This means that $y \in X$ and there is an isotone bijection $g \in U$ such that $\operatorname{dom}(g) = X_y$ and $\operatorname{rng}(g) = \beta$. Since β is an ordinal number, we conclude that α is also an ordinal number and $\alpha \subset \beta$. Consider $x \equiv g^{-1}(\alpha) \in X_y$. If $t \in X_x \subset X_y$, then $g(t) < g(x) = \alpha$, i. e. $g(t) \in \alpha$. If $\gamma \in \alpha$, then $\gamma \in \beta$ and we can take an element $z \equiv g^{-1}(\gamma) \in X_y$. It follows from $g(z) = \gamma < \alpha = g(x)$ that $z < x$, i. e. $z \in X_x$. Moreover, $g(z) = \gamma$. This implies that the function $f \equiv g|X_x$ maps X_x on α. Clearly, it is bijective and isotone. Since $f \subset g \in U$ and U is quasitransitive, we get $f \in U$. Hence, $\alpha = H(x) \in c$. This means that the set c is transitive.

Let $\alpha, \beta \in c$. Then, $\alpha = H(x)$ and $\beta = H(y)$ for some $\alpha, \beta \in \mathbf{On}$ and $x, y \in Z$ such that $\varphi^U(x, \alpha; X)$ and $\varphi^U(y, \beta; X)$. This means that $x, y \in X$ and there are isotone bijections $f, g \in U$ such that $\operatorname{dom}(f) = X_x$, $\operatorname{dom}(g) = X_y$, $\operatorname{rng}(f) = \alpha$, and $\operatorname{rng}(g) = \beta$. Since α and β are ordinal numbers, we see that $\alpha \in \beta$ or $\beta \in \alpha$ or $\alpha = \beta$. Therefore, the set c linearly ordered with respect to the binary relation $\in \cup =$.

Let $\varnothing \neq \alpha \subset c$. By the regularity axiom there is $r \in \alpha$ such that $r \cap \alpha = \varnothing$. Take any $s \in \alpha$ such that $s \in r$ or $s = r$. It follows from $r \cap \alpha = \varnothing$ that $s \notin r$. Hence, $s = r$. This means that r is a minimal element in α. Consequently, c is well-ordered.

Thus, we have proven that c is an ordinal number.

Check that the function H is bijective and isotone. Let $x, y \in Z$ and $x < y$. Then, for ordinal numbers $a \equiv H(x)$ and $b \equiv H(y)$ there are isotone bijections $f, g \in U$ such that $dom(f) = X_x$, $dom(g) = X_y$, $\operatorname{rng}(f) = a$, and $\operatorname{rng}(g) = b$. Consider the ordinal number $a' \equiv g(x) \in b$. If $t \in X_x$, then $t < x < y$ implies $g(t) < g(x) \equiv a'$, i. e. $g(t) \in a'$. If $\alpha \in a' \subset b$, then for the element $s \equiv g^{-1}(\alpha)$ it follows from $\alpha < a'$ that $s < x$, i. e. $s \in X_x$ and $g(s) = \alpha$. Therefore, the function $f' \equiv g|X_x$ is an isotone bijection from X_x onto a'. IT follows from the uniqueness proven above that $f' = f$. Thus, $f \subset g$ implies $a \subset b$. Suppose that $a = b$; then, $X_x = f^{-1}[a] = f'^{-1}[a] = g^{-1}[a] = g^{-1}[b] = X_y$. This contradicts the inequality $x < y$. Hence, $a \in b$, i. e. $a < b$. Conversely, let $x, y \in Z$ and $a < b$. Since $a \in b$, we can take the element $x'' \equiv g^{-1}(a) \in X_y$. If $t \in X_{x''}$, then $t < x'' < y$ implies $g(t) < g(x'') = a$, i. e. $g(t) \in a$. If $\alpha \in a \subset b$, then for the element $s \equiv g^{-1}(\alpha)$ it follows from $\alpha < a$ that $s = g^{-1}(\alpha) < g^{-1}(a) = x''$, i. e. $s \in X_{x''}$ and $g(s) = \alpha$. Therefore, the function $f'' \equiv g|X_{x''}$ is an isotone bijection from $X_{x''}$ onto a. Consider the isotone bijections $p \equiv f^{-1} : a \rightarrowtail\!\!\!\twoheadrightarrow X_x$ and $p'' \equiv f''^{-1} : a \rightarrowtail\!\!\!\twoheadrightarrow X_{x''}$. We prove, as above, that $p = p''$. Consequently, $X_x = X_{x''}$ implies $x = x'' < y$.

Thus, the surjective function H is isotone. Therefore, H is an isotone bijection from $Z \subset X$ onto $c \in A$. Assume that $Z \neq X$; then, the set $X \backslash Z$ has a minimal element y. Consider the initial interval X_y. If $x \in X_y$, then $x \in Z$, i. e. $X_y \subset Z$. Conversely, let $x \in Z$. Suppose that $y \leqslant x$. Consider the ordinal number $a \equiv H(x)$. For it there is a bijection $f \in U$ such that $dom(f) = X_x$ and $\operatorname{rng}(f) = a$. If $y = x$, then $y \in Z$, which is impossible.

Suppose $y < x$. Consider the ordinal number $b \equiv f(y) \in a$ and the isotone bijection $g \equiv f|X_y$ from X_y onto b. Since $g \subset f \in U$ and U is quasitransitive, we obtain $g \in U$. Hence, $b = H(y)$ and $y \in Z$, which is impossible. This contradiction implies $x < y$, i. e. $x \in X_y$. As a result, we obtain $X_y = Z$.

Consider the set $Y \equiv Z \cup \{y\}$ and define the function $f : Y \twoheadrightarrow a \equiv c + 1$ setting $f|Z \equiv H$ and $f(y) \equiv c$. Let $x, x' \in Y$ and $x < x'$. If $x, x' \in Z$, then $f(x) = H(x) < H(x') = f(x')$. If $x \in Z$ and $x' = y$, then $f(x) = H(x) \in c$ implies $f(x) < c = f(x')$. Therefore, f is strictly monotone. Conversely, let $f(x) < f(x')$ for $x, x' \in Y$. If $x, x' \in Z$, then $H(x) < H(x')$ implies $x < x'$. If $x \in Z$ and $x' = y$, then $x < y = x'$. If $x' \in Z$ and $x = y$, then $f(x') = H(x') \in c$. Consequently, $f(x') < c = f(x)$. This contradicts the condition. As a result, we obtain $x < x'$. Hence, f is isotone, and therefore, f is bijective.

Assume that $X \backslash Z \neq \{y\}$. Then, the non-empty set $X \backslash Y$ has a minimal element x. If $x = y$, then $x \in Y$, which is impossible. If $x < y$, then $x \notin X \backslash Z$, i. e. $x \in Z \subset Y$, which is also impossible. Hence, $y < x$. Let $t \in Y$. If $t \in Z = X_y$, then $t < y < x$, i. e. $t \in X_x$. If $t = y < x$, then we again get $t \in X_x$. Therefore, $Y \subset X_x$. Conversely, if $t \in X_x$, then $t < x$ implies $t \notin X \backslash Y$, i. e. $t \in Y$. As a result, we obtain $Y = X_x$. Consequently, f is an isotone bijection from X_x onto a. It follows from $y \in X \in U$ that $y \in U$. Hence, $\langle y, c\rangle \in U$ and $\{\langle y, c\rangle\} \in U$. Further, $H \subset Z * c \in U$ implies $H \in U$ by vitrue of the quasitransitivity of U, and therefore, $f = H \cup \{\langle y, c\rangle\} \in U$. This means that $a = H(x) \in c \in a$, which is impossible. This contradiction implies $X = Y$. Thus, f is an isotone bijection from X onto a. Since $a = c \cup \{c\} \in U$, we get $a \in A$.

If $Z = X$, then put $a \equiv c$ and $f \equiv H$.

Thus, in any case, we have constructed the isotone bijection $f \in U$ from X onto $a \in A$. Since $|X| \subset a \in U$, the quasitransitivity of U implies $|X| \in U$. □

Now, let us show that in a scheme-universal set, as in a universal set, the $\in$-induction principle similar to the $\in$-induction principle in ZF holds (see Lemma 4 (A.2.2) and Lemma 5 (A.4.1)).

Lemma 3. *Let U be a scheme-universal set, $C \subset U$, and $\forall x \in U(x \subset C \Rightarrow x \in C)$. Then, $C = U$.*

Proof. The proof is similar to the proof of Lemma 5 (A.4.1) except its central part changing as follows. Denote R_x^x by R_x. Consider the following formula of ZF: $\varphi(x, y) \equiv (x \in \omega \wedge y = R_x)$. Consider also the formula $\varphi^U(x, y) \equiv ((x \in \omega)^U \wedge (y = R_x)^U)$ for $x, y \in U$. Since $x, \omega, y, R_x \in U$, using the transitivity of the set U, one can prove that $(x \in \omega)^U \Leftrightarrow x \in \omega$ è $(y = R_x)^U \Leftrightarrow y = R_x$. Hence, $\varphi^U(x, y) \Leftrightarrow \varphi(x, y)$ for $x, y \in U$. Consider the function

$$
\begin{aligned}
[\varphi|U] \equiv &\{z \in U * U \mid \exists x, y \in U(z = \langle x, y\rangle \wedge \varphi^U(x, y))\} = \\
&\{z \in U * U \mid \exists x, y \in U(z = \langle x, y\rangle \wedge \varphi(x, y))\} = \\
&\{z \in U * U \mid \exists x, y \in U(z = \langle x, y\rangle \wedge x \in \omega \wedge y = R_x)\} \subset U * U.
\end{aligned}
$$

It is clear that $\mathrm{dom}[\varphi|U] = \omega \grave{e} A \equiv \mathrm{rng}[\varphi|U] = \{q \in U \mid \exists p \in \omega(q = R_p)\}$. Since $[\varphi|U] \leftrightarrows \omega \to U$, properties 5, 4, and 2 from the definition of a scheme-universal set guarantees that $A \in U$ and $Q \equiv \cup A \in U$. Evidently, $Q = \{y \mid \exists n \in \omega(y \in R_n)\}$. Hence, $R_n \subset Q$ for every $n \in \omega$, and therefore, $P = R_0 \subset Q$. It follows from the uniqueness property mentioned above that $u(m) = u(n)|(m+1)$ for all $m \leqslant n$, i. e. $R_k^m = R_k^n$ for every $k \in m+1$. Therefore, $\cup R_k = \cup R_m^m = \cup R_m^{m+1} = R_{m+1}^{m+1} \equiv R_{m+1}$. □

For a scheme-universal set, as well as for a universal set, the following analogue of the von Neumann identity holds (see Lemma 7 (A.3.3) and Lemma 6 (A.4.1)).

Lemma 4. *Let U be a scheme-universal set. Then,*
1) $V_\alpha \in U$ *for every* $\alpha \in \mathbf{On} \cap U$;
2) $U = \cup\{V_\alpha \subset U \mid \alpha \in \mathbf{On} \cap U\}$.

Proof. 1. Consider the sets $A \equiv \mathbf{On} \cap U$ and $C' \equiv \{\alpha \in A \mid V_\alpha \in U\}$ and the classes $\mathbf{C}'' \equiv \mathbf{On} \backslash U$ and $\mathbf{C} \equiv C' \cup \mathbf{C}''$. By Lemma 2 $0 = V_0 = \varnothing \in U$. Hence, $0 \in \mathbf{C}$. Let $\alpha \in \mathbf{C}$. Suppose that $\alpha+1 \in A$. Since $\alpha \in \alpha+1 \in U$, by property 1, we get $\alpha \in U$, and therefore, $\alpha \in A \cap \mathbf{C} = C'$. Then, the condition $V_\alpha \in U$ implies $V_{\alpha+1} = V_\alpha \cup \mathcal{P}(V_\alpha) \in U$ by virtue of properties 2 and 3. Therefore, $\alpha+1 \in C' \subset \mathbf{C}$. If $\alpha+1 \notin A$, then we immediately get $\alpha+1 \in \mathbf{C}'' \subset \mathbf{C}$.

Let α be a limit ordinal number and $\alpha \subset \mathbf{C}$. Suppose that $\alpha \in A$. If $\beta \in \alpha$, then $\beta \in \alpha \in U$ implies $\beta \in A \cap \mathbf{C} = C'$.

Consider the functional formula $\varphi(x,y) \equiv (x \in \alpha \Rightarrow y = V_x) \wedge (x \notin \alpha \Rightarrow y = \varnothing)$. Then, $[\varphi|U] = \{z \mid \exists x \in U \exists y \in U(z = \langle x,y\rangle \wedge (x \in \alpha \Rightarrow y = V_x)^U \wedge (x \notin \alpha \Rightarrow y = \varnothing)^U) \wedge \alpha \in U\}$. Since, by condition, $\alpha \in U$ and $x \in \alpha \Rightarrow v_x \in U$, this formula is equivalent to the formula $\{z \mid \exists x \in U \exists y \in U(z = \langle x,y\rangle \wedge (x \in \alpha \Rightarrow y = V_x) \wedge (x \notin \alpha \Rightarrow y = \varnothing))\}$. Evidently, in this case, $[\varphi|U] \leftrightarrows \alpha \to U$ and $\mathrm{rng}[\varphi|U] = \{V_y \mid y \in \alpha\}$. By property 4, $\{V_y \mid y \in \alpha\} \in U$, and therefore, $V_\alpha = \cup\{V_y \mid y \in \alpha\} \in U$. Hence, $\alpha \in C' \subset \mathbf{C}$. If $\alpha \notin A$, then we immediately get $\alpha \in \mathbf{C}'' \subset \mathbf{C}$.

By the transfinite induction principle we conclude that $\mathbf{C} = \mathbf{On}$, and therefore, $C' = A$.

2. It follows from the above that $V_\alpha \subset U$ for every $\alpha \in A$. Hence, $P \equiv \cup\{V_\alpha \mid \alpha \in A\} \subset U$. Check that P satisfies the $\in$-induction principle from Lemma 3. Consider the formula $\varphi(u,z) \equiv (u \in P \Rightarrow z = \mathrm{sm}\{\alpha \in A \mid p \in V_\alpha\}) \wedge (u \notin P \Rightarrow z = \varnothing)$.

Let $x \in U$ and $x \subset P$. If $x = \varnothing$, then $x \in P$. In what follows, we assume that $x \neq \varnothing$. If $y \in x \subset P$, then $y \in V_\alpha$ for some $\alpha \in A$. Hence, $\varphi(y) \leqslant \alpha \in U$ implies $\varphi(y) \in A$ by virtue of Lemma 1. Therefore, we can consider the functional formula $\psi \equiv \varphi|x$. It is easy to prove that $[\psi|U] \leftrightarrows x \to A$. By property 4, we get $R \equiv \mathrm{rng}[\psi|A] \in U$, and by property 2, we get $\rho \equiv \cup R \in U$. Since $\varnothing \neq R \subset \mathbf{On}$, Lemma 2 (A.2.2) implies that ρ is an ordinal number. Consequently, $\rho \in A$.

If $y \in x$, then $[\psi|U](y) \subset \rho$ implies $y \in V_{[\psi|U](y)} \subset V_\rho$ in view of Lemma 1 (A.3.2). Then, by virtue of Lemma 2 (A.3.2), it follows from $x \subset V_\rho \in V_{\rho+1}$ that $x \in V_{\rho+1}$. By property 3, it follows from $\rho+1 = \rho \cup \{\rho\} \in U$ that $\rho+1 \in A$. Consequently, $x \in P$.

Now, Lemma 3 implies $P = U$. □

Theorem 2. *Let U be a scheme-universal set. Then,*
1) $U = V_\varkappa$ *for* $\varkappa \equiv \sup(\mathbf{On} \cap U) = \cup(\mathbf{On} \cap U) \subset U$;
2) $\varkappa$ *is a scheme-inaccessible cardinal number;*
3) *the correspondence* $\mathbf{q} : U \mapsto \varkappa$ *such that* $U = V_\varkappa$ *is an isotone injective mapping from the class* $\mathbf{U}'$ *of all scheme-universal sets into the class* $\mathbf{In}'$ *of all scheme-inaccessible cardinal numbers.*

Proof. 1. Since $A \equiv \mathbf{On} \cap U$ contains the element ω by property 5, it is non-empty. Then, by Lemma 2 (A.2.2) $\varkappa$ is an ordinal number.

Let $\varkappa \in U$. Then, according to the properties of a scheme-universal set, we get $\varkappa + 1 = \varkappa \cup \{\varkappa\} \in U$. Since $\varkappa + 1 \in \mathbf{On}$, we get $\varkappa + 1 \in (\mathbf{On} \cap U)$, i. e. $\varkappa + 1 \leqslant \varkappa$, which is impossible. Hence, $\varkappa \notin U$.

Suppose that $\varkappa = \alpha + 1$ for some ordinal number α; then, $\alpha \in U$ since $\varkappa \subset U$ and $\alpha \in \varkappa$. Since $\varkappa = \alpha \cup \{\alpha\}$, the properties of a scheme-universal set imply $\varkappa \in U$, which is impossible.

Thus, $\varkappa$ is a limit ordinal number.

Therefore, $V_\varkappa = \cup\{V_\beta \mid \beta \in \varkappa\}$. By Lemma 4, we get $U = \cup\{V_\alpha \mid \alpha \in A\}$. If $\alpha \in A$, then $\alpha \leqslant \varkappa$ implies $V_\alpha \subset V_\varkappa$. Hence, $U \subset V_\varkappa$. If $\beta \in \varkappa = \cup A$, then $\beta \in \alpha \in A$ for some α. Property 1 guarantees that $\beta \in A$. Therefore, $V_\varkappa \subset U$.

Thus, $U = V_\varkappa$.

2. Obviously, $\varkappa \neq 0$.

Suppose that the ordinal number $\varkappa$ is not scheme-regular; then $\exists \alpha(\alpha \in \varkappa \wedge [\varphi|U] \leftrightarrows \alpha \to \varkappa \wedge \cup \operatorname{rng}[\varphi|U] = \varkappa)$ for some formula $\varphi(x, y, \vec{p})$. But $\alpha \in U$ and $\varkappa \subset U$ imply $\operatorname{rng}[\varphi|U] \in U$ in view of property 4 of a scheme-universal set, and therefore, $\cup \operatorname{rng}[\varphi|U] \in U$. Hence, $\cup \operatorname{rng}[\varphi|U] \neq \varkappa$. This contradiction provides that the ordinal number $\varkappa$ is scheme-regular.

Let λ be an ordinal number such that $\lambda < \varkappa$. Since $\lambda \in \varkappa \subset U$, property 2 implies $\mathcal{P}(\lambda) \in U$. By Theorem 1, we get $|\mathcal{P}(\lambda)| \in U$. Hence, $|\mathcal{P}(\lambda)| \leqslant \varkappa$. Assuming that $\varkappa = |\mathcal{P}(\lambda)| \in U$, as above, we arrive at a contradiction. Therefore, $|\mathcal{P}(\lambda)| < \varkappa$.

3. Using Lemma 1 (A.3.2), we conclude that $\varkappa$ is unique. Therefore, we can define the mapping $\mathbf{q} : \mathbf{U}' \to \mathbf{In}'$ such that $\mathbf{q}(U) = \varkappa$, where $U = V_\varkappa$. Lemma 1 (A.3.2) also guarantees that q is isotone. □

Corollary 1. *If U is a scheme-universal set, then $|U|$ is a scheme-inaccessible cardinal number,* $|U| = \sup(\mathbf{On} \cap U)$, *and* $U = V_{|U|}$.

Proof. By Theorem 2 $U = V_\varkappa$ for the scheme-inaccessible cardinal number $\varkappa \equiv \sup(\mathbf{On} \cap U)$. By Lemma 3 (A.8.1) $\varkappa = |V_\varkappa| = |U|$. □

Theorem 3. *For any set U the following conclusions are equivalent:*
1) *U is a scheme-inaccessible cumulative set;*
2) *U is a scheme-universal set.*

Proof. (1) $\vdash$ (2). Let $U = V_\varkappa$ for a scheme-inaccessible cardinal number $\varkappa > \omega$. Prove that the set U is scheme-universal. The property $x \in U \Rightarrow x \subset U$ follows from Lemma 3 (A.3.2).

The property $x \in U \Rightarrow \mathcal{P}(x) \in U$ follows from Lemma 6 (A.3.2). The property $x \in U \wedge y \in U \Rightarrow x \cup y \in U$ follows from Lemma 5 (A.3.2). The property $x \in U \wedge y \in U \Rightarrow \{x, y\}, \langle x, y\rangle, x * y \in U$ follows from Corollaries 1 and 2 to Lemma 5 (A.3.2). The property $\omega \in U$ follows from Lemma 7 (A.3.2). The property $x \in U \Rightarrow \cup x \in U$ follows from Lemma 5 (A.8.1). The property $x \in U \wedge [\varphi|U] \leftrightharpoons x \to U \Rightarrow \mathrm{rng}[\varphi|U] \in U$ follows from Corollary 1 to Lemma 4 (A.8.1). Thus, the set U is scheme-universal.

(2) $\vdash$ (1). This deduction follows directly from Theorem 2. □

A.8.3 Supertransitive standard models of the ZF set theory in the ZF set theory

In this subsection, we consider supertransitive standard models of the ZF set theory in the ZF set theory.

Proposition 1. *In the ZF set theory, the following conclusions are equivalent for a set U:*
1) *U is a supertransitive standard model set for ZF;*
2) *U is a scheme-universal.*

Proof. Consider an arbitrary sequence $s \equiv x_0, \dots, x_q, \dots$ of elements of the set U and translations of some axioms and axiom schemes of the ZF theory with respect to the standard interpretation $M \equiv (U, I)$ on the sequence s.

Instead of $\theta_M[s]$ and $M \models \varphi[s]$, we write θ^t and φ^t for terms θ and formulas φ, respectively.

To simplify the further presentation, we first consider the translations of certain simple formulas. Let u and v be some sets.

The formula $u \in v$ translates into the formula $(u \in v)^t = (\langle u^t, v^t\rangle \in B)$. Denote the last formula by γ. By definition, this formula is equivalent to the formula $(\exists x \exists y(x \in U \wedge y \in U \wedge \langle u^t, v^t\rangle = \langle x, y\rangle \wedge x \in y))$. Using the property of an ordered pair, we conclude that $u^t = x$ and $v^t = y$. Therefore, it is deduced from γ that $\delta \equiv (u^t \in v^t)$. By the deduction theorem, $\gamma \Rightarrow \delta$. Conversely, consider the formula δ. It was proven in ZF that for sets u^t and v^t there is a set z such that $z = \langle u^t, v^t\rangle$. By virtue of logical axiom scheme LAS3 we deduce from the formula δ the formula $(z = \langle u^t, v^t\rangle \Rightarrow u^t \in U \wedge v^t \in U \wedge z = \langle u^t, v^t\rangle \wedge u^t \in v^t)$. Since the formula $z = \langle u^t, v^t\rangle$ is deduced from the axioms, the formula $(u^t \in U \wedge v^t \in U \wedge z = \langle u^t, v^t\rangle \wedge u^t \in v^t)$ is also deduced. By LAS13, we deduce the formula $\exists x \exists y(x \in U \wedge y \in U \wedge z = \langle x, y\rangle \wedge x \in y)$ equivalent to the formula $z \in B$, and therefore, to the formula γ. By the deduction theorem, $\delta \Rightarrow \gamma$. Thus, the first equivalence $(u \in v)^t \Leftrightarrow u^t \in v^t$ holds.

The formula $v \subset w$ translates into the formula $(v \subset w)^t$. Denote the last formula by ε. By the first equivalence proven above is equivalent to the formula $\varepsilon' \equiv \forall u \in U(u \in$

$v^t \Rightarrow u \in w^t$). According to LAS11, from the formula ε' we deduce the formula $\varepsilon'' \equiv (x \in U \Rightarrow (x \in v^t \Rightarrow x \in w^t))$. If $x \in v^t$, then $v^t \in U$ and transitivity of U imply $x \in U$. Then, the formula ε'' implies $x \in v^t \Rightarrow x \in w^t$. Consequently, by the deduction theorem we deduce $(\varepsilon \Rightarrow (x \in v^t \Rightarrow x \in w^t))$. By the rule of generalization (Gen) the formula $\forall x(\varepsilon \Rightarrow (x \in v^t \Rightarrow x \in w^t))$ is deduced. By LAS12, we deduce the formula $(\varepsilon \Rightarrow \forall x(x \in v^t \Rightarrow x \in w^t))$, i. e. the formula $(\varepsilon \Rightarrow v^t \subset w^t)$.

Conversely, let the formula $v^t \subset w^t$ be given. Using the logical axioms, we sequentially deduce from it the formulas $(u \in v^t \Rightarrow u \in w^t)$ and $(u \in U \Rightarrow (u \in v^t \Rightarrow u \in w^t))$. By (Gen) we deduce the formula ε'. Hence, by the deduction theorem, we get the formula $(v^t \subset w^t \Rightarrow \varepsilon)$. Thus, the second equivalence $(v \subset w)^t \Leftrightarrow v^t \subset w^t$ holds.

We obtain the third equivalence $(u = v)^t \Leftrightarrow u^t = v^t$ in exactly the same way as the first equivalence.

In what follows, we will write not literal transformations of axioms but their equivalent variants obtained by using the mentioned equivalences.

The extensionality axiom A1 translates into the formula $A1^t \Leftrightarrow A1^U = \forall X \in U \forall Y \in U(\forall u \in U(u \in X \Leftrightarrow u \in Y) \Rightarrow X = Y)$.

The pair axiom A2 translates into the formula $A2^t \Leftrightarrow A2^U = \forall u \in U \forall v \in U \exists x \in U \forall z \in U(z \in x \Leftrightarrow z = u \vee z = v)$.

The union axiom A4 translates into the formula $A4^t \Leftrightarrow A4^U = \forall X \in U \exists Y \in U \forall u \in U(u \in X \Leftrightarrow \exists z \in U(u \in z \wedge z \in X))$.

The power set axiom A5 translates into the formula $A5^t \Leftrightarrow A5^U = \forall X \in U \exists Y \in U \forall u \in U(u \subset X \Leftrightarrow u \in Y)$.

The replacement axiom scheme AS6 translates into the formula scheme

$$AS6^t \Leftrightarrow \forall x \in U \forall y \in U \forall y' \in U(\varphi^\tau(x,y) \wedge \varphi^\tau(x,y') \Rightarrow y = y') \Rightarrow$$
$$\Rightarrow \forall X \in U \exists Y \in U \forall x \in U(x \in X \Rightarrow \forall y \in U(\varphi^\sigma(x,y) \Rightarrow y \in Y)),$$

where φ^τ and φ^σ are denotations of the formulas $M \models \varphi[s^\tau]$ and $M \models \varphi[s^\sigma]$ and s^τ and s^σ denote the corresponding changes of the sequence s under translation of the quantifier overformulas indicated above. Denote the last formula scheme by $\alpha \Rightarrow \beta$.

The empty set axiom A7 translates into the formula $A7^t \Leftrightarrow A7^U = \exists x \in U \forall z \in U(z \notin x)$.

The infinity axiom A8 translates into the formula $A8^t \Leftrightarrow A8^\tau \equiv \exists Y \in U(\varnothing^t \in Y \wedge \forall y \in U(y \in Y \Rightarrow (y \cup \{y\})^\tau \in Y))$, where

- the set $\varnothing^t$ is determined by the formula $A7^U$;
- the set $Z_1 \equiv Z_1(y) \equiv (y \cup \{y\})^\tau$ is determined by the formula $\exists Z_1 \in U \forall u \in U(u \in Z_1 \Leftrightarrow \exists z \in U(u \in z \wedge z \in \{y, \{y\}\}^\sigma))$;
- the set $Z_2 \equiv Z_2(y) \equiv \{y, \{y\}\}^\sigma$ is determined by the formula $\exists Z_2 \in U \forall u \in U(u \in Z_2 \Leftrightarrow u = y \vee u = \{y\}^\rho)$;
- the set $Z_3 \equiv Z_3(y) \equiv \{y\}^\rho$ is determined by the formula $\exists Z_3 \in U \forall u \in U(u \in Z_3 \Leftrightarrow u = y)$.

Since M is a model of the ZF theory, all the translations written above are deducible formulas in the ZF theory.

Therefore, the formula $A7^U$ asserts the existence of some $x \in U$ denoted by $\varnothing^t$. If $z \in U$, then $A7^U$ implies $z \notin x$. Now, suppose that $z \notin U$ and $z \in x$. Then, by virtue of transitivity of U we obtain $z \in U$, but it contradicts the condition. Hence, $z \notin x$. Thus, we deduce $z \notin x$. By (Gen) the formula $\forall z(z \notin x)$ meant $x = \varnothing$ is deduced. Thus, $\varnothing^t = \varnothing$ and $\varnothing \in U$.

Check now that if $y \in U$, then $Z_3 = \{y\}$. Let $u \in Z_3$. Since $Z_3 \in U$ and U is transitive, we get $u \in U$. If $u \in U$, then the formula for Z_3 presented above implies $u = y$, where $u \in \{y\}$. Therefore, $Z_3 \subset \{y\}$. Conversely, suppose $u \in \{y\}$. Then, $u = y$. Since $y \in U$, we get $u \in U$, where, by the same formula, we obtain $u \in Z_3$. Consequently, $\{y\} \subset Z_3$, which implies the required equality. This equality eliminates the index ρ in the formula for Z_2.

Using this equality, show that $Z_2 = \{y, \{y\}\}$. Let $u \in Z_2$. Then, as above, $u \in U$. Therefore, the formula for Z_2 presented above implies $u = y$ or $u = \{y\}$, where $u \in \{y, \{y\}\}$. Consequently, $Z_2 \subset \{y, \{y\}\}$. Conversely, suppose $u \in \{y, \{y\}\}$. Then, $u = y \in U$ or $u = \{y\} = Z_3 \in U$, i. e. $u \in U$ in both cases. Hence, by the same formula we get $u \in Z_2$, where $\{y, \{y\}\} \subset Z_2$. This implies the required equality. This equality eliminates the index σ in the formula for Z_1.

Finally, we verify that if $y \in U$, then $Z_1 = y \cup \{y\}$. Let $u \in Z_1$. Since $Z_1 \in U$ and U is transitive, we get $u \in U$. It follows from the formula for Z_1 that there exists $z \in U$ such that $u \in z$ and $z \in \{y, \{y\}\}$. Therefore, $u \in \cup\{y, \{y\}\} \equiv Z$, i. e. $Z_1 \subset Z$. Conversely, suppose $u \in Z$. Then, there exists $z \in \{y, \{y\}\}$ such that $u \in z$. It follows from $z = y \in U$ or $z = \{y\} = Z_3 \in U$ that $z \in U$. Then, the formula presented above implies $u \in Z_1$. Hence, $Z \subset Z_1$, which implies the required equality. This equality eliminates the index τ in the formula for $A8^\tau$.

All said above implies $A8^\tau = \exists Y \in U(\varnothing \in Y \wedge \forall y \in U(y \in Y \Rightarrow y\cup\{y\} \in Y))$. If $y \in Y$, then it follows from $Y \in U$ and transitivity of U that $y \in U$. Then, $y\cup\{y\} \in Y$ deduced from this formula. By the deduction theorem, we deduce $y \in Y \Rightarrow y \cup \{y\} \in Y$. By the generalization rule we deduce $\forall y \in Y(y \cup \{y\} \in Y)$. Thus, we deduce from $A8^t$ the formula $\exists Y \in U(\varnothing \in Y \wedge \forall y \in Y(y \cup \{y\} \in Y))$ almost coinciding with the infinity axiom and asserting the existence of an inductive set $Y \in U$.

Using the obtained translations, let us prove that the set U is scheme-universal.

Consider the formula $A2^U$. According to it, for any $u, v \in U$ there is a corresponding set $x \in U$. If $z \in x$, then by transitivity of U we get $z \in U$. Therefore, the formula $z = u \vee z = v$ is deduced from it. If $z = u \vee z = v$, then $z \in U$, and therefore, it is deduced from $A2^U$ that $z \in x$. Since $A2^U$ is deducible in ZF, by the deduction theorem and the generalization rule, the formula $\forall z(z \in x \Leftrightarrow z = u \vee z = v)$ is deduced. This formula means that $x = \{u, v\}$. Hence, $\{u, v\} \in U$. By the deduction theorem, we deduce the formula $u, v \in U \Rightarrow \{u, v\} \in U$. This implies $\{u\} \in U$ and $\langle u, v\rangle \in U$.

Consider the formula $A4^U$. According to it, for any $X \in U$ there is a corresponding set $Y \in U$. As above, transitivity of U implies $Y = \cup X$. Consequently, $\cup X \in U$, and by the deduction theorem, we deduce the formula $X \in U \Rightarrow \cup X \in U$. This implies that it follows from $X, Y \in U$ that $X \cup Y \equiv \cup\{X, Y\} \in U$.

Consider the formula $A5^U$. According to it, for any $X \in U$ there is a corresponding set $Y \in U$. Clearly, $Y \subset \mathcal{P}(X)$. Let $y \in \mathcal{P}(X)$. Then, $y \subset X \in U$ implies $y \in U$ in view of quasitransitivity of U. Hence, $Y = \mathcal{P}(X)$. Therefore, $\mathcal{P}(X) \in U$, and by the deduction theorem, we deduce $X \in U \Rightarrow \mathcal{P}(X) \in U$.

If $X, Y \in U$, then $X * Y \subset \mathcal{P}(\mathcal{P}(X \cup Y)) \in U$ implies $X * Y \in U$ in view of quasitransitivity of U.

Consider the inductive set $Y \in U$, whose existence was proven above. Since ω is the smallest among all inductive sets, we get $\omega \subset Y$. By the quasitransitivity property, this implies $\omega \in U$.

Property 4 from the definition of a scheme-universal set holds automatically.

Thus, we have proven that (1) $\vdash$ (2).

(2) $\vdash$ (1). Let U be a scheme-universal set. According to A.8.2, it is supertransitive. Consider the standard interpretation $M \equiv (U, I)$ of the theory ZF. We have translated above some axioms and axiom schemes of ZF under the interpretation M on the sequence s. Prove that they are deducible in ZF.

Consider the formula $\mathrm{A1}^U$. Let $X, Y \in U$ and $\chi \equiv \forall u \in U(u \in X \Leftrightarrow u \in Y)$. Take an arbitrary set u. If $u \in X$, then by transitivity of U, we obtain $u \in U$, and therefore, the formula $u \in Y$ is deduced. Similarly, we deduce $u \in Y$ from $u \in X$. Then, by the deduction theorem, the formula $u \in X \Leftrightarrow u \in Y$ is deduced, and by the generalisation rule (Gen), the formula $\forall u(u \in X \Leftrightarrow u \in Y)$ is deduced. According to the extensionality axiom A1, the equality $X = Y$ is deduced. By the deduction theorem, in ZF, the formula $\chi \Rightarrow X = Y$ is deduced. Further, by logical tools we deduce $\mathrm{A1}^t$.

Consider the formula $\mathrm{A2}^U$. Let $u, v \in U$. By the property of a universal set $\{u, v\} \in U$. It follows from the pair axiom A2 that $\forall z \in U(z \in \{u, v\} \Leftrightarrow z = u \vee z = v)$. Then, by LAS13, we deduce $\exists x \in U \forall z \in U(z \in x \Leftrightarrow z = u \vee z = v)$. Further, by logical tools we deduce $\mathrm{A2}^t$.

The separation axiom scheme AS3 translates into the formula scheme $AS3^t \Leftrightarrow \forall X \in U \exists Y \in U \forall u \in U(u \in Y \Leftrightarrow u \in X \wedge \varphi^\tau(u))$, where Y is not a free variable in $\varphi(u)$ and φ^τ denotes the formula $M \vDash \varphi[s^\tau]$, where s^τ denote the corresponding changes of the sequence s under translation of the quantifier overformulas $\forall x(\dots)$, $\exists Y(\dots)$, and $\forall u(\dots)$ indicated above. According to AS3 for $X \in U$ there is Y such that $\forall u \in U(u \in Y \Leftrightarrow u \in X \wedge \varphi^\tau(u))$. Since $Y \subset X \in U$, by Lemma 1 (A.8.2), we get $Y \in U$. Therefore, $AS3^t$ is deduced in ZF.

Similar to the deducibility of $\mathrm{A2}^t$, we verify the deducibility of $\mathrm{A4}^t$ and $\mathrm{A5}^t$.

Let us verify the deducibility of $\mathrm{AS6}^t$. Suppose that the formula α holds. Consider the set $X \in U$. According to the separation axiom scheme AS3, the set $F \equiv \{z \in U \mid \exists x, y \in U(z = \langle x, y\rangle \wedge \varphi^\sigma(x, y))\}$ exists. Clearly, $F \subset U * U$. It follows from transitivity of U that $X \subset U$. Therefore, there is a set $Z \equiv F[X] \subset U$. Consider the set $G \equiv \{z \in U \mid \exists x, y \in U(z = \langle x, y\rangle \wedge \varphi^\sigma(x, y) \wedge x \in X)\} = F|X \subset X * Z$. Let $x \in X \subset U$. If $x \notin \operatorname{dom} G$, then $G\langle x\rangle = \varnothing \in U$. Let $x \in \operatorname{dom} G$, i. e. $G\langle x\rangle \neq \varnothing$. If $y, y' \in G\langle x\rangle \subset U$, then the formula

$\varphi^\sigma(x, y) \wedge \varphi^\sigma(x, y')$ or, more precisely, the formula $\varphi^\sigma(x, y, X, Y) \wedge \varphi^\sigma(x, y', X, Y)$ holds (since X and Y can be free variables of the formula φ^σ). Since $\varphi^\tau(x, y) = \varphi^\sigma(x, y, X \parallel X_M[s], Y \parallel Y_M[s])$ and, similarly, for y', by virtue of LAS11 we obtain $\varphi^\tau(x, y) \wedge \varphi^\tau(x, y')$. Hence, the formula α implies $y = y'$. Therefore, $G\langle x\rangle = \{y\} \in U$. Thus, $G\langle x\rangle \in U$ for every $x \in X$. By properties 4 and 2 of a scheme-universal set, we get $Y_0 \equiv \operatorname{rng} G = \cup\{G\langle x\rangle \mid x \in X\} \in U$.

If $x \in X \subset U$, $y \in U$, and $\varphi^\sigma(x, y)$, then $\langle x, y\rangle \in G$ implies $y \in Y_0$. This means that the formula β deduced from the formula α. By the deduction theorem, the formula $\alpha \Rightarrow \beta$ is deduced, and therefore, the scheme $AS6^t$ is deduced.

According to Lemma 2 (A.8.2), $\varnothing \in U$. Then, we deduce $A7^t$ from this and A7.

Consider the formula $A8^\tau$ and the set $\omega \in U$. It follows from the above that $\varnothing^t = \varnothing \in \omega$. Let $y \in U$ and $y \in \omega$. Then, as above, we check that $Z_3 = \{y\}$, $Z_2 = \{y, \{y\}\}$ and $Z_1 = y \cup \{y\} \in \omega$. By the deduction theorem, we deduce $(y \in \omega \Rightarrow Z_1 \in \omega)$. Further, by logical tools we deduce $(\varnothing^t \in \omega \wedge \forall y \in U(y \in \omega \Rightarrow (y \cup \{y\})^\tau \in \omega))$, and therefore, the formula $A8^t$.

The regularity axiom translates into the formula $A9^t \Leftrightarrow A9^\tau \equiv \forall X \in U(X \neq \varnothing^t \Rightarrow \exists x \in U(x \in X \wedge (x \cap X)^\tau = \varnothing^t))$, where
- the set $\varnothing^t$ is determined by $A7^U$ and, as was proven above, it coincides with the empty set $\varnothing$,
- the set $Z \equiv (x \cap X)^\tau$ is determined by the formula $\exists Z \in U \forall u \in U(u \in Z \Leftrightarrow u \in x \wedge u \in X)$.

Check now that if $X \in U$ and $x \in U$, then $Z = x \cap X$. Let $u \in Z$. Since $Z \in U$ and U is transitive, we get $u \in U$. Therefore, it follows from the formula for Z that $u \in x \wedge u \in X$, i. e. $u \in x \cap X$. Hence, $Z \subset x \cap X$. Conversely, suppose $u \in x \cap X$, i. e. $u \in x \wedge u \in X$. Then, by virtue of transitivity we get $u \in U$ and the mentioned formula implies $u \in Z$. Thus, $x \cap X \subset Z$, which implies the required equality. This equality eliminates the index τ in the formula $A9^\tau$.

Let $X \in U$ and $X \neq \varnothing^t = \varnothing$. By the regularity axiom there is $x \in X$ such that $x \cap X = \varnothing$. By virtue of transitivity we get $x \in U$. Further, by logical tools we deduce $A9^t$.

Finally, the choice axiom A10 translates into the formula

$$A10^t \Leftrightarrow A10^\tau \equiv$$

$$\equiv \forall X \in U(X \neq \varnothing^t \Rightarrow \exists z \in U((z \leftrightharpoons \mathcal{P}(X)\backslash\{\varnothing\} \to X)^\tau \wedge$$

$$\wedge \forall Y \in U(Y \in (\mathcal{P}(X)\backslash\{\varnothing\})^\sigma \Rightarrow z(Y)^\sigma \in Y))),$$

where
- the set $Z_1 \equiv Z_1(X) \equiv (\mathcal{P}(X)\backslash\{\varnothing\})^\sigma$ is determined by the formula $\exists Z_1 \in U \forall u \in U(u \in Z_1 \Leftrightarrow u \in \mathcal{P}(X)^\rho \wedge u \notin \{\varnothing\}^\rho)$,
- the set $Z_2 \equiv z(Y)^\sigma$ is determined by the formula $\langle Y, Z_2\rangle^\rho \in z$,

and $\varphi^\tau \equiv (z \leftrightarrows \mathcal{P}(X)\backslash\{\varnothing\} \to X)^\tau$ denotes the formula $M \models \varphi[s^\tau]$, where s^τ denote the corresponding changes of the sequence s under translation of the quantifier overformulas $\forall X(\dots)$ and $\exists z(\dots)$ indicated above.

Fix the conditions $X \in U$ and $X \neq \varnothing^t = \varnothing \in U$. As was shown above, this implies $\mathcal{P}(X)^\rho = \mathcal{P}(X)$ and $\{\varnothing\}^\rho = \{\varnothing\}$. This equality eliminates the index ρ in the formula for Z_1.

Check that $Z_1 = \mathcal{P}(X)\backslash\{\varnothing\} \equiv Z$. Let $u \in Z_1$. Since $Z_1 \in U$ and U is transitive, we get $u \in U$. Then, the formula for Z_1 implies $u \in Z$. Hence, $Z_1 \subset Z$. Conversely, suppose $u \in Z$. Since $\mathcal{P}(X) \in U$ and U is transitive, we get $\mathcal{P}(X) \subset U$. This implies $u \in U$. Consequently, the mentioned formula implies $u \in Z_1$. Therefore, $Z \subset Z_1$, which implies the required equality. This guarantees that Z_1 is replaced by Z in the formula A10$^\tau$.

Consider the formula $\varphi \equiv (z \leftrightarrows Z \to X)$. It is the conjunction of the following three formulas: $\varphi_1 \equiv (z \subset Z * X)$, $\varphi_2 \equiv (\operatorname{dom} z = Z)$, and $\varphi_3 \equiv (\forall x(x \in Z \Rightarrow \forall y(y \in X \Rightarrow \forall y'(y' \in X \Rightarrow (\langle x, y\rangle \in z \wedge \langle x, y'\rangle \in z \Rightarrow y = y')))))$.

Then, $\varphi^\tau = \varphi_1^\tau \wedge \varphi_2^\tau \wedge \varphi_3^\tau$. Since $\varphi_1 = (\forall u(u \in z \Rightarrow \exists x\exists y(x \in Z \wedge y \in X \wedge u = \langle x, y\rangle)))$, we obtain $\varphi_1^\tau \Leftrightarrow (\forall u \in U(u \in z \Rightarrow \exists x \in U\exists y \in U(x \in Z \wedge y \in X \wedge u = \langle x, y\rangle^\sigma)))$. Similarly, it follows from $\varphi_2 = (\forall x(x \in Z \Rightarrow \exists y(y \in X \wedge \langle x, y\rangle \in z)))$ that $\varphi_2^\tau \Leftrightarrow (\forall x \in U(x \in Z \Rightarrow \exists y \in U(y \in X \wedge \langle x, y\rangle^\sigma \in z)))$.

Finally, $\varphi_3^\tau \Leftrightarrow (\forall x \in U(x \in Z \Rightarrow \forall y \in U(y \in X \Rightarrow \forall y' \in U(y' \in X \Rightarrow (\langle x, y\rangle^\sigma \in z \wedge \langle x, y'\rangle^\sigma \in z \Rightarrow y = y')))))$.

By the transitivity property for x, y, and y' in the formulas φ_1^τ, φ_2^τ è φ_3^τ, we have $x, y, y' \in U$. Therefore, as was shown above, the equalities $\langle x, y\rangle^\sigma = \langle x, y\rangle$ and $\langle x, y'\rangle^\sigma = \langle x, y'\rangle$ hold in these formulas. This implies that the formulas φ_1^τ, φ_2^τ, and φ_3^τ differ from the formulas φ_1, φ_2, and φ_3, respectively, only by bounded quantifier prefixes $\forall \dots \in U$ and $\exists \dots \in U$. For X by the choice axiom A10 there is z such that $\chi \equiv ((z \leftrightarrows Z \to X) \wedge \forall Y(Y \in Z \Rightarrow z(Y) \in Y))$.

Hence, the formula $\varphi = \varphi_1 \wedge \varphi_2 \wedge \varphi_3$ is deduced, and therefore, the formulas φ_1, φ_2, and φ_3 are also deduced.

Let $u \in U$ and $u \in z$. Then, it is deduced from the formula φ_1 that there are $x \in Z$ and $y \in X$ such that $u = \langle x, y\rangle$. Since $x \in Z \in U$, $y \in X \in U$, and U is transitive, we get $x, y \in U$. This means that for the given conditions $u \in U$ and $u \in z$ the formula $\exists x \in U\exists y \in U(x \in Z \wedge y \in X \wedge u = \langle x, y\rangle^\sigma)$ is deduced. Applying the deduction theorem and the deduction rules twice, we deduce the formula φ_1^τ.

Let $x \in U$ and $x \in Z$. Then, we deduce from the formula φ_2 that for x, there is $y \in X$ such that $\langle x, y\rangle \in z$. It follows from $y \in X \in U$ that $y \in U$. This means that for the given conditions $x \in U$ and $x \in Z$ the formula $\exists y \in U(y \in X \wedge \langle x, y\rangle^\sigma \in z)$ is deduced. Therefore, as above, we deduce the formula φ_2^τ.

Let $x \in U$, $x \in Z$, $y \in U$, $y \in X$, $y' \in U$, $y' \in X$, $\langle x, y\rangle \in z$, and $\langle x, y'\rangle \in z$. Then, it is deduced from φ_3 that $y = y'$. Applying alternately the deduction theorem and the deduction rules several times, we deduce the formula φ_3^τ.

Thus, the formula φ^τ is deduced.

Since $z \leftrightharpoons Z \to X$, we get $z\langle Y\rangle = \{z(Y)\}$. If $Y \in U$ and $Y \in Z_1 = Z$, then $Z_2 \in U$ implies $\langle Y, Z_2\rangle^\rho = \langle Y, Z_2\rangle$. Then, $\langle Y, Z_2\rangle \in z$ implies $Z_2 \in z\langle Y\rangle$, where $Z_2 = z(Y)$. Therefore, for the function z, the conditions $Y \in U$ and $Y \in Z_1$ imply $Z_2 = z(Y) \in Y$.

Since $Z = Z_1 \in U$ and $X \in U$, we get $Z * X \in U$. It follows from $z \subset Z * X$ by Lemma 1 (A.8.2) that $z \in U$.

Thus, we see that it is deduced from axiom A10 that there exists the object $z \in U$ satisfying the formula χ, implying the formula $\xi \equiv (\varphi^\tau \wedge \forall Y \in U(Y \in Z_1 \Rightarrow Z_2 \in Y))$. Consequently, we deduce the formula $\exists z \in U\ \xi$ from the fixed conditions. Applying alternately the deduction theorem and the generalization rule several times, we, as a result, deduce the formula $\mathrm{A10}^t$.

Thus, M is a supertransitive standard model of the ZF set theory. □

Corollary 1. *Any uncountable scheme-inaccessible cumulative set $V_\varkappa$ is a supertransitive standard model set for the ZF set theory.*

Proof. The assertion follows from Proposition 1 and Theorem 3 (A.8.2). □

Using Theorems 2 and 3 (A.8.2) and Proposition 1, we infer the following theorem.

Theorem 1. *In the ZF set theory, the following conclusions are equivalent for a set U:*
1) *$U = V_\varkappa$ for the scheme-inaccessible cardinal number $\varkappa = |U| = \sup(\mathbf{On} \cap U)$;*
2) *U is a supertransitive standard model set for the ZF set theory.*

Proof. (1) $\vdash$ (2). By Theorem 3 (A.8.2), the set $U = V_\varkappa$ is scheme-universal. By Proposition 1, the set U is a supertransitive standard model set.

(2) $\vdash$ (1). By Proposition 1 U is scheme-universal. By Theorem 2 (A.8.2) $U = V_\varkappa$ and $\varkappa = \sup(\mathbf{On} \cap U)$. By Corollary 1 to Theorem 2 (A.8.2) $\varkappa = |U|$. □

This theorem yields the *canonical form of supertransitive standard model sets for the ZF set theory*. Thus, we have described all natural models of the ZF set theory.

A.8.4 Tarski scheme sets. Characterization of all natural models of the ZF set theory

A set U in the ZF set theory will be called a *scheme Tarski set* if:
1) $x \in U \Rightarrow x \subset U$ (the *transitivity property*);
2) $x \in U \Rightarrow \mathcal{P}(x), \cup x \in U$;
3) $\forall \vec{p}, \vec{u} \in U(([\varphi(x, y, \vec{p})|U] \leftrightharpoons \langle\sigma(x; \vec{u})|U\rangle \rightarrowtail \varepsilon) \wedge \varepsilon \in |U| \Rightarrow \langle\sigma(x; \vec{u})|U\rangle \in U)$, where φ and σ are metavariables denoting arbitrary formulas of ZF;
4) $\omega \in U$ and $|U| \subset U$.

It follows from A.7.1 that any Tarski set of uncountable cardinality is a scheme Tarski set.

Lemma 1. *If U is a scheme Tarski set and $x \in U$, then $|x| \in |U|$.*

The proof is completely the same as the proof of Lemma 2 (A.7.1).

Lemma 2. *Any scheme Tarski set is supertransitive.*

The proof is completely the same as the proof of Lemma 1 (A.7.1).

Lemma 3. *If U is a scheme Tarski set and $x, y \in U$, then $\{x\}, \{x, y\}, \langle x, y\rangle \in U$.*

Proof. Consider the formulas $\sigma_1(s; u) \equiv (s = u)$, $\sigma_2(s; u, v) \equiv (s = u \vee s = v)$, $\varphi_1(s, t; u) \equiv (s = u \Rightarrow t = 0)$, and $\varphi_2(s, t; u, v) \equiv (s = u \Rightarrow t = 0) \wedge (s = v \wedge v = u \Rightarrow t = 0) \wedge (s = v \wedge v \neq u \Rightarrow t = 1)$. Then, $X_1 \equiv \langle \sigma_1(s; x)|U\rangle = \{x\}$ and $X_2 \equiv \langle \sigma_2(s; x, y)|U\rangle = \{x, y\}$.

Consider the correspondences $f_1 \equiv [\varphi_1(s, t; x)|U]$ and $f_2 \equiv [\varphi_2(s, t; x, y)|U]$. If $s \in X_1$ and $\langle s, t\rangle \in f_1$, then $s = x$ and $t = 0$. Therefore, f_1 is an injective mapping from X_1 into $\{0\} \equiv 1 \in |U|$. By property 3, $X_1 \in U$.

Now, let $s \in X_2$ and $\langle s, t\rangle \in f_2$. If $s = x$, then $t = 0$. If $s = y \wedge y = x$, then $t = 0$. If $s = y \wedge y \neq x$, then $t = 1$. Therefore, f_2 is an injective mapping from X_2 into $\{0, 1\} = 2 \in |U|$. By property 3, $X_2 \in U$. Thus, we conclude that $\langle x, y\rangle \in U$. □

Corollary 1. *If U is a scheme Tarski set and $x, y \in U$, then $x \cup y \in U$.*

Proof. Lemma 3 and property 2 imply $x \cup y = \cup\{x, y\} \in U$. □

Corollary 2. *If U is a scheme Tarski set and $x, y \in U$, then $x * y \in U$.*

Proof. Since $x * y \subset \mathcal{P}(\mathcal{P}(x \cup y)) \in U$, by Lemma 2, we get $x * y \in U$. □

Lemma 4. *Let U be a scheme Tarski set, $\varphi(a, b; \vec{r})$ be a formula in the ZF, and $x \in U$. If $\vec{r} \in U$ and $[\varphi(a, b; \vec{r})|U] \leftrightharpoons x \to U$, then $\operatorname{rng}[\varphi(a, b; \vec{r})|U] \in U$.*

Proof. Denote $[\varphi(a, b; \vec{r})|U]$ and $\operatorname{rng}[\varphi|U]$ for a given $\vec{r} \in U$ by f and R, respectively. Consider the formula $\rho(b; \vec{r}, y) \equiv \exists a \in y\varphi(a, b; \vec{r})$. Then, $\langle \rho(b; \vec{r}, x)|U\rangle = \{b \in U \mid \exists a \in U(a \in x \wedge \varphi^U(a, b; \vec{r}))\} = R$ for given $\vec{r}, x \in U$.

Consider the formula $\psi(b, c; \vec{r}, y) \equiv \forall a \in c(a \in y \wedge \varphi(a, b; \vec{r})) \wedge \forall a \in y(\varphi(a, b; \vec{r}) \Rightarrow a \in c)$ and the correspondence $[\psi(b, c; \vec{r}, y)|U] = \{t \in U*U \mid \exists b, c \in U(t = \langle b, c\rangle \wedge \forall a \in c(a \in y \wedge \varphi^U(a, b; \vec{r})) \wedge \forall a \in y(\varphi^U(a, b; \vec{r}) \Rightarrow a \in c))\}$. It is easily proven that the correspondence $g \equiv [\psi(b, c; \vec{r}, x)|U]$ is an injective mapping from R into $S \equiv \mathcal{P}(x)$ such that $g(b) = f^{-1}(b)$ for every $b \in R$.

Properties 2 and 4 and Lemma 1, we get $S \in U$, $|S| \in |U|$, and $|S| \in U$. Consider some bijection $h : S \rightarrowtail\!\!\!\!\!\twoheadrightarrow |S|$. By Corollary 2 to Lemma 3 $S * |S| \in U$. It follows from $h \subset S * |S|$ that $h \in U$ by virtue of Lemma 2. Consider the formula $\chi(s, t; e) \equiv (\langle s, t\rangle \in e)$. Then, for value of the parameter e equal to h, we have $[\chi(s, t; h)|U] \equiv$

$\{z \in U * U \mid \exists s, t \in U(z = \langle s, t\rangle \wedge \langle s, t\rangle \in h)\} = h$. It remains to take the composition of the mappings g and h. For this purpose consider the formula $\zeta(b, t; \vec{r}, y, e) \equiv \exists s \in \mathcal{P}(y)(\forall a \in s(a \in y \wedge \varphi(a, b; \vec{r})) \wedge \forall a \in y(\varphi(a, b; \vec{r}) \Rightarrow a \in s) \wedge \langle s, t\rangle \in e)$ and the correspondence $[\zeta(b, t; \vec{r}, y, e)|U] = \{z \in U * U \mid \exists b, t \in U(z = \langle b, t\rangle \wedge \exists s \in \mathcal{P}(y)(\forall a \in s(a \in y \wedge \varphi^U(a, b; \vec{r})) \wedge \forall a \in y(\varphi^U(a, b; \vec{r}) \Rightarrow a \in s) \wedge \langle s, t\rangle \in e)\}$. It is clear that $F \equiv [\zeta(b, t; \vec{r}, x, h)|U] = h \circ g$. Consequently, F is an injective mapping from R to $|S| \in U$. By property 3, $R \in U$. □

Proposition 1. *Any scheme Tarski set is scheme-universal.*

Proof. The assertion follows from properties 2 and 4, Lemma 3, Corollaries 1 and 2 to Lemma 3, and Lemma 4. □

Theorem 1 (the Zakharov theorem on the characterization of natural models of the ZF set theory). *In the ZF set theory, the following conclusions are equivalent for a set U:*
1) *U is a scheme-inaccessible cumulative set, i. e. $U = V_\varkappa$ for some scheme-inaccessible cardinal number $\varkappa$;*
2) *U is a scheme-universal set;*
3) *U is a supertransitive standard model set for the ZF set theory;*
4) *U is a scheme Tarski set.*

Proof. The equivalence of (1) and (2) follows from Theorem 3 (A.8.2).

The equivalence of (2) and (3) follows from Theorem 1 (A.8.3). The deduction (4) $\vdash$ (2) follows from Proposition 1 (A.8.3). (1) $\vdash$ (4). Let $U = V_\varkappa$ for some scheme-inaccessible cardinal number $\varkappa > \omega$. Show that U is a scheme Tarski set. The property $x \in U \Rightarrow x \subset U$ follows from Lemma 3 (A.3.2). The property $x \in U \Rightarrow \mathcal{P}(x) \in U$ follows from Lemma 6 (A.3.2). The property $x \in U \Rightarrow \cup x \in U$ follows from Lemma 5 (A.8.1). Property 3 follows from Lemma 6 (A.8.1). The property $\omega \in U$ follows from Lemma 7 (A.3.2). Finally, the property $|U| \subset U$ follows from Lemma 1 (A.3.2) and Lemma 1 (A.8.2). Thus, U is a scheme Tarski set. □

B Local theory of sets as a foundation for category theory and its connection with the Zermelo – Fraenkel set theory

Introduction

The crises that arose in the naive set theory at the beginning of the 20th century brought to the origin of some strict axiomatic theories of mathematical totalities.

The most widely used of them are the *theory of sets in Zermelo – Fraenkel's axiomatics* (ZF) (see A.2 and also [*Kuratowski and Mostowski*, 1967; *Tourlakis*, 2003b]) and the *theory of classes and sets in Neumann – Bernays – Gödel's axiomatics* (NBG) (see 1.1 and also [*Kelley*, 1975; *Mendelson*, 1997]).

These axiomatic theories eliminated all the known paradoxes of naive set theory at the expense of the sharp restriction of possible expressive means. At the same time, they gave the opportunity to include almost all then-existing mathematical objects and constructions within the framework of these theories.

In 1945, the new mathematical notion of a *category* was introduced by Eilenberg and MacLane in their initial paper [*Eilenberg and MacLane*, 1945]. Henceforth, the category theory became an independent branch of mathematics. But from the very beginning, the category theory unfortunately not did not go within the framework of the theory of sets in Zermelo – Fraenkel's axiomatics but even within the framework of the theory of classes and sets in Neumann – Bernays – Gödel's axiomatics NBG (see [*Eilenberg and MacLane*, 1945]).

By this reason, S. MacLane [1961] put the *general problem of constructing a new and more flexible axiomatic set theory that could serve as an adequate logical foundation for all the naive category theory*.

Different variants of new axiomatic theories of mathematical totalities, adjusted for ones or others needs of category theory, were proposed by C. Ehresmann [1957], P. Dedecker [1959], J. Sonner [1962], A. Grothendieck [*Gabriel*, 1962], N. da Costa [1965, 1967], J. Isbell [1966], S. MacLane [1969, 1971], S. Feferman [1969], H. Herrlich, and G. Strecker [1979], and others.

C. Ehresmann, P. Dedecker, J. Sonner, and A. Grothendieck introduced the important notion of a (*categorical*) *universe* U, i. e. such totality of objects, which satisfies the following properties of closedness:

1) $X \in U \Rightarrow X \subset U$ (Ehresmann – Dedecker did not propose this property);
2) $X \in U \Rightarrow \mathcal{P}(X), \cup X \in U$;
3) $X, Y \in U \Rightarrow X \cup Y, \{X, Y\}, \langle X, Y\rangle, X \times Y \in U$;
4) $X \in U \wedge (F \in U^X) \Rightarrow \operatorname{rng} F \in U$;
5) $\omega \in U$ ($\omega = \{0, 1, 2, \dots\}$ is here the set of all finite ordinal numbers).

https://doi.org/10.1515/9783110550948-003

Within the framework of such a universe, it is possible to develop a quite rich category theory, in particular, the theory of categories with direct and inverse limits. To satisfy all needs of category theory, these authors proposed to strengthen the ZF or the NBG set theory by the strong *axiom of universality*, postulating that *every set belongs to some universal set*. In MacLane's axiomatics, the *existence of at least one universal set* is postulated. Similar versions were proposed by J. Isbell and S. Feferman. Herlich – Srecker's axiomatics have dealings with objects of three types: *sets, classes, and conglomerations*.

Within the framework of each of these axiomatics, some definitions of a category and a functor are given. But the notions of a category given in [*Isbell*, 1966; *MacLane*, 1969; 1971; *Feferman*, 1969; *Herrlich and Strecker*, 1979] are not closed with respect to such important operations of naive category theory as "*the category of categories*" and "*the category of functors*" (see [*Hatcher*, 1982, 8.4]).

Within the framework of the axiomatic theories from [*Ehresmann*, 1957; *Dedecker*, 1959; *Sonner*, 1962; *Gabriel*, 1962], the definition of a *$\mathcal{U}$-category, $\mathcal{U}$-functor* and *natural $\mathcal{U}$-tranformation* consisting of subsets of a universal set $\mathcal{U}$ is given. This notion of a category is closed with respect to such operations as the *$\mathcal{V}$-category of $\mathcal{U}$-categories* and the *$\mathcal{V}$-category of $\mathcal{U}$-functors*, where $\mathcal{V}$ is some universal set containing the universal set $\mathcal{U}$ as an element.

Axiomatics from [*Da Costa*, 1965; 1967] also give definitions of a category and a functor closed under mentioned operations of naive category theory. But N. da Costa uses logic with non-constructive rule of deduction $\varphi(V_1), \varphi(V_2), \varphi(V_3), \dots \vdash \forall t\, \varphi(t)$, where $V_1, V_2, \dots, V_n, \dots$ is an infinite sequence of constants which denotes universes like the NBG-universe (see [*Da Costa*, 1967]).

In connection with logical difficulties of constructing a set-theoretical foundation for all naive category theory, different attempts to construct purely arrow-axiomatic foundations were undertaken (see [*Lawvere*, 1966; *Blanc and Preller*, 1975; *Blanc and Donnadieu*, 1976]). But in attempts of arrow-axiomatic description of indexed categories and fiber categories, their own logical difficulties appeared (see [*Hatcher*, 1982, 8.4]).

Consider now more precisely what the mentioned above formulation of MacLane's problem means. To do it, we need to have the strict definition of a category. This definition became possible after the *elementary* (≡*first-order*) *category theories* T_c took their shapes. There are many such theories, two-sorted and one-sorted (see, for example, [*MacLane*, 1971, I.3, I.8], [*Hatcher*, 1982, 8.2], [*Goldblatt*, 1979, 2.3, 11.1]). But for every set theory S, there exist canonical one-to-one correspondences between the totalities of models of these theories in the theory S. Therefore, for the strict definition of a category, one can use any elementary theory T_c. According to the definition of Lawvere [1966] (see also [*Hatcher*, 1982, 8.2]), a *category in some set theory S* is any model of the theory T_c in the theory S. *The complete [a partial] formalization of naive category theory C in the set theory S* is an adequate translation of all [some] notions and constructions of naive category theory C into strict notions and constructions for categories (as models of the theory T_c) in the set theory S.

Along with the notion of a category in the set theory S, there exists also the notion of an *abstract category in S*. An *abstract category in the set theory S* is any abstract model of the theory T_c in the set theory S. *The complete [a partial] abstract formalization of naive category theory C in the set theory S* is an adequate translation of all [some] notions and constructions of naive category theory C into strict notions and constructions for abstract categories (as abstract models of the theory T_c) in the set theory S.

According to these definitions, abstract categories in the ZF set theory can be considered also on *classes* (as abstracts of the ZF set theory), and abstract categories in the set theory NBG can be considered also on *assemblies* (as abstracts of the theory NBG). However, the complete abstract formalization of naive category theory C in any set theory S is impossible because it is impossible to take abstracts of abstracts. Only a partial abstract formalization of C in S is possible. It means that the notions of an abstract category in S and a partial abstract formalization of C in S have only an auxiliary value with respect to the notions of a category in S and the complete formalization of C in S.

Therefore, more precisely, the MacLane's problem is understood as *constructing a set theory that admits the complete formalization of naive category theory C in this set theory*.

According to the definitions, mentioned above categories in the ZF set theory can be considered only on sets, but categories in the set theory NBG can be considered also on classes. But the complete formalization of naive category theory C in these set theories is impossible because it is impossible to define the operations "the category of categories" and "the category of functors" (at least nowadays we have no methods of approach to this formalization).

By this reason, C. Ehresman, P. Dedecker, J. Sonner and A. Grothendieck proposed the idea of formalization of naive category theory C within the framework of the ZF+AU set theory with the axiom of universality AU stronger than the ZF set theory. Using the totality of universal sets, we can make the complete adequate translation of all notions and constructions of naive theory C to the strict notions and constructions for categories in ZF+AU. It holds also for the sets theories of N. da Costa.

For other mentioned set theories, such a complete translation is impossible. Therefore, the theory ZF+AU and the sets theories of da Costa are the most adequate with respect to MacLane's problem. Moreover, by virtue of the deficiency of da Costa's set theories mentioned above the set theory ZF+AU is more preferable.

However, the theory ZF+AU is too strong for the complete formalization of category theory by virtue of redundancy of the totality of all universal sets, because for formalization of the operations "the category of categories" and "the category of functors" it is sufficient to have only a countable totality of universal sets as it is done in da Costa's axiomatics.

The ZF+AU(ω) set theory with the axiom of ω-universality AU(ω) postulating the existence of an infinite totality of universal sets is weaker than ZF+AU and satisfies all needs of the theory C. But it leaves behind the limits of categorical consideration such mathematical systems that are not elements of universal sets from this infinite totality.

Therefore, the necessity arose to create a set theory S having an infinite totality **U** of some objects of the theory S, called *universes* and satisfying the following conditions:

1) S has to be in some sense weaker than the redundant ZF+AU;
2) S has to satisfy all needs of the theory C in such an extent as the ZF+AU(ω) does it;
3) in the contrary to ZF+AU(ω) the theory S has not to have objects laying outside of the totality of universes **U**.

In 2000, V. K. Zakharov proposed in the capacity of more adequate foundation of category theory the *local theory of sets* (*LTS*), satisfying all these conditions, and in 2003, he proposed its equiconsistent strengthening: the *locally minimal theory of sets* (*LMTS*) (see [*Zakharov*, 2005b; *Zakharov et al.*, 2006]). The main idea of the LTS and the the LMTS consist in that *for the construction of a set theory satisfying conditions* 1 – 3 *it is not necessary to assign a global set-theoretical structure, but it is sufficient to assign only local variants of this structure in each universe U.*

The local theory of sets tries to remain all positive that is contained in the globally internal concept of Ehresmann – Dedecker – Sonner – Grothendieck.

The locally external ideology of the LTS states that it is necessary to take the NBG-universe as the basic one and *to duplicate externally its local copies*, and to get some hierarchy of universes with the following properties:

1) every class belongs as a set to some universal class, which is a usual NBG-universe;
2) all subclasses of a given universal class are sets of any larger universal class (*the property of value change*);
3) there exists the least universal class (≡ the *infra-universe*), belonging to all other universal classes.

The first of these properties is similar to the axiom of universality mentioned above.

Thus, in the LTS the notion of a big totality becomes relative: totalities that are "*big*" in one universe become "*small*" in any larger universe.

This appendix is devoted to rigorous development of the expressed ideas.

In the first section, all proper axioms and axiom schemes of the LTS are stated and the important set-theoretical constructions are defined.

In the second section, such key categorical constructions as "the category of categories" and "the category of functors" are formalized in the LTS. As in the globally internal concept, in the LTS, all categorical notions and constructions are defined only within the framework of local NBG-universes. Therefore, categories under consideration are called *local*.

In the third section, the notions of ordinals, cardinals, and inaccessible cardinals in the LTS are introduced, cumulative Mirimanov – Neumann classes are constructed, and the connection between universal classes and cumulative classes with indices that are inaccessible cardinals are stated. Through that, it was proven that the assembly of all universal classes in the LTS is well-ordered with respect to the order by inclusion $U \subset V$ and has some more complicated structures. Here, we also carry out the *globalization*, i. e. for the assembly of all classes we define almost all local set-theoretical constructions excluding the most important construction of the full union, which is basic for the construction by transfinite induction.

In the fourth section, the relative consistency between the LTS and the ZF set theory with some additional axioms is stated. It is also shown there that the LTS satisfies conditions 1 – 3 on a set theory S indicated above and, therefore, gives the solution of the MacLane problem of constructing an adequate foundation for the naive category theory. The stricter version of the solution is considered in B.6.3.

In the fifth section, the *method of abstract interpretation* is considered. Further, with its help, the independence of the introduced additional axioms and the undeducibility in the LTS of the global axiom scheme of replacement are stated.

Finally, in the seventh section, the finite axiomatizability of the LTS and the NBG set theory is proven.

For the reader's convenience, this appendix contains all necessary notions as was done in the whole book. Proofs are all detailed, making them useful for young mathematicians.

B.1 The local theory of sets

B.1.1 Proper axioms and axiom schemes of the local theory of sets

The local theory of sets is a first-order theory with two predicate symbols: a binary predicate symbol of *belonging* $\in$ (write $A \in B$) and an unary predicate symbol of *universality* $\bowtie$ (write $A \bowtie$), and also with two constants $\varnothing$ (*the empty class*) and $\mathfrak{a}$ (*the infra-universe*). The set of functional letters in the LTS is empty. By this reason, terms in the LTS are constant symbols and variables.

Objects of the LTS are called *classes*.

A notation $\varphi(\vec{u})$ is used for the formula $\varphi(u_0, \dots, u_{n-1})$, where $u_0, \dots, u_{n-1}$ are free variables of the formula φ.

By technical reasons, it is useful to consider the totality $\mathbf{C}$ of all classes A satisfying a given formula $\varphi(x)$. This totality $\mathbf{C}$ is called the *assembly defined by the formula* φ. The totality $\mathbf{C}$ of all classes A, satisfying the formula $\varphi(x, \vec{u})$, is called the *assembly defined by the formula* φ *through the parameter* $\vec{u}$. Along with these, we will use the notations

$$A \in \mathbf{C} \equiv \varphi(A), \ A \in \mathbf{C} \equiv \varphi(A, \vec{u}) \ \text{and} \ \mathbf{C} \equiv \{x \mid \varphi(x)\}, \ \mathbf{C} \equiv \{x \mid \varphi(x, \vec{u})\}.$$

If $\mathbf{C} \equiv \{x \mid \varphi(x)\}$ and φ contains only one free variable x, then the assembly $\mathbf{C}$ is called *well-defined by the formula* φ. Assemblies will be usually denoted by semibold Latin letters.

Every class A can be considered as the assembly $\{x \mid x \in A\}$.

The *universal assembly* is the assembly of all classes $\overline{\mathbf{V}} \equiv \{x \mid x = x\}$.

An assembly $\mathbf{C} \equiv \{x \mid \varphi(x)\}$ is called a *subassembly of an assembly* $\mathbf{D} \equiv \{x \mid \psi(x)\}$ (in notation, $\mathbf{C} \subset \mathbf{D}$) if $\forall x(\varphi(x) \Rightarrow \psi(x))$. Assemblies $\mathbf{C}$ and $\mathbf{D}$ are called *equal* if $(\mathbf{C} \subset \mathbf{D}) \wedge (\mathbf{D} \subset \mathbf{C})$ (in notation, $\mathbf{C} = \mathbf{D}$).

We will use the notation $\{x \in A \mid \varphi(x)\} \equiv \{x \mid x \in A \wedge \varphi(x)\}$.

A1. (The *extensionality axiom.*) $\forall y \forall z ((y = z) \Rightarrow (\forall X (y \in X \Leftrightarrow z \in X)))$.

Let α be some fixed class. A class A will be called a *class of the class* α ($\equiv$ *α-class*) if $A \subset \alpha$. A class A is called a *set of the class* α ($\equiv$ *α-set*) if $A \in \alpha$.

A formula φ is called *α-predicative* (see [*Mendelson*, 1997, 4.1]) if for all variables x all symbol-strings $\forall x$ and $\exists x$, occurring in the formula φ, are situated only in the following positions: $\forall x(x \in \alpha \Rightarrow \ldots)$ and $\exists x(x \in \alpha \wedge \ldots)$.

In what follows, we use symbol-strings of the form $\vec{x}, \vec{p} \equiv x_1, \ldots, x_m, p_1, \ldots, p_n$ with $m \geqslant 1$ and $n \geqslant 0$, assuming that the case $n = 0$ corresponds to the symbol-string $x_1, \ldots, x_m$. The variables $x_1, \ldots, x_m$ will be called *basic* and the variables $p_1, \ldots, p_n$ will be called *auxiliary* (or *parameters for* φ). If all variables of φ occur among the symbol-string $\vec{x}, \vec{p}$ only, we shall write $\varphi[\vec{x}, \vec{p}]$.

The short symbol-strings $\forall \vec{x}$ and $\exists \vec{x}$ are the designations for the symbol-strings $\forall x_1 \ldots \forall x_m$ and $\exists x_1 \ldots \exists x_m$, respectively.

AS2. (The *full comprehension axiom scheme.*) *Let $\varphi[x, \vec{p}]$ be an X-predicative formula such that the substitution $\varphi[x \parallel y, \vec{p}]$ is admissible. Then,*

$$\forall X(\exists Y(\forall y((y \in Y) \Leftrightarrow (y \in X \wedge \varphi[y, \vec{p}])))).$$

This axiom scheme postulates that for each class X and any X-predicative formula $\varphi[x, \vec{p}]$, there exists a unique class defined as $\{x \in X \mid \varphi[x, \vec{p}]\}$.

Let A be a class and an assembly $\mathbf{C} \equiv \{x \mid \varphi(x)\}$ is defined by A-predicative formula φ. If $\mathbf{C} \subset A$, then the assembly $\mathbf{C}$ is a class. By AS2, there exists a class $B \equiv \{x \in A \mid \varphi(x)\}$. If $x \in \mathbf{C}$, then $\forall x(\varphi(x) \Rightarrow x \in A)$ implies $x \in B$. Therefore, $\mathbf{C} \subset B$. Conversely, if $x \in B$, then, by AS2, $x \in A \wedge \varphi(x)$, i. e. $x \in \mathbf{C}$. Hence, $\mathbf{C} = B$.

A3. (The *empty class axiom.*) $\forall Z ((\forall x(x \notin Z)) \Leftrightarrow Z = \varnothing)$.

Lemma 1. $\forall X (\varnothing \subset X)$.

Proof. Denote the formulas $x \in \varnothing$ and $x \in X$ by φ and ψ, respectively. Since it follows from A3 that $\neg\varphi$, by rules of deduction we obtain $\varphi \Rightarrow \psi$. Applying the rule of generalization, we get $\forall x (x \in \varnothing \Rightarrow x \in X)$. □

A class α will be called *universal* if $\alpha \bowtie$.

A4. (The *axiom of equiuniversality.*) $\forall U\, \forall V\, ((U = V) \Rightarrow (U \bowtie \Leftrightarrow V \bowtie))$.

This axiom postulates that equal classes are simultaniously universal or not universal.

A5. (The *infra-universality axiom.*) $\mathfrak{a} \bowtie \ \wedge \forall U\, (U \bowtie \ \Rightarrow \mathfrak{a} \subset U)$.

This axiom postulates that the class $\mathfrak{a}$ is the "smallest" universal class. We will call it *infra-universal* or the *infra-universe*.

A6. (The *universality axiom.*) $\forall X\, \exists U\, (U \bowtie \ \wedge\ X \in U)$.

This axiom postulates that every class A is an element of some universal class.

The following axioms explain what the notion "universality" means.

A7. (The *transitivity axiom.*) $\forall U\, (U \bowtie \ \Rightarrow \forall X\, (X \in U \Rightarrow X \subset U))$.

This axiom postulates that if α is a universal class, then every α-set is an α-class.

A8. (The *subset* (or *quasitransitivity*) *axiom.*)

$$\forall U\, (U \bowtie \ \Rightarrow \forall X\, \forall Y\, (X \in U \wedge Y \subset X \Rightarrow Y \in U)).$$

This axiom postulates that if α is a universal class then every subclass of every α-set is an α-set.

Within the framework of every class α we can define all basic set-theoretical constructions.

For every class A the α-class $\mathcal{P}_\alpha(A) \equiv \{x \in \alpha \mid x \subset A\}$ is called the *full α-ensemble of the class A.*

A9. (The *full ensemble axiom.*) $\forall U\, (U \bowtie \ \Rightarrow \forall X\, (X \in U \Rightarrow \mathcal{P}_U(X) \in U))$.

This axiom postulates that if α is an universal class and A is an α-set, then $\mathcal{P}_\alpha(A)$ is an α-set.

For classes A and B the α-class $A \bigcup_\alpha B \equiv \{x \in \alpha \mid x \in A \vee x \in B\}$ is called the *α-union of the classes A and B*; the α-class $A \bigcap_\alpha B \equiv \{x \in \alpha \mid x \in A \wedge x \in B\}$ is called the *α-intersection of the classes A and B.*

A10. (The *binary union axiom.*)

$$\forall U\, (U \bowtie \ \Rightarrow \forall X\, \forall Y\, (X \in U \wedge Y \in U \Rightarrow X \cup_U Y \in U)).$$

This axiom postulates that if α is a universal class then the binary α-union of α-sets is an α-set. Axioms A10 and A8 imply that the same holds also for the binary α-intersection.

For a class A consider the *solitary α-class* $\{A\}_\alpha \equiv \{x \in \alpha \mid x = A\}$.

Call the α-class $\{A, B\}_\alpha \equiv \{A\}_\alpha \bigcup_\alpha \{B\}_\alpha$ the *unordered α-pair*, and the α-class $\langle A, B\rangle_\alpha \equiv \{\{A\}_\alpha, \{A, B\}_\alpha\}_\alpha$ the *coordinate α-pair of the classes A and B.*

Lemma 2. *Let α be a universal class and $a, b \in \alpha$. Then, $\{a\}_\alpha$, $\{a, b\}_\alpha$ and $\langle a, b\rangle_\alpha$ are α-sets.*

Proof. If a is an α-set, then by A9 $\mathcal{P}_\alpha(a) \in \alpha$. From $\{a\}_\alpha \subset \mathcal{P}_\alpha(a)$ by A8, it follows that $\{a\}_\alpha$ is an α-set.

From A10, now, we have that $\{a, b\}_\alpha \in \alpha$. This fact together with $\{a\}_\alpha \in \alpha$ according to the proven property implies $\langle a, b\rangle_\alpha \in \alpha$. □

Corollary 1. *Let α be a universal class and $a, a', b, b' \in \alpha$ and $\langle a, b\rangle_\alpha = \langle a', b'\rangle_\alpha$. Then, $a = a'$ and $b = b'$.*

For classes A and B, the α-class $A *_\alpha B \equiv \{x \in \alpha \mid \exists y\, \exists z\, (y \in A \wedge z \in B \wedge x = \langle y, z\rangle_\alpha)\}$ will be called the *coordinate α-product of classes A and B.*

Lemma 3. *Let α be a universal class and $A, B \in \alpha$. Then, $A *_\alpha B \in \alpha$.*

Proof. Let $a \in A$ and $b \in B$. Then, $\{a\}_\alpha \subset A \cup_\alpha B$ and $\{b\}_\alpha \subset A \cup_\alpha B$ implies $\{a, b\}_\alpha \subset A \cup_\alpha B$. By A10, $A \cup_\alpha B \in \alpha$. According to Lemma 2, $\{a\}_\alpha \in \mathcal{P}_\alpha(A \cup_\alpha B)$ and $\{a, b\}_\alpha \in \mathcal{P}_\alpha(A \cup_\alpha B)$. By the same reason, $\langle a, b\rangle_\alpha = \{\{a\}_\alpha, \{a, b\}_\alpha\}_\alpha \subset \mathcal{P}_\alpha(A \cup_\alpha B)$. Hence, $\langle a, b\rangle_\alpha \in \mathcal{P}_\alpha(\mathcal{P}_\alpha(A \cup_\alpha B))$. Therefore, $A *_\alpha B \subset \mathcal{P}_\alpha(\mathcal{P}_\alpha(A \cup_\alpha B)) \in \alpha$. By A8, we have $A *_\alpha B \in \alpha$. □

Further, A and B will denote some fixed α-classes.

An α-subclass u of the α-class $A *_\alpha B$ will be called an *α-correspondence from the α-class A into the α-class B* and will be denoted also by $u : A \longrightarrow\!\!\!\!<_\alpha B$. The formula $u \subset A *_\alpha B$ will be denoted also by $u \leftrightharpoons A \longrightarrow\!\!\!\!<_\alpha B$. For the α-correspondense $u : A \longrightarrow\!\!\!\!<_\alpha B$ consider the α-classes

$$\operatorname{dom}_\alpha u \equiv \{x \in \alpha \mid x \in A \wedge ((\exists y\, (y \in B \wedge \langle x, y\rangle_\alpha \in u))\} \text{ and}$$
$$\operatorname{rng}_\alpha u \equiv \{y \in \alpha \mid y \in B \wedge ((\exists x\, (x \in A \wedge \langle x, y\rangle_\alpha \in u))\}.$$

The α-subclass $B_a \equiv u\langle a\rangle \equiv \{y \in \alpha \mid y \in B \wedge \langle a, y\rangle_\alpha \in u\}$ of the α-class B will be called the *α-class of values of the α-correspondence u on the element $a \in A$*, the α-subclass $u[A'] \equiv \{y \in \alpha \mid y \in B \wedge (\exists x\, (x \in A' \wedge \langle x, y\rangle_\alpha \in u)\}$ of the class B the *image of the subclass A' of the class A with respect to the α-correspondence u*. It is clear that $u[\{a\}_\alpha] = u\langle a\rangle$ for each $a \in A$ and $u[A] = \operatorname{rng}_\alpha u$.

If $u\langle a\rangle$ contains a single element $b \in B$ (in such sense that $\exists y\, (y \in B \wedge u\langle a\rangle = \{y\}_\alpha)$), then this single element b is called a *value of the α-correspondence u on the element $a \in A$* and is denoted by $u(a)$ or by b_a.

An α-correspondence u will be called *total* if $\operatorname{dom}_\alpha u = A$ and *single-valued* if $u\langle a\rangle = \{u(a)\}_\alpha$ for every $a \in \operatorname{dom}_\alpha u$. The single-valued α-correspondence is called also the *α-mapping* (≡ *α-function*).

The total single-valued α-correspondence $u : A \longrightarrow\!\!\!\!<_\alpha B$ is called the *α-mapping* (≡ *α-function*) *from the α-class A into the α-class B* and is denoted by $u : A \to_\alpha B$. A formula, expressing the property for the α-class u to be an α-mapping from the α-class A into the α-class B, will be denoted by $u \leftrightharpoons A \to_\alpha B$.

An α-mapping $u : A \to_\alpha B$ is called:
- *injective* if $\forall x, y \in A(u(x) = u(y) \Rightarrow x = y)$ (it is denoted by $u : A \rightarrowtail_\alpha B$);
- *surjective* if $\mathrm{rng}_\alpha u = B$ (it is denoted by $u : A \twoheadrightarrow_\alpha B$);
- *bijective* (≡*one-to-one*) if it is injective and surjective (it is denoted by $u : A \rightarrowtail\!\!\!\twoheadrightarrow_\alpha B$).

The α-class $\{x \in \alpha \mid x \rightleftharpoons A \to_\alpha B\}$ of all α-mappings from the α-class A into the α-class B which are α-sets will be denoted by $B^A_{(\alpha)}$ or by $Map_\alpha(A, B)$.

A11. (The *full union axiom.*)

$$\forall U(U\bowtie \Rightarrow \forall X \forall Y \forall z(X \in U \wedge Y \subset U \wedge (z \subset X *_U Y) \wedge \\ \wedge (\forall x(x \in X \Rightarrow z\langle x\rangle \in U))) \Rightarrow (\mathrm{rng}_U z \in U))).$$

An α-correspondence u from α-class A into α-class B will also be called a (*multivalued*) *α-collection of α-subclasses B_a of the α-class B, indexed by the α-class A*. In this case, the class u and the formula $u \leftrightharpoons A \longrightarrow\!\!\!\!<_\alpha B$ will be denoted also by $(B_a \subset B \mid a \in A)_\alpha$ and $u \leftrightharpoons (B_a \subset B \mid a \in A)_\alpha$, respectively. An α-mapping u from A into B will also be called a *simple α-collection of the elements b_α of the α-class B, indexed by the α-class A*. In this case, the class u, the class $\mathrm{rng}\, u$, and the formula $u \leftrightharpoons A \to_\alpha B$ are denoted also by $(b_a \in B \mid a \in A)_\alpha$, $\{b_a \in B \mid a \in A\}$, and $u \leftrightharpoons (b_a \in B \mid a \in A)_\alpha$, respectively.

The α-class $\{y \in \alpha \mid \exists x \in A(y \in B_x)\}$ is called the *α-union of α-collection* $(B_a \subset B \mid a \in A)_\alpha$ and is denoted by $\cup_\alpha(B_a \subset B \mid a \in A)_\alpha$. The α-class $\{y \in \alpha \mid \forall x \in A(y \in B_x)\}$ is called the *α-intersection of α-collection* $(B_a \subset B \mid a \in A)_\alpha$ and is denoted by $\cap_\alpha(B_a \subset B \mid a \in A)_\alpha$.

In these terms and notations, the axiom of full union means that if α is a universal class and $(B_a \subset B \mid a \in A)_\alpha$ is an α-collection of α-subsets B_a of the α-class B, indexed by the α-set A, then its α-union $\cup_\alpha(B_a \subset B \mid a \in A)_\alpha$ is an α-set.

With α-class A, it is associated in the canonical way the *α-collection* $(a \subset \alpha \mid a \in A)_\alpha$ *of one-element α-sets of the α-class A* (according to axiom A7, $a \in A \subset \alpha$ implies $a \subset \alpha$). The α-union of this α-collection $(a \subset \alpha \mid a \in A)_\alpha$ is called the *α-union* (≡ *α-sum*) of the *α-class A* and is denoted by $\cup_\alpha A$. If α is a universal class and A is an α-set, then $\cup_\alpha A$ is also an α-set. With every α-class A, it is associated in the canonical way the *simple α-collection* $(a \in A \mid a \in A)_\alpha$ *of elements of the α-class A*. It is clear that $\{a \in A \mid a \in A\}_\alpha = A$.

The next axiom serves, in particular, to exclude the possibility for a set to be its own element.

A12. (The *regularity axiom.*)

$$\forall U\,(U\bowtie \Rightarrow \forall X\,(X \subset U \wedge X \neq \varnothing \Rightarrow \exists x\,(x \in X \wedge x \cap_U X = \varnothing))).$$

The next axiom postulates the existence of the infinite set.

A13. (The *infra-infinity axiom.*)

$$\exists X\,(X \in \mathfrak{a} \wedge \varnothing \in X \wedge \forall x\,(x \in X \Rightarrow (x \cup_{\mathfrak{a}} \{x\}_{\mathfrak{a}} \in X))).$$

Denote the postulated $\mathfrak{a}$-set by π.

This axiom implies that the class $\varnothing$ is an $\mathfrak{a}$-set. By axiom A5, $\varnothing$ is an α-set for every universe α.

Consider the $\mathfrak{a}$-class $\varkappa \equiv \{Y \in \mathfrak{a} \mid Y \subset \pi \wedge \varnothing \in Y \wedge \forall y\,(y \in Y \Rightarrow (y \subset \mathfrak{a} \wedge y \bigcup_{\mathfrak{a}} \{y\}_{\mathfrak{a}} \in Y))\}$. Since $\varkappa \subset \mathcal{P}_{\mathfrak{a}}(\pi)$, axioms A9 and A8 imply that $\varkappa$ is an α-set.

Consider the $\mathfrak{a}$-class $\omega \equiv \{y \in \mathfrak{a} \mid \forall Y\,(Y \in \varkappa \Rightarrow y \in Y)\}$. Since $\omega \subset \pi$, we infer by axiom A8 that ω is an $\mathfrak{a}$-set. Call it the $\mathfrak{a}$-*set of natural numbers*. By axiom A5, ω is an α-set for every universe α.

Consider the *initial natural numbers* $0 \equiv \varnothing$, $1 \equiv 0 \bigcup_{\mathfrak{a}} \{0\}_{\mathfrak{a}}$, $2 \equiv 1 \bigcup_{\mathfrak{a}} \{1\}_{\mathfrak{a}}, \ldots$. From the definitions of $\varkappa$ and ω, it follows that $0, 1, 2, \cdots \in \omega$. By axiom A7, $0, 1, 2, \cdots \in \alpha$ for every universe α.

The last axiom postulates the existence of a choice function.

A14. (The *choice axiom.*) $\forall U\,(U\bowtie \Rightarrow \forall X\,(X \in U \wedge X \neq \varnothing \Rightarrow \exists z\,((z \leftrightharpoons \mathcal{P}_U(X) \setminus \{\varnothing\}_U \rightarrow_U X) \wedge \forall Y\,(Y \in \mathcal{P}_U(X) \setminus \{\varnothing\}_U \Rightarrow z(Y) \in Y))))$.

The description of the list of mathematical axioms and axiom schemes of the LTS is finished. It was proposed by V. K. Zakharov [2005b].

B.1.2 Some constructions in the local theory of sets

Almost all modern mathematics (except naive category theory and naive theory of mathematical systems) can be formalized within the framework of the infra-universe $\mathfrak{a}$. Only the mentioned naive theories require using other higher universes.

To show that all naive category theory can be formalized within the framework of the local theory of sets we need to introduce an analogue of the coordinate α-pair $\langle A, B\rangle_\alpha$ working also with α-classes A and B not only with α sets a and b (see Corollary 1 to Lemma 2 (B.1.1)).

Now, let α be some fixed universal class.

Let $A, A', A'', \ldots,$ be α-classes, where the prime symbol (′) is used only for the sake of uniformity of notations.

The α-collection $[\alpha_i \subset \alpha \mid i \in 2]_\alpha$, such that $\alpha_0 \equiv A$ and $\alpha_1 \equiv A'$ will be called the (*multivalued*) *sequential α-pair of α-classes A and A'* and will be denoted by $[A, A']_\alpha$. The α-collection $[\alpha_i \subset \alpha \mid i \in 3]_\alpha$, such that $\alpha_0 \equiv A$, $\alpha_1 \equiv A'$, and $\alpha_2 \equiv A''$, will be called the (*multivalued*) *sequential α-triplet of α-classes A, A', and A''* and will be denoted by $[A, A', A'']_\alpha$, and so on.

Let now $a, a', a'', \ldots$ be α-sets.

The simple α-collection $(a_i \in \alpha \mid i \in 2)_\alpha$, such that $a_0 \equiv a$ and $a_1 \equiv a'$ will be called the *simple sequential α-pair of α-sets a and a'* and will be denoted by $(a, a')_\alpha$.

The simple α-collection $(a_i \in \alpha \mid i \in 3)_\alpha$, such that $a_0 \equiv a$, $a_1 \equiv a'$, and $a_2 \equiv a''$, will be called the *simple sequential α-triplet of α-sets a, a', and a''* and will be denoted by $(a, a', a'')_\alpha$, and so on.

If A, A', B, and B' are α-classes and $(A, A']_\alpha = (B, B']_\alpha$, then $A = B$ and $A' = B'$. If a, a', b, and b' are α-sets and $(a, a')_\alpha = (b, b')_\alpha$, then $a = b$ and $a' = b'$. The similar properties are valid also for every finite α-collections. Thus, the α-pairs $(A, A']_\alpha$ and $(a, a')_\alpha$ possess the mentioned property of the Kuratowski α-pair $\langle a, a'\rangle_\alpha$ (see Corollary 1 to Lemma 2 (B.1.1)). However, in contrast to the latter one, the α-pair $(A, A']_\alpha$ works also for α-classes not only for α-sets.

Let some α-collection $u \equiv (A_i \subset \alpha \mid i \in I]_\alpha$ be indexed by α-class $I \neq \varnothing$. The α-class $\prod_\alpha (A_i \subset \alpha \mid i \in I]_\alpha \equiv \{z \in \alpha \mid (z : I \to_\alpha \alpha) \wedge (\forall x\,(x \in I \Rightarrow z(x) \in A_x))\}$ will be called the *α-product of the α-collection u*. In the particular case $A, A', A'', \ldots$ are α-classes, then the α-classes $\prod_\alpha (A, A']_\alpha$, $\prod_\alpha (A, A', A'']_\alpha, \ldots$ will be called the *α-product of the α-pair $(A, A']_\alpha$, the α-triplet $(A, A', A'']_\alpha, \ldots$* and will be denoted by $A \times_\alpha A'$, $A \times_\alpha A' \times_\alpha A'', \ldots$

One can check that $A \times_\alpha A' = \{x \in \alpha \mid (\exists y \exists y' (y \in A \wedge y' \in A' \wedge x = (y, y')_\alpha))\}$. It is seen from this equality that the α-product $A \times_\alpha A'$ is similar to the coordinate α-product $A *_\alpha A'$, but in contrast to the latter one, it is a partial case of the general product $\prod_\alpha (A_i \subset \alpha \mid i \in I]_\alpha$.

If $A = A' = A'' = \ldots$, then $A \times_\alpha A = A^2_{(\alpha)} \equiv Map_\alpha(2, A)$, $A \times_\alpha A \times_\alpha A = A^3_{(\alpha)} \equiv Map_\alpha(3, A), \ldots$. At the same time, $A *_\alpha A \neq A^2_{(\alpha)}$, between the α-classes $A *_\alpha A$ and $A^2_{(\alpha)}$, there exists only a bijective α-mapping of the canonical form $\langle a, a'\rangle \mapsto (a, a')$. Namely, this stipulates the necessity of introducing the non-coordinate α-product $A \times_\alpha A'$, $A \times_\alpha A' \times_\alpha A'', \ldots$

If $n \in \omega$, then an α-subclass R of the α-class $A^n_{(\alpha)} \equiv \mathrm{Map}_\alpha(n, A)$ is called *n-placed α-correspondence on α-class A*. α-mapping $O : A^n_{(\alpha)} \to_\alpha A$ is called *n-placed α-operation on α-class A*. Note that $O \subset A^n_{(\alpha)} *_\alpha A \neq A^{n+1}_{(\alpha)}$. Therefore, an n-placed operation O cannot be considered as an $(n+1)$-placed correspondence.

B.2 The MacLane problem on a set-theoretical foundation for the naive category theory. The solution of this problem within the framework of the local theory of sets

By the naive notion of a category, we mean the notion of *metacategory*, given by S. MacLane in [1971]. According to [*MacLane*, 1971], *a metacategory* consists of a *objects* $a, b, c, \ldots$, *arrows* $f, g, h, \ldots$, and four *operations* $f \mapsto \mathrm{dom}\, f$, $f \mapsto \mathrm{codom}\, f$, $a \mapsto i_a$, and $f, g \mapsto g \circ f$, satisfying some additional conditions. Unfortunately, even such a pathological object as the *metacategory of all metacategories* satisfies this definition.

The *MacLane problem* appeared because of the internal contradictoriness of the notion of metacategory. The aim of any formalization of the naive category theory is 1) to construct some axiomatic theory (set-theoretical, arrow or mixed) and 2) to give in it a strict definition of some fundamental notion that (a) corresponds to the naive understanding of a category, (b) is closed with respect to all operations and constructions of naive category theory, (c) includes in itself all known important concrete examples of categories, and (d) cuts off such naive pathological examples as the metacategory of all metacategories.

B.2.1 The definition of a local category in the local theory of sets

Using all notations of the previous section we can formalize the naive notion of a category in the following way. For a universe α we define an α-category as a big two-sorted algebraic system with two relations and one operation (see [*Bourbaki*, 1954; 1957; *Zakharov and Mikhalev*, 2000c]). For this purpose we will use the notion of an α-collection of α-classes and an α-pair of α-classes introduced in the previous section.

Consider fixed $\mathfrak{a}$-set Ω_c, consisting of three elements of the class α, denoted by the signs #, $\circ$, and $\leftrightarrow$, called the *symbol of partition*, the *symbol of composition*, and the *symbol of identification*, respectively. The set Ω_c is called a *signature of the category*. Since $\mathfrak{a} \subset \alpha$, we infer that Ω_c is also an α-set for every universe α.

Fix some universe α.

Consider an α-pair $A \equiv (Obj, Arr)_\alpha$, containing two α-classes Obj and Arr, and the α-collection $s_c \equiv (\omega_A \subset \alpha \mid \omega \in \Omega_c)_\alpha$ with the three components $\zeta \equiv s_c\langle\#\rangle \equiv \#_A$, $\eta \equiv s_c\langle\circ\rangle \equiv \circ_A$, and $\vartheta \equiv s_c\langle\leftrightarrow\rangle \equiv \leftrightarrow_A$.

An α-class $\mathcal{C} \equiv (A, s_c)_\alpha$ will be called an *α-category* (≡ *a category of the class* α) if the α-classes Obj, Arr, ζ, η, and ϑ occur in the sequential conjuction of the following formulas (written in informal way):

$P_c1.$ $(\zeta \subset (Obj \times_\alpha Obj) *_\alpha Arr) \wedge (\eta \subset (Arr \times_\alpha Arr) *_\alpha Arr) \wedge (\vartheta \subset Obj *_\alpha Arr)$;

This formula postulates that the partition assignes to every pair of elements of Obj some α-class of elements of Arr; the composition assignes to every pair of elements of Arr some third element Arr, and the identification assignes to every element of Obj some element of Arr.

$P_c2.$ $(\mathrm{rng}_\alpha \zeta = Arr) \wedge (\forall x, y \in Obj \times_\alpha Obj\, (x \neq y \Rightarrow \zeta\langle x\rangle \bigcap_\alpha \zeta\langle y\rangle = \varnothing))$; ζ is usually written in the form of the α-collection $\zeta \equiv (Arr(\pi, \varkappa) \subset Arr \mid (\pi, \varkappa)_\alpha \in Obj \times_\alpha Obj)_\alpha$; in this notation the indicated property means that the α-class Arr is equal to the α-union of this pairwise disjoint α-collection.

$P_c3.$ $(\mathrm{dom}_\alpha \eta = \{x \in \alpha \mid \exists u\, \exists v\, \exists w\, \exists v'\, \exists w'\, ((u, v, w \in Obj) \wedge (v', w' \in Arr) \wedge (v' \in Arr(u, v)) \wedge (w' \in Arr(v, w)) \wedge (x = (v', w')_\alpha))\}) \wedge (\eta : \mathrm{dom}_\alpha \eta \to Arr)$;

$P_c4.$ $(\vartheta : Obj \to_\alpha Arr) \wedge (\forall X \in Obj(\vartheta(X) \in Arr(X, X))) \wedge (\forall f \in Obj(\vartheta(f) \in Arr(f, f)))$; ϑ is usually written in the form of the simple α-collection $\vartheta \equiv (i_\pi \in Arr \mid \pi \in Obj)_\alpha$, where $i_\pi \equiv \vartheta(\pi)$;

$P_c5.$ $\eta[Arr(\pi, \varkappa) \times_\alpha Arr(\varkappa, \rho)] \subset Arr(\pi, \rho)$ for every elements $\pi, \varkappa, \rho \in Obj$;

$P_c6.$ $\eta(\eta(F, G), H)) = \eta(F, \eta(G, H))$ for every elements $\pi, \varkappa, \rho, \sigma \in Obj$ and every elements $F \in Arr(\pi, \varkappa)$, $G \in Arr(\varkappa, \rho)$ and $H \in Arr(\rho, \sigma)$;

$P_c7.$ $\eta(F, i_\pi) = F$ and $\eta(i_\pi, G) = G$ for every elements $\pi, \varkappa, \rho \in Obj$ and every elements $F \in Arr(\varkappa, \pi)$ and $G \in Arr(\pi, \rho)$.

α-Categories defined in such a way can be called *local.*

Elements F of the class $Arr(\pi, \varkappa)$ are called *arrows from the object π to the object $\varkappa$.* The formula $F \in Arr(\pi, \varkappa)$ is also denoted by $F : \pi \dot{\rightarrow} \varkappa$. The correspondence η is called the *composition* and is usually denoted simply by $\circ$; in this case along with $\circ(F, G)$ we write also $G \circ F$.

An α-category $\mathcal{C}$ is called *small,* if $\mathcal{C}$ is an α-set. An α-Category $\mathcal{C}$ is called *locally small* if every α-class $Arr(\pi, \varkappa)$ is an α-set.

Remark. In category theory, the refusal of the term "*a morphism F from an object π to an object $\varkappa$*" took its place by the following reason. For algebraic systems U and V, the notion of *homomorphism f from U to V* is usual; for smooth manifolds U and V, the notion of a *diffeomorphism f from U to V* is usual, and so on; the generalization of all these notions is the notion of a *morphism of mathematical systems* (see [*Bourbaki*, 1954; *Zakharov and Mikhalev*, 2000b]). However, mathematical systems $U, V, \ldots$ of a type $\mathfrak{C}$ and morphisms f from U to V of a status $\mathfrak{S}$ do not form a category because the morphism f does not define uniquely the system V; therefore, for some systems, the property $(U, V) \neq (U', V') \Rightarrow Mor(U, V) \bigcap Mor(U', V') \neq \varnothing$ is possible. But this property contradicts the property P_c2. To form the corresponding category, it is necessary to take not morphisms f from U into V, but triplets (f, U, V), which can be naturally called *arrows from the system U into the system V defined by the morphisms f* (see [*MacLane*, 1971, I.8]).

B.2.2 Functors and natural transformations and generated by them "the category of categories" and "the category of functors" in the local theory of sets

Let $\mathcal{C}$ and $\mathcal{D}$ be α-categories. An α-class $\Phi \equiv (\Phi_O, \Phi_T)_\alpha$ will be called a (*covariant*) *α-functor* (≡ a *functor of the class α*) *from the α-category $\mathcal{C}$ to the α-category $\mathcal{D}$* if:

1) Φ_O is an α-mapping from the α-class $Obj\mathcal{C}$ into the α-class $Obj\mathcal{D}$;
2) Φ_T is an α-mapping from the α-class $Arr\mathcal{C}$ into the α-class $Arr\mathcal{D}$;
3) $\Phi_T(F) \in Arr_{\mathcal{D}}(\Phi_O(\pi), \Phi_O(\varkappa))$ for every objects $\pi, \varkappa \in Obj\mathcal{C}$ and every arrow $F \in Arr_{\mathcal{C}}(\pi, \varkappa)$;
4) $\Phi_T(G \circ F) = \Phi_T(G) \circ \Phi_T(F)$ for every objects $\pi, \varkappa, \rho \in Obj\mathcal{C}$ and every arrows $F \in Arr_{\mathcal{C}}(\pi, \varkappa)$ and $G \in Arr_{\mathcal{C}}(\varkappa, \rho)$;
5) $\Phi_T(i_\pi) = i_{\Phi_O(\pi)}$ for every object $\pi \in Obj\mathcal{C}$.

Usually, α-mappings Φ_O and Φ_T are denoted by one symbol Φ.

α-Functors are exactly homomorphisms between α-categories considered as algebraic systems.

The *composition* $\Psi \circ \Phi$ *of an* α*-functor* $\Phi \equiv (\Phi_O, \Phi_T]_\alpha$ *from* $\mathcal{C}$ *to* $\mathcal{D}$ *and an* α*-functor* $\Psi \equiv (\Psi_O, \Psi_T]_\alpha$ *from* $\mathcal{D}$ *to* $\mathcal{E}$ is the α-functor $(\Psi_O \circ \Phi_O, \Psi_T \circ \Phi_T]_\alpha$ from $\mathcal{C}$ to $\mathcal{E}$.

The *identity* α*-functor* $I_\mathcal{C}$ *for the* α*-category* $\mathcal{C}$ is the α-functor $(id_{Obj\mathcal{C}}, id_{Arr\mathcal{C}}]_\alpha$, containing the two identical mappings for the α-classes $Obj\mathcal{C}$ and $Arr\mathcal{C}$, respectively.

Now, we will formalize the operation of naive category theory known as "the category of categories".

Take any universe β such that $\alpha \in \beta$.

Consider the β-class $Cat_\alpha^\beta \equiv \{X \in \beta \mid X \text{ is a } \alpha\text{-category}\}$ of all α-categories $\mathcal{C}$.

Consider also the β-class ${}^f Arr_\alpha^\beta \equiv \{X \in \beta \mid \exists x \exists Y \exists Z ((Y, Z \text{ are } \alpha\text{-categories}) \wedge (x$ is an α-functor from Y to $Z) \wedge (X = (x, Y, Z]_\alpha))\}$ of all α*-functorial arrows* $F \equiv (\Phi, \mathcal{C}, \mathcal{D}]_\alpha$.

For every simple β-pair $(\mathcal{C}, \mathcal{D})_\beta$ of α-categories $\mathcal{C}$ and $\mathcal{D}$, consider the β-class ${}^f Arr_\alpha^\beta(\mathcal{C}, \mathcal{D}) \equiv \{X \in \beta \mid \exists x \exists Y \exists Z ((Y, Z$ are α-categories$) \wedge (x$ is an α-functor from Y to $Z) \wedge (X = (x, Y, Z]_\alpha))\}$. Consider the β-collection $\zeta' \equiv ({}^f Arr_\alpha^\beta(\mathcal{C}, \mathcal{D}) \subset {}^f Arr_\alpha^\beta \mid (\mathcal{C}, \mathcal{D})_\beta \in Cat_\alpha^\beta \times_\beta Cat_\alpha^\beta]_\beta$.

Consider the β-correspondence η' from ${}^f Arr_\alpha^\beta \times_\beta {}^f Arr_\alpha^\beta$ to ${}^f Arr_\alpha^\beta$, generated by the composition of α-functors and consider the β-mapping ϑ' from Cat_α^β to ${}^f Arr_\alpha^\beta$ such that $\vartheta'(\mathcal{C}) = (I_\mathcal{C}, \mathcal{C}, \mathcal{C}]_\alpha$.

These β-classes give us the opportunity to consider the β-correspondence s'_c from Ω_c into β such that $s'_c\langle\#\rangle \equiv \zeta'$, $s'_c\langle\circ\rangle \equiv \eta'$, and $s'_c\langle\leftrightarrow\rangle \equiv \vartheta'$.

As a result, we get the β-category $\mathbb{C}_\alpha^\beta \equiv ((Cat_\alpha^\beta, {}^f Arr_\alpha^\beta]_\beta, s'_c]_\beta$. It will be called the β*-category of all* α*-categories and all* α*-functorial arrows between them.*

Now, let $\mathcal{C}$ and $\mathcal{D}$ be fixed α-categories. Suppose that Φ and Ψ are α-functors from $\mathcal{C}$ to $\mathcal{D}$. A simple α-collection $T = (t_\pi \in Arr\mathcal{D} \mid \pi \in Obj\mathcal{C})_\alpha$ will be called a (*natural*) α*-transformation from the* α*-functor* Φ *to the* α*-functor* Ψ if:
1) $t_\pi \in Arr_\mathcal{D}(\Phi(\pi), \Psi(\pi))$ for every object π from $Obj\mathcal{C}$;
2) $\Psi(F) \circ t_\pi = t_\varkappa \circ \Phi(F)$ for every objects π, $\varkappa$ from $Obj\mathcal{C}$ and every arrow F from $Arr_\mathcal{C}(\pi, \varkappa)$.

The *composition* $U \circ T$ *of an* α*-transformation* $T = (t_\pi \in Arr\mathcal{D} \mid \pi \in Obj\mathcal{C})_\alpha$ *from* Φ *to* Ψ *and an* α*-transformation* $U = (u_\pi \in Arr\mathcal{D} \mid \pi \in Obj\mathcal{C})_\alpha$ *from* Ψ *to* Ω is the α-transformation $(u_\pi \circ t_\pi \in Arr\mathcal{D} \mid \pi \in Obj\mathcal{C})_\alpha$ from Φ to Ω.

The *identity* α*-transformation* I_Φ *from the* α*-functor* Φ *to the* α*-functor* Φ is the α-transformation $(I_{\Phi(\pi)} \in Arr\mathcal{D} \mid \pi \in Obj\mathcal{C})_\alpha$ from Φ to Φ.

Finally, we will formalize the operation of naive category theory known as "the category of functors".

Consider the β-class $Funct_\alpha^\beta(\mathcal{C}, \mathcal{D})$ of all α-functors from the α-category $\mathcal{C}$ to the α-category $\mathcal{D}$. Also consider the β-class ${}^c Arr_\alpha^\beta \equiv \{X \in \beta \mid \exists x \exists Y \exists Z ((Y, Z$ are α-functors from $\mathcal{C}$ to $\mathcal{D}) \wedge (x$ is an α-transformation from Y to $Z) \wedge (X = (x, Y, Z]_\alpha))\}$ of all α*-transformational arrows* $F \equiv (T, \Phi, \Psi]_\alpha$.

For every simple β-pair $(\Phi, \Psi)_\beta$ of α-functors Φ and Ψ from $\mathcal{C}$ to $\mathcal{D}$, consider the β-class ${}^c Arr_\alpha^\beta(\Phi, \Psi) \equiv \{X \in \beta \mid \exists x\,((x \text{ is an } \alpha\text{-transformation from } \Phi \text{ to } \Psi) \wedge (X = (x, \Phi, \Psi)_\alpha))\}$. Consider the β-collection $\zeta'' \equiv ({}^c Arr_\alpha^\beta(\Phi, \Psi) \subset {}^c Arr_\alpha^\beta \mid (\Phi, \Psi)_\beta \in Funct_\alpha^\beta(\mathcal{C}, \mathcal{D}) \times_\beta Funct_\alpha^\beta(\mathcal{C}, \mathcal{D}))_\beta$.

Consider the β-correspondence η'' from ${}^c Arr_\alpha^\beta \times_\beta {}^c Arr_\alpha^\beta$ to ${}^c Arr_\alpha^\beta$ generated by the composition of α-transformations.

Consider the β-mapping ϑ'' from $Funct_\alpha^\beta(\mathcal{C}, \mathcal{D})$ to ${}^c Arr_\alpha^\beta$, such that $\vartheta''(\Phi) = (I_\Phi, \Phi, \Phi)_\alpha$.

These β-classes give us the opportunity to consider the β-correspondence s_c'' from Ω_c into β such that $s_c''\langle\#\rangle \equiv \zeta''$, $s_c''\langle\circ\rangle \equiv \eta''$, and $s_c''\langle\leftrightarrow\rangle \equiv \vartheta''$.

As a result, we get the β-category $\mathbb{F}_\alpha^\beta(\mathcal{C}, \mathcal{D}) \equiv ((Funct_\alpha^\beta(\mathcal{C}, \mathcal{D}),\ {}^c Arr_\alpha^\beta)_\beta, s_c'')_\beta$. It will be called the *β-category of all α-functors from the α-category $\mathcal{C}$ to the α-category $\mathcal{D}$ and all α-transformational arrows between α-functors.*

The constructions $\mathbb{C}_\alpha^\beta$ and $\mathbb{F}_\alpha^\beta(\mathcal{C}, \mathcal{D})$ show that the notion of an α-category is closed with respect to such important operations of naive category theory as "the category of categories" and "the category of functors". Thus, the notion of an *α-category* has all the good properties of the notion of an $\mathcal{U}$-category.

B.3 Universal classes, ordinals, cardinals, and cumulative classes in the local theory of sets

B.3.1 The relativization of formulas of the LTS to universal classes. The interpretation of the ZF set theory in universal classes

We use the abbreviations $\forall x \in X(\varphi)$ for $\forall x(x \in X \Rightarrow \varphi)$ and $\exists x \in X(\varphi)$ for $\exists x(x \in X \wedge \varphi)$. As in A.6.1 by φ^U, we denote the formula (the *relativization of the formula φ to the class U*) received by changing in φ all subformulas of the form $\forall x(\varphi')$ and $\exists x(\varphi')$ to $\forall x \in U\varphi'$ and $\exists x \in U\varphi'$, respectively.

Statement 1. *Let U be a universal class in the LTS. Consider the interpretation $M \equiv (U, I)$ of the ZF set theory in the LTS in which the correspondence I assigns to the predicate symbols $\in_{ZF}$ and $=_{ZF}$ the binary U-correspondences $B \equiv \{z \in U \mid \exists x \in U \exists y \in U(z = (x, y)_U \wedge x \in_{LTS} y)\}$ and $E \equiv \{z \in U \mid \exists x \in U \exists y \in U(z = (x, y)_U \wedge x =_{LTS} y)\}$ on the class U. Then, the interpretation M is a model of the ZF set theory in the LTS.*

Proof. We need to check that in the LTS there exists a deduction of the formula or the scheme of formulas $M \models \varphi[s]$ for every proper axiom or axiom scheme φ of the ZF set theory and every sequence $s = x_0, \dots, x_q, \dots$ of elements of the class U.

On s, axiom A1 is translated into the formula $M \models A1[s] = A1^U \equiv \forall X \in U \forall Y \in U(\forall u \in U(u \in X \Leftrightarrow u \in Y) \Rightarrow X = Y)$. By the definition of equality in the LTS, this formula is evidently deducible.

On s, axiom A2 is translated into the formula $M \models A2[s] = A2^U \equiv \forall u \in U(\forall v \in U \exists x \in U \forall z \in U(z \in x \Leftrightarrow z = u \vee z = v)$.

For the U-sets u and v, consider the unordered U-pair $x \equiv \{u, v\}_U$. By Lemma 2 (B.1.1) $x \in U$. From the corresponding definitions, we infer that $x = \{u\}_U \cup \{v\}_U = \{y \in U \mid y \in \{u\}_U \vee y \in \{v\}_U\} = \{y \in U \mid y = u \vee y = v\}$. By axiom scheme AS2 (LTS), we have $\forall z \in U(z \in x \Leftrightarrow z = u \vee z = v)$, and we get the desired deducibility.

On s, the axiom scheme of separation AS3 is translated into the scheme $M \models AS3[s] \equiv \forall X \in U \exists Y \in U \forall u \in U(u \in Y \Leftrightarrow u \in X \wedge \varphi^U(u, \vec{p}_M[s]))$, where Y is not a free variable of the formula $\varphi(u, \vec{p})$.

By AS2 (LTS), for the U-predicative formula $\varphi^U(u, \vec{p}_M[s])$ and U-set X, there exists an U-class $Y \equiv \{u \in U \mid \varphi^U(u, \vec{p}_M[s]) \wedge u \in X\}$ such that $u \in Y \Leftrightarrow (u \in U \wedge u \in X \wedge \varphi^U(u, \vec{p}_M[s]))$. Since $Y \subset X \in U$, we have, by subset axiom A8 (LTS), $Y \in U$, and it gives us the desired deducibility.

On s, the axiom of union A4 is translated into the formula $M \models A4[s] = A4^U \equiv \forall X \in U \exists Y \in U \forall z \in U \forall u \in U(u \in z \wedge z \in X \Rightarrow u \in Y)$.

For the U-set X and the corresponding U-predicative formula, by AS2 (LTS), there exists the U-class $Z \equiv \{w \in U \mid \exists x, y \in U(x \in X \wedge y \in x \wedge w = \langle x, y\rangle_U)\} \subset X *_U U$. Since $Z\langle x\rangle = x \in U$ for any $x \in X$, by the axiom of full union A11 (LTS), $Y \equiv \cup_U\{x \subset U \mid x \in X\}_U \equiv \mathrm{rng}_U Z \in U$. If $u \in z \in X$, then $u \in Y$, and we get the desired deducibility.

On s, the axiom of power set A5 is translated into the formula $M \models A5[s] = A5^U \equiv \forall X \in U \exists Y \in U \forall u \in U(u \subset X \Rightarrow u \in Y)$.

For the U-set X, by AS2 (LTS), there exists an U-class $Y \equiv \mathcal{P}_U(X) \equiv \{x \in U \mid x \subset X\}$. By axiom A8 (LTS), $Y \in U$. If $u \in U$ and $u \subset X$, then, by AS2, $u \in Y$, and we get the desired deducibility.

On the sequence s, the axiom scheme of replacement AS6 is translated into the scheme

$$\begin{aligned}M \models AS6[s] \equiv \forall x \in U \forall y \in U \forall y' \in U(\varphi^U(x, y, \vec{p}_M[s]) \wedge \\ \wedge \varphi^U(x, y', \vec{p}_M[s]) \Rightarrow y = y') \Rightarrow \forall X \in U \exists Y \in U(\forall x \in U(x \in X \Rightarrow \\ \Rightarrow \forall y \in U(\varphi^U(x, y, \vec{p}[s]) \Rightarrow y \in Y)))))\end{aligned}$$

where $\vec{p}_M[s]$ denotes the line of values of the terms $p_0, \ldots, p_{m-1}$ on s under the interpretation M.

By AS2 (LTS), for the U-predicative formula $\varphi^U(x, y, \vec{p}[s])$, there exists the U-class $F \equiv \{z \in U \mid \exists x, y \in U(z = \langle x, y\rangle_U \wedge \varphi^U(x, y, \vec{p}_M[s]))\}$. From the formula scheme cited above, we infer that the U-classes F is a U-function.

Consider any U-set X and the U-class $Y \equiv F[X]$. Consider the U-class $G \equiv \{z \in U \mid \exists x, y \in U(z = \langle x, y\rangle_U \wedge \varphi^U(x, y, \vec{p}_M[s]) \wedge x \in X\} = F|X \subset X *_U Y$. If $x \in X$, then $G\langle x\rangle = \varnothing \in U$ for $x \notin \mathrm{dom}_U F$ and $G\langle x\rangle = \{F(x)\}_U$ for $x \in \mathrm{dom}_U F$. Therefore, by the axiom of full union A11 (LTS), $Y = \mathrm{rng}_U G \in U$.

If $x \in X, y \in U$, and $\varphi^U(x, y, \vec{p}_M[s])$, then $\langle x, y\rangle_U \in F$. Thus, $y \in F[X] \equiv Y$. It proves the formula scheme $M \models AS6[s]$.

On s, the axiom of empty set A7 is translated into the formula $M \models A7[s] = A7^U \equiv \exists x \in U \forall z \in U(\neg(z \in x))$.

The empty U-set $\varnothing_{LTS}$, by axiom A3 (LTS), possesses the necessary property $\forall z \in U(z \notin \varnothing_{LTS})$.

On s, the axiom of infinity A8 is translated into the formula $M \models A8[s] \equiv \exists Y \in U(\varnothing \in Y \wedge \forall y \in U(y \in Y \Rightarrow y \cup_U \{y\}_U \in Y))$.

Consider the $\mathfrak{a}$-set π postulated by axiom of infra-infinity A13 (LTS). By this axiom, $\varnothing \in \pi$ and, if $y \in U$ and $y \in \pi$, then $y \cup_{\mathfrak{a}} \{y\}_{\mathfrak{a}} \in \pi$. Check that $A \equiv y \cup_{\mathfrak{a}} \{y\}_{\mathfrak{a}} = y \cup_U \{y\}_U \equiv B$. Let $x \in A$. Then, $x \in \mathfrak{a}$ and $x \in y \vee x = y$. Since by axiom A5 (LTS) $\mathfrak{a} \subset U$, then $x \in U$. Therefore, $x \in B$. Conversely, let $x \in B$, i.ė., $x \in U$ and $x \in y \vee x = y$. Since $y \in \pi \in \mathfrak{a}$, we have, by axiom A7 (LTS), $y \in \mathfrak{a}$. If $x \in y$, then by the same reason, $x \in \mathfrak{a}$. If $x = y$, then again $x \in \mathfrak{a}$. Thus, in each case $x \in \mathfrak{a}$. Therefore, $x \in A$. From the equality which was proven above we infer that $y \cup_U \{y\}_U \in \pi$. It means that the translation of axiom A8 (ZF) is deduced in the LTS.

On s, the axiom of regularity A9 is translated into the formula $M \models A9[s] \equiv \forall X \in U(X \neq \varnothing \Rightarrow \exists x \in U(x \in X \wedge x \cap_U X = \varnothing))$. This formula is evidently deduced from the axioms of transitivity A7 (LTS) and regularity A12 (LTS).

Finally, on s, the choice axiom A10 is translated into the formula $M \models A10[s] \equiv \forall X \in U(X \neq \varnothing \Rightarrow \exists z \in U((z \leftrightharpoons \mathcal{P}_U(X) \setminus \{\varnothing\}_U \rightarrow_U X) \wedge \forall Y \in U(Y \in \mathcal{P}_U(X) \setminus \{\varnothing\}_U \Rightarrow z(Y) \in Y)))$.

If $\varnothing \neq X \in U$, then by choice axiom A14 (LTS), there exists the class z such that $(z \leftrightharpoons \mathcal{P}_U(X) \setminus \{\varnothing\}_U \rightarrow_U X) \wedge \forall Y(Y \in \mathcal{P}_U(X) \setminus \{\varnothing\}_U \Rightarrow z(Y) \in Y)$. By axiom A9, $\mathcal{P}_U(X) \in U$, and, by axiom A8, $A \equiv \mathcal{P}_U(X) \setminus \{\varnothing\}_U \in U$. Therefore, by Lemma 3 (B.1.1) $B \equiv A *_U X \in U$. From $z \subset B$ according to axiom A8, we infer that $z \in U$, and that gives us the required deducibility. □

According to Statement 1, we can use in every universal class U all assertions for U-classes and U-sets which can be proven in ZF for the classes and sets.

B.3.2 The globalization of local constructions

In the same manner as in ZF for classes (see A.2.1), we define in the LTS for assemblies $\mathbf{A}$ and $\mathbf{B}$ and classes A and B the following assemblies:

1) $\mathcal{P}(\mathbf{A}) \equiv \{x \mid x \subset \mathbf{A}\}$;
2) $\mathbf{A} \cup \mathbf{B} \equiv \{x \mid x \in \mathbf{A} \vee x \in \mathbf{B}\}$;
3) $\mathbf{A} \cap \mathbf{B} \equiv \{x \mid x \in \mathbf{A} \wedge x \in \mathbf{B}\}$;
4) $\{A\} \equiv \{x \mid x = A\}$;
5) $\{A, B\} \equiv \{A\} \cup \{B\} = \{x \mid x = A \vee x = B\}$;
6) $\langle A, B\rangle \equiv \{A, \{A, B\}\}$;

7) $\mathbf{A} * \mathbf{B} \equiv \{x \mid \exists y \in \mathbf{A} \exists z \in \mathbf{B}(x = \langle y, z\rangle)\}$;
8) $\cup\mathbf{A} \equiv \{x \mid \exists y \in \mathbf{A}(x \in y)\}$.

In the same manner as in ZF changing the word "class" to the word "assembly" and the word "set" to the word "class", we define in the LTS *a correspondence* $\mathbf{C}$ *with the domain* $\operatorname{dom}\mathbf{C}$ and *the class of values* $\operatorname{rng}\mathbf{C}$, *a function* (≡ *a mapping*) $\mathbf{F}$, a *correspondence* $\mathbf{C} : \mathbf{A} \longrightarrow\!\!\!\!\!< \mathbf{B}$, *a function* $\mathbf{F} : \mathbf{A} \to \mathbf{B}$, a (*multivalued*) *collection* $(\mathbf{B}_a \subset \mathbf{B} \mid a \in \mathbf{A})$ *with the union* $\cup(\mathbf{B}_a \subset \mathbf{B} \mid a \in \mathbf{A})$ *and the intersection* $\cap(\mathbf{B}_a \subset \mathbf{B} \mid a \in \mathbf{A})$, *a simple collection* $(b_a \in \mathbf{B} \mid a \in \mathbf{A})$ *with the assembly of members* $\{b_a \in \mathbf{B} \mid a \in \mathbf{A}\}$, the (*multivalued*) *sequential pair* $(\mathbf{A}, \mathbf{A}')$, *triplet* $(\mathbf{A}, \mathbf{A}', \mathbf{A}'')$, ... *of assemblies* $\mathbf{A}, \mathbf{A}', \mathbf{A}'', \dots$, *the simple sequential pair* (a, a'), *triplet* (a, a', a''), ... *of classes* $a, a', a'', \dots$, *the product* $\prod(\mathbf{A}_i \subset \mathbf{A} \mid i \in \mathbf{I})$ *of a collection* $(\mathbf{A}_i \subset \mathbf{A} \mid i \in \mathbf{I})$, *the product* $\mathbf{A} \times \mathbf{A}'$, $\mathbf{A} \times \mathbf{A}' \times \mathbf{A}''$, ... *of the pair* $(\mathbf{A}, \mathbf{A}')$, *triplet* $(\mathbf{A}, \mathbf{A}', \mathbf{A}'')$, ... *of assemblies* $\mathbf{A}, \mathbf{A}', \mathbf{A}'', \dots$, *an n-placed relation* $\mathbf{R} \subset \mathbf{A}^n \equiv \operatorname{Map}(n, \mathbf{A})$ *on an assembly* $\mathbf{A}$, *an n-placed operation* $\mathbf{O} : \mathbf{A}^n \to \mathbf{A}$ *on an assembly* $\mathbf{A}$, and so on.

One can check that the pairs $\langle a, b\rangle$ and (a, b) possesses the usual property: $\langle a, b\rangle = \langle a', b'\rangle \Leftrightarrow a = a' \wedge b = b'$ and $(a, b) = (a', b') \Leftrightarrow a = a' \wedge b = b'$ for every classes a and b.

With every assembly $\mathbf{A}$, it is associated in the canonical way the *collection* $(a \subset \overline{\mathbf{V}} \mid a \in \mathbf{A})$ *of element classes of the assembly* $\mathbf{A}$ and the *simple collection* $(a \in \mathbf{A} \mid a \in \mathbf{A})$ *of elements of the assembly* $\mathbf{A}$. The equalities $\cup\mathbf{A} = \cup(a \in \overline{\mathbf{V}} \mid a \in \mathbf{A})$ and $\mathbf{A} = \{a \in \mathbf{A} \mid a \in \mathbf{A}\}$ are valid for them.

Now, we will state the connection between local notions and constructions and corresponding global ones.

Lemma 1. *Let α and β be universal classes, $A \in \alpha$, and $A \in \beta$. Then, $\mathcal{P}_\alpha(A) = \mathcal{P}_\beta(A) = \mathcal{P}(A)$.*

Proof. Let $x \in \mathcal{P}_\alpha(A)$, i. e. $x \in \alpha$ and $x \subset A$. Since $A \in \beta$, by axiom A8, $x \in \beta$. Therefore, $x \in \mathcal{P}_\beta(A)$. Hence, $\mathcal{P}_\alpha(A) \subset \mathcal{P}_\beta(A)$. The converse implication is checked analogously.

It is clear that $\mathcal{P}_\alpha(A) \subset \mathcal{P}(A)$. The inclusion $\mathcal{P}(A) \subset \mathcal{P}_\alpha(A)$ can be checked as above. □

Corollary 1. *For every class A the assembly $\mathcal{P}(A)$ is a class.*

Proof. By the axiom of universality A6, for A, there exists a universal class α such that $A \in \alpha$. Then, by the proven lemma, $\mathcal{P}(A) = \mathcal{P}_\alpha(A)$. But by axiom scheme AS2, $\mathcal{P}_\alpha(A)$ is a class. □

Lemma 2. *Let α and β be universal classes, $A, B \subset \alpha$, and $A, B \subset \beta$. Then, $A \cup_\alpha B = A \cup_\beta B = A \cup B$ and $A \cap_\alpha B = A \cup_\beta B = A \cap B$.*

Proof. Let $x \in A \cup_\alpha B$, i. e. $x \in \alpha$, and $x \in A \vee x \in B$. Then, $x \in \beta$, and therefore, $x \in A \cup_\beta B$. Thus, $A \cup_\alpha B \subset A \cup_\beta B$. The converse inclusion can be checked in the same way.

It is clear that $A \cup_\alpha B \subset A \cup B$. The inclusion $A \cup B \subset A \cup_\alpha B$ is checked as was done above. □

Lemma 3. *Let α and β be universal classes, $A, B \in \alpha$, and $A, B \in \beta$. Then, $\{A\}_\alpha = \{A\}_\beta = \{A\}$, $\{A, B\}_\alpha = \{A, B\}_\beta = \{A, B\}$ and $\langle A, B\rangle_\alpha = \langle A, B\rangle_\beta = \langle A, B\rangle$.*

Proof. If $y \in \{A\}_\alpha \equiv \{x \in \alpha \mid x = A\}$, then $y = A \in \beta$, and therefore, $y \in \{x \in \beta \mid x = A\} \equiv \{A\}_\beta$. Thus, $\{A\}_\alpha \subset \{A\}_\beta$. The converse inclusion is checked in the same way. It is clear that $\{A\}_\alpha \subset \{a\}$. The inclusion $\{A\} \subset \{A\}_\alpha$ is checked as was done above.

Now, according to the proven assertions and Lemma 2, $\{A, B\}_\alpha \equiv \{A\}_\alpha \cup_\alpha \{B\}_\alpha = \{A\}_\beta \cup_\alpha \{B\}_\beta = \{A\}_\beta \cup_\beta \{B\}_\beta \equiv \{A, B\}_\alpha$. Similarly, $\{A, B\}_\alpha = \{A, B\}$.

Finally, by Lemma 2 (B.1.1) $\{A\}_\alpha \in \alpha$, $\{A\}_\beta \in \beta$, $\{A, B\}_\alpha \in \alpha$ and $\{A, B\}_\beta \in \beta$. Therefore, by the properties proven above, $\{A\}_\beta \in \alpha$ and $\{A, B\}_\beta \in \alpha$.

Consequently, applying the equality proven above, we get

$$\langle A, B\rangle_\alpha \equiv \{\{A\}_\alpha, \{A, B\}_\alpha\}_\alpha = \{\{A\}_\beta, \{A, B\}_\beta\}_\alpha = \{\{A\}_\beta, \{A, B\}_\beta\}_\beta \equiv \langle A, B\rangle_\beta.$$

Similarly, $\langle A, B\rangle_\alpha = \langle A, B\rangle$. □

Corollary 1. *For every class A, the assembly $\{A\}$ is a class.*

Proof. By the axiom of universality A6, for A, there exists a universal class α such that $A \in \alpha$. Then, by the proven lemma $\{A\} = \{A\}_\alpha$. But, by axiom scheme AS2, $\{A\}_\alpha$ is a class. □

Lemma 4. *Let α and β be universal classes, $A, B \subset \alpha$, and $A, B \subset \beta$. Then, $A *_\alpha B = A *_\beta B = A * B$.*

Proof. Let $x \in A *_\alpha B$, i. e. $x \in \alpha$ and $\exists y \exists z(y \in A \wedge z \in B \wedge x = \langle y, z\rangle_\alpha)$. Since $y \in A \subset \beta$, we have $y \in \beta$. Similarly, $z \in \beta$. By Lemmas 2 (B.1.1) and 3 (B.3.2), $x = \langle y, z\rangle_\alpha = \langle y, z\rangle_\beta \in \beta$. Therefore, $x \in \beta$ and $\exists y \exists z(y \in A \wedge z \in B \wedge x = \langle y, z\rangle_\beta)$, i. e. $x \in A *_\beta B$. Therefore, $A *_\alpha B \subset A *_\beta B$. The inverse inclusion is checked in the same way.

It is clear that $A *_\alpha B \subset A * B$. The inclusion $A * B \subset A *_\alpha B$ can be checked as above. □

Lemma 5. *Let α and β be universal classes, $A, B \subset \alpha$ and $A, B \subset \beta$. Then, for every class u, the following assertions are equivalent:*

1) $u \fallingdotseq A \relbar\joinrel\prec_\alpha B$ [*, respectively,* $u \fallingdotseq A \to_\alpha B$];
2) $u \fallingdotseq A \relbar\joinrel\prec_\beta B$ [*, respectively,* $u \fallingdotseq A \to_\beta B$];
3) $u \fallingdotseq A \relbar\joinrel\prec B$ [*, respectively,* $u \fallingdotseq A \to B$].

Besides, $\operatorname{dom}_\alpha u = \operatorname{dom}_\beta u = \operatorname{dom} u$ *and* $\operatorname{rng}_\alpha u = \operatorname{rng}_\beta u = \operatorname{rng} u$.

Proof. (1) $\vdash$ (2). By Lemma 4, $u \subset A *_\alpha B = A *_\beta B$. Therefore, $\operatorname{dom}_\beta u \subset A$. If $x \in A$, then, by (1), $x \in A = \operatorname{dom}_\alpha u$. Therefore, $x \in \alpha$ and $\langle x, y\rangle_\alpha \in u$ for some $y \in B$. By Lemma 3, $\langle x, y\rangle_\beta \in u$. Hence, $x \in \operatorname{dom}_\beta u$. Thus, $A \subset \operatorname{dom}_\beta u$. As a result, $\operatorname{dom}_\beta u = A$.

Let $\langle x, y\rangle \in u$ and $\langle x, y'\rangle_\beta \in u$ for some $x \in A$. Then, $y, y' \in \operatorname{rng}_\beta u \subset B$. Since $x, y, y' \in \alpha$ and $x, y, y' \in \beta$, we infer by Lemma 3, that $\langle x, y\rangle_\alpha = \langle x, y\rangle_\beta \in u$ and $\langle x, y'\rangle_\alpha = \langle x, y'\rangle_\beta \in u$. From (1), we now infer $y = y'$. It means that $u \leftrightharpoons A \rightarrow_\beta B$.

All other deducibilities are proven in the same manner.

The equalities $\operatorname{dom}_\alpha u = \operatorname{dom}_\beta u = \operatorname{dom} u$ and $\operatorname{rng}_\alpha u = \operatorname{rng}_\beta u = \operatorname{rng} u$ are checked with the help of Lemma 3 in an obvious way. □

Lemma 6. *Let α and β be universal classes, $A \subset \alpha$, and $A \subset \beta$. Then, $\cup_\alpha A = \cup_\beta A = \cup A$.*

Proof. By definition, $\cup_\alpha A \equiv \cup_\alpha(a \subset \alpha \mid a \in A)_\alpha \equiv \{z \in \alpha \mid \exists y \in A(z \in y)\}$. Therefore, if $x \in \cup_\alpha A$, then $x \in y \in A \subset \beta$, by axiom A7, implies $x \in \beta$. Thus, $x \in \cup_\beta A$. Thus, $\cup_\alpha A \subset \cup_\beta A$. The inverse inclusion is checked similarly.

It is clear that $\cup_\alpha A \subset \cup A$. The inclusion $\cup A \subset \cup_\alpha A$ is checked in the same way as above. □

Corollary 1. *For every class A, the assembly $\cup A$ is a class.*

Proof. By the axiom of universality A6, for A, there exists a universal class α such that $A \in \alpha$. Then, by the previous lemma, $\cup A = \cup_\alpha A$. But, $\cup_\alpha A$, by axiom scheme AS2, is a class. □

Unfortunately, for classes A and B, we cannot yet prove that the assemblies $A \cup B$, $A \cap B$, $\{A, B\}$, $\langle A, B\rangle$, and $A * B$ are classes. It will be done in B.3.5 (see Corollary 3 to Theorem 1 (B.3.5)).

B.3.3 Ordinals and cardinals in the local theory of sets

In the same manner as in ZF, changing the word "class" to the word "assembly" and the word "set" to the word "class" in the LTS, we can define *ordered* and *well-ordered assemblies, ordinals* and *ordinal numbers*.

In the same manner as in ZF, changing the word "class" to the word "U-class", the word "set" to the word "U-set", and the word "relation" to the word "U-relation", in the LTS for every universal class U we can define *U-ordered* and *well-U-ordered U-classes, U-ordinals,* and *U-ordinal numbers* with the following change of the definition of a well-U-ordered U-class. Namely, an U-ordered U-class P is called *well-U-ordered*, if $\forall Q(Q \subset U \wedge \varnothing \neq Q \subset P \Rightarrow \exists x \in U(x \in Q \wedge \forall y \in U(y \in Q \Rightarrow x \leqslant y)))$, what means that every non-empty U-subclass of the U-class P has the smallest element. From $x, y \in Q \subset P \subset U$, we infer that this formula is equivalent to the formula $\forall Q(\varnothing \neq Q \subset P \Rightarrow \exists x \in Q(\forall y \in Q(x \leqslant y)))$, cited in condition 5 from the definition of a well-ordered class in ZF. But in the LTS, this formula has a wider sense, namely, it means that every non-empty subclass of the U-class P has the smallest element (compare with Lemma 1 (A.2.2) in ZF).

It implies that the following lemma is fulfilled.

Lemma 1. *Let U be a universal class and $\alpha \in U$. Then, the following assertions are equivalent:*
1) *α is an ordinal number;*
2) *α is a U-ordinal number.*

Now, we will infer from this lemma that the assembly $\overline{\mathbf{On}} \equiv \{x \mid On(x)\}$ of all ordinal numbers in the LTS is well-ordered by the relation $\in \cup =$.

Lemma 2. *The assembly $\overline{\mathbf{On}}$ is well-ordered by the relation $\in \cup =$.*

Proof. Let α and β be ordinal classes and $\alpha \neq \beta$. By the axiom of universality A6, $\alpha \in U$ and $\beta \in V$ for some universal classes U and V. Then, either $\alpha \not\subset \beta$ or $\beta \not\subset \alpha$. Let for certainty $\beta \not\subset \alpha$. In this case, by subset axiom A8, the non-empty V-set $\beta \setminus \alpha = \{\eta \in V \mid \eta \in \beta \wedge \eta \notin \alpha\} \in V$ has the smallest element $\gamma \in V$. We have $\gamma \notin \alpha$, by the definition of the V-set $\beta \setminus \alpha$. Since every element γ by virtue of the minimality of γ is an element of α, we have $\gamma \subseteq \alpha$. By the subset axiom, $\gamma \in U$. From $\gamma \notin \alpha$ and $\gamma \leqslant \alpha$, it follows that $\gamma = \alpha$, i. e. $\alpha \in \beta$.

We proved that $\in \cup =$ is a linear order on the assembly $\overline{\mathbf{On}}$. Show that this assembly is well-ordered with respect to the given order. Suppose that we have some non-empty class S of ordinal numbers. Consider a universe U, containing S. Then, by axiom A7, $S \subset U$. By Lemma 1, S is a U-class of U-ordinal numbers. Since by Statement 1 (B.3.1), the universe U is a model of the ZF set theory, we infer that the U-class S of U-ordinal numbers has the smallest element. □

The next lemma is similar to Lemma 1 (A.2.2), but it has completely another proof.

Lemma 3. *Let $\mathbf{A}$ be a non-empty subassembly of the assembly $\overline{\mathbf{On}}$. Then, $\mathbf{A}$ has the smallest element.*

Proof. By condition, there exists some ordinal number $\alpha \in \mathbf{A}$. By the axiom of universality A6 (LTS), there exists a universal class U such that $\alpha \in U$. Consider the assembly $\mathbf{B} \equiv \{x \in U \mid x \in \mathbf{A} \wedge x \in \alpha \cup_U \{\alpha\}_U\}$. By full comprehension axiom scheme AS2 (LTS), this assembly is an U-class. Since $\alpha \in \mathbf{B} \subset \overline{\mathbf{On}}$ and the assembly $\overline{\mathbf{On}}$ is well-ordered, we infer that the class $\mathbf{B}$ has the smallest element β. Take an arbitrary element $\gamma \in \mathbf{A}$.

If $\gamma < \alpha$, then $\gamma \in \alpha \in U$, where, by the axiom of transitivity A7 (LTS), $\gamma \in U$. Therefore, in this case, $\gamma \in \mathbf{B}$ and hence $\gamma \geqslant \beta$. If $\gamma = \alpha$, then again $\gamma \in \mathbf{B}$ and so $\gamma \geqslant \beta$. Finally, if $\gamma > \alpha$, then $\gamma > \beta$. Therefore, β is the smallest element of the assembly $\mathbf{A}$. □

Lemma 4. *Let α be an ordinal number. Then:*
1) *the assembly $\alpha^+ \equiv \alpha + 1 \equiv \alpha \cup \{\alpha\}$ is an ordinal number:*
2) *α^+ is the smallest of all ordinal numbers which are greater than the number α.*

Proof. 1. By axiom A6, $\alpha \in U$ for some U⋈. Let $x \in \alpha_U^+ \equiv (\alpha + 1)_U \equiv \alpha \cup_U \{\alpha\}_U$. Then, $x \in U$ and either $x \in \alpha$ or $x = \alpha$. Therefore, $x \in \alpha^+$. Let $y \in \alpha^+$. In this case, either $y \in \alpha$ or $y \in \alpha$. In both the cases, $y \in U$. It means that $y \in \alpha_U^+$. Thus, $\alpha^+ = \alpha_U^+$ and α^+ is a U-ordinal number and therefore an ordinal number. It is clear that $\alpha^+ > \alpha$.

2. Let β be an ordinal number such that $\beta > \alpha$. Suppose that $\alpha^+ > \beta$. Then, $\beta \in \alpha^+$, i. e. either $\beta \in \alpha$ or $\beta = \alpha$, but it contradicts to the condition $\beta > \alpha$. From this contradiction, we infer that $\beta \geqslant \alpha^+$. □

An ordinal number α^+ will be called the *successor of the ordinal number α*.

Lemma 5. *If A is a non-empty class of ordinal numbers, then:*
1) *$\cup A$ is an ordinal number;*
2) *$\cup A = \sup A$ in the ordered assembly $\overline{\mathbf{On}}$.*

Proof. By axiom A6, $A \in U$ for some universal class U. Then, by axiom A7, $A \subset U$. Consider the assembly $X \equiv \cup A$ and the U-class $Y \equiv \cup_U A$. By Lemma 6 (B.3.2), $X = Y$. By Lemma 2 (A.2.2) from ZF and Statement 1 (B.3.1), Y is an U-ordinal number; and therefore, by Lemma 1, it is an ordinal number. Thus, X is also an ordinal number.

Let $a \in A$. If $X < a$, then $X \in a \in A$ implies $X \in \cup A = X$, but it is impossible. From which $a \leqslant X$, and so X is an upper bound of the class A. Let α be an ordinal number and $\alpha \geqslant a$ for every $a \in A$. Suppose that $X > \alpha$. Then, $\alpha \in X = \cup A$ implies $\alpha \in a$ for some $a \in A$. Hence, $\alpha < a$, but it is impossible. Therefore, $X \leqslant \alpha$. Thus, $X = \sup A$ in $\overline{\mathbf{On}}$. □

A *limit ordinal* is an ordinal which is not equal to α^+ for any ordinal number α.

As in A.2.2, classes A and B are called *equivalent* ($A \sim B$) if there exists a one-to-one ($\equiv$ bijective) function $u : A \rightarrowtail\!\!\!\twoheadrightarrow B$.

An ordinal number α will be called *cardinal*, if for every ordinal number β the conditions $\beta \leqslant \alpha$ and $\beta \sim \alpha$ imply $\beta = \alpha$. The assembly of all cardinal numbers will be denoted by $\overline{\mathbf{Cn}}$. The assembly $\overline{\mathbf{Cn}}$ with the order, induced from the assembly $\overline{\mathbf{On}}$, is well-ordered.

Let U be a universal class. U-Classes A and B are called *U-equivalent* ($A \sim_U B$) if there exists a bijective U-function $u : A \rightarrowtail\!\!\!\twoheadrightarrow_U B$.

A U-ordinal number α is called *U-cardinal* if for every U-ordinal number β the conditions $\beta \leqslant \alpha$ and $\beta \sim_U \alpha$ imply $\beta = \alpha$.

Proposition 1. *Let U be a universal class and $\alpha \in U$. Then, the following assertions are equivalent:*
1) *α is a cardinal number;*
2) *α is an U-cardinal number.*

Proof. (1) $\vdash$ (2). By Lemma 1, α is a U-ordinal number. Let β be a U-ordinal number such that $\beta \leqslant \alpha$ and $\beta \sim_U \alpha$. It means that there exists a bijective U-mapping $f : \beta \rightarrowtail\!\!\!\twoheadrightarrow_U \alpha$. By axiom of transitivity A7, $\beta \subset \alpha \subset U$. Therefore, by Lemma 5 (B.3.2), $f \leftrightharpoons \beta \rightarrowtail\!\!\!\twoheadrightarrow \alpha$. Hence, $\beta \sim \alpha$. By condition 1, we get $\beta = \alpha$.

(2) $\vdash$ (1). By Lemma 1, α is an ordinal number. Let β be an ordinal number such that $\beta \leqslant \alpha$ and $\beta \sim \alpha$. It means that there exists a bijective mapping $f : \beta \rightarrowtail\!\!\!\twoheadrightarrow \alpha$. By axiom A7, $\beta \subset \alpha \subset U$. Therefore, by Lemma 5 (B.3.2) $f \leftrightharpoons \beta \rightarrowtail\!\!\!\twoheadrightarrow_U \alpha$, i. e. $\beta \sim_U \alpha$. By condition 2, we infer $\beta = \alpha$. □

The *power* $\operatorname{card}_U A$ *of a set* $A \in U$ in a universe U is a U-cardinal $\alpha \in U$ such that there exists a one-to-one U-mapping $f : A \rightarrowtail\!\!\!\twoheadrightarrow_U \alpha$. The *power* $\operatorname{card} A$ of a *class* A is a cardinal α such that $A \sim \alpha$.

Proposition 2. *Suppose that $A \in U \in V$, $U\bowtie$, and $V\bowtie$. Then,* $\operatorname{card} A = \operatorname{card}_U A = \operatorname{card}_V A < \operatorname{card}_V U$.

Proof. Let $\operatorname{card}_U A = \alpha$, $\alpha \in U$, Then, $\alpha \in V$. By definition, there exists a one-to-one function $f : A \rightarrowtail\!\!\!\twoheadrightarrow_U \alpha$. By Lemma 5 (B.3.2), $f \leftrightharpoons A \rightarrowtail\!\!\!\twoheadrightarrow_V \alpha$.

By Proposition 1, α is a U-cardinal number. Therefore, $\alpha = \operatorname{card}_V A$.

Similarly, by Lemma 5 (B.3.2) $f \leftrightharpoons A \rightarrowtail\!\!\!\twoheadrightarrow \alpha$, and by Proposition 1, α is a cardinal number. Thus, $\alpha = \operatorname{card} A$.

Show now that $\operatorname{card}_V A < \operatorname{card}_V U$. According to Statement 1 (B.3.1), $\operatorname{card}_V A \leqslant \operatorname{card}_V U$. Suppose that $\operatorname{card}_V A = \operatorname{card}_V U = \alpha, \quad \alpha \in V$.

From the assertions proven above, we infer that $\operatorname{card}_U A = \operatorname{card}_V A = \alpha$ implies $\alpha \in U$. By axiom A7, $\alpha \subset U$, $\alpha \subset V$, and $U \subset V$. By the definition of V-power, there exists a bijective V-function $f : \alpha \rightarrowtail\!\!\!\twoheadrightarrow_V U$. According to Lemma 5 (B.3.2), $f \leftrightharpoons \alpha \rightarrowtail\!\!\!\twoheadrightarrow_U U$. By the axiom of full union A11, $U = \operatorname{rng}_U f \in U$.

We infer that $U \in U$, but it is impossible. □

An inaccessible cardinal number is defined in the LTS as was done in the ZF set theory.

A U-cardinal number $\varkappa$ will be called *U-regular* if for every U-ordinal number β, for which there exists a U-function $f : \beta \to_U \varkappa$ such that $\cup_U \operatorname{rng}_U f = \varkappa$, it is valid $\varkappa \leqslant \beta$. A U-cardinal number $\varkappa > \omega_0 \equiv \omega$ will be called *U-inaccessible* if $\varkappa$ is U-regular and for every U-ordinal number λ from $\lambda \in \varkappa$ it follows that $\mathcal{P}_U(\lambda) \in \varkappa$.

Proposition 3. *Let U be a universal class and $\alpha \in U$. Then, the following conditions are equivalent:*
1) *α is an inaccessible cardinal number;*
2) *α is a U-inaccessible U-cardinal number.*

Proof. (1) $\vdash$ (2). By Proposition 1, α is an U-cardinal number. Let $\beta \in U$, β be a U-ordinal number, and there exist a U-mapping $f : \beta \to_U \alpha$ such that $\cup_U \operatorname{rng}_U f = \alpha$. By Lemma 1, β is an ordinal number. Since $\beta, \alpha \in U$, then, by the axiom of transitivity $\beta, \alpha \subset U$. Therefore, by Lemma 5 (B.3.2), $f \leftrightharpoons \beta \to \alpha$ and $\operatorname{rng}_U f = \operatorname{rng} f$, where $\alpha = \cup_U \operatorname{rng} f$. By Lemma 6 (B.3.2), $\alpha = \cup \operatorname{rng} f$. Since α is a regular cardinal number, we infer that $\alpha \leqslant \beta$. Hence, α is a U-regular U-cardinal number.

Let $\beta \in U$, β is an U-ordinal number and $\beta \in \alpha$. By Lemma 1, β is an ordinal number. Since α is inaccessible, then $\mathcal{P}(\beta) \in \alpha$. By Lemma 1 (B.3.2), $\mathcal{P}_U(\beta) = \mathcal{P}(\beta) \in \alpha$. Therefore, α is a U-inaccessible U-cardinal number.

(2) $\vdash$ (1). By Proposition 1, α is a cardinal number. Let β be an ordinal number and let $f : \beta \to \alpha$ be a mapping such that $\cup \operatorname{rng} f = \alpha$. Suppose that $\beta \in \alpha \in U$. By the axiom of transitivity A7, $\beta \in U$. By Lemma 1, β is a U-ordinal number. By A8, $\beta, \alpha \subset U$. Therefore, by Lemma 1 $f \leftrightharpoons \beta \to_U \alpha$ and $\operatorname{rng} f = \operatorname{rng}_U f$. By Lemma 6 (B.3.2) $\alpha = \cup \operatorname{rng}_U f = \cup_U \operatorname{rng}_U f$. Since α is a U-regular U-cardinal number, we infer that $\alpha \leqslant \beta < \alpha$, but it is impossible. From this contradiction, we infer that the case $\beta \in \alpha$ is impossible. Thus, $\alpha \leqslant \beta$. Hence, α is a regular cardinal number.

Let β be an ordinal number and $\beta \in \alpha$. By axiom A7, $\beta \in U$. By Lemma 1 β is a U-ordinal number. Since α is U-inaccessible, then $\mathcal{P}_U(\beta) \in \alpha$. By Lemma 1 (B.3.2) and Proposition 2 $\operatorname{card} \mathcal{P}(\beta) = \operatorname{card} \mathcal{P}_U(\beta) = \operatorname{card}_U \mathcal{P}_U(\beta) \in \alpha$. Therefore, α is an inaccessible cardinal number. □

In the LTS, we can use the principle of transfinite induction by virtue of the well-ordering of the assembly of all ordinals. Show that

Theorem 1 (the principle of transfinite induction in the LTS). *Let* **C** *be an assembly of ordinal numbers such that:*

1) $\varnothing \in \mathbf{C}$;
2) $\alpha \in \mathbf{C} \Rightarrow \alpha + 1 \in \mathbf{C}$;
3) (α *is a limit ordinal number* $\wedge\ \alpha \subset \mathbf{C}) \Rightarrow \alpha \in \mathbf{C}$.

Then, $\mathbf{C} = \overline{\mathbf{On}}$.

Proof. Suppose that it is not true. Consider the subassembly $\mathbf{D} \equiv \overline{\mathbf{On}} \setminus \mathbf{C}$.

Since the assembly **D** is not empty, we infer by Lemma 3 that it has the smallest element.

Then, $\gamma \neq \varnothing$, because $\varnothing \in \mathbf{C}$. Therefore, γ is either β^+ for some ordinal number β or a limit ordinal number. Suppose that $\gamma = \beta + 1$. Since $\beta \in \gamma$, it follows that $\beta \notin \mathbf{D}$ and so $\gamma \in \mathbf{C}$. By condition 2 of the theorem, $\gamma = \beta + 1 \in \mathbf{C}$. From this contradiction, we infer that $\gamma \neq \beta + 1$. Now, the case remains when γ is a limit ordinal number. In this case, $\gamma = \sup \gamma$ and $\gamma \subset \mathbf{C}$; thus, by the condition 3 of the theorem, $\gamma = \sup \gamma \in \mathbf{C}$. Therefore, the assembly **D** is empty and so $\mathbf{C} = \overline{\mathbf{On}}$. □

Theorem 2 (the principle of natural induction in the LTS). *Let* **C** *be some assembly in LTS such that:*
1) $\varnothing \in \mathbf{C}$;
2) *for all* $n \in \omega$ *the condition* $n \in \mathbf{C}$ *implies* $n + 1 \in \mathbf{C}$.

Then, $\omega \subseteq \mathbf{C}$.

Proof. Consider the assembly $\widetilde{\mathbf{C}} = \mathbf{C} \cap \omega$. This assembly is not empty because $\varnothing \in \mathbf{C} \wedge \varnothing \in \omega \Rightarrow \varnothing \in \widetilde{\mathbf{C}}$, and besides, it contains only ordinal numbers. Suppose now that the assertion of the theorem is not fulfilled. In this case, the assembly $\overline{\mathbf{C}} = \{x \mid x \in \omega \wedge x \notin \widetilde{\mathbf{C}}\}$ is not empty because it is a subassembly of $\overline{\mathbf{On}}$, and therefore contains the smallest element $\alpha \in \omega$. We know that $\alpha \neq \varnothing$, because $\varnothing \in \widetilde{\mathbf{C}}$. Since $\alpha \in \omega$ and $\alpha \neq \varnothing$, then there exists β such that $\alpha = \beta + 1$. In this case, $\beta \in \widetilde{\mathbf{C}}$ because α is the smallest ordinal number in $\overline{\mathbf{C}}$. By condition 2 of the theorem, in this case, $\beta + 1 \in \widetilde{\mathbf{C}}$, i. e. $\alpha \in \widetilde{\mathbf{C}}$, and we get the contradiction with our assumption. □

B.3.4 Cumulative classes in the LTS and their connection with universal classes

Using all previous material, we will construct cumulative classes in the LTS.

Consider an arbitrary universe U. Since it is a model of ZF, we can, using the construction by transfinite induction, define cumulative U-sets V_α^U for every U-ordinal number $\alpha \in U$. By the axiom of universality A6 and Lemma 1, for every ordinal number α we define some cumulative U-set V_α^U for every universe U such that $\alpha \in U$.

Lemma 1. *Let* U *and* W *be universal classes and* α *be an ordinal number such that* $\alpha \in U$ *and* $\alpha \in W$. *Then,* $V_\alpha^U = V_\alpha^W$.

Proof. We will prove it by the transfinite induction. Consider the subassembly **C** of the assembly $\overline{\mathbf{On}}$, consisting of all ordinal numbers α such that either $\alpha \in U$, $\alpha \in W$ and $V_\alpha^U = V_\alpha^W$, or $\alpha \notin U$, or $\alpha \notin W$.

Since $\varnothing \in U$, $\varnothing \in W$, $V_\varnothing^U = \varnothing$ and $V_\varnothing^W = \varnothing$, it follows that $V_\varnothing^U = V_\varnothing^W$. Therefore, $\varnothing \in \mathbf{C}$.

Let $\alpha \in \mathbf{C}$. If $\alpha \notin U$ or $\alpha \notin W$, then respectively $\alpha+1 \notin U$ or $\alpha+1 \notin W$, i. e. $\alpha+1 \in \mathbf{C}$. Let $\alpha \in U$, $\alpha \in W$. By the facts proven in section A.2, $V_{\alpha+1}^U = V_\alpha^U \cup_U \mathcal{P}_U(V_\alpha^U)$ and $V_{\alpha+1}^W = V_\alpha^W \cup_W \mathcal{P}_W(V_\alpha^W)$. By assumption, $V_\alpha^U = V_\alpha^W \equiv V_\alpha$. Since $V_\alpha \in U$ and $V_\alpha \in W$, it follows by Lemma 1 (B.3.2) that $\mathcal{P}_U(V_\alpha) = \mathcal{P}_W(V_\alpha)$. Hence, $V_{\alpha+1}^U = V_\alpha \cup_U \mathcal{P}_U(V_\alpha) \equiv P$ and $V_{\alpha+1}^W = V_\alpha \cup_W \mathcal{P}_U(V_\alpha) \equiv Q$. By axiom A9 (LTS), $\mathcal{P}_U(V_\alpha) \in U$ and $\mathcal{P}_U(V_\alpha) \in W$. Using axiom A7, we can easily check that $P = Q$. Thus, $V_{\alpha+1}^U = V_{\alpha+1}^W$. Therefore, $\alpha + 1 \in \mathbf{C}$.

Let α be a limit ordinal number such that $\alpha \subset \mathbf{C}$. If $\alpha \notin U$ or $\alpha \notin W$, then $\alpha \in \mathbf{C}$. Consider the case $\alpha \in U$ and $\alpha \in W$. By axiom A7, $\alpha \subset U$ and $\alpha \subset W$. Since $\alpha \subset \mathbf{C}$, we have

$V_\beta^U = V_\beta^W \equiv V_\beta$ for all $\beta \in \alpha$. According to assertions from A.3.1, we get $V_\alpha^U = \cup_U(V_\beta^U \mid \beta \in \alpha)_U$ and $V_\alpha^W = \cup_W(V_\beta^W \mid \beta \in \alpha)_W$. Since $V_\beta^U = V_\beta^W = V_\beta$ and $V_\beta \in U$, $V_\beta \in W$ for all $\beta \in \alpha$, we can write that $V_\alpha^U = \cup_U(V_\beta \mid \beta \in \alpha)_U$, and $V_\alpha^W = \cup_W(V_\beta \mid \beta \in \alpha)_W$. Show that $R \equiv \cup_U(V_\beta \mid \beta \in \alpha)_U = \cup_W(V_\beta \mid \beta \in \alpha)_W \equiv S$.

Let $x \in R$. Then, $x \in U$ and $x \in V_\beta$ for some $\beta \in \alpha$. Since $x \in V_\beta \in W$, we have, by axiom A7, $x \in W$. Therefore, $x \in S$. Thus, $R \subset S$. The converse implication is checked similarly. Thus, we infer that $V_\alpha^U = R = S = V_\alpha^W$. It means that $\alpha \in \mathbf{C}$.

By Theorem 1 (A.2.2), $\mathbf{C} = \overline{\mathbf{On}}$. □

Using this lemma, we can for every $\alpha \in \overline{\mathbf{On}}$ define in the LTS the *cumulative class* $\overline{V_\alpha}$ (we draw a line over V_α to differ these classes from the corresponding classes in ZF) as the U-class V_α^U for every universe U, satisfying the condition $\alpha \in U$. In the result, we get the *Mirimanov – Neumann collection* $(\overline{V_\alpha} \subset \overline{\mathbf{V}} | \alpha \in \overline{\mathbf{On}})$ in the LTS. It satisfies properties 1–3 of the Mirimanov – Neumann collection in ZF, listed in section A.2.

Lemma 2. *The collection* $(\overline{V_\alpha} \subset \overline{\mathbf{V}} \mid \alpha \in \overline{\mathbf{On}})$ *possesses the following properties:*
1) $\alpha = \beta \Longleftrightarrow \overline{V_\alpha} = \overline{V_\beta}$;
2) $\alpha < \beta \Longleftrightarrow \overline{V_\alpha} \in \overline{V_\beta}$.

Proof. At the beginning we will show that $\alpha < \beta$ implies $\overline{V_\alpha} \in \overline{V_\beta}$. Suppose that $\beta \in U$ for some universe U. Then, $\alpha \in U$ implies $\overline{V_\alpha} = V_\alpha^U$ and $\overline{V_\beta} = V_\beta^U$. By Statement 1 (B.3.1), our assertion follows now from the fact that in ZF, by Lemma 1 (A.3.2) $\alpha < \beta \Rightarrow V_\alpha \in V_\beta$. Now, we will prove all assertions of the lemma. The assertion $\alpha = \beta \Rightarrow \overline{V_\alpha} = \overline{V_\beta}$ is proven above. If $\overline{V_\alpha} = \overline{V_\beta}$, then either $\alpha < \beta$ or $\alpha = \beta$, or $\beta < \alpha$. If $\alpha < \beta$, then $\overline{V_\alpha} \in \overline{V_\beta}$; if $\beta < \alpha$, then $\overline{V_\beta} \in \overline{V_\alpha}$, therefore $\beta = \alpha$.

The assertion $\alpha < \beta \Rightarrow \overline{V_\alpha} \in \overline{V_\beta}$, is already proven. If $\overline{V_\alpha} \in \overline{V_\beta}$, then $\alpha < \beta$, because for $\alpha = \beta$, we have $\overline{V_\alpha} = \overline{V_\beta}$, and for $\beta < \alpha$, we have $\overline{V_\beta} \in \overline{V_\alpha}$. □

The following theorem shows that all universal classes in the LTS are cumulative sets for inaccessible cardinal indices.

Theorem 1. *Let U be an arbitrary universal class. Then:*
1) $\varkappa \equiv \sup(\overline{\mathbf{On}} \cap U) = \cup(\overline{\mathbf{On}} \cap U) \subset U$ *is an inaccessible cardinal number;*
2) $U = \overline{V_\varkappa}$;
3) *the correspondence* $q : U \mapsto \varkappa$ *such that* $U = \overline{V_\varkappa}$ *is an injective isotone mapping from the assembly* $\mathbf{U}$ *of all universal classes into the assembly* $\overline{\mathbf{In}}$ *of all inaccessible cardinal numbers.*

Proof. 1. Since $A \equiv \overline{\mathbf{On}} \cap U$ is a non-empty class, because it contains the element $0 \equiv \varnothing$, we infer, by Lemma 5 (B.3.3) that $\varkappa$ is an ordinal number.

Suppose that $\varkappa$ is not a cardinal number. In this case, there exist an ordinal number $\alpha < \varkappa$ and a bijective function $f : \alpha \rightarrowtail\!\!\!\!\rightarrow \varkappa$. Since $\varkappa \subset U$ and $\alpha \subset U$, we have, by Lemma 4 (B.3.2) $\alpha * \varkappa = \alpha *_U \varkappa$. Therefore, f is a U-function $f : \alpha \rightarrowtail_U \varkappa$. Since $\alpha \in \varkappa \subset U$ and $f(x) \in \varkappa \subset U$ for every $x \in \beta$, by the axiom of full union, for the universal class U we infer that $\varkappa = \operatorname{rng}_U f \in U \cap \overline{\mathbf{On}}$ and therefore, by the axiom of binary union, $\varkappa_U^+ \equiv \varkappa \cup_U \{\varkappa\}_U \in U$. By Lemma 4 (B.3.3) $\varkappa_U^+ = \varkappa^+ \in \overline{\mathbf{On}}$. Thus, $\varkappa_U^+ \leqslant \varkappa < \varkappa_U^+$, but it is impossible. From this contradiction, we infer that $\varkappa$ is a cardinal number.

Suppose that the cardinal $\varkappa$ is not regular. Then, $\alpha \equiv cf(\varkappa) < \varkappa$. By definition, there exists a function $f : \alpha \to \varkappa$ such that $\sup f[\alpha] = \varkappa$. As above, f is a U-function $f : \alpha \to_U \varkappa$ and $\operatorname{rng}_U f \in U$. It is clear that $\operatorname{rng}_U f \subset f[\alpha]$. Conversely, if $y \in f[\alpha]$, then $y = f(x)$ for some $x \in \alpha$. Since $f(x) \in \varkappa \subset U$, we have $y \in U$. Consequently, $y \in \operatorname{rng}_u f$. As a result, $f[\alpha] = \operatorname{rng}_U f \in U$. By the axiom of full union, $\varkappa = \sup f[\alpha] = \cup f[\alpha] = \cup_U (y \subset U \mid y \in f[\alpha])_U \in U$. Similarly, as was done before, the property $\varkappa \in U$ brings us to the contradiction. Therefore, $\varkappa$ is a regular cardinal.

Let λ be a cardinal number such that $\lambda < \varkappa$. Since $\lambda \in \varkappa \subset U$, we have, by the axiom of full ensemble and by Lemma 1 (B.3.2) $\mathcal{P}(\lambda) = \mathcal{P}_U(\lambda) \in U$. Consequently, $\alpha \equiv \operatorname{card} \mathcal{P}(\lambda) = \operatorname{card} \mathcal{P}_U(\lambda)$. According to Proposition 2 (B.3.3), this last number is equal to the number $\operatorname{card}_U \mathcal{P}_U(\lambda) \in U$. Thus, $\alpha \in U \cap \overline{\mathbf{On}}$. Therefore, $\alpha \leqslant \varkappa$. Suppose that $\varkappa = \alpha$. Then, $\varkappa \in U$. But as above this property leads us to the contradiction. As a result, we infer that $\alpha < \varkappa$.

Now, it remains only to show that $\varkappa > \omega$. Since $\omega \in \mathfrak{a}$ (see after axiom A13 in section B.1), we have $\omega \in U$ and so $\omega + 1 = \omega \cup \{\omega\} \in U$. Therefore, $\omega \in \omega + 1 \in A$ implies $\omega \in \cup A = \varkappa$.

Assertion 1 is proven.

2. From (1) it follows that $\varkappa$ is a limit ordinal number.

Therefore, $\overline{V_\varkappa} = \cup(\overline{V_\beta} | \beta \in \varkappa)$. Since $\beta \in \varkappa \subset U$, we have, by the definition, $\overline{V_\beta} = V_\beta^U \subset U$. Consequently, $\overline{V_\varkappa} \subset U$. Conversely, let $x \in U$. By Lemma 7 (A.3.3) $\Pi = \mathbf{V}$ in ZF. Similarly, $\cup_U (V_\alpha^U \mid \alpha \in \overline{\mathbf{On}} \cap U) = U$ in the LTS. Therefore, $x \in V_\alpha^U$ for some $\alpha \in A \subset U$. Since $V_\alpha^U = \overline{V_\alpha}$, we have $x \in \overline{V_\alpha} \subset \overline{V_\varkappa}$. Therefore, $U \subset \overline{V_\varkappa}$. As a result, we infer that $U = \overline{V_\varkappa}$.

3. From Lemma 2, we infer that $\varkappa$ is unique. Therefore, we can define a mapping $q : \mathbf{U} \to \overline{\mathbf{In}}$ such that $q(U) = \varkappa$, where $U = \overline{V_\varkappa}$. From Lemma 2 we also infer that q is isotone. □

Corollary 1. *If U is a universal class, then $\varkappa \equiv \operatorname{card} U$ is an inaccessible cardinal number and $U = \overline{V_\varkappa}$.*

Proof. According to Theorem 1, we only need to show that for any inaccessible cardinal number $\varkappa$, we have $\operatorname{card} \overline{V_\varkappa} = \varkappa$.

Consider some universal class W such that $\varkappa \in W$. In this case, $\overline{V_\varkappa} = V_\varkappa^W \in W$. By Proposition 2 (B.3.3), $\operatorname{card} \overline{V_\varkappa} = \operatorname{card}_W \overline{V_\varkappa}$. Since the universe W is a model of ZF, by Lemma 2 (A.3.3) the property $\varkappa = \operatorname{card}_W V_\varkappa^W = \operatorname{card} \overline{V_\varkappa}$ is fulfilled in it.

Therefore, if U is a universal class, then, by Theorem 1, $U = \overline{V_\varkappa}$, where $\varkappa$ is an inaccessible cardinal number. Our assertion now follows directly from the property card U = card $\overline{V_\varkappa} = \varkappa$. □

Corollary 2. *In the LTS, the equality* $\cup(\overline{V_\alpha} \mid \alpha \in \overline{\mathbf{On}}) = \overline{\mathbf{V}}$ *is valid.*

Proof. We need to show that for an arbitrary class x in the LTS the assertion $x \in \cup(\overline{V_\alpha} \mid \alpha \in \overline{\mathbf{On}})$ is valid, i. e. there exists $\alpha \in \overline{\mathbf{On}}$ such that $x \in \overline{V_\alpha}$.

By the axiom of universality, there exists a universe U such that $x \in U$, and, by Theorem 1, $U = \overline{V_\varkappa}$ for some $\varkappa \in \overline{\mathbf{On}}$. Therefore, $x \in \overline{V_\varkappa}$, i. e. our assertion is true. □

Theorem 1 allows to make the following conclusions about the structure of the assembly $\mathbf{U} \equiv \{U \mid U \bowtie\}$ of all universal classes.

The relation $\in \cup =$ is a relation of order on the assembly $\mathbf{U}$. We will denote it by $\leqslant$, i. e. $U \leqslant V$, if $U \in V$ or $U = V$. By axiom A7, the assembly $\mathbf{U}$ is transitive. Therefore, $U \in V$ implies $U \subset V$. Thus, $U \leqslant V$ implies $U \subset V$. We will prove now that these relations are equivalent.

Proposition 1. *Let U and V be universal classes. Then, the relation $U \leqslant V$ is equivalent to the relation $U \subset V$.*

Proof. We only need to check that $U \subset V$ implies $U \leqslant V$. By Theorem 1, $U = \overline{V_\pi}$ and $V = \overline{V_\varkappa}$ for some inaccessible cardinals π and $\varkappa$. If $\pi = \varkappa$, then $U = \overline{V_\pi} = \overline{V_\varkappa} = V$. If $\pi < \varkappa$, then, by Lemma 2 $U = \overline{V_\pi} \in \overline{V_\varkappa} = V$. Finally if $\pi > \varkappa$, then, by the same lemma, $V = \overline{V_\varkappa} \in \overline{V_\pi} = U \subset V$, but it is impossible. Therefore, $U \leqslant V$. □

Corollary 1. *The infra-universe $\mathfrak{a}$ is the smallest element in the assembly* **U** *of all universal classes.*

Corollary 2. *If U is a universal class, then either $U = \mathfrak{a}$ or $\mathfrak{a} \in U$.*

Corollary 3. *With the universal class $\mathfrak{a}$ it is associated a unique inaccessible cardinal number $\varkappa^*$ such that $\mathfrak{a} = \overline{V_{\varkappa^*}}$. This number is the smallest in the assembly* $\overline{\mathbf{In}}$ *of all inaccessible cardinal numbers.*

Thus, in the LTS, there exists at least one inaccessible cardinal number.

Prove now that in the LTS there exists more than one inaccessible cardinal number.

B.3.5 The structure of the assemblies of all universal classes and all inaccessible cardinals in the local theory of sets

Proposition 1. *In the LTS for every ordinal number α, there exists an inaccessible cardinal number $\varkappa$ such that $\alpha < \varkappa$.*

Proof. By the axiom of universality, $\alpha \in U$ for some universal class U. By Theorem 1 (B.3.4), $U = \overline{V_\varkappa}$ for some inaccessible cardinal $\varkappa$. By definition, $\overline{V_\alpha} = V_\alpha^U \subset U = \overline{V_\varkappa}$. By Lemma 2 (B.3.4) $\alpha \leqslant \varkappa$. Suppose that $\varkappa = \alpha$. Then, $\varkappa \in U$. But this property leads to the contradiction, as it was shown in the proof of Theorem 1 (B.3.4). Therefore, $\alpha < \varkappa$. □

This property is similar to the axiom of universality, which postulates that every class in the LTS is an element of some universal class.

The parallelism between properties of universal classes and inaccessible cardinals in the LTS is confirmed also by the following assertions.

Theorem 1. *The assembly* $\mathbf{U}$ *of all universal classes with respect to the order* $\subset$ *is well-ordered. Furthermore, every subassembly of the assembly* $\mathbf{U}$ *has the smallest element.*

Proof. Let $\varnothing \neq \mathbf{A} \subset \mathbf{U}$. Using the injective and strictly monotone mapping $q : U \mapsto \varkappa$ from the assembly $\mathbf{U}$ into the assembly $\overline{\mathbf{On}}$ of the form $U = \overline{V_\varkappa}$ from Theorem 1 (B.3.4), we can consider for the assembly $\mathbf{A}$ the subassembly $\mathbf{B} \equiv q[\mathbf{A}] \equiv \{x \mid x \in \overline{\mathbf{On}} \wedge \exists U \in \mathbf{A}(z = q(\mathbf{U}))\}$ of the assembly $\overline{\mathbf{On}}$. By Lemma 3 (B.3.3), it has the smallest element π, which is an inaccessible cardinal. Since $\pi \in \mathbf{B}$, we have $\pi = q(U)$ for some $U \in \mathbf{A}$, i. e. $U = \overline{V_\pi}$. Since the mapping q is injective and strictly monotone, it follows that U is the smallest element in the assembly $\mathbf{A}$. □

Corollary 1. *For every class A, there exists a universe $U(A)$ which is the smallest in the assembly of all universes U such that $A \in U$.*

Corollary 2. *The intersection* $\cap\mathbf{A} \equiv \{x \mid \forall U \in \mathbf{A}(x \in U)\}$ *of any non-empty assembly* $\mathbf{A}$ *of universal classes is a universal class.*

Proof. By Theorem 1, the subassembly $\mathbf{A}$ of the assembly $\mathbf{U}$ has the smallest element $\mathbf{U}$. It is clear that $\cap\mathbf{A} \subset U$. If $V \in \mathbf{A}$, then $U \leqslant V$ implies $U \subset V$. Therefore, $U \subset \cap\mathbf{A}$. Thus, $\cap\mathbf{A} = U$. □

Theorem 1 allows to finish the globalization of local constructions, which was started in B.3.2.

Corollary 3. *For every class A and B, the assemblies $A \cup B$, $A \cap B$, $\{A, B\}$, $\langle A, B\rangle$, and $A * B$ are classes.*

Proof. By the axiom of universality A6, for A and B, there exist universal classes β and γ such that $A \in \beta$ and $B \in \gamma$. By Theorem 1, either $\beta \subset \gamma$ or $\gamma \subset \beta$. Therefore, there exists a universal class α ($\alpha = \beta$ or $\alpha = \gamma$) such that $\beta, \gamma \subset \alpha$. Thus, $A, B \in \alpha$. By the axiom of transitivity A7, $A, B \subset \alpha$. Therefore, by Lemmas 2–5 (B.3.2) $A \cup B = A \cup_\alpha B$, $A \cap B = A \cap_\alpha B$, $\{A, B\} = \{A, B\}_\alpha$, $\langle A, B\rangle = \langle A, B\rangle_\alpha$, and $A * B = A *_\alpha B$. By axiom scheme AS2, the right parts of all these equalities are classes, because they are defined by α-predicative formulas. □

Corollary 4. *Let $n \in \omega \setminus 1$ and $\mathbf{F}$ be an assembly such that $\mathbf{F}$ is a mapping from the class n into the assembly $\overline{\mathbf{V}}$. Then, the assembly $\mathbf{F}$ is a class.*

Proof. Consider the assembly $\mathbf{K}' \subset n$, consisting of all natural numbers $k \in n$ such that the assembly $\mathbf{F}|(k+1)$ is a class. Consider the assembly $\mathbf{K}'' \equiv \omega \setminus n$ and the assembly $\mathbf{K} \equiv \mathbf{K}' \cup \mathbf{K}''$.

Since $\mathbf{F}|1 = \langle 0, \mathbf{F}(0)\rangle$, we infer by Corollary 1 to Theorem 1 that $\mathbf{F}|1$ is a class.

Let $k \in \omega$ and $k \in \mathbf{K}$. If $k < n - 1$, then $k + 1 \in n$. Since in this case $k \in \mathbf{K}'$, the assembly $\mathbf{F}|(k+1)$ is a class. By Corollary 3 to Theorem 1, the assembly $\langle k+1, \mathbf{F}(k+1)\rangle$ is also a class. Now, from the equality $\mathbf{F}|(k+2) = \mathbf{F}|(k+1) \cup \{\langle k+1, \mathbf{F}(k+1)\rangle\}$ by the mentioned corollary we infer that the assembly $\mathbf{F}|(k+2)$ also is a class. It means that $k+1 \in \mathbf{K}' \subset \mathbf{K}$.

If $k \geqslant n-1$, then $k+1 \geqslant n$ implies $k+1 \in \mathbf{K}'' \subset \mathbf{K}$.

Applying Theorem 2 (B.3.3), we conclude that $\omega \subset \mathbf{K} \subset \omega$. Therefore, $\mathbf{K}' = n$. Consequently, $n-1 \in \mathbf{K}'$ means that the assembly $\mathbf{F} = \mathbf{F}|((n-1)+1)$ is a class. □

Corollary 5. *Let $A, A', A'', \ldots$ be classes. Then, the assemblies (A, A'), (A, A', A''), ... are classes.*

Proof. By definition, the assemblies (A, A'), (A, A', A''), ... are the particular cases of sequences $(A_0, \ldots, A_{n-1})$ when $n \in \omega \setminus 2$. But, by the previous corollary, the sequences $(A_0, \ldots, A_{n-1})$ are classes. □

Thus, the assembly $\overline{\mathbf{V}}$ of all classes in the LTS allows us to produce almost all set-theoretical constructions which are possible in a universal class or in the NBG-universe, except the construction of full union, which is basic for the construction by transfinite induction. The fact that the construction of full union is impossible in the LTS follows from Statement 1 (B.6.1). It means that the global assembly of all classes in the LTS with respect to its constructive possibilities is much more poor than local universal classes in it.

The next theorem describes the structure of the assembly $\mathbf{U}$ of all universes in the LTS. It is proven with the help of Theorem 1.

Theorem 2. *In the LTS, for every $n \in \omega$, there exist a unique universal class U and a unique U-sequence of universal classes $u(n) \equiv (U_k \in U | k \in n+1)_U$ such that $U_0 = \mathfrak{a}$, $U_k \in U_l$ for every $k \in l \in n+1$ and if V is an universal class and $U_0 \subset V \in U$, then $V = U_k$ for some $k \in n+1$ (the property of incompressibility).*

From the property of uniqueness, it follows that $u(n) \mid m+1 = u(m)$ for all $m \leqslant n$, i. e. these finite sequences continue each other.

Proof. Consider the set N, consisting of all $n \in \omega$, for which there exist a unique universal class U and a unique sequence of universal classes $u \equiv u(n) \equiv (U_k \in U \mid k \in n+1)_U$ such that $U_0 = \mathfrak{a}$, $U_k \in U_l$ for any $k \in l \in n+1$ and if V is a universal class and $U_0 \subset V \in U$, then $V = U_k$ for some $k \in n+1$.

By Corollary 1 to Theorem 1, for the infra-universe $\mathfrak{a}$, there exists a universe U^* which is the smallest from all universes U such that $\mathfrak{a} \in U$. Since the universe $U \equiv U^*$ and the sequence $(U_k \in U \mid k \in 1)$ such that $U_0 \equiv \mathfrak{a}$, satisfy all mentioned properties, we have $0 \in N$. Let $n \in N$. Consider the assembly $\mathbf{V} \equiv \{i \mid i \bowtie \wedge u \neq U \wedge \forall k \in n+1(u \neq U_k)\}$. By axiom A6, for U, there exists a universal class K such that $U \in K$. Therefore, the assembly $\mathbf{V}$ is non-empty and consequently, by Theorem 1, it contains the smallest element V. Clear that $V \neq U$ and $V \geqslant U^* > U_0$. Suppose that $V \in U$. Then, $U_0 \leqslant V \in U$, by supposition, implies $A = U_k$ for some $k \in n+1$, but it is impossible. Therefore, $U \in V$.

Thus, in the universe V we can define a V-sequence $v \equiv (V_k \in V \mid k \in n+2)_V$, setting $V_k \equiv U_k$ for every $k \in n+1$ and $V_{n+1} \equiv U$. It is clear that $V_0 = \mathfrak{a}$ and $V_k \in V_l$ for every $k \in l \in n+2$. Let W be a universe and $V_0 \leqslant W \in V$. Then, $U_0 \subset W \in V$. If $W = U$, then $W = V_{n+1}$. If $W \in U$, then $U_0 \subset W \in U$ implies $W = U_k = V_k$ for some $k \in n+1$. Finally, if $U \in W$, then $U_0 \in U_1 \in \cdots \in U_n \in U \in W$ implies $W \in \mathbf{V}$. Therefore, $V \subset W$. But this case is impossible. From the two previous cases, we infer that $W = V_k$ for some $k \in n+2$. It means that the pair V and v possesses all necessary properties. Check its uniqueness. Suppose that there exist a universe W and a W-sequence of universes $w \equiv (W_k \in W \mid k \in n+2)|_W$ such that $W_0 = \mathfrak{a}$, $W_k \in W_l$ for all $k \in l \in n+2$, and if K is a universe and $W_0 \subset K \in W$, then $K = W_k$ for some $k \in n+2$.

Consider the universe $U' \equiv W_{n+1}$ and the U'-sequence $u' \equiv (U_k{}' \in U' \mid k \in n+1)_{U'}$ such that $U_k{}' \equiv W_k$ for every $k \in n+1$. It is clear that $U_0' = W_0 = \mathfrak{a}$ and $U_k{}' \equiv W_k \in W_k \equiv U_l{}'$ for every $k \in l \in n+1$. If K is a universe and $U_0{}' \leqslant K \in U'$, then $U_0' \leqslant K \in U'$ and $W_0 \leqslant K \in W_{n+1} \in W$. By the axiom of transitivity A7, $W_0 \subset K \in W$. Therefore, by assumption, $K = W_k$ for some $k \in n+2$. Since $K \in W_{n+1}$, we have $K = W_k = U_k'$ for some $k \in n+1$. Thus, the pair U' and u' possesses all the properties for n mentioned above. Therefore, by virtue of the uniqueness of this pair, we infer that $U = U' = W_{n+1}$ and $u = u' = w|n+1$, i. e. $W_{n+1} = U \equiv V_{n+1}$ and $W_k = U_k \equiv V_k$ for all $k \in n+1$. Thus, $w = v$. If $W \in V$, then $V_0 = W_0 \subset W \in V$, by the above, implies $W = V_k = W_k$ for some $k \in n+2$, but it is impossible. If $V \in W$, then $W_0 = V_0 \subset V \in W$ in a similar manner implies $V = W_k = V_k$ for some $k \in n+2$, but this is impossible. Thus, $W = V$ and the uniqueness of the universe V and the sequence v is proven. Therefore, $n+1 \in N$. By the principle of natural induction, $N = \omega$. Thus, for any $n \in \omega$, there exists the mentioned unique pair V and u. □

Unfortunately, in the LTS, in contrast to ZF, where there is the axiom scheme of replacement, we have no means to componate all these finite sequences into one infinite sequence of universal classes.

With the help of Proposition 1 (B.3.5), we can prove the following

Theorem 3. *In the LTS, for every* $n \in \omega$, *there exists a unique sequence* $c(n) \equiv (\varkappa_k \in \mathbf{I} \mid k \in n+1)$ *of inaccessible cardinals such that* $\varkappa_0 = \varkappa^*$, $\varkappa_k \in \varkappa_l$ *for every* $k \in l \in n+1$ *and if* π *is an inaccessible cardinal and* $\varkappa_0 \leqslant \pi < \varkappa_n$, *then* $\pi = \varkappa_k$ *for some* $k \in n$ *(the property of incompressibility).*

From the property of uniqueness, it follows that $c(n)|m+1 = c(m)$ *for all* $m \leqslant n$, *i. e. these finite sequences continue each other.*

The proof is similar to the proof of the previous theorem.

The remark, which was made after Theorem 2, is valid also in this case: in the LTS, there are no means to construct from these finite sequences continuing each other one infinite sequence of inaccessible cardinals. In the next section, we show how to do it in the ZF set theory.

B.4 Relative consistency between the LTS and the ZF set theory

B.4.1 Additional axioms on inaccessible cardinals in the ZF set theory

To prove the relative consistency, we need to write the axioms on inaccessible cardinals in more formal way. Therefore, we are forced to adduce the following formal definitions in the ZF set theory (some of these notions, notations, and axioms were also considered in A.2, A.4, and A.5):

- α *is an ordinal number* $\equiv On(\alpha) \equiv (\forall x(x \in \alpha \Rightarrow \forall y(y \in x \Rightarrow y \in \alpha))) \wedge (\forall x, y, z (x, y, z \in \alpha \wedge x \in y \wedge y \in z \Rightarrow x \in z)) \wedge (\forall x, y(x, y \in \alpha \Rightarrow x \in y \vee x = y \vee y \in x)) \wedge \forall z(\varnothing \neq z \subset \alpha \Rightarrow \exists x(x \in z \wedge \forall y(y \in z \Rightarrow x \in y)))$;
- f is a *function* $\equiv func(f) \equiv \forall x \forall y \forall y'(\langle x, y\rangle \in f \wedge \langle x, y'\rangle \in f \Rightarrow y = y')$;
- $f \leftrightharpoons A \to B \equiv func(f) \wedge \operatorname{dom} f = A \wedge \operatorname{rng} f \subset B$;
- $\varkappa$ *is a cardinal number* $\equiv Cn(\varkappa) \equiv On(\varkappa) \wedge \forall x(On(x) \wedge (x = \varkappa \vee x \in \varkappa) \wedge \exists u(u \leftrightharpoons x \rightarrowtail\!\!\!\twoheadrightarrow \varkappa) \Rightarrow x = \varkappa)$;
- $\varkappa$ *is a regular cardinal number* $\equiv Rcn(\varkappa) \equiv Cn(\varkappa) \wedge \forall x(On(x) \wedge \exists u(u \leftrightharpoons x \to \varkappa \wedge \cup \operatorname{rng} u = \varkappa) \Rightarrow (\varkappa = x \vee x \in \varkappa))$;
- $\varkappa$ *is an inaccessible cardinal number* $\equiv Icn(\varkappa) \equiv Rcn(\varkappa) \wedge \forall x(On(x) \wedge x \in \varkappa \Rightarrow \operatorname{card} \mathcal{P}(x) \in \varkappa)$;

AIC. (The *axiom of inaccessible cardinal.*) $\exists \varkappa(Icn(\varkappa))$ (see A.5.2).

AI(ω). (The *ω-inaccessibility axiom.*) $\exists X(\forall x \in X(Icn(x)) \wedge X \neq \varnothing \wedge \forall x \in X \exists y \in X(x \in y))$. (see A.5.1).

AI($\omega + \omega$). $\equiv \exists X \exists Y(\forall x \in X \forall y \in Y(Icn(x) \wedge Icn(y) \wedge x \in y) \wedge X \neq \varnothing \wedge Y \neq \varnothing \wedge \forall x \in X \exists x' \in X(x \in x') \wedge \forall y \in Y \exists y' \in Y(y \in y'))$;

AI. (The *inaccessibility axiom.*) $\forall \alpha(On(\alpha) \Rightarrow \exists \varkappa(Icn(\varkappa) \wedge \alpha \in \varkappa))$ (see A.4.3).

Consider the class (possibly empty) $\mathbf{In} \equiv \{x \mid Icn(x)\}$ of all inaccessible cardinal numbers in the ZF set theory.

The adduced list will allow us later in process of investigation of corresponding interpretations to look accurately what values some derivative terms such as $\operatorname{rng} u$, $\cup \operatorname{rng} u$, $\mathcal{P}(x)$, $\{x\}$, $x \cup \{x\}$, $\langle x, y\rangle$, $\operatorname{dom} f$, $\operatorname{rng} f$, and others take under the interpretation, and also what formulas such formulas as $z \subset \alpha$, $u \leftrightharpoons x \to \varkappa$, and others are translated into.

Theorem 1. *In the ZF set theory, the following assertions are equivalent:*
1) *$AI(\omega)$;*
2) *for every $n \in \omega$, there exists a finite set of inaccessible cardinals of the power $n + 1$;*
3) *for every $n \in \omega$, there exists a finite sequence $u \equiv (\iota_k \mid k \in n+1)$ of inaccessible cardinals such that $\iota_k < \iota_l$ for all $k \in l \in n+1$, i. e. the sequence u is strictly increasing;*
4) *there exists an inaccessible cardinal $\varkappa^*$ and for every $n \in \omega$, there exists a unique finite strictly increasing sequence $u(n) \equiv (\iota_k^n \mid k \in n+1)$ of inaccessible cardinals such that $\iota_0^n = \varkappa^*$ and from the fact that $\varkappa$ is an inaccessible cardinal and $\iota_0^n \leqslant \varkappa \leqslant \iota_n^n$, it follows that $\varkappa = \iota_k^n$ for some $k \in n+1$ (the property of incompressibility);*
5) *there exists a denumerable set of inaccessible cardinal;*
6) *there exists an infinite sequence $u \equiv (\iota_n \mid n \in \omega)$ of inaccessible cardinals such that $\iota_k < \iota_l$ for every $k \in l \in \omega$, i. e. the sequence u is strictly increasing;*
7) *there exists an infinite strictly increasing sequence $u \equiv (\iota_n \mid n \in \omega)$ of inaccessible cardinals such that from $n \in \omega$, $\varkappa$ is an inaccessible cardinal, and $\iota_0 \leqslant \varkappa \leqslant \iota_n$ it follows that $\varkappa = \iota_k$ for some $k \in n+1$ (the property of incompressibility);*
8) *there exists an infinite set of inaccessible cardinals.*

Proof. (1) $\vdash$ (4). Let I be a non-empty set, the existence of which is postulated by axiom AI(ω). Consider the non-empty class $\mathbf{I} \equiv \{x \mid Icn(x) \wedge \exists y \in I(x \leqslant y)\}$. If $x \in \mathbf{I}$, then $x \leqslant y$ for some $y \in I$. By AI(ω), for $y \in I$, there exists $z \in I$ such that $y < z$. Therefore, $x < z \in \mathbf{I}$. Thus, the class $\mathbf{I}$ possesses all the properties, listed in formula AI(ω).

Since $\varnothing \neq \mathbf{I} \subset \mathbf{In}$, by Lemma 1 (A.2.2) in $\mathbf{I}$ there exists the smallest element $\varkappa^*$. From $\varkappa^* \leqslant y$ for every $y \in I$, we infer that $\varkappa^* \in \mathbf{I}$. The class $\mathbf{I}$ possesses the following property: if $z \in \mathbf{In}$ and $z \leqslant y$ for some $y \in \mathbf{I}$, then $z \in \mathbf{I}$.

Consider the set N, consisting of all $n \in \omega$, for which there exists a unique sequence $u \equiv u(n) \equiv (\iota_k \in \mathbf{I} \mid k \in n+1)$ such that $\iota_0 = \varkappa^*$, $\iota_k < \iota_l$ for every $k \in l \in n+1$ and $\varkappa \in \mathbf{In}$ and $= \iota_0 \leqslant \varkappa < \iota_n$ imply $\varkappa = \iota_k$ for some $k \in n$.

Since the sequence $(\iota_k \in \mathbf{I} \mid k \in 1)$ such that $\iota_0 \equiv \varkappa^*$ possesses all the mentioned properties, we infer that $0 \in N$. Let $n \in N$. By the property of the class $\mathbf{I}$, for $\iota_n \in \mathbf{I}$, there exists $z \in \mathbf{I}$ such that $\iota_n < z$. Therefore, the class $\mathbf{J} \equiv \{x \in \mathbf{I} \mid \iota_n < x\}$ is not empty. Hence, by Lemma 1 (A.2.2), it contains the smallest element α.

Therefore, we can define a sequence $v \equiv (\pi_k \in \mathbf{I} \mid k \in n+2)$, setting $\pi_k \equiv \iota_k$ for every $k \in n+1$ and $\pi_{n+1} \equiv \alpha$, i.e. $v = u \cup \{\langle n+1, \alpha\rangle\}$. It is clear that $\pi_0 = \varkappa^*$ and $\pi_k < \pi_l$ for all $k \in l \in n+2$. Let $\varkappa \in \mathbf{In}$ and $\pi_0 \leqslant \varkappa < \pi_{n+1}$. Then, $\varkappa \in \mathbf{I}$ and $\iota_0 \leqslant \varkappa < \alpha$. If $\varkappa = \iota_n$, then $\varkappa = \pi_n$. If $\varkappa < \iota_n$, then $\iota_0 \leqslant \varkappa < \iota_n$ implies $\varkappa = \iota_k = \pi_k$ for some $k \in n$. Finally, if $\varkappa > \iota_n$, then $\varkappa \in \mathbf{J}$. Thus, $\alpha \leqslant \varkappa$, but it contradicts the property $\varkappa < \alpha$. Therefore, this case is impossible. It follows from the two previous cases that $\varkappa = \pi_k$ for some $k \in n+1$. It means that the sequence v possesses all the necessary properties. Check its uniqueness. Suppose that there exists a sequence $w \equiv (\varkappa_k \in \mathbf{I} \mid k \in n+2)$ such that $\varkappa_0 = \varkappa^*$, $\varkappa_k < \varkappa_l$ for all $k \in l \in n+2$, and $\varkappa \in \mathbf{In}$ and $\varkappa_0 \leqslant \varkappa < \varkappa_{n+1}$ imply $\varkappa = \varkappa_k$ for some $k \in n+1$. Since the sequence $w|n+1 \equiv (\varkappa_k \in \mathbf{I} \mid k \in n+1)$ possesses all the mentioned properties for n, by virtue of the uniqueness of the sequence u we infer that $u = w|(n+1)$, i.e. $\varkappa_k = \iota_k \equiv \pi_k$ for all $k \in n+1$. If $\varkappa_{n+1} < \pi_{n+1}$, then $\pi_0 = \varkappa_0 \leqslant \varkappa_{n+1} < \pi_{n+1}$ by the above implies $\varkappa_{n+1} = \pi_k = \varkappa_k$ for some $k \in n+1$, but it is impossible. If $\pi_{n+1} < \varkappa_{n+1}$, then $\varkappa_0 = \pi_0 \leqslant \pi_{n+1} < \varkappa_{n+1}$ in a similar way implies $\pi_{n+1} = \varkappa_k = \pi_k$ for some $k \in n+1$, but it is also impossible. Therefore, $\varkappa_{n+1} = \pi_{n+1}$. Hence, the uniqueness of the sequence v is proven. Consequently, $n+1 \in N$. By the principle of natural induction, $N = \omega$. Thus, for every $n \in \omega$, there exists the unique mentioned sequence $u(n)$. By virtue of its uniqueness, we can denote it by $(\iota_k^n \mid k \in n+1)$.

(4) ⊢ (7). Consider the following formula of the ZF set theory: $\varphi(x, y) \equiv (x \in \omega \Rightarrow y = \iota_x^x) \wedge (x \notin \omega \Rightarrow y = \varnothing)$. By the axiom scheme of replacement AS6 (ZF), for ω, there exists a set Y such that $\forall x \in \omega(\forall y(\varphi(x, y) \Rightarrow y \in Y))$. If $n \in \omega$, then $\varphi(n, \iota_n^n)$ implies $\iota_n^n \in Y$. Therefore, we can in the set $\omega \times Y$ define an infinite sequence $u \equiv (\iota_n \in Y \mid n \in \omega)$, setting $u \equiv \{z \in \omega \times Y \mid \exists x \in \omega(z = \langle x, \iota_x^x\rangle)\}$. It follows from the mentioned above property of uniqueness that $u(m) = u(n)|m+1$ for all $m \leqslant n$. Hence, $u|n+1 = u(n)$. It is clear that the sequence u possesses all the necessary properties.

(6) ⊢ (1). Consider the next formula of the ZF set theory: $\varphi(x, y) \equiv (x \in \omega \Rightarrow y = \iota_x) \wedge (x \notin \omega \Rightarrow y = \varnothing)$. By axiom scheme AS6 (ZF), for ω, there exists a set Y such that $\forall x \in \omega(\forall y(\varphi(x, y) \Rightarrow y \in Y))$. If $n \in \omega$, then $\varphi(n, \iota_n)$ implies $\iota_n \in Y$. By axiom scheme AS3 (ZF), a class $X \equiv \{\iota_n \mid n \in \omega\} \equiv \{y \mid \exists x \in \omega(y = \iota_x)\} = \{y \mid y \in Y \wedge \exists x \in \omega(y = \iota_x)\}$ i a set. Since the sequence u is strictly increasing, then the set X satisfies axiom AI(ω).

Deducibilities (7) ⊢ (6) ⊢ (5) ⊢ (2) are evident.

Deducibilities (4) ⊢ (3) ⊢ (2) are also evident.

(2) ⊢ (3) and (2) ⊢ (6). Consider the non-empty class $\mathbf{A}$ of all finite sets of inaccessible cardinals. Then, the class $\mathbf{I} \equiv \cup\mathbf{A}$ is also non-empty, and therefore, by Lemma 1 (A.2.2) in $\mathbf{I}$ there exists the smallest element $\varkappa^*$.

Consider the set N, consisting all $n \in \omega$, for which there exists a unique sequence $u \equiv u(n) \equiv (\iota_k \in \mathbf{I} \mid k \in n+1)$ such that $\iota_0 = \varkappa^*$, $\iota_k < \iota_l$ for all $k \in l \in n+1$ and $\varkappa \in \mathbf{I}$ and $\iota_0 \leqslant \varkappa < \iota_n$ imply $\varkappa = \iota_k$ for some $k \in n$ (*the property of* **I**-*incompressibility*). Since the sequence $(\iota_k \in \mathbf{I} \mid k \in 1)$ such that $\iota_0 \equiv \varkappa^*$, possesses all the listed properties, we have $0 \in N$. Let $n \in N$, i.e. for n the sequence $u \equiv (\iota_k \in \mathbf{I} \mid k \in n+1)$ is constructed. Consider the finite set $A \equiv \{\iota_k \in \mathbf{I} \mid k \in n+1\}$ of the power $n+1$. By condition 2, for $n+2$, there exists a finite set $B \in \mathbf{A}$ of power $n+2$. Take in B the smallest element a and the greatest element b. By definition, $a \geqslant \varkappa^*$. Suppose that $b \leqslant \iota_n$. Then, for every $c \in B$, the inequality $\iota_0 = \varkappa^* \leqslant a \leqslant c \leqslant b \leqslant \iota_n$ is valid. If $c < \iota_n$, then from $c \in \mathbf{I}$, by the property of **I**-incompressibility, we infer that $c = \iota_k$ for some $k \in n$, i. e. $c \in A$. If $c = \iota_n$, then again $c \in A$. In the result, we come to the inclusion $B \subset A$, but it is impossible. From this contradiction, we infer that $\iota_n < b$. Since $b \in \mathbf{I}$, the class $\mathbf{J} \equiv \{x \in \mathbf{I} \mid \iota_n < \varkappa\}$ is not empty. Therefore, by Lemma 1 (A.2.2), it contains the smallest element α.

Consequently, we can define sequence $v \equiv (\pi_k \in \mathbf{I} \mid k \in n+2)$, setting $\pi_k \equiv \iota_k$ for every $k \in n+1$ and $\pi_{n+1} \equiv \alpha$, i. e. $v = u \cup \{\langle n+1, \alpha\rangle\}$. Then, in almost the same manner as in the deduction (1) ⊢ (4) changing **In** by **I**, we make sure that the sequence v possesses all the necessary properties and that it is unique. Therefore, $n+1 \in N$. By the principle of natural induction, $N = \omega$. Thus, for every $n \in \omega$, there exists the unique mentioned sequence $u(n)$. By virtue of its uniqueness, we can denote it by $(\iota_k^n \mid k \in n+1)$. Thus, deduction (2) ⊢ (3) is finished.

Now, as in deducibility (4) ⊢ (7) using the sequences $(\iota_k^n \mid k \in n+1)$, we construct an infinite strictly increasing sequence $u \equiv (\iota_n \mid n \in \omega)$ of inaccessible cardinals. It gives us deduction (2) ⊢ (6).

Thus, the following deducibilities and equivalences are proven: (1) ⊢ (4) ⊢ (7) ⊢ (6) ⊢ (1) and (6) ⊢ (5) ⊢ (2) ⊢ (6) and (2) ~ (3). This implies immediately the equivalence of all assertions 1–7.

(8) ⊢ (6). Let I be an infinite set of inaccessible cardinals. By Lemma 1 (A.2.2) in I, there exists the smallest $\varkappa^*$.

Consider the set N, consisiting of all $n \in \omega$, for which there exists a unique sequence $u \equiv u(n) \equiv (\iota_k \in I \mid k \in n+1)$ such that $\iota_0 = \varkappa^*$, $\iota_k < \iota_l$ for every $k \in l \in n+1$, and $\varkappa \in I$ and $\iota_0 \leqslant \varkappa < \iota_n$ imply $\varkappa = \iota_k$ for some $k \in n$ (*the property of I-incompressibility*).

Since a sequence $(\iota_k \in I \mid k \in 1)$ such that $\iota_0 = \varkappa^*$, possesses all the listed properties, we infer that $0 \in N$. Let $n \in N$. Consider the set $J \equiv I \setminus \{\iota_k \mid k \in n+1\}$. It is not empty because in the opposite case, the set I has to be finite; therefore, it contains the smallest element α. It is clear that $\alpha \neq \iota_n$ and $\alpha \geqslant \varkappa^* = \iota_0$. Suppose that $\alpha < \iota_n$. Then, $\iota_0 \leqslant \alpha < \iota_n$ by the condition $n \in N$ implies $\alpha = \iota_k$ for some $k \in n$, but it is impossible. Thus, $\iota_n < \alpha$.

Therefore, we can define a sequence $v \equiv (\pi_k \in I \mid k \in n+2)$, setting $\pi_k \equiv \iota_k$ for every $k \in n+1$ and $\pi_{n+1} \equiv \alpha$, i. e. $v = u \cup \{\langle n+1, \alpha\rangle\}$. It is clear that $\pi_0 = \varkappa^*$ and $\pi_k < \pi_l$ for all $k \in l \in n+2$. Let $\varkappa \in I$ and $\pi_0 \leqslant \varkappa < \pi_{n+1}$. Then, $\iota_0 \leqslant \varkappa < \alpha$. If $\varkappa = \iota_n$, then $\varkappa = \pi_n$. If $\varkappa < \iota_n$, then $\iota_0 \leqslant \varkappa < \iota_n$ implies $\varkappa = \iota_k = \pi_k$ for some $k \in n$. Finally, if $\varkappa > \iota_n$, then $\varkappa > \iota_k$ for all $k \in n+1$. Hence, $\varkappa \in J$. Therefore, $\alpha \leqslant \varkappa$, but it contradicts the property $\varkappa < \alpha$. Therefore, this case is impossible. It follows from the two previous cases that $\varkappa = \pi_k$ for some $k \in n+1$. It means that the sequence v possesses all the necessary properties.

Check its uniqueness. Suppose that there exists a sequence $w \equiv (\varkappa_k \in I \mid k \in n+2)$ such that $\varkappa_0 = \varkappa^*$, $\varkappa_k \in \varkappa_l$ for all $k \in l \in n+2$, and $\varkappa \in I$ and $\varkappa_0 \leqslant \varkappa < \varkappa_{n+1}$ imply $\varkappa = \varkappa_k$ for some $k \in n+1$. Since the sequence $w|n+1 \equiv (\varkappa_k \in I \mid k \in n+1)$ possesses all properties for n mentioned above, by virtue of the uniqueness of the sequence u, we infer that $u = w|n+1$, i. e. $\varkappa_k = \iota_k \equiv \pi_k$ for all $k \in n+1$. If $\varkappa_{n+1} < \pi_{n+1}$, then $\pi_0 = \varkappa_0 \leqslant \varkappa_{n+1}$

$< \pi_{n+1}$, by the above, implies $\varkappa_{n+1} = \pi_k = \varkappa_k$ for some $k \in n+1$, but it is impossible. If $\pi_{n+1} < \varkappa_{n+1}$, then $\varkappa_0 = \pi_0 \leqslant \pi_{n+1} < \varkappa_{n+1}$ in a similar way implies $\pi_{n+1} = \varkappa_k = \pi_k$ for some $k \in n+1$, but it is also impossible. Therefore, $\varkappa_{n+1} = \pi_{n+1}$. Thus, the uniqueness of the sequence v is proven. Consequently, $n+1 \in N$. By the principle of natural induction, $N = \omega$. Hence, for every $n \in \omega$, there exists the unique mentioned sequence $u(n)$. By virtue of its uniqueness, we can denote it by $(\iota_k^n \mid k \in n+1)$. Further, as in the deduction (4) $\vdash$ (7), using the sequences $(\iota_k^n \mid k \in n+1)$ we construct the infinite strictly increasing sequence $u \equiv (\iota_n \mid n \in \omega)$ of inaccessible cardinals.

(6) $\vdash$ (8). In the same manner as in the proof of deducibility (6) $\vdash$ (1) for the sequence u, consider the set $X \equiv \{\iota_n \mid n \in \omega\}$ of its members. Suppose that the set X is finite. Then, X contains the greatest element $\varkappa$, but it contradicts the fact that the sequence U is strictly increasing. □

Proposition 1. *In the ZF set theory, the following assertions are equivalent:*
1) *$AI(\omega+\omega)$;*
2) *there exist infinite sequences $u \equiv (\iota_m \mid m \in \omega)$ and $v \equiv (\varkappa_n \mid n \in \omega)$ of inaccessible cardinals such that $\iota_k < \iota_m < \varkappa_l < \varkappa_n$ for every $k \in m \in \omega$ and $l \in n \in \omega$, i.e. the sequences u and v are strictly increasing and continue each other.*

Proof. (1) $\vdash$ (2). By $\mathrm{AI}(\omega+\omega)$, there exist sets X and Y, satisfying axiom $\mathrm{AI}(\omega)$. Therefore, by Theorem 1, there exist strictly increasing infinite sequences $u \equiv (\iota_m \in X \mid m \in \omega)$ and $v \equiv (\varkappa_n \in Y \mid n \in \omega)$. By $\mathrm{AI}(\omega+\omega)$, $\iota_k < \varkappa_l$ for all $k, l \in \omega$.

(2) $\vdash$ (1). Similarly to the proof of the deducibility (6) $\vdash$ (1) from Theorem 1, we check that the classes $X \equiv \{\iota_m \mid m \in \omega\}$ and $Y \equiv \{\varkappa_n \mid n \in \omega\}$ are sets. These sets satisfy axiom $\mathrm{AI}(\omega+\omega)$. □

It becomes clear from assertion 2 of this proposition that ZF+$\mathrm{AI}(\omega+\omega)$ ensures the existence of $\omega+\omega$ different inaccessible cardinals.

Now, we will clear up the correlation between the axioms on inaccessible cardinals:
- AIC $\equiv$ *there exists at least one inaccessible cardinal*;
- $\mathrm{AI}(\omega) \sim$ *there exists an infinite set of inaccessible cardinals*;
- $\mathrm{AI}(\omega+\omega) \sim$ *there exist two following one after another infinite sets of inaccessible cardinals*;
- AI $\equiv$ *for every ordinal α, there exists an inaccessible cardinal which is greater than α.*

Proposition 2. *In the ZF set theory, the deducibilities $AI \vdash AI(\omega+\omega) \vdash AI(\omega) \vdash AIC$ are valid.*

Proof. Prove that from AI we can infer property 2 of Proposition 1. The arguments repeats completely the proof of deducibility (1) $\vdash$ (2) from Theorem 1.

Using now the equivalence of (2) and $\mathrm{AI}(\omega + \omega)$, we get the proof of deducibility $\mathrm{AI} \vdash \mathrm{AI}(\omega + \omega)$. All other deducibilities are evident. □

In Appendix A, it is proven that in the ZF set theory, the axiom of inaccessibility AI is equivalent to the *axiom of universality*

$$AU \equiv \forall X \exists U (U \bowtie \wedge X \in U),$$

where $U\bowtie$ denotes the property of a set U to be universal (see Theorem 1 (A.4.3)). Hence, the set theories ZF+AI and ZF+AU are equivalent.

It that appendix, it is also proven that in ZF the axiom of ω-inaccessibility $\mathrm{AI}(\omega)$ is equivalent to the *axiom of ω-universality*

$$AU(\omega) \equiv \exists X(\forall U \in X(U\bowtie) \wedge X \neq \varnothing \wedge \forall U \in X \exists V \in X(U \in V)),$$

postulating the existence of an infinite set of universal sets (Proposition 2 (A.5.1)). Therefore, the set theories ZF+AI(ω) and ZF+AU(ω) are also equivalent.

B.4.2 “Forks” of relative consistency

Using globalization of local constructions in the LTS, which was made above, we can prove the following statement.

Statement 1. *All axioms of the ZF set theory, except the axiom scheme of separation* (AS3) *and the axiom scheme of replacement* (AS6), *are deducible in the LTS as formulas of the LTS.*

Proof. For every formula φ in the first-order theory, the formula scheme $\varphi \Rightarrow \varphi$ is deduced. By the definition of equality in the LTS, it gives us the formula $\forall u(u \in X \Leftrightarrow u \in Y) \Rightarrow X = Y$. Formula A1 is inferred from it by the rule Gen.

For classes u and v, by Corollary 3 to Theorem 1 (B.3.5), there exists the class $\{u, v\}$. By definition, $z \in \{u, v\} \equiv z = u \vee z = v$. Using the scheme $\varphi \Rightarrow \varphi$, we get the formula $(z \in \{u, v\}) \Leftrightarrow z = u \vee z = v$. By the rule Gen, logical axiom scheme LAS12, and the rule MP, we can infer from it the formula $\exists x \forall z(z \in x \Leftrightarrow z = u \vee z = v)$. Formula A2 (ZF) is inferred from it by the rule Gen.

For a class X, by Corollary 1 to Lemma 6 (B.3.2), there exists a class $\cup X$. By definition, $u \in \cup X \equiv \exists z(u \in z \wedge z \in X)$. Using the formula scheme $\varphi \Rightarrow \varphi$, we get the formula $(u \in \cup X) \Leftrightarrow \exists z(u \in z \wedge z \in X)$. By the rule Gen, logical axiom scheme LAS12, and the rule MP, the formula $\exists Y \forall u(u \in Y \Leftrightarrow \exists z(u \in z \wedge z \in X))$ is inferred from it. Formula A4 (ZF) is inferred from it by the rule Gen.

For a class X, by Corollary to Lemma 1 (B.3.2), there exists the class $\mathcal{P}(X)$. By definition, $u \in \mathcal{P}(X) \equiv u \subset X$. As it was done above, we consecutively infer the formulas $u \in \mathcal{P}(X) \Leftrightarrow u \subset X$, $\forall u(u \in \mathcal{P}(X) \Leftrightarrow u \subset X)$, $\exists Y \forall u(u \in Y \Leftrightarrow u \subset X)$, and A5 (ZF).

Consider the class π, which exists by axiom of infra-infinity A13 (LTS). Let $u \in \pi$. Then, $u \in \pi \in \mathfrak{a}$ by the axiom of transitivity A7 implies $u \in \mathfrak{a}$. Therefore, by Lemma 3 (B.3.2) $\{u\}_{\mathfrak{a}} = \{u\}$. By Lemma 2 (B.1.1), $\{u\} \in \mathfrak{a}$. By A7 u, $\{u\} \subset \mathfrak{a}$. Hence, by Lemma 2 (B.3.2) $u \cup \{u\} = u \cup_{\mathfrak{a}} \{u\} = u \cup_{\mathfrak{a}} \{u\}_{\mathfrak{a}}$. By virtue of A13, we infer $u \cup \{u\} \in \pi$. Since we did not apply the rule of generalization, then by the deduction theorem and by the generalization rule, the formula $\forall u(u \in \pi \Rightarrow u \cup \{u\} \in \pi)$ is deduced. Since, by A13, $\pi \neq \varnothing$, we infer that by the derivative rule of conjuction, the formula $\pi \neq \varnothing \wedge \forall u(u \in \pi \Rightarrow u \cup \{u\} \in \pi)$ is deduced. Using logical axiom scheme LAS12 and the rule MP, we infer from it formula A7 (ZF).

For every non-empty class X, by the axiom of universality A6, there exists a universal class α such that $X \in \alpha$. By the axiom of transitivity A7, $X \subset \alpha$. From the axiom of regularity A12 (LTS), the formula $\exists x(x \in X \wedge x \cap_\alpha X = \varnothing)$ is deduced. Since $x \in X \subset \alpha$, we have, by A7, $x \subset \alpha$. Hence, by Lemma 2 (B.3.2) $x \cap_\alpha X = x \cap X$. Thus, the formula $\exists x(x \in X \wedge x \cap X = \varnothing)$ is deduced. By the theorem of deduction, the formula $X \neq \varnothing \Rightarrow \exists x(x \in X \wedge x \cap X = \varnothing)$ is deduced. By the rule Gen, we infer from it formula A8 (ZF).

Consider in the LTS the empty class $\varnothing$. From A3 (LTS), the formula $\forall z(z \notin \varnothing)$ is deduced. Using LAS12 and MP, we infer formula A9 (ZF) from it.

For every non-empty class X, by A6, there exists a universal class α such that $X \in \alpha$. From choice axiom A14 (LTS), we infer the formula $\exists z((z \leftrightharpoons \mathcal{P}_\alpha(X) \setminus \{\varnothing\}_\alpha \to_\alpha X) \wedge \forall Y(Y \in \mathcal{P}_\alpha(X) \setminus \{\varnothing\}_\alpha \Rightarrow z(Y) \in Y))$. By Lemmas 1 and 3 (B.3.2), $\mathcal{P}_\alpha(X) = \mathcal{P}(X)$ and $\{\varnothing\}_\alpha = \{\varnothing\}$. Therefore, $z \leftrightharpoons \mathcal{P}(X) \setminus \{\varnothing\} \to_\alpha X$. By the axiom of transitivity A7, $X \subset \alpha$ and $\varnothing \subset \alpha$. Consequently, by Lemma 5 (B.3.2) $z \leftrightharpoons \mathcal{P}(X) \setminus \{\varnothing\} \to X$. Thus, the formula $\exists z(z \leftrightharpoons \mathcal{P}(X) \setminus \{\varnothing\} \to X) \wedge \forall Y(Y \in \mathcal{P}(X) \setminus \{\varnothing\} \Rightarrow z(Y) \in Y))$ is deduced. By the theorem of deduction and the rule Gen, formula A10 (ZF) is deduced from it. □

Undeducibility in the LTS of the axiom scheme of replacement will be proven later.

The existence of inaccessible cardinal numbers in the LTS proven in B.3.4 and B.3.5 allows to prove the following statement.

Statement 2. *If the LTS is consistent, then the theory ZF+AIC is consistent.*

Proof. By axiom A6, for a fixed universe U_0, there exists a universe U such that $U_0 \in U$. Consider the interpretation $M \equiv (U, I)$ of the ZF set theory in the LTS, described in the proof of Statement 1 (B.3.1). Prove that this interpretation is a model of ZF+AIC. According to the proof of Statement 1 (B.3.1), we need only to consider the translation of axiom AIC and prove its deducibility in the LTS.

We will use all notations of the proof of Statement 1 (B.3.1).

On the sequence s axiom AIC is translated into the formula $\varphi_0 \equiv M \models AIC[s] = AIC^U \equiv \exists x' \in U(Icn(x'))^U$, where

- $\varphi_1(x') \equiv (Icn(x'))^U \equiv (Rcn(x'))^U \ \wedge\ \forall x'' \in U((On(x''))^U \ \wedge\ x'' \in x' \Rightarrow \mathrm{card}_U\, \mathcal{P}_U (x'') \in x'$), where
- $\varphi_2(x') \equiv (Rcn(x'))^U \equiv (Cn(x'))^U \wedge \forall x''' \in U((On(x'''))^U \wedge \exists u \in U(u \leftrightharpoons x''' \to_U x' \wedge \cup_U \mathrm{rng}_U\, u = x') \Rightarrow (x' = x''' \vee x' \in x'''))$, where
- $\varphi_3(x') \equiv (Cn(x'))^U \equiv (On(x'))^U \wedge \forall x''' \in U((On(x'''))^U \wedge (x''' = x' \vee x''' \in x') \wedge \exists u \in U(u \leftrightharpoons x''' \rightarrowtail\!\!\!\!\twoheadrightarrow_U x') \Rightarrow x''' = x')$, where
- $\varphi_4(x'') \equiv (On(x''))^U \ \equiv\ \forall x \in U(x \in x'' \Rightarrow \forall y \in U(y \in x \Rightarrow y \in x''))) \wedge (\forall x, y, z \in U(x, y, z \in x'' \ \wedge\ x \in y \ \wedge\ y \in z \Rightarrow x \in z)) \ \wedge\ (\forall x, y \in U(x, y \in x'' \Rightarrow x \in y \vee x = y \vee y \in x)) \wedge \forall z \in U(z \neq \varnothing \wedge \forall x \in U(x \in z \Rightarrow x \in x'') \Rightarrow \exists x \in U(x \in z \wedge \forall y \in U(y \in z \Rightarrow x \in y)))$,
- $(On(x'))^U = \varphi_4(x'' \parallel x') \equiv \varphi_4(x')$, and
- $((On(x'''))^U = \varphi_4(x'' \parallel x''') \equiv \varphi_4(x''')$.

The comparison of the formula $\varphi_4(x'')$ with the definition shows that the subformula $\psi \equiv \forall x \in U(x \in z \Rightarrow x \in x'')$ is unusual in it. But in that place, where it is staying, it is equivalent to the formula $\psi' \equiv z \subset x''$. If $x \in z$, then from $z \in U$, by the axiom of transitivity A7 (LTS), it follows $x \in U$ and therefore $x \in x''$. Thus, we can substitute ψ' for ψ. Under this substitution we see that the formula $\varphi_4(x'')$ means that x'' is a U-ordinal number in the universal class U. Consequently, $\varphi_4(x')$ and $\varphi_4(x''')$ mean that x' and x''' are U-ordinal numbers.

It leads to the following form of the formula $\varphi_3(x')$:

$$\varphi_3(x') = (x' \text{ is an } U\text{-ordinal number}) \wedge \forall x''' \in U((x''' \text{ is an } U\text{-ordinal number}) \wedge \\ \wedge (x''' \leqslant x') \wedge \exists u \in U(u \leftrightharpoons x''' \rightarrowtail\!\!\!\!\twoheadrightarrow_U x') \Rightarrow x''' = x').$$

In this formula, the subformula $\chi \equiv \exists u \in U(u \leftrightharpoons x''' \rightarrowtail\!\!\!\!\twoheadrightarrow_U x')$ is unusual. Since x''', $x' \in U$, we infer that $u \subset x''' *_U x' \in U$, by the axiom of subset A8, implies $u \in U$. Therefore, χ is equivalent to the formula $\chi' \equiv \exists u(u \leftrightharpoons x''' \to_U x')$. Substituting χ' for χ, we see that $\varphi_3(x')$ means that x' is a U-cardinal number.

It leads to the following form of the formula $\varphi_2(x')$:

$$\varphi_2(x') = (x' \text{ is an } U\text{-cardinal number}) \wedge \forall x''' \in U((x''' \text{ is an } U\text{-ordinal number}) \wedge \\ \wedge \exists u \in U(u \leftrightharpoons x''' \to_U x' \wedge \cup_U \mathrm{rng}_U\, u = x') \Rightarrow (x' \leqslant x''')).$$

By the same reasons as above, in φ_2, the quantifier prefix $\exists u \in U$ can be replaced by $\exists u$. But then the formula $\varphi_2(x')$ means that x' is a U-regular U-cardinal number.

It leads to the following form of the formula $\varphi_1(x')$:

$$\varphi_1(x') = (x' \text{ is a } U\text{-regular } U\text{-cardinal number}) \wedge \\ \wedge \forall x'' \in U((x'' \text{ is a } U\text{-ordinal number}) \wedge x'' \in x' \Rightarrow \operatorname{card}_U \mathcal{P}_U(x'') \in x').$$

This means that x' is a U-inaccessible U-cardinal number in the universe U. Thus, axiom AIC has been translated into the formula $\varphi_0 \equiv M \models AIC[s] = \exists x' \in U(x'$ is a U-inaccessible U-cardinal number).

Infer this formula in the LTS. By Corollary 3 to Proposition 1 (B.3.4), there exists an inaccessible cardinal number $\varkappa$ such that $\mathfrak{a} = \overline{V_\varkappa}$. Since $\varkappa \subset \mathfrak{a} \subset U_0 \in U$, we have, by the axiom of subset A8, $\varkappa \in U$. By Proposition 3 (B.3.3), $\varkappa$ is an U-inaccessible U-cardinal number. As a result, we deduced the desired formula. □

Statement 2 was proven with the help of constructing a model of the ZF set theory+AIC in the LTS. It follows from Theorem 3 (B.3.5) that to construct a model of the LTS in the theory ZF it is necessary to have in ZF at least the same as in Theorem 3 (B.3.5) "metasequence" $c(0), c(1), \dots, c(n), \dots$ of finite incompressible sequences $c(n) \equiv (\varkappa_k^n \mid k \in n+1)$ of inaccessible cardinals. But in ZF, this metasequence can be globalized by the axiom scheme of replacement into the usual unfinite sequence $c \equiv (\varkappa_n \equiv \varkappa_n^n \mid n \in \omega)$. The existence of such an infinite sequence of inaccessible cardinals, as Theorem 1 (B.4.1) shows, is equivalent to axiom AI(ω).

Using Theorem 1 (B.4.1), we can prove the following statement.

Statement 3. *If the theory ZF+AI(ω) is consistent, then the LTS is consistent.*

Proof. Consider the sequence $(\iota_n \mid n \in \omega)$ from Theorem 1 (B.4.1) and the set $A \equiv \{\iota_n \mid n \in \omega\}$. By Lemma 2 (A.2.2), $\alpha \equiv \cup A = \sup A$ is an ordinal number. Further, instead of V_{ι_n} we will write W_n. Since $\iota_n \leqslant \alpha$, we have $W_n \subset V_\alpha$. Therefore, $D \equiv \cup[W_n \mid n \in \omega] \subset V_\alpha$. By AS3, D is a set. Since $0 < \iota_n < \iota_n{+}1 \leqslant \iota_{n+1}$, we obtain, by Lemma 1 (A.3.2) $\varnothing = V_0 \in W_n \in W_{n+1} \subset D$ for every $n \in \omega$.

The set D is transitive. If $y \in D$, then $y \in W_m$ for some m, and Lemma 3 (A.3.2) implies that $y \subset W_m \subset D$. Similarly, with the help of Lemma 2 (A.3.2) we can check that if $x \subset y \in D$, then $x \in D$. We will often use later these two properties.

Choose the set D in the capacity of the domain of interpretation of the LTS in the theory ZF+AI(ω). Consider in D the subset $R \equiv \{x \in D \mid \exists n \in \omega(x = W_n)\}$. Define a correspondence I, assigning to the predicate symbol $\in$ in the LTS the two-placed relation $B \equiv \{z \in D \times D \mid \exists x, y \in D(z = (x, y) \wedge x \in y)\}$, to the symbol $\bowtie$ in LTS the one-placed relation R, and to the constant symbols $\varnothing$ and $\mathfrak{a}$ in the LTS the elements $\varnothing$ and W_0 of the set D, respectively.

Let s be some sequence $x_0, \dots, x_q, \dots$ of elements of D. We will consider translations $M \models \varphi[s]$ of axioms and axiom schemes of the LTS on the sequence s under the interpretation M and will prove their deducibility in the theory ZF+AI(ω). Instead of $\theta_M[s]$ and $M \models \varphi[s]$ we will write θ^t and φ^t for terms θ and formulas φ, respectively.

To simplify our account further, consider at first the translations of some basic formulas. Let u and v be some classes in the LTS.

The formula $u \in v$ is translated into the formula $(u \in v)^t = ((u^t, v^t) \in B)$. Denote this last formula by γ. By definition, this formula is equivalent to the formula $(\exists x \exists y(x \in D \wedge y \in D \wedge (u^t, v^t) = (x, y) \wedge (x \in y))$. Using the property of a sequential pair, we conclude that $u^t = x$ and $v^t = y$. Consequently, the formula $\delta \equiv (u^t \in v^t)$ is deduced from γ. By the theorem of deduction, $\gamma \Rightarrow \delta$. Conversely, consider the formula δ. In the ZF set theory, one can prove that for sets u^t and v^t, there exists the set z such that $z = (u^t, v^t)$. By virtue of LAS3, the formula $(z = (u^t, v^t) \Rightarrow u^t \in D \wedge v^t \in D \wedge z = (u^t, v^t) \wedge u^t \in v^t)$ is deduced from δ. Since the formula $z = (u^t, v^t)$ is deduced from axioms, we infer that the formula $(u^t \in D \wedge v^t \in D \wedge z = (u^t, v^t) \wedge u^t \in v^t)$ is also deduced. By LAS12, the formula $\exists x \exists y(x \in D \wedge y \in D \wedge z = (x, y) \wedge x \in y)$ is deduced, and it is equivalent to the formula $z \in B$ and so to the formula γ. By the theorem of deduction, $\delta \Rightarrow \gamma$. Thus, the first equivalence $(u \in v)^t \Leftrightarrow u^t \in v^t$ is valid.

The formula $v \subset w$ is translanted into the formula $(v \subset w)^t$. Denote this last formula by ε. By the equivalence proven above, it is equivalent to the formula $\varepsilon' \equiv \forall u \in D(u \in v^t \Rightarrow u \in w^t)$. According to LAS11, the formula $\varepsilon'' \equiv (x \in D \Rightarrow (x \in v^t \Rightarrow x \in w^t))$ is deduced from the formula ε'. If $x \in v^t$, then $v^t \in D$ and the transitivity of the set D imply $x \in D$. Then, the formula ε'' implies $x \in v^t \Rightarrow x \in w^t$. Consequently, by the theorem of deduction, the formula $(\varepsilon \Rightarrow (x \in v^t \Rightarrow x \in w^t))$ is deduced. By the rule of generalization, the formula $\forall x(\varepsilon \Rightarrow (x \in v^t \Rightarrow x \in w^t))$ is deduced. By LAS13, we infer the formula $(\varepsilon \Rightarrow \forall x(x \in v^t \Rightarrow x \in w^t))$, i. e. the formula $(\varepsilon \Rightarrow v^t \subset w^t)$.

Conversely, let the formula $v^t \subset w^t$ be given. With the help of logical axioms we can consecutively infer from it the formulas $(u \in v^t \Rightarrow u \in w^t)$ and $(u \in D \Rightarrow (u \in v^t \Rightarrow u \in w^t))$. By the rule Gen, the formula ε' is deduced. Therefore, by the theorem of deduction, the formula $(v^t \subset w^t \Rightarrow \varepsilon)$ is deduced. Thus, the second equivalence $(v \subset w)^t \Leftrightarrow v^t \subset w^t$ is valid.

It follows from this equivalence that we get the equivalence $(v =_{LTS} w)^t \Leftrightarrow (v^t \subset w^t) \wedge (w^t \subset v^t)$. By the axiom of extensionality A1 (ZF), the last formula implies $v^t =_{ZF} w^t$. By the theorem of deduction in the ZF set theory, the formula $((v^t \subset w^t) \wedge (w^t \subset v^t) \Rightarrow v^t = w^t)$ is deduced. Conversely, if $v^t =_{ZF} w^t$, then, by the replacement of equals principle (see the beginning of section A.2), the formula $(u \in v^t \Rightarrow u \in w^t)$ is deduced. By the rule Gen, the formula $v^t \subset w^t$ is deduced. Similarly, the formula $w^t \subset v^t$ is deduced. Therefore, the formula $(v^t \subset w^t) \wedge (w^t \subset v^t)$ is also deduced. By the deduction theorem, the formula $(v^t =_{ZF} w^t \Rightarrow (v^t \subset w^t) \wedge (w^t \subset v^t))$ is deduced. Thus, we get the third equivalence $(v =_{LTS} w)^t \Leftrightarrow v^t =_{ZF} w^t$.

Further on, we will write not literal translations of axioms and axiom schemes, but their equivalent variants, which are received by using the proven equivalences $(u \in_{LTS} v)^t \Leftrightarrow u^t \in_{ZF} v^t$, $(v \subset_{LTS} w)^t \Leftrightarrow v^t \subset_{ZF} w^t$, and $(v =_{LTS} w)^t \Leftrightarrow v^t =_{ZF} w^t$. We will denote these equivalent variants, using the sign "~" over them.

$\widetilde{A^t1} \equiv \forall y \in D \forall z \in D((y = z) \Rightarrow (\forall X \in D(y \in X \Leftrightarrow z \in X)))$.

In ZF, the formula $y = z \Rightarrow z = y$ can be proven in the following way. By the property of changing equals, we have $(y = z \Rightarrow (y = y \Rightarrow z = y))$. Since the formula $y = y$ is valid for any y, we obtain $y = z \Rightarrow z = y$. Besides, by the property of changing equals, $(y = z) \Rightarrow (y \in X \Rightarrow z \in X)$ and $(z = y) \Rightarrow (z \in X \Rightarrow y \in X)$ are deduced. With $y = z \Rightarrow z = y$, then $(y = z) \Rightarrow (y \in X \Leftarrow z \in X)$ and $(y = z) \Rightarrow (y \in X \Leftrightarrow z \in X)$ are deduced. Now, by LAS1 and the rule of generalization $\forall X \in D((y = z) \Rightarrow (y \in X \Leftrightarrow z \in X))$ is deduced. From LAS13 and the rule of generalization, the formula $\widetilde{A^t1}$ is deduced.

$\widetilde{AS^t2}$: if $\varphi(x)$ be an X-predicative formula of the LTS such that the substitution $\varphi(x \parallel y)$ is admissible and φ does not contain Z as a free variable, then $\forall X \in D(\exists Z \in D(\forall y \in D((y \in Z) \Leftrightarrow (y \in X \wedge \varphi^\tau(y)))))$, where φ^τ denotes the formula $M \models [s^\tau]$, in which by s^τ we denote the corresponding change of the sequence s under the translation of the quantifier over-formulas $\forall X(\dots)$, $\exists Y(\dots)$ and $\forall y(\dots)$, indicated above.

Since φ^τ is a formula of ZF, we infer that $\widetilde{AS^t2}$ is deduced from AS3 (ZF). By AS3 (ZF), for $X \in D$, there exists Z such that $\forall y(y \in Z \Leftrightarrow y \in X \wedge \varphi^\tau(y))$. Therefore, $Z \subset X \in D$. By the definition of D, there exists $n \in \omega$ such that $X \in W_n$. By Lemma 2 (A.3.2), $Z \in W_n \subset D$. Thus, for $X \in D$, there exists $Z \in D$ such that $\forall y \in D(y \in Z \Leftrightarrow y \in X \wedge \varphi^\tau(y))$.

$\widetilde{A^t3} \equiv \forall Z \in D((\forall x \in D(x \notin Z)) \Leftrightarrow Z = \varnothing)$.

Fix the condition $Z \in D$. Consider the formula $\chi \equiv \forall x(x \in D \Rightarrow x \notin Z)$. If $x \in Z$ then, by condition, $x \in D$ and then χ implies $x \notin Z$. If $x \notin Z$, then evidently χ implies $x \notin Z$. Consequently, under our condition from χ it is deduced $x \notin Z$. By the rule of generalization from χ, it is deduced $\forall x(x \notin Z)$. By axioms A1 and A7 (ZF), $\forall x(x \notin Z)$ implies $Z = \varnothing$. Thus, from the totality $Z \in D$ and χ it is deduced $Z = \varnothing$. By the theorem of deduction, $Z \in D$ implies the formula $\chi \Rightarrow Z = \varnothing$. Conversely, by A1 and A7, $Z = \varnothing$ implies $\forall x(x \notin Z)$. Therefore, from the totality $Z \in D$ and $Z = \varnothing$ we can infer χ. By the theorem of deduction, from $Z \in D$, we can infer the formula $Z = \varnothing \Rightarrow \chi$. Therefore, from the condition $Z \in D$ the formula $\chi \Leftrightarrow Z = \varnothing$ is deduced. By the theorem of deduction, the formula $Z \in D \Rightarrow (\chi \Leftrightarrow Z = \varnothing)$ is deduced. Thus, by the rule of generalization the formula $\widetilde{A^t3}$ is deduced.

$\widetilde{A^t4}$. $\forall U \in D \forall V \in D((U = V) \Rightarrow (U \in R \Leftrightarrow V \in R))$. By the property of changing equals, we have $(U = V) \Rightarrow (U \in R \Rightarrow V \in R)$. By the proof above, $(U = V) \Rightarrow (V = U)$, and therefore $(U = V) \Rightarrow (V \in R \Rightarrow U \in R)$. It follows that $(U = V) \Rightarrow (U \in R \Leftrightarrow V \in R)$, and by the rule of generalization, we infer the formula $\widetilde{A^t4}$.

$\widetilde{A^t5}$. $W_0 \in R \wedge \forall U \in D(U \in R \Rightarrow W_0 \subset U)$.

The formula $W_0 \in R$ is deduced in ZF+AI(ω) by virtue of AS3. Therefore, we need only to deduce the formula $\forall U \in D(U \in R \Rightarrow W_0 \subset U)$. In another form, we can write this formula as $\forall U \in D(\exists n \in \omega(U = W_n) \Rightarrow W_0 \subset U)$. By virtue of Corollary 1 to Lemma 3 (A.3.2), $\iota_0 \leqslant \iota_n$ implies $W_0 \subset W_n$ for every $n \in \omega$. It follows from this assertion that the mentioned formula is deduced in ZF+AI(ω).

$\widetilde{A^t6}$. $\forall X \in D \exists U \in D(U \in R \wedge X \in U)$.

It follows from $X \in D$ that $X \in W_n$ for some $n \in \omega$. Consider $U = W_n$. Then, $X \in D$ implies $U \in D \wedge U \in R \wedge X \in U$, and as a result, we obtain $\widetilde{A^t 6}$.

$\widetilde{A^t 7} \equiv \forall U \in D(U \in R \Rightarrow \forall X \in D(X \in U \Rightarrow X \subset U))$.

It follows from $U \in R$ that $U = W_n$ for some $n \in \omega$, and $X \in W_n$ implies $X \subset W_n$ by Lemma 3 (A.3.2). This gives the desired formula.

$\widetilde{A^t 8} \equiv \forall U \in D(U \in R \Rightarrow \forall X \in D \forall Y \in D(X \in U \wedge Y \subset X \Rightarrow Y \in U))$.

Since from $U \in R$, it follows that $U = W_n$ for some $n \in \omega$, we need only to prove that for any X, Y from D it is valid $(X \in W_n \wedge Y \subset X \Rightarrow Y \in W_n)$. But it directly follows from Lemma 2 (A.3.2).

$\widetilde{A^t 9} \equiv \forall U \in D(U \in R \Rightarrow \forall X \in D(X \in U \Rightarrow \mathcal{P}_U(X)^\tau \in U))$, where the set $Z \equiv \mathcal{P}_U(X)^\tau$ is defined by the formula $\exists Z \in D(\forall y \in D((y \in Z) \Leftrightarrow (y \in U \wedge y \subset X)))$.

At first, check that if $U \in R$, $X \in D$, and $X \in U$, then $Z = \mathcal{P}(X)$. Let $y \in Z$. Since $Z \in D$, it follows from the property of transitivity that $Z \subset D$. Consequently, $y \in D$. But in this case $y \in D$ and $y \in Z$ imply $y \subset X$, i. e. $y \in \mathcal{P}(X)$. Conversely, let $y \in \mathcal{P}(X)$, i. e. $y \subset X$. From $X \in D$, by the property of the set D proven above, we get $y \in D$. It follows from $U \in R$ that $U = W_n$ for some $n \in \omega$. Consequently, $y \subset X \in W_n$ implies by Lemma 2 (A.3.2) that $y \in W_n = U$. But then $y \in D$, $y \in U$, and $y \subset X$ imply $y \in Z$, and it proves the desired equality.

By Lemma 6 (A.3.2), $X \in W_n$ implies $Z = \mathcal{P}(X) \in W_n = U$. From here, the formula $\widetilde{A^t 9}$ is deduced by logical means.

$\widetilde{A^t 10} \equiv \forall U \in D(U \in R \Rightarrow \forall X \in D \forall Y \in D(X \in U \wedge Y \in U \Rightarrow (X \cup_U Y)^\tau \in U))$, where the set $Z \equiv (X \cup_U Y)^\tau$ is determined from the formula $\exists Z \in D(\forall y \in D((y \in Z) \Leftrightarrow (y \in U \wedge (y \in X \vee y \in Y))))$.

In the same manner as at the deduction of formula $\widetilde{A^t 9}$, we check that the conditions $U \in R$, $X \in D$, $Y \in D$, $X \in U$ and $Y \in U$ imply the equality $Z = X \cup Y$, and also $U = W_n$ for some $n \in \omega$. By Lemma 5 (A.3.2), $X, Y \in W_n$ implies $Z = X \cup Y \in W_n = U$. From here the formula $\widetilde{A^t 10}$ is deduced.

$\widetilde{A^t 11} \equiv \forall U \in D(U \in R \Rightarrow \forall X \in D \forall Y \in D \forall z \in D((X \in U \wedge Y \subset U \wedge (z \subset (X *_U Y)^\sigma) \wedge (\forall x \in D(x \in X \Rightarrow z\langle x\rangle^\tau \in U))) \Rightarrow ((\mathrm{rng}_U\, z)^\sigma \in U)))$, where

- the set $Z_1 \equiv (X *_U Y)^\sigma$ is determined from the formula $\exists Z_1 \in D(\forall y \in D((y \in Z_1) \Leftrightarrow (y \in U \wedge (\exists u \in D \exists v \in D(u \in X \wedge v \in Y \wedge y = \langle u, v\rangle_U^*)))))$;
- the set $Z_2 \equiv Z_2(x) \equiv z\langle x\rangle^\tau$ is determined from the formula $\exists Z_2 \in D(\forall y \in D((y \in Z_2) \Leftrightarrow (y \in U \wedge y \in Y \wedge \langle x, y\rangle_U^* \in z)))$;
- the set $Z_3 \equiv (\mathrm{rng}_U\, z)^\sigma$ is determined from the formula $\exists Z_3 \in D(\forall y \in D((y \in Z_3) \Leftrightarrow (y \in U \wedge y \in Y \wedge (\exists x \in D(x \in X \wedge \langle x, y\rangle_U^* \in z)))))$.

As before, we check that the conditions $U \in R$, $u \in D$, $v \in D$, $u \in U$, and $v \in U$ imply consecutively the equalities $\{u\}_U^* = \{u\}$, $\{u, v\}_U^* = \{u, v\}$ and $\langle u, v\rangle_U^* = \langle u, v\rangle$, where $U = W_n$ for some $n \in \omega$. By Corollary 1 to Lemma 6 (A.3.2), $u, v \in W_n$ implies $\langle u, v\rangle_U^* = \langle u, v\rangle \in W_n = U$.

Thus, in its turn, we infer that the conditions $U \in R$, $X \in D$, $Y \in D$, $X \in U$, $Y \subset U$, $x \in D$, and $x \in X$ imply the equalities $Z_1 = X * Y$, $Z_2 = z\langle x\rangle$ and $Z_3 = \mathrm{rng}\, z$. Let we have also the condition $\forall x \in D(x \in X \Rightarrow z\langle x\rangle^{\tau} \in U)$. Since z is a correspondence from X into $Y \subset U$, it follows that z is a correspondence from X into W_n. If $x \in X \in D$, then from the transitivity of D it follows $x \in D$. Therefore, this additional condition implies $z\langle x\rangle = z\langle x\rangle^{\tau} \in W_n$. Since $X \in W_n$, it follows by Lemma 4 (A.3.3) that $(\mathrm{rng}_U\, z)^{\sigma} = \mathrm{rng}\, z \in W_n = U$. From here, by logical means, we infer the formula $\widetilde{A^t 11}$.

$\widetilde{A^t 12} \equiv \forall U \in D(U \in R \Rightarrow \forall X \in D(X \subset U \wedge X \neq \varnothing \Rightarrow \exists x \in D(x \in X \wedge (x \cap_U X)^{\tau} = \varnothing)))$, where the set $Z \equiv (x \cap_U X)^{\tau}$ is determined from the formula $\exists Z \in D(\forall y \in D((y \in Z) \Leftrightarrow (y \in U \wedge (y \in x \wedge y \in X))))$.

If we fix some $U \in R$, i. e. $U = W_n$ for $n \in \omega$, then we need to prove the formula $\forall X \in D(X \subset W_n \wedge X \neq \varnothing \Rightarrow \exists x \in D(x \in X \wedge (x \cap_{W_n} X)^{\tau} = \varnothing))$. Since $X \subset W_n$ implies $X \in D$, and $x \in X$ implies $x \in D$, it follows that this formula can be transformed to the formula $\forall X(X \subset W_n \wedge X \neq \varnothing \Rightarrow \exists x(x \in X \wedge (x \cap_{W_n} X)^{\tau} = \varnothing))$. For $x, X \in W_n$ we have $((x \cap_{W_n} X)^{\tau} = x \cap X)$, therefore we only need to prove that $\forall X \subset W_n(X \neq \varnothing \Rightarrow \exists x(x \in X \wedge x \cap X = \varnothing))$. But it is the direct consequence from the axiom of regularity in ZF.

$\widetilde{A^t 13} \equiv \exists X \in D(X \in W_0 \wedge \varnothing \in X \wedge \forall x \in D(x \in X \Rightarrow ((x \cup_{W_0} \{x\}_{W_0})^{\tau} \in X)))$, where

- the set $Z_1 \equiv (x \cup_{W_0} \{x\}_{W_0})^{\tau}$ is determined from the formula $\exists Z_1 \in D(\forall y \in D((y \in Z_1) \Leftrightarrow (y \in W_0 \wedge (y \in x \vee y \in \{x\}^*_{W_0}))))$;
- the set $Z_2 \equiv \{x\}^*_{W_0}$ is determined from the formula $\exists Z_2 \in D(\forall y \in D((y \in Z_2) \Leftrightarrow (y \in W_0 \wedge y = x)))$.

From the conditions $X \in D$, $X \in W_0$, $x \in D$, and $x \in X$, it follows that $Z_2 = \{x\}$ and therefore $Z_1 = x \cup \{x\}$. Consider in ZF the set $X \equiv \omega$. This set evidently satisfies the condition $\varnothing \in X \wedge \forall x \in X(x \cup \{x\} \in X)$. We need to prove that $\omega \in W_0$. Since $W_0 = V_{\iota_0}$ and $\iota_0 > \omega$, by the definition of an inaccessible cardinal, it follows by Lemma 7 (A.3.2) that $\omega \in W_0$.

$\widetilde{A^t 14} \equiv \forall U \in D(U \in R \Rightarrow \forall X \in D(X \in U \wedge X \neq \varnothing \Rightarrow \exists z \in D((z \coloneqq \mathcal{P}_U(X) \setminus \{\varnothing\}_U \rightarrow_U X)^{\sigma} \wedge \forall Y \in D(Y \in (\mathcal{P}_U(X) \setminus \{\varnothing\}_U)^{\tau} \Rightarrow z(Y)^{\tau} \in Y))))$, where

- the set $Z_1 \equiv (\mathcal{P}_U(X) \setminus \{\varnothing\}_U)^{\tau}$ is determined from the formula $\exists Z_1 \in D(\forall y \in D((y \in Z_1) \Leftrightarrow (y \in U \wedge (y \in \mathcal{P}_U(X)^* \wedge y \notin \{\varnothing\}^*_U))))$;
- the set $Z_2 \equiv z(Y)^{\tau}$ is defined from the formula $Z_2 \in U \wedge \langle Y, Z_2\rangle^{\tau}_U \in z$;
- η^{τ} denotes the formula $M \models \eta[s^{\tau}]$, in which s^{τ} denotes the corresponding change of the sequence s under the translation of the quantifier over-formulas $\forall U(\dots)$, $\forall X(\dots)$, and $\exists z(\dots)$, indicated above.

Fix the conditions $U \in D$, $U \in R$, $X \in D$, and $X \in U$. Denote $\mathcal{P}_U(X) \setminus \{\varnothing\}_U$ by S and $\mathcal{P}(X) \setminus \{\varnothing\}$ by T. We have proven above that $X \in U$ and $\varnothing \in U$ imply $\mathcal{P}_U(X)^* = \mathcal{P}(X)$ and $\{\varnothing\}^*_U = \{\varnothing\}$. Since $U \in R$, it follows that $U = W_n$ for some $n \in \omega$. Therefore, by Lemma 6 (A.3.2) $X \in U$ implies $\mathcal{P}(X) \in U$, and, by Lemma 2 (A.3.2) it implies $T \in U$. By Lemma 3 (A.3.2), $y \in T \in U$ implies $y \in U$. All these properties imply $Z_1 = T$.

If $Y \in D$ and $Y \in Z_1$, then $Y \in T \in U$ implies $Y \in U$. As it was stated above, $Z_2 \in U$ and $Y \in U$ imply $\langle Y, Z_2\rangle^{\tau}_U = \langle Y, Z_2\rangle$. Then, $\langle Y, Z_2\rangle \in z$ implies $Z_2 \in z\langle Y\rangle$. From here and from the previous conditions, we cannot yet infer that $Z_2 = z(Y)$.

Consider the formula $\varphi \equiv (z \leftrightharpoons \mathcal{P}_U(X) \setminus \{\varnothing\}_U \to_U X)$. It is the conjuction of the three following formulas: $\varphi_1 \equiv (z \subset S *_U X)$, $\varphi_2 \equiv (\mathrm{dom}_U\, z = S)$, and $\varphi_3 \equiv (\forall x(x \in S \Rightarrow \forall y(y \in X \Rightarrow \forall y'(y' \in X \Rightarrow (\langle x, y\rangle_U \in z \wedge \langle x, y'\rangle_U \in z \Rightarrow y = y')))))$.

Therefore, $\varphi^\sigma = \varphi_1^\sigma \wedge \varphi_2^\sigma \wedge \varphi_3^\sigma$. Since $\varphi_1 = (\forall u(u \in z \Rightarrow u \in U \wedge \exists x \exists y(x \in S \wedge y \in X \wedge u = \langle x, y\rangle_U)))$, it follows that $\varphi_1^\sigma \Leftrightarrow (\forall u \in D(u \in z \Rightarrow u \in U \wedge \exists x \in D \exists y \in D(x \in Z_1 \wedge y \in X \wedge u = \langle x, y\rangle_U^*)))$. Similarly, $\varphi_2 = (\forall x(x \in S \Rightarrow x \in U \wedge x \in S \wedge \exists y(y \in X \wedge \langle x, y\rangle_U \in z)))$ implies $\varphi_2^\sigma \Leftrightarrow (\forall x \in D(x \in Z_1 \Rightarrow x \in U \wedge x \in Z_1 \wedge \exists y \in D(y \in X \wedge \langle x, y\rangle_U^* \in z)))$.

Finally, $\varphi_3^\sigma \Leftrightarrow (\forall x \in D(x \in Z_1 \Rightarrow \forall y \in D(y \in X \Rightarrow \forall y' \in D(y' \in X \Rightarrow (\langle x, y\rangle_U^* \in z \wedge \langle x, y'\rangle_U^* \in z \Rightarrow y = y')))))$.

For X, by choice axiom A10 (ZF), there exists z such that $\chi \equiv (z \leftrightharpoons \mathcal{P}(X) \setminus \{\varnothing\} \to X) \wedge \forall Y(Y \in \mathcal{P}(X) \setminus \{\varnothing\} \Rightarrow z(Y) \in Y)$.

Consider the formula $\psi \equiv (z \leftrightharpoons \mathcal{P}(X) \setminus \{\varnothing\} \to X)$. As above $\psi = \psi_1 \wedge \psi_2 \wedge \psi_3$, where $\psi_1 = (\forall u(u \in z \Rightarrow \exists x \exists y(x \in T \wedge y \in X \wedge u = \langle x, y\rangle)))$, $\psi_2 = (\forall x(x \in T \Rightarrow x \in T \wedge \exists y(y \in \wedge \langle x, y\rangle \in z)))$ and $\psi_3 = (\forall x(x \in T \Rightarrow \forall y(y \in X \Rightarrow \forall y'(y' \in X \Rightarrow (\langle x, y\rangle \in z \wedge \langle x, y'\rangle \in z \Rightarrow y = y')))))$.

Since $T \in U$, by Corollary 2 to Lemma 6 (A.3.2), we get $T * X \in U$. The formula ψ_1 means that $z \subset T * X$. Therefore, $z \in U$ and $z \in D$.

Deduce from these properties the formula φ_1^σ. Let $u \in D$ and $u \in z$. Then, $u \in z \in U$, by Lemma 3 (A.3.2) implies $u \in U$. From the formula ψ_1, it follows that for u, there exist $x \in T$ and $y \in X$ such that $u = \langle x, y\rangle$. By Lemma 3 (A.3.2), $x, y \in U \subset D$. We have stated above that in this case $\langle x, y\rangle_U^* = \langle x, y\rangle$. Since $x \in T$ and $T = Z_1$, it follows $x \in Z_1$. Thus, from $u \in z$ it is deduced the formula $(u \in U \wedge \exists x \in D \exists y \in D(x \in Z_1 \wedge y \in X \wedge u = \langle x, y\rangle_U^*))$. Applying the theorem of deduction and the rules of deduction, we deduce the formula φ_1^σ.

Let $x \in D$ and $x \in Z_1 = T$. Since $T \in U$, it follows by Lemma 3 (A.3.2) that $x \in U$. From the formula ψ_2, we infer that for x, there exists $y \in X$ such that $\langle x, y\rangle \in z$. The condition $y \in X \in D$, by the transitivity of D, implies $y \in D$. Similarly, $y \in X \in U$ by Lemma 3 (A.3.2) implies $y \in U$. But in this case $\langle x, y\rangle = \langle x, y\rangle_U^*$. Thus, from $x \in Z_1$, it is deduced the formula $(x \in U \wedge x \in Z_1 \wedge \exists y \in D(y \in X \wedge \langle x, y\rangle_U^* \in z))$. From here, as above, we deduce the formula φ_2^σ.

Let $x \in Z_1 = T, y \in D, y \in X, y' \in D, y' \in X, \langle x, y\rangle_U^* \in z$, and $\langle x, y'\rangle_U^* \in z$. As above, these conditions imply $\langle x, y\rangle \in z$ and $\langle x, y'\rangle \in z$. But then, the formula ψ_3 implies $y = y'$. Applying several times in turn the theorem of deduction and the rules of deduction, we deduce from the formula ψ the formula φ_3^σ.

Thus, the formula φ^σ is deduced.

Since $z \leftrightharpoons T \to X$, it follows that $z\langle Y\rangle = \{z(Y)\}$. Consequently, from $Z_2 \in \{z(Y)\}$, we infer that $Z_2 = z(Y)$. Therefore, for the function z the conditions $Y \in D$ and $Y \in Z_1 = T$ imply $Z_2 = z(Y) \in Y$.

All this means that from axiom A10 (ZF), the existence of an object z is deduced, satisfying the formula χ, from which the formula $\xi \equiv \varphi^\sigma \wedge \forall Y \in D(Y \in Z_1 \Rightarrow Z_2 \in Y)$ is deduced. Therefore, in ZF from the fixed conditions, it is deduced the formula

$\exists z \in D(\xi)$. Applying several times the deduction theorem and the generalization rule, we deduce, as a result, the formula $\widetilde{A^t 14}$.

Since all the translations of the axioms of LTS turned out to be deducible formulas of ZF+AI(ω), it follows that the LTS is consistent. □

An absence in the LTS of an axiom scheme like the axiom scheme of replacement in the ZF set theory apparently renders the interpretation of the ZF set theory+AI(ω) impossible in the LTS. But this interpretation becomes possible, if to strengthen the LTS by the following axiom.

AU(ω). (The *ω-universality axiom.*) $\exists X(\forall U \in X(U\bowtie) \wedge X \neq \varnothing \wedge \forall U \in X \exists V \in X (U \in V))$.

Consider also the following axiom in the LTS.

ATU(ω). (The *axiom of transitive ω-universality.*) *There exists a set Y such that:*

a) $Y \neq \varnothing$;
b) $\forall U \in Y(U\bowtie)$;
c) $\forall U \forall V(U \bowtie \wedge U \in V \wedge V \in Y \Rightarrow U \in Y)$ (*the transitivity property with respect to universal sets*);
d) $\forall V \in Y \exists W \in Y(V \in W)$ (*the unboundedness property*).

Lemma 1. *In the LTS, the following assertions are equivalent:*
1) *AU(ω);*
2) *ATU(ω).*

Proof. (1) $\vdash$ (2). Denote by D the class, the existence of which is postulated by axiom AU(ω). It follows from the axiom of universality A6 that $D \in \alpha$ for some universal class α. By the axiom of transitivity, A7, $\forall V \in D(V \in \alpha)$. By Corollary 2 to Proposition 1 (B.3.4), $\mathfrak{a} \in \alpha$ or $\mathfrak{a} = \alpha$. It is clear that the second case contradicts AU(ω). Therefore, $\mathfrak{a} \in \alpha$.

Consider the class $E \equiv \{U \in \alpha \mid U \bowtie \wedge \exists V \in D(U \in V)\}$. If $U \in D$, then, by AU(ω), $\exists V \in D(U \in V)$. Thus, $D \subset E$. The class E is universally transitive. If $U\bowtie$ and $U \in V \in E$, then $U \in V \in W \in D$ for some W. Using axiom A7, we get by turns $U \in W \in D \in \alpha$, $U \in W \in \alpha$, and $U \in \alpha$. Therefore, $U \in E$.

If $V \in E$, then, by definition, $V \in W \in D \subset E$ for some W. Therefore, E has property (d).

(2) $\vdash$ (1). This deduction is obvious. □

Lemma 2. *Let E be a non-empty class of universes with the property of transitivity with respect to universes, i. e. E possesses properties* a) – c) *from Lemma* 1. *Then,* $\mathfrak{a} \in E$.

Proof. Let $V \in E$. By Corollary 2 to Proposition 1 (B.3.4), $V = \mathfrak{a}$ or $\mathfrak{a} \in V$. In the first case $\mathfrak{a} \in E$. In the second case, $\mathfrak{a} \in V \in E$ implies $\mathfrak{a} \in E$. □

Statement 4. *If LTS +AU(ω) is consistent, then ZF+AI(ω) is consistent.*

Proof. Consider the class C, the existence of which is postulated by axiom AU(ω). According to axiom of universality A6, there exists a universal class D such that $C \in D$. Consider the interpretation $M \equiv (D, I)$ of the ZF set theory in the LTS, described in the proof of Statement 1 (B.3.1). In the proof of Statement 1 (B.3.1), it was established that the interpretation M is a model of ZF in the LTS, and so a model of ZF in the LTS+AU(ω). Check the deducibility of the translation of axiom AI(ω) under the interpretation M on an arbitrary infinite sequence $s \equiv x_0, \dots, x_q, \dots$ of elements of the class D. This translation has the form $\varphi \equiv M \models AI(\omega)[s] = \exists X \in D(\forall x \in D(x \in X \Rightarrow Icn(x)^D) \wedge X \neq \varnothing \wedge \forall y \in D(y \in X \Rightarrow \exists z \in D(z \in X \wedge y \in z)))$. In the proof of Statement 2, it was established that the formula $Icn(x)^D$ means that x is a D-inaccessible D-cardinal number in the LTS.

By Theorem 1 (B.3.4), there exists the injective mapping $q : \mathbf{U} \rightarrowtail \overline{\mathbf{In}}$ such that $q(\alpha) \subset \alpha$ and $\alpha = \overline{V_{q(\alpha)}}$. If $\alpha \in C$, then $q(\alpha) \subset \alpha \in C \in D$, by axioms A7 and A8, implies $q(\alpha) \in D$. Therefore, $p \equiv q|C$ is an injective mapping from C into D. Since $C \subset D$, it follows by Lemma 5 (B.3.2) that $p \leftrightharpoons C \rightarrowtail_D D$ and $K \equiv \operatorname{rng} p = \operatorname{rng}_D p \subset D$. Hence, $p \leftrightharpoons C \rightarrowtail\!\!\!\!\twoheadrightarrow_D K$. Since $p(\alpha) \in D$, it follows, by the axiom of the full union A11, that $K \in D$. From $C \neq \varnothing$, it follows that $K \neq \varnothing$. If $\varkappa \in K$, then $\varkappa \in D$. By Proposition 3 (B.3.3), $\varkappa$ is a D-inaccessible D-cardinal. Consequently, K consists of D-inaccessible D-cardinals.

By axiom AU(ω), for $\varkappa \in K$ and $\beta \equiv \overline{V_\varkappa} \in C$, there exists $\gamma \in C$ such that $\beta \in \gamma$. Consider the inaccessible cardinal $\lambda \equiv p(\gamma) \in K \subset D$. Since q is strictly monotone, it follows that $\varkappa \in \lambda$. Consequently, the formula $\psi(K) \equiv (\forall x \in D(x \in K \Rightarrow (x$ is a D-inaccessible D-cardinal$)) \wedge K \neq \varnothing \wedge \forall y \in D(y \in K \Rightarrow \exists z \in D(z \in K \wedge y \in z))))$ is deduced in the LTS+AU(ω). Thus, the formula $\varphi = \exists X \in D\psi(X)$ is also deduced. □

Axiom AU(ω) allows to componate the continuing each other finite sequences of universal classes from Theorem 2 (B.3.5) into one infinite sequence.

Proposition 1. *In the LTS, the following assertions are equivalent:*

1) *AU(ω);*
2) *there exist a universal class X and an infinite strictly increasing X-sequence $u \equiv (U_n \in X \mid n \in \omega)_X$ of universal classes such that from $n \in \omega$, V is a universal class, and $U_0 \subset V \subset V_n$ it follows that $V = V_k$ for some $k \in n+1$ (the property of incompressibility);*
3) *there exists a universal class X such that $U^n_k \in X$ for any $k \in n+1$ and any $n \in \omega$, where $u(n) \equiv (U^n_k \in U(n) \mid k \in n+1)_{U(n)}$ are the continuing each other strictly increasing incompressible $U(n)$-sequences of universal classes from Theorem 2 (B.3.5).*

Proof. (1) ⊢ (2). Consider the class A, the existence of which is postulated by axiom AU(ω). By the axiom of universality, there exists a universal class X such that $A \in X$. By Theorem 1 (B.3.5) in the class A, there exists the smallest element α.

In the same way as in the proof of Theorem 2 (B.3.5), it is proven that for every $n \in \omega$, there exists a unique X-sequence of universal classes $u(n) \equiv (U_k \in A \mid k \in n+1)_X$ such that $U_0 = \alpha$, $U_k \in U_l$ for every $k \in l \in n+1$, and if V is a universal class and $U_0 \subset V \subset U_n$, then $V = U_k$ for some $k \in n+1$ (*the property of incompressibility*). It follows from the property of uniqueness that $u(n)|(m+1) = u(m)$ for all $m \leqslant n$, i. e. these finite sequences continue each other. By virtue of this uniqueness, we can denote the sequence $u(n)$ by $(U_k^n \mid k \in n+1)$.

Consider the X-class $\omega *_X A$ and the X-correspondence $u \equiv \{z \in \omega *_X A \mid \exists x \in \omega(z = \langle x, U_x^x \rangle_X)\}$. Since $u\langle n\rangle = \{U_n^n\}_X \in X$ for every $n \in \omega$, it follows that $u \leftrightarrows \omega \to_X A$. By the axiom of transitivity, $U_n \equiv U_n^n \in A \in X$ implies $U_n \in X$. Therefore, u is an X-sequence $(U_n \in X \mid n \in \omega)$.

(2) ⊢ (3). From the property of uniqueness of the $U(n)$-sequence $u(n)$ and the class $U(n)$ from Theorem 2 (B.3.5), it follows that $u|n+1 = u(n)$, i. e. $U_{n+1} = U(n)$ and $U_k = U_k^n$ for any $n \in \omega$ and any $k \in n+1$. Therefore, $U_k^n \in X$.

(3) ⊢ (1). Consider the X-class $Y \equiv \{y \in X \mid \exists x \in \omega(y = U_x^x)\} = \{U_n^n \in X \mid n \in \omega\}_X$. Since the $U(n)$-sequences $u(n)$ are strictly increasing and continue each other, it follows that the class Y satisfies axiom AU(ω). □

It follows from assertion 2 of this statement that the LTS+AU(ω) resembles the axiomatics of N. da Costa [1965, 1967] with the denumerable set of constants $\mathfrak{U}_1, \dots, \mathfrak{U}_n, \dots$ for universes with axioms like the axiom $X \subset \mathfrak{U}_n \Rightarrow X \in \mathfrak{U}_{n+1}$. However, the theory of da Costa uses the non-constructive rule of deduction (namely, the ω-rule of Carnap) and some properties of natural numbers.

Corollary 1. *In the theory LTS+AU(ω), there exist a universal class X and an infinite strictly increasing incompressible X-sequence of universal classes $u \equiv (U_m \in X \mid m \in \omega)_X$, and also for every $n \in \omega$ there exist a unique universal class V and a unique finite strictly increasing incompressible V-sequence of universal classes $v(n) \equiv (V_k \in V \mid k \in n+1)_V$ such that $V_0 = X$.*

From the property of uniqueness, it follows that $v(n)|(l+1) = v(l)$ for all $l \leqslant n$, i. e. these finite sequences continue each other.

Proof. By Proposition 1, there exist the corresponding X and $u \equiv (U_m \in X \mid m \in \omega)_X$. Further, similarly to the proof of Theorem 2 (B.3.5), it is deduced the existence of corresponding V and $v(n) \equiv (V_k \in V \mid k \in n+1)_V$. □

Roughly speaking, LTS+AU(ω) ensures the existence of $\omega + \forall n$ different universal classes.

It follows from this corollary and the remark, which was done before Theorem 1 (B.4.1) that to construct a model of the LTS+AU(ω) in the ZF set theory, it is necessary to have in ZF at least two infinite sequences $u \equiv (\iota_m \mid m \in \omega)$ and $v \equiv (\varkappa_n \mid n \in \omega)$ of inaccessible cardinals such that $\iota_k < \iota_m < \varkappa_l < \varkappa_n$ for every $k \in m \in \omega$ and $l \in n \in \omega$. But the existence of such infinite sequences, as Proposition 1 (B.4.1) shows, is equivalent to axiom AI($\omega + \omega$).

With the help of Proposition 1 (B.4.1), we can prove the following statement.

Statement 5. *If the theory ZF+AI($\omega+\omega$) is consistent, then the theory LTS+AU(ω) is also consistent.*

Proof. In the same way, as in the proof of Statement 3, consider the sequences $(\iota_m \mid m \in \omega)$ and $(\varkappa_n \mid n \in \omega)$ from Proposition 1 (B.4.1). Consider the sets $A \equiv \{\iota_m \mid m \in \omega\}$ and $B \equiv \{\varkappa_n \mid n \in \omega\}$. Further, instead of V_{ι_m} and $V_{\varkappa_n}$ we will write $W_m{}'$ and $W_n{}''$, respectively. Consider the ordinal numbers $\alpha \equiv \cup A = \sup A$ and $\beta \equiv \cup B = \sup B$ and the sets $D \equiv \cup(W_m{}' \mid m \in \omega) \subset V_\alpha$ and $E \equiv D \cup \cup(W_n{}'' \mid n \in \omega) \subset V_\beta$. It is clear that $\varnothing = V_0 \in W_0{}' \in W_k{}' \in W_m{}' \subset D \subset V_\alpha \subset W_0{}'' \in W_l{}'' \in W_n{}'' \subset E \subset V_\beta$ for every $0 \in k \in m \in \omega$ and $0 \in l \in n \in \omega$. Therefore, by Lemma 2 (A.3.2) we infer that $W_k{}', W_l{}'' \in E$ for all $k, l \in \omega$.

Choose the set E as the domain of interpretation of LTS+AU(ω) in the set theory ZF+AI($\omega + \omega$). Consider in E the subsets $R \equiv \{x \in E \mid \exists m \in \omega(x = W_m{}')\}$ and $S \equiv \{x \in E \mid \exists n \in \omega(x = W_n{}' \vee x = W_n{}'')\}$. Define a correspondence J, assigning to the predicate symbol $\in$ in the LTS the two-placed relation $C \equiv \{z \in E \times E \mid \exists x, y \in E(z = (x, y) \wedge x \in y)\}$, to the symbol $\bowtie$ in the LTS, the one-placed relation $S \subset E$, and to the constant symbols $\varnothing$ and $\mathfrak{a}$ in LTS, the elements $\varnothing$ and $W_0{}'$ of the domain E. Consider the interpretation $N \equiv (E, J)$ of the LTS in the theory ZF+AI($\omega + \omega$), which is similar to the interpretation $M \equiv (D, I)$, described in the proof of Statement 3. Prove that this interpretation is a model of the LTS+AU(ω). According to the proof of Statement 3, we need only to consider the translation of axiom AU(ω) and to prove its deducibility in ZF+AI($\omega + \omega$).

Let $s \equiv x_0, \ldots, x_q, \ldots$ be an arbitrary sequence of elements of the set E. The translation of axiom AU(ω) under the interpretation N on the sequence s has the form $\varphi \equiv N \models AU(\omega)[s] \equiv \exists X \in E(\forall U \in E(U \in X \Rightarrow U \in R) \wedge X \neq \varnothing \wedge \forall V \in E(V \in X \Rightarrow \exists W \in E(W \in X \wedge V \in W)))$.

Since $\iota_m < \varkappa_0 < \varkappa_1$ for every $m \in \omega$, it follows that $W_m{}' \in W_0{}'' \in W_1{}''$. By Lemma 2 (A.3.2), $R \in W_1{}''$, and therefore, $R \in E$. From $W_0{}' \in R$, it follows that $R \neq \varnothing$. If $U \in E$ and $U \in R$, then $U = W_m{}'$ for some $m \in \omega$. Since $\iota_m < \iota_m + 1 \leqslant \iota_{m1}$, by Lemma 1 (A.3.2) we infer that $U = W_m{}' \in W_{m+1}{}' \equiv V \in R \subset E$. Consequently, for the set R in ZF+AI($\omega + \omega$) the formula $\psi(R) \equiv (\forall U \in E(U \in R \Rightarrow U \in R) \wedge R \neq \varnothing \wedge \forall V \in E(V \in R \Rightarrow \exists W \in E(W \in R \wedge V \in W)))$ is deduced. Thus, the formula $\varphi \equiv \exists X \in E\psi(X)$ is also deduced. □

Thus, we have proven the following chain of interpretations:

$$ZF + AIC \prec LTS \prec ZF + AI(\omega) \prec LTS + AU(\omega) \prec ZF + AI(\omega + \omega) \prec ZF + AI,$$

where $T \prec S$ denotes the interpretability of the theory T in the set theory S. Denoting by $cons(T)$ the consistency of the theory T, we get the converse chain of relative consistency:

$$cons(ZF + AI) \Rightarrow cons(ZF + AI(\omega + \omega)) \Rightarrow cons(LTS + AU(\omega)) \Rightarrow$$
$$\Rightarrow cons(ZF + AI(\omega)) \Rightarrow cons(LTS) \Rightarrow cons(ZF + AIC).$$

These chains were proven by V. K. Zakharov and E. I. Bunina in [2006]. The similar chains remain valid, apparently, if we change the theory ZF by the theory NBG.

It follows from Proposition 2 (B.4.1) that inside the theory ZF+AI, which is the strongest of the mentioned ones, the chain of mutual interpretations can be continued further, if we take as the next steps theories LTS+AU(ω+ω), ZF+AI(ω+ω+ω), and so on.

Since, according to Theorem 1 (A.4.3), the theories ZF+AU and ZF+AI are equivalent, it follows from Proposition 2 (B.4.1) that the theory ZF+AI(ω) is weaker than the theory ZF+AU. It follows from Statement 3 that the LTS is weaker than ZF+AI(ω). Hence, the LTS is weaker than the theory ZF+AU. Thus, the LTS satisfies condition 1 formulated in the introduction.

It follows from Theorem 2 (B.3.5) that the LTS, so as the theories ZF+AU(ω) and ZF+AI(ω) (see B.4), has a countable totality of different universes, and therefore, satisfies all needs of category theory in such an extent as the theory ZF+AU(ω) does it. Therefore, the LTS satisfies condition 2.

Finally, axiom of universality A6 from B.1.1 asserts that in the LTS there are no objects, which are not elements of this countable totality of universes. By the same token, the LTS satisfies condition 3.

As a result, we obtain that *the LTS is more adequate foundation for category theory than the theories ZF+AU and ZF+AU(ω)*. Moreover, the consistency of the theory ZF+AU(ω) implies the consistency of the LTS. Besides, the LTS is a more adequate foundation for category theory than the set theory of da Costa, because the existence of a countable totality of universes in the LTS is deduced, but is not postulated beforehand as in the axiomatics of da Costa. This saves from the necessity to attract externally the natural numbers and their properties.

Note also that in B.6, we will show that in the LTS we cannot prove that the constructed in Theorem 2 (B.3.5) assemblies' sequence of finite $U(n)$-sequences $u(n) \equiv (U_k^n \in U(n) \mid k \in n+1)_{U(n)}$ of universal classes U_k^n can be continued as was done in the assertion 3 of Proposition 1. It means that in the LTS, we have only countable assembly of universes from Theorem 2 (B.3.5).

B.5 The proof of relative consistency by the method of abstract interpretation

The method of abstract interpretation, going back to K. Gödel (see [*Jech*, 1971, 10]), is the direct generalization of the method of interpretation stated in the section A.1.

B.5.1 Abstracts of a set theory

Let S be some set theory (see A.1). We will consider that either S is a theory with the equality or, in S, the equality is introduced by the formula $(A \subset B) \wedge (B \subset A)$. We introduced classes in the ZF set theory (A.1.1) and assemblies in the LTS (B.1.1). Now in exactly the same way we introduce the abstracts $\mathbf{C} \equiv \{x \mid \varphi(x)\}$ and $\mathbf{C}(\vec{u}) \equiv \{x \mid \varphi(x, \vec{u})\}$ in an arbitrary set theory S. For the abstracts $\mathbf{C} \equiv \{x \mid \varphi(x)\}$ and $\mathbf{D} \equiv \{x \mid \psi(x)\}$ as in B.1.2, we define the formulas $\mathbf{C} \subset \mathbf{D}$ and $\mathbf{C} = \mathbf{D}$. Define the formula $\mathbf{C} \in y$ as a notation for the formula $\exists z(z \in y \wedge z = \mathbf{C})$.

As in B.3.2 for abstracts $\mathbf{A}$ and $\mathbf{B}$ and objects A and B of the set theory S, we introduce the abstracts $\mathcal{P}(\mathbf{A})$, $\mathbf{A} \cup \mathbf{B}$, $\mathbf{A} \cap \mathbf{B}$, $\{A\}$, $\{A, B\}$, $\langle A, B\rangle$, $\mathbf{A} * \mathbf{B}$, and $\cup\mathbf{A}$.

Similarly, we define in S a *correspondence* $\mathbf{C}$ *with the domain of definition* dom $\mathbf{C}$ and *the range of values* rng $\mathbf{C}$, *a function* (≡ *a mapping*) $\mathbf{F}$, *a correspondence* $\mathbf{C} : \mathbf{A} \longrightarrow\!\!\!< \mathbf{B}$, *a function* $\mathbf{F} : \mathbf{A} \to \mathbf{B}$, a (*multivalued*) *collection* $[\mathbf{B}_a \subset \mathbf{B} \mid a \in \mathbf{A}]$ *with the union* $\cup[\mathbf{B}_a \subset \mathbf{B} \mid a \in \mathbf{A}]$ and *the intersection* $\cap[\mathbf{B}_a \subset \mathbf{B} \mid a \in \mathbf{A}]$, *a simple collection* $(b_a \in \mathbf{B} \mid a \in \mathbf{A})$ *with the abstract of members* $\{b_a \in \mathbf{B} \mid a \in \mathbf{A}\}$, the (*multivalued*) *sequential pair* $[\mathbf{A}, \mathbf{A}']$, *triplet* $[\mathbf{A}, \mathbf{A}', \mathbf{A}'']$, ... *of abstracts* $\mathbf{A}, \mathbf{A}', \mathbf{A}'', \ldots$, *the simple sequential pair* (a, a'), *triplet* (a, a', a''),... *of objects* $a, a', a'', \ldots$, *the product* $\prod[\mathbf{A}_i \subset \mathbf{A} \mid i \in \mathbf{I}]$ *of a collection* $[\mathbf{A}_i \subset \mathbf{A} \mid i \in \mathbf{I}]$, *the product* $\mathbf{A} \times \mathbf{A}'$, $\mathbf{A} \times \mathbf{A}' \times \mathbf{A}''$, ... *of the pair* $[\mathbf{A}, \mathbf{A}']$, *triplet* $[\mathbf{A}, \mathbf{A}', \mathbf{A}'']$, ... *of abstracts* $\mathbf{A}, \mathbf{A}', \mathbf{A}'', \ldots$, *an n-placed relation* $\mathbf{R} \subset \mathbf{A}^n \equiv \mathrm{Map}(n, \mathbf{A})$ *on an abstract* $\mathbf{A}$, *an n-placed operation* $\mathbf{O} : \mathbf{A}^n \to \mathbf{A}$, and so on.

The abstract $\mathfrak{U} \equiv \{x \mid x = x\}$, consisting of all objects of the theory S, is called *universal.*

Note that, operating with abstracts, we are always staying within the framework of formulas of the set theory S.

B.5.2 The abstract interpretation of a first-order theory in a set theory

A set theory S will be called *finitely closed* (≡ *closed up to finite collections*) if in the theory S some formula $on(x)$ defines natural numbers as the objects of this theory and for every natural number $n \geqslant 1$ every abstract $\mathbf{F}$ that is a mapping from the object n into the universal abstract $\mathfrak{U}$ is an object of the theory S.

In a finitely closed set theory S, we define the abstract $\omega \equiv \{x \mid on(x)\}$, consisting of all natural numbers. Therefore, in such a theory abstracts $\mathbf{s} : \omega \to \mathbf{A}$ are defined. They can be called *abstract infinite sequences of elements of the abstract* $\mathbf{A}$.

Suppose that in a set theory S we have selected by means of this theory some abstract $\mathbf{D} \equiv \{x \mid \varphi(x)\}$.

Let S be some fixed finitely closed set theory with some selected abstract $\mathbf{D}$.

An abstract interpretation of a first-order theory T *in the finitely closed set theory* S *with the selected abstract* $\mathbf{D}$ is a pair $\mathbf{M}$, consisting of the abstract $\mathbf{D}$ and some correspondence I, assigning to every predicate letter P_i^n some n-placed relation $I(P_i^n)$ in $\mathbf{D}$, every functional letter F_i^n some n-placed operation $I(F_i^n)$ in $\mathbf{D}$, and every constant symbol a_i some element $I(a_i)$ of $\mathbf{D}$.

Let $\mathbf{s}$ be some abstract infinite sequence $x_0, \dots, x_q, \dots$ of elements of the abstract $\mathbf{D} \equiv \{x \mid \varphi(x)\}$.

Define the *value of a term t of the theory T on the sequence* $\mathbf{s}$ *under the abstract interpretation* $\mathbf{M}$ *of the theory T in the set theory S* (in notation $t_{\mathbf{M}}[\mathbf{s}]$) by induction in the following way:

- if $t \equiv v_i$, then $t_{\mathbf{M}}[\mathbf{s}] \equiv x_i$;
- if $t \equiv a_i$, then $t_{\mathbf{M}}[\mathbf{s}] \equiv I(a_i)$;
- if $t \equiv F(t_0, \dots, t_{n-1})$, where F is an n-placed functional symbol and $t_0, \dots, t_{n-1}$ are terms, then $t_{\mathbf{M}}[\mathbf{s}] \equiv I(F)(t_{0\mathbf{M}}[\mathbf{s}], \dots, t_{n-1\mathbf{M}}[\mathbf{s}])$.

Since $I(F)$ is an operation from $\mathbf{D}^n$ into $\mathbf{D}$, it follows that for the term $t \equiv F(t_0, \dots, t_{n-1})$, we have $t_{\mathbf{M}}[\mathbf{s}] \in \mathbf{D}$. Consequently, the value of a term is always an element of the abstract $\mathbf{D}$, i. e. it is some object of the theory S.

Define *the translation of the formula φ on the sequence* $\mathbf{s}$ *under the abstract interpretation* $\mathbf{M}$ *of the theory T in the finitely closed set theory S* (in notation $\mathbf{M} \vDash \varphi[\mathbf{s}]$) by induction in the following way:

- if $\varphi \equiv (P(t_0, \dots, t_{n-1})$, where P is a n-placed predicate symbol and $t_0, \dots, t_{n-1}$ are terms, then $\mathbf{M} \vDash \varphi[\mathbf{s}] \equiv ((t_{0\mathbf{M}}[\mathbf{s}], \dots, t_{n-1\mathbf{M}}[\mathbf{s}]) \in I(P))$;
- if $\varphi \equiv (\neg\theta)$, then $\mathbf{M} \vDash \varphi[\mathbf{s}] \equiv (\neg\mathbf{M} \vDash \theta[\mathbf{s}])$;
- if $\varphi \equiv (\theta_1 \Rightarrow \theta_2)$, then $\mathbf{M} \vDash \varphi[\mathbf{s}] \equiv (\mathbf{M} \vDash \theta_1[\mathbf{s}] \Rightarrow \mathbf{M} \vDash \theta_2[\mathbf{s}])$;
- if $\varphi \equiv (\forall v_i \theta)$, then

$$\mathbf{M} \vDash \varphi[\mathbf{s}] \equiv (\forall x(x \in \mathbf{D} \Rightarrow \mathbf{M} \vDash \theta[x_0, \dots, x_{i-1}, x, x_{i+1}, \dots, x_q, \dots])).$$

On other formulas the translation is continued similarly to A.1.3.

This definition needs in some explanation. For the formula $\varphi \equiv (P(t_0, \dots, t_{n-1}))$ the symbol-string $((t_{0\mathbf{M}}[\mathbf{s}], \dots, t_{n-1\mathbf{M}}[\mathbf{s}]) \in I(P))$ is a formula of the theory S. We have mentioned above that the values of the terms $t_{i\mathbf{M}}[\mathbf{s}]$ are objects of the theory S. Since the theory S is finitely closed, it follows that the abstract $v \equiv (t_{0\mathbf{M}}[\mathbf{s}], \dots, t_{n-1\mathbf{M}}[\mathbf{s}])$ is an object of the theory S. By definition, $I(P)$ is a subabstract of the abstract $\mathbf{D}^n$. Therefore, the symbol-string $(v \in I(P))$ is, in fact, a formula of the theory S. It follows from the other items of this definition that as a result of the translation we always get formulas of the theory S. Thus, $\mathbf{M} \vDash \varphi[\mathbf{s}]$ is a formula of the theory S.

An abstract interpretation $\mathbf{M}$ is called an *abstract model of the axiomatic theory* (T, Φ_a) *in the axiomatic finitely closed set theory* (S, Ξ_a) *with the selected abstract* $\mathbf{D}$, if for every abstract sequence $\mathbf{s}$ of elements of $\mathbf{D}$ the translation $\mathbf{M} \vDash \varphi[\mathbf{s}]$ of every axiom φ of the theory T is a deducible formula in the theory (S, Ξ_a).

All other definitions and assertions from A.1.3 are transferred to the case of abstract interpretation under the corresponding insignificant changes.

B.6 Undeducibility of some axioms in the LTS

In this section, all proofs are given by the method of abstract interpretation described in the previous section.

B.6.1 The undeducibility of the axiom scheme of replacement

In B.3, we have made the globalization of almost all main constructions used in naive set theory.

Now, we need only to know if in the LTS one can deduce the global axiom scheme of replacement.

ASR. (*The axiom scheme of replacement.*) Let $\varphi(x, y)$ be a formula of the LTS and φ does not contain Y as a free variable. Then, $\forall x \forall y \forall y'(\varphi(x, y) \wedge \varphi(x, y') \Rightarrow y = y') \Rightarrow \forall X \exists Y \forall x \in X \forall y(\varphi(x, y) \Rightarrow y \in Y)$.

Statement 1.
1) *If the LTS is consistent, then LTS + ¬ASR is consistent.*
2) *If the LTS is consistent, then axiom scheme ASR is not deducible in the LTS.*

Proof. 1. Consider for $n \in \omega$ the class U and the U-sequence $u(n)$ from Theorem 2 (B.3.5). By virtue of their uniqueness, we can denote them by $U(n)$ and $u(n) \equiv (U_k^n \in U(n) \mid k \in n + 1)_{U(n)}$.

Consider the assembly $\mathbf{C} \equiv \{z \mid \exists x \exists y(x \in \omega \wedge y \in U(x) \wedge z = \langle x, y\rangle)\}$. It is clear that $\mathbf{C}$ is a correspondence from the class ω into the assembly $\overline{\mathbf{V}}$ such that $\mathbf{C}\langle n\rangle = U_n$ for every $n \in \omega$. Therefore, $\mathbf{C}$ can be written as the collection $\mathbf{C} \equiv [U(n) \in \overline{\mathbf{V}} \mid n \in \omega]$. Consider the assemblies $\mathbf{D} \equiv \cup[U(n) \mid n \in \omega] = \{y \mid \exists x \in \omega(y \in U(x))\}$ and $\mathbf{R} \equiv \{y \mid \exists x \in \omega(y = U(x))\}$. By Theorem 2 (B.3.5), $U(m) = U_{m+1}^{m+1} = U_{m+1}^{n+1} \in U_{n+1}^{n+1} = U(n) \subset \mathbf{D}$ for any $m \in n \in \omega$. Therefore, $\mathbf{R} \subset \mathbf{D}$. Besides, $\varnothing \in U_0 \subset \mathbf{D}$ implies $\varnothing \in \mathbf{D}$.

By virtue of Corollaries 4 and 1 to Theorem 1 (B.3.5), the LTS is a finitely closed set theory.

Choose the assembly $\mathbf{D}$ as in the capacityof the domain of abstract interpretation of the theory $T \equiv LTS + \neg ASR$ in the finitely closed set theory $S \equiv LTS$.

Define the correspondence I, assigning to the predicate symbol $\in$ in T the two-placed relation $\mathbf{B} \equiv \{z \mid \exists x \exists y(x \in \mathbf{D} \wedge y \in \mathbf{D} \wedge z = (x, y) \wedge x \in y)\}$ on $\mathbf{D}$, to the predicate symbol $\bowtie$ in T the one-placed relation $\mathbf{R}$ on $\mathbf{D}$, and to the constants $\varnothing$ and $\mathfrak{a}$ in T the elements $\varnothing$ and U_0 of the domain $\mathbf{D}$, respectively. Consider the abstract interpretation $\mathbf{M} \equiv [\mathbf{D}, I]$.

Let $\mathbf{s}$ be an abstract sequence $x_0, \ldots, x_q, \ldots$ of elements of $\mathbf{D}$. We will consider the translations $\mathbf{M} \vDash \varphi[\mathbf{s}]$ of axioms and axiom schemes of the theory T on the sequence $\mathbf{s}$ under the interpretation $\mathbf{M}$ and will prove their deducibility in the set theory S.

Instead of $\theta_{\mathbf{M}}[\mathbf{s}]$ and $\mathbf{M} \models \varphi[\mathbf{s}]$, we will write θ^t and φ^t for terms θ and formulas φ, respectively.

The assembly $\mathbf{D}$ is transitive. If $y \in \mathbf{D}$, then $y \in U(m)$ for some m. By axiom A7, $y \subset U(m) \subset \mathbf{D}$. Besides, if $x \subset y \in \mathbf{D}$, then $x \in \mathbf{D}$. By axiom A8, $x \subset y \in U(m)$ implies $x \in U(m) \subset \mathbf{D}$. We will often use these two properties in our proofs further.

To further simplify the account, consider first the translations of some simple formulas. Let u and v be some classes in the LTS.

The formula $u \in v$ is translated into the formula $(u \in v)^t = ((u^t, v^t) \in \mathbf{B})$. By Corollary 5 to Theorem 1 (B.3.5), the assembly (u^t, v^t) is a class, i. e. there exists the class z such that $z = (u^t, v^t)$. Therefore, similarly to the arguments from the proof of Statement 3 (B.4.2) we establish the first equivalence $(u \in v)^t \Leftrightarrow u^t \in v^t$.

Further, similarly to the proof of Statement 3 (B.4.2), we establish the second equivalence $(v \subset w)^t \Leftrightarrow v^t \subset w^t$.

This equivalence immediately implies the third equivalence $(v = w)^t \Leftrightarrow v^t = w^t$.

We will now write not literal translations of axioms and axiom schemes, but their equivalent variants which are received by using the proven equivalences $(u \in v)^t \Leftrightarrow u^t \in v^t$, $(v \subset w)^t \Leftrightarrow v^t \subset w^t$ and $(v = w)^t \Leftrightarrow v^t = w^t$. We will denote these equivalent variants, using the sign "~" over them.

$\widetilde{A^t 1} \equiv \forall y \in \mathbf{D} \forall z \in \mathbf{D}((y = z) \Rightarrow (\forall X \in \mathbf{D}(y \in X \Leftrightarrow z \in X)))$.

This formula is directly deduced from the axiom of extensionality A1 in the LTS.

$\widetilde{AS^t 2}$: if $\varphi(x)$ be an X-predicative formula in the LTS such that the substitution $\varphi(x \parallel y)$ is admissible and φ does not contain Y as a free variable, then $\forall X \in \mathbf{D}(\exists Y \in \mathbf{D}(\forall y \in \mathbf{D}((y \in Y) \Leftrightarrow (y \in X \wedge \varphi^\tau(y)))))$, where φ^τ denotes the formula $\mathbf{M} \models \varphi[\mathbf{s}^\tau]$, in which by $\mathbf{s}^\tau$ we denote the corresponding change of the sequence $\mathbf{s}$ under the translation of the quantifier over-formulas $\forall X(\dots)$, $\exists Y(\dots)$, and $\forall y(\dots)$, indicated above.

Let $X \in \mathbf{D}$. Check that the formula φ^τ is also X-predicative.

Suppose that the subformula $\psi \equiv \exists x(x \in X \wedge \dots)$ occurs in the formula φ. Then, the subformula ψ^τ occurs in the formulas φ^τ. By the proven equivalence, taking into account the external quantification by X, the formula ψ^τ is equivalent to the formula $\psi' \equiv \exists x(x \in \mathbf{D} \wedge x \in X \wedge \dots)$. If $x \in X$, then it follows from the transitivity of $\mathbf{D}$ that $x \in \mathbf{D}$. Consequently, the formula ψ' is equivalent to the formula $\psi^* \equiv \exists x(x \in X \wedge \dots)$. Substituting in the formula φ^τ the subformula ψ^τ by the equivalent formula ψ^*, we get the formula φ^*, which is equivalent to the formula φ^τ and contains the subformula $\psi^* \equiv \exists x(x \in X \wedge \dots)$. Making this substitution with all subformulas of the form ψ in the formula φ, we get the formula $\varphi^\times$, which is equivalent to the formula φ^τ and contains only subformulas of the form ψ^*.

Now, suppose that the formula φ contains the subformula $\chi \equiv \forall x(x \in X \Rightarrow \dots)$. Then, the subformula χ^τ occurs in the formula φ^τ. Therefore, it occurs also in the formula $\varphi^\times$. As above, the formula χ^τ is equivalent to the formula $\chi' \equiv \forall x(x \in \mathbf{D} \Rightarrow (x \in X \Rightarrow \dots))$.

Consider the formula χ'. We deduce from it the formula $\sigma \equiv (x \in \mathbf{D} \Rightarrow (x \in X \Rightarrow \ldots)$. If $x \in X$, then $x \in X \in \mathbf{D}$, by the transitivity of $\mathbf{D}$, implies $x \in \mathbf{D}$. It means that χ' and $x \in X$ imply the subformula $(x \in X \Rightarrow \ldots$ of the formula σ and so the formula $\chi^* \equiv \forall x(x \in X \Rightarrow \ldots)$. By the theorem of deduction in the LTS, we deduce the formula $\chi' \Rightarrow \chi^*$.

Conversely, consider the formula χ^*. By LAS12, it implies the subformula $(x \in X \Rightarrow \ldots)$. By LAS1, the formula σ is deduced. By the theorem of deduction in the LTS, the formula $\chi^* \Rightarrow \sigma$ is deduced. By the rule of generalization, the formula $\forall x(\chi^* \Rightarrow \sigma)$ is deduced. By virtue of LAS13, the formula $\chi^* \Rightarrow \chi'$ is deduced. Thus, in the LTS, we deduce the equivalence of the formulas χ' and χ^*.

Substituting in the formula $\varphi^\times$, the subformula χ^τ by the equivalent formula χ^*, we get the formula $\varphi^{\times *}$, which is equivalent to the formula $\varphi^\times$ and contains the subformula $\chi^* \equiv \forall x(x \in X \Rightarrow \ldots)$. Making this substitution with all subformulas of the form χ in the formula φ, we get the formula $\varphi^{\times\times}$, which is equivalent to the formula φ^τ and contains only subformulas of the form ψ^* and χ^*. Consequently, the formula $\varphi^{\times\times}$ is X-predicative.

Therefore, by AS2 from the LTS, $X \in \mathbf{D}$ implies the formula $\pi \equiv (\exists Y(\forall y(y \in Y \Leftrightarrow (y \in X \wedge \varphi^{\times\times}(y)))))$. Consider the formula $\alpha \equiv (\forall y \in Y \Leftrightarrow (y \in X \wedge \varphi^{\times\times}(y))))$. The formula α, by LAS11, implies the formula $\beta \equiv (y \in Y \Leftrightarrow (y \in X \wedge \varphi^{\times\times}(y)))$. By LAS1, from α, we infer the formula $\gamma \equiv (y \in \mathbf{D} \Rightarrow \beta)$. Consequently, from $X \in \mathbf{D}$ and α, we infer the formula $\delta \equiv (y \in \mathbf{D} \Rightarrow (y \in Y \Leftrightarrow (y \in X \wedge \varphi^\tau(y))))$. By the rule of generalization, we infer the formula $\varepsilon \equiv (\forall y \in \mathbf{D}(y \in Y \Leftrightarrow (y \in X \wedge \varphi^\tau(y))))$.

Besides, from α, we infer the formula $Y \subset X$. Since $X \in \mathbf{D}$, by the second property of the totality $\mathbf{D}$ proven above, we get $Y \in \mathbf{D}$. Therefore, from $X \in \mathbf{D}$ and α, we infer the formula $Y \in \mathbf{D} \wedge \varepsilon$. By LAS12, it implies the formula $\varkappa \equiv \exists Y \in \mathbf{D}\varepsilon$.

Thus, by the theorem of deduction, from $X \in \mathbf{D}$, we infer the formula $\alpha \Rightarrow \varkappa$. By the rule of generalization, we infer the formula $\forall Y(\alpha \Rightarrow \varkappa)$. Then, by LAS14, from $X \in \mathbf{D}$, we infer the formula $\pi \Rightarrow \varkappa$.

Since we have already deduced above the formula π under the condition $X \in \mathbf{D}$, by the rule of implication, we infer the formula $\varkappa$. By the theorem of deduction in the LTS, the formula $(X \in \mathbf{D} \Rightarrow \varkappa)$ is deduced. Thus, by the rule of generalization, the formula $\widetilde{AS^t2}$ is deduced.

$\widetilde{A^t3} \equiv \forall Z \in \mathbf{D}((\forall x \in \mathbf{D}(x \notin Z)) \Leftrightarrow z = \varnothing)$.

Fix the condition $Z \in \mathbf{D}$. Consider the formula $\chi \equiv \forall x(x \in \mathbf{D} \Rightarrow x \notin Z)$. If $x \in Z$, then, by the property of transitivity, $x \in \mathbf{D}$ and then χ implies $x \notin Z$. If $x \notin Z$, then evidently, χ implies $x \notin Z$. Thus, under our condition, we infer that χ implies $x \notin Z$. By the rule of generalization, χ implies $\forall x(x \notin Z)$. By axiom A3, χ implies $Z = \varnothing$. By the theorem of deduction, $Z \in \mathbf{D}$ implies the formula $\chi \Rightarrow Z = \varnothing$. Conversely, $Z = \varnothing$, by A3, implies $\forall x(x \notin Z)$. Therefore, $Z \in \mathbf{D}$ and $Z = \varnothing$ imply the formula $(Z = \varnothing \Rightarrow \chi)$. Thus, the condition $Z \in \mathbf{D}$ implies the formula $(\chi \Leftrightarrow Z = \varnothing)$. By the theorem of deduction, we infer the formula $Z \in \mathbf{D} \Rightarrow (\chi \Leftrightarrow Z = \varnothing)$. So, by the rule of generalization, we infer the formula $\widetilde{A^t3}$.

$\widetilde{A^t4} \equiv \forall U \in \mathbf{D}\forall V \in \mathbf{D}((U = V \Rightarrow (U \in \mathbf{D} \Leftrightarrow V \in \mathbf{R}))$.

Let $U = V$. If $U \in \mathbf{R}$, then $U = U_n$ for some $n \in \omega$. Then, $V = U_n$ implies $V \in \mathbf{R}$, and vice versa. Therefore, $U \in \mathbf{R} \Leftrightarrow V \in \mathbf{R}$.

$\widetilde{A^t5} \equiv U_0 \in \mathbf{R} \wedge \forall U \in \mathbf{D}(U \in \mathbf{R} \Rightarrow U_0 \subset U))$.

Consider the formulas $\varphi(x, y) \equiv (x \in \omega \wedge y = U(x))$ and $\psi \equiv \exists x\, \varphi(x, y)$. Since $0 \in \omega \wedge U(0) = U(0)$, by LAS12 in the LTS, we infer the formula $\exists x(x \in \omega \wedge U(0) = U(x))$, i. e. the formula $\psi(y \parallel U(0))$. By definition, it means that $U(0) \in \mathbf{R}$.

If $U \in \mathbf{R}$, then $U = U(n)$ for some $n \in \omega$. If $n = 0$, then $U(0) = U$. If $n > 0$, then, as it was indicated at the beginning of the proof, $U(0) \in U(n)$. By the axiom of transitivity A7, $U(0) \subset U(n) = U$. By the theorem of deduction in the LTS, we infer the formula $\alpha \equiv (U \in \mathbf{R} \Rightarrow U(0) \subset U)$. By LAS1, we infer the formula $(U \in \mathbf{D} \Rightarrow \alpha)$ and, by the rule of generalization, we infer the formula $\forall U \in \mathbf{D}\alpha$. Thus, we infer the formula $\widetilde{A^t5}$.

$\widetilde{A^t6} \equiv \forall X \in \mathbf{D}\exists U \in \mathbf{D}(U \in \mathbf{R} \wedge X \in U)$.

From $X \in \mathbf{D}$, it follows that $X \in U(n)$ for some $n \in \omega$. In the same way as in the deduction of A^t5, we prove the deducibility of the formula $U(n) \in \mathbf{R} \subset \mathbf{D}$. Consequently, from $X \in \mathbf{D}$, we infer the formula $\alpha \equiv (U(n) \in \mathbf{D} \wedge U(n) \in \mathbf{R} \wedge X \in U(n))$. By LAS12, we infer the formula $\beta \equiv \exists U \in \mathbf{D}(U \in \mathbf{R} \wedge X \in U)$. By the theorem of deduction in the LTS, we infer the formula $\gamma \equiv (X \in \mathbf{D} \Rightarrow \beta)$. By the rule of generalization, we infer $\widetilde{A^t6}$.

$\widetilde{A^t7} \equiv \forall U \in \mathbf{D}(U \in \mathbf{R} \Rightarrow \forall X \in \mathbf{D}(x \in U \Rightarrow x \subset U))$.

This formula is deduced from axiom A7 in the LTS.

$\widetilde{A^t8} \equiv \forall U \in \mathbf{D}(U \in \mathbf{R} \Rightarrow \forall X \in \mathbf{D}\forall Y \in \mathbf{D}(X \in U \wedge Y \subset X \Rightarrow Y \in U))$.

This formula is deduced from subset axiom A8 in the LTS.

$\widetilde{A^t9} \equiv \forall U \in \mathbf{D}(U \in \mathbf{R} \Rightarrow \forall X \in \mathbf{D}(X \in U \Rightarrow \mathcal{P}_U(X)^\tau \in U))$, where the U-class $Z \equiv \mathcal{P}_U(X)^\tau$ is determined from the formula $\exists Z \in \mathbf{D}(\forall y \in \mathbf{D}((y \in Z) \Leftrightarrow (y \in U \wedge y \subset X)))$.

First check that if $U \in \mathbf{R}$, $X \in \mathbf{D}$, and $X \in U$, then $Z = \mathcal{P}_U(X) \equiv Y$. Let $y \in Z$. Since $Z \in \mathbf{D}$, by the proven above transitivity, $y \in \mathbf{D}$. But then, $y \in \mathbf{D}$ and $y \in Z$ imply $y \in U \wedge y \subset X$, i. e. $y \in Y$. Conversely, let $y \in Y$, i. e. $y \in U \wedge y \subset X$. Since $X \in \mathbf{D}$, by the proven above second property of the assembly $\mathbf{D}$, we get $y \in \mathbf{D}$. From $U \in \mathbf{R}$, it follows that U is a universal class. Therefore, by subset axiom A8, $y \subset X \in U$ implies $y \in U$. But then, $y \in \mathbf{D}$, $y \in U$ and $y \subset X$ implies $y \in Z$, which prove the required equality.

By axiom A9, $X \in U$ implies $Z = Y \in U$. From here, by logical means, we infer the formula $\widetilde{A^t9}$.

$\widetilde{A^t10} \equiv \forall U \in \mathbf{D}(U \in \mathbf{R} \Rightarrow \forall X \in \mathbf{D}\forall Y \in \mathbf{D}(X \in U \wedge Y \in U \Rightarrow (X \cup_U Y)^\tau \in U))$, where the U-class $Z \equiv (X \cup_U Y)^\tau$ is determined from the formula $\exists z \in \mathbf{D}(\forall y \in \mathbf{D}((y \in z) \Leftrightarrow (y \in U \wedge (y \in X \vee y \in Y))))$.

In the same way as in the deduction of the formula $\widetilde{A^t9}$, we check that the conditions $U \in \mathbf{R}$, $X \in \mathbf{D}$, $Y \in \mathbf{D}$, $X \in U$, and $Y \in U$ imply the equality $Z = X \cup_U Y$, where

U is a universal class. By axiom A10, $Z = X \cup_U Y \in U$. From here, we infer the formula $\widetilde{A^t 10}$.

$\widetilde{A^t 11} \equiv \forall U \in \mathbf{D}(U \in \mathbf{R} \Rightarrow \forall X \in \mathbf{D} \forall Y \in \mathbf{D} \forall z \in \mathbf{D}((X \in U \wedge Y \subset U \wedge (z \subset (X *_U Y)^\sigma) \wedge (\forall x \in \mathbf{D}(x \in X \Rightarrow z\langle x \rangle^\tau \in U))) \Rightarrow ((\mathrm{rng}_U\, z)^\sigma \in U)))$, where

- the U-class $Z_1 \equiv (X *_U Y)^\sigma$ is determined from the formula $\exists Z_1 \in \mathbf{D}((\forall y \in \mathbf{D}((y \in Z_1) \Leftrightarrow (y \in U \wedge (\exists u \in \mathbf{D} \exists v \in \mathbf{D}(u \in X \wedge v \in Y \wedge y = \langle u, v \rangle_U^*))))))$;
- the U-class $Z_2 \equiv Z_2(x) \equiv z\langle x \rangle^\tau$ is determined from the formula $\exists Z_2 \in \mathbf{D}(\forall y \in \mathbf{D}((y \in Z_2) \Leftrightarrow (y \in U \wedge y \in Y \wedge \langle x, y \rangle_U^* \in z)))$;
- the U-class $Z_3 \equiv (\mathrm{rng}_U\, z)^\sigma$ is determined from the formula $\exists Z_3 \in \mathbf{D}(\forall y \in \mathbf{D}((y \in Z_3) \Leftrightarrow (y \in U \wedge y \in Y \wedge (\exists x \in \mathbf{D}(x \in X \wedge \langle x, y \rangle_U^t \in z)))))$.

We check, as above, that the conditions $U \in \mathbf{R}$, $u \in \mathbf{D}$, $v \in \mathbf{D}$, $u \in U$, and $v \in U$ imply successively the equalities $\{u\}_U^* = \{u\}_U$, $\{u, v\}_U^* = \{u, v\}_U$, and $\langle u, v \rangle_U^* = \langle u, v \rangle_U$, where U is a universal class. By Lemma 2 (B.1.1), $u, v \in U$ implies $\langle u, v \rangle_U^* = \langle u, v \rangle_U \in U$.

From here in its turn, we infer that the conditions $U \in \mathbf{R}$, $X \in \mathbf{D}$, $Y \in \mathbf{D}$, $X \in U$, $Y \subset U$, $x \in \mathbf{D}$, and $x \in X$ imply the equalities $Z_1 = X *_U Y$, $Z_2 = z\langle x \rangle$, and $Z_3 = \mathrm{rng}_U\, z$. Let use have one more condition $\forall x \in \mathbf{D}(x \in X \Rightarrow z\langle x \rangle^\tau \in U)$. Since z is a U-correspondence from X into $Y \subset U$, it follows that z is a U-correspondence from X into U. If $x \in X \in \mathbf{D}$, then from the transitivity of $\mathbf{D}$ we infer that $x \in \mathbf{D}$. Therefore, the additional condition implies $z\langle x \rangle = z\langle x \rangle^\tau \in U$. Since $X \in U$, it follows by the axiom of full union A11 that $Z_3 = \mathrm{rng}_U\, z \in U$. From here, by logical means, we infer the formula $\widetilde{A^t 11}$.

$\widetilde{A^t 12} \equiv \forall U \in \mathbf{D}(U \in \mathbf{R} \Rightarrow \forall X \in \mathbf{D}(X \subset U \wedge X \neq \varnothing \Rightarrow \exists x \in \mathbf{D}(x \in X \wedge (x \cap_U X)^\tau = \varnothing)))$, where the U-class $Z \equiv (x \cap_U X)^\tau$ is determined from the formula $\exists Z \in \mathbf{D}(\forall y \in \mathbf{D}((y \in Z) \Leftrightarrow (y \in U \wedge (y \in x \wedge y \in X))))$.

Check that the conditions $U \in \mathbf{R}$ and $X \in \mathbf{D}$ imply the equality $Z = x \cap_U X \equiv Y$. Let $y \in Z$. Since $X \in \mathbf{D}$, it follows that $y \in \mathbf{D}$. But in this case $y \in \mathbf{D}$ and $y \in Z$ imply $y \in U \wedge y \in x \wedge y \in X$, i. e. $y \in Y$. Conversely, let $y \in Y$, i. e. $y \in U \wedge y \in x \wedge y \in X$. Since $y \in X \in \mathbf{D}$, it follows that $y \in \mathbf{D}$. Consequently, $y \in Z$, which proves the required equality. From $U \in \mathbf{R}$, it follows that U is a universal class.

By the axiom of regularity A12, for $\varnothing \neq X \subset U$, there exists $x \in X$ such that $Z = Y = \varnothing$. Since $x \in X \in \mathbf{D}$, it follows $x \in \mathbf{D}$. From here, by logical means, we infer the formula $\widetilde{A^t 12}$.

$\widetilde{A^t 13} \equiv \exists X \in \mathbf{D}(X \in U_0 \wedge \varnothing \in X \wedge \forall x \in \mathbf{D}(x \in X \Rightarrow ((x \cup_{U_0} \{x\}_{U_0})^\tau \in X)))$, where

- the U_0-class $Z_1 \equiv Z_1(x) \equiv (x \cup_{U_0} \{x\}_{U_0})^\tau$ is determined from the formula $\exists Z_1 \in \mathbf{D}(\forall y \in \mathbf{D}((y \in Z_1) \Leftrightarrow (y \in U_0 \wedge (y \in x \vee y \in \{x\}_{U_0}^*))))$;
- the U_0-class $Z_2 \equiv Z_2(x) \equiv \{x\}_{U_0}^*$ is determined from the formula $\exists Z_2 \in \mathbf{D}(\forall y \in \mathbf{D}((y \in Z_2) \Leftrightarrow (y \in U_0 \wedge y = x)))$.

From the conditions $X \in \mathbf{D}$, $X \in U_0 = \mathfrak{a}$, $x \in \mathbf{D}$, and $x \in X$ it follows that $Z_2 = \{x\}_\mathfrak{a}$ and therefore, $Z_1 = x \cup_\mathfrak{a} \{x\}_\mathfrak{a}$.

Consider the $\mathfrak{a}$-set π from the axiom of infra-infinity A13. It is clear that $\pi \in \mathfrak{a} = U_0 \subset \mathbf{D}$. Since π possesses the property $\pi \in U_0 \wedge \varnothing \in \pi \wedge \forall x \in \pi(Z_1(x) \in \pi)$, it follows that in the LTS the formula $\widetilde{A^t13}$ is deduced.

$\widetilde{A^t14} \equiv \forall U \in \mathbf{D}(U \in \mathbf{R} \Rightarrow \forall X \in \mathbf{D}(X \in U \wedge X \neq \varnothing \Rightarrow \exists z \in \mathbf{D}((z \leftrightharpoons \mathcal{P}_U(X)\backslash\{\varnothing\}_U \to_U X)^\sigma \wedge \forall Y \in \mathbf{D}(Y \in \mathcal{P}_U(X) \setminus \{\varnothing\}_U)^\tau \Rightarrow z(Y)^\tau \in Y))))$, where

- the U-class $Z_1 \equiv Z_1(X) \equiv (\mathcal{P}_U(X) \setminus \{\varnothing\}_U)^\tau$ is determined from the formula $\exists Z_1 \in \mathbf{D}(\forall y \in \mathbf{D}((y \in Z_1) \Leftrightarrow (y \in U \wedge (y \in \mathcal{P}_U(X)^* \wedge y \notin \{\varnothing\}^*_U))))$;
- the U-class $Z_2 \equiv Z_2(Y) \equiv z(Y)^\tau$ is determined from the formula $Z_2 \in U \wedge \langle Y, Z_2\rangle^\tau_U \in z$;
- η^τ denotes the formula $\mathbf{M} \vDash \eta[\mathbf{s}^\tau]$, in which $\mathbf{s}^\tau$ denotes the corresponding change of the sequence $\mathbf{s}$ under the translation of the quantifier over-formulas $\forall U(\dots)$, $\forall X(\dots)$, $\exists z(\dots)$, and $\forall Y(\dots)$, indicated above.

Fix the conditions $U \in \mathbf{U}$, $U \in \mathbf{R}$, $X \in \mathbf{D}$, and $X \in U$. We established above that under these conditions $\mathcal{P}_U(X)^* = \mathcal{P}_U(X)$ and $\{\varnothing\}^*_U = \{\varnothing\}_U*$. From here, by virtue of the transitivity of $\mathbf{D}$, as above, we infer $Z_1 = \mathcal{P}_U(X) \setminus \{\varnothing\}_U \equiv T$. From $U \in \mathbf{R}$, it follows that U is a universal class. Consequently, by A9, $\mathcal{P}_U(X) \in U$. Therefore, by A8, $T \in U$.

If $Y \in \mathbf{D}$ and $Y \in Z_1$, then $Y \in T \in U$ implies $Y \in U$. As it was established above, $Z_2 \in U$ and $Y \in U$ imply $\langle Y, Z_2\rangle^\tau_U = \langle Y, Z_2\rangle_U$. Then, $\langle Y, Z_2\rangle_U \in z$ implies $Z_2 \in z\langle Y\rangle$. From here and from the previous conditions, we cannot yet infer that $Z_2 = z(Y)$.

Consider the formula $\varphi \equiv (z \leftrightharpoons T \to_U X)$. It is the conjuction of the three following formulas: $\varphi_1 \equiv (z \subset T *_U X)$, $\varphi_2 \equiv (\mathrm{dom}_U\, z = T)$, and $\varphi_3 \equiv (\forall x(x \in T \Rightarrow \forall y(y \in X \Rightarrow \forall y'(y' \in X \Rightarrow (\langle x, y\rangle_u \in z \wedge \langle x, y'\rangle_U \in z \Rightarrow y = y')))))$.

Therefore, $\varphi^\sigma = \varphi_1^\sigma \wedge \varphi_2^\sigma \wedge \varphi_3^\sigma$. Since $\varphi_1 = (\forall u(u \in z \Rightarrow u \in U \wedge \exists x \exists y(x \in T \wedge y \in X \wedge u = \langle x, y\rangle_U)))$, it follows that $\varphi_1^\sigma \Leftrightarrow (\forall u \in \mathbf{D}(u \in z \Rightarrow u \in U \wedge \exists x \in \mathbf{D} \exists y \in \mathbf{D}(x \in Z_1 \wedge y \in X \wedge u = \langle x, y\rangle^*_U)))$. Similarly, $\varphi_2 = (\forall x(x \in T \Rightarrow x \in U \wedge x \in T \wedge \exists y(y \in X \wedge \langle x, y\rangle_U \in z)))$ implies $\varphi_2^\sigma \Leftrightarrow (\forall x \in \mathbf{D}(x \in Z_1 \Rightarrow x \in U \wedge x \in Z_1 \wedge \wedge \exists y \in \mathbf{D}(y \in X \wedge \langle x, y\rangle^*_U \in z)))$.

Finally, $\varphi_3^\sigma \Leftrightarrow (\forall x \in \mathbf{D}(x \in Z_1 \Rightarrow \forall y \in \mathbf{D}(y \in X \Rightarrow \forall y' \in \mathbf{D}(y' \in X \Rightarrow (\langle x, y\rangle^*_U \in z \wedge \langle x, y'\rangle^*_U \in z \Rightarrow y = y')))))$.

By virtue of the properties of transitivity for x, y and y' in the formulas φ_1^σ, φ_2^σ, and φ_3^σ, we have $x, y, y' \in U$. Therefore, by the proof above (see the proof of deducibility of $\widetilde{A^t11}$), in these formulas, we have the equalities $Z_1 = T$, $\langle x, y\rangle^*_U = \langle x, y\rangle_U$, and $\langle x, y'\rangle^*_U = \langle x, y'\rangle_U$. It follows from here that the formulas φ_1^σ, φ_2^σ, and φ_3^σ differ from the formulas φ_1, φ_2, and φ_3, respectively, only by the bounded quantifier prefixes $\forall \ldots \in \mathbf{D}$ and $\exists \ldots \in \mathbf{D}$.

For X, by choice axiom A14, there exists z such that $\chi \equiv (z \leftrightharpoons \mathcal{P}_U(X) \setminus \{\varnothing\}_U \to_U X) \wedge \forall Y(Y \in \mathcal{P}_U(X) \setminus \{\varnothing\}_U \Rightarrow z(Y) \in Y)$.

Thus, in the LTS, we infer the formula $\varphi = \varphi_1 \wedge \varphi_2 \wedge \varphi_3$, and consequently, the formulas φ_1, φ_2, and φ_3.

Let $u \in \mathbf{D}$ and $u \in z$. Then, from the formula φ_1, we infer that there exist $x \in T$ and $y \in X$ such that $u = \langle x, y\rangle_U$. Since $x \in T \in U$ and $y \in X \in U$, it follows that, by the property of transitivity, $x, y \in U \subset \mathbf{D}$. It means that under the given conditions $u \in \mathbf{D}$ and $u \in z$ in the LTS, we infer the formula $(u \in U \wedge \exists x \in \mathbf{D}\exists y \in \mathbf{D}(x \in T \wedge y \in X \wedge u = \langle x, y\rangle_U))$. Applying two times the theorem of deduction and the rules of deduction, we infer the formula φ_1^σ.

Let $x \in \mathbf{D}$ and $x \in Z_1 = T$. Then, from the formula φ_2, we infer that for x, there exists $y \in X$ such that $\langle x, y\rangle_U \in z$. From $y \in X \in \mathbf{D}$, by the transitivity of $\mathbf{D}$, it follows that $y \in \mathbf{D}$. It means that under the given conditions $x \in \mathbf{D}$ and $x \in T$ in the LTS, we infer the formula $(x \in U \wedge x \in T \wedge \exists y \in \mathbf{D}(y \in X \wedge \langle x, y\rangle_U \in z))$. From here, as in the previous indentation, we infer the formula φ_2^σ.

Let $x \in \mathbf{D}$, $x \in Z_1 = T$, $y \in \mathbf{D}$, $y \in X$, $y' \in \mathbf{D}$, $y' \in X$, $\langle x, y\rangle_U \in z$, and $\langle x, y\rangle_U \in z$. Then, from the formula φ_3, we infer that $y = y'$. Applying several times in turn the theorem of deduction and the rule of deduction, we infer the formula φ_3^σ.

Thus, the formula φ^σ is deduced.

Since $z \leftrightharpoons T \to_U X$, it follows that $z\langle Y\rangle = \{z(Y)\}_U$.

Consequently, from $Z_2 \in U\{z(Y)\}_U$, we conclude that $Z_2 = z(Y)$. Therefore, for the U-mapping z, the conditions $Y \in \mathbf{D}$ and $Y \in Z_1 = T$ imply $Z_2 = z(Y) \in Y$.

Since $T \in U$ and $X \in U$, it follows that by Lemma 3 (B.1.1) $T *_U X \in U$. From $z \subset T *_U X$ by axiom A8, it follows that $z \in U \subset \mathbf{D}$.

All this means that from axiom A14, we deduce the existence of an object z, satisfying the formula χ, from which we infer the formula $\xi \equiv \varphi^\sigma \wedge \forall Y \in \mathbf{D}(Y \in Z_1 \Rightarrow Z_2 \in Y)$. By the same token, in the LTS from the fixed conditions, we infer the formula $\exists z \in \mathbf{D}\xi$. Applying several times in turn the theorem of deduction and the rule of generalization, we infer, as a result, the formula $\widetilde{A^t14}$.

Consider now the translation of axiom scheme of replacement ASR.

$\widetilde{ASR^t}$: if $\varphi(x, y)$ be a formula of the theory T that does not contain Y as a free variable, then $\forall x \in \mathbf{D}\forall y \in \mathbf{D}\forall y' \in \mathbf{D}(\varphi^\vee(x, y) \wedge \varphi^\vee(x, y') \Rightarrow y = y') \Rightarrow \forall X \in \mathbf{D}\exists Y \in \mathbf{D}\forall x \in \mathbf{D}(x \in X \Rightarrow \forall y \in \mathbf{D}(\varphi^\wedge(c, y) \Rightarrow y \in Y))$, where $\varphi^\vee$ and $\varphi^\wedge$ denote the formulas $\mathbf{M} \vDash \varphi[\mathbf{s}^\vee]$ and $\mathbf{M} \vDash \varphi[\mathbf{s}^\wedge]$, respectively, in which by $\mathbf{s}^\vee$, we denote the corresponding change of the sequence $\mathbf{s}$ under the translation of the indicated above quantifier over-formulas $\forall x(\ldots)$, $\forall y(\ldots)$, and $\forall y'(\ldots)$, and by $\mathbf{s}^\wedge$, we denote the corresponding change of the sequence $\mathbf{s}$ under the translation of the indicated above quantifier over-formulas $\forall X(\ldots)$, $\exists Y(\ldots)$, $\forall x(\ldots)$, and $\forall y(\ldots)$.

Denote the first part of this scheme by α and the second part by β. Then, $\widetilde{ASR^t} = (\alpha \Rightarrow \beta)$. Therefore, the equivalence $(\neg(\alpha \Rightarrow \beta)) \Leftrightarrow (\alpha \wedge \neg\beta)$ implies

$$(\neg\widetilde{ASR})^t = (\forall y \in \mathbf{D}\forall y \in \mathbf{D}\forall y' \in \mathbf{D}(\varphi^\vee(x, y) \wedge \varphi^\wedge(x, y') \Rightarrow y = y')) \wedge$$
$$\wedge (\exists X \in \mathbf{D}\forall Y \in \mathbf{D}\exists x \in \mathbf{D}(x \in X \wedge \exists y \in \mathbf{D}(\varphi^\wedge(x, y) \wedge y \notin Y))).$$

Further on, the first part of this scheme, we will denote by α', and the second part by β'. For the adduced concrete formula φ below, the symbol-strings α' and β' will be formulas of the LTS.

Consider the following formulas of the theory T: $\psi(x, y, z) \equiv (x \in \omega \wedge \mathfrak{a} \in y \wedge y \bowtie \wedge (z \leftrightharpoons x + 1 \to_y y) \wedge \forall k \in x + 1(z(k) \bowtie) \wedge z(0) = \mathfrak{a} \wedge \forall k \in x + 1(\forall l \in x + 1(k \in l \Rightarrow z(k) \in z(l))) \wedge \forall V((V \bowtie \wedge z(0) \subset V \wedge V \in y) \Rightarrow \exists k \in x + 1(V = z(k))))$ and $\varphi(x, y) \equiv \exists z \psi(x, y, z)$, where $x + 1$ denotes the class $x \cup_{\mathfrak{a}} \{x\}_{\mathfrak{a}}$.

For the formula φ, we have the following translations:

- $\varphi^{\vee}(x, y) \Leftrightarrow \exists z \in \mathbf{D}(x \in \omega^{\sigma} \wedge \mathfrak{a} \in y \wedge y \in \mathbf{R} \wedge (z \leftrightharpoons x+1 \to_y y)^{\sigma} \wedge \forall k \in \mathbf{D}(k \in (x+1)^{\sigma} \Rightarrow z(k)(k \in (x + 1)^{\sigma} \Rightarrow \forall l \in \mathbf{D}(l \in (x + 1)^{\sigma} \Rightarrow (k \in l \Rightarrow z(k)^{\sigma} \in z(l)^{\sigma}))) \wedge \forall V \in \mathbf{D}((V \in \mathbf{R} \wedge (z(0)^{\sigma} \subset V) \wedge V \in y) \Rightarrow \exists k \in \mathbf{D}(k \in (x + 1)^{\sigma} \wedge (V = z(k)^{\sigma})));$
- $\varphi^{\wedge}(x, y) \Leftrightarrow \exists z \in \mathbf{D}(x \in \omega^{\tau} \wedge \mathfrak{a} \in y \wedge y \in \mathbf{R} \wedge (z \leftrightharpoons x+1 \to_y y)^{\tau} \wedge \forall k \in \mathbf{D}(k \in (x+1)^{\tau} \Rightarrow z(k)^{\tau} \in \mathbf{R}) \wedge z(0)^{\tau} = \mathfrak{a} \wedge \forall k \in \mathbf{D}(k \in (x + 1)^{\tau} \Rightarrow \forall l \in \mathbf{D}(l \in (x + 1)^{\tau} \Rightarrow (k \in l \Rightarrow z(k)^{\tau} \in z(l)^{\tau}))) \wedge \forall V \in \mathbf{D}((V \in \mathbf{R} \wedge (z(0)^{\tau} \subset V) \wedge V \in y) \Rightarrow \exists k \in \mathbf{D}(k \in (x + 1)^{\tau} \wedge (V = z(k)^{\tau})))),$

where $\theta^{\sigma}, \theta^{\tau}, \eta^{\sigma}$, and η^{τ} denote the terms $\theta_{\mathbf{M}}[\mathbf{s}^{\sigma}]$ and $\theta_{\mathbf{M}}[\mathbf{s}^{\tau}]$ and the formulas $\mathbf{M} \vDash \eta[\mathbf{s}^{\sigma}]$ and $\mathbf{M} \vDash \eta[\mathbf{s}^{\tau}]$, in which $\mathbf{s}^{\sigma}$ and $\mathbf{s}^{\tau}$ denote the corresponding changes of the sequences $\mathbf{s}^{\vee}$ and $\mathbf{s}^{\wedge}$ under the translation of the quantifier over-formula $\exists z \in \mathbf{D}(\dots)$.

Check that $\omega^{\sigma} = \omega$ and $\omega^{\tau} = \omega$. By means of the formula $On(x)$, the class ω is assigned by the formula $\nu \equiv \exists! z(On(z) \wedge z \neq \varnothing \wedge \forall x(On(x) \Rightarrow z \neq x \cup_{\mathfrak{a}} \{x\}_{\mathfrak{a}}) \wedge \forall y ((On(y) \wedge y \neq \varnothing \wedge \forall x(On(x) \Rightarrow y \neq x \cup_{\mathfrak{a}} \{x\}_{\mathfrak{a}})) \Rightarrow z \subset y))$, which is deduced in the theory T. Therefore, with regard to the proven above three equivalences, the value ω^{σ} is defined from the formula $\nu^{\sigma} \Leftrightarrow \exists! z \in \mathbf{D}(On^{\mathbf{D}}(z) \wedge z \neq \varnothing \wedge z \in \mathfrak{a} \wedge \forall x \in \mathbf{D}(On^{\mathbf{D}}(x) \wedge x \in \mathfrak{a} \Rightarrow z \neq (x \cup_{\mathfrak{a}} \{x\}_{\mathfrak{a}})^{*} \wedge \forall y \in \mathbf{D}((On^{\mathbf{D}}(y) \wedge y \neq \varnothing \wedge y \in \mathfrak{a} \wedge \forall x \in \mathbf{D}(On^{\mathbf{D}}(x) \wedge x \in \mathfrak{a} \Rightarrow y \neq (x \cup_{\mathfrak{a}} \{x\}_{\mathfrak{a}})^{*})) \Rightarrow z \subset y))$ because $\mathfrak{a}$ is translated into $\mathfrak{a}$.

Since the assembly $\mathbf{D}$ is transitive, we can prove by direct check, using the definition of the formula $On(z)$ from the beginning of B.4, that for $z \in \mathbf{D}$, we have the equivalence $on^{\mathbf{D}}(z) \Leftrightarrow On(z)$.

When we checked the deducibility of formula $\widetilde{A^{t}13}$, we established that the conditions $x \in \mathbf{D}$ and $x \in \mathfrak{a}$ imply the equality $(x \cup_{\mathfrak{a}} \{x\}_{\mathfrak{a}})^{*} = x \cup_{\mathfrak{a}} \{x\}_{\mathfrak{a}}$. Therefore, the formula ν^{σ} is equivalent to the formula ν', which differs from the formula ν only by bounded quantifier prefixes $\exists \dots \in \mathbf{D}$ and $\forall \dots \in \mathbf{D}$. But since the formula ν and the formula ν' contain the subformulas $z \in \mathfrak{a}$, $x \in \mathfrak{a}$, and $y \in \mathfrak{a}$, which immediately imply the restrictions $z \in \mathbf{D}$, $x \in \mathbf{D}$, and $y \in \mathbf{D}$, it follows that in the LTS we infer the equivalence $\nu \Leftrightarrow \nu' \Leftrightarrow \nu^{\sigma}$. From the equivalence $\nu \Leftrightarrow \nu^{\sigma}$, it follows that $\omega^{\sigma} = \omega$. In the same way, it is checked that $\omega^{\tau} = \omega$.

We have checked above that under the conditions $x \in \omega$, $\omega \in \mathfrak{a}$, and $\mathfrak{a} \in \mathbf{D}$ by virtue of the transitivity of $\mathbf{D}$ it is valid the equality $(x+1)^{\sigma} = x \cup_{\mathfrak{a}} \{x\}_{\mathfrak{a}} = x+1$. We also checked that under the condition $y \in \mathbf{R}$ the equivalence $(z \leftrightharpoons x+1 \to_y y)^{\sigma} \Leftrightarrow (z \leftrightharpoons x+1 \to_y y)$ is valid. Therefore, $z(i)^{\sigma} = z(i)$ for every $i \in x+1$. The same is valid for the variant with the sign τ.

Therefore, $\varphi^{\vee}(x, y)$ and $\varphi^{\wedge}(x, y)$ are equivalent to the same formula $\varphi^{*}(x, y) \equiv \exists z \in \mathbf{D}(x \in \omega \wedge \mathfrak{a} \in y \wedge y \in \mathbf{R} \wedge (z \leftrightharpoons x+1 \to_y y) \wedge \forall k \in \mathbf{D}(k \in x+1 \Rightarrow z(k) \in \mathbf{R}) \wedge z(0) =$

$\mathfrak{a} \wedge \forall k \in \mathbf{D}(k \in x+1 \Rightarrow \forall l \in \mathbf{D}(l \in x+1 \Rightarrow (k \in l \Rightarrow z(k) \in z(l)))) \wedge \forall V \in \mathbf{D}((V \in \mathbf{D} \wedge (z(0) \subset V) \wedge V \in y) \Rightarrow \exists k \in \mathbf{D}(k \in x+1 \wedge V = z(k))))).$

In the same way as in the proof of Theorem 1 (B.3.5) in the LTS, we infer the formula

$$\forall x \forall y \forall y' \forall z \forall z' (\psi(x, y, z) \wedge \psi(x, y', z') \Rightarrow y = y' \wedge z = z'),$$

which means that y and z are defined uniquely by x. We also infer the formula

$$\forall x \forall x' \forall y \forall y' \forall z \forall z' ((\psi(x, y, z) \wedge \psi(x', y', z') \wedge x \in x') \Rightarrow$$
$$\Rightarrow \forall k \in k+1(z'(k) = z(k)) \wedge (z'(x+1) = y'))),$$

which means that for $m \in n$, the sequence $u(n)$ continues the sequence $u(m)$ and $U^n_{m+1} = U(m)$.

Deduce now the formula $\forall x \forall y \forall z(\psi(x, y, z) \Rightarrow \forall k \in ((x+1) \setminus 1) \exists x' \exists y' \exists z'(x' \in x \wedge \varphi(x', y', z') \wedge z(k) = y'))$, which means that all members of the sequence $u(n)$, beginning from the first one, are constructed from the previous classes $U(m)$ for $m \in n$.

Denote the formula $\forall k \in ((x+1) \setminus 1) \exists x' \exists y' \exists z'(x' \in x \wedge \psi(x', y', z') \wedge z(k) = y')$ by $\eta(x, y, z)$. We will infer this formula under the condition $\psi(x, y, z)$. The formula $\psi(x, y, z)$ postulates that x is a natural number, y is a universal class, greater than $\mathfrak{a}$, and there exists a finite y-sequence of universal classes $u(x) \equiv (y_k \in y \mid k \in x+1)_y$ such that $y_0 = \mathfrak{a}$, $y_k \in y_l$ for all $k \in l \in x+1$, and if V is a universal class and $y_0 \subset V \in y$, then $V = y_k$ for some $k \in x+1$. By Theorem 2 (B.3.5), for every $x \in \omega$ such a sequence exists and is unique, and, besides, $u(x)|m+1 = u(m)$ for all $m \leqslant x$.

The formula $\eta(x, y, z)$ has the form $\forall k(k \in (x+1) \setminus 1 \Rightarrow \exists x' \exists y' \exists z'(x' \in x \wedge \psi(x', y', z') \wedge z(k) = y')$. Show that from our conditions and the condition $k \in (x+1) \setminus 1$, the formula $\eta'(x, y, z, k) \equiv \exists x' \exists y' \exists z'(x' \in x \wedge \psi(x', y', z') \wedge z(k) = y')$ is deduced.

Consider x' such that $x'+1 = k$. This is possible because $1 \subseteq k$. By Theorem 2 (B.3.5) for the given x', there exist y' and z' such that $\psi(x', y', z')$. Since $x' \in k \subseteq x$, it follows that $x' \in x$. It remains to show that $z(k) = y'$, i. e. $y' = y_k$ in the sequence $u(x)$. Since $\psi(x', y', z')$, it follows that $\mathfrak{a} = y_0 \in y'$, i. e. $y_0 \in y'$. Since $y' \bowtie$, there are only three possibilities: 1. $y \in y'$; 2) $y = y'$; 3) $y' \in y$. In the first case, $y = z'(l)$ for some $l \leqslant x' < x$, but this is impossible; in the second case, the sequences $u(x)$ and $u(x')$ have to coincide, but this it impossible because $x' < x$. As a result, only case 3 $y' \in y$ is possible. Thus, $y_0 \subset y' \in y$, and consequently, by the condition, $y' = y_m$ for some $m \in x+1$. Show that $m = k$. If $m < k$, then $y_m = z'(m)$, but it is impossible because $z'(m) \in y'$ for all $m < k$. If $m > k$, then, taking $V \equiv y_k$, we get the condition $y_0{}' \subset V \in y'$, which imply $V = y_l{}'$ for $l < k$, but it is also impossible. Therefore, $y' = y_k$, i. e. $z(k) = y'$. Thus, under the conditions $\psi(x, y, z)$ and $k \in (x+1) \setminus 1$, we infer the formula $\eta'(x, y, z, k)$, and so, by the theorem of deduction, under the condition $\psi(x, y, z)$ we infer the formula $k \in (x+1) \setminus 1 \Rightarrow \eta'(x, y, z, k)$, from here by the rule of generalization, the formula $\eta(x, y, z)$ is deduced. Then, applying the theorem of deduction and the rule of generalization one more time, we get the required formula.

Consequently, $z(k) \in \mathbf{R}$ for all $k \in x + 1$. Take in the capacity of value of the variable X the class ω. Since $\omega \in \mathfrak{a} = U(0) \subset \mathbf{D}$, it follows that $\omega \in \mathbf{D}$. Take any $Y \in \mathbf{D}$. Then, $Y \in U(x_0)$ for some $x_0 \in \omega$. This means that in the LTS the formula $\exists x \exists y (\exists! z(\psi(x, y, z) \wedge Y \in y)$ is deduced. Denote the unique values of the variables y and z, corresponding to the value x_0, by y_0 and z_0. From all this, it follows that the formula $\psi(x_0, y_0, z_0) \wedge Y \in y_0$ is deduced.

By definition, $y_0 \in \mathbf{R} \subset \mathbf{D}$. From the previous arguments, it follows that $z_0(k) \in \mathbf{R}$ for any $k \in x_0 + 1$. It is clear that $k \in \mathbf{D}$ and $l \in \mathbf{D}$.

If $V \in \mathbf{D}$ and $V \in \mathbf{R}$, then $V\bowtie$, and therefore, the conditions $z_0(0) \subset V$ and $V \in y_0$, implies $\exists k \in \mathbf{D}(k \in x_0 + 1 \wedge V = z_0(k))$. Thus, in the LTS, the following formula is deduced:

$$\delta(x_0, y_0, z_0) \equiv (x_0 \in \omega \wedge \mathfrak{a} \in y_0 \wedge y_0 \in \mathbf{R} \wedge (z_0 \leftrightharpoons x_0 + 1 \rightarrow_{y_0} y_0) \wedge \forall k \in \mathbf{D}(k \in x_0 + 1 \Rightarrow z_0(k) \in \mathbf{R}) \wedge z_0(0) = \mathfrak{a} \wedge \forall k \in \mathbf{D}(k \in x_0 + 1 \Rightarrow \forall l \in \mathbf{D}(l \in x_0 + 1 \Rightarrow (k \in l \Rightarrow z_0(k) \in z_0(l)))) \wedge \forall V \in \mathbf{D}((V \in \mathbf{R} \wedge z_0(0) \subset V \wedge V \in y_0) \Rightarrow \exists k \in \mathbf{D}(k \in x_0 + 1 \wedge V = z_0(k))))).$$

Since $z_0 \leftrightharpoons x_0 + 1 \rightarrow_{y_0} y_0$, it follows that, by axiom A11, $P \equiv \mathrm{rng}_{y_0} z_0 \in y_0$. Therefore, $z_0 \subset ((x_0 + 1)^*_{y_0} P) \equiv Q$. Besides, $x_0 + 1 \in \omega \in \mathfrak{a} \in y_0$ implies $x_0 + 1 \in y_0$. By Lemma 3 (B.1.1), $Q \in y_0$. Consequently, by axiom A8, $z_0 \in y_0 \in \mathbf{D}$. Hence, we get $z_0 \in \mathbf{D}$. Thus, in the LTS, the formula $\varphi^*(x_0, y_0) = \exists z \in \mathbf{D}\delta(x_0, y_0, z_0)$ is deduced. Besides, the formula $Y \in y_0$ was deduced. By the axiom of regularity A12, the formula $y_0 \notin Y$ is deduced.

By the same token, we deduced the formula $\varphi \wedge (x_0, y_0) \wedge y_0 \notin Y$. Since $x_0 \in \mathbf{D}$, $x_0 \in \omega$, and $y_0 \in \mathbf{D}$, further, by logical means, we infer the formula $\exists x \in \mathbf{D}(x \in \omega \wedge \exists y \in \mathbf{D}(\varphi^\wedge(x, y) \wedge y \in Y)$. Since $\omega \in \mathbf{D}$ and $Y \in \mathbf{D}$, then further by logical means, we infer the formula β'.

Let now $x, y, y' \in \mathbf{D}$, $x \in \omega$, $\in y$, $\mathfrak{a} \in y'$, $y \in \mathbf{R}$, and $y' \in \mathbf{R}$.

Infer now the formula $\varphi^*(x, y) \wedge \varphi^*(x, y') \Rightarrow y = y'$. Consider the condition $\mu(x, y, y') \equiv \varphi^*(x, y) \wedge \varphi^*(x, y')$. According to this condition, for a natural number x, there exist universal classes y and y' from $\mathbf{R}$, greater than $\mathfrak{a}$ and finite sequences of universal classes $u(x) \equiv (y_k \in y \mid k \in x + 1)_y$, and $u'(x) \equiv (y_k{}' \in y' \mid k \in x + 1)_{y'}$ such that $y_0 = y_0{}' = \mathfrak{a}, y_k \in y_l$ and $y_k{}' \in y_l{}'$ for any $k \in l \in x+1$, and if V and W are universal classes from $\mathbf{R}$ and $y_0 \subset V \in y, y_0{}' \subset W \in y'$, then $V = y_m, W = y_l{}'$ for some $m, l \in x+1$. Suppose that $y \neq y'$. Since y and y' are universal classes, it follows that in this case either $y \in y'$ or $y' \in y$. Suppose, for example, that $y \in y'$. Set $W \equiv y$. We get $W \in \mathbf{R} \wedge \mathfrak{a} \subset W \wedge W \in y'$, $W = y_l{}'$ for some $l \in x + 1$. Thus, $y = y_l{}'$ for $l \in x + 1$. Similarly, for any $k < x + 1$, there exists $l(k) < x + 1$ such that $y_k = y_{l(k)}{}'$. Since $\forall k \in x + 1(y_k \in y)$, it follows that $\forall k \in x+1(y_{l(k)}{}' \in y_l{}')$, i. e. $l(k) \in l$ for all $k \in x+1$. Besides, if $k, m \in x+1$ and $k \neq m$, then $y_{l(k)}{}' \neq y_{l(m)}{}'$. Consequently, there exists an inclusion of the set $x+1$ into the set $l \in x+1$, but it is impossible. The case $y' \in y$ is checked in just the same way. Consequently, $y = y'$. Applying the theorem of deduction, we deduce the formula $\mu(x, y, y') \Rightarrow y = y'$.

Thus, the formula $\varphi^{\vee}(x, y) \wedge \varphi^{\vee}(x, y') \Rightarrow y = y'$ is deduced. From here, by logical means, the formula α' is deduced.

As a result, in the theory S, the formula $\alpha' \wedge \beta'$ is deduced, which is equal to the formula $(\widetilde{\neg ASR})^t$ with the given concrete formula φ.

Since all the translations of the axioms of the theory T turned out to be deducible formulas of the theory S, it follows that the theory T is consistent.

2. We will argue in naive propositional logic with the symbol of implication $\supset$.

Denote by Φ_a and Ξ_a the totalities of axioms of the theories $T \equiv LTS + \neg ASR$ and $S \equiv LTS$, respectively.

Consider the propositions $A \equiv cons(S) \supset \neg(\Xi_a \vdash ASR)$ and $B \equiv cons(S) \wedge (\Xi_a \vdash ASR)$. Then, $\neg A = cons(S) \wedge \neg\neg(\Xi_a \vdash ASR)$. Using LAS10, we get $\neg A \supset B$.

It is clear that $B \supset (\Phi_a \vdash ASR)$ and $\Phi_a \vdash \neg ASR$. Thus, the proposition $B \supset (\Phi_a \vdash ASR) \wedge (\Phi_a \vdash \neg ASR)$, i. e. the expression $B \supset \neg cons(T)$ is valid.

By the rule of deduction, $\neg A \supset \neg cons(T)$.

According to item 1 of our statement, the proposition $cons(S) \supset cons(T)$ is valid. Therefore, $B \supset cons(T)$ is valid. By the rule of deduction, we have $\neg A \supset cons(T)$.

Thus, the proposition $(\neg A \supset cons(T)) \wedge (\neg A \supset \neg cons(T))$ is deduced. Applying the tautology $(\neg A \supset C) \wedge (\neg A \supset \neg C) \supset A$ (see [*Kolmogorov and Dragalin*, 1982, I, 7]), we infer the proposition A. □

With the help of more complicated abstract interpretation, one can prove also that if the LTS is consistent, then axiom scheme ASR is not deducible in the LTS+AU(ω).

In B.3 we have proven the closeness of the assembly $\overline{\mathbf{V}}$ of all classes in the LTS with respect to all basic finite set-theoretical operations. Therefore, we can, in the assembly $\overline{\mathbf{V}}$, define on classes such basic mathematical systems as groups, topological spaces, automats, and other, and also define morphisms between them. By the same token, we can in $\overline{\mathbf{V}}$ consider abstract categories of all such mathematical systems and morphisms between them.

From the undeducibility in the LTS of the global axiom scheme of replacement, it follows that the assembly $\overline{\mathbf{V}}$ does not possess the fourth of the five properties of the Eresmann – Dedecker – Sonner – Grothendieck universe listed in the introduction and necessary for developing in this universe valuable category theory. Therefore, the theory of abstract categories in the LTS will be essentially poorer than the theory of (local) categories in the LTS.

In particular, these abstract categories of mathematical systems will not be closed with respect to such infinite operations as the sum $\cup[A_i \mid i \in I]$ and the product $\prod[A_i \mid i \in I]$ of the collection $[A_i \mid i \in I]$ of objects A_i of these categories and will consequently be abstract categories without direct and inverse limits (see [*Bucur and Deleanu*, 1972, ch. 2, 1; ch. 3, 2]).

On the contrary, the set theories ZF or NBG with the axiom of universality abstract categories of mathematical systems with the point of view of naive category theory do not differ absolutely from local U-categories of mathematical systems.

B.6.2 The independence of axiom AU(ω) of the axioms of the LTS

In section B.4, we compared the LTS with the ZF set theory with some additional axioms. The fact that axiom AIC postulating the existence of an inaccessible cardinal is independent of the axioms of the ZF set theory is well-known (see [*Jech*, 1971, 13]). Consequently, axioms AI(ω) and AI($\omega + \omega$), postulating the existence of infinite sets of inaccessible cardinals, are also independent of the axioms of the ZF set theory.

It remained only to clarify the independence of axiom AU(ω) postulating the existence of an infinite class of universal classes of the axioms of the local theory of sets.

Statement 1.
1) *If the LTS is consistent, then the LTS+AU(ω) is consistent.*
2) *If the LTS is consistent, then axiom AU(ω) is not deducible in the LTS.*

Proof. 1. By Lemma 1 (B.4.2), axiom AU(ω) is equivalent to axiom ATU(ω). Therefore, we will consider the equivalent theory $T \equiv LTS{+}\neg ATU(\omega)$. Consider the abstract interpretation $\mathbf{M} \equiv (\mathbf{D}, I)$ of the theory T in the finitely closed set theory $S \equiv LTS$, described in the proof of Statement 1 (B.6.1). In the proof of Statement 1 (B.6.1), we established that the interpretation $\mathbf{M}$ is an abstract model of the LTS in the set theory S.

By virtue of Lemma 2 (B.4.2) in the conjunctive kernel of axiom ATU(ω), we can insert one more formula $\mathfrak{a} \in Y$. Therefore, consider the formula

$\varphi \equiv ATU(\omega) \equiv \exists Y(\forall U(U \in Y \Rightarrow U\bowtie) \wedge \mathfrak{a} \in Y \wedge \forall U \forall V(U \bowtie \ \wedge U \in V \wedge V \in Y \Rightarrow U \in Y) \wedge \forall V(V \in Y \Rightarrow \exists W(W \in Y \wedge V \in W)))$.

The translation of this formula on some abstract sequence $\mathbf{s}$ of elements of the assembly $\mathbf{D}$ under the interpretation $\mathbf{M}$ has the form of the formula

$\psi \equiv \mathbf{M} \vDash \varphi[\mathbf{s}] = \exists Y \in \mathbf{D}(\forall U \in \mathbf{D}(U \in Y \Rightarrow U \in \mathbf{R}) \wedge \mathfrak{a} \in Y \wedge \forall U \in \mathbf{D} \forall V \in \mathbf{D}(U \in \mathbf{R} \wedge U \in V \wedge V \in Y \Rightarrow U \in Y) \wedge \forall V \in \mathbf{D}(V \in Y \Rightarrow \exists W \in \mathbf{D}(W \in Y \wedge V \in W)))$.

Suppose that the condition ψ is fixed and consider a class $E \in \mathbf{D}$, the existence of which follows from this condition. Consider the classes $A_n \equiv \{x \in U(n) \mid \exists k \in n + 1(x = U_k^n)\}$, consisting of all members of $U(n)$-sequences $u(n)$ from Theorem 2 (B.3.5). Further, along with U_n^n, we will write U_n.

Prove by the natural induction that $A_n \subset E$ for every $n \in \omega$. Consider the assembly $\mathbf{X} \equiv \{x \mid x \in \omega \wedge Ax \subset E\}$. If $n = 0$, then $U_0 \in E$ implies $A_0 \equiv \{x \in U(0) \mid \exists k \in 1(x = U_k^0)\} \subset E$. Thus, $0 \in \mathbf{X}$.

Let $n \in \mathbf{X}$, i.e. $U_k^n \in E$ for every $k \in n + 1$. By the property of the class E, there exists $V \in \mathbf{D}$ such that $V \in E$ and $U_n \in V$. Besides, $U_n \equiv U_n^n = U_n^{n+1} \in U_{n+1}^{n+1} \equiv U_{n+1}$. If $U_{n+1} = V$, then $U_{n+1} \in E$. Let $U_{n+1} \in V$. In the proof of Theorem 2 (B.3.5), we established that $U_{n+1} = U(n) \in \mathbf{R}$. Since $V \in E$, by the property of the class E, we conclude that $U_{n+1} \in E$. Finally, let $V \in U_{n+1}$. Then, $V \in \mathbf{D}$ and $V \in E$ imply $V \in \mathbf{R}$. Consequently, V is a universal class. From $V \in U_{n+1} \in U(n + 1)$, by the axiom of universality, we infer that $V \in U(n + 1)$. Therefore, the condition $U_0^{n+1} = \mathfrak{a} \subset V \in U(n + 1)$, by the property of incompressibility from Theorem 2 (B.3.5) implies the equality $U = U_k^{n+1}$ for some $k \in n+2$. From $V \in U_{n+1}^{n+1}$, it follows that $k < n+1$. Hence, $U_n^{n+1} = U_n^n \equiv U_n \in V = U_k^{n+1}$ implies $n < k < n+1$, but this is impossible. From the obtained contradiction, we infer that the third case $V \in U_{n+1}$ is impossible. By virtue of Proposition 1 (B.3.4) and Theorem 1 (B.3.5), there are no other possibilities except the considered three cases. In the first and the second cases. we get $U_{n+1} \in E$. Besides, $U_k^{n+1} = U_k^n \in E$ for all $k \in n+1$. So $A_{n+1} \subset E$ implies $n+1 \in \mathbf{X}$. By the principle of natural induction in the LTS (Theorem 2 (B.3.3)) $\omega \subset \mathbf{X}$.

Thus, from the formula ψ, we deduced the formula $\chi \equiv \forall x \in \omega(U_x \in E)$. Besides, from ψ one can infer $E \in \mathbf{D}$. Consequently, $E \in U(m) = U_{m+1}$ for some $m \in \omega$. Therefore, $U_{m+1} \notin E$. This means that from ψ the formula $\neg\chi$ is inferred. By the theorem of deduction, in the theory S, we deduce the formulas $(\psi \Rightarrow \chi)$ and $(\psi \Rightarrow \neg\chi)$.

Applying now LAS9 and the logical explicit axiom $(\psi \Rightarrow \chi) \Rightarrow ((\psi \Rightarrow \neg\chi) \Rightarrow \neg\psi)$, we consecutively deduce the formulas $(\psi \Rightarrow \neg\chi) \Rightarrow \neg\psi$ and $\neg\psi$. The last formula is equal to the formula $\mathbf{M} \models (\neg\varphi)[\mathbf{s}]$. Thus, $\mathbf{M}$ is an abstract model of T in S.

2. The proof is similar to the proof of assertion 2 from Statement 1 (B.6.1). □

Thus, axiom AU(ω) does not depend of axioms of the LTS, i. e. it is a new axiom for the LTS.

Corollary 1. *If the LTS is consistent, then the assertion* 3) *from Proposition* 1 (B.4.2) *is not deducible in the LTS.*

It follows from here that in the LTS, there is only countable assembly of universes, constructed in Theorem 2 (B.3.5), and there are no means to continue it further on like this was done in assertion 3 from Proposition 1 (B.4.2).

B.6.3 The locally minimal theory of sets

The local theory of sets with additional axiom, which states that infinite class of universes does not exist, will be called the *locally minimal theory of sets* (LMTS), i. e. LMTS≡ LTS+¬AU(ω). Statement 1 (B.6.2) implies that the LTS and the LMTS are mutually consistent.

In the LMTS, the assembly $\mathbf{U}$ of all universal classes has the following complete description.

Theorem 1. *In the LMTS, an assembly sequence* $(u(n) \mid n \in \omega)$ *of finite* $U(n)$*-sequences* $u(n) \equiv (U_k^n \in U(n) \mid k \in n+1)_{U(n)}$ *of universal classes from Theorem* 2 (B.3.5) *includes all universal classes; more strict,* $\forall U(U\bowtie \Rightarrow \exists n \in \omega(U = U_n^n))$.

Proof. Consider an assembly $\mathbf{U}^* \equiv \{U \mid U\bowtie \wedge \exists n \in \omega(U = U_n^n)\}$. Suppose that $\mathbf{U}^* \neq \mathbf{U}$, i. e. there exists a universal class X such that $X \notin \mathbf{U}^*$.

Fix numbers $n \in \omega$ and $k \in n+1$. Suppose that $X = U^n_k$. Then, by Theorem 2 (B.3.5) $X = U^n_k = U^k_k \in \mathbf{U}^*$, but it is not true. Therefore, $X \neq U^n_k$ for all $n \in \omega$ and $k \in n+1$. By Theorem 1 (B.3.5) and Proposition 1 (B.3.4), either $X \in U^n_k$ or $U^n_k \in X$. Suppose that $X \in U^n_k$. By Corollary 2 to Proposition 1 (B.3.4), $U^n_0 = \mathfrak{a} \subset X \in U^n_k$. By axiom A8 from LTS, we have $U^n_0 \in U^n_k$. Since $k \leqslant n$, we have that $n \geqslant 1$. Thus, $U^{n-1}_0 = \mathfrak{a} \subset X \in U^n_k \subset U^n_n = U(n-1)$. By the property of incompressibility from Theorem 2 (B.3.5), $X = U^n_l$ for some $k \in n+1$. But we have proven above that it is impossible. This contradiction implies that $U^n_k \in X$ for all $n \in \omega$ and $k \in n+1$.

Therefore, according to Proposition 1 (B.4.2), we deduce axiom AU(ω), but it is impossible if the LMTS is consistent. Consequently, our assumption is not true and $\mathbf{U}^* = \mathbf{U}$. □

We have proven that the LMTS satisfies the two following properties: (1) it has the property of universal comprehension, which is expressed as universality axiom A6 ($\forall x \exists U(x \in U \wedge U\bowtie)$); (2) it has countable metasequence (assembly) of all universes. Neither ZF+AU nor ZF+AU(ω) (see the Introduction to this Appendix and subsection A.4.3 and A.5.1 in Appendix A) does not satisfy these two properties.

Besides, Statement 1 (B.6.1) implies that the LMTS is strictly weaker than the theories ZF+AU and ZF+AU(ω). Therefore, the LMTS satisfies conditions 1–3, given in introduction. Thus, this theory is more natural for category theory than the theories ZF+AU and ZF+AU(ω). In comparison with the LTS, the LMTS satisfies condition 3 in a stricter form.

B.7 The finite axiomatizability of the LTS and the NBG set theory

The local theory of sets has 13 axioms (A1, A3 – A14) and one axiom scheme AS2 (see B.1.1). In this section, we show that this axiom scheme can be replaced by finitely many axioms that are special cases of the comprehension axiom scheme. This means that the LTS and the LMTS (see B.6.3), as well as the NBG set theory, are finitely axiomatizable.

Note that the finite axiomatizability cannot be proven by using the sketch of the proof of the finite axiomatizability of NBG given in [*Mendelson*, 1997] because of the condition that the formula φ in axiom scheme AS2 of the LTS must be X-predicative. Thus, we introduce two more axioms, which look rather unusual for those familiar with the proof given in Mendelson's book.

In contrast to [*Mendelson*, 1997, 4.1], we present the detailed proof of the finite axiomatizability theorem for the LTS (see Theorem 2 (B.7.2)). This gives us the pleasant opportunity to present the abridged proof of Bernays' outstanding result on the finite axiomatizability of the very NBG set theory in B.7.3.

B.7.1 Replacement of the full comprehension axiom scheme by finitely many axioms

Associate axiom scheme AS2 with explicit axioms A2.1 – A2.15 stated below. Denote by LTS* the theory determined by axioms A1, A2.1 – A2.15, A3, A6, A7, and A12.

A2.1. (The *pair axiom.*) $\forall A, B \exists Y \forall y (y \in Y \Leftrightarrow (y = A \vee y = B))$.

According to this axiom, in LTS* for any classes A and B, the *unordered pair* $\{A, B\}$ exists.

Define in LTS* the *ordered pair* $\langle A, B\rangle \equiv \{\{A\}, \{A, B\}\}$, where $\{A\} \equiv \{A, A\}$.

Put $\langle A\rangle \equiv A$ and define the *ordered suits* $\langle A_0, \dots, A_{n+1}\rangle \equiv \langle\langle A_0, \dots, A_n\rangle, A_{n+1}\rangle$ for $n \geqslant 1$ by induction.

Lemma 1. *If* $\langle A, B\rangle = \langle A', B'\rangle$, *then* $A = A'$ *and* $B = B'$.

Lemma 2. $U \bowtie \ \wedge \langle A, B\rangle \in U \Rightarrow A, B \in U$.

Corollary 1. $U \bowtie \ \wedge \langle A_0, \dots, A_{n+1}\rangle \in U \Rightarrow A_0, \dots, A_{n+1} \in U$.

A2.2. (The *local pair axiom.*) $\forall U (U \bowtie \Rightarrow \forall A, B \in U(\{A, B\} \in U))$.

According to the local pair axiom A2.2, $A, B \in U$ implies $\langle A, B\rangle \in U$. It is easily proven by induction that $A_1, \dots, A_n \in U$ implies $\langle A_1, \dots, A_n\rangle \in U$.

A2.3. (The *axiom of specification of local universal elements.*)

$$\forall U (U \bowtie \Rightarrow \exists Y \forall y (y \in Y \Leftrightarrow y \in U \wedge y \bowtie)).$$

A2.4. (The *axiom of specification of a local restriction.*)

$$\forall A \forall U (U \bowtie \Rightarrow \exists Y \forall y (y \in Y \Leftrightarrow y \in U \wedge y \in A)).$$

A2.5. (The *axiom of specification of a local frame.*)

$$\forall A \forall U (U \bowtie \Rightarrow \exists Y \forall y (y \in Y \Leftrightarrow y \in U \wedge A \in y)).$$

A2.6. (The *axiom of specification of a local product.*)

$$\forall U (U \bowtie \Rightarrow \forall A \exists Y \forall y (y \in Y \Leftrightarrow y \in U \wedge \exists a, v (a, v \in U \wedge a \in A \wedge y = \langle a, v\rangle)))).$$

Check that in axiom A2.6 and other axioms, the subformula $y = \langle a, v\rangle$ is U-predicative. This subformula has the form $\eta \equiv \forall z (z \in y \Leftrightarrow z \in \langle a, v\rangle)$. Consider the formula $\zeta \equiv \forall z \in U (z \in y \Leftrightarrow z \in \langle a, v\rangle)$. It is clear that $\eta \Rightarrow \zeta$. Let ζ and take an arbitrary z. If $z \subset y$, then $y \in U$ implies $z \in U$. Therefore, ζ and $z \in y$ imply $z \in \langle a, v\rangle$. By the deduction theorem, we deduce from ζ the formula $z \in y \Rightarrow z \in \langle a, v\rangle$. Conversely, if $z \in \langle a, v\rangle$, then by axiom A7, it follows from $\langle a, v\rangle \in U$ that $z \in U$. Hence, ζ implies $z \in y$.

By the deduction theorem, we deduce from ζ the formula $z \in \langle a, v\rangle \Rightarrow z \in y$. Hence, by the generalization rule, we deduce from ζ the formulas $y \subset \langle a, v\rangle \equiv \forall z(z \in y \Rightarrow z \in \langle a, v\rangle)$ and $\langle a, v\rangle \subset y \equiv \forall z(z \in \langle a, v\rangle \Rightarrow z \in y)$. Therefore, $\zeta \vdash y = \langle a, v\rangle$ (see [*Kolmogorov and Dragalin*, 1982, III, 2]). By the deduction theorem, $\zeta \Rightarrow \eta$. Thus, we get $\eta \Leftrightarrow \zeta$.

A2.7. (The *first axiom of specification of a local permutation.*)

$$\forall U(U\bowtie \Rightarrow \forall A\exists Y\forall y(y \in Y \Leftrightarrow y \in U \wedge \exists u, v(u, v \in U \wedge y = \langle u, v\rangle \wedge \langle v, u\rangle \in A))).$$

A2.8. (The *second axiom of specification of a local permutation.*)

$$\forall U(U\bowtie \Rightarrow \forall A\exists Y\forall y(y \in Y \Leftrightarrow$$
$$\Leftrightarrow y \in U \wedge \exists u, v, w(u, v, w \in U \wedge y = \langle u, v, w\rangle \wedge \langle v, w, u\rangle \in A))).$$

As above, we can check that the subformula $y = \langle u, v, w\rangle$ in axioms A2.8 and A2.9 is U-predicative.

A2.9. (The *third axiom of specification of a local permutation.*)

$$\forall U(U\bowtie \Rightarrow \forall A\exists Y\forall y(y \in Y \Leftrightarrow$$
$$\Leftrightarrow y \in U \wedge \exists u, v, w(u, v, w \in U \wedge y = \langle u, v, w\rangle \wedge \langle u, w, v\rangle \in A))).$$

A2.10. (The *axiom of specification of local membership.*)

$$\forall U(U\bowtie \Rightarrow \exists Y\forall y(y \in Y \Leftrightarrow y \in U \wedge \exists u, v(u, v \in U \wedge y = \langle u, v\rangle \wedge u \in v))).$$

A2.11. (The *axiom of specification of a local complement.*)

$$\forall U(U\bowtie \Rightarrow \forall A\exists Y\forall y(y \in Y \Leftrightarrow y \in U \wedge y \notin A)).$$

A2.12. (The *axiom of specification of a local binary union.*)

$$\forall A\forall B\forall U(U\bowtie \Rightarrow \exists Y\forall y(y \in Y \Leftrightarrow y \in U \wedge (y \in A \vee y \in B))).$$

It follows from this axiom that in LTS* for any A, B, and $U\bowtie$, there exists a U-class $A \cup_U B$.

A2.13. (The *axiom of specification of a local domain of definition.*)

$$\forall A\forall U(U\bowtie \Rightarrow \exists Y\forall y(y \in Y \Leftrightarrow y \in U \wedge \exists z(z \in U \wedge \langle y, z\rangle \in A))).$$

Unfortunately, unlike the proof of the finite axiomatizability theorem for the NBG set theory given in [1997], the condition of X-predicativity of the formula φ in axiom scheme AS2 prevents the reduction of the elementary subformulas $p_k \in p_l$ and $p_k\bowtie$ for $k, l \in n + 1 \setminus 1$ of the formula $\varphi(x_1, \dots, x_m; p_1, \dots, p_n)$ in the proof of Proposition 1 below to the equivalent formulas

$$\exists x_j(\forall x_i(x_i \in x_j \Leftrightarrow x_i \in p_k) \wedge x_j \in p_l) \text{ and } \exists x_j(\forall x_i(x_i \in x_j \Leftrightarrow x_i \in p_k) \wedge x_j\bowtie)$$

because these equivalent formulas are not X-predicative. Thus, we are forced to introduce two more axioms, which look rather unusual for the reader familiar with the NBG set theory.

A2.14. (The *axiom of specification of external universal elements.*)

$$\forall A \forall U(U\bowtie \Rightarrow \exists Y \forall y(y \in Y \Leftrightarrow y \in U \wedge A\bowtie)).$$

A2.15. (The *axiom of specification of external membership.*)

$$\forall A \forall B \forall U(U\bowtie \Rightarrow \exists Y \forall y(y \in Y \Leftrightarrow y \in U \wedge A \in B)).$$

The theory determined by axioms A1, A2.1 – A2.15, and A3 – A14 will be called the *finite axiomatizable local theory of sets* and will be denoted by LTS^f.

B.7.2 The deductive equivalence of the theories LTS and LTS^f

Lemma 1. *The formula*

$$\forall A \forall B \forall U(U\bowtie \Rightarrow \exists Y \forall y(y \in Y \Leftrightarrow y \in U \wedge y \in A \wedge y \in B))$$

is deducible in the LTS.*

Proof. By axiom A2.11, there exist Y' and Y'' such that $\forall y(y \in Y' \Leftrightarrow y \in U \wedge y \notin A)$ and $\forall y(y \in Y'' \Leftrightarrow y \in U \wedge y \notin B)$. By A2.1, there is Y''' such that $\forall y(y \in Y''' \Leftrightarrow y \in U \wedge (y \in Y' \vee y \in Y''))$. By axiom A2.11, again, there is Y such that $\forall y(y \in Y \Leftrightarrow y \in U \wedge y \notin Y''')$.

Suppose $y \in Y$; then $y \in U \wedge y \notin Y'''$. Hence, $y \notin Y' \wedge y \notin Y''$, and therefore, $y \in A \wedge y \in B$. By the deduction theorem the formula $y \in Y \Rightarrow y \in U \wedge y \in A \wedge y \in B$ is deducible in the LTS*.

Conversely, suppose $y \in U \wedge y \in A \wedge y \in B$; then $y \notin Y'$ and $y \notin Y''$. Hence, $y \notin Y'''$, and therefore, $y \in Y$. By the deduction theorem, the formula $y \in U \wedge y \in A \wedge y \in B \Rightarrow y \in Y$ is deducible. Further, by logical tools, we deduce the required formula. □

This lemma means that in the LTS* for any A, B, and $U\bowtie$ the class $A \cap_U B$ exists.

Prove that in the LTS* for $U\bowtie$, there exists a U-class determined by the elementary formula $y \in y$.

Lemma 2. *The formula $\forall X(X \notin X)$ is deducible in the LTS*.*

Proof. By the universality axiom A6, $X \in U$ for some universal class U. By axiom A2.1, there is the class $\{X\} \equiv \{X, X\} \subset U$ such that $X \in \{X\}_U$. By A3, $\{X\} \neq \varnothing$. Then, the regularity axiom A12 guarantees that there exists $x \in \{X\}$ such that $x \cap_U \{X\} = \varnothing$. By A2.1, $x \in U \wedge x = X$. Suppose that $X \in X$. Since $X \in U$, Lemma 1 (B.7.1) implies $X \in X \cap_U \{X\} = \varnothing$. But this contradicts axiom A3. Therefore, $X \notin X$. □

Corollary 1. *The formula $\forall U \forall y(y \in \varnothing \Leftrightarrow y \in U \wedge y \in y)$ is deducible in the LTS*.*

Proof. Let $y \in \varnothing$. By empty class axiom A3, $y \in \varnothing \wedge y \notin \varnothing$. By the law of non-contradiction, $\neg(y \in \varnothing \wedge y \notin \varnothing)$. By the falsehood elimination rule, $y \in U \wedge y \in y$. According to the deduction theorem, the formula $y \in \varnothing \Rightarrow y \in U \wedge y \in y$ is deduced in the LTS*. Conversely, let $y \in U \wedge y \in y$. By Lemma 2, $y \in y \wedge y \notin y$. As above, we obtain the formula $y \in U \wedge y \in y \Rightarrow y \in \varnothing$. The required formula is deduced by the generalization rule. □

Corollary 2. *The formula $\forall U(U\bowtie \Rightarrow \exists Y \forall y(y \in Y \Leftrightarrow y \in U \wedge y \in y)$ is deducible in the LTS*.*

Proof. Verify that the empty class $\varnothing$ can be taken as Y. Let $y \in \varnothing$. By the empty class axiom, we get $y \notin \varnothing$. Then, the falsehood elimination rule yields $y \in U \wedge y \in y$. Conversely, let $y \in U \wedge y \in y$. Corollary 1 implies $y \in \varnothing$. □

Corollary 3. *The formula $\forall z \forall U \forall y(y \in \varnothing \Leftrightarrow y \in U \wedge z \in z))$ is deducible in the LTS*.*

Proof. The proof is quite similar to that of Corollary 1. □

Corollary 4. *The formula $\forall z \forall U(U\bowtie \Rightarrow \exists Y \forall y(y \in Y \Leftrightarrow y \in U \wedge z \in z))$ is deducible in the LTS*.*

Lemma 3. *The formula*

$$\forall U(U\bowtie \Rightarrow \forall A \forall B \exists Y \forall y(y \in Y \Leftrightarrow y \in U \wedge \exists a, b(a, b \in U \wedge a \in A \wedge b \in B \wedge y = \langle a, b \rangle)))$$

is deducible in the LTS.*

Proof. By axiom A2.6, there are classes Y' and Z such that

$$\xi \equiv \forall y(y \in Y' \Leftrightarrow y \in U \wedge \exists a, w(a, w \in U \wedge a \in A \wedge y = \langle a, w \rangle)) \text{ and}$$
$$\zeta \equiv \forall y(y \in Z \Leftrightarrow y \in U \wedge \exists b, w(b, w \in U \wedge a \in A \wedge y = \langle b, w \rangle)).$$

By axiom A2.7, there is a class Y'' such that $\eta \equiv \forall y(y \in Y'' \Leftrightarrow y \in U \wedge \exists u, v(u, v \in U \wedge y = \langle u, v \rangle \wedge \langle v, u \rangle \in Z))$.

Suppose $y \in Y''$; then the formula η implies $y \in U \wedge \exists u, v(u, v \in U \wedge y = \langle u, v \rangle \wedge \langle v, u \rangle \in Z))$. According to the formula ζ, we get $\exists b, w(b, w \in U \wedge b \in D \wedge \langle v, u \rangle = \langle b, w \rangle)$. By Lemma 1 (B.7.1), $v = b$ and $u = w$. Hence, $y = \langle u, v \rangle = \langle u, b \rangle$. Thus, we obtain the formula $\theta \equiv y \in U \wedge \exists u, b(u, b \in U \wedge b \in B \wedge y = \langle u, b \rangle)$. By the deduction theorem, we deduce the formula $y \in Y'' \Rightarrow \theta$.

Conversely, suppose θ; then $y \in U \wedge y = \langle u, b \rangle_U$ for some $u, b \in U$ such that $b \in B$. Consider $z = \langle b, u \rangle$. Since $u, b \in U$, axiom A2.2 provides $z \in U$. According to the formula ζ, we get $z \in Z$. Thus, $y \in U \wedge \exists u, b(u, b \in U \wedge y = \langle u, b \rangle \wedge \langle b, u \rangle \in Z)$. Applying the formula η, we conclude that $y \in Y''$. By the deduction theorem, the formula $\theta \Rightarrow y \in Y''$ is deducible in the LTS*. Consequently, the formula $\xi \equiv \forall y(y \in Y'' \Leftrightarrow \theta)$ is deducible as well.

According to Lemma 1, there exists the class $Y \equiv Y' \cap_U Y''$; for this class, we have $\forall y(y \in Y \Leftrightarrow y \in U \wedge y \in Y' \wedge y \in Y'')$. The formula ξ implies $\exists a, w(a, w \in U \wedge a \in A \wedge y = \langle a, w \rangle)$. The formula χ implies $\exists u, b(u, b \in U \wedge b \in B \wedge y = \langle u, b \rangle)$. By Lemma 1 (B.7.1), $u = a$ and $w = b$. Hence, $\exists a, b(a, b \in U \wedge a \in A \wedge b \in B \wedge y = \langle a, b \rangle_U)$, from where we obtain the required assertion. □

Corollary 1. *The formula* $\forall U(U\bowtie \Rightarrow \forall A \forall B \exists Y \forall u, v(\langle u, v \rangle \in Y \Leftrightarrow \langle u, v \rangle \in U \wedge u \in A \wedge v \in B))$ *is deducible in the LTS*.*

Proof. Consider the class Y from Lemma 3. Let u and v be arbitrary classes. Let $y \equiv \langle u, v \rangle \in Y$. Then, Lemma 3 implies $y \in U \wedge \exists a, b \in U(a \in A \wedge b \in B \wedge y = \langle a, b \rangle)$. Therefore, $u = a \in A$ and $v = b \in B$, where $u, v \in U$. By the deduction theorem, the formula $\langle u, v \rangle \in Y \Rightarrow u, v \in U \wedge \langle u, v \rangle \in U \wedge u \in A \wedge v \in B$ is deducible in LTS*.

Conversely, let $y \equiv \langle u, v \rangle \in Y \wedge u \in A \wedge v \in B \Rightarrow \langle u, v \rangle \in Y$. By Lemma 3, $y \in Y$. By the deduction theorem, we deduce the formula $u, v \in U \wedge \langle u, v \rangle \in U \wedge u \in A \wedge v \in B \Rightarrow \langle u, v \rangle \in Y$. Further, by logical tools, we deduce the required formula. □

Corollary 2. *Let* $n \geqslant 2$. *Then, the formula*

$$\forall A_1 \ldots \forall A_n \forall U(U\bowtie \rightarrow \exists Y \forall u_1 \ldots \forall u_n(\langle u_1, \ldots, u_n \rangle \in Y \Leftrightarrow$$
$$\Leftrightarrow \langle u_1, \ldots, u_n \rangle \in U \wedge u_1 \in A_1 \wedge \ldots \wedge u_n \in A_n))$$

is deducible in the LTS.*

Proof. The proof is by induction on n. Suppose that the assertion holds for n. Fix $A_1, \dots, A_{n+1}$ and $U\bowtie$. By the inductive hypothesis, there is a class Z such that $\zeta \equiv \forall z_1 \cdots \forall z_n(\langle z_1, \dots, z_n\rangle \in Z \Leftrightarrow \langle z_1, \dots, z_n\rangle \in U \wedge z_1 \in A_1 \wedge \dots \wedge z_n \in A_n)$. For the classes Z and A_{n+1}, consider the class Y from Corollary 1.

Let $\langle u_1, \dots, u_{n+1}\rangle \equiv \langle z, v\rangle \in Y$, where $z \equiv \langle u_1, \dots, u_n\rangle$ and $v \equiv u$. By Corollary 1, $z, v \in U \wedge \langle z, v\rangle \in U \wedge z \in Z \wedge v \in A_{n+1}$. By the formula ζ, $u_1, \dots, u_n \in U \wedge z \in U \wedge u_1 \in A_1 \wedge \dots \wedge u_n \in A_n$. Then, by the deduction theorem, we deduce the formula $\langle u_1, \dots, u_{n+1}\rangle \in Y \Rightarrow u_1, \dots, u_{n+1} \in U \wedge \langle u_1, \dots, u_{n+1}\rangle \in U \wedge u_1 \in A_1 \wedge \dots \wedge u_{n+1} \in A_{n+1}$. Denote the condition and the conclusion in this formula by η and θ, respectively. Thus, we have deduced the formula $\eta \Rightarrow \theta$.

Conversely, let θ. By Lemma 3 (B.3.2), $z \equiv \langle u_1, \dots, u_{n+1}\rangle_U \in U$. Then, using the formula ζ, we get $z \in Z$. Since $z, u_{n+1} \in U \wedge \langle z, u_{n+1}\rangle_U \in U \wedge z \in Z \wedge u_{n+1} \in A_{n+1}$, by Corollary 1, we get $\langle z, u_{n+1}\rangle_U \in Y$, i. e. the formula η is deduced. By the deduction theorem, we deduce the formula $\theta \Rightarrow \eta$. Further, by logical tools, we deduce the required formula for $n + 1$. □

The proofs of the following three lemmas are similar to that of Corollary 1 to Lemma 3.

Lemma 4. *In the LTS*, axiom A2.8 is equivalent to the formula* $\forall A \forall U(U\bowtie \Rightarrow \exists Y \forall u, v, w(\langle u, v, w\rangle \in Y \Leftrightarrow \langle u, v, w\rangle \in U \wedge \langle v, w, u\rangle \in A))$.

Lemma 5. *In the LTS*, axiom A2.9 is equivalent to the formula* $\forall A \forall U(U\bowtie \Rightarrow \exists Y \forall u, v, w(\langle u, v, w\rangle \in Y \Leftrightarrow \langle u, v, w\rangle \in U \wedge \langle u, w, v\rangle \in A))$.

Lemma 6. *In the LTS*, axiom A2.10 is equivalent to the formula* $\forall U(U\bowtie \Rightarrow \exists Y \forall u, v(\langle u, v\rangle \in Y \Leftrightarrow \langle u, v\rangle \in U \wedge u \in V))$.

We prove the following assertion by a method similar to that used in [1997] to prove Proposition 4.4 but with a number of substantial changes related to the U-predicativity of the formula φ.

Proposition 1. *Let* $\varphi[\vec{x}, \vec{p}]$ *be a U-predicative formula such that the substitution* $\varphi[\vec{x} \parallel \vec{y}, \vec{p}]$ *is admissible. Then, the formula* $\forall U(U\bowtie \Rightarrow \exists Y \forall \vec{y}(\langle \vec{y}\rangle \in Y \Leftrightarrow \langle \vec{y}\rangle \in U \wedge \varphi[\vec{y}, \vec{p}]))$ *is deducible in the LTS*.*

Proof. We shall prove by induction on the number $s \in \omega$ of logical connectives and quantifiers in the formula φ. Denote by $A(s)$ the following assertion:

for every $m \geqslant 1$ *and* $n \geqslant 0$, *for every variable* Y, *for every formula of the LTS* having the form* $\varphi(x_1, \dots, x_m; p_1, \dots, p_n)$ *such that:*

1. φ *is composed of variables, the predicate symbols* $\in$ *and* $\bowtie$, *and the logical symbols* $\exists$, $\neg$, *and* $\vee$ *;*
2. φ *contains precisely s logical symbols* $\exists$, $\neg$, *and* $\vee$ *;*
3. φ *is U-predicative;*

and for every symbol-string $\vec{y} \equiv y_1, \ldots, y_m$ *such that the substitution* $\varphi[\vec{x} \parallel \vec{y}, \vec{p}]$ *is admissible; the formula* $\forall U(U\bowtie \Rightarrow \exists Y \forall \vec{y}(\langle\vec{y}\rangle \in Y \Leftrightarrow \langle\vec{y}\rangle \in U \wedge \varphi(\vec{y};\vec{p})))$ *is deducible in LTS*.*

We shall prove this assertion, using the complete induction principle $A(0) \wedge \forall s (\forall r < s(A(r)) \Rightarrow A(s)) \Rightarrow \forall s(A(s))$.

Let $s = 0$. Then, the formula φ has one of the following forms: $x_i\bowtie$, $p_k\bowtie$, $x_i \in p_k$, $p_k \in x_i$, $p_k \in p_l$, or $x_i \in x_j$ for $i, j \in (m+1) \setminus 1$ and $k, l \in n \setminus (n+1) \setminus 1$.

If $x_i\bowtie$, then by axiom A2.3, there is a class Y such that $\forall y_i(y_i \in Y \Leftrightarrow y_i \in U \wedge y_i\bowtie)$.

If $x_i \in p_k$, then by axiom A2.4, there is a class Y such that $\forall y_i(y_i \in Y \Leftrightarrow y_i \in U \wedge y_i \in p_k)$.

If $p_k \in x_i$, then by axiom A2.5, there is a class Y such that $\forall y_i(y_i \in Y \Leftrightarrow y_i \in U \wedge p_k \in y_i)$.

If $x_i \in x_i$, then by Corollary 2 to Lemma 2, there is a class Y such that $\forall y_i(y_i \in Y \Leftrightarrow y_i \in U \wedge y_i \in y_i)$.

If $p_k\bowtie$, then by axiom A2.14, there is a class Y such that $\forall y_i(y_i \in Y \Leftrightarrow y_i \in U \wedge p_k\bowtie)$.

If $p_k \in p_l$ for $k \neq l$, then by axiom A2.15, there is a class Y such that $\forall y_i(y_i \in Y \Leftrightarrow y_i \in U \wedge p_k \in p_l)$.

Finally, if $p_k \in p_k$, then by Corollary 2 to Lemma 2, there is a class Y such that $\forall y_i(y_i \in Y \Leftrightarrow y_i \in U \wedge p_k \in p_k)$.

By Corollary 2 to Lemma 3 for $A_1 = U, \ldots, A_{i-1} = U, A_i = Y, A_{i+1} = U, \ldots, A_m = U$, there is Z such that $\forall \vec{y}\ (\langle\vec{y}\rangle \in Z \Leftrightarrow \langle\vec{y}\rangle \in U \wedge y_i \in Y)$. From the formulas deduced above, we conclude that $\forall \vec{y}(\langle\vec{y}\rangle \in Z \Leftrightarrow \langle\vec{y}\rangle \in U \wedge \varphi[\vec{y}, \vec{p}])$. This means that the assertion is proven in the cases considered above.

It remains to consider the case $x_i \in x_j$ for $i \neq j$. It can be assumed that $i < j$. By axiom A2.10, there is a class Y such that $\zeta \equiv \forall u(u \in Y \Leftrightarrow u \in U \wedge \exists y_i, y_j(y_i, y_j \in U \wedge u = \langle y_i, y_j\rangle_U \wedge y_i \in y_j))$.

If $i = 1$, then put $X_i = Y$.

Let $i > 1$. By Corollary 1 to Lemma 3 for $A = Y$ and $B = U$, there is a class X_1 such that $\eta \equiv \forall u, y_1(\langle u, y_1\rangle_U \in X_1 \Leftrightarrow u, y_1 \in U \wedge \langle u, y_1\rangle_U \in U \wedge u \in Y)$. Let $\langle y_i, y_j, y_1\rangle_U \in X_1$. Then, it follows from η and ζ that $\langle y_i, y_j\rangle_U, y_1 \in U \wedge \langle y_i, y_j, y_1\rangle_U \in U \wedge y_i, y_j \in U \wedge y_i \in y_j$. Hence, we get the formula $\theta \equiv y_i, y_j, y_1 \in U \wedge \langle y_i, y_j, y_1\rangle_U \in U \wedge y_i \in y_j$. Conversely, suppose θ. By axiom A2.2, $u \equiv \langle y_i, y_j\rangle_U \in U$. Then, ζ implies $u \in Y$. Further, η implies $\langle y_i, y_j, y_1\rangle_U \in X_1$. By the deduction theorem, the formula $\langle y_i, y_j, y_1\rangle_U \in X_1 \Leftrightarrow \theta$ is deduced in LTS*. By the generalization rule, $\xi \equiv \forall y_i, y_j, y_1(\langle y_i, y_j, y_1\rangle_U \in X_1 \Leftrightarrow \theta)$.

By Lemma 4, for $A \equiv X_1$, there is a class X_2 such that

$$\chi \equiv \forall y_1, y_i, y_j(\langle y_1, y_i, y_j\rangle_U \in X_2 \Leftrightarrow$$
$$\Leftrightarrow y_1, y_i, y_j \in U \wedge \langle y_1, y_i, y_j\rangle_U \in U \wedge \langle y_i, y_j, y_1\rangle_U \in X_1).$$

We deduce from the formulas ξ and χ the formula $\forall y_1, y_i, y_j(\langle y_1, y_i, y_j\rangle_U \in X_2 \Leftrightarrow y_1, y_i, y_j \in U \wedge \langle y_1, y_i, y_j\rangle_U \in U \wedge y_i \in y_j)$. If $i = 2$, then put $X_i \equiv X_2$. If $i > 2$, then it is proven in a similar way, by induction, that in the LTS* there is a class X_i such that $\gamma \equiv \forall y_1, \dots, y_i, y_j(\langle y_1, \dots, y_i, y_j\rangle_U \in X_i \Leftrightarrow y_1, \dots, y_i, y_j \in U \wedge \langle y_1, \dots, y_i, y_j\rangle_U \in U \wedge y_i \in y_j)$.

If $j = i + 1$, then put $Y_{ij} \equiv X_i$. Let $i + 1 < j$. By Corollary 1 to Lemma 3 for $A = X_i$ and $B = U$, there is a class X_{i1} such that $\delta \equiv \forall w, y_{i+1}(\langle w, y_{i+1}\rangle_U \in X_{i1} \Leftrightarrow w, y_{i+1} \in U \wedge \langle w, y_{i+1}\rangle_U \in U \wedge w \in X_i)$.

Denote $\langle y_1, \dots, y_i\rangle_U$ by u. Let $\langle u, y_j, y_{i+1}\rangle_U \in X_{i1}$. Then, δ for $w \equiv \langle u, y_j\rangle_U$ implies $\langle u, y_j\rangle, y_{i+1} \in U \wedge \langle u, y_j, y_{i+1}\rangle_U \in U \wedge \langle u, y_j\rangle_U \in U$. Now, γ implies $\langle u, y_j\rangle, y_{i+1} \in U \wedge \langle u, y_j, y_{i+1}\rangle_U \in U \wedge y_1, \dots, y_i, y_j \in U \wedge y_i \in y_j$. Hence, $\varkappa \equiv u, y_j, y_{i+1} \in U \wedge \langle u, y_j, y_{i+1}\rangle_U \in U \wedge y_i \in y_j$. Conversely, let $\varkappa$. By axiom A2.2, $w \equiv \langle u, y_j\rangle_U \in U$. Therefore, it follows from γ that $w \in X_i$. Further, δ implies $\langle u, y_j, y_{i+1}\rangle_U \in X_{i1}$. By the deduction theorem, the formula $\langle u, y_j, y_{i+1}\rangle_U \in X_{i1} \Leftrightarrow \varkappa$ is deduced in the LTS*. By the generalization rule, we deduce the formula $\varepsilon \equiv \forall u, y_j, y_{i+1}(\langle u, y_j, y_{i+1}\rangle_U \in X_{i1} \Leftrightarrow \varkappa)$.

By Lemma 5 for $A \equiv X_{i1}$, there is a class X_{i2} such that

$$\pi \equiv \forall u, y_{i+1}, y_j(\langle u, y_{i+1}, y_j\rangle_U \in X_{i2} \Leftrightarrow \\ \Leftrightarrow u, y_{i+1}, y_j \in U \wedge \langle u, y_{i+1}, y_j\rangle_U \in U \wedge \langle u, y_j, y_{i+1}\rangle_U \in X_{i1}).$$

The formulas ε and π imply $\forall u, y_{i+1}, y_j(\langle u, y_{i+1}, y_j\rangle_U \in X_{i2} \Leftrightarrow u, y_{i+1}, y_j \in U \wedge \langle u, y_{i+1}, y_j\rangle_U \in U \wedge y_i \in y_j)$. This implies the formula

$$\forall y_1, \dots, y_{i+1}, y_j(\langle y_1, \dots, y_{i+1}, y_j\rangle_U \in X_{i2} \Leftrightarrow \\ \Leftrightarrow y_1, \dots, y_{i+1}, y_j \in U \wedge \langle y_1, \dots, y_{i+1}, y_j\rangle_U \in U \wedge y_i \in y_j).$$

If $j = i + 2$, then put $Y_{ij} \equiv X_{i2}$. If $j > i + 2$, then it is proven in a similar way, by induction, that in the LTS*, there is a class Y_{ij} such that $\forall y_1, \dots, y_j(\langle y_1, \dots, y_j\rangle_U \in Y_{ij} \Leftrightarrow y_1, \dots, y_j \in U \wedge \langle y_1, \dots, y_j\rangle_U \in U \wedge y_i \in y_j)$.

Further, starting at Y_{ij} and using the induction, we prove that there is a class Z, such that $\forall y_1, \dots, y_m(\langle y_1, \dots y_m\rangle_U \in Z \Leftrightarrow y_1, \dots, y_m \in U \wedge \langle y_1, \dots, y_m\rangle_U \in U \wedge y_i \in y_j)$. This means that $\forall \vec{y}(\langle \vec{y}\rangle_U \in Z \Leftrightarrow \langle \vec{y}\rangle_U \in U \wedge \varphi[\vec{y}, \vec{p}])$. Thus, we have $A(0)$.

Suppose that the required assertion is proven for any $r < s$ and $\varphi[\vec{x}, \vec{p}]$ contains s logical connectivities and quantifiers.

Let φ be $\neg\psi$. By inductive hypothesis, there is a class Z such that $\zeta \equiv \forall \vec{y}(\langle \vec{y}\rangle_U \in Z \Leftrightarrow \langle \vec{y}\rangle \in U \wedge \psi[\vec{y}, \vec{p}])$.

Consider for the classes Z and U the class Y from axiom A2.11 such that $\chi \equiv \forall y(y \in Y \Leftrightarrow y \in U \wedge y \notin Z)$. Let $\langle \vec{y}\rangle_U \in Y$. Then, χ implies $\langle \vec{y}\rangle \in U \wedge \langle \vec{y}\rangle_U \notin Z$. But it deduced from ζ that $\langle \vec{y}\rangle \notin Z \Rightarrow (\langle \vec{y}\rangle \notin U \vee \neg\psi[\vec{y}, \vec{p}])$. Therefore, the formula $\langle \vec{y}\rangle \in U \wedge (\langle \vec{y}\rangle_U \notin U \vee \neg\psi[\vec{y}, \vec{p}])$ is deduced. It is equivalent to the formula $(\langle \vec{y}\rangle \in U \wedge \langle \vec{y}\rangle \notin U) \vee (\langle \vec{y}\rangle \in U \wedge \neg\psi[\vec{y}, \vec{p}])$. By the falsehood elimination rule, we get $\langle \vec{y}\rangle \in U \wedge \neg\psi[\vec{y}, \vec{p}]$. By the deduction theorem, the formula $\langle \vec{y}\rangle_U \in Y \Rightarrow \langle \vec{y}\rangle_U \in U \wedge \varphi[\vec{y}, \vec{p}]$ is deduced in the propositional calculus.

Let $\langle \vec{y}\rangle_U \in U \wedge \varphi[\vec{y}, \vec{p}]$. Then, $(\langle \vec{y}\rangle_U \in U \wedge \neg\psi[\vec{y}, \vec{p}]) \vee (\langle \vec{y}\rangle \in U \wedge \langle \vec{y}\rangle_U \notin U)$. It is equivalent to the formula $\langle \vec{y}\rangle_U \in U \wedge (\neg\psi[\vec{y}, \vec{p}] \vee \langle \vec{y}\rangle_U \notin U)$. The last formula, together with the formula ζ, implies the formula $\langle\vec{y}\rangle_U \in U \wedge \langle\vec{y}\rangle_U \notin Z$. Using χ, we obtain $\langle\vec{y}\rangle_U \in Y$. By the deduction theorem, the formula $\langle\vec{y}\rangle_U \in U \wedge \varphi[\vec{y},\vec{p}] \Rightarrow \langle\vec{y}\rangle_U \in Y$ is deduced in the propositional calculus.

Applying the generalization rule, we deduce the formula $\forall\vec{y}(\langle\vec{y}\rangle_U \in U \wedge \varphi[\vec{y},\vec{p}] \Leftrightarrow \langle\vec{y}\rangle_U \in Y)$.

Let φ be $\psi \Rightarrow \theta$, which is equivalent to the formula $\neg\psi \vee \theta$. By inductive hypothesis, there are classes Z_1 and Y_2 such that $\zeta \equiv \forall\vec{y}(\langle\vec{y}\rangle_U \in Z_1 \Leftrightarrow \langle\vec{y}\rangle_U \in U \wedge \psi[\vec{y},\vec{p}])$ and $\eta \equiv \forall\vec{y}(\langle\vec{y}\rangle_U \in Y_2 \Leftrightarrow \langle\vec{y}\rangle_U \in U \wedge \theta[\vec{y},\vec{p}])$. By A2.11, for Z_1, there is a class Y_1 such that $\xi \equiv \forall\vec{y}(\langle\vec{y}\rangle_U \in Y_1 \Leftrightarrow \langle\vec{y}\rangle_U \in U \wedge \langle\vec{y}\rangle_U \notin Z_1)$. By A2.12, for Y_1 and Y_2, there is Y such that $\chi \equiv \forall\vec{y}(\langle\vec{y}\rangle_U \in Y \Leftrightarrow \langle\vec{y}\rangle_U \in U \wedge (\langle\vec{y}\rangle_U \in Y_1 \vee \langle\vec{y}\rangle_U \in Y_2))$.

Let $\langle\vec{y}\rangle \in Y$. Then, it follows from χ, η, and ξ that $\langle\vec{y}\rangle_U \in U \wedge ((\langle\vec{y}\rangle_U \in U \wedge \langle\vec{y}\rangle_U \notin Z_1) \vee (\langle\vec{y}\rangle_U \in U \wedge \theta[\vec{y},\vec{p}])$, which is equivalent to the formula $\langle\vec{y}\rangle_U \in U \wedge (\langle\vec{y}\rangle_U \in U \vee \theta[\vec{y},\vec{p}])$. Using ζ, we get $\langle\vec{y}\rangle_U \in U \wedge ((\langle\vec{y}\rangle_U \notin U \vee \neg\psi) \vee \theta)$. It is equivalent to $((\langle\vec{y}\rangle_U \in U \wedge \langle\vec{y}\rangle_U \notin U) \vee (\langle\vec{y}\rangle_U \in U \wedge (\neg\psi \vee \theta))$. By the falsehood elimination rule, we obtain $\langle\vec{y}\rangle_U \in U \wedge (\neg\psi \vee \theta)$. By the deduction theorem, the formula $\langle\vec{y}\rangle_U \in Y \Rightarrow \langle\vec{y}\rangle_U \in U \wedge (\neg\psi \vee \theta)$ is deduced in the propositional calculus.

Conversely, let $\langle\vec{y}\rangle_U \in U \wedge (\neg\psi \vee \theta)$. Then, we get $((\langle\vec{y}\rangle_U \in U \wedge \langle\vec{y}\rangle_U \notin U) \vee ((\langle\vec{y}\rangle_U \in U \wedge \neg\psi) \vee ((\langle\vec{y}\rangle_U \in U \wedge \theta))$, where $\langle\vec{y}\rangle_U \in U \wedge ((\langle\vec{y}\rangle_U \notin U \vee \neg\psi \vee ((\langle\vec{y}\rangle_U \in U \wedge \theta))$, which is equivalent to $\langle\vec{y}\rangle_U \in U \wedge (\neg(\langle\vec{y}\rangle_U \in U \wedge \psi) \vee (\langle\vec{y}\rangle_U \in U \wedge \theta))$. Applying ζ and η, we obtain $\langle\vec{y}\rangle_U \in U \wedge (\langle\vec{y}\rangle_U \notin Z_1 \vee \langle\vec{y}\rangle_U \in Y_2)$, which is equivalent to $\langle\vec{y}\rangle_U \in U((\langle\vec{y}\rangle_U \in U \wedge \langle\vec{y}\rangle_U \in Z_1) \vee \langle\vec{y}\rangle_U \in Y_2)$. Applying ξ, we get $\langle\vec{y}\rangle_U \in U \wedge (\langle\vec{y}\rangle_U \in Y_1 \vee \langle\vec{y}\rangle_U \in Y_2)$. Applying χ, we get $\langle\vec{y}\rangle_U \in Y$. By the deduction theorem, the formula $\langle\vec{y}\rangle_U \in U \wedge (\psi \vee \neg\theta) \Rightarrow \langle\vec{y}\rangle_U \in Y$ is deduced in the propositional calculus. By the generalization rule, we deduce the formula $\forall\vec{y}(\langle\vec{y}\rangle_U \in Y \Leftrightarrow \langle\vec{y}\rangle_U \in U \wedge \varphi[\vec{y},\vec{p}])$.

Finally, let $\varphi[\vec{x},\vec{p}]$ be $\exists z(z \in U \wedge \psi[\vec{y},z,\vec{p}])$. By the inductive hypothesis, for the formula $\psi[\vec{y},z,\vec{p}]$, there is a class Z such that $\zeta \equiv \forall\vec{y}\forall z(\langle\vec{y},z\rangle_U \in Z \Leftrightarrow \langle\vec{y},z\rangle_U \in U \wedge \psi[\vec{y},z,\vec{p}])$.

By axiom A2.13, there is class Y such that $\chi \equiv \forall\vec{y}(\langle\vec{y}\rangle_U \in U \Leftrightarrow \langle\vec{y}\rangle_U \in U \wedge \exists z(z \in U \wedge \langle\vec{y},z\rangle_U \in Z))$.

Let $\langle\vec{y}\rangle_U \in Y$. Then, χ implies $\langle\vec{y}\rangle_U \in U \wedge \exists z(z \in U \wedge \langle\vec{y},z\rangle_U \in Z)$. Applying ζ, we obtain $\langle\vec{y}\rangle_U \in U \wedge \exists z(z \in U \wedge \langle\vec{y},z\rangle_U \in U \wedge \psi[\vec{y},z,\vec{p}])$, where $\langle\vec{y}\rangle_U \in U \wedge \exists z(z \in U \wedge \psi[\vec{y},z,\vec{p}])$. By the deduction theorem, the formula $\langle\vec{y}\rangle_U \in Y \Rightarrow \langle\vec{y}\rangle_U \in U \wedge \exists z(z \in U \wedge \psi[\vec{y},z,\vec{p}])$ is deduced in the propositional calculus.

Conversely, let $\langle\vec{y}\rangle_U \in U \wedge \exists z(z \in U \wedge \psi[\vec{y},z,\vec{p}])$. By axiom A2.2, $\langle\vec{y},z\rangle_U \equiv \langle\langle\vec{y}\rangle_U, z\rangle_U \in U$. Consequently, $\langle\vec{y}\rangle_U \in U \wedge \exists z(z \in U \wedge \langle\vec{y},z\rangle_U \in U \wedge \psi[\vec{y},z,\vec{p}])$. Applying ζ, we get $\langle\vec{y}\rangle_U \in U \wedge \exists z(z \in U \wedge \langle\vec{y},z\rangle_U \in Z)$. Using χ, we obtain $\langle\vec{y}\rangle_U \in Y$. By the deduction theorem, the formula $\langle\vec{y}\rangle_U \in Y \wedge \exists z(z \in U \wedge \psi[\vec{y},z,\vec{p}]) \Rightarrow \langle\vec{y}\rangle_U \in Y$ is deduced in the propositional calculus.

By the generalization rule, we deduce the formula $\forall\vec{y}(\langle\vec{y}\rangle_U \in Y \Leftrightarrow \langle\vec{y}\rangle_U \in U \wedge \varphi[\vec{y},\vec{p}])$.

Thus, the assertion is proven for s. □

Corollary 1. *Let $\varphi[x, \vec{p}]$ be a U-predicative formula such that the substitution $\varphi[x \parallel y, \vec{p}]$ is admissible. Then, the formula $\forall U(U\bowtie \Rightarrow \exists Y \forall y(y \in Y \Leftrightarrow y \in U \wedge \varphi[y, \vec{p}]))$ is deducible in the LTS*.*

Theorem 1. *Let $\varphi[x, \vec{p}]$ be an X-predicative formula such that the substitution $\varphi[x \parallel y, \vec{p}]$ is admissible. Then, the formula $\forall X \exists Y \forall y(y \in Y \Leftrightarrow y \in X \wedge \varphi[y, \vec{p}])$ is deducible in the LTS*.*

Proof. Instead of $\varphi[x, \vec{p}]$, we shall write simply $\varphi[x]$.

By the universality and transitivity axioms A6 and A7, there is a universal class U such that $X \subset U$. Check that the formula $\varphi[x]$ is equivalent to some U-predicative formula $\varphi'[x]$.

Suppose that φ contains the subformula $\xi \equiv \forall x \varepsilon$ or $\eta \equiv \exists x \varepsilon$. By assumption, $\xi = \forall x(x \in X \Rightarrow \gamma)$ and $\eta = \exists x(x \in X \wedge \delta)$. Consider the formulas $\xi' \equiv \forall x(x \in U \Rightarrow (x \in X \Rightarrow \gamma))$ and $\eta' \equiv \exists x(x \in U \wedge x \in X \wedge \delta)$. It is clear that the formula $\xi \Rightarrow \xi'$ is deducible in LTS*. Conversely, suppose that ξ' holds and $x \in X$. Since $X \subset U$, we get $x \in U$. It follows from ξ' that γ holds. Hence, by the deduction theorem, the formula $x \in X \Rightarrow \gamma$ is deduced in LTS*. By the generalization rule, the formula ξ is deduced. Again, by the deduction theorem we deduce $\xi' \Rightarrow \xi$. Hence,the equivalence $\xi \Leftrightarrow \xi'$ is deduced in the LTS*. Similarly, the equivalence $\eta \Leftrightarrow \eta'$ is deduced. These equivalences imply that the formula φ is equivalent to some U-predicative formula φ'.

Consider the U-predicative formula $x \in X \wedge \varphi'[x]$. By Corollary 1 to Proposition 1, there exists a class Y such that $\chi \equiv \forall y(y \in Y \Leftrightarrow y \in U \wedge y \in X \wedge \varphi'[y])$. It is clear that χ implies the formula $y \in Y \Rightarrow y \in X \wedge \varphi[y]$. Let $y \in X \wedge \varphi[y]$. Since $X \subset U$, using χ, we deduce $y \in Y$. By the deduction theorem, the formula $y \in X \wedge \varphi[y] \Rightarrow y \in Y$ is deduced in the LTS*. Hence, the formula χ implies the formula $\psi \equiv \forall y(y \in Y \Leftrightarrow y \in X \wedge \varphi[y])$. Finally, by logical tools we deduce the formula $\forall X \exists Y \psi$. □

Theorem 2 (The Zakharov theorem on the finite axiomatization of the LTS). *The theories LTS and LTS^f are deductively equivalent.*

Proof. It follows from Theorem 1 that the LTS is weaker than the LTS^f.

On the other hand, it is clear that axioms A2.3 – A2.15 are specific cases of axiom scheme AS2. Axiom A2.1 is deduced in the LTS by virtue of Corollary 3 to Theorem 1 (B.3.5). If U is a universal class and $A, B \in U$, then by Lemma 2 (B.1.1) $\{A, B\}_U \in U$. But by Lemma 3 (B.3.2) $\{A, B\} = \{A, B\}_U$. This means that axiom A2.2 is deduced in the LTS. Therefore, the LTS^f is weaker than the LTS. □

B.7.3 The finite axiomatization of the NBG set theory by P. Bernays

Here, we shortly consider the finite axiomatization of the NBG set theory discovered and proven by P. Bernays [1968].

Further, as in A.6.2, the formula $\exists X(A \in X)$ meaning that the class A is a set is denoted by $S(A)$. All proper axioms and axioms schemes of NBG was listed also in A.6.2.

Associate axiom scheme AS2 with explicit axioms A2.1 – A2.9 stated below.

A2.1. (The *unordered pair axiom.*)

$$\forall a, b(S(a) \wedge S(b) \Rightarrow \exists Y(S(Y) \wedge \forall y(S(y) \Rightarrow (y \in Y \Leftrightarrow (y = a \vee y = b))))).$$

A2.2. (The *specification axiom for a product.*)

$$\forall A \exists Y \forall y(S(y) \Rightarrow (y \in Y \Leftrightarrow \exists a, v(S(a) \wedge S(v) \wedge a \in A \wedge y = \langle a, v\rangle))).$$

A2.3. (The *first specification axiom for a permutation.*)

$$\forall A \exists Y \forall y(S(y) \Rightarrow (y \in Y \Leftrightarrow \exists u, v(S(u) \wedge S(v) \wedge y = \langle u, v\rangle \wedge \langle v, u\rangle \in A))).$$

A2.4. (The *second specification axiom for a permutation.*)

$$\forall A \exists Y \forall y(S(y) \Rightarrow (y \in Y \Leftrightarrow \exists u, v, w(S(u) \wedge S(v) \wedge S(w) \wedge y = \langle u, v, w\rangle \wedge \langle v, w, u\rangle \in A))).$$

A2.5. (The *third specification axiom for a permutation.*)

$$\forall A \exists Y \forall y(S(y) \Rightarrow (y \in Y \Leftrightarrow \exists u, v, w(S(u) \wedge S(v) \wedge S(w) \wedge y = \langle u, v, w\rangle \wedge \langle u, w, v\rangle \in A))).$$

A2.6. (The *specification axiom for membership.*)

$$\exists Y \forall y(S(y) \Rightarrow (y \in Y \Leftrightarrow \exists u, v(S(u) \wedge S(v) \wedge y = \langle u, v\rangle \wedge u \in v))).$$

A2.7. (The *specification axiom for a complement.*) $\forall A \exists Y \forall y(S(y) \Rightarrow (y \in Y \Leftrightarrow y \notin A))$.

A2.8. (The *specification axiom for binary intersection.*)

$$\forall A \forall B \exists Y \forall y(S(y) \Rightarrow (y \in Y \Leftrightarrow (y \in A \wedge y \in B))).$$

A2.9. (The *specification axiom for a domain of definition.*)

$$\forall A \exists Y \forall y(S(y) \Rightarrow (y \in Y \Leftrightarrow \exists z(S(z) \wedge \langle y, z\rangle \in A))).$$

The theory of classes and sets determined by axioms A1, A2.1 – A2.9 will be denoted by NBG*. Introduce in the NBG* some basic notions.

The set Y in axiom A2.1 is unique by A1. It is called the *unordered pair of the sets a and b* and is denoted by $\{a, b\}$. It has a and b as its only members.

If a is a set, then the *singleton* $\{a\}$ is the set $\{a, a\}$ consisting of exactly one element a.

The *ordered pair of a and b* is defined by Kuratovski's formula $\langle a, b\rangle \equiv \{\{a\}, \{a, b\}\}$. Following the proof of Proposition 2 (1.1.6) and using in it axiom A2.1 instead of axiom scheme AS2, one can prove that $\langle a, b\rangle = \langle a', b'\rangle$ implies $a = a'$ and $b = b'$.

The *ordered triple of a, b, and c* is defined as $\langle a, b, c\rangle \equiv \langle\langle a, b\rangle, c\rangle$. Similarly, the *ordered suits of sets $a_1, \dots, a_n$* can be defined by induction as $\langle a_1, \dots, a_n\rangle \equiv \langle\langle a_1, \dots, a_{n-1}\rangle, a_n\rangle$.

Lemma 1. *In the NBG*, the formula $\exists Y\forall y(S(y) \Rightarrow (y \in Y \Leftrightarrow \exists A(y \in A)))$ is deduced.*

Proof. Consider the class X from axiom A2.6. By axiom A2.9 for X, there is the class Y such that $\xi \equiv \forall y(S(y) \Rightarrow (y \in Y \Leftrightarrow \exists z(S(z) \wedge \langle y, z\rangle \in X)))$. Then, for every $y \in Y$, we have $\langle y, z\rangle \in X$ and $\langle y, z\rangle = \langle u, v\rangle$ for some sets u and v. This implies $y = u$ and $z = v$, and therefore, $y \in z$. By the deduction theorem, we deduce the formula $y \in Y \Rightarrow \exists A(y \in A)$.

Conversely, let $y \in A$ for some class A. Then, $\{y\}$ is a set and $y \in \{y\}$. Denote $\{y\}$ by z. Consider the set $x \equiv \langle y, z\rangle$. Then, using A2.6, we infer that $x \in X$. By the deduction theorem, we deduced the formula $\exists A(y \in A) \Rightarrow y \in Y$.

Thus, we have the equivalence $y \in Y \Leftrightarrow \exists A(y \in A)$ and the formula $\eta \equiv S(y) \Rightarrow (y \in Y \Leftrightarrow \exists A(y \in A))$. Applying the generalization rule, we get the formula $\forall y(\eta)$. As a result, we deduce the necessary formula $\exists Y\forall y(\eta)$. □

The class Y in this lemma is unique by A1. It is called the *universe of all sets* and is denoted by $\mathbb{U}$.

Lemma 2. *In the NBG*, the formula $\exists Z\forall z(S(z) \Rightarrow (z \in Z \Leftrightarrow \forall A(z \notin A)))$ is deduced.*

Proof. According to axiom A2.7, for the class $\mathbb{U}$, there is a class Z such that $\zeta \equiv \forall z(S(z) \Rightarrow (z \in Z \Leftrightarrow z \notin \mathbb{U}))$.

Let $S(z)$ and $z \in Z$. Then, $z \notin \mathbb{U}$ implies $\forall A(z \notin A)$. By the deduction theorem, we deduce the formula $z \in Z \Rightarrow \forall A(z \notin A)$.

Conversely, let $S(z)$ and $\forall A(z \notin A)$. Then, $z \notin \mathbb{U}$. Hence, by ζ we infer that $z \in Z$. By the deduction theorem, we obtain the formula $\forall A(z \notin A) \Leftrightarrow z \in Z$.

Thus, we have the equivalence $z \in Z \Leftrightarrow \forall A(z \notin A)$ and the formula $\theta \equiv S(z) \Rightarrow (z \in Z \Leftrightarrow \forall A(z \notin A))$. Applying the generalization rule, we get the formula $\forall z(\theta)$. As a result, we deduce the necessary formula $\exists Z\forall z(\eta)$. □

The class Z in this lemma is unique by A1. It is called the *empty class* and is denoted by $\varnothing$.

Thus, axioms A1, A2.1 – A2.9 gave the opportunity to prove the existence of the empty class $\varnothing$. Unfortunately, they do not give the opportunity to prove that $\varnothing$ is a set. Therefore, we need some other axioms, in particular, the infinity axiom (A7), to assert that $\varnothing$ is a set. These detailed reasoning about the empty set $\varnothing$ allows us to avoid using the explicit *empty set axiom* presented in [1997, 4.1].

But explicit axioms A1, A2.1 – A2.9 allow us to eliminate axiom scheme AS2.

For $m \geqslant 1$ and $n \geqslant 0$, consider the list of variables $\vec{x}, \vec{p} \equiv x_1, \ldots, x_m, p_1, \ldots, p_n$. If φ is a formula whose variables occur among the symbol-string $\vec{x}, \vec{p}$ only, then we shall write $\varphi[\vec{x}, \vec{p}]$.

The following proposition was proven by P. Bernays.

Proposition 1. *Let $\varphi[\vec{x}, \vec{p}]$ be a predicative formula such that the substitution $\varphi[\vec{x} \parallel \vec{u}, \vec{p}]$ is admissible. Then, the formula $\exists Y \forall y (S(y) \Rightarrow (y \in Y \Leftrightarrow \exists u_1, \ldots, u_m (S(u_1) \wedge \ldots \wedge S(u_m) \wedge y = \langle u_1, \ldots, u_m \rangle \wedge \varphi[\vec{u}, \vec{p}])))$ is deducible in the NBG*.*

The proof of this remarkable result may be carried out in a manner similar to the proof of Proposition 1 (B.7.2).

Corollary 1. *Let $\varphi[x, \vec{p}]$ be a predicative formula such that the substitution $\varphi[x \parallel y, \vec{p}]$ is admissible. Then, the formula $\exists Y \forall y (y \in Y \Leftrightarrow \exists Z (y \in Z) \wedge \varphi[y, \vec{p}])$ is deducible in the NBG*.*

The theory of classes and sets determined by axioms A1, A2.1 – A2.9, A3 – A8 will be denoted by NBG^f.

Theorem 1 (The Bernays theorem on the finite axiomatization of the NBG set theory). *The theories NBG and NBG^f are deductively equivalent.*

Proof. Corollary 1 to Proposition 1 shows that the NBG is weaker than the NBG^f.

On the other hand, it is clear that axioms A2.2 – A2.9 are specific cases of axiom scheme AS2. Axiom A2.1 is deduced in the NBG by virtue of Lemma 4 (1.1.6). Therefore, the NBG^f is weaker than the NBG. □

C Compactness theorem for generalized second-order language

Introduction

For the first-order language, the *compactness theorem* was proven by K. Gödel and A. I. Maltsev (see e.g. [*Tourlakis*, 2003a, 1.5.42]). It was also proven by J. Loś [1955] by means of the *method of ultraproducts* (see also [*Ershov and Palyutin*, 1984, § 17], [*Maltsev*, 1973, 8.3], [*Mendelson*, 1997, 2.14]).

Unfortunately, for the usual second-order language (see e.g. [*Maltsev*, 1973, § 6], [*Mendelson*, 1997, Appendix], [*Takeuti*, 2013, § 16]) the compactness theorem does not hold (see e.g. [*Mendelson*, 1997, Appendix], [*Boolos et al.*, 2007, §18]). Moreover, the method of ultraproducts is also inapplicable to second-order models.

A possible way out of this situation is to refuse the most vulnerable place in the construction of ultraproducts connected with the factorization with respect to an ultrafilter, i. e. to stay working with the ordinary non-factorized product. This refusal compels us instead of the single usual set–theoretical equality = to use several *generalized equalities* $\approx_{\text{first}}$ and $\approx_{\text{second}}$ for first and second orders, and instead of the single usual set-theoretical belonging $\in$ to use several *generalized belongings* $\lessdot_\tau$. By this reason, it is necessary to refuse the usual set-theoretical interpretation $(\gamma(x_0), \dots, \gamma(x_k)) \in \gamma(u)$ of the second basic (after equality) atomic formula $(x_0, \dots, x_k)u$ and to replace it by the generalized interpretation $(\gamma(x_0), \dots, \gamma(x_k)) \lessdot_\tau \gamma(u)$, where $x_i^{\tau_i}$ are variables of the first-order types τ_i, u^τ is a variable of the second-order type $\tau = [\tau_0, \dots, \tau_k]$ (i. e. predicate), and γ is some evaluation of variables on some mathematical system U.

This appendix is devoted to rigorous development of the expressed general idea. A short presentation of this idea was announced in [*Zakharov*, 2008b]; the complete proof was given in [*Zakharov and Yashin*, 2014].

In the capacity of initial formulas, the formulas of the following two forms were taken: the formula $y^\sigma \delta_\sigma z^\sigma$ for the *generalized equality* δ_σ and the formula $(x_0^{\tau_0}, \dots, x_k^{\tau_k}) \varepsilon_\tau u^\tau$ for the *generalized belonging* ε_τ, where y^σ and z^σ are the variables of the first- or second-order type σ and $x_i^{\tau_i}$ and u^τ are the variables of the first-order types τ_i and the second-order type $\tau \equiv [\tau_0, \dots, \tau_k]$, respectively.

These atomic formulas are interpreted on an evaluated mathematical system $\langle U, \gamma \rangle$ (with an evaluation γ of variables on U) in the following generalized way: $\gamma(y) \approx_\sigma \gamma(z)$ and $(\gamma(x_0), \dots, \gamma(x_k)) \lessdot_\tau \gamma(u)$, where $\approx_\sigma$ is a *generalized ratio of equality* and $\lessdot_\tau$ is a *generalized ratio of belonging*. Generalized equalities and generalized belongings are connected with each other by the *initial principle of change of equals* (see axiom E4 from C.1.3).

More exactly, we introduce a *generalized second-order signature* Σ_2^g containing, in addition to individual and predicate constants and variables, the symbols δ_τ and ε_τ.

https://doi.org/10.1515/9783110550948-004

With respect to this signature formulas φ in the language $L(\Sigma_2^g)$ are defined by common induction, when we start from the above-mentioned atomic formulas.

To give a semantics of the language $L(\Sigma_2^g)$, we define *mathematical systems U of the signature* Σ_2^g. The *satisfaction of a formula* φ *on a system U with respect to an evaluation of variables* γ is defined according to the above-mentioned generalized interpretation of the atomic formulas (in notation $U \models \varphi[\gamma]$).

The semantics for the language $L(\Sigma_2^g)$ presented in this appendix differs both from the standard semantics (see [*Mendelson*, 1997, Appendix], [*Takeuti*, 2013, §16]) and from the Henkin semantics (see [*Mendelson*, 1997, Appendix], [*Takeuti*, 2013, §21], [*Rossberg*, 2004; *Shapiro*, 1991; *Väänänen*, 2001]), which restricts the range of values of the evaluation $\gamma(x^\tau)$ for a variable x^τ of a second-order type τ by some subset of the set $\mathcal{P}(\tau(X))$ of the *terminal* $\tau(X)$ of the mathematical system $U \equiv (X, S)$.

In this appendix, the following *generalized compactness theorem* is proven:

Let Φ *be a set of formulas of the language* $L(\Sigma_2^g)$. *Let for every finite subset f of the set* Φ *there exist a mathematical system* U_f *of the signature* Σ_2^g *and an evaluation of variables* γ_f *on the system* U_f *such that* $U_f \models \varphi[\gamma_f]$ *for every formula* $\varphi \in f$. *Then, there exist a mathematical system U of the signature* Σ_2^g *and an evaluation of variables* γ *on the system U such that* $U \models \varphi[\gamma]$ *for every formula* $\varphi \in \Phi$ (see Theorem 1 (C.3.3)).

This system U is constructed with the help of some ultrafilter starting from the systems U_f by means of the *method of infraproducts* based on the refusal of the Łós factorization.

The most delicate point in the proof of the generalized compactness theorem is the demonstration of the *property of infrafiltration* for a quantified formula $\exists x^\tau \psi$ for a variable x^τ of a second-order type $\tau = [\tau_0 \dots, \tau_k]$, which requires some preliminary assertions (see Propositions 2 (C.2.4) and 1 (C.3.2)).

To enlarge the area of possible applications of the above-mentioned generalized compactness theorem, it is proven in a polygrade language with *basic* and *auxiliary grades*. Therefore, interpretations are defined on polygrade domains of the form $[A_0, \dots, A_m, K_0, \dots, K_{n-1}]$, where $K_0, \dots, K_{n-1}$ are the fixed *auxiliary sets* (which are absent when $n = 0$). It allows to consider in capacity of models modules A_K over the fixed ring K.

The presence of the suite $H \equiv [K_0, \dots, K_{n-1}]$ of the fixed auxiliary sets requires introducing the additional condition of *H-concordance of mathematical systems* $U \equiv (X, S)$ *and* $V \equiv (Y, T)$, where S and T are the polygrade superstructures over the supports $X \equiv [A_0, \dots, A_m, K_0, \dots, K_{n-1}]$ and $Y \equiv [B_0, \dots, B_m, K_0, \dots, K_{n-1}]$. This condition means the similarity of the systems U and V with respect to all elements of the signature Σ_2^g connnected with the fixed auxiliary suite H. Also, we use the similar condition of *H-concordance of an evaluation* γ *on the system U and an evaluation* δ *on the system V*. In turn, this entails the necessity of introducing the additional condition of *H-concordance* in defining the satisfactions $U \models (\exists x^\tau \varphi)[\gamma]$ and $U \models (\forall x^\tau \varphi)[\gamma]$, which is not required for $n = 0$, i. e. when the auxiliary suite is absent.

At the end of the appendix, the method of infraproducts is applied for the construction of models of the second-order generalized Peano – Landau arithmetic. The supports of these models are the generalized Baire sets $\mathbb{N}_0^F$, which are uncountable in general.

In this appendix we fix any rich axiomatic set theory ST such as ZF, NBG, LTS, and so on. Therefore, the general term *set* in ST can mean the *set* in ZF, the *class* in NBG and LTS, and so on.

C.1 Types, formations, terminals, signatures, and formulas

C.1.1 Types

For fixed integers $m, n \in \omega$, define by induction the *semitypes* and the *types*:
1) for any $i \in m + 1$, the symbol-string $\langle i, 1\rangle$ is the *semitype* and the *type*;
2) for any $j \in n$ the symbol-string, $\langle j, 0\rangle$ is the *semitype* and the *type*;
3) if τ is a type, then τ is the *semitype*:
4) if τ is a semitype, then $[\tau]$ is the *type*;
5) if $\tau_0, \dots, \tau_k$ are semitypes and $k \geqslant 1$, then $(\tau_0, \dots, \tau_k)$ is the *semitype*.

This definition is a generalization of the corresponding definition from [*Takeuti*, 2013, § 20].

Further, instead of $[(\tau_0, \dots, \tau_k)]$, we shall write simply $[\tau_0, \dots, \tau_k]$. Thus, the notation $[\tau_0, \dots, \tau_k]$ may be used for $k \geqslant 0$.

Semantics of semitypes and types will be explained in the next section.

Types $\langle i, 1\rangle$ and $\langle j, 0\rangle$ will be called the *first-order types*. If $\tau_0, \dots, \tau_k$ are first-order types and $k \geqslant 0$ then $[\tau_0, \dots, \tau_k]$ will be called the *second-order type.*

For a type $\tau \equiv [\tau_0, \dots, \tau_k]$ with $k \geqslant 0$, the types $\tau_0, \dots, \tau_k$ will be called the *parents of the type* τ and will be denoted by $p_0\tau, \dots, p_k\tau$, respectively. Consider *the set* $P(\tau) \equiv \{p_0\tau, \dots, p_k\tau\}$ *of all parents of* the type τ.

For any first-order type τ, put formally $p\tau \equiv \tau$ and $P(\tau) \equiv \{p\tau\} = \{\tau\}$.

With any type τ, we associate *the semitype* $\check{\tau}$ *of the type* τ as follows:
1) if τ is a first-order type, then $\check{\tau} \equiv \tau$;
2) if $\tau = [\tau_1]$ and τ_1 is a semitype, then $\check{\tau} \equiv \tau_1$.

In other words, the semitype of a type is obtained by omitting the square brackets.

Auxiliary types are defined by induction in the following way:
1) any type of the form $\langle j, 0\rangle$ is an *auxiliary type* for every $j \in n$;
2) if τ is an auxiliary type, then $[\tau]$ is the auxiliary type;
3) if $\tau_0, \dots, \tau_k$ are auxiliary types and $k \geqslant 1$, then $[\tau_0, \dots, \tau_k]$ is an *auxiliary type.*

A type will be called *basic* if it is not auxiliary.

Thus, for a second-order type $\tau \equiv [\tau_0, \dots, \tau_k]$, the index set $k + 1$ is decomposed on two subsets $M(\tau)$ and $N(\tau)$, so that for any $\mu \in M(\tau)$ the type τ_μ is basic and for any $\nu \in N(\tau)$ the type τ_ν is auxiliary.

C.1.2 Formations and terminals

Further in the appendix, $K_0, \dots, K_{n-1}$ are fixed *auxiliary sets*. If $n = 0$, then all the fixed sets are absent.

Define the *formation* $G \equiv [P_0, \dots, P_{l-1}]$ *of the rank* $l \in \omega$ in the following way:

1) $G \equiv [P_0, \dots, P_{-1}] \equiv \varnothing$ for $l = 0$;
2) $G \equiv [P_0, \dots, P_0] \equiv P$ for $l = 1$;
3) $G \equiv [P_0, \dots, P_{l-1}] \equiv (P_i \mid i \in l) \equiv (P_0, \dots, P_{l-1})$ for $l \geqslant 2$.

Further on, we fix the *auxiliary formation* $H \equiv [K_0, \dots, K_{n-1}]$ of the rank $n \in \omega$.

Define the *formation* $X \equiv [A_0, \dots, A_m, K_0, \dots, K_{n-1}]$ *of the rank* $m+1|n$ *over the set* H in the following way:

1) $X \equiv [A_0, \dots, A_m, K_0, \dots, K_{n-1}] \equiv [A_0, \dots, A_m]$ for $n = 0$ and $m \in \omega$;
2) $X \equiv [A_0, \dots, A_m, K_0, \dots, K_{n-1}] \equiv \langle [A_0, \dots, A_m], [K_0, \dots, K_{n-1}] \rangle$ for $n \geqslant 1$ and $m \in \omega$.

The sets $A_0, \dots, A_m$ are called *basic in* X. A formation X may be without auxiliary sets but should contain at least one basic set.

Define the *terminals* $\tau(X)$ *of the semitypes* τ *over the formation* X by induction:

1) $\langle i, 1 \rangle(X) \equiv A_i$;
2) $\langle j, 0 \rangle(X) \equiv K_j$;
3) if τ is a semitype, then $[\tau](X) \equiv \mathcal{P}(\tau(X))$, where $\mathcal{P}$ denotes taking the set of all parts of the intended set (see 1.1.6);
4) if $\tau_0, \dots, \tau_k$ are semitypes and $k \geqslant 1$, then $(\tau_0, \dots, \tau_k)(X) \equiv \tau_0(X) \times \dots \times \tau_k(X)$.

Thus, for semitypes $\tau_0, \dots, \tau_k$ with $k \geqslant 1$, for the type $\tau \equiv [\tau_0, \dots, \tau_k]$, and for its semitype $\check{\tau} = (\tau_0, \dots, \tau_k)$, the following equalities $\tau(X) = \mathcal{P}(\tau_0(X) \times \dots \times \tau_k(X))$ and $\check{\tau}(X) = \tau_0(X) \times \dots \times \tau_k(X)$ are fulfilled.

C.1.3 Signatures and formulas

A non-empty set Θ of types τ will be called the *type domain of rank* $m + 1|n$ if $\tau \in \Theta$ implies $p\tau \in \Theta$ for every parent $p\tau$ of the type τ. In the type domain Θ, select the *belonging type subdomain* $\Theta_b \equiv \{\tau \in \Theta \mid \exists k \in \omega \exists \tau_0, \dots, \tau_k \in \Theta(\tau = [\tau_0, \dots, \tau_k])\}$.

A collection $\Sigma_c \equiv (\Sigma_c^\tau \mid \tau \in \Theta)$ of *collections* $\Sigma_c^\tau \equiv (\sigma_\omega^\tau \mid \omega \in \Omega_\tau)$ *of constants* σ_ω^τ *of the types* τ will be called the *signature of constants of the type domain* Θ. Sets Ω_τ may be empty, and then $\Sigma_c^\tau = \varnothing$.

The constants σ_ω^τ of the first-order type τ are called *individual* or *objective*. The constants of other types are called *predicate*.

A collection $\Sigma_e \equiv (\delta_\tau \mid \tau \in \Theta)$ of *binary predicate symbols of (generalized) equalities* δ_τ *of the types* τ will be called the *signature of (generalized) equalities of the type domain* Θ. It follows from the definition of the type domain that for every equality symbol δ_τ, the collection Σ_e contains necessarily the equality symbols $\delta_{p\tau}$ for every parent $p\tau$ of the type τ.

A collection $\Sigma_b \equiv (\epsilon_\tau \mid \tau \in \Theta)$ of binary *predicate symbols of (generalized) belongings ϵ_τ of the types τ* will be called the *signature of (general) belongings of the type domain* Θ.

A collection $\Sigma_v \equiv (\Sigma_v^\tau \mid \tau \in \Theta)$ of denumerable *sets Σ_v^τ of variables $x^\tau, y^\tau, \ldots$ of the types τ* will be called the *signature of variables of the type domain* Θ. The sets Σ_v^τ may be empty. The variables $x^\tau, y^\tau, \ldots$ of the first-order types τ are called *individual* or *objective*. The variables of other types are called *predicate*.

Further, we shall always assume that for every type $\tau \in \Theta$ there are either constants or variables of this type.

The quadruple $\Sigma^g \equiv \Sigma_c|\Sigma_e|\Sigma_b|\Sigma_v$ will be called the *(polygrade) generalized signature of the rank $m+1|n$* or the *signature with generalized equalities and belongings*.

The *language $L(\Sigma^g)$ of the generalized signature Σ^g* consists of:
1) all types τ from the type domain Θ;
2) all members of all signatures from Σ^g;
3) logical symbols $\neg$, $\vee$, $\wedge$, $\Rightarrow$, $\forall$, and $\exists$;
4) parenthesis.

If the type domain Θ contains first- and second-order types only and at least one second-order type, then we shall say that the signature Σ^g and the language $L(\Sigma^g)$ have the *second order* (see [*Mendelson*, 1997, Appendix], [*Dalen*, 1997, 4]). In this case, the notations Σ_2^g and $L(\Sigma_2^g)$ will be used.

Constants and variables of a type τ are called *terms of the type τ of the language* $L(\Sigma^g)$.

Atomic formulas of the language $L(\Sigma^g)$ are defined in the following way:
1) if q and r are terms of a type $\tau \in \Theta$, then $q\delta_\tau r$ is an *atomic formula*;
2) if $\tau_0, \ldots, \tau_k$ are types from Θ for $k \geqslant 0$, $\tau \equiv [\tau_0, \ldots, \tau_k] \in \Theta_b$, $q_0^{\tau_0}, \ldots, q_k^{\tau_k}$ are terms of the types $\tau_0, \ldots, \tau_k$, respectively, and r^τ is a term of the type τ, then $(q_0^{\tau_0}, \ldots, q_k^{\tau_k})\epsilon_\tau r^\tau$ is an *atomic formula*; in particular, for $k = 0$, the symbol-string $q_0^{\tau_0}\epsilon_{[\tau_0]} r^{[\tau_0]}$ is an *atomic formula*.

The *formulas of the language $L(\Sigma^g)$* are constructed from atomic ones with the use of connectives $\vee$, $\wedge$, $\neg$, $\Rightarrow$, quantifiers $\exists x^\tau$ and $\forall x^\tau$ with respect to variables x^τ, and parenthesis.

The *logical axiom schemes of the (polygrade) type theory in the language $L(\Sigma^g)$ of the generalized signature Σ^g* are the schemes of the predicate calculus (see 1.1.4), where variables and terms substituting each other must be of the same type $\tau \in \Theta$.

In addition to these axiom schemes, consider the following *equality axioms for types* $\tau \in \Theta$.

E1. $\forall x^\tau(x\delta_\tau x)$.
E2. $\forall x^\tau, y^\tau(x\delta_\tau y \Rightarrow y\delta_\tau x)$.
E3. $\forall x^\tau, y^\tau, z^\tau(x\delta_\tau y \wedge y\delta_\tau z \Rightarrow x\delta_\tau z)$.
E4. (The *initial principle of change of equals.*) $\forall x_0^{\tau_0}, y_0^{\tau_0}, \dots, x_k^{\tau_k}, y_k^{\tau_k}, u^\tau, v^\tau(x_0\delta_{\tau_0} y_0 \wedge \dots \wedge x_k\delta_{\tau_k} y_k \wedge u\delta_\tau v \Rightarrow ((x_0, \dots, x_k)\epsilon_\tau u \Leftrightarrow (y_0, \dots, y_k)\epsilon_\tau v)))$, where $\tau \equiv [\tau_0, \dots, \tau_k]$.

The *inference rules* in the depicted type theory are:

$$\frac{\varphi,\ \varphi \Rightarrow \psi}{\psi}\ (MP) \qquad \text{and} \qquad \frac{\varphi(x^\tau)}{\forall x^\tau \varphi(x^\tau)}\ (Gen).$$

If there are non-logical axioms or axiom schemes written by second-order formulas of the language $L(\Sigma_2^g)$, we shall say that a (*mathematical*) *generalized second-order theory* is given.

C.2 Mathematical systems of the signature Σ^g with generalized equalities and belongings

C.2.1 The definition of a mathematical system of the generalized signature Σ^g

Let Σ^g be a fixed signature of the rank $m+1|n$ defined in C.1.3. Fix also a formation $X \equiv [A_0, \dots, A_m, K_0, \dots, K_{n-1}]$ of the rank $m+1|n$.

For the formation X and the signature Σ^g, consider the following collections:

1) the collection $S_c \equiv (S_c^\tau \mid \tau \in \Theta)$ of collections $S_c^\tau \equiv (s_\omega^\tau \mid \omega \in \Omega_\tau)$ of *constant structures $s_\omega^\tau \in \tau(X)$ of the types τ*;
2) the collection $S_e \equiv (\approx_\tau \mid \tau \in \Theta)$ of *generalized ratios of equality* $\approx_\tau \subset \tau(X) \times \tau(X)$ *of the types τ on the sets $\tau(X)$*, containing the usual set-theoretic ratios of equality $=$ on the sets $\tau(X)$, i. e. such ratios $\approx_\tau$ that for every elements $r, s \in \tau(X)$ the equality $r = s$ implies the generalized equality $r \approx_\tau s$;
3) the collection $S_b \equiv (<_\tau \mid \tau \in \Theta_b)$ of *generalized ratios of belonging* $<_\tau \subset \check{\tau}(X) \times \tau(X)$ *of the types τ*, containing the usual set-theoretic ratios of belonging $\in$ from the sets $\check{\tau}(X)$ into the sets $\tau(X)$, i. e. such ratios $<_\tau$ that for every elements $p \in \check{\tau}(A)$ and $P \in \tau(X)$ the belonging $p \in P$ implies the generalized belonging $p <_\tau P$;
4) the collection $S_v \equiv (\tau(X) \mid \tau \in \Theta)$ of the terminals $\tau(X)$ of the types τ over the formation X.

The quadruple $S \equiv (S_c, S_e, S_b, S_v)$ of the above-mentioned collections will be called a (*polygrade*) *superstructure of the signature Σ^g over the formation X*.

The pair $U \equiv (X, S)$ will be called a *mathematical system of the generalized signature Σ^g with the support* (*carrier*) *X and the superstructure S*. This notion is a generalization of the notion of an *algebraic system of the signature* Σ_1 (see [*Ershov and Palyutin*, 1984, § 15]).

The mathematical system $U \equiv (X, S)$ will be called also an *interpretation of the signature Σ^g on the support X*.

Further, for a type $\tau = [\tau_0, \dots, \tau_k]$ and elements $p \equiv (p(0), \dots, p(k))$, $q \equiv (q(0), \dots, q(k)) \in \check{\tau}(X) = \tau_0(X) \times \cdots \times \tau_k(X)$ along with $p(0) \approx_{\tau_0} q(0) \wedge \cdots \wedge p(k) \approx_{\tau_k} q(k)$, we shall also write $p \approx_{\check{\tau}} q$.

C.2.2 Concordance of mathematical systems of the generalized second-order signature

Two mathematical systems $U \equiv (X, S)$ and $V \equiv (Y, T)$ of the signature Σ_2^g will be called *H-concordant* if:

1) for every auxiliary type $\tau \in \Theta$ and every $\omega \in \Omega_\tau$, the constants $s_\omega^\tau \in \tau(X)$ and $t_\omega^\tau \in \tau(Y) = \tau(X)$ coincide, where by the definition of terminals $\tau(X) = \tau(Y)$;
2) for every auxiliary type $\tau \in \Theta$, the equalities $\approx_\tau \subset \tau(X) \times \tau(X)$ and $\approx_\tau \subset \tau(Y) \times \tau(Y)$ coincide, where as above $\tau(Y) \times \tau(Y) = \tau(X) \times \tau(X)$;
3) for every auxiliary type $\tau \in \Theta_b$, the belongings $<_\tau \subset \check{\tau}(X) \times \tau(X)$ and $<_\tau \subset \check{\tau}(Y) \times \tau(Y)$ coincide, where by the same reason, $\check{\tau}(Y) \times \tau(Y) = \check{\tau}(X) \times \tau(X)$;
4) for every suite $p \equiv (p(0), \dots, p(k)) \in s_\omega^\tau \subset \check{\tau}(X) = \tau_0(X) \times \cdots \times \tau_k(X)$, there exists a suite $q \equiv (q(0), \dots, q(k)) \in t_\omega^\tau \subset \check{\tau}(Y) = \tau_0(Y) \times \cdots \times \tau_k(Y)$ such that $q(\nu) = p(\nu)$, and for every q, there exists p such that $p(\nu) = q(\nu)$ for every $\nu \in N(\tau)$ and every type $\tau \equiv [\tau_0, \dots, \tau_k]$, such that $M(\tau) \neq \varnothing$ and $N(\tau) \neq \varnothing$.

The property of H-concordance means the identity of the systems U and V with respect to all elements connected with the auxiliary set H.

The generalized equalities $\approx_\tau$ and the generalized belongings $<_\tau$ admit some additional conditions.

A system U will be called *balanced* if $\forall P, Q \in \tau(X)(P \approx_\tau Q \Leftrightarrow \forall p \in P \exists q \in Q(q \approx_{\check{\tau}} p) \wedge \forall q \in Q \exists p \in P(p \approx_{\check{\tau}} q))$, where $\tau_0, \dots, \tau_k \in \Theta$, $k \geqslant 0$, and $\tau \equiv [\tau_0, \dots, \tau_k] \in \Theta$.

A system U will be called *regular* if $\forall p \in \check{\tau}(X) \forall P \in \tau(X)(p <_\tau P \Leftrightarrow \exists q \in P(p \approx_{\check{\tau}} q))$, where $\tau_0, \dots, \tau_k \in \Theta$, $k \geqslant 0$, and $\tau \equiv [\tau_0, \dots, \tau_k] \in \Theta$.

A system U will be called *normal* if $\forall p, q \in \sigma(X)(p \approx_\sigma q \Leftrightarrow p = q) \wedge \forall p \in \check{\tau}(X) \forall P \in \tau(X)(p <_\tau P \Leftrightarrow p \in P)$.

A system U will be called *extensional* if $\forall P, Q \in \tau(X)(P \approx_\tau Q \Leftrightarrow \forall p(p <_\tau P \Rightarrow p <_\tau Q) \wedge \forall q(q <_\tau Q \Rightarrow q <_\tau P))$, where $\tau \in \Theta_b$.

C.2.3 Evaluations and models

An *evaluation on a system $U \equiv (X, S)$ of the signature Σ^g* is a mapping γ, defined on the set of all variables of the signature Σ^g and associating with the variable x^τ of the type $\tau \in \Theta$, the element $\gamma(x^\tau)$ of the terminal $\tau(X)$ (see [*Ershov and Palyutin*, 1984, § 16], [*Takeuti*, 2013, 16.17]). The pair (U, γ) consisting of the system U of the signature Σ^g

and the evaluation γ on U will be called an *evaluated mathematical system of the signature* Σ^g.

Evaluated mathematical systems (U, γ) and (V, δ) of the signature Σ_2^g will be called *H-concordant* if:

1) the systems U and V are H-concordant;
2) for every auxiliary type $\tau \in \Theta$ the evaluations $\gamma(x^\tau) \in \tau(X)$ and $\delta(x^\tau) \in \tau(Y) = \tau(X)$ coincide, i. e. $\gamma(x^\tau) = \delta(x^\tau)$ (see C.2.1);
3) for every suite $p \equiv (p(0), \ldots, p(k)) \in \gamma(x^\tau) \subset \check{\tau}(X) = \tau_0(X) \times \cdots \times \tau_k(X)$ there exists a suite $q \equiv (q(0), \ldots, q(k)) \in \delta(x^\tau) \subset \check{\tau}(Y) = \tau_0(Y) \times \cdots \times \tau_k(Y)$ such that $q(v) = p(v)$ and, for every q there exists p such that $p(v) = q(v)$ for every $v \in N(\tau)$ and every type $\tau = [\tau_0, \ldots, \tau_k]$ such that $M(\tau) \neq \varnothing$ and $N(\tau) \neq \varnothing$.

The property of H-concordance means the identity of the evaluated systems (U, γ) and (V, δ) with respect to all elements connected with the auxiliary set H.

An evaluation γ on a system U and an evaluation δ on a system V will be called *H-concordant* if they satisfy conditions 2) and 3) from the previous definition.

Define the *value* $q[\gamma]$ *of a term* q *with respect to the evaluation* γ *on the system* U in the following way (see [*Ershov and Palyutin*, 1984, § 16], [*Maltsev*, 1973, § 6], [*Mendelson*, 1997, 2.2], [*Shoenfield*, 2001, 2.5]):

- if σ_ω^τ is a constant of a type $\tau \in \Theta$, then $\sigma_\omega^\tau[\gamma] \equiv s_\omega^\tau$;
- if x^τ is a variable of a type $\tau \in \Theta$, then $x^\tau[\gamma] \equiv \gamma(x^\tau)$.

Define the *satisfaction* (*translation* as in A.1.3) *of a formula* φ *of the language* $L(\Sigma_2^q)$ *on a system* U *of the signature* Σ_2^q *with respect to an evaluation* γ (in notation, $U \vDash \varphi[\gamma]$) by induction in the following way (see [*Mendelson*, 1997, 2.2], [*Shoenfield*, 2001, 2.5], [*Takeuti*, 2013, 16.17]):

1) if q and r are terms of a type $\tau \in \Theta$ and $\varphi \equiv (q\delta_\tau r)$, then $U \vDash \varphi[\gamma]$ is equivalent to $q[\gamma] \approx_\tau r[\gamma]$;
2) if $\tau_0, \ldots, \tau_k$ are types from Θ for $k \geqslant 0$, $\tau \equiv [\tau_0, \ldots, \tau_k] \in \Theta$, $q_0, \ldots, q_k$ are terms of the types $\tau_0, \ldots, \tau_k$, respectively, r is a term of the type τ, and $\varphi \equiv (q_0, \ldots, q_k)\epsilon_\tau r$, then $U \vDash \varphi[\gamma]$ is equivalent to $(q_0[\gamma], \ldots, q_k[\gamma]) \lessdot_\tau r[\gamma]$;
3) if $\varphi \equiv \neg\psi$, then $U \vDash \varphi[\gamma]$ iff $U \vDash \psi[\gamma]$ is not true;
4) if $\varphi \equiv (\psi \vee \xi)$, then $U \vDash \varphi[\gamma]$ iff $U \vDash \psi[\gamma]$ or $U \vDash \xi[\gamma]$;
5) if $\varphi \equiv (\psi \wedge \xi)$, then $U \vDash \varphi[\gamma]$ iff $U \vDash \psi[\gamma]$ and $U \vDash \xi[\gamma]$;
6) if $\varphi \equiv (\psi \Rightarrow \xi)$, then $U \vDash \varphi[\gamma]$ iff that $U \vDash \psi[\gamma]$ implies $U \vDash \xi[\gamma]$;
7) if $\varphi \equiv \exists x^\tau \psi$, then $U \vDash \varphi[\gamma]$ is equivalent to $U \vDash \psi[\gamma']$ for some evaluation γ' H-concordant with γ and such that $\gamma'(y^\sigma) = \gamma(y^\sigma)$ for every variable $y^\sigma \neq x^\tau$;
8) if $\varphi \equiv \forall x^\tau \psi$, then $U \vDash \varphi[\gamma]$ is equivalent to $U \vDash \psi[\gamma']$ for every evaluation γ' H-concordant with γ and such that $\gamma'(y^\sigma) = \gamma(y^\sigma)$ for every variable $y^\sigma \neq x^\tau$.

Note that bringing into use in points 7) and 8) of this definition the additional (in comparison with [*Ershov and Palyutin*, 1984, § 16], [*Mendelson*, 1997, 2.2], [*Shoenfield*,

2001, 3.2]) property of H-concordance of the evaluations γ and γ' is stipulated by the initial polygrade structure of considered mathematical systems and by the presence of the fixed auxiliary formation $H \equiv [K_0, \ldots, K_{n-1}]$.

Let Φ be a set of formulas of the language $L(\Sigma_2^q)$. An evaluated mathematical system (U, γ) of the signature Σ_2^q will be called a *model for the set* Φ if $U \vDash \varphi[\gamma]$ for every formula $\varphi \in \Phi$ (see [*Ershov and Palyutin*, 1984, § 17]).

A model (U, γ) will be called *balanced, regular, normal, extensional*, etc. if the system U is the same.

A model (U, γ) for a set Φ will be called *second-order* if at least one formula from Φ contains at least one second-order variable.

Remark that if a system $U \equiv (X, S)$ is considered in an axiomatic set theory, then the satisfaction of a closed formula φ of the language $L(\Sigma_2^g)$ with respect to any evaluation γ is reduced to correctness of the relativization φ^r of φ on the corresponding terminals of the support X in this set theory.

In particular, since equality axioms E1 – E4 are closed formulas, their relativizations E1^r – E4^r take the following forms:

$$E1^r \equiv \forall x \in \tau(X)(x \approx_\tau x);$$
$$E2^r \equiv \forall x, y \in \tau(X)(x \approx_\tau y \Rightarrow y \approx_\tau x);$$
$$E3^r \equiv \forall x, y, z \in \tau(X)(x \approx_\tau y \wedge y \approx_\tau z \Rightarrow x \approx_\tau z);$$
$$E4^r \equiv \forall x_0, y_0 \in \tau_0(X) \ldots \forall x_k, y_k \in \tau_k(X) \forall u, v \in \tau(X)(x_0 \approx_{\tau_0} y_0 \wedge \ldots \wedge x_k \approx_{\tau_k} y_k \wedge$$
$$\wedge u \approx_\tau v \Rightarrow ((x_0, \ldots, x_k) \lessdot_\tau u \Leftrightarrow (y_0, \ldots, y_k) \lessdot_\tau v)), \text{ where } \tau \equiv [\tau_0, \ldots, \tau_k],$$
$$k \geqslant 0, \text{ and all types are in } \Theta.$$

The satisfaction of formulas E1^r – E3^r means that all generalized equalities $\approx_\tau$ are equivalence relations on corresponding sets $\tau(X)$, and the satisfaction of formula E4^r means the initial principle of change of equals in the atomic formula with the generalized belonging $\lessdot_\tau$.

Further on, we shall say that a system U of the signature Σ_2^g has *true generalized equalities and belongings* if axioms E1 – E4 from C.1.3 are satisfied on U with respect to some (and, consequently, to any) evaluation γ. This means that formulas E1^r – E4^r are correct for the system U in the used set theory.

C.2.4 The generalized equality of values of evaluations and satisfiability

For every formula φ of the language $L(\Sigma_2^g)$ we define the formula φ^* by induction:

1) $\varphi^* \equiv \varphi$ for every atomic formula φ;
2) $(\psi \wedge \xi)^* \equiv \psi^* \wedge \xi^*$;
3) $(\neg\psi)^* \equiv \neg\psi^*$;
4) $(\exists x^\tau \psi)^* \equiv \exists x^\tau \psi^*$;

5) $(\psi \vee \xi)^* \equiv \neg(\neg\psi^* \wedge \neg\xi^*)$;
6) $(\psi \Rightarrow \xi)^* \equiv \neg(\psi^* \wedge \neg\xi^*)$;
7) $(\forall x^\tau \psi)^* \equiv \neg(\exists x^\tau(\neg\psi^*))$.

A formula φ is said to be *normalizable* if for every mathematical Σ_2^g-system U and every evaluation γ on U the following condition holds: $U \vDash \varphi[\gamma] \Leftrightarrow U \vDash \varphi^*[\gamma]$.

Lemma 1. *Let formulas ψ and ξ be normalizable. Then, formulas $\psi \wedge \xi$, $\neg\psi$, $\psi \vee \xi$, $\psi \Rightarrow \xi$, $\forall x^\tau \psi$, and $\exists x^\tau \psi$ are normalizable too.*

The proof of this lemma uses the definition of satisfiability and some well known tautologies only, so it is omitted.

Proposition 1. *Every formula of the language $L(\Sigma_2^g)$ of the generalized second-order signature Σ_2^g is normalizable.*

Proof. Denote by Φ the set of all formulas of the language $L(\Sigma_2^g)$. The subset of the set Φ consisting of formulas containing at most $n \in \omega$ logical symbols $\neg$, $\wedge$, $\Rightarrow$, $\vee$, $\exists$, $\forall$, denote by Φ_n. It is clear that $\Phi = \bigcup(\Phi_n \mid n \in \omega)$.

Prove by the complete induction principle the following assertion $A(n)$: *every formula $\varphi \in \Phi$ is normalizable.*

If $n = 0$, then the formula φ is atomic, and so by the definition of the operation $\varphi \mapsto \varphi^*$ we have $\varphi^* \equiv \varphi$. Consequently, the assertion $A(0)$ is true.

Suppose that for all $m < n$ the assertion $A(m)$ is true. Let $\varphi \in \Phi_n$. If $\varphi \equiv \psi \wedge \xi$, $\varphi \equiv \neg\psi$, $\varphi \equiv \exists x^\tau \psi$, $\varphi \equiv \psi \vee \xi$, $\varphi \equiv \psi \Rightarrow \xi$, or $\varphi \equiv \forall x^\tau \psi$, then $\psi, \xi \in \Phi_{n-1}$. Therefore, by the induction hypothesis, the formulas ψ and ξ are normalizable. By Lemma 1 the formula φ is normalizable. Hence the assertion $A(n)$ is true. □

Proposition 2. *Let U be a mathematical system of the second-order signature Σ_2^g with true generalized equalities and belongings. Then, for every formula φ of the language $L(\Sigma_2^g)$ and every H-concordant evaluations γ and δ on the system U such that $\gamma(x^\tau) \approx_\tau \delta(x^\tau)$ for every variable x^τ of every type $\tau \in \Theta$ the properties $U \vDash \varphi[\gamma]$ and $U \vDash \varphi[\delta]$ are equivalent.*

Proof. The set of all formulas φ of the language $L(\Sigma_2^g)$ constructed by induction from the atomic formulas with the use of connectives $\neg$ and $\wedge$ and quantifier $\exists$ denote by Ψ. The subset of the set Ψ consisting of formulas containing at most $n \in \omega$ logical symbols $\neg$, $\wedge$, and $\exists$ denote by Ψ_n. It is clear that $\Psi = \bigcup(\Psi_n \mid n \in \omega)$.

Prove by the complete induction principle the assertion $A(n)$: *for every formula $\varphi \in \Psi_n$ and every mentioned evaluations γ and δ the assertion of the Proposition holds.*

Let $n = 0$ and $\varphi \in \Psi_0$. Then, φ is an atomic formula. At first consider the atomic formula φ of the form $q^\tau \delta_\tau r^\tau$. Suppose that $q^\tau = x^\tau$ and $r^\tau = \sigma^\tau_\omega$. Then, $U \models \varphi[\gamma]$ is equivalent to $\gamma(x) \approx_\tau s^\tau_\omega$ and $U \models \varphi[\delta]$ is equivalent to $\delta(x) \approx_\tau s^\tau_\omega$. Since, by our condition $\gamma(x) \approx_\tau \delta(x)$, assuming $U \models \varphi[\gamma]$ and using axioms E2$'$ and E3$'$ we infer $U \models \varphi[\delta]$. The inverse inference is checked in the same way. For the terms q^τ and r^τ of other forms the reasons are quite similar.

Now, consider the atomic formula φ of the form $(q_0^{\tau_0}, \ldots, q_k^{\tau_k})\varepsilon_\tau r^k$ for $\tau \equiv [\tau_0, \ldots, \tau_k] \in \Theta_b$. Assume that $q_\lambda^{\tau_\lambda} = x_\lambda^{\tau_\lambda}$ and $r^\tau = u^\tau$ for some variables x_λ and u. Then, $U \models \varphi[\gamma]$ is equivalent to $(\gamma(x_0), \ldots, \gamma(x_k)) \lessdot_\tau \gamma(u)$ and $U \models \varphi[\delta]$ is equivalent to $(\delta(x_0), \ldots, \delta(x_k)) \lessdot_\tau \delta(u)$.

Suppose $U \models \varphi[\gamma]$. Since, by our condition, $\gamma(x_\lambda^{\tau_\lambda}) \approx_{\tau_\lambda} \delta(x_\lambda^{\tau_\lambda})$, using axiom E4$'$ we infer $U \models \varphi[\delta]$. The inverse inference is checked in the same way. For the terms $q_\lambda^{\tau_\lambda}$ and r^τ of other kinds the reasons are quite similar.

Assume that assertion $A(m)$ is true for every $m < n$. Let $\varphi \equiv \exists x^\tau \psi$. Then, $\psi \in \Psi_{n-1}$. Let be given some H-concordant evaluations γ and δ such that $\gamma(x^\tau) \approx_\tau \delta(x^\tau)$.

Suppose $U \models \varphi[\gamma]$. It is equivalent to $U \models \psi[\gamma']$ for some evaluation γ', H-concordant with γ and such that $\gamma'(y) = \gamma(y)$ for any $y^\sigma \neq x^\tau$.

Define an evaluation δ' on U setting $\delta'(y) \equiv \delta(y)$ for every $y^\sigma \neq x^\tau$ and $\delta'(x) \equiv \gamma'(x)$. Then, $\delta'(y) = \delta(y) \approx_\sigma \gamma(y) = \gamma'(y)$ and $\delta'(x) = \gamma'(x)$, i. e. $\delta'(x) \approx_\tau \gamma'(x)$.

Check that the evaluations δ' and γ' are H-concordant. If σ is an auxiliary first-order type, then $\delta'(y^\sigma) \equiv \delta(y) = \gamma(y) = \gamma'(y)$. If τ is an auxiliary first-order type, then $\delta'(x^\tau) = \gamma'(x^\tau)$.

Let σ and τ be second-order types. Let $p \in \delta'(y^\sigma) = \delta(y)$. Since δ and γ are H-concordant, for p there exists $q \in \gamma(y)$ such that $q(v) = p(v)$ for every $v \in N(\sigma)$. Since γ and γ' are H-concordant, there exists $r \in \gamma'(y)$ such that $r(v) = q(v)$. Thus, for p, there is $r \in \gamma'(y)$ such that $r(v) = p(v)$ for every $v \in N(\sigma)$. The inverse property can be established in the same way. The property of H-concordancy for x^τ holds automatically because $\delta'(x^\tau) \equiv \gamma'(x^\tau)$.

Since δ' and γ' are H-concordant in the above mentioned sense and $\delta'(x^\tau) \approx \gamma'(x^\tau)$, then, by our condition, $U \models \psi[\gamma'] \Leftrightarrow U \models \psi[\delta']$. Consequently, we obtain the property $U \models \psi[\delta']$. By construction, $\delta'(y) = \delta(y)$ for every $y^\sigma \neq x^\tau$.

Check that the evaluations δ and δ' are H-concordant. If σ is an auxiliary first-order type, then $\delta'(y^\sigma) = \delta(y)$. If τ is an auxiliary first-order type, then $\delta'(x^\tau) = \gamma'(x) = \gamma(x) = \delta(x)$.

Let σ and τ bee second-order types. Since $\delta'(y^\sigma) = \delta(y)$, the property of H-concordancy obviously holds. Let $p \in \delta(x)$. Since δ and γ are H-concordant, we see that there exists $q \in \gamma(x)$ such that $q(v) = p(v)$. Since γ and γ' are H-concordant, there exists $r \in \gamma'(x) = \delta'(x)$ such that $r(v) = q(v)$. Thus, $r(v) = p(v)$ for every $v \in N(\tau)$. The inverse property is established in the same way.

By the definition of satisfiability, we conclude that $U \models \varphi[\delta]$. The inverse inference of $U \models \varphi[\gamma]$ from $U \models \varphi[\delta]$ is established quite analogously.

Now, let $\varphi \equiv \psi \wedge \xi$. Then, $\psi, \xi \in \Psi_{n-1}$. Consequently, $U \models \psi[\gamma] \Leftrightarrow U \models \psi[\delta]$ and $U \models \xi[\gamma] \Leftrightarrow U \models \xi[\delta]$. From here, $(U \models \psi[\gamma] \wedge U \models \xi[\gamma]) \Leftrightarrow (U \models \psi[\delta] \wedge U \models \xi[\delta])$. Thus, $U \models \varphi[\gamma] \Leftrightarrow U \models \varphi[\delta]$.

Finally, let $\varphi \equiv \neg\psi$. Then, $\psi \in \Psi_{n-1}$. Consequently, $U \models \psi[\gamma] \Leftrightarrow U \models \psi[\delta]$. From here, $U \models \varphi[\gamma] \Leftrightarrow \neg(U \models \psi[\gamma]) \Leftrightarrow \neg(U \models \psi[\delta]) \Leftrightarrow U \models \varphi[\delta]$.

This proves that the assertion $A(n)$ is true. By the complete induction principle, the assertion $A(n)$ is true for every natural number $n \in \omega$, i. e. the assertion of the Proposition holds for every formula $\varphi \in \Psi$.

Now, let φ be an arbitrary formula of the language $L(\Sigma_2^g)$. In virtue of Proposition 1 we have $U \models \varphi[\gamma] \Leftrightarrow U \models \varphi^*[\gamma]$ and $U \models \varphi[\delta] \Leftrightarrow U \models \varphi^*[\delta]$. By the definition of the operation $\varphi \mapsto \varphi^*$, we have $\varphi^* \in \Psi$. As was shown above, $U \models \varphi^*[\gamma] \Leftrightarrow U \models \varphi^*[\delta]$. As a result, we obtain the equivalence $U \models \varphi[\gamma] \Leftrightarrow U \models \varphi[\delta]$. □

C.2.5 An example of a good model for the second-order equality axioms

Construct for axioms E1 – E4 a regular, balanced, extensional, second-order model.

Take $m = 0$, $n = 0$, $\rho \equiv \langle 0, 1\rangle$, $\sigma \equiv [\rho]$, $\Theta \equiv \{\rho, \sigma\}$, $\Omega_\rho = \varnothing$, $\Omega_\sigma = \varnothing$, $\Sigma_c^\rho = \varnothing$, and $\Sigma_c^\sigma = \varnothing$. Then, $\Sigma_e \equiv (\delta_\rho, \delta_\sigma)$, $\Theta_b = \{\sigma\}$, $\Sigma_b \equiv (\varepsilon_\tau \mid \tau \in \Theta_b)$, i. e. Σ_b consists of the symbol $\varepsilon_\sigma = \varepsilon_{[\rho]}$ only, and the collection $\Sigma_v \equiv (\Sigma_v^\tau \mid \tau \in \Theta)$ consists of a denumerable set Σ_v^ρ of variables $x^\rho, y^\rho, \ldots$ of the first-order type ρ and a denumerable set Σ_v^σ of variables $u^\sigma, v^\sigma, \ldots$ of the second-order type σ.

Consider the one-grade signature $\Sigma_0 \equiv \Sigma_c \mid \Sigma_e \mid \Sigma_b \mid \Sigma_v$ of the rank 1|0 and its language $L(\Sigma_0)$. This language contains the three atomic formulas: $x^\rho \delta_\rho y^\rho$, $u^\sigma \delta_\sigma v^\sigma$, and $x^\rho \varepsilon_\sigma u^\sigma$.

Take the set of all closed segments of straight lines on the plane as the set $A \equiv A_0$. Then, $X = A$. Since $\Omega_\rho = \Omega_\sigma = \varnothing$, there are no constants. For segments $p, q \in A$, put $p \approx_\rho q$ if q is obtained from p by some parallel transfer. For sets $P, Q \in \mathcal{P}(A)$ of segments put $P \approx_\sigma Q$ if $(\forall p \in P \exists q \in Q(p \approx_\rho q)) \wedge (\forall q \in Q \exists p \in P(q \approx_\rho p))$. For a segment $p \in A$ and a set of segments $P \in \mathcal{P}(A)$ put $p \in_\sigma P$ if and only if $\exists q \in A(q \approx_\rho p \wedge q \in P)$, i. e. the segment p can be transferred into the set P by some parallel transfer.

The collection of terminals $S_v \equiv [\tau(X) \mid \tau \in \Theta]$ consists of the terminal $\rho(X) = A$ and the terminal $\sigma(X) = \mathcal{P}(A)$.

The constructed collections form the one-grade superstructure S over the set $X = A$. Consider the mathematical system $U \equiv [A, S]$ of the signature Σ_0.

Proposition 1. *The above-constructed mathematical system U together with any evaluation γ of variables of the language $L(\Sigma_0)$ on the system U forms the regular, balanced, extensional, second-order model for equality axioms* E1 – E4.

Proof. The correctness of the equality axioms is evident. The regularity follows from the definition. The same is true for the balance property.

Check the extensionality property. Let $P, Q \in \sigma(A) = \mathcal{P}(A)$. Assume $p \in P$. Then, $p \lessdot_\sigma P$. Suppose the right side of the extensionality formula. By condition we conclude $p \lessdot_\sigma Q$. By the regularity property, there exists an element $q \in Q$ such that $q \approx_\rho p$. The inverse finding of an element $p \in P$ for a given element $q \in Q$ such that $p \approx_\rho q$ is established quite similarly. In accordance with the definition of the equality $\approx_\sigma$, we conclude that $P \approx_\sigma Q$. Thus, we have inferred the left side of the extensionality formula. It follows from the correctness of axiom E4$'$ that the left side implies the right one. □

C.3 Infraproducts, infrafiltration, and generalized compactness theorem

C.3.1 Infraproducts of collections of evaluated mathematical systems of the generalized second-order signature Σ_2^g

Let F be a set and $(U_f \mid f \in F)$ be a pairwise H-concordant collection of mathematical systems of the second-order signature Σ_2^g with true generalized equalities and belongings.

By definition, $U_f \equiv (X_f, S_f)$, where $X_f \equiv [A_{0f}, \dots, A_{mf}, K_0, \dots, K_{n-1}]$.

Consider the sets $A_i \equiv \prod(A_{if} \mid f \in F)$ and the formation $X \equiv \operatorname{prod}(X_f \mid f \in F) \equiv [A_0, \dots, A_m, K_0, \dots, K_{n-1}]$.

Let $\tau \equiv [\tau_0, \dots, \tau_k]$ be a second-order type and $k \geqslant 0$.

If $\mu \in M(\tau)$, then $\tau_\mu = \langle i, 1\rangle$ for some i. Thus, $\tau_\mu(X) = A_i = \prod(A_{if} \mid f \in F) = \prod(\tau_\mu(X_f) \mid f \in F)$. If $\nu \in N(\tau)$ then $\tau_\nu = \langle j, 0\rangle$ for some j. Thus, $\tau_\nu(X) = K_j = \tau_\nu(X_f)$ for every $f \in F$. This means that the terminals of different parent types of the type τ over the formation X have quite different constitutions. Therefore, it is convenient to introduce the following notation. For elements $p \in \check{\tau}(X) = \tau_0(X) \times \dots \times \tau_k(X)$ and $f \in F$, define the element $p(f) \in \check{\tau}(X_f) = \tau_0(X_f) \times \dots \times \tau_k(X_f)$ setting $p(f)(\mu) \equiv p(\mu)(f)$ for every $\mu \in M(\tau)$ and $p(f)(\nu) \equiv p(\nu)$ for every $\nu \in N(\tau)$).

For elements $P \subset \check{\tau}(X)$ and $f \in F$, define the element $P\langle f\rangle \subset \check{\tau}(X_f)$ setting $P\langle f\rangle \equiv \{\xi \in \check{\tau}(X_f) \mid \exists p \in P(p(f) = \xi)\}$.

Let $\mathcal{D}$ be a subset of the set $\mathcal{P}(F)$, i. e. an ensemble on F. Define some superstructure S of the signature Σ_2^g over the formation X.

First, define some constant structures $s_\omega^\tau \in \tau(X)$ for $\tau \in \Theta$ and $\omega \in \Omega_\tau$.

If τ is a basic first-order type, then $\tau(X) = \prod(\tau(X_f) \mid f \in F)$. Therefore, define $s_\omega^\tau \in \tau(X)$ setting $s_\omega^\tau(f) \equiv s_{\omega f}^\tau$ for every $f \in F$. If τ is an auxiliary first-order type, then $\tau(X) = \tau(X_f)$ and $s_{\omega f}^\tau$ does not depend on the index f. Therefore, put $s_\omega^\tau \equiv s_{\omega f}^\tau$ for some (and then for every) $f \in F$.

If $\tau = [\tau_0, \dots, \tau_k]$ is a second-order type, then put $s_\omega^\tau \equiv \{p \in \check{\tau}(X) \mid \forall f \in F(p(f) \in s_{\omega f}^\tau)\}$.

As a result, we obtain the collections $S_c^\tau \equiv (s_\omega^\tau \mid \omega \in \Omega_\tau)$ and the collection $S_c \equiv (S_c^\tau \mid \tau \in \Theta)$.

Now, define generalized equalities $\approx_\tau \subset \tau(X) \times \tau(X)$.

If τ is a basic first-order type, then for $p, q \in \tau(X)$ put $p \approx_\tau q$ iff $\exists G \in \mathcal{D} \forall g \in G(p(g) \approx_{\tau,g} q(g))$. If τ is an auxiliary first-order type, then the equality $\approx_{\tau,f}$ does not depend on the index f. Therefore, for $p, q \in \tau(X)$ put $p \approx_\tau q$ iff $p \approx_{\tau,f} q$ for some (and then for every) $f \in F$.

If $\tau = [\tau_0, \ldots, \tau_k]$ is a second-order type, then for $P, Q \subset \check{\tau}(X)$ put $P \approx_\tau Q$ iff $\exists G \in \mathcal{D} \forall g \in G(P\langle g\rangle \approx_{\tau,g} Q\langle g\rangle)$.

As a result, we obtain the collection $S_e \equiv (\approx_\tau \mid \tau \in \Theta)$.

Now, define the generalized belongings $\lessdot_\tau \subset \check{\tau}(X) \times \tau(X)$. Let $\tau \in \Theta_b$. By definition, $\tau = [\tau_0, \ldots, \tau_k]$ for some $\tau_0, \ldots, \tau_k \in \Theta$. For $p \in \check{\tau}(X)$ and $P \subset \check{\tau}(X)$ put $p \lessdot_\tau P$ iff $\exists G \in \mathcal{D} \forall g \in G(p(g) \lessdot_{\tau,g} P\langle g\rangle)$[1].

Thus, we obtain the collection $S_b \equiv (\lessdot_\tau \mid \tau \in \Theta_b)$.

Consider also the collection $S_v \equiv (\tau(X) \mid \tau \in \Theta)$ consisting of the τ-terminals of the formation X.

The constructed collections compose the superstructure $S \equiv (S_c, S_e, S_b, S_v)$ over the formation X. Therefore, we can consider the mathematical system $U \equiv (X, S)$ of the signature Σ_2^g. It will be called the *infra-$\mathcal{D}$-product of the collection of mathematical systems $(U_f \mid f \in F)$ of the generalized second-order signature* Σ_2^g and will be denoted by infra-$\mathcal{D}$-prod $(U_f \mid f \in F)$.

An ensemble D on F is called a *filter on* F if it has the following properties:

1) $\forall G, H \in \mathcal{D}\ (G \cap H \in \mathcal{D})$;
2) $\forall G \in \mathcal{D}\ \forall H \in \mathcal{P}(F)\ (G \subset H \Rightarrow H \in \mathcal{D})$.

A filter $\mathcal{D}$ is called *proper* if $\mathcal{D} \neq \mathcal{P}(F)$, and a proper filter $\mathcal{D}$ is called an *ultrafilter* if for any proper filter $\mathcal{E}$ on F such that $\mathcal{D} \subset \mathcal{E}$ we have $\mathcal{D} = \mathcal{E}$, i. e. $\mathcal{D}$ is a maximal element in the set of all proper filters on F (see 1.1.15).

Further on, we assume that $\mathcal{D}$ is a filter.

Now, let $((U_f, \gamma_f) \mid f \in F)$ be a pairwise H-concordant collection of evaluated mathematical systems of the second-order signature Σ_2^g with true generalized equalities and belongings.

Define an evaluation γ on the system $U \equiv$ infra-$\mathcal{D}$-prod $(U_f \mid f \in F)$ in the following way.

Let x be a variable of a type τ. If τ is a first-order basic type, then define $\gamma(x) \in \tau(X)$ setting $\gamma(x)(f) \equiv \gamma_f(x)$ for every $f \in F$. If τ is an auxiliary first-order type, then put $\gamma(x) \equiv \gamma_f(x)$ for some (and then for every) $f \in F$.

If $\tau = [\tau_0, \ldots, \tau_k]$ is a second-order type, then put $\gamma(x) \equiv \{p \in \check{\tau}(X) \mid \forall f \in F(p(f) \in \gamma_f(x))\}$.

The evaluation γ will be called the *crossing of the collection of evaluations* $(\gamma_f \mid f \in F)$ and will be denoted by $\bowtie (\gamma_f \mid f \in F)$.

1 Note that the usage of a generalized belonging was explored in the forcing method in the form $x \in_p y$ (see e. g., [*Shoenfield*, 2001, 9.8]).

Lemma 1. *Let $((U_f, \gamma_f) \mid f \in F)$ be a pairwise H-concordant collection of evaluated mathematical systems of the second-order signature Σ_2^g and let every evaluated mathematical system (U_f, γ_f) be a model for equality axioms* E1 – E4. *Then, the pair (infra-$\mathcal{D}$-*prod $(U_f \mid f \in F)$, $\bowtie(\gamma_f \mid f \in F))$ *is also a model for axioms* E1 – E4.

Proof. Let $t_0, t_0' \in \tau_0(X), \ldots, t_k, t_k' \in \tau_k(X)$, $P, P' \subset \breve{\tau}(X) = \tau_0(X) \times \ldots \times \tau_k(X)$, $p \equiv (t_0, \ldots, t_k)$, $p' \equiv (t_0', \ldots, t_k')$, $p \approx_{\breve{\tau}} p'$, and $P \approx_\tau P'$.

Assume that $p \lessdot_\tau P$. By the definition of the belonging, $\exists G_1 \in \mathcal{D} \forall g \in G_1(p(g) \lessdot_{\tau,g} P\langle g\rangle)$. By the definition of the first-order equalities, $\exists G_2 \in \mathcal{D} \forall g \in G_2(p(g) \approx_{\breve{\tau},g} p'(g))$. Finally, by the definition of the second-order equalities $\exists G_3 \in \mathcal{D} \forall g \in G_3(P\langle g\rangle \approx_{\tau,g} P'\langle g\rangle)$. Since every system (U_g, γ_g) satisfies E4, we see that $p'(g) \lessdot_{\tau,g} P'\langle g\rangle$ for every $g \in G \equiv G_1 \cap G_2 \cap G_3$. Thus, $p' \lessdot_\tau P'$. Hence, $p \lessdot_\tau P \Rightarrow p' \lessdot_\tau P'$. The inverse implication is checked quite similarly. This proves axiom E4. The validity of axioms E1, E2, E3 is obvious. □

Further, for a formula $\varphi \in L(\Sigma)$ the set $\{f \in F \mid U_f \vDash \varphi[\gamma_f]\}$ will be denoted by G_φ.

Lemma 2. *Let $\tau = [\tau_0, \ldots, \tau_k]$ be a second-order type. Let s_ω^τ be the constants constructed above for the support $X \equiv$ *prod$(X_f \mid f \in F)$. *Then, $s_\omega^\tau\langle f\rangle = s_{\omega f}^\tau$ for every $f \in F$.*

Proof. Let $\xi \in s_\omega^\tau\langle f\rangle$, i. e. $\xi = p(f)$ for some $p \in s_\omega^\tau$. By definition, $\xi = p(f) \in s_{\omega f}^\tau$. Consequently, $s_\omega^\tau\langle f\rangle \subset s_{\omega f}^\tau$.

Conversely, let $\xi_f \in s_{\omega f}^\tau$. Since the collection of systems $(U_f \mid f \in F)$ is H-concordant, using the axiom of choice, we can find a collection $(\xi_g \mid g \in F \setminus \{f\})$ such that $\xi_g \in s_{\omega g}^\tau$ and $\xi_g(\nu) = \xi_f(\nu)$ for every $\nu \in N(\tau)$. Define the element $p \in \breve{\tau}(X)$ setting $p(\mu)(g) \equiv \xi_g(\mu)$ for every $g \in F$ and every $\mu \in M(\tau)$ and $p(\nu) \equiv \xi_f(\nu)$ for every $\nu \in N(\tau)$. Then, $p(g) = \xi_g \in s_{\omega g}^\tau$ for every $g \in F$ implies $p \in s_\omega^\tau$. Since $\xi_f = p(f)$, we have $\xi_f \in s_\omega^\tau\langle f\rangle$. Hence, $s_{\omega f}^\tau \subset s_\omega^\tau\langle f\rangle$. □

Lemma 3. *Let $\tau = [\tau_0, \ldots, \tau_k]$ be a second-order type. Let x be a variable of the type τ and $\gamma(x)$ be the evaluation constructed above for the system $U \equiv (X, S)$. Then, $\gamma(x)\langle f\rangle = \gamma_f(x)$ for every $f \in F$.*

The proof is completely similar to the proof of the previous lemma.

C.3.2 Infrafilteration of formulas of the second-order language $L(\Sigma_2^g)$ of the generalized second-order signature Σ_2^g

Consider a non-empty set F and a filter $\mathcal{D}$ on F.

By analogy with the first-order language (see [*Ershov and Palyutin*, 1984, § 17], [*Maltsev*, 1973, 8.2]), a formula φ of the language $L(\Sigma_2^g)$ of the second-order signature Σ_2^g with generalized equalities and belongings will be called *infrafiltrated with respect to the filter $\mathcal{D}$* if for every pairwise H-concordant collection $((U_f, \gamma_f) \mid f \in F)$ of evaluated mathematical systems of the second-order signature Σ_2^g with true generalized equalities and belongings the property infra-$\mathcal{D}$-prod$(U_f \mid f \in F) \vDash \varphi[\bowtie(\gamma_f \mid f \in F)]$ is equivalent to the property $\{g \in F \mid U_g \vDash \varphi[\gamma_g]\} \in \mathcal{D}$.

Lemma 1. *Every atomic formula is infrafiltrated with respect to any filter $\mathcal{D}$ on the set F.*

Proof. First, consider an atomic formula φ of the form $q^\tau \delta_\tau r^\tau$. Assume that $q^\tau = x^\tau$ and $r^\tau = \sigma_\omega^\tau$. Then, $U \vDash \varphi[\gamma]$ is equivalent to $\gamma(x) \approx_\tau s_\omega^\tau$, and analogously for the pair (U_f, γ_f).

Let τ be a first-order type. Let $G_\varphi \in \mathcal{D}$, i. e. $\gamma_g(x) \approx_{\tau,g} s_{\omega g}^\tau$ for every $g \in G_\varphi \in \mathcal{D}$. If τ is a basic type, then $\gamma_g(x) = \gamma(x)(g)$ and $s_{\omega g}^\tau$ implies $\gamma(x)(g) \approx_{\tau,g} s_\omega^\tau(g)$ for every $g \in G_\varphi \in \mathcal{D}$. Thus, $\gamma(x) \approx_\tau s_\omega^\tau$. If τ is an auxiliary type, then $\gamma_g(x) = \gamma(x)$ and $s_{\omega g}^\tau = s_\omega^\tau$. Besides, $\approx_{\tau,g}$ coincides with $\approx_\tau$. Hence, $\gamma(x) \approx_\tau s_\omega^\tau$. In both cases, we have obtained the property $U \vDash \varphi[\gamma]$.

Conversely, let $U \vDash \varphi[\gamma]$, i. e. $\gamma(x) \approx_\tau s_\omega^\tau$. If τ is a basic type, then there exists $G \in \mathcal{D}$ such that $\gamma(x)(g) \approx_{\tau,g} s_\omega^\tau(g)$ for every $g \in G$. But it means that $\gamma_g(x) \approx_{\tau,g} s_{\omega g}^\tau$, i. e. $U_g \vDash \varphi[\gamma_g]$ for every $g \in G \in \mathcal{D}$. Since $G \subset G_\varphi$, we have $G_\varphi \in \mathcal{D}$. If τ is an auxiliary type then $\gamma_f(x) \approx_{\tau,f} s_{\omega f}^\tau$ for every $f \in F$. Consequently, $G_\varphi \in \mathcal{D}$ again.

Now, let $\tau \equiv [\tau_0, \dots, \tau_k]$ be a second-order type. Let $G_\varphi \in \mathcal{D}$, i. e. $\gamma_g(x) \approx_{\tau,g} s_{\omega g}^\tau$ for every $g \in G_\varphi \in \mathcal{D}$. According to Lemmas 2 and 3 (C.3.1), the equalities $s_{\omega g}^\tau = s_\omega^\tau\langle g\rangle$ and $\gamma_g(x) = \gamma(x)\langle g\rangle$ are correct. Therefore, $\gamma(x)\langle g\rangle \approx_{\tau,g} s_\omega^\tau\langle g\rangle$ for every $g \in G_\varphi$. Consequently, $\gamma(x) \approx_\tau s_\omega^\tau$, i. e. $U \vDash \varphi[\gamma]$.

Conversely, let $U \vDash \varphi[\gamma]$, i. e. $\gamma(x) \approx_\tau s_\omega^\tau$. By the definition of the second-order equality, $\gamma(x)\langle g\rangle \approx_{\tau,g} s_\omega^\tau\langle g\rangle$ for some $G \in \mathcal{D}$ and every $g \in G$. Using Lemmas 2 and 3 (C.3.1), we obtain $\gamma_g(x) \approx_{\tau,g} s_{\omega g}^\tau$, i. e. $U_g \vDash \varphi[\gamma_g]$ for every $g \in G$. Since $G \subset G_\varphi$, we infer that $G_\varphi \in \mathcal{D}$.

For the terms q^τ and r^τ of other forms, the reasons are quite similar.

Now, consider an atomic formula φ of the form $(q_0^{\tau_0}, \dots, q_k^{\tau_k})\varepsilon_\tau r^\tau$ for $\tau \equiv [\tau_0, \dots, \tau_k] \in \Theta_b$. Assume that $q_\lambda^{\tau\lambda} = x_\lambda^{\tau\lambda}$ and $r^\tau = u^\tau$ for some variables x_λ and u. Then, $U \vDash \varphi[\gamma]$ is equivalent to $(\gamma(x_0), \dots, \gamma(x_k)) \lessdot_\tau \gamma(u)$ and analogously for the pair (U_f, γ_f).

Let $G_\varphi \in \mathcal{D}$, i. e. $(\gamma_g(x_0), \dots, \gamma_g(x_k)) \lessdot_{\tau,g} \gamma_g(u)$ for every $g \in G_\varphi \in \mathcal{D}$. Consider the elements $\xi_f \equiv (\gamma_f(x_0), \dots, \gamma_f(x_k))$ and $p \equiv (\gamma(x_0), \dots, \gamma(x_k)) \in \check{\tau}(X)$. Let $f \in F$. Then, $p(f)(\mu) \equiv p(\mu)(f) = \gamma(x_\mu)(f) = \gamma_f(x_\mu) = \xi_f(\mu)$ for every $\mu \in M(\tau)$ and $p(f)(\nu) \equiv p(\nu) = \gamma(x_\nu) = \gamma_f(x_\nu) = \xi_f(\nu)$ for every $\nu \in N(\tau)$. Consequently, $p(f) = \xi_f$. By Lemma 2 (C.3.1), $\gamma_f(u) = \gamma(u)\langle f\rangle$. As a result, we obtain $p(g) \lessdot_{\tau,g} \gamma(u)\langle g\rangle$ for every $g \in G_\varphi \in \mathcal{D}$. By definition, it means that $p \lessdot_\tau \gamma(u)$, i. e. $U \vDash \varphi[\gamma]$.

Conversely, let $U \vDash \varphi[\gamma]$, i. e. $(\gamma(x_0), \dots, \gamma(x_k)) \lessdot_\tau \gamma(u)$. By the definition of the second-order belonging, for $p \equiv (\gamma(x_0), \dots, \gamma(x_k))$ there exists $G \in \mathcal{D}$ such that $p(g) \lessdot_{\tau,g} \gamma(u)\langle g\rangle$ for every $g \in G$. By Lemma 3 (C.3.1) $\gamma(u)\langle g\rangle = \gamma_g(x)$. By the previous subsection, $\xi_g = p(g)$. Consequently, $\xi_g \lessdot_{\tau,g} \gamma_g(u)$, i. e. $U_g \vDash \varphi[\gamma_g]$ for every $g \in G$. Since $G \subset G_\varphi$, we infer that $G_\varphi \in \mathcal{D}$.

For the terms $q_\lambda^{\tau\lambda}$ and r^τ of other forms, the reasons are quite similar. □

A proof of the property of infrafiltration for the quantified formula $\exists x^\tau \varphi$ for the polygrade language $L(\Sigma_2^g)$ of the generalized second-order signature Σ_2^g is more delicate than for the first-order language. Therefore, we begin it with a subsidiary proposition.

Let $((U_f, \gamma_f) \mid f \in F)$ be a pairwise H-concordant collection of evaluated mathematical systems of the second-order signature Σ_2^g with true generalized equalities and belongings. Let β be an evaluation on the system $U \equiv$ infra-$\mathcal{D}$-prod$(U_f \mid f \in F)$, H-concordant with the evaluation $\gamma \equiv \bowtie(\gamma_f \mid f \in F)$.

For the evaluation β and for every $f \in F$ define the evaluation δ_f on the system U_f in the following way. Let x be a variable of a type τ. If τ is a basic first-order type, then put $\delta_f(x) \equiv \beta(x)(f)$. If τ is an auxiliary first-order type, then put $\delta_f(x) \equiv \beta(x)$. If τ is a second-order type, then put $\delta_f(x) \equiv \beta(x)\langle f\rangle$.

Proposition 1.
1) *The collection $((U_f, \delta_f) \mid f \in F)$ of the evaluated mathematical systems (U_f, δ_f) of the second-order signature Σ_2^g with true generalized equalities and belongings is pairwise H-concordant;*
2) *the evaluated systems (U_f, γ_f) and (U_f, δ_f) are H-concordant;*
3) *for the evaluation $\delta \equiv \bowtie(\delta_f \mid f \in F)$, the equalities $\delta(x^\tau) \approx_\tau \beta(x^\tau)$ hold for any variable x^τ;*
4) *the evaluations δ and β are H-concordant.*

Proof. 1. Let x be a variable of a type τ. If τ is an auxiliary first-order type, then $\delta_f(x) = \beta(x) = \delta_g(x)$ for every $f, g \in F$.

Let $\tau \equiv [\tau_0, \dots, \tau_k]$ be a second-order type. Fix some $f, g \in F$. Consider an arbitrary element $\xi \in \delta_f(x) = \beta(x)\langle f\rangle$. By definition, $\xi = p(f)$ for some $p \in \beta(x) \subset \check{\tau}(X) = \tau_0(X) \times \dots \times \tau_k(X)$ Consider the element $\eta \equiv p(g) \in \beta(x)\langle g\rangle = \delta_g(x)$. Then, $\eta(\nu) = p(g)(\nu) = p(\nu)$ and $\xi(\nu) = p(f)(\nu) = p(\nu)$ implies $\eta(\nu) = \xi(\nu)$ for every $\nu \in N(\tau)$. The inverse finding the element ξ corresponding to the given element η is realized in the similar manner.

2. If τ is an auxiliary first-order type, then $\delta_f(x^\tau) \equiv \beta(x^\tau)$ and $\gamma(x^\tau) \equiv \gamma_f(x^\tau)$. By condition, $\beta(x^\tau) = \gamma(x^\tau)$. Consequently, $\delta_f(x^\tau) = \gamma_f(x^\tau)$.

Let $\tau \equiv [\tau_0, \dots, \tau_k]$ be a second-order type. Consider an arbitrary element $\xi \in \gamma_f(x)$. By virtue of Lemma 3 (C.3.1) we have $\gamma_f(x) = \gamma(x)\langle f\rangle$. Since $\xi \in \gamma(x)\langle f\rangle$, by definition, there exists $p \in \gamma(x)$ such that $\xi = p(f)$. By condition, for $p \in \gamma(x)$, there is $q \in \beta(x)$ such that $q(\nu) = p(\nu)$ for any $\nu \in N(\tau)$. Consider the element $\eta \equiv q(f) \in \beta(x)\langle f\rangle = \delta_f(x)$. Then, $\eta(\nu) = q(f)(\nu) = q(\nu) = p(\nu) = p(f)(\nu) = \xi(\nu)$. The inverse condition is checked in the same way.

3. Let x be a variable of a type τ. If τ is a basic first-order type, then by the definition of the evaluations δ and δ_f, we obtain $\delta(x)(f) \equiv \delta_f(x) = \beta(x)(f)$ for any $f \in F$, i. e. $\delta(x) = \beta(x)$. If τ is an auxiliary first-order type, then $\delta(x) = \delta_f(x) = \beta(x)$ for some $f \in F$.

Let τ be a second-order type. By virtue of Lemma 3 (C.3.1), we get $\delta(x)\langle f\rangle = \delta_f(x) = \beta(x)\langle f\rangle$ for any $f \in F$. By the definition of the second-order equality, we conclude that $\delta(x) \approx_\tau \beta(x)$.

4. Let τ be an auxiliary first-order type. Then, $\delta(x^\tau) = \delta_f(x^\tau)$ for some $f \in F$. By definition, $\delta_f(x^\tau) = \beta(x^\tau)$. Consequently, $\delta(x^\tau) = \beta(x^\tau)$.

Let $\tau = [\tau_0, \dots, \tau_k]$ be a second-order type. Let $p \in \beta(x)$. By the definition of the cut, $p(f) \in \beta(x)\langle f\rangle = \delta_f(x)$ for every $f \in F$. By the definition of the crossing, $p \in \delta(x)$. Thus, for $p \in \beta(x)$, there exists $q = p \in \delta(x)$ such that $q(\nu) = p(\nu)$ for every $\nu \in N(\tau)$.

Conversely, let $q \in \delta(x^\tau)$. By the definition of the crossing, $q(f) \in \delta_f(x) = \beta(x)\langle f\rangle$ for every $f \in F$. Fix some element $f_0 \in F$. By the definition of the cut, there is $p \in \beta(x)$ such that $p(f_0) = q(f_0)$. If $\nu \in N(\tau)$, then $p(f_0)(\nu) = q(f_0)(\nu)$. However $p(f_0)(\nu) = p(\nu)$ and $q(f_0)(\nu) = q(\nu)$. Therefore, $p(\nu) = q(\nu)$ for every $\nu \in N(\tau)$. □

Proposition 2. *Let a formula ψ be infrafiltrated with respect to the filter $\mathcal{D}$. Then, the formula $\exists x^\tau \psi$ is infrafiltrated with respect to $\mathcal{D}$ as well.*

Proof. Denote the formula $\exists x^\tau \psi$ by φ. Let $G_\varphi \in \mathcal{D}$, i. e. $U_g \vDash \varphi[\gamma_g]$ for every $g \in G_\varphi \in \mathcal{D}$. Further, we shall write simply G instead of G_φ.

The presented satisfaction property means that $U_g \vDash \psi[\gamma'_g]$ for some evaluation γ'_g, H-concordant with the evaluation γ_g and such that $\gamma'_g(y) = \gamma_g(y)$ for every $y^\sigma \neq x^\tau$. For every $f \in F$, define the evaluation δ_f setting $\delta_f \equiv \gamma_f$ if $f \in F\backslash G$ and $\delta_f \equiv \gamma'_f$ if $f \in G$. Check that the evaluated systems (U_f, δ_f) and (U_g, δ_g) are H-concordant for every $f, g \in F$. If $f, g \in F\backslash G$, then $\delta_f = \gamma_f$ and $\delta_g = \gamma_g$. Since the evaluations γ_f and γ_g are H-concordant, our assertion is true. Let $f, g \in G$. Then, $\delta_f = \gamma'_f$ and $\delta_g = \gamma'_g$. Let x be a variable of a type τ.

For an auxiliary first-order type τ, we have $\gamma'_f(x) = \gamma_f(x)$ and $\gamma'_g(x) = \gamma_g(x)$. Since the evaluations γ_f and γ_g are H-concordant, we infer that $\gamma_f(x) = \gamma_g(x)$. Consequently, $\delta_f(x) = \gamma'_f(x) = \gamma'_g(x) = \delta_g(x)$.

Let τ be a second-order type. Let $p \in \delta_f(x) = \gamma'_f(x)$. Then, there exists $q \in \gamma_f(x)$ such that $q(\nu) = p(\nu)$ for every $\nu \in N(\tau)$. Since the evaluations γ_f and γ_g are H-concordant, there is $r \in \gamma_g(x)$ such that $r(\nu) = q(\nu)$. Since the evaluations γ_g and γ'_g are H-concordant as well, there exists $s \in \gamma'_g(x) = \delta_g(x)$ such that $s(\nu) = r(\nu) = q(\nu) = p(\nu)$ for any $\nu \in N(\tau)$. The inverse condition is checked in the same way.

In the cases when $f \in F\backslash G$ and $g \in G$, or conversely, the arguments are similar.

Thus, the collection $((U_f, \delta_f) \mid f \in F)$ of evaluated mathematical systems of the signature Σ_2^g with true generalized equalities and belongings is pairwise H-concordant. Consider the evaluation $\delta \equiv \bowtie(\delta_f \mid f \in F)$.

Check that $\delta(y) = \gamma(y)$ for every $y^\sigma \neq x^\tau$. Let σ be a basic first-order type. If $g \in G$, then $\delta(y)(g) = \delta_g(y) = \gamma'_g(y) = \gamma_g(y) = \gamma(y)(g)$. If $f \in F\backslash G$, then $\delta(y)(f) = \delta_f(y) = \gamma_f(y) = \gamma(y)(f)$. Consequently, $\delta(y) = \gamma(y)$.

Let σ be an auxiliary first-order type. Then, $\delta(y) = \delta_f(y) = \gamma_f(y) = \gamma(y)$ for some $f \in F\backslash G$.

Let σ be a second-order type. If $f \in G$, then $\delta_f(y) = \gamma'_f(y) = \gamma_f(y)$. If $f \in F\backslash G$, then $\delta_f(y) = \gamma_f(y)$. Let $p \in \delta(y)$. By the definition of the crossing, $p(f) \in \delta_f(y)$ for every $f \in F$. By the above, $p(f) \in \gamma_f(y)$ for every $f \in F$. This means that $p \in \gamma(y)$, whence $\delta(y) \subset \gamma(y)$. The inverse inclusion is checked in the same way. Consequently, $\delta(y) = \gamma(y)$.

Thus, for every $y \neq x$, we have $\delta(y) = \gamma(y)$.

Check that the evaluations γ and δ are H-concordant. Let $y^\sigma \neq x^\tau$.

If σ in an auxiliary first-order type, then $\delta(y) = \gamma(y)$.

Let σ be a second-order type. It was proven above that $\delta(y) = \gamma(y)$. Consequently, for every $p \in \delta(y)$, there is $q \equiv p \in \gamma(y)$ such that $q(\nu) = p(\nu)$ for any $\nu \in N(\sigma)$.

If τ is an auxiliary first-order type, then $\delta(x) = \delta_f(x) = \gamma_f(x) = \gamma(x)$ for some $f \in F\backslash G$.

Let τ be a second-order type. Let $p \in \gamma(x^\tau)$. By the definition of the crossing, $p(f) \in \gamma_f(x)$ for every $f \in F$. If $f \in F\backslash G$, then $\delta_f = \gamma_f$. If $g \in G$, then $\delta_g = \gamma'_g$ and the evaluations γ'_g and γ_g are H-concordant.

Consider the non-empty set $A \equiv \bigcup(\gamma'_g(x) \mid g \in G)$. Define the mapping $\alpha\colon G \to \mathcal{P}(A) \setminus \{\varnothing\}$ setting $\alpha(g) \equiv \{\eta \in \gamma'_g(x) \subset A \mid \forall \nu \in N(\tau)(\eta(\nu) = p(g)(\nu))\}$. According to the point 3 of the definition of H-concordant systems, the set $\alpha(g)$ is non-empty.

By the axiom of choice, there exists a function $ch\colon \mathcal{P}(A) \setminus \{\varnothing\} \to A$ such that $chP \in P$. Consider the function $\beta \equiv ch \circ \alpha\colon G \to A$ and the corresponding collection $\beta = (\eta_g \in A \mid g \in G)$. Since $\eta_g = \beta(g) = ch(\alpha(g)) \in \alpha(g)$, then we have $\eta_g(\nu) = p(g)(\nu) = p(\nu)$ for every $\nu \in N(\tau)$.

Define the element $q \in \check{\tau}(X)$ setting $q(\mu)(f) \equiv p(\mu)(f)$ for every $f \in F\backslash G$, $q(\mu)(g) \equiv \eta_g(\mu)$ for every $g \in G$ and every $\mu \in M(\tau)$, and $q(\nu) \equiv \eta_g(\nu) = p(\nu)$ for every $\nu \in N(\tau)$ and every $g \in G$.

Then, $q(f)(\mu) = q(\mu)(f) = p(\mu)(f) = p(f)(\mu)$ for every $\mu \in M(\tau)$ and $q(f)(\nu) = q(\nu) = p(\nu) = p(f)(\nu)$ for every $\nu \in N(\tau)$ implies $q(f) = p(f) \in \gamma_f(x) = \delta_f(x)$ for every $f \in F\backslash G$. If $g \in G$, then $q(g)(\mu) = q(\mu)(g) = \eta_g(\mu)$ for every $\mu \in M(\tau)$ and $q(g)(\nu) = q(\nu) = \eta_g(\nu)$ for every $\nu \in N(\tau)$ implies $q(g) = \eta_g \in \gamma'_g(x) = \delta_g(x)$. Consequently, by the definition of the crossing, $q \in \delta(x^\tau)$. Besides, $q(\nu) = p(\nu)$ for every $\nu \in N(\tau)$. The inverse finding the element p corresponding to the given element q is realized in the similar manner.

Thus, the evaluations γ and δ are really H-concordant.

By condition and construction, $U_g \vDash \psi[\delta_g]$ for every $g \in G \in \mathcal{D}$. Since the formula ψ is infrafiltered, the obtained property implies the property $U \vDash \psi[\delta]$. Since the evaluation δ is H-concordant with the evaluation γ and $\delta(y^\sigma) = \gamma(y^\sigma)$ for every $y^\sigma \neq x^\tau$, we obtain the property $U \vDash \varphi[\gamma]$.

Conversely, let $U \vDash \varphi[\gamma]$. It is equivalent to $U \vDash \psi[\beta]$ for some evaluation β, H-concordant with the evaluation γ and such that $\beta(y) = \gamma(y)$ for every $y^\sigma \neq x^\tau$.

Consider the evaluation $\delta \equiv \bowtie(\delta_f \mid f \in F)$ from Proposition 1, corresponding to the evaluation β. According to Proposition 1, the evaluations δ and β are H-concordant and $\delta(z^\rho) \approx_\rho \beta(z^\rho)$ for every variable z^ρ. It follows from Proposition 2 (C.2.4) that the property $U \models \psi[\beta]$ is equivalent to the property $U \models \psi[\delta]$. Since the formula ψ is infrafiltrated, the property $U \models \psi[\delta]$ is equivalent to the property $G \equiv \{g \in F \mid U_g \models \psi[\delta_g]\} \in \mathcal{D}$.

By Proposition 1, the evaluations δ_g and γ_g are H-concordant. Let $y^\sigma \neq x^\tau$. If σ is a basic first-order type, then $\delta_g(y) = \beta(y)(g) = \gamma(y)(g) = \gamma_g(y)$. If σ is an auxiliary first-order type, then $\delta_g(y) = \beta(y) = \gamma(y) = \gamma_g(y)$. Finally, if σ is a second-order type, then $\delta_g(y) = \beta(y)\langle g\rangle = \gamma(y)\langle g\rangle$. Since by Lemma 3 (C.3.1) $\gamma(y)\langle g\rangle = \gamma_g(y)$, we have $\delta_g(y) = \gamma_g(y)$. Consequently, in all the cases $\delta_g(y) = \gamma_g(y)$ for every $y^\sigma \neq x^\tau$. Therefore, the property $U_g \models \psi[\delta_g]$ is equivalent to the property $U_g \models \varphi[\gamma_g]$. Thus, $\{g \in F \mid U_g \models \varphi[\gamma_g]\} = G \in \mathcal{D}$. This implies $G_\varphi \in \mathcal{D}$. □

The following two lemmas are the same as ones for the first-order language.

Lemma 2. *Let formulas ψ and ξ be infrafiltrated with respect to the filter $\mathcal{D}$. Then, the formula $\psi \wedge \xi$ is infrafiltrated with respect to $\mathcal{D}$ as well.*

Proof. Denote the formula $\psi \wedge \xi$ by φ. Let $G_\varphi \in \mathcal{D}$, i. e. $U_g \models \varphi[\gamma_g]$ for all $g \in G_\varphi \in \mathcal{D}$. This property is equivalent to the conjunction of the properties $U_g \models \psi[\gamma_g]$ and $U_g \models \xi[\gamma_g]$. Since these formulas are infrafiltrated, it is equivalent to the conjunction of the properties $U \models \psi[\gamma]$ and $U \models \xi[\gamma]$, but it is equivalent to the property $U \models \varphi[\gamma]$.

Conversely, let $U \models \varphi[\gamma]$. It is equivalent to the conjunction of the properties $U \models \psi[\gamma]$ and $U \models \xi[\gamma]$. Then, $G_\psi \in \mathcal{D}$ and $G_\xi \in \mathcal{D}$. Consider $G \equiv G_\psi \cap G_\xi$. Then, $U_g \models \psi[\gamma_g]$ and $U_g \models \xi[\gamma_g]$ implies $U_g \models \varphi[\gamma_g]$ for every $g \in G \in \mathcal{D}$. Hence, $G_\varphi \in \mathcal{D}$. □

Lemma 3. *Let a formula ψ be infrafiltrated with respect to the ultrafilter $\mathcal{D}$. Then, the formula $\neg\psi$ is infrafiltrated with respect to $\mathcal{D}$ as well.*

Proof. Denote the formula $\neg\psi$ by φ. By assumption, the properties $G_\psi \in \mathcal{D}$ and $U \models \psi[\gamma]$ are equivalent.

By definition, $F \setminus G_\varphi = \{g \in F \mid$ the property $U_g \models \varphi[\gamma_g]$ does not hold$\}$. But $U_g \models \varphi[\gamma_g]$ is equivalent to the assertion that the property $U_g \models \psi[\gamma_g]$ does not hold. Consequently, the property $U_g \models \psi[\gamma_g]$ is equivalent to the assertion that the property $U_g \models \varphi[\gamma_g]$ does not hold. It implies $F \setminus G_\varphi = G_\psi$.

Let $G_\varphi \in \mathcal{D}$. Since $\mathcal{D}$ is an ultrafilter, we have $G_\psi = F \setminus G_\varphi \notin \mathcal{D}$. Thus, the property $U \models \psi[\gamma]$ does not hold. By the definition of the satisfiability, it means that $U \models \varphi[\gamma]$.

Conversely, let $U \models \varphi[\gamma]$. Then, the property $U \models \psi[\gamma]$ does not hold. Therefore, $G_\psi \notin \mathcal{D}$. Since $\mathcal{D}$ is an ultrafilter, we have $G_\varphi = F \setminus G_\psi \in \mathcal{D}$. □

Theorem 1 (Zakharov). *Every formula φ of the language $L(\Sigma_2^g)$ of the second-order signature Σ_2^g with generalized equalities and belongings is infrafiltrated with respect to any ultrafilter $\mathcal{D}$ on the set F.*

Proof. The set of all formulas φ of the language $L(\Sigma_2^g)$, constructed by induction from atomic formulas by means of the connectives $\neg$ and $\wedge$ and the quantifier $\exists$, will be denoted by Ψ. The subset of the set Ψ, consisting of all formulas containing at most n logical symbols $\neg$, $\wedge$, and $\exists$, will be denoted by Ψ_n. Obviously, $\Psi = \bigcup(\Psi_n \mid n \in \omega)$.

Using the complete induction principle, we shall prove the following assertion $A(n)$: *every formula $\varphi \in \Psi_n$ is infrafiltrated.*

If $n = 0$, then φ is an atomic formula. By Lemma 1 (C.3.2), it is infrafiltrated. Consequently, $A(0)$ holds.

Assume that for every $m < n$ the assertion $A(m)$ holds. Let $\varphi \in \Psi_n$. If $\varphi = \neg\psi$, then $\psi \in \Psi_{n-1}$. Therefore, ψ is infrafiltrated. By Lemma 3, the formula φ is infrafiltrated as well. If $\varphi = \psi \wedge \xi$, then $\psi, \xi \in \Psi_{n-1}$. Therefore, by the inductive assumption, the formulas ψ and ξ are infrafiltered. By Lemma 2, the formula φ is infrafiltrated as well. Finally, if $\varphi = \exists x^\tau \psi$, then $\psi \in \Psi_{n-1}$. Consequently, as above, the formula ψ is infrafiltrated. By Proposition 2, the formula φ is infrafiltrated as well. Thus, the assertion $A(n)$ holds.

By the complete induction principle, the assertion $A(n)$ holds for every $n \in \omega$. This means that any formula $\varphi \in \Psi$ is infrafiltrated.

Let φ be an arbitrary formula of the language $L(\Sigma_2^g)$. Consider for φ the accompanying formula φ^* defined in C.2.4. By the definition of the operation $\varphi \mapsto \varphi^*$, we have $\varphi^* \in \Psi$. By the proven above, the formula φ^* is infrafiltrated, i. e. $\{g \in F \mid U_g \vDash \varphi^*[\gamma_g]\} \in \mathcal{D} \Leftrightarrow U \vDash \varphi^*[\gamma]$. Proposition 1 (C.2.4) implies the equivalences $U \vDash \varphi^*[\gamma] \Leftrightarrow U \vDash \varphi[\gamma]$ and $U_g \vDash \varphi^*[\gamma_g] \Leftrightarrow U_g \vDash \varphi[\gamma_g]$. As a result, we get the following chain of equivalences: $\{g \in F \mid U_g \vDash \varphi[\gamma_g]\} \in \mathcal{D} \Leftrightarrow \{g \in F \mid U_g \vDash \varphi^*[\gamma_g]\} \in \mathcal{D} \Leftrightarrow U \vDash \varphi^*[\gamma] \Leftrightarrow U \vDash \varphi[\gamma]$. It means that the formula φ is infrafiltrated. □

This theorem has one important corollary. Let Φ be some set of formulas of the language $L(\Sigma_2^g)$ of the generalized second-order signature Σ_2^g. Let the set Φ has a model (U_0, γ_0) of the signature Σ_2^g with true generalized equalities and belongings. Take an arbitrary set F and an arbitrary ultrafilter $\mathcal{D}$ on F. Consider the collection of the models $((U_f, \gamma_f) \mid f \in F)$ such that $(U_f, \gamma_f) \equiv (U_0, \gamma_0)$. The infra-$\mathcal{D}$-product infra-$\mathcal{D}$-prod $(U_f \mid f \in F)$ of the collection $(U_f \mid f \in F)$ will be called the *infra-$\mathcal{D}$-power of the system U_0 with the exponent F* and will be denoted by infra-$\mathcal{D}$-power(U_0, F). The crossing $\bowtie(\gamma_f \mid f \in F)$ of the collection $(\gamma_f \mid f \in F)$ will be called the *crossing of the evaluation γ_0 in the quantity F* and will be denoted by $\bowtie(\gamma_0, F)$.

Corollary 1. *Let Φ be some set of formulas of the language $L(\Sigma_2^g)$. If the set Φ has a model (U_0, γ_0) of the signature Σ_2^g with true generalized equalities and belongings, then for every set F and every ultrafilter $\mathcal{D}$ on F, the set Φ has also the model* (infra-$\mathcal{D}$-power (U_0, F), $\bowtie(\gamma_0, F)$) *of the signature Σ_2^g with true generalized equalities and bolongingc.*

This implies that if a set Φ of formulas of the language $L(\Sigma_2^g)$ has a model with true generalized equalities and belongings, then it has the same model of an arbitrary large power. Therefore, the *generalized second-order logic has the upper*

Lövengame – Skolem property, in contrast with the standard second-order logic, which does not have this property (see [*Mendelson*, 1997, Appendix, (III)]).

C.3.3 Compactness theorem for formulas of the language $L(\Sigma_2^g)$ of the generalized second-order signature

Theorem 1 (the Zakharov compactness theorem for generalized second-order language). *Let Φ and Ψ be some sets of formulas of the language $L(\Sigma_2^g)$ of the generalized second-order signature Σ_2^g. Let for every finite subset f of the set Φ, the set of formulas f+(E1 – E4)+Ψ has a model (U_f, γ_f) of the signature Σ_2^g such that collection $((U_f, \gamma_f) \mid f \in F)$ is pairwise H-concordant. Then, the set of formulas Φ+(E1 – E4)+Ψ has a model (U, γ) of the signature Σ_2^g.*

Proof. Consider the set $F \equiv \{f \subset \Phi \mid 0 < |f| < \omega\}$ of all finite non-empty subsets from Φ.

For an element $f \in F$, consider the set $F_f \equiv \{g \in F \mid f \subset g\}$. Since $f \in F_f$, we have $F_f \neq \varnothing$. The ensemble $\mathfrak{C} \equiv \{F_f \mid f \in F\}$ has the finite intersection property, i. e. it is multiplicative. Hence, there is some ultrafilter $\mathcal{D}$ on the set F including the set $\mathfrak{C}$.

Consider the system $U \equiv$ infra-$\mathcal{D}$-prod $(U_f \mid f \in F)$ and the evaluation $\gamma \equiv \bowtie(\gamma_f \mid f \in F)$ on the system U constructed in C.3.1. By Lemma 1 (C.3.1), U is a system with the true generalized equalities and belongings.

Prove that the evaluated system (U, γ) is a model for the set Φ.

Suppose $\varphi \in \Phi$. Consider the set $F_{\{\varphi\}}$. By condition, $U_{\{\varphi\}} \vDash \varphi[\gamma_{\{\varphi\}}]$. Consider the set $G_\varphi \equiv \{g \in F \mid U_g \vDash \varphi[\gamma_g]\}$. If $g \in F_{\{\varphi\}}$, then $\{\varphi\} \subset g$ implies $\varphi \in g$. Therefore, $U_g \vDash \varphi[\gamma_g]$. Consequently, $F_{\{\varphi\}} \subset G_\varphi$. Since $F_{\{\varphi\}} \in \mathcal{D}$, we have $G_\varphi \in \mathcal{D}$.

By Theorem 1 (C.3.2), we infer the property $U \vDash \varphi[\gamma]$. Thus, (U, γ) is a model for the set Φ. The fact that (U, γ) is a model for the set Ψ follows immediately from Theorem 1 (C.3.2). □

C.3.4 Uncountable models of the second-order generalized Peano – Landau arithmetic

First, we describe the Peano – Landau arithmetic in the generalized second-order language of the one-grade signature of the rank 1|0.

Put $m = 0$ and $n = 0$, i. e. we shall consider the single basic first-order type of the form $\pi \equiv \langle 0, 1\rangle$ without auxiliary first-order types. Consider the second-order types $\varkappa \equiv [\pi]$ and $\rho \equiv [\pi, \pi]$ and the type domain $\Theta \equiv \Theta_{Ar2}^g \equiv \{\pi, \varkappa, \rho\}$ of the rank 1|0 with the belonging type subdomain $\Theta_b \equiv \{\varkappa, \rho\}$.

Put $\Omega_\pi \equiv 1$, $\Omega_\varkappa \equiv \varnothing$, $\Omega_\rho \equiv 1$, and consider the collections $\Sigma_c^\pi \equiv (\sigma_\omega^\pi \mid \omega \in \Omega_\pi) = \sigma_0^\pi$, $\Sigma_c^\varkappa \equiv (\sigma_\omega^\varkappa \mid \omega \in \Omega_\varkappa) = \varnothing$, and $\Sigma_c^\rho \equiv \left(\sigma_\omega^\rho \mid \omega \in \Omega_\rho\right) = \sigma_0^\rho$. They compose the signature of constants of the type domain Θ of the form $\Sigma_c \equiv (\Sigma_c^\tau \mid \tau \in \Theta) = (\sigma_0^\pi, \varnothing, \sigma_0^\rho)$ containing the constant σ_0^π, which is an objective first-order constant for denoting the natural number 0, and the constant σ_0^ρ, which is a predicate second-order constant for expressing the *succession relation of Peano* between a natural number a and its *successor* $a + 1$.

Further, along with σ_0^π and σ_0^ρ, we shall simply write 0 and σ, respectively.

Take the signature of the generalized equalities of the type domain Θ of the form $\Sigma_e \equiv (\delta_\tau \mid \tau \in \Theta) = (\delta_\pi, \delta_\varkappa, \delta_\rho)$ containing the first-order equality δ_π and the second-order equalities $\delta_{[\pi]}$ and $\delta_{\pi,\pi}$.

Take the signature of the generalized belongings of the type domain Θ of the form $\Sigma_b \equiv (\varepsilon_\tau \mid \tau \in \Theta_b) = (\varepsilon_\varkappa, \varepsilon_\rho)$.

Finally, take a denumerable set Σ_v^π of objective variables $x^\pi, y^\pi, \ldots$ of the first-order type π and denumerable sets $\Sigma_v^\varkappa$ and Σ_v^ρ of predicate variables $u^\varkappa, v^\varkappa, \ldots$ and $u^\rho, v^\rho, \ldots$ of the second-order types $\varkappa$ and ρ, respectively.

They form the signature $\Sigma_v \equiv (\Sigma_v^\tau \mid \tau \in \Theta) = (\Sigma_v^\pi, \Sigma_v^\varkappa, \Sigma_v^\rho)$ of variables of the type domain Θ.

Consider the one-grade generalized signature $\Sigma_{Ar2}^g \equiv \Sigma_c|\Sigma_e|\Sigma_b|\Sigma_v$ of the rank 1|0 and its language $L(\Sigma_{Ar2}^g)$. Terms $p, q, r, \ldots$ of this language are constants and variables only, the atomic equality formulas have the forms $q^\pi\delta_\pi r^\pi$, $q^\varkappa\delta_\varkappa r^\varkappa$, and $q^\rho\delta_\rho r^\rho$. Respectively, the atomic belonging formulas have the forms $q^\pi\varepsilon_\varkappa r^\varkappa$ and $(p^\pi, q^\pi)\varepsilon_\rho r^\rho$.

Further, along with $x^\pi, y^\pi, \ldots$ and δ_π, we shall simply write $x, y, \ldots$ and δ.

The *axioms of the second-order generalized Peano – Landau arithmetic* are the following ones.

A1. $\forall x_1, x_2, y((x_1, y)\varepsilon_\rho\sigma \wedge (x_2, y)\varepsilon_\rho\sigma \Rightarrow x_1\delta x_2)$.
A2. $\forall x, y_1, y_2((x, y_1)\varepsilon_\rho\sigma \wedge (x, y_2)\varepsilon_\rho\sigma \Rightarrow y_1\delta y_2)$.
A3. $\forall x, y((x, y)\varepsilon_\rho\sigma \Rightarrow \neg(y\delta 0))$.
A4. $\forall u^\varkappa(0\varepsilon_\varkappa u^\varkappa \wedge \forall x, y(x\varepsilon_\varkappa u^\varkappa \wedge (x, y)\varepsilon_\rho\sigma \Rightarrow y\varepsilon_\varkappa u^\varkappa) \Rightarrow \forall z(z\varepsilon_\varkappa u^\varkappa))$.

Consider the following generalized *extensionality properties*.

PE1. $\forall u^\varkappa, v^\varkappa(u^\varkappa\delta_\varkappa v^\varkappa \Leftrightarrow \forall x(x\varepsilon_\varkappa u^\varkappa \Leftrightarrow x\varepsilon_\varkappa v^\varkappa))$.
PE2. $\forall u^\rho, v^\rho(u^\rho\delta_\rho v^\rho \Leftrightarrow \forall x, y((x, y)\varepsilon_\rho u^\rho \Leftrightarrow (x, y)\varepsilon_\rho v^\rho))$.

Consider the set $\mathbb{N}_0 \equiv \omega$ of all natural numbers constructed in the NBG set theory (see 1.2.6) or in any set theory ST mentioned in Introduction.

For the formation $\mathbb{N}_0$ of the rank 1|0 and the signature Σ_{Ar2}^g, consider the following collections $S_c^\pi \equiv [s_\omega^\pi \mid \omega \in \Omega_\pi] = s_0^\pi$, $S_c^\varkappa \equiv [s_\omega^\varkappa \mid \omega \in \Omega_\varkappa] = \varnothing$, and $S_c^\rho \equiv [s_\omega^\rho \mid \omega \in \Omega_\rho] = s_0^\rho$. They compose the collection of constant structures $S_c \equiv [S_c^\tau \mid \tau \in \Theta] = [s_0^\pi, \varnothing, s_0^\rho]$, containing the constant structure $s_0^\pi \in \pi(\mathbb{N}_0) - \mathbb{N}_0$, which is the initial natural number, and the constant structure $s_0^\rho \in \rho(\mathbb{N}_0) = \mathcal{P}(\mathbb{N}_0 \times \mathbb{N}_0)$, which is the set of all pairs of natural numbers of the form $\langle a, a + 1\rangle$.

Further, along with s_0^π and s_0^ρ, we shall write simply 0 and s, respectively.

Consider the collection of the equality ratios of the form $S_e \equiv (\approx_\tau \mid \tau \in \Theta) = (\approx_\pi, \approx_\varkappa, \approx_\rho) \equiv (= |\mathbb{N}_0^2, = |\mathcal{P}(\mathbb{N}_0)^2, = |\mathcal{P}(\mathbb{N}_0 \times \mathbb{N}_0)^2)$ containing in the capacity of the first-order equality ratio $\approx_\pi$ and of the second-order equality ratios $\approx_\varkappa$ and $\approx_\rho$ the restrictions on the indicated sets one and the same set-theoretical equality $=$ in the NBG set theory.

Consider the collection of the belonging ratios of the form $S_b \equiv (\lessdot_\tau \mid \tau \in \Theta) = (\lessdot_\varkappa, \lessdot_\rho) \equiv (\in |\mathbb{N}_0 \times \mathcal{P}(\mathbb{N}_0), \in |(\mathbb{N}_0 \times \mathbb{N}_0) \times \mathcal{P}(\mathbb{N}_0 \times \mathbb{N}_0))$ containing in the capacity of the belonging ratio $\lessdot_\varkappa$ and $\lessdot_\rho$ the restrictions on the indicated sets one and the same set-theoretical belonging $\in$ in the NBG set theory.

Finally, take the collection of the terminals over the formation $\mathbb{N}_0$ of the form $S_v \equiv (\tau(\mathbb{N}_0) \mid \tau \in \Theta) = (\pi(\mathbb{N}_0), \varkappa(\mathbb{N}_0), \rho(\mathbb{N}_0)) = (\mathbb{N}_0, \mathcal{P}(\mathbb{N}_0), \mathcal{P}(\mathbb{N}_0 \times \mathbb{N}_0))$.

These collections compose the one-grade superstructure $S_{Ar2} \equiv (S_c, S_e, S_b, S_v)$ of the signature Σ^g_{Ar2} of the rank 1|0 over the formation $\mathbb{N}_0$.

The system $Ar2 \equiv (\mathbb{N}_0, S_{Ar2})$ of the signature Σ^g_{Ar2} can be called the *natural series of Peano – Landau of the second order in the NBG set theory*, because it models in NBG the following *Peano – Landau postulates*.

P1. $\forall a_1, a_2, b(\langle a_1, b\rangle \in s \wedge \langle a_2, b\rangle \in s \Rightarrow a_1 = a_2)$.
P2. $\forall a, b_1, b_2(\langle a, b_1\rangle \in s \wedge \langle a, b_2\rangle \in s \Rightarrow b_1 = b_2)$.
P3. $\forall a, b(\langle a, b\rangle \in s \Rightarrow b \neq 0)$.
P4. $\forall P(0 \in P \wedge \forall a, b(a \in P \wedge \langle a, b\rangle \in s \Rightarrow b \in P) \Rightarrow \forall c(c \in P))$.

Consider an evaluation γ on the system $Ar2$ such that $\gamma(x) \in \pi(\mathbb{N}_0) = \mathbb{N}_0$, $\gamma(u^\varkappa) \in \varkappa(\mathbb{N}_0) = \mathcal{P}(\mathbb{N}_0)$, and $\gamma(u^\rho) \in \rho(\mathbb{N}_0) = \mathcal{P}(\mathbb{N}_0 \times \mathbb{N}_0)$.

For the evaluated system $(Ar2, \gamma)$ the following assertion holds.

Lemma 1. *The evaluated system $(Ar2, \gamma)$ is the standard model for the set of formulas* E1, E2, E3, E4, A1, A2, A3, A4, PE1, *and* PE2 *of the language* $L(\Sigma^g_{Ar2})$.

Proof. The satisfactions $Ar2 \models A1[\gamma]$, $Ar2 \models A2[\gamma]$, $Ar2 \models A3[\gamma]$, and $Ar2 \models A4[\gamma]$ follow from the correctness for the system $Ar2$ of Peano – Landau postulates $P1$, $P2$, $P3$, and $P4$, respectively. The other satisfactions are checked immediately. □

Therefore, the evaluated system $(Ar2, \gamma)$ and also the mathematical system $Ar2$ (see A.1.3) can be called the *generalized natural series of Peano – Landau of the second order in the NBG set theory*. Note that along with the NBG set theory any set theory ST mentioned in Introduction can be used.

Now, construct an uncountable model. Take an arbitrary set F and an arbitrary ultrafilter $\mathcal{D}$ on F. Consider the system infra-$\mathcal{D}$-power$(Ar2, F)$ and the evaluation $\bowtie(\gamma, F)$ defined in C.3.2.

Theorem 1. *The evaluated system* (infra-$\mathcal{D}$-power$(Ar2, F)$, $\bowtie(\gamma, F)$) *is the generalized model for the set of formulas* E1, E2, E3, E4, A1, A2, A3, A4, PE1, *and* PE2 *of the language*

$L(\Sigma^g_{Ar2})$. The support of the model is the generalized Baire set $\mathbb{N}_0^F$ (see [*Engelking*, 1977, 4.3.G]). *If* $|F| \geqslant \omega$, *then the support is uncountable.*

Proof. The assertion follows from Lemma 1 and Theorem 1 (C.3.2). □

Index of terms

2 In this index, any articles (a, an, the) are omitted.

3 In this index, for terms having different values in different set theories, the abbreviations in brackets indicate subsections where the term is explained within the framework of the corresponding theory.

https://doi.org/10.1515/9783110550948-005

4 In this index, the words “with respect to” are abbreviated to “w. r. t.”.

Index of notations

5 In this index, for notations having different values in different set theories, the abbreviations in brackets indicate subsections where the notation is explained within the framework of the corresponding theory.

https://doi.org/10.1515/9783110550948-006

Digits

Arrows

Other symbols of operations and relations

Brackets

Infinities

Ascenders

Special cases of using upper and lower indices

Miscellaneous

Bibliography

Akilov, G. P. and V. N. Dyatlov

[1983] *Fundamentals of mathematical analysis*, Nauka, Novosibirsk, 1983. (in Russian).

Alexandrov, A. D.

[1940] *Additive set functions on abstract spaces. I*, Matem. Sborn. **8** (1940), no. 2, 303–348.

[1941] *Additive set functions on abstract spaces. II*, Matem. Sborn. **9** (1941), no. 2–3, 563–628.

[1943] *Additive set functions on abstract spaces. III*, Matem. Sborn. **13** (1943), no. 3, 169–238.

Alexandrov, P. S.

[1977] *Introduction to set theory and general topology*, Nauka, Moscow, 1977. (in Russian).

Anger, B. and C. Portenier

[1992] *Radon integrals: an abstract approach to integration and Riesz representation through function cones*, Progress in Math., vol. 103, Birkhäuser, Boston, 1992.

Arens, Richard

[1950] *Representation of functionals by integrals*, Duke Math. J. **17** (1950), 499–506.

[1963] *Representation of moments by integrals*, Illinois J. Math. **7** (1963), 609–614.

Baire, R.

[1899] *Sur les fonctions de variables réelles*, Ann. Mat. Pura Appl. Ser. IIIa **3** (1899), 1–122.

[1905] *Leçons sur les fonctions discontinues*, Gauthier–Villar, Paris, 1905.

Banach, S.

[1931] *Über analytisch darstellbare Operationen in abstrakten Räumen*, Fundam. Math. **17** (1931), 283–295.

[1932] *Théorie des opérations linéares*, Monografie Matematyczne, vol. 1, PWA, Warszawa, 1932.

[1937] *The Lebesgue integral in abstract spaces* (1937). (Note II in [*Saks*, 1937, p. 320–330]).

Bernays, P.

[1968] *Axiomatic set theory*, North-Holland, Amsterdam, 1968.

[1976] *A system of axiomatic set theory*, Sets and classes: on the work of Paul Bernays, Studies in Logic and the Foundations of Mathematics, vol. 84, North-Holland, Amsterdam, 1976, pp. 1–117.

Benedetto, J. J. and C. Wojciech

[2009] *Integration and modern analysis*, Birkhäuser, Basel, 2009.

Blanc, G. and A. Preller

[1975] *Lawvere's basic theory of the category of categories*, J. Symbolic Logic **40** (1975), 14–18.

Blanc, G. and M. R. Donnadieu

[1976] *Axiomatisation de la catégorie des catégories*, Cahiers Topol. Géom. Diff. **17** (1976), 135–170.

Bogachev, V. I.

[2007] *Measure theory. Vol. I-II*, Springer–Verlag, Berlin, 2007.

Boolos, G., J. P. Burgess, and R. Jeffrey

[2007] *Computability and logic*, Cambridge University Press, Cambridge, 2007.

Borel, E.

[1898] *Leçons sur la théorie des fonctions*, Gauthier–Villars, Paris, 1898.

[1905] *Leçons sur les fonctions de variables réelles*, Gauthier–Villars, Paris, 1905.

[1922] *Méthodes et problèmes de la théorie des fonctions*, Gauthier–Villar, Paris, 1922.

Bourbaki, Nicolas

[1951] *Eléments de Mathématique. Livre II. Algebre. Chapitre 1. Structures algébriques*, 2nd ed., Actualities sci. Ind., vol. 1111, Hermann, Paris, 1951.

https://doi.org/10.1515/9783110550948-007

[1952] *Eléments de Mathématique. Livre II. Algebre. Chapitre 6. Groupes et corps ordonné*, Actualities sci. Ind., vol. 1179, Hermann, Paris, 1952.

[1953] *Eléments de Mathématique. Livre V. Espaces vectoriels topologiques. Chapitre 1. Espaces vectoriels topologiques sur un corps value. Chapitre 2. Ensembles convexes et espaces localement convexes*, Actualities sci. Ind., vol. 1189, Hermann, Paris, 1953.

[1954] *Eléments de Mathématique. Livre I. Théorie des ensembles. Chapitre 1. Description de la mathéique formelle. Chapitre 2. Théorie des ensembles*, Actualities sci. Ind., vol. 1212, Hermann, Paris, 1954.

[1956] *Eléments de Mathématique. Livre VI. Intégration. Chapitre 5. Intégration des mesures*, Actualities sci. Ind., vol. 1244, Hermann, Paris, 1956.

[1957] *Eléments de Mathématique. Livre I. Théorie des ensembles. Chapitre 4. Structures*, Actualities sci. Ind., vol. 1258, Hermann, Paris, 1957.

[1958] *Eléments de Mathématique. Livre III. Topologie générale. Chapitre 9. Utilisation des nombres réels en topologie générale*, 2nd ed., Actualities sci. Ind., vol. 1045, Hermann, Paris, 1958.

[1959] *Eléments de Mathématique. Livre VI. Intégration. Chapitre 6. Intégration vectorielle*, Actualities sci. Ind., vol. 1281, Hermann, Paris, 1959.

[1960] *Eléments de mathématique. Livre III. Topologie générale. Chapitre 3. Groupes topologiques (théorie élémentaire). Chapitre 4. Nombres réels*, 3rd ed., Actualities Sci. Ind., vol. 1143, Hermann, Paris, 1960.

[1961a] *Eléments de mathématique. Livre III. Topologie générale. Chapitre 10. Espaces fonctionels*, 2nd ed., Actualities sci. Ind., vol. 1084, Hermann, Paris, 1961.

[1961b] *Eléments de mathématique. Livre III. Topologie générale. Chapitre 1. Structures topologiques. Chapitre 2. Structures uniformes*, 3rd ed., Actualities sci. Ind., vol. 1142, Hermann, Paris, 1961.

[1963] *Eléments de mathématique. Livre I. Théorie des ensembles. Chapitre 3. Ensembles ordonnés. Cardinaux. Nombres entiers*, 2nd ed., Actualities sci. Ind., vol. 1243, Hermann, Paris, 1963.

[1965] *Eléments de mathématique. Livre VI. Intégration. Chapitre 1. Inégalités de convexité. Chapitre 2. Espaces de Riesz. Chapitre 3. Mesures sur les espaces localement compact. Chapitre 4. Prolongement d'une mesure. Espaces L^p*, 2nd ed., Actualities sci. Ind., vol. 1175, Hermann, Paris, 1965.

[1969] *Eléments de mathématique. Livre VI. Intégration. Chapitre 9. Intégration sur les espaces topologiques separes*, Actualities sci. Ind., vol. 1343, Hermann, Paris, 1969.

Brudno, A. L.

[1971] *Theory of functions of a real variable*, Nauka, Moscow, 1971. (in Russian).

Bunina, E. I. and V. K. Zakharov

[2003] *A canonical form for supertransitive standard models in Zermelo–Fraenkel set theory*, Russian Math. Surveys **58** (2003), no. 4, 782–783.

[2005] *Canonical form of Tarski sets in Zermelo–Fraenkel set theory*, Math. Notes **77** (2005), no. 3, 297–306.

[2006] *Characterization of model Mirimanov–von Neumann cumulative sets*, J. Math. Sci. **138** (2006), no. 4, 5830–5891.

[2007] *Formula-inaccessible cardinals and a characterization of all natural models of Zermelo–Fraenkel set theory*, Izv. Math. **71** (2007), no. 2, 219–245.

Bucur, I. and A. Deleanu

[1972] *Introduction to the theory of categories and functors*, Wiley, 1972.

Cohn, D. L.

[2013] *Measure theory*, Birkhäuser, Basel, 2013.

Da Costa, N.
[1965] *On two systems of set theory*, Proc. Koninkl. Nederl. Akad. Wet. Ser. A. **68** (1965), no. 1, 95–98.
[1967] *Two formal systems of set theory*, Proc. Koninkl. Nederl. Akad. Wet. Ser. A. **70** (1967), no. 1, 45–51.
Da Costa, N. and A. de Caroli
[1967] *Remarques sur les universe d'Ehresmann–Dedecker*, C. R. Acad. Sci. Paris. Sér. A. **265** (1967), 761–763.
Dalen, D. van
[1997] *Logic and Structure*, Springer, Berlin, 1997.
Daniell, P. J.
[1918] *A general form of integral*, Annals of Math. Ser. 2 **19** (1918), no. 4, 279–294.
Dashiell, F., A. Hager, and M. Henriksen
[1980] *Order–Caushy completions of rings and vector lattices of continuous functions*, Can. J. Math. **XXXII** (1980), no. 3, 657–685.
Day, M. M.
[1958] *Normed linear spaces*, Springer Verlag, Berlin, 1958.
Dedecker, P.
[1959] *Introduction aux structures locales*, Colloque Géom. Diff. Globale (Bruxelles, 1958), Centre Belge Rech. Math., Louvain, 1959, pp. 103–135.
De Jonge, E. and A. C. M. Van Rooij
[1977] *Introduction to Riesz spaces*, Math. Centre Tracts, vol. 78, Matematisch Centrum, Amsterdam, 1977.
Dinculeanu, N.
[1967] *Vector measures*, Internat. Ser. Monogr. Pure Appl. Math., vol. 95, Pergamon Press, Oxford, 1967.
Dorogovtsev, A. Ya.
[1989] *Elements of general theory of measure and integral*, Vysha shkola, Kiev, 1989. (in Russian).
Dunford, N. and J. T. Schwartz
[1958] *Linear operators. Part I. General theory*, Wiley, 1958.
Edwards, R. E.
[1953] *A theory of Radon measures on locally compact spaces*, Acta Math. **89** (1953), 133–164.
[1965] *Functional analysis: theory and applications*, Dover Publications, New York, 1965.
Ehresmann, C.
[1957] *Guttungen von lokalen Strukturen*, J. Deutschen Math.-Verein. **60** (1957), no. 2, 49–77.
Eilenberg, S. and S. MacLane
[1945] *General theory of natural equivalences*, Trans. Am. Math. Soc. **58** (1945), 231–294.
Engelking, R.
[1977] *General topology*, Monografie Matematyczne, vol. 60, PWN, Warszawa, 1977.
Ershov, Yu. L. and E. A. Palyutin
[1984] *Mathematical logic*, Mir, Moscow, 1984.
Feferman, S.
[1963] *The number systems. Foundations of algebra and analysis*, Addison – Wesley Publishing, Reading, 1963.
[1969] *Set-theoretical foundations of category theory*, Reports of the Midwest Category Seminar. III, Lecture Notes in Math., vol. 106, Springer, Berlin, 1969, pp. 201–247.
Fine, N. T., L. Gillman, and J. Lambek
[1965] *Rings of quotients of rings of functions*, McGill University Press, Montreal, 1965.

Fichtenholz, G. and L. Kantorovich
[1934] *Sur les opérations linéaires dans l'èspaces des founctions bornées*, Studia Math. **5** (1934), 69–98.
Fisher, C. A.
[1915] *Linear functionals of N-spreads*, Annals of Math. Ser. 2 **19** (1915), no. 4, 37–43.
Forster, T.
[1995] *Set theory with a universal set*, Oxford University Press, New York, 1995.
Fraenkel, A.
[1922] *Zu den Grandlagen der Cantor–Zermeloschen Mengenlehre*, Math. Ann. **86** (1922), 230–237.
Fréchet, M.
[1904] *Sur les opérations linéaires*, Trans. AMS **5** (1904), 493–499.
[1910] *Extension au cas des intégrales multiples d'une definition de l'intégrale due à Stieltjes*, Nouvelles Annales de Math. Ser. 4. **10** (1910), no. 5, 241–256.
[1915a] *Sur l'integrale d'une fonctionnelle étendue á un ensemble abstrait*, Bull. Soc. Math. de France **XLIII** (1915), 248–265.
[1915b] *Sur les fonctionnelles bilinéaires*, Trans. AMS **16** (1915), no. 3, 215–234.
Fremlin, D. H.
[1974] *Topological Riesz spaces and measure theory*, Cambridge University Press, London–New York, 1974.
[2003a] *Measure theory. Vol. 1. The irreducible minimum*, Torres Fremlin, Colchester, 2003.
[2003b] *Measure theory. Vol. 2. Broad foundations*, Torres Fremlin, Colchester, 2003.
[2004] *Measure theory. Vol. 3. Measure algebras*, Torres Fremlin, Colchester, 2004.
[2006] *Measure theory. Vol. 4. Topological measure spaces*, Torres Fremlin, Colchester, 2006.
[2008] *Measure theory. Vol. 5. Set-theoretic measure theory*, Torres Fremlin, Colchester, 2008.
Gabriel, P.
[1962] *Des catégories abéliennes*, Bull. Soc. Math. France **90** (1962), 323–448.
Gavrilov, V. I. and A. I. Shtern
[2006] *Introducing of the exponential, logarithmical, and trigonometrical functions ...*, Moscow State University, Moscow, 2006. (in Russian).
Gödel, K.
[1940] *The consistency of the axiom of choice and of the generalized continuum hypothesis with the axioms of set theory*, Princeton Univ. Press, Princeton, NJ, 1940.
Goldblatt, Robert
[1979] *Topoi. The categorial analysis of logic*, North-Holland, Amsterdam, 1979.
Gordon, Hugh
[1971] *Ring of functions determined by zero-sets*, Pacific J. Math. **36** (1971), no. 1, 133–157.
Hadamard, J.
[1903] *Sur les opérations fonctionelles*, C. R. Acad. Sci. Paris **136** (1903), 351–354.
Halmos, Paul R.
[1950] *Measure theory*, Chelsea Publishing Company, New York, 1950.
Hansell, R. W.
[2001] *Descriptive sets and the topology of nonseparable Banach spaces*, Serdica Math. J. **27** (2001), 1–66.
Hatcher, W. S.
[1982] *The logical foundations of mathematics*, Pergamon, Oxford, 1982.
Hausdorff, F.
[1914] *Grundzüge der Mengenlehre*, Vien, Leipzig, 1914.
[1915] *Über halbstetige Functionen und deren Verallgemeinerung*, Math. Zeit. **4** (1915), 292–309.

[1927] *Mengenlehre*, 2nd ed., Walter De Gruyter, Berlin, 1927; English transl. in [2005].
[2005] *Set Theory*, Am. Math. Soc., Providence, RI, 2005.
Helly, Edward
[1912] *Über lineare Funktionaloperationen*, Sitzungsberichten der Kaiserl. Akad. der Wissen. in Wien. Math.-naturw. Klasse **121** (1912), no. 2, 265–297.
Henstock, R.
[1988] *Lectures on the theory of integration*, Series in Real Analysis, vol. 1, World Scientific, Singapore, 1988.
Herrlich, H. and G. Strecker
[1979] *Category theory. An introduction*, Heldermann, Berlin, 1979.
Hewitt, E.
[1952] *Integration on locally compact spaces. I*, Univ. of Washington Publ. in Math. **3** (1952), 71–75.
Hewitt, E. and K. Stromberg
[1965] *Real and abstract analysis. A modern treatment of the theory of functions of a real variable*, Springer–Verlag, New York, 1965.
Holmes, M. S.
[1998] *Elementary set theory with universal set*, Academia-Bruylant, Louvain-la-Neuve, 1998.
Isbell, J.
[1966] *Structure of categories*, Bull. Am. Math. Soc. **72** (1966), 619–655.
Jacobs, K.
[1978] *Measure and integral*, Probab. Math. Stat., Academic Press, New York, 1978.
Jech, Thomas J.
[1971] *Lectures in set theory with particular emphasis on the method of forcing*, Lecture notes in Math., vol. 217, Springer, Berlin, 1971.
[2003] *Set theory*, Springer, New York, 2003.
Johnson, K. G.
[1962] *Algebras of Riemann integrable functions*, Proc. of Am. Math. Soc. **13** (1962), no. 3, 437–441.
Kakutani, S.
[1941] *Concrete representation of abstract (M)-spaces*, Ann. Math. Ser. 2 **42** (1941), 994–1024.
Kamke, Dr. E. von
[1956] *Das Lebesgue–Stieltjes integral*, Teumner Verlagsgesellschaft, Leipzig, 1956.
Kanamori, Akihiro
[2003] *The higher infinite: Large cardinals in set theory from their beginnings*, 2nd ed., Springer Monographs in Mathematics, Springer, Berlin–Heidelberg, 2003.
Kanovei, V. G. and V. A. Lubetsky
[2010] *Modern set theory. Borel and projective sets*, MCCME, Moscow, 2010. (in Russian).
Kelley, J. L.
[1975] *General topology*, Springer, New York, 1975.
Kolmogorov, A. N. and A. G. Dragalin
[1982] *Introduction to mathematical logic*, Moscow State University, Moscow, 1982. (in Russian).
Kolmogorov, A. N. and S. V. Fomin
[1999] *Elements of theory of functions and functional analysis*, Dover Publications, Mineola, NY, 1999.
König, H.
[1995] *The Danielle–Stone–Riesz representation theorem*, Operator Theory: Advances Appl. **75** (1995), 191–222.

[1997] *Measure and integration: an advanced course in basis procedures and applications*, Springer–Verlag, Berlin, 1997.
[2000] *On the inner Daniell–Stone and Riesz representation theorems*, Documenta Math. **5** (2000), 301–315.
[2002] *Measure and integration: an attempt at unified systematization*, Rend. Istit. Mat. Univ. Trieste **34** (2002), 155–214.
[2008] *New version of the Danielle–Stone–Riesz representation theorem*, Positivity **12** (2008), no. 1, 105–118.

Kubrusly, C. S.
[2015] *Essentials of measure theory*, Springer, New York, 2015.

Kuratowski, K.
[1966] *Topology*, Academic Press, New York, 1966.

Kuratowski, K. and A. Mostowski
[1967] *Set theory*, North-Holland, Amsterdam, 1967.

Kurzweil, J.
[2000] *Henstock–Kurzweil integration: its relation to topological vector spaces*, Series in Real Analysis, vol. 7, World Scientific, Singapore, 2000.

Lawvere, F. W.
[1966] *The category of categories as a foundation for mathematics*, Proceedings of the Conference on Categorical Algebra, Springer, Berlin, 1966, pp. 1–20.

Lebesgue, H.
[1902] *Integral, longeur, aire*, Ann. Math.(3) **VII** (1902), 231–359.
[1903] *Sur les séries triginométriques*, Ann. Ecole Norm. Super. (3) **XX** (1903), 453–485.
[1904] *Leçons sur la l'integration et recherche des fonctions primitives*, Gauthier–Villar, Paris, 1904.
[1905] *Sur les fonctions représentables analytiquement*, J. Math. Ser. 6 **1** (1905), 139–216.

Lee, T. Y.
[2011] *Henstock–Kurzweil integration on Euclidean spaces*, Series in Real Analysis, vol. 12, World Scientific, Singapore, 2011.

Loś, J.
[1955] *Quelques remarques, théorems et problèms sur les classes définisables d'algèbres*, Mathematical interpretations of formal systems, North-Holland, Amsterdam, 1955, pp. 98–113.

Lukashenko, T. P., V. A. Skvortsov, and A. P. Solodov
[2011] *Generalized integrals*, 2nd ed., Librocom, Moscow, 2011. (in Russian).

MacLane, S.
[1961] *Locally small categories and the foundations of set theory*, Proc. Symp. Foundation Math., Pergamon, Oxford, 1961, pp. 25–43.
[1969] *One universe as a foundation for category theory*, Reports of the Midwest Category Seminar. III, Lecture Notes in Math., vol. 106, Springer, Berlin, 1969, pp. 192–200.
[1971] *Categories for working mathematician*, Springer, Berlin, 1971.

Maltsev, A. I.
[1973] *Algebraic systems*, Springer, Berlin, 1973.

Marcus, S.
[1959] *La mesure de Jordan et l'integrale de Riemann dans un espace mesuré topologique*, Acta Sci. Math. **20** (1959), no. 2–3, 156–163.

Mendelson, E.
[1964] *Introduction to mathematical logic*, Van Nostrand, New York, 1964.
[1997] *Introduction to mathematical logic*, 4th ed., Chapman&Hall / CRC, London, 1997.

Mikhalev, A. V., A. A. Seredinskii, and V. K. Zakharov

[2002] *Purely ring characteristics of rings of continuous functions on compact topological spaces*, J. Math. Sci. **122** (2002), no. 6, 4583–4597.

[2007] *Characterization of the space of Riemann integrable functions by means of cuts of the space of continuous functions. I*, Moscow Univ. Math. Bull. **62** (2007), no. 5, 173–180.

[2008] *Characterization of the space of Riemann integrable functions by means of cuts of the space of continuous functions. II*, Moscow Univ. Math. Bull. **63** (2008), no. 5, 183–192.

Mirimanov, D.

[1917] *Les antinomies de Russel et de Burali–Forti et le probléme fondamentale de la théorie des ensembles*, L'Enseignment Math. **19** (1917), 37–52.

Montague, R. and L. Vaught

[1959] *Natural models of set theory*, Fundam. Math. **47** (1959), 219–242.

Naimark, M. A.

[1959] *Normed rings*, Noordhoff N. V., Groningen, 1959.

Nakano, H. and T. Shimogaki

[1962] *A note on the cut extensions of C-spaces*, Proc. Japan Acad. **38** (1962), no. 8, 473–477.

Natanson, I. P.

[1955] *Theory of functions of a real variable*, Frederick Ungar Publishing Co., New York, 1955.

Neumann, John von

[1929] *Über eine Widerspruchfreiheitsfrage in der axiomatischen Mengenlehre*, J. Reine Angew. Math. **160** (1929), 227–241.

Neveu, J.

[1964] *Bases mathématiques du calcul des probabilités*, Masson et Cie, Paris, 1964.

Nikol'skii, S. M.

[1990] *Course in mathematical analysis*, 4th ed., Vol. 1,2, Nauka, Moscow, 1990. (in Russian).

Oxtoby, John C.

[1971] *Measure and category*, Graduate Texts in Math., Springer, New York, 1971.

Pollard, D. and F. Topsøe

[1975] *A unified approach to Riesz type representation theorems*, Studia Math. **54** (1975), no. 2, 173–190.

Prokhorov, Yu. V.

[1956] *Convergence of random processes and limit theorems in probability theory*, Theory Probab. Appl. **1** (1956), 157–214.

Radon, J.

[1913] *Theorie und Anwendangen der absolut additiven Mengenfunktionen*, Sitzungsberichten der Kaiserl. Akad. der Wissen. in Wien. Math.-naturw. Klasse **122** (1913), no. 7, 1295–1438. (reprinted in [*Radon*, 1987, p. 45–188]).

[1987] *Gesammelte Abhandlungen (Collected works)*, Vol. 1, Verlag der Österreich. Akad. der Wissen. & Birkhäuser, Wien–Basel–Boston, 1987.

Raja, M.

[2002] *On some class of Borel measurable maps and absolute Borel topological spaces*, Topol. Appl. **123** (2002), 267–282.

Regoli, G.

[1977] *Some characterization of sets of measurable functions*, Am. Math. Monthly **84** (1977), no. 6, 455–458.

Riesz, F.

[1909] *Sur les opérations fonctionelles linéaires*, C. R. Acad. Sci. Paris **149** (1909), 974–977.

[1911] *Sur systèmes singuliers d'équations intégrales*, Annales Sci. de l'École Norm. Super. Ser. 3 **28** (1911), no. 1, 33–62.

[1914] *Démonstration nouvelle d'un théorème concernant les opérations fonctionelles linéaires*, Annales Sci. de l'École Norm. Super. Ser. 3 **31** (1914), no. 1, 9–14.

Robertson, A. P. and W. Robertson

[1964] *Topological vector spaces*, Cambridge University Press, Cambridge, 1964.

Rossberg, M.

[2004] *First-order and second-order logic and completeness*, First-order logic revisited, Logos, Berlin, 2004, pp. 303–322.

Sadovnichii, V. A.

[1999] *Theory of operators*, 3rd ed., Vyshaya shkola, Moscow, 1999. (in Russian).

Saks, Stanisław

[1937] *Theory of the integral*, 2nd ed., Monografie Matematyczne, vol. 7, PWA – G.E.Stechert, Warszawa – Lwów – New York, 1937.

[1938] *Integration in abstract metric spaces*, Duke Math. J. **4** (1938), no. 2, 408–411.

Schaefer, Helmut H.

[1966] *Topological vector spaces*, Macmillan, New York, 1966.

Schwartz, L.

[1973] *Radon measures on arbitrary topological spaces and cylindrical measures*, Oxford Univ. Press, London–New York, 1973.

Semadeni, Z.

[1971] *Banach spaces of continuous functions*, Monografie Matematyczne, vol. 55, PWA, Warszawa, 1971.

Shapiro, S.

[1991] *Foundations without foundationalism: a case for second-order logic*, Oxford University Press, Oxford, 1991.

Shatery, H. R. and J. Zafarani

[2004] *The equality between Borel and Baire classes*, Real Analysis Exchange **30** (2004), no. 1, 373–384.

Shepherdson, J.

[1951] *Inner models for set theory. I*, J. Symbolic Logic **16** (1951), 161–190.

[1952] *Inner models for set theory. II*, J. Symbolic Logic **17** (1952), 225–237.

[1953] *Inner models for set theory. III*, J. Symbolic Logic **18** (1953), 145–167.

Shilov, G. E. and B. L. Gurevich

[1967] *Integral, measure, and derivative (general theory)*, Nauka, Moscow, 1967. (in Russian).

Shoenfield, J. R.

[2001] *Mathematical logic*, Natick, MS, Assoc. Symbol. Logic, 2001.

Sierpiński, W.

[1921] *Sur les fonctions développables en séries absolument convergentes de fonctions continues*, Fundam. Math. **2** (1921), 15–27.

Sierpiński, W. and A. Tarski

[1930] *Sur une propriété caractéristique des nombres inaccessibles*, Fundam. Math. **15** (1930), 292–300.

Skvortsov, V. A. and F. Tulone

[2013] *Generalized Hake property for integrals of Henstock type*, Moscow Univ. Math. Bull. **68** (2013), no. 6, 270–274.

Sonner, J.

[1962] *The formal definition of categories*, Math. Zeit. **80** (1962), 163–176.

Spurný, J.

[2010] *Borel sets and functions in topological spaces*, Acta Math. Hungar. **129** (2010), no. 1–2, 47–69.

Srivastava, S. M.
[1998] *A course on Borel sets*, Graduate Texts in Mathematics, vol. 180, Springer, 1998.
Stone, M. N.
[1937] *Applications of the theory of Boolean rings to general topology*, Trans. AMS **41** (1937), 375–481.
[1948a] *Notes on integration. I*, Proc. Natl. Acad. Sci. USA **34** (1948), no. 7, 336–342.
[1948b] *Notes on integration. II*, Proc. Natl. Acad. Sci. USA **34** (1948), no. 9, 447–455.
[1948c] *Notes on integration. III*, Proc. Natl. Acad. Sci. USA **34** (1948), no. 10, 483–490.
[1949] *Notes on integration. IV*, Proc. Natl. Acad. Sci. USA **35** (1949), no. 1, 50–58.
Takeuti, G.
[2013] *Proof theory*, Dover Publications, Mineola, NY, 2013.
Tarski, A.
[1938] *Über unerreichbare Kardinalzahlen*, Fundam. Math. **30** (1938), 68–89.
Topsøe, F.
[1970] *Topology and measure*, Lecture notes in Math., vol. 133, Springer–Verlag, Berlin, 1970.
[1976] *Further results on integral representations*, Studia Math. **55** (1976), no. 3, 239–245.
[1983] *Radon measures, some basic constructions*, Measure theory and its applications (Sherbrooke, QC 1982), Lecture notes in Math., vol. 1033, Springer–Verlag, Berlin, 1983, pp. 303–311.
Tourlakis, George
[2003a] *Lectures in logic and set theory. Vol. 1: Mathematical Logic*, Cambridge studies in advanced math., vol. 82, Cambridge University Press, Cambridge, 2003.
[2003b] *Lectures in logic and set theory. Vol. 2: Set theory*, Cambridge studies in advanced math., vol. 83, Cambridge University Press, Cambridge, 2003.
Väänänen, J.
[2001] *Second order logic and foundations of mathematics*, Bull. Symbolic Logic **7** (2001), no. 4, 504–520.
Voronov, M. V. and V. K. Zakharov
[2008] *Basic mathematical concepts. Part 1*, PPI, Pskov, 2008. (in Russian).
Vulikh, B. Z.
[1965] *Course in theory of function of a real variable*, Nauka, Moscow, 1965. (in Russian).
[1967a] *Introduction to functional analysis*, Nauka, Moscow, 1967. (in Russian).
[1967b] *Introduction to the theory of partially ordered vector spaces*, Wolters–Noordhoff Scientific Publ., Groningen, 1967.
Yosida, K.
[1965] *Functional analysis*, Springer Verlag, Berlin, 1965.
Young, W. H.
[1905] *On upper and lower integration*, Proc. London Math. Soc. Ser. 2 **II** (1905), 52–66.
[1911] *A new method in the theory of integration*, Proc. London Math. Soc. Ser. 2 **9** (1911), 15–50.
[1913] *On functions and their associated sets of points*, Proc. London Math. Soc. Ser. 2 **12** (1913), 260–277.
[1914] *On integration with respect to a function o bounded variation*, Proc. London Math. Soc. Ser. 2 **13** (1914), 109–150.
Zakharov, V. K.
[1980] *The functional representation of the uniform completion of the maximal and of the countably dense modules of fractions of the module of continuous functions*, Russian Math. Surveys **35** (1980), no. 4, 200–201.
[1981] *Functional characterization of absolute and Dedekind completion*, Bull. Acad. Polonais Sci. Ser. math. **39** (1981), no. 5–6, 293–297.

[1982] *On the hyperstonean cover of a compact*, Math. Nachr. **107** (1982), 175–186.
[1984] *Functional characterization of the absolute, vector lattices of functions with the Baire property and of quasinormal functions and modules of particular*[6] *continuous functions*, Trans. Mosc. Math. Soc. (1984), no. 1, 69–105.
[1985] *Some perfect preimages connected with extensions of the family of continuous functions*, Topology and applications (Eger, Hungary, 1983), Colloq. Math. Soc. János Bolyai, vol. 41, North-Holland, 1985, pp. 703–728.
[1987a] *On functions connected with absolute, Dedekind completion, and divisible envelope*, Periodica Math. Hungar. **18** (1987), no. 1, 17–26.
[1987b] *On functions connected with sequential absolute, Cantor completion, and classical ring of quotients*, Periodica Math. Hungar. **19** (1987), no. 2, 113–133.
[1988a] *Hyperstonean cover and second dual extension*, Acta Math. Hungar. **51** (1988), no. 1–2, 125–149.
[1988b] *Lebesgue cover and Lebesguean extension*, Studia Sci. Math. Hungar. **23** (1988), 343–368.
[1989a] *Alexandrovian cover and Sierpińskian extension*, Studia Sci. Math. Hungar. **24** (1989), no. 2–3, 93–117.
[1989b] *Borel cover and Borel extension*, Studia Sci. Math. Hungar. **24** (1989), no. 4, 407–428.
[1990] *The connection between the complete ring of quotients of the ring of continuous functions, regular completion, Hausdorff–Sierpiński extensions*, Russian Math. Surveys **45** (1990), no. 6, 177–178.
[1991] *Connections between the Lebesgue extension and the Borel first class extension, and between the preimages corresponding to them*, Math. USSR Izvestya **37** (1991), no. 2, 273–302.
[1993a] *The Gordon preimage of the Aleksandrov space as an enclosed covering*, Russian Acad. Sci. Izv. Math. **40** (1993), no. 2, 405–424.
[1993b] *Arens extension of a ring of continuous functions*, St. Petersburg Math. J. **4** (1993), no. 1, 131–148.
[1994] *Countably divisible and Baire's extensions as divisible envelopes of a ring and Banach algebra of continuous functions*, St. Petersburg Math. J. **5** (1994), no. 6, 1141–1156.
[1995a] *Connection between the classical ring of quotients of the ring of continuous functions and Riemann integrable functions*, Fundam. Prikl. Mat. **1** (1995), no. 1, 161–176. (in Russian).
[1995b] *Extensions of a ring of continuous functions generated by regular, countably divisible, and complete rings of quotients, and their corresponding pre-images*, Izv. Math. **59** (1995), no. 4, 677–720.
[1995c] *The Kaplan extension of the ring and Banach algebra of continuous functions as a divisible hull*, Izv. Math. **45** (1995), no. 3, 477–493.
[1995d] *Extensions of the ring of continuous functions generated by the classical, rational, and regular rings of fractions as divisible hulls*, Sb. Math. **186** (1995), no. 12, 1773–1809.
[1995e] *Classical extensions of the ring of continuous functions and the corresponding preimages of a completely regular space*, J. Math. Sci. **73** (1995), no. 1, 114–139.
[1996] *Relationships between the Riemann extension and the classical ring of quotients and between the Semadeni preimage and the sequential absolute*, Trans. Moscow Math. Soc. **57** (1996), 223–243.
[2002a] *Classification of Borel sets and functions for an arbitrary space*, Doklady Math. **66** (2002), no. 1, 99–101.

6 A mistake of translators of the article: "quotients of" should be instead of "particular".

[2002b] *The problem of the characterization of Radon integrals*, Doklady Math. **66** (2002), no. 1, 118–120.
[2002c] *Classification of Borel sets and functions in the general case*, Russian Math. Survey **57** (2002), no. 4, 822–823.
[2002d] *Divisible envelopes connected with the regular and rational rings of quotients and the Hausdorff–Sierpinski extension of continuous functions*, J. Math. Sci. **110** (2002), no. 3, 2660–2682.
[2005a] *The Riesz–Radon problem of the characterization of integrals and the weak compactness of Radon measures*, Proc. Steklov Inst. Math. **248** (2005), no. 1, 101–110.
[2005b] *Local set theory*, Math. Notes **77** (2005), no. 1–2, 177–193.
[2005c] *Description of extension of families of continuous functions by means of order boundaries*, Doklady Math. **71** (2005), no. 1, 80–83.
[2006a] *New classes of functions related to general families of sets*, Doklady Math. **73** (2006), no. 2, 197–201.
[2006b] *Characterization of the classical extensions of the family of continuous functions as Dedekind hulls*, Doklady Math. **74** (2006), no. 3, 849–853.
[2008a] *Compactness theorem for generalized second-order language*, Contemporary problems of fundamental and applied mathematics, MPhTI, Dolgoprudnii, 2008, pp. 11–31. (in Russian).
[2008b] *Hausdorff theorems on measurable functions and a new class of uniform functions*, Moscow Univ Math. Bull. **63** (2008), no. 1, 1–6.

Zakharov, V. K., E. I. Bunina, A. V. Mikhalev, and P. V. Andreev

[2006] *Local theory of sets as a foundation for the category theory and its connection with Zermelo–Frenkel set theory*, J. Math. Sci. **138** (2006), no. 4, 5763–5829.

Zakharov, V. K. and A. V. Koldunov

[1980] *Sequential absolute and its characterisations*, Soviet Math. Dokl. **22** (1980), no. 1, 70–74.
[1982a] *Characterization of the σ-cover of a compact*, Sibir. Math. J. **XXIII** (1982), no. 6, 91–99. (in Russian).
[1982b] *Characterization of the σ-cover of a compact*, Math. Nachr. **107** (1982), 7–16.

Zakharov, V. K. and A. V. Mikhalev

[1997a] *Radon problem for regular measures on an arbitrary Hausdorff space*, Fundam. Prikl. Mat. **3** (1997), no. 3, 801–808. (in Russian).
[1997b] *Integral representation for Radon measures on an arbitrary Hausdorff space*, Fundam. Prikl. Mat. **3** (1997), no. 4, 1135–1172. (in Russian).
[1998a] *The problem of integral representation for Radon measures on a Hausdorff space*, Doklady Math. **57** (1998), no. 3, 337–339.
[1998b] *On the concept of mathematical system*, Fundam. Prikl. Mat. **4** (1998), no. 3, 927–935. (in Russian).
[1999] *The problem of general Radon representation for an arbitrary Hausdorff space*, Izv. Math. **63** (1999), no. 5, 881–921.
[2000a] *The MacLane problem about a set theoretical foundation for the category theory*, J. Math. Sci. **102** (2000), no. 3, 4018–4031.
[2000b] *Mathematical systems*, Algebra. Proceedings of the International Algebraic Conference (Moscow, 1998), Walter De Gruyter, Berlin, 2000, pp. 370–383.
[2000c] *Big mathematical systems*, J. Math. Sci. **102** (2000), no. 6, 4679–4695.
[2001] *Connections between the integral Radon representations for locally compact and Hausdorff spaces*, Fundam. Prikl. Mat. **7** (2001), no. 1, 33–46. (in Russian).
[2002] *The problem of the general Radon representation for an arbitrary Hausdorff space. II*, Izv. Math. **66** (2002), no. 6, 1087–1101.

[2003a] *The MacLane problem about a set theoretical foundation for the category theory. II*, J. Math. Sci. **114** (2003), no. 2, 1067–1085.
[2003b] *Riesz–Radon problem of characterization of Radon integrals*, Kolmogorov and contemporary mathematics, Marcel Dekker, New York, 2003, pp. 260–261.
[2004] *A local two-sorted theory of classes and sets: MacLane's problem and Mostowski's program*, J. Math. Sci. **123** (2004), no. 2, 3906–3920.

Zakharov, V. K., A. V. Mikhalev, and T. V. Rodionov
[2010a] *The problem of characterization of general Radon integrals*, Doklady Math. **82** (2010), no. 1, 613–616.
[2010b] *Riesz–Radon–Fréchet problem of characterization of integrals*, Russian Math. Surveys **65** (2010), no. 4, 741–765.
[2012a] *Characterization of Radon integrals as linear functionals*, J. Math. Sci. **85** (2012), no. 2, 233–281.
[2012b] *Characterization of integrals with respect to arbitrary Radon measures by the boundedness indices*, J. Math. Sci. **85** (2012), no. 3, 417–429.
[2015] *Postclassical families of functions proper to descriptive and prescriptive spaces*, Doklady Math. **92** (2015), no. 2, 259–262.
[2016] *Descriptive spaces and proper classes of functions*, J. Math. Sci. **213** (2016), no. 2, 163–200.
[2017] *Postclassical families of functions proper for descriptive and prescriptive spaces*, J. Math. Sci. **221** (2017), no. 3, 360–383.

Zakharov, V. K. and T. V. Rodionov
[2008a] *Classification of Borel sets and functions for an arbitrary space*, Sborn. Math. **199** (2008), no. 6, 833–869.
[2008b] *A class of uniform functions and its relationship with the class of measurable functions*, Math. Notes **84** (2008), no. 6, 756–770.
[2014a] *A fine correlation between Baire and Borel functional hierarchies*, Acta Math. Hungar. **142** (2014), no. 2, 384–402.
[2014b] *Naturalness of the class of Lebesgue–Borel–Hausdorff measurable functions*, Math. Notes **95** (2014), no. 4, 500–508.

Zakharov, V. K. and A. A. Seredinskii
[2006a] *A new characterization of Riemann-integrable functions*, J. Math. Sci. **139** (2006), no. 4, 6708–6714.
[2006b] *A new characterization of Riemann integral and Riemann-integrable functions*, Moscow Univ. Math. Bull. **61** (2006), no. 2, 16–23.

Zakharov, V. K. and A. D. Yashin
[2011] *Finite axiomatizability of the local set theory*, Math. Notes **90** (2011), no. 1, 64–78.
[2014] *Compactness theorem for some generalized second-order language*, J. Math. Research **6** (2014), no. 3, 21–38.

Zermelo, E.
[1908] *Untersuchungen über die Grundlagen der Mengenlehre. I*, Math. Ann. **65** (1908), 261–281.
[1930] *Über Grenzzahlen und Mengenbereiche*, Fundam. Math. **16** (1930), 29–47.

De Gruyter Studies in Mathematics

Volume 68/2
Valeriy K. Zakharov, Timofey V. Rodionov, Alexander V. Mikhalev
Set, Functions, Measures. Volume 2: Fundamentals of Functions and Measure Theory
ISBN 978-3-11-055009-2, e-ISBN 978-3-11-055096-2, Set-ISBN 978-3-11-055097-9

Volume 67
Alexei Kulik
Ergodic Behavior of Markov Processes. With Applications to Limit Theorems
ISBN 978-3-11-045870-1, e-ISBN 978-3-11-045893-0, Set-ISBN 978-3-11-045894-7

Volume 66
Igor V. Nikolaev
Noncommutative Geometry. A Functorial Approach, 2017
ISBN 978-3-11-054317-9, e-ISBN 978-3-11-054525-8, Set-ISBN 978-3-11-054526-5

Volume 65
Günter Mayer
Interval Analysis and Automatic Result Verification, 2017
ISBN 978-3-11-050063-9, e-ISBN 978-3-11-049946-9, Set-ISBN 978-3-11-049947-6

Volume 64
Dorina Mitrea, Irina Mitrea, Marius Mitrea, Michael Taylor
The Hodge-Laplacian. Boundary Value Problems on Riemannian Manifolds, 2016
ISBN 978-3-11-048266-9, e-ISBN 978-3-11-048438-0, Set-ISBN 978-3-11-048439-7

Volume 63
Evguenii A. Rakhmanov
Orthogonal Polynomials, 2016
ISBN 978-3-11-031385-7, e-ISBN 978-3-11-031386-4, Set-ISBN 978-3-11-038325-6

Volume 62
Ulrich Krause
Positive Dynamical Systems in Discrete Time, 2015
ISBN 978-3-11-036975-5, e-ISBN 978-3-11-036569-6, Set-ISBN 978-3-11-036571-9

Volume 61
Francesco Altomare, Mirella Cappelletti Montano, Vita Leonessa, Ioan Rasa
Markov Operators, Positive Semigroups and Approximation Processes, 2015
ISBN 978-3-11-037274-8, e-ISBN 978-3-11-036697-6, Set-ISBN 978-3-11-036698-3

Bei Fragen zur Produktsicherheit wenden Sie sich bitte an:
If you have any questions regarding product safety,
please contact:

Walter de Gruyter GmbH
Genthiner Straße 13
10785 Berlin
productsafety@degruyterbrill.com